AF749586

Surface Microscopy with Low Energy Electrons

Ernst Bauer

Surface Microscopy with Low Energy Electrons

Springer

Ernst Bauer
Department of Physics
Arizona State University
Tempe, AZ, USA

ISBN 978-1-4939-0934-6 ISBN 978-1-4939-0935-3 (eBook)
DOI 10.1007/978-1-4939-0935-3
Springer New York Heidelberg Dordrecht London

Library of Congress Control Number: 2014938038

© Springer Science+Business Media New York 2014
This work is subject to copyright. All rights are reserved by the Publisher, whether the whole or part of the material is concerned, specifically the rights of translation, reprinting, reuse of illustrations, recitation, broadcasting, reproduction on microfilms or in any other physical way, and transmission or information storage and retrieval, electronic adaptation, computer software, or by similar or dissimilar methodology now known or hereafter developed. Exempted from this legal reservation are brief excerpts in connection with reviews or scholarly analysis or material supplied specifically for the purpose of being entered and executed on a computer system, for exclusive use by the purchaser of the work. Duplication of this publication or parts thereof is permitted only under the provisions of the Copyright Law of the Publisher's location, in its current version, and permission for use must always be obtained from Springer. Permissions for use may be obtained through RightsLink at the Copyright Clearance Center. Violations are liable to prosecution under the respective Copyright Law.
The use of general descriptive names, registered names, trademarks, service marks, etc. in this publication does not imply, even in the absence of a specific statement, that such names are exempt from the relevant protective laws and regulations and therefore free for general use.
While the advice and information in this book are believed to be true and accurate at the date of publication, neither the authors nor the editors nor the publisher can accept any legal responsibility for any errors or omissions that may be made. The publisher makes no warranty, express or implied, with respect to the material contained herein.

Printed on acid-free paper

Springer is part of Springer Science+Business Media (www.springer.com)

To Anastassia

About the Author

Professor Ernst Bauer is a distinguished German-American physicist and surface scientist who has made fundamental contributions to the understanding of epitaxial growth and to the development of microscopy techniques. He is one of the founders of surface physics and the physics of thin films.

In 1958 he derived the classification of the thin film growth mechanisms that provides the theoretical framework of epitaxy which is used worldwide to this day. In 1962 he invented LEEM (Low Energy Electron Microscopy), which came to fruition in 1985. In the late 1980s/early 1990s he extended the LEEM technique in two important directions by developing Spin-Polarized Low Energy Electron Microscopy (SPLEEM) and Spectroscopic Photo Emission and Low Energy Electron Microscopy (SPELEEM). The combination of these methods now allows a comprehensive (structural, chemical, magnetic, and electronic) characterization of surfaces and thin films on the 10 nm scale.

Ernst Bauer's interest in the development of synchrotron radiation microscopy techniques and his involvement with the Synchrotron source Elettra in Trieste, Italy resulted in the development of the Nanospectroscopy beamline, which is today one of the leading synchrotron radiation microscopy facilities worldwide.

His work directly or indirectly impacts many areas of modern materials science: surfaces, thin films, electronic materials, and instrumentation. The invention and development of surface microscopy with slow electrons has revolutionized the study of surface science and thin film science.

Ernst Bauer has authored or coauthored more than 450 publications (among them 85 review papers and book chapters) and one book ("Electron Diffraction: Theory, Practice and Applications," 1958, in German). His papers are widely cited.

Numerous LEEM instruments are now installed and are operating in many laboratories and synchrotron radiation facilities around the world (USA, Europe, Asia, and Australia). An important recognition for Ernst Bauer's efforts in the field of surface microscopy is the increasing number of the scientists involved in LEEM research, which is reflected in the organization of biannual LEEM/PEEM workshops, the first of which was organized by Ernst Bauer and Anastassia Pavlovska in Arizona in 1998.

Broad international collaboration is typical of Ernst Bauer's research. He had longstanding scientific cooperations with NASA, University of Pretoria (South Africa), Synchrotron Radiation Source in Trieste (Italy), Poland, Ukraine, Bulgaria, and Czech Republic. About 80 visiting scientists had a possibility to perform high-quality research in his group in Germany. Presently he has collaborations with Japan, Poland, Italy, Germany, and Hong Kong.

The scientific achievements of Ernst Bauer have been multiply honored. He was the recipient of the E.W. Muller Award in 1985, the Gaede Prize of the German Vacuum Society in 1988, the Medard W. Welch Award of the American Vacuum Society in 1992, the Niedersachsenpreis for Science (Germany) in 1994, the BESSY Innovation Award on Synchrotron Radiation in 2004, and the very prestigious Davisson-Germer Prize of the American Physical Society in 2005. In 2003 Ernst Bauer received the first Award of the Japan Society of Promotion of Science's 141st Committee on Microbeam Analysis and was made an honorary member of this organization. He was elected a Member of the Goettingen Academy of Sciences in 1989, Fellow of the American Physical Society in 1991, and Fellow of the American Vacuum Society in 1994. In 2008 he was honored with a Humboldt Research Prize and Doctor Honoris Causa at the Marie Sklodovska-Curie University, Lublin, Poland. In 2012 he was appointed Fellow of Elettra Sincrotrone Trieste and in 2014 he received the Doctor Honoris Causa title from the University of Wrocław, Poland.

More information can be found:
on Ernst Bauer's website at Arizona State University: http://ernstbauer.physics.asu.edu/ and in Wikipedia: Ernst G. Bauer.

Preface

"Seeing is believing" has not always been true and still is not. When Galileo saw more than 400 years ago that there were some moons circling around Jupiter many astronomers and philosophers initially did not believe this. Four hundred years later most scientists see in the steady temperature rise and extreme weather events clear signs of human-caused climate change but some scientists and many people do not believe this. Fortunately exact science is fact-based and if something is seen and independently confirmed it is generally believed. Since Galileo's worldview-changing discoveries, seeing has expanded our worldview immensely beyond the narrow range that our eyes can recognize without instruments. Seeing with telescopes using electromagnetic waves ranging from X-rays to radio waves has given us deep insight into the universe. Microscopes using electromagnetic waves ranging from the infrared to X-rays have also opened our eyes to the microworld in and around us. Their wavelength-limited access to the nanoworld has been overcome by microscopes using electron waves, which now allow us to see down into the sub-nanometer world, an ability of utmost importance in modern technology, medicine, biology, and other disciplines.

The dimensions encountered in these fields are increasingly in the sub-micrometer range, which has the consequence that the ratio of surface to volume becomes important to such an extent that the surface either increasingly determines the properties of the material or *is* the material as in graphene and other so-called two-dimensional structures. Understanding these properties calls for methods which allow us to "see" the surface with all its properties, not only its geometric structure. Numerous methods have been developed in the last several decades with this goal, including field electron microscopy, field ion microscopy, and various scanning probe microscopies, which have given deep insight into the surface nanoworld.

This book describes one of these methods, cathode lens or immersion lens electron microscopy, which was born already in the early 1930s as a twin of its faster maturing brother, high energy electron microscopy. Because its aim is to image surfaces, which tend to be covered with a wide variety of surface contaminants, it did not come out of adolescence until ultrahigh vacuum and related cleaning methods became available. This opened the door to a new scientific discipline, surface science, which in turn stimulated the development of methods for seeing surfaces better than was possible using ultraviolet light in photoelectron emission microscopy (PEEM). Thus Low Energy Electron Microscopy (LEEM) was born in the early 1960s, motivated by the desire to see and believe what Low Energy Electron Diffraction (LEED) suggested. Later results of these developments are synchrotron radiation and pulsed laser excited photoemission electron microscopy, which now play an increasingly dominant role in cathode lens microscopy.

Just as we experience our environment multimodally by seeing, smelling, and hearing, we not only want to see surfaces but also to "smell" and "hear" them. What do they consist of and what are their properties? The smelling wish has been fulfilled by combining imaging and spectroscopy, primarily X-ray photoelectron spectroscopy, with the original LEEM instrument in the Spectroscopic Photo Emission and Low Energy Electron Microscope (SPELEEM). The hearing wish has been satisfied to some degree, too, for example in the understanding of magnetic microstructures, by banging on them with the angular momentum of photons in magnetic dichroism PEEM and electrons in Spin-Polarized Low Energy Electron Microscope (SPLEEM) and seeing their response. Thus by combining microscopy and diffraction with spectroscopy, and making full use of the properties of the electrons and photons, surface microscopy with slow electrons has grown far beyond what the fathers of cathode lens electron microscopy could have imagined in the 1930s. Today's instruments are no longer pure imaging systems but small laboratories for the analysis of the properties of surfaces, thin films, and nanostructures. As a result, full field cathode lens microscopy occupies a unique position in this field of materials science.

This book tries to lead the reader through the world of surface microscopy with slow electrons, starting with a brief recount of the history of the field before the advent of surface science (Chap. 1). This introduction shows the high level that the field had achieved in instrumentation before it faded away because of the lack of sufficiently good vacuum. Before continuing with the evolution of the instrumentation in the ultrahigh vacuum age, Chap. 2 describes the fundamental interactions of photons and electrons with matter, which are necessary to the understanding of the methods used in imaging. Chapter 3 is an overview of the wide variety of instruments and their components that have been developed, but not all of them completely. Since most researchers in surface microscopy work with only one type of instrument, this chapter hopefully will give them some ideas about other instruments of potential interest. In any case, many components such as the objective lens are common to all instruments and the user should be aware of their possibilities and limitations. The fundamental understanding of resolution and contrast in imaging with slow electrons is the subject of Chap. 4. It is necessarily based on wave optics, where coherence plays a fundamental role which distinguishes reflection from emission microscopy. This presentation demonstrates that, presently, the image detection system rather than the optics limits the resolution.

With this background the remaining chapters in the book describe the wide variety of applications of modern cathode lens electron microscopy. Applications occupy a large fraction of the book because only they can justify the human efforts and costs of developing the instruments as described in Chap. 3. These results should also give the reader not currently working in this field an idea of the capabilities and limitations of imaging with slow electrons. In the early years the emphasis was on surface science and work in this field is still continuing. Therefore Chap. 5, which covers this field, is by far the longest and most detailed. Applications to younger fields such as graphene and plasmonics, which are described in Chap. 6, are still evolving so that the book presents only their early phases. A case study at the end of this chapter illustrates that the cathode lens microscope is not only powerful for studying known materials but also for identifying unknown materials by making full use of its seeing and smelling capabilities. Magnetic imaging, largely based on X-ray photoemission, is not so young but has grown to such an extent that it deserved a separate treatment in Chap. 7. While most of the surface science studies in Chap. 5 make full use of the in situ capabilities of the microscopes, most of the work described in Chaps. 6 and 7 uses ex situ prepared samples and is combined with many complementary methods. Chapter 8 discusses briefly other surface imaging methods with electrons, which complement and/or compete with cathode lens electron microscopy and ends the book with some concluding remarks.

A few comments on the presentation of the material included in this book should be made. While the original intention was to cover all published work in the field, an intention still partially realized in Chap. 5, it soon turned out that this was not feasible, in particular because of the rapid growth of the number of publications in the fields covered in Chaps. 6 and 7. Nevertheless the book should give a good presentation of the state of the art in late 2013 shortly before the completion of the manuscript in early 2014, with the support of references to related reviews. The book does not intend to explain the scientific problems studied with cathode lens electron microscopy but only to show what it can contribute to their solution. For the science aspect of the problems the reader is referred to the references, which include the names of all authors and the complete titles of the publications. This should make it easier to decide which publications to read. No quality or importance criteria were used in their selection.

This book would never have been completed without the support of many people, foremost of my wife and coworker for more than 30 years, Anastassia Pavlovska. Without her encouragement it would never have been started and its completion would have been impossible without her strong steady support. She not only took care of all the figures, references, and permissions but also did a lot of editorial work, including correction of typographical and grammar errors, and making passages, which were difficult to read, more understandable. The amount and quality of work she did would justify including her as coauthor but then I could not have dedicated the book to her. Of all the other people whom I want to thank, I should first mention Ruud Tromp, Lothar Fritsche and John Spence, who helped me to understand a number of theoretical problems. There are many others who in some way or other contributed to this book. For figures I want to thank (listed alphabetically) M.S. Altman, R. Belkhou, J. Feng, K. Grzelakowski, E. Ilkova, M. Kiskinova, T. Koshikawa, M. Louwers (Philips), A. Locatelli, O. Menteş,

H. Ohldag, T. Osada (JEOL), A. Pang, T. Schmidt, S. Schramm and T. Yasue, some of whom have also contributed to the understanding of the background of the figures. In addition my thanks go to all the colleagues who have sent me their publication lists, preprints, and clarifications of their work. Finally I want to acknowledge the very important help in computer matters by J. Blackburn and the outstanding support by the library of my university and to thank David Packer and Sara Kate Heukerott (Springer), for their friendly and effective editorial guidance.

Tempe, AZ, USA
February, 2014

Ernst Bauer

Contents

List of Abbreviations

AEM	Auger electron microscopy
AES	Auger electron spectroscopy
AFC	Antiferromagnetic coupling
AFM	Antiferromagnetic, Atomic force microscopy (depending on context)
ARPES	Angle-resolved photoelectron spectroscopy
ARUPS	Angle-resolved ultraviolet photoelectron spectroscopy
CCD	Charge-coupled device
CFE	Cold field emission
CHA	Concentric hemispherical analyzer
CMOS	Complementary metal oxide semiconductor
CTF	Contrast transfer function
CVD	Chemical vapor deposition
CW	Continuous wave
DLD	Delay line detector
DQE	Detective quantum efficiency
EM	Electromagnetic
ESCA	Electron spectroscopy for chemical analysis
ESD	Electron-stimulated desorption
EXAFS	Extended X-ray absorption spectroscopy
FC	Ferromagnetic coupling
FEL	Free electron laser
FELPEEM	Free electron laser PEEM
FM	Ferromagnetic
FWHM	Full width at half maximum
HRTEM	High-resolution transmission electron microscopy
IMFP	Inelastic mean free path
ITR	Interferometric time resolution

LEED	Low energy electron diffraction
LEEM	Low energy electron microscopy
LEETS	Low energy electron transmission spectroscopy
LITD	Laser-induced thermal desorption
LSP	Localized surface plasmon
MCD	Magnetic circular dichroism
MCDPEEM	Magnetic circular dichroism photoemission electron microscopy
MCP	Multichannel plate
MEED	Medium energy electron diffraction
MEM	Mirror electron microscopy
MFM	Magnetic force microscopy
MIEEM	Metastable impact electron emission microscopy
MIES	Metastable impact electron spectroscopy
MLD	Magnetic linear dichroism
MTF	Modulation transfer function
NEXAFS	Near edge X-ray absorption fine structure
*n*PPE ($n = 2, 3, \ldots$)	*n*-Photon photoemission
*n*PPEEM	*n*-Photon photoemission electron microscopy
NPS	Noise power spectrum
OFET	Organic field effect transistor
PE	Photoelectron
PED	Photoelectron diffraction
PEEM	Photoemission electron microscopy
PFM	Piezoelectric force microscopy, piezoresponse force microscopy
PM	Paramagnetic
PSF	Point spread function
QE	Quantum efficiency
QSE	Quantum size effect
REM	Reflection electron microscopy
RHEED	Reflection high energy electron diffraction
RLD	Richardson-Laue-Dushman
SE	Secondary electron
SEEM	Secondary electron emission microscopy
SEM	Scanning electron microscopy
SEMPA	Scanning electron microscopy with polarization analysis
SLEEM	Scanning low energy electron microscopy
SP	Surface plasmon
SPELEEM	Spectroscopic photoemission and low energy electron microscopy

SPEM	Scanning photoemission microscopy
SPLEED	Spin-polarized low energy electron diffraction
SPLEEM	Spin-polarized low energy electron microscopy
SPP	Surface plasmon polariton
SREM	Scanning reflection electron microscopy
SRT	Spin reorientation transition
STEM	Scanning transmission electron microscopy
STM	Scanning tunneling microscopy
STXM	Scanning transmission X-ray microscopy
TCS	Total current spectroscopy
TE	Thermionic electron
TEEM	Thermionic electron emission microscopy
TEM	Transmission electron microscopy
TR-PEEM	Time-resolved photoemission microscopy
UHV	Ultrahigh vacuum
UPS	Ultraviolet photoelectron spectroscopy
UVMCDPEEM	Ultraviolet magnetic circular dichroism photoemission microscopy
UVPEEM	Ultraviolet photoemission microscopy
VLEED	Very low energy electron diffraction
VUV	Vacuum ultraviolet
XANES	X-ray absorption near edge structure
XAS	X-ray absorption spectroscopy
XLD	X-ray linear dichroism
XLDPEEM	X-ray linear dichroism photoemission microscopy
XMCD	X-ray magnetic circular dichroism
XMCDPEEM	X-ray magnetic circular dichroism photoemission microscopy
XMLD	X-ray magnetic linear dichroism
XMLDPEEM	X-ray magnetic linear dichroism photoemission microscopy
XNLD	X-ray natural linear dichroism
XNLDPEEM	X-ray natural linear dichroism photoemission microscopy
XPEEM	X-ray photoemission microscopy
XPS	X-ray photoemission spectroscopy
XSPEM	X-ray scanning photoemission microscopy
XUV	Extreme ultraviolet

Chapter 1
Introduction: History

Abstract This introductory chapter discusses the evolution of surface imaging with slow electrons from the beginnings to the time when imaging under ultrahigh vacuum (UHV) conditions was developed. It started in the early 1930s with thermionic electron emission microscopy (TEEM) and photo emission electron microscopy (PEEM), followed by mirror electron microscopy (MEM). World War II caused a considerable delay in the evolution of the field, but thereafter it flourished due to the development of sophisticated instruments and new methods such as ion bombardment-induced secondary electron emission microscopy (SEEM). The parallel evolution of transmission electron microscopy (TEM) and scanning electron microscopy (SEM), which gave higher resolution, led to a decline of emission and mirror microscopy. In addition the non-UHV conditions in the instruments used at this time were not compatible with the emerging field of surface science in the 1960s. The UHV requirements finally led to a new age in surface imaging with slow electrons. This is discussed in the subsequent chapters.

1.1 The Early Years

The late 1920s saw two important events in the physics of free electrons: (1) the discovery by Davisson and Germer [1] that slow electrons are diffracted by crystals similar to X-rays, confirming deBrogie's theory that electrons have wave nature and (2) Busch's calculations [2] of the magnetic fields of coils, which showed that coils act like lenses for electron beams. Had these two observations been combined at that time, it would have been the birth of electron microscopy. However, this had to wait for a few more years and came from a completely different side: from efforts to improve cathode ray oscillographs and thermionic electron emitters. Gábor, who received later (in 1971) the Nobel Prize in Physics for his invention and development of holography, had already developed earlier in his Ph.D. thesis in the

E. Bauer, *Surface Microscopy with Low Energy Electrons*,
DOI 10.1007/978-1-4939-0935-3_1, © Springer Science+Business Media New York 2014

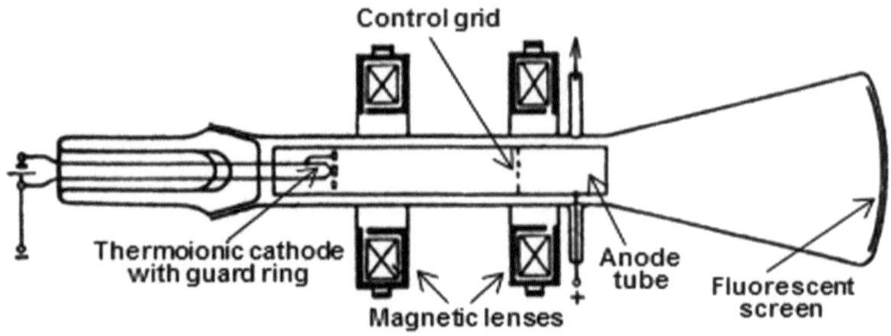

Fig. 1.1 Schematic of the magnetic thermionic electron emission microscope of Knoll et al. Adapted from Ref. [6] with permission from Springer Science + Business Media

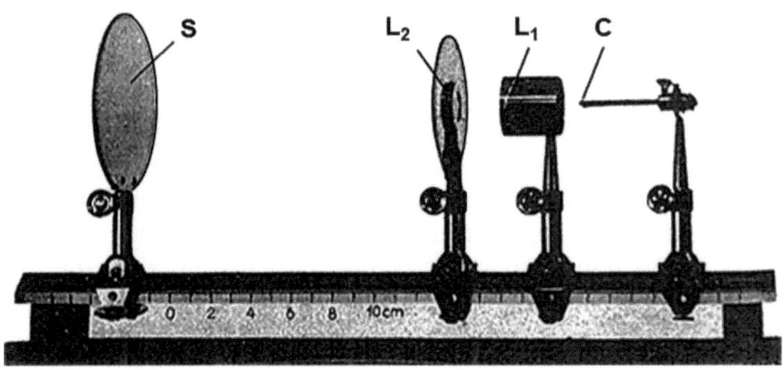

Fig. 1.2 Optical bench with Brüche's electrostatic emission microscope with the vacuum enclosure removed. Adapted from Ref. [8] with permission from Springer Science + Business Media

Electrotechnical Laboratory of the Technical University Berlin iron-encapsulated coils to bundle electrons [3]. Stimulated by Busch's work Knoll and Ruska [4, 5] used in the same institute such coils to image a "cold" cathode from which electrons were emitted upon bombardment with ions from a gas discharge and accelerated to high energies. Knoll et al. [6] also used such lenses (Fig. 1.1) to image hot cathodes with thermionic electron emission. Independently Brüche and Johannson [7, 8], who worked in an industrial research laboratory (AEG) and were interested in the understanding and improvement of thermionic electron emitters, were imaging such cathodes too. They used simple axially symmetric electric fields similar to those whose focusing properties Davisson and Calbick had calculated earlier [9, 10]. Figure 1.2 [8] shows a photo of their experimental setup (without vacuum enclosure) and Fig. 1.3 [8] the geometry and potential distribution, which they had determined in an electrolytic tank. In the early studies the vacuum was produced by mercury diffusion pumps and the vacuum chambers were made of glass. While Knoll and Ruska's work may be considered as the origins of transmission electron microscopy (TEM) with fast electrons, that of Brüche is the beginning of surface microscopy with

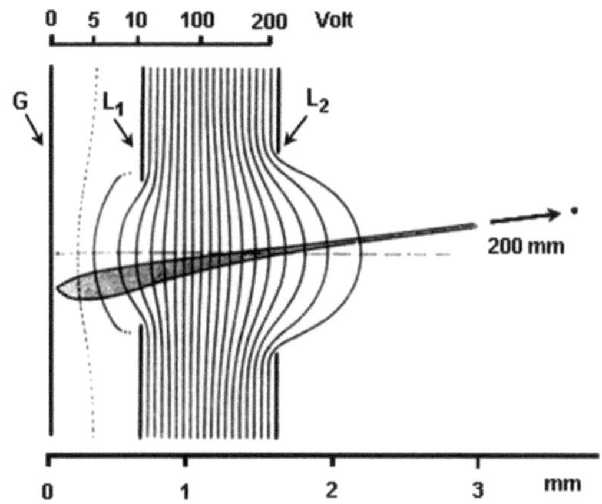

Fig. 1.3 Potential distribution of an electrostatic triode immersion lens and electron bundle emitted from an object feature, which is imaged on the fluorescent screen in 200 mm distance from the object G. The electrodes L_1 and L_2 are at the potentials shown on *top*. Adapted from Ref. [8] with permission from Springer Science + Business Media

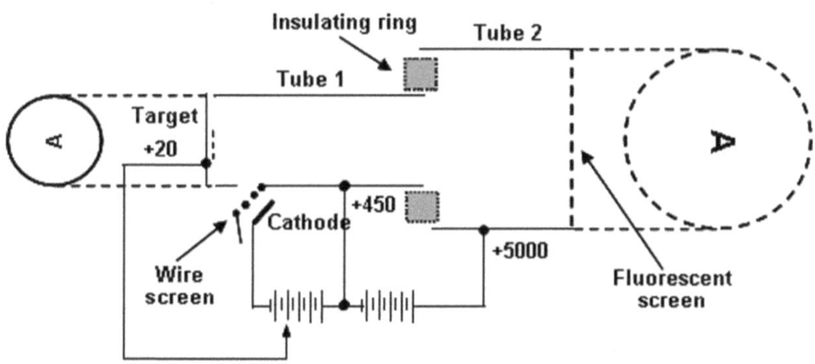

Fig. 1.4 Schematic of Zworykin's "secondary electron" emission microscope. The lens consists of two tubes at the indicated potentials. The A's in the full and dashed circles indicate the object on the target and the image on the fluorescent screen. Adapted with permission from Ref. [12]. Copyright 1933 Elsevier

slow electrons, the subject of this book. Of course, in the early years the magnification and resolution of the images were far below that of light-optical microscopes and it was not until 10 years later that electron microscopy surpassed light microscopy in these respects. In the following we sketch this evolution.

In 1933 Brüche demonstrated that a surface could also be imaged by photoelectric emission [11] and Zworykin showed that imaging was also possible by electron bombardment of the surface [12]. Zworykin used the system shown in Fig. 1.4 and a bombardment energy of 20 eV. Although he called it secondary electron imaging we know today that at this energy the secondary electron yield is still small compared to the reflectivity so that the contrast in his image was caused by the different reflectivity of the materials on the cathode (C and Ni). Also in 1933 Johannson published the first systematic studies of three-electrode immersion objective lenses and

Fig. 1.5 Thermionic electron emission image of a Ba-Sr oxide-activated polycrystalline Ni surface after annealing at 1150 °C. Reproduced from Ref. [17] with permission from Springer Science + Business Media

discussed their aberrations [13]. These studies showed already the importance of high field strength at the cathode for good resolution. In the following year he complemented these studies with measurements of four-electrode immersion objectives [14]. These measurements, which were made on thin plates with concentric round holes, were important for later more sophisticated lens designs which allowed higher fields at the cathode. Using such an electrostatic immersion lens, Behne [15] demonstrated that imaging was possible not only with near-zero energy electrons as in thermionic and photoelectric emission but also with electrons as fast as 500 eV, transmitted through a holy film cathode. This was possible because the electrons had a very narrow energy distribution and were nearly parallel to the optical axis. The state of the art 2 years after the invention of emission electron microscopy [16, 17] is illustrated in Fig. 1.5.

In 1934 [18] and 1935 [19] the first calculations of the chromatic aberration of simple immersion systems, consisting of a homogeneous acceleration field and an aberration-free thin lens, appeared. A few years later Langmuir calculated the spherical aberration of the homogeneous acceleration field [20]. In 1936 Boersch [21] introduced wave-optics into the discussion of image formation, though for transmission microscopy, and found that insertion of an aperture into the back focal plane of the imaging lens increased the image sharpness. He attributed this to the reduction of the influence of the spherical and chromatic aberration of the lens by the aperture. In the same year Recknagel [22, 23] published his theoretical studies of the properties of electron mirrors and showed that their imaging properties were reverse to those of lenses. This will be discussed in Chap. 3. Imaging with an electron mirror was experimentally demonstrated in 1937 by Hottenroth [24] by replacing the electrostatic einzel lens M shown in Fig. 1.6 with a charged plate surrounded by a charged cylinder. The first systematic study of metal and semiconductor surfaces with mirror electron microscopy (MEM) was made in 1939 by Orthuber but published much later [25]. He compared contrast formation in MEM with that in thermionic electron emission and used MEM with a semiconducting cathode as an image converter for infrared. Mahl [26] showed that

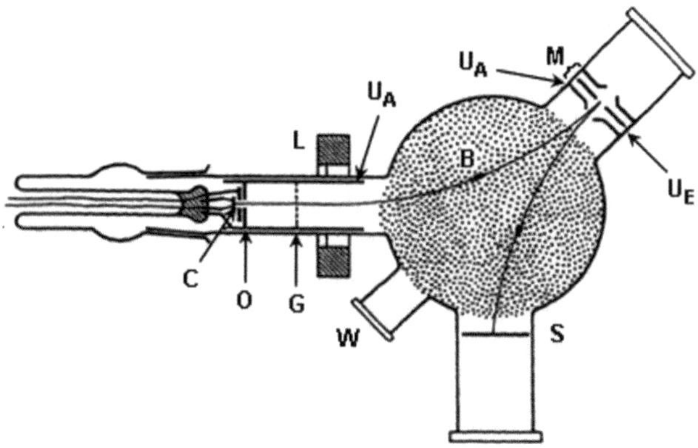

Fig. 1.6 Schematic of Hottenroth's mirror microscope. C: cathode, O: immersion objective lens, G: grid to be imaged, L: magnetic lens, B: magnetic cross field, M: mirror, S: fluorescent screen, W: observation window, U_A: acceleration potential, U_E: mirror potential. Adapted from Ref. [24] with permission from Wiley-VCH Verlag GmbH & Co. KGaA, Weinheim

the emission distribution from tunnel (also called field emission) cathodes could be imaged with an immersion lens. In addition to the cited work there have been more studies of the electron optics and many applications, mainly using thermionic emission in the years up to the breakthrough through the light microscope resolution barrier. Most of them came from Brüche's AEG Forschungsinstitut in Berlin but some also from other groups. Examples are Refs. [27–30]. In Ref. [30] true secondary electron emission excited by electron bombardment was used for the first time. Much of this work is cited in early reviews [17, 31] and in a more recent one [32].

In 1941 Recknagel [33] published a rigorous geometrical optical theory of the immersion lens for arbitrary axially symmetric potential distribution $U(z,r)$ and emission energies ε small compared to the final energy eU_o of the electrons. For the specific case of the homogeneous acceleration field and assuming that all emitted electrons contribute to the image he obtained a resolution of $4\varepsilon/eF$, where F is the field strength. For a mean thermionic emission energy of 0.1 eV and field strength of 10 kV/cm this gives a resolution of 400 nm, a value much lower than what had been achieved at that time. It was obvious that the resolution could be improved considerably compared to the low micrometer range achieved previously with fields of 1–2 kV/cm by significantly increasing F. Indeed, in 1942 three authors, Boersch [34], Kinder [35] and Mecklenburg [36], succeeded in reaching with thermionic emission a resolution below the theoretical predictions: 70 nm, 90 nm and 40 nm with field strengths of 30 kV/cm, 100 kV/cm and 40 kV/cm and improved three-electrode lenses (electrostatic "triodes") [34, 36] and a magnetic lens [35], respectively. In the same year Boersch [37] explained theoretically the improvement of the resolution in emission microscopy by insertion of an aperture in the back focal plane of the lens, which he had observed 6 years earlier in transmission microscopy [21], by the reduction of the spherical aberration and by diffraction at the aperture. He also demonstrated for the first time dark-field imaging with diffracted beams (in transmission microscopy) [21, 38]. Brüche summarized the understanding of the resolution of the emission electron microscope in 1942 [39]. In 1943 Recknagel [40] solved the Schrödinger equation for the homogeneous acceleration field and determined the resolution wave-optically assuming an aberration-free lens following the acceleration field, no emission angle limitation, and thermal emission with energy width $\Delta\varepsilon << \varepsilon$. The results gave a limiting resolution proportional to $F^{-1/4}$ and to $\varepsilon^{-1/8}$ or $\lambda^{1/4}$. For Mecklenburg's instrument he predicted a resolution of 5 nm at $F = 80$ kV/cm. Finally a theoretical paper by Artsimovich [41] marked the end of the first phase of immersion lens electron microscopy because the increasing intensity of World War II stopped all nonmilitary research and development including the work in the center of surface electron microscopy in Berlin.

1.2 The Postwar Revival

Recknagel's expression dominated emission microscopy for more than 40 years, with prefactors ranging from the original 4 down to 0.5, depending upon the definition of point resolution, location of the image plane, smaller aberrations due to improvements in the lens design, and the use of an angle-limiting aperture. Immersion lens microscopy resurged again only in the early 1950s when research and development resumed in several groups in France, the Netherlands, Germany, Russia and the USA. In the meantime high energy TEM, which had advanced much more than low energy surface microscopy already before the war, had reached a high level of sophistication, so that low-energy surface microscopy could build on all the experience of the TEM in electron optics, vacuum, and mechanical design. Indeed, one of the first instruments was produced by a company that built TEM instruments, Philips Industries in Eindhoven [42, 43]. It was for some time the most sophisticated instrument and could be operated in a pure electrostatic triode mode to a nearly pure magnetic diode mode (Fig. 1.7). The early glass systems were replaced by metal systems, mercury diffusion pumps by oil diffusion pumps, picein wax and vacuum grease by rubber gaskets, etc., and samples could be exchanged by airlocks. Independent developments went on in the early 1950s in several groups: in France in Septier's group in Paris [44–46], in Germany in Möllenstedt's group in Tübingen [47], both with ion bombardment-induced secondary electron emission, in Paris by Huguenin [48–50] with photoelectron emission, and in the USA in the Bell Laboratories by Heidenreich [51] with thermionic emission. In Russia Spivak's group in Moscow still used simple glass systems with which it nevertheless succeeded to obtain images of the magnetic stray fields above ferromagnetic cathodes using electron bombardment-induced secondary electron emission [52, 53]. His group also used the first tetrode objective lens and the first pulsed system [54].

In the second half of the 1950s and in the 1960s other groups entered the field and the existing groups developed additional imaging modes. In Toulouse Fert's

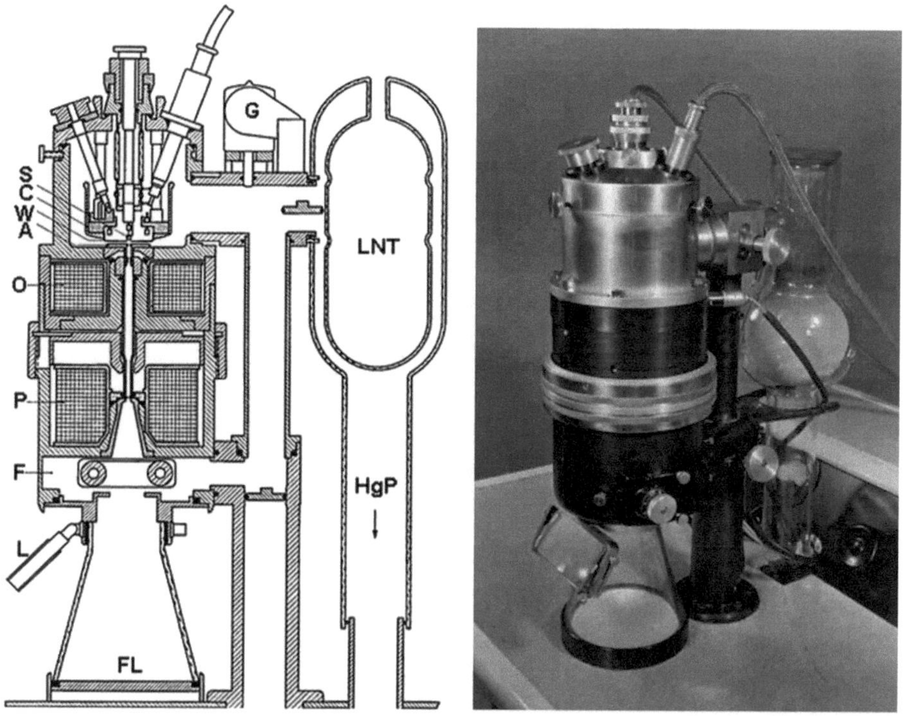

Fig. 1.7 The Philips magnetic thermionic emission electron microscope. S: specimen, C: cathode enclosure, W: Wehnelt cylinder, A: anode, O: objective lens, P: projective lens, F: film camera, L: lens for observation of fluorescent screen FL, G: Penning gauge, LNT: liquid nitrogen trap, HgP: mercury diffusion pump. Adapted from Ref. [43] with permission from Philips Company Archives

group built an instrument with a magnetic diode objective lens and a contrast aperture, which allowed a point resolution of 25 nm in ion bombardment-induced secondary electron (SE) imaging [55–57]. The groups of Bethge in Halle [58], Bas in Zürich [59, 60], Drahoš in Brno [61, 62], Ruska in Berlin [63, 64], and Maruse in Nagoya [65] built increasingly better instruments, and several companies, such as Leitz in Wetzlar [66], VEB Carl Zeiss in Jena (EF-Z6) [67, 68], Balzers in Liechtenstein [69, 70], JEOL in Tokyo [71], Ford Motor Company in Dearborn [72], or KYKY in Beijing [73], developed emission microscopes, some of them commercially quite successful and long-lived, for example the Balzers KE-3 (Fig. 1.8) photoemission microscope. Wegmann's group [74] in particular used this instrument to demonstrate the possibilities of photoemission electron microscopy (PEEM) in materials science, including insulating materials, whose charging they suppressed by coating with carbon. Möllenstedt's group improved imaging with ion bombardment-induced secondary electron emission [75–77] and extended it to electron bombardment-induced secondary electron imaging [78], X-ray photoemission imaging [79], and UV photoemission imaging [80]. Some other well-designed instruments built in this period can be found in Ref. [81]. Parallel to these

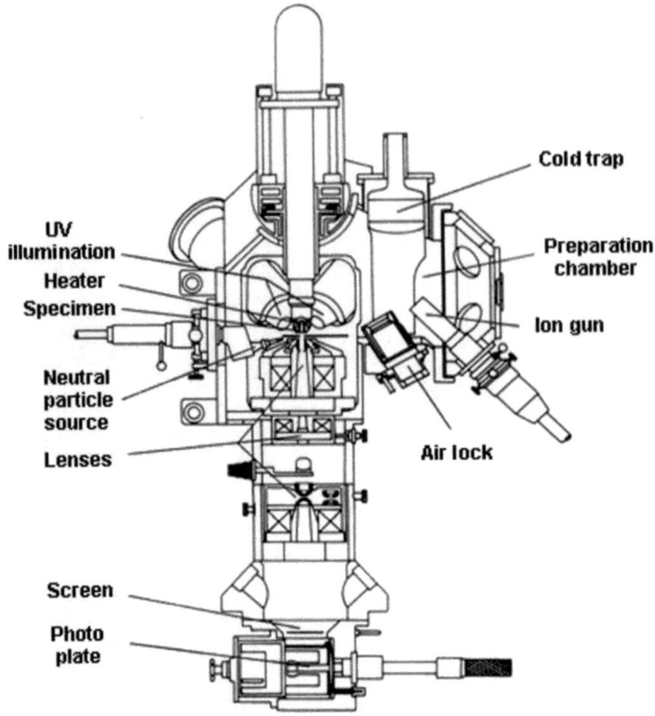

Fig. 1.8 Schematic of the Balzers Metioscope KE 3 photoelectron emission microscope. Adapted from Ref. [74] with permission from Wiley-Blackwell

developments also new imaging methods were developed such as field emission or tunneling microscopy through thin insulating layers [82], and the theory of image formation was significantly extended [83–86].

While the instrument developments just described led to considerable improvements of the resolution, for example to 10–15 nm line resolution in thermionic and ion bombardment-induced secondary emission imaging [67, 68] or 12 nm point resolution in photoemission [63, 64], other limitations became increasingly apparent. One was the wide energy distribution in ion bombardment-induced secondary electron imaging, the most frequently used method in order to keep the surface clean by simultaneous sputtering. This problem was attacked by energy filtering, not only with the contrast aperture, as already introduced by Boersch [34], but also a filter lens [87], and a biprisma-mirror analyzer [88] (Fig. 1.9) were used. While the first one eliminates the low-energy part of the spectrum, the second one eliminates the high-energy part. The second problem was that fluorescent screens were not sensitive enough and had too large grains so that sensitive in-vacuum fine-grained photographic plates or film had to be used to achieve the high resolution of the electron optics. Their outgassing limited the vacuum and contributed to the specimen contamination. The introduction of Channeltron electron multiplier arrays

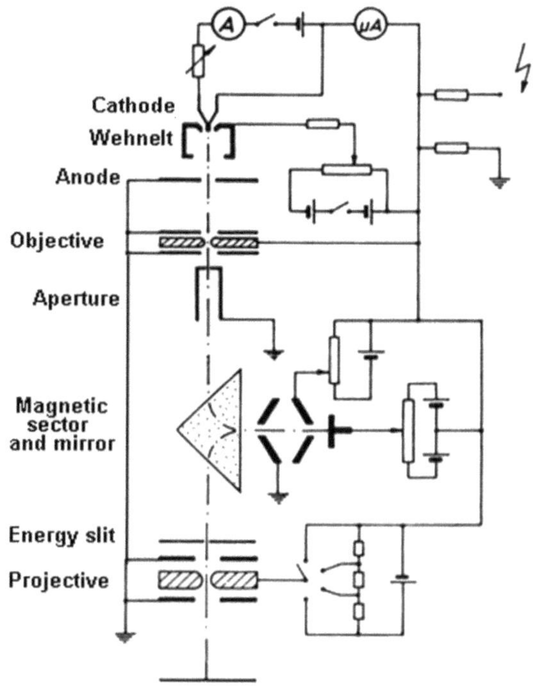

Fig. 1.9 Schematic of high-pass energy filtering with a Castaing-Henry filter (*dotted triangle*) and an electron mirror (*wide dark lines*) used by Gruner et al. to monochromatize thermally emitted electrons. Adapted with permission from Ref. [88]. Copyright 1971 Urban and Fischer (Elsevier) (The Netherlands)

(CEMA), now usually called multichannel plates (MCP), with gains of 10^7, coupled to fiber optics windows, solved this problem [89, 90]. The major problem was however the high residual gas pressure in the instruments which caused rapid specimen contamination. It could be avoided only by continuous sputtering during imaging which made ion bombardment-induced secondary electron microscopy the most popular imaging method at that time. First steps to alleviate this problem were liquid nitrogen cooling to trap condensable gases such as the hydrocarbons from the oil diffusion pumping system or greased gaskets, and Ti sublimation, which reduced the base pressure into the 10^{-8} Torr range without baking the system [91].

In the early 1960s clean all-metal ultrahigh vacuum (UHV) systems with oil-free pumps such as the electrostatic Orbion pump, magnetic sputter ion pumps, or Ti sublimation pumps and Cu gaskets became available and were increasingly used in microscopes. After bakeout pressures in the 10^{-10} Torr range could be obtained and the specimens could be kept clean during imaging [92, 93]. The most sophisticated PEEM of this type was however not used for surface studies but for the study of biological specimens [94, 95] (Fig. 1.10). The main driving force towards UVH was however the new field of surface science that emerged in the early 1960s. The author started with the development of low energy electron microscopy (LEEM) in 1962, a subject that will be discussed later. In the early 1970s others combined emission microscopy with laterally averaging surface analysis methods in order to

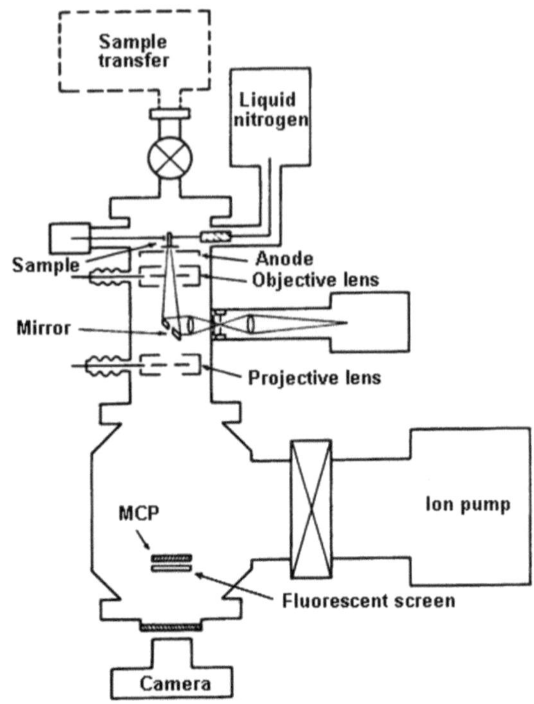

Fig. 1.10 Schematic diagram of the ultrahigh vacuum electrostatic photoelectron emission microscope of Griffith et al. The sample is illuminated through the objective lens. Adapted from Ref. [94] with permission from O.H. Griffith

characterize the imaged surfaces crystallographically and chemically via low energy electron diffraction (LEED) and Auger electron spectroscopy (AES). Delong et al. [96, 97] added LEED to their emission microscope and discussed the possibility of AES in their instrument [98]. Recknagel et al. combined their microscope with LEED and reflection high energy electron diffraction [99]. These efforts represent the transition to true UHV surface microscopy of well-defined surfaces, which did not become viable until the early 1980s and will be discussed in later chapters. More references to pre-UHV emission microscopy with instrument descriptions and discussions of the image formation and many applications can be found in Refs. [32, 100–102].

Also mirror microscopy flourished in the 1960s and 1970s, starting already in the early 1950s in the USA [103], Germany [104, 105], and Russia [106] with instruments of the beam separator type used by Hottenroth [24] and Orthuber [25] before World War II. Although this type of instrument was built and used for many years (see, e.g., Refs. [107–109]), it was soon displaced by straight beam systems, first by Mayer [110] and subsequently by other groups [111–121]. This was done in part for their simplicity and in part in order to avoid the image distortions caused by the simple magnetic deflection fields used to separate incident and reflected beam. The incorporation of beam separators with low astigmatism [122] into mirror microscopes [123, 124] (Fig. 1.11 [124]) revived the interest in separator-type instruments again. The most sophisticated separator-type instrument, built in the 1960s

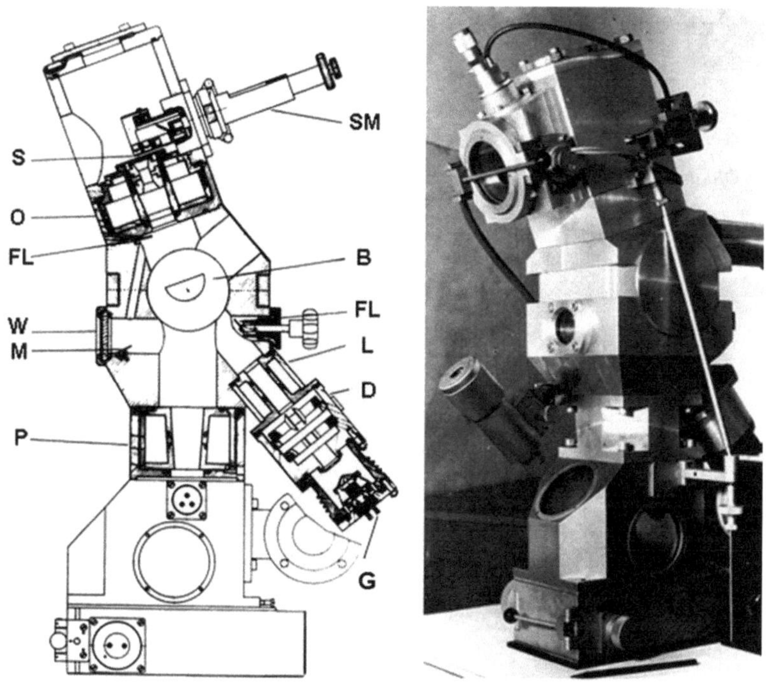

Fig. 1.11 Mirror electron microscope of Bethge and Heidenreich. G: electron gun, D: alignment deflectors, L: condenser lens, FL: fluorescence screens, B: beam separator, SM: specimen manipulator, S: specimen, O: objective lens, W: window, M: mirror, P: projective lens. Modified from Ref. [124]

by Bok at Delft University, reduced the astigmatism by using smaller deflection angles and several deflections to obtain minimum chromatic aberration [125, 126].

With these improved separator-type instruments focused mirror images could now be obtained by focusing the incident beam into the back focal plane of the objective lens and inserting there a contrast aperture. This was first demonstrated in 1964 by deflecting the beam in a straight beam (projection)-type system [127] and subsequently in more detail in a beam separator system [128]. All previous separator-type instruments had no contrast aperture and insufficient focusing so that only out-of-focus, point projection or shadow or "Schlieren" images could be obtained, just like in the straight beam instruments. Nevertheless a large amount of information was obtained on electrical and magnetic stray fields from magnetic and ferroelectric surfaces and potential variations on semiconductor surfaces, e.g., at p–n junctions. In particular, Mayer's application examples such as to recorded patterns on magnetic tape [129] made mirror microscopy as popular as emission microscopy, despite the fact, realized already early by geometrical optics calculations [130] and experimentally [131], that it was in general difficult to extract quantitative information on surface topography from the image contrast. A wave-optical treatment of the

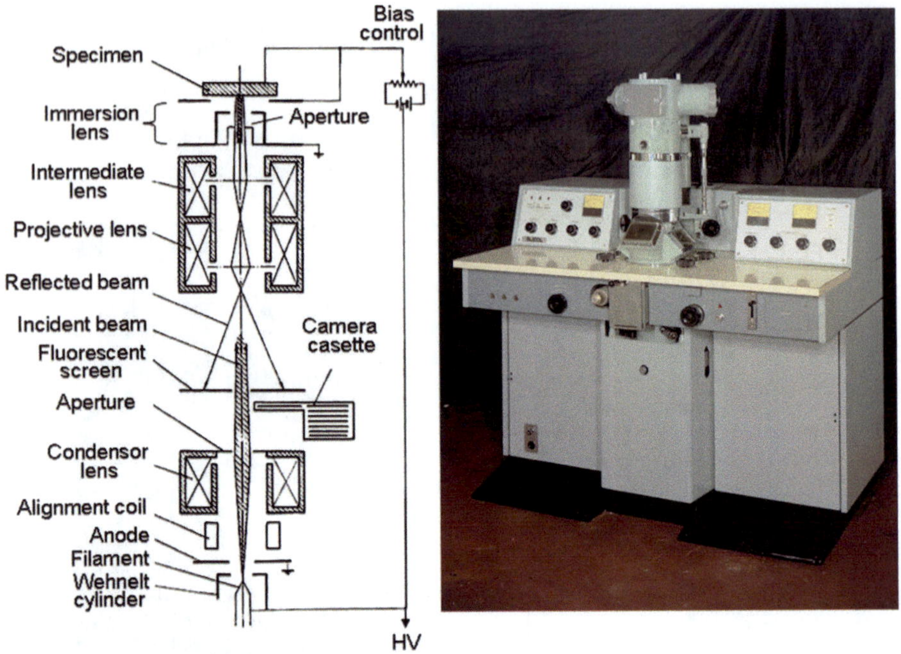

Fig. 1.12 JEOL JEM-M1 electron mirror microscope. Adapted from Ref. [134] with permission from JEOL and Courtesy of T. Osada (JEOL)

image formation for the case of parallel illumination, that is for focused imaging, predicted a resolution of 8 nm at field strength of 10 kV/mm [132], a value which, however, has not been demonstrated experimentally. Wave-optical calculations for the case of projection imaging gave for the same field strength a resolution of 84 nm [133], in approximate agreement with the experimental value of 70 nm reported in Ref. [117]. The interest in mirror microscopy stimulated also commercial instruments, a straight beam system built by JEOL in Japan [134, 135] and illustrated in Fig. 1.12 and a separator-type instrument manufactured by the Polish Optical Factory PZO [136].

Beginning in the 1970s the evolution of surface science led to the development of UHV systems which combined mirror imaging with other surface analysis methods such as LEED, work function, and reflectivity measurement at the expense of electron-optical sophistication [137–150]. Finally, also a sophisticated mirror interference microscope with a field emission gun and a Castaing-type biprism beam separator was developed and used to measure surface roughness and potential steps. For a sinus-shaped height distribution with a period of 1 μm a height resolution of 0.2 nm was reported and for an array of 2.5 nm high Au patterns a height resolution of 0.05 nm, and potential steps of 1 mV could be resolved [151, 152]. By the end of the 1970s the interest in mirror microscopy had disappeared except in connection with surface science studies until it reemerged around 1985 as a subsidiary method in LEEM. However in the about 20 years in which this method was applied it reached a high level of sophistication and made important contributions to materials science and technology despite the generally poor vacuum. These contributions and more details on the instrumentation and image formation can be found in a number of reviews [32, 153–158]. The proceedings of the first (and last) international conference on emission electron microscopy [159] contain an extensive list of references up to 1979, with emphasis on PEEM.

References

1. Davisson, C.J., Germer, L.H.: Diffraction of electrons by a crystal of nickel. Phys. Rev. **30**, 705–740 (1927)
2. Busch, H.: Rechnung der Bahn von Elektronenstrahlen im axialsymmetrischen elektromagnetischen Feld. Ann. Phys. **81**, 974–993 (1926)
3. Gábor, D.: Oszillographieren von Wanderwellen mit dem Kathodenoszillographen. In: Matthias, A. (ed.) Forschungshaefte der Studiengesellschaft für Höchspannungsanlagen, vol. 1, pp. 7–24. Vereinigung Elektrizitätswerke, Berlin (1927)
4. Knoll, M., Ruska, R.: Beitrag zur geometrischen Elektronenoptik. Ann. Phys. **12**, 607–661 (1932)
5. Knoll, M., Ruska, R.: Das Elektronenmikroskop. Z. Physik **78**, 318–335 (1932)
6. Knoll, M., Houtermans, F.G., Schulze, W.: Untersuchungen der Emissionsverteilung an Glühkathoden mit dem magnetischen Elektronenmikroskop. Z. Physik **78**, 340–362 (1932)
7. Brüche, E.: Elektronenmikroskop. Naturwissenschaften **20**, 49 (1932)
8. Brüche, E., Johannson, H.: Elektronenoptik und Elektronenmikroskopie. Naturwissenschaften **20**, 353–358 (1932)
9. Davisson, C.J., Calbick, C.J.: Electron lenses. Phys. Rev. **38**, 585 (1931)
10. Davisson, C.J., Calbick, C.J.: Electron lenses. Phys. Rev. **42**, 580 (1932)
11. Brüche, E.: Elektronenmikroskopische Abbildung mit lichtelektrischen Elektronen. Z. Physik **86**, 448–450 (1933)
12. Zworykin, V.K.: On electron optics. J. Franklin Inst. **215**, 535–555 (1933)
13. Johannson, H.: Über das Immersionsobjektiv der geometrischen Elektronenoptik. Ann. Phys. **18**, 385–413 (1933)
14. Johannson, H.: Über das Immersionsobjektiv der geometrischen Elektronenoptik II. Ann. Phys. **21**, 274–284 (1934)
15. Behne, R.: Die Eigenschaften des Immersionsobjektivs für die Abbildung mit schnellen Elektronen. Ann. Phys. **26**, 372–384 (1936)
16. Knecht, W.: Das kombinierte Licht- und Elektronenmikroskop, seine Eigenschaften und seine Anwendung. Ann. Phys. **20**, 161–184 (1934)
17. Ramsauer, C.: Elektronenmikroskopie. Bericht über Arbeiten des AEG Forschungs-Instituts 1930 bis 1942. Springer, Berlin (1943)
18. Brüche, E., Scherzer, O.: Geometrische Elektronenoptik. Springer, Berlin (1934)
19. Henneberg, W., Recknagel, A.: Der chromatische Fehler bei elektronenoptischen Anwendungen, insbesondere beim Bildwandler. Z. Techn. Phys. **16**, 230–235 (1935)
20. Langmuir, D.B.: Theoretical limitations of cathode-ray tubes. Proc. Inst. Radio Eng. **25**, 977–991 (1937)
21. Boersch, H.: Über das primäre and sekundäre Bild im Elektronenmikroskop. I. Eingriffe in das Beugungsbild und ihr Einfluss auf die Abbildung. Ann. Phys. **26**, 631–644 (1936)
22. Recknagel, A.: Zur Theorie des Elektronenspiegels. Z. Techn. Physik **17**, 643–645 (1936)
23. Recknagel, A.: Zur Theorie des Elektronenspiegels. Z. Physik **104**, 384–394 (1936)
24. Hottenroth, G.: Untersuchungen über Elektronenspiegel. Ann. Phys. **422**, 689–712 (1937)

25. Orthuber, R.: Über die Anwendung des Elektronenspiegels zum Abbilden der Potentialverteilung auf metallischen und Halbleiteroberflächen. Z. Angew. Phys. **1**, 79–89 (1948)
26. Mahl, H.: Feldemission aus geschichteten Kathoden bei Elektronbestrahlung. Z. Techn. Phys. **18**, 559–563 (1937)
27. Heinze, W., Wagener, S.: Die Vorgänge bei der Aktivierung von Oxydkathoden. Z. Techn. Phys. **17**, 645–653 (1936)
28. McMillen, J.H., Scott, G.H.: A magnetic electron microscope of simple design. Rev. Sci. Instrum. **8**, 228–290 (1937)
29. Burgers, W.G., Ploos van Amstel, J.J.A.: Electronoptical observation of metal surfaces. Physica **4**, 5–14 (1937)
30. Meschter, E.: An electron microscope for studying thermal and secondary electron emission. Rev. Sci. Instrum. **9**, 12–15 (1938)
31. Brüche, E., Recknagel, A.: Elektronengeräte. Prinzipien und Systematik. Springer, Berlin (1941)
32. Griffith, O.H., Engel, W.: Historical perspective and current trends in emission microscopy, mirror electron microscopy, and low energy electron microscopy. Ultramicroscopy **36**, 1–28 (1991)
33. Recknagel, A.: Theorie des elektrischen Elektronenmikroskops für Selbststrahler. Z. Physik **117**, 689–708 (1941)
34. Boersch, H.: Erhöhung der Auflösung im Emissions-Elektronenmikroskop. Naturwissenschaften **30**, 120 (1942)
35. Kinder, E.: Emissions-Übermikroskopie mit magnetischen Linsen. Naturwissenschaften **30**, 591–592 (1942)
36. Mecklenburg, W.: Über das elektrostatische Emissions-Übermikroskop. Z. Physik **120**, 21–32 (1942)
37. Boersch, H.: Die Verbesserung des Auflösungsvermögens im Emissions-Elektronenmikroskop. Z. Techn. Phys. **23**, 129–130 (1942)
38. Boersch, H.: Über das primäre and sekundäre Bild im Elektronenmikroskop. II. Strukturuntersuchung mittels Elekktronenbeugung. Ann. Phys. **27**, 75–80 (1936)
39. Brüche, E.: Die Auflösungsgrenze des Emissions-Elektronenmikroskops. Kolloid Z. **100**, 192–206 (1942)
40. Recknagel, A.: Das Auflösungsvermögen des Elektronenmikroskops für Selbststrahler. Z. Physik **120**, 331–362 (1943)
41. Artsimovich, L.A.: Electron-optical properties emission-systems. Izv. Akad. Nauk. SSSR Ser. Fiz. **8**, 313–329 (1944) (in Russian)
42. Rathenau, G.W., Baas, G.: Grain growth in a texture. Studied by means of electron emission microscopy. Physica **17**, 117–128 (1951)
43. Baas, G., Rathenau, G.W.: An emission electron microscope for research at high temperatures. Philips Tech. Rev. **18**, 1–10 (1956/1957)
44. Gauzit, M., Septier, A.: Le microsope électronique a émission en metallographie. Bull. Microsc. Appl. **1**, 109–118 (1951)
45. Septier, A.: Sur l'obtention d'images électroniques de surfaces bombardées par des ions. C. R. Acad. Sci. Paris **237**, 231–233 (1953)
46. Septier, A.: Quelque types d'objectifs électrostatiques a immersion a fort grandissement. Ann. Radioélectricité **9**, 374–410 (1954)
47. Möllenstedt, G., Düker, H.: Emissionsmikroskopische Oberflächenabbildung mit Elektronen, die durch schrägen Ionenbeschuß ausgelöst wurden. Optik **10**, 192–205 (1953)
48. Huguenin, E.L.: Un essai de microscopie électronique à photoémission. C. R. Acad. Sci. Paris **239**, 404–406 (1954)
49. Huguenin, E.L.: Observation des structures cristallines au microscope électronique à photoémission. C. R. Acad. Sci. Paris **241**, 307–309 (1955)
50. Huguenin, E.L.: Sur le microscope électronique à photoémission. Ann. Phys. **2**, 214–266 (1957)
51. Heidenreich, R.D.: Thermionic emission microscopy of metals. J. Appl. Phys. 26, 757–765 (Part I) and 879–889 (Part II) (1955)

52. Spivak, G.V., Kanavina, N.G., Chernishev, I.N., Sbitnikova, I.S.: Electron-optical method for imaging heterogeneous magnetic objects. Dokl. Akad. Nauk. SSSR (in Russian) **92** (541–543) (1953)
53. Spivak, G.V., Kanavina, N.G., Sbitnikova, I.S., Dombrovskaya, T.N.: Imaging ferromagnetic domains by electron-optical method. Dokl. Akad. Nauk. SSSR (in Russian) **105**(706–708) (1955)
54. Prilejaeva, I.N., Lifshits, V.V., Spivak, G.V.: Electron optical study of nonstationary emission of oxydcathode in vacuum and in gas. Zh. Tech. Fiz. (in Russian) **25**, 97–107 (1955)
55. Fert, C., Simon, R.: Amélioration du pouvoir de resolution du microscope électronique à émission. C. R. Acad. Sci. Paris **243**, 1300–1303 (1956)
56. Fert, C., Simon, R.: Pouvoir séparateur d'un microscope électronique à émission en presence d'un diaphragme, et spectre de vitesse des électrons transmis. C. R. Acad. Sci. Paris **244**, 1177–1179 (1957)
57. Fert, C., Simon, R.: Progrès récents en microscopie électronique à emisión. J. Phys. Radium **18**, 87–89 (1957)
58. Bethge, H., Eggert, H., Herbold, K.: Über einige anwendungstechnsche Erfahrungen mit einem Emissions-Mikroskop. In: Bargmann, W., et al. (eds.) Proceedings of the 4th International Conference on Electron Microscopy, Berlin 1958, pp. 217–222. Springer, Berlin (1960)
59. Bas, E.B.: De la microscopie à émission d'électrons, et de son application en métallurgie. Le Vide **96**, 303–316 (1961)
60. Aeschlimann, R., Bas, E.B.: Ein universal-emissions-elektronenmikroskop. In: Sydney, J., Breese, S.S. (ed.) Proceedings of the 5th International Congress Electron Microscopy, Philadephia 1962, pp. D-9. Academic Press, New York (1962)
61. Delong, A., Drahoš, V.: Ein Kombiniertes Emissions-Elektronenmikroskop. In: Titlbach, M. (ed.) Proceedings of the 3rd European Regional Conference on Electron Microscopy, Prague 1964, pp. 25–26. Academia, Prague (1964)
62. Delong, A., Drahoš, V.: On some features of a new electrón emisión microscope. J. Sci. Instrum. **1**, 397–400 (1968)
63. Engel, W.: Emission microscopy with different kinds of electron emission. In: Ueda, R. (ed.) Proceedings of the 6th International Congress on Electron Microscopy, Kyoto 1966, pp. 217–218. Maruzen, Tokyo (1966)
64. Engel, W.: Entwicklung eines Emisssionsmikroskops höher Auflösung. Ph.D. thesis, Freie Universität Berlin (1968)
65. Uchikawa, Y., Kojima, M., Ichihashi, M., Maruse, S.: The ion-induced emission electron microscope and an image contrast due to specimen contamination. Jpn. J. Appl. Phys. **8**, 436–442 (1969)
66. Bartz, G.: Über ein Elektronenmikroskop und ein Verfahren zur direkten Sichtbarmachung von isolierten Oberflächen. In: Bargmann, W., et al. (eds.) Proceedings of the 4th International Conference on Electron Microscopy, Berlin 1958, pp. 201–207. Springer, Berlin (1960)
67. Soa, E.-A.: Entwicklung eines Emissionsmikroskops und Anwendung auf einige Fragen der Phasengrenzvorgänge bei der Metalloxydation. In: Jenaer Jahrbuch 1963, vol. II. pp. 427–486. VEB Gustav Fischer, Jena (1964)
68. Soa, E.-A.: Ein neues Emissionselektronenmikroskop. Optik **22**, 66–76 (1965)
69. Wegmann, L.: Auf dem Weg zum Metall-Elektronenmikroskop (Progress toward metallurgical electron microscope). Prakt. Metallogr. **5**, 241–263 (1968)
70. Wegmann, L.: Photoemissions-Elektronenmikroskopie. In: Müller, E.A.W. (ed.) Handbuch der Zerstörungsfreien Werksstoffprüfung, vol. 3, pp. 1–23. Oldenbourg Verlag, München (1969)
71. Ion Bombardement Secondary Emission Electron Microscope, Model JEM-E3 In: JEOL News, vol. 3, No. 1. pp. 3–4 (1965)
72. Eichen, E., Forgacs, R.L., Parafin, B.A.: New thermionic emission microscope. Rev. Sci. Instrum. **37**, 438–444 (1966)
73. Huang, L.Y.: The construction of commercial electron microscopes in China. Adv. Imag. Electron Phys. **96**, 805–847 (1996)

74. Wegmann, L.: The photo-emission electron microscope: its technique and applications (invited review). J. Microsc. **96**, 1–23 (1972)
75. Bayh, W.: Direkte Sichtbarmachung von Metalloberflächen mit ionenausgelösten Elektronen. Z. Physik **151**, 281–295 (1958)
76. Möllenstedt, G., Düker, H.: Ein Elektronen-Emissions-Mikroskop für metallkundische Anwendungen. In: Bargmann, W., et al. (eds.) Proceedings of the 4th International Conference on Electron Microscopy, Berlin 1958, pp. 212–215. Springer, Berlin (1960)
77. Düker, H.: Das Elektronen-Emissions-Mikroskop als Hilfsmittel in der Metallkunde. Z. Metallkunde **51**, 314–321 (1960)
78. Bayh, W.: Emissionsmikroskopie mit Sekundärelektronen (15 keV-Primärelektronen). Z. Physik **150**, 10–15 (1958)
79. Huang, L.Y.: Röntgen-Bildwandler-Mikroskopie. Z. Physik **149**, 225–253 (1957)
80. Koch, W.: Ein hochauflösendes Emissions-Mikroskop zur Sichtbarmachung von Oberflächen mit UV-ausgelösten Elektronen. Z. Physik **152**, 1–18 (1958)
81. Von Ardenne, M.: Tabellen zur Angewandten Physik, vol. 1. VEB Deutscher Verlag der Wissenschaften, Berlin (1962)
82. Delong, A., Drahoš, V.: Emissions-Elektronenmikroskopie ebener Oberflächen mittels Feldelektronen. In: Paty, L. (ed.) Proceedings of the 3rd Czechoslovakia Conference on Electronics Vacuum Physics Transactions, Prag 1965, pp. 633–644. Academia, Prag (1967)
83. Storbeck, F.: Über die Auflösungsgrenze des Emissions-Elektronenmikroskops. Ann. Phys. **475**, 272–291 (1967)
84. Storbeck, F.: Zur Kontrastübertragung im Emissions-Elektronenmikroskop. Ann. Phys. **478**, 209–227 (1969)
85. Uchikawa, Y., Maruse, S.: Theory of chromatic aberrations of the cathode lens. J. Electron Microsc. **18**, 118–127 (1969)
86. Uchikawa, Y., Maruse, S.: Resolving power of the cathode lens. J. Electron Microsc. **19**, 12–16 (1970)
87. Nevzorov, A.N., Sedov, N.N., Dükov, V.G.: Filtration des vitesses des électrons dans un microsope à émission. In: Favard, P. (ed.) Proceedings of the September Congress International Microscopie Électronique, Grenoble 1970, pp. 191–192. Société Francaise de Microscopie Électronique, Paris (1970)
88. Gruner, H., Gauckler, K.H., Möllenstedt, G.: Monochromatisierung von Elektronenstrahlen mittels Spiegelung und ihre Anwendung in der Emissions-Mikroskopie. Optik **33**, 255–269 (1971)
89. McCune, R.C.: Image enhancement in the emission electron microscope. In: Bailey, G.W. (ed.) Proceedings of the 31th Annual EMSA Meeting, New Orleans 1973, pp. 90–91. Claitor's Publishing Div, Baton Rouge (1973)
90. Baxter, W.J., Rouze, S.R.: A photoemission electron microscope using an electron multiplier array. In: Bailey, G.W. (ed.) Proceedings of the 31th Annual EMSA Meeting, New Orleans 1973, pp. 138–139. Claitor's Publishing Div, Baton Rouge (1973)
91. Koch, W.: Ein Photoelektronen-Emissionsmikroskop für den 10^{-8} Torr-Bereich. Optik **25**, 523–534 (1967)
92. Turner, G.H., Bauer, E.: An Ultrahigh Vacuum Electron Microscope and Its Application to Work Function Studies. In: Ueda, R. (ed.) Proceedings of the 6th International Congress on Electron Microscopy, Kyoto 1966, pp. 163–164. Maruzen, Tokyo (1966)
93. Druzhinin, A.V., Dükov, V.G., Isaev, A.A.: Ultrahigh vacuum emission microscope. In: Favard, P. (ed.) Proceedings of the September Congress International Microscopie Électronique, Grenoble 1970, pp. 187–188. Société Francaise de Microscopie Électronique, Paris (1970)
94. Griffith, O.H., Lesch, G.H., Rempfer, G.F., Birrell, G.B., Burke, C.A., Schlosser, D.W., Mallon, M.H., Lee, G.B., Stafford, R.G., Jost, P.C., Marriott, T.B.: Photoelectron microscopy: a new approach to mapping organic and biological surfaces. Proc. Nat. Acad. Sci. U.S.A. **69**, 561–565 (1972)

95. Griffith, O.H., Rempfer, G.F., Lesch, G.H.: A high vacuum photoelectron microscope for the study of biological specimens. In: Johari, O. (ed.) Scanning Electron Microscopy/1981 Meetings, Dallas 1981, pp. 123–130. SEM Inc AMF O'Hare, Chicago, IL (1981)
96. Delong, A., Drahoš, V.: Low energy electron diffraction in an emission microscope. Nat. Phys. Sci. **230**, 196–197 (1971)
97. Drahoš, V., Delong, A., Kolařik, V., Lenč, M.: On a mirror-type selected area LEED. J. Microsc. **18**, 135–146 (1973)
98. Delong, A., Drahoš, V., Kolarik, V., Lenc, M.: On the possibility of Auger electron spectroscopy in the electron emission microscope. In: Proceedings of the 5th Czechoslovakia Conference on Electronics and Vacuum Physics, Berlin, p. c 6 (1972)
99. Recknagel, A., Mahnert, H., Brückner, J.: Über ein Gerät, das Beugung mit langsamen und schnellen Elektronen und emissionselektronenmikroskopische Abbildung von Festkörperoberflächen gestattet. Optik **35**, 376–390 (1972)
100. Möllenstedt, G., Lenz, F.: Electron emission microscopy. Adv. Electron. Electron Phys. **18**, 251–329 (1963)
101. Schwarzer, R.A.: Physical aspects of emission electron microscopy. In: Pfefferkorn, G., Schur, K. (eds.) Proceedings of the First International Conference on Emission Electron Microscopy: Beiträge zur elektronenmikroskopischen Direktabbildung von Oberflächen, Tubingen/Germany 1979, pp. 3–38. Verlag R.A. Remy, Münster (1979)
102. Schwarzer, R.A.: Emission electron microscopy – a review. Microsc. Acta **84**, 51–84 (1981)
103. Mayer, L.: Electron optical image of an electron beam. J. Appl. Phys. **24**, 105–106 (1953)
104. Bartz, G., Weissenberg G., Wiskott, D.: Ein Auflichtelektronenmikroskop. In: Ross, R. (ed.) International Conference on Electron Microscopy, London 1954, pp. 395–404 (1954)
105. Bartz, G., Weissenberg, G., Wiskott, D.: Ein Auflicht Elektronenmikroskop. Radex-Rundschau **4/5**, 163–172 (1956)
106. Spivak, G.V., Prilejaeva, I.N., Azovcev, V.K.: Magnetic contrast in electron mirror and observation ferromagnetic domains. Dokl. Akad. Nauk. SSSR (in Russian) **105**(965–967) (1955)
107. Kuehler, J.D.: A new electron mirror design. IBM J. Res. Dev. **4**, 202–204 (1960)
108. Mayer, L., Rickett, R. Steinemann, H.: Design features of a new electron mirror microscope. In: Sydney, j., Breese S.S., (ed.) Proceedings of the 5th International Congress Electron Microscopy, Philadephia 1962, pp. D-10. Academic Press, New York (1962)
109. Kuhlmann, F., Hein, U., Burka, C., Knick, K.-D., Schneidereit, A.: Zur Elektronenspiegelmikroskopie in Schattenabbildung mit monochromatisierten Elektronen. Exp. Tech. Phys. **26**, 285–295 (1978)
110. Mayer, L.: On electron mirror microscopy. J. Appl. Phys. **26**, 1228–1230 (1955)
111. Spivak, G.V., Igras, É., Pryamkova, I.A., Zheludev, I.S.: Detection of the domain structure in barium titanate with an electron reflector. Sov. Phys. Cryst. (in Russian) **4**(115–117) (1959)
112. Bethge, H., Hellgardt, J., Heydenreich, J.: Zum Aufbau einfacher Anordnungen für Untersuchngen mit dem Elektronenspiegel. Exp. Tech. Phys. **8**, 49–61 (1960)
113. Spivak, G.V., Pryamkova, I.A., Fetisov, D.V., Kabanov, A.N., Lazareva, L.V., Shilina, A.I.: Mirror electron microscope for studying the structure of surfaces. Izv. Akad. Nauk. SSSR Ser. Fiz. (in Russian) **25**, 698–705 (1961)
114. Bacquet, G., Santouil, A.: Un microscope électronique à miroir. C. R. Acad. Sci. Paris **255**, 1263–1265 (1962)
115. Spivak, G.V., Luk'yanov, A.E., Toshev, S.D., Koptschik, V.A.: Observation of the domain structure of triglyceride sulfate by means of an electron mirror. Bull. Acad. Sci. USSR **27**, 1177–1179 (1963)
116. Igras, E., Warminski, T.: Electron mirror observations of semiconductor surfaces at low temperatures. Phys. Status Solidi A **9**, 79–85 (1965)
117. Ivanov, R.D., Abalmazova, M.G.: Universal high-resolution mirror electron microscope. Bull. Acad. Sci. USSR **30**, 813–815 (1966)
118. Barnett, M.E., Nixon, W.C.: A mirror electron microscope using magnetic lenses. J. Sci. Instrum. **44**, 893–898 (1967)

119. Someya, T., Watanabe, M.: Development of electron mirror microscope. In: Bocciarelli, D.S. (ed.) Proceedings of the 4th European Regional Conference on Electron Microscopy, Rome 1968, pp. 97–98. Tipografia Poliglotta Vaticana, Rome (1968)
120. Szentesi, O.I., Bernett, M.E.: Electron mirror observation of an interdigital electrode structure. J. Phys. E Sci. Instrum. **2**, 855–858 (1969)
121. Becherer, G., Kuhlmann, F.: Untersuchungen zum Aufbau eines Elektronenspiegel-Mikroskops im Druckbereich von 10^{-7}–10^{-8} Torr (Teil 1). Exp. Tech. Phys. **19**, 119–131 (1971)
122. Archard, G.D., Mulvey, T.: Magnetic deflexion of electron beams without astigmatism. J. Sci. Instrum. **35**, 279–283 (1958)
123. Schwartze, W.: Ein Elektronenspiegel-Oberflächenmikroskop. Exp. Tech. Phys. **14**, 293–302 (1966)
124. Bethge, H., Heydenreich, J.: Ein Elektronenspiegel-Mikroskop mit Umlenkfeld und seine Einsatzmöglichkeiten. Exp. Tech. Phys. **14**, 375–382 (1971)
125. Bok, A.B.: A mirror electron microscope. Ph.D. thesis, Technical University of Delft (1968)
126. Bok, A.B.: Scanning mirror electron microscopy. In: Amelinckx, S., Gevers, R., Remaut, G., van Landuyt, J. (eds.) Modern Diffraction and Imaging Techniques in Material Science, pp. 655–682. North Holland, Amsterdam (1970)
127. Forst, G., Wende, B.: Zur Bildentstehung im Elektronenspiegelmikroskop. Z. Angew. Phys. **17**, 479–481 (1964)
128. Schwartze, W.: Elektronenspiegel-Oberflächenmikroskopie mit fokussierter Abbildung. Optik **25**, 260–272 (1967)
129. Mayer, L.: Electron mirror microscopy of patterns recorded on magnetic tape. J. Appl. Phys. **29**, 658–660 (1958)
130. Wiskott, D.: Zur Theorie des Auflicht-Elektronenmikroskops. I. Geometrische Elektronenoptik in der Umgebung des Objects. Optik **13**, 463–478 (1956)
131. Brand, U., Schwartze, W.: Modellversuche zur Bildentstehung im Elekronenspiegel-Oberflächenmikroskop. Exp. Tech. Phys. **11**, 18–25 (1963)
132. Wiskott, D.: Zur Theorie des Auflicht-Elektronenmikroskops. II. Wellenmechanische Elektronenoptik in der Umgebung des Objects. Optik **13**, 481–493 (1956)
133. Gvosdover, R.S., Zel'dovich, Y.B.: The quantum theory of the image contrast formation of electrical and magnetic microfields in mirror electron microscopy. J. Microsc. **17**, 107–130 (1973)
134. JEM-M1 Electron Mirror Microscope. In: JEOL News, vol. 6, No 1. pp. 2–3 (1968)
135. Someya, T., Kobayashi, J.: Quantitative application of electron-mirror microscopy to the determination of pure shear of ferroelectric Gd2(MoO4)3. Phys. Status Solidi A **26**, 325–336 (1974)
136. Warminski, T.: Sensitivity of the electron mirror microscope for surface electrical inhomogeneities. In: Proceedings of the 5th European Congress Electron Microscopy 1972, pp. 28–29. Institute of Physics, Manchester (1972)
137. Guittard, C., Babout, M., Pernoux, E.: Mesure des discontinutés de potentiel dans un microscope à miroir. C. R. Acad. Sci. Paris **261**, 5358–5361 (1965)
138. Guittard, C., Babout, M., Pernoux, E.: Microscopie electronique a miroir: mesure des variations locales de potentiel. In: Bocciarelli, D.S. (ed.) Proceedings of the 4th European Regional Conference on Electron Microscopy, Rome 1968, pp. 107–108. Tipografia Poliglotta Vaticana, Rome (1968)
139. Guittard, C., Guittard, S., Bernard, R.: Faisceaux Électroniques et surfaces. Perspectives de regroupement des principales techniques en vue d'une corrélation plus efficace des resultants. Bull. Soc. Chim. Fr. **Numero Special**, 137–143 (1970)
140. Babout, M., Laydevant, L., Guittard, C.: Microscopie Electronique a Miroir. Images et mesure des potentiels locaux. In: Favard, P. (ed.) Proceedings of the September Congress International de Microscopie Électronique, Grenoble 1970, pp. 271–272. Société Française de Microscopie Électronique, Paris (1970)
141. Guittard, C., Bernard, R.: Microscope-Volumetre Électronique pour measures superficielle ponctuelles. In: Favard, P. (ed.) Proceedings of the September Congress International de Microscopie Électronique, Grenoble 1970, pp. 275–276. Société Francaise de Microscopie Électronique, Paris (1970)

142. Laydevant, L., Guittard, C., Bernard, R.: Diffraction d'electrons tres lents en microscopie electronique a miroir. In: Proceedings of the 5th European Congress Electron Microscopy, Manchester 1972, pp. 662–663. Institute of Physics (1972)
143. Bernard, R., Laydevant, L., Dupuy, J. C.: Groupe de pompage automatisable en ultra-vide: Application à un microscope élecronique à miroir. Le Vide, Les Couches Minces **169**, 111–116 (1974)
144. Berger, C., Laydevant, L., Bernard, J.: Utilisation de l'objectif de Johannson pour l'obtention de diagrammes de diffraction d'électrons lents. J. Microsc. Biol. Cell **22**, 1–14 (1975)
145. Pantel, R., Bujor, M., Bardolle, J.: Continuous measurement of surface potential variations during oxygen adsorption on the (100), (110) and (111) faces of niobium using mirror electron microscope (with LEED, AES). Surf. Sci. **62**, 589–609 (1977)
146. Pantel, R., Bujor, M., Bardolle, J.: Oxygen adsorption on various vicinal faces close to the (0001) basal plane of rhenium. Surf. Sci. **83**, 228–242 (1979)
147. Chang, C.C.: LEED optics as electron mirror microscope. Rev. Sci. Instrum. **42**, 189–191 (1971)
148. Dupuy, J.C., Sibai, A., Vilotitch, B.: Mirror electron microscopy (MEM): work function and imaging of an electron beam biased junction of silicon (100). Surf. Sci. **147**, 191–202 (1984)
149. Forster, M.S., Campuzano, J.C., Willis, R.F., Dupuy, J.C.: A mirror electron microscope for surface analysis. J. Microsc. **140**, 395–403 (1985)
150. Shern, C.S., Unertl, W.N.: A simple mirror electron microscope-low-energy electron diffraction apparatus for surface studies. J. Vac. Sci. Technol. A **5**, 1266–1270 (1987)
151. Lichte, H.: Ein Elektronen-Auflicht-Interferenzmikroskop zur Präzisionsmessung von Unebenheiten und Potentialunterschieden auf Oberflächen. Optik **57**, 35–67 (1980)
152. Lichte, H.: Coherent electron optical experiments using an electron mirror. In: International Symposium on Foundations of Quantum Mechanics, Tokyo 1983, pp. 29–38 (1983)
153. Bok, A.B., Le Poole, J.B., Roos, J., De Lang, H.: Mirror electron microscopy. In: Adv. Optical Electron Microscopy, vol. 4. pp. 161–261. Academic Press, London (1971)
154. Luk'yanov, A.E., Spivak, G.V., Gvozdover, R.S.: Mirror electron microscopy. Sov. Phys. Usp. **16**, 529–552 (1974)
155. Bethge, H., Heydenreich, J.: Special methods for direct imaging of surfaces: emission electron microscopy, reflection electron microscopy and mirror electron microscopy. In: Bethge, H., Heydenreich, J. (eds.) Electron Microscopy in Solid State Physics, pp. 229–235. Elsevier, Amsterdam (1987)
156. Godenhardt, R.: Mirror electron microscopy. In: Hawkes, P. (ed.) Advantages of Imaging in Electron Physics, vol. 94, pp. 81–150. Academic, London (1995)
157. Eichen, E.: Thermionic emission microscopy and its metallurgical applications. In: Bunshah, R.F. (ed.) Techniques in Metals Research Part I: Techniques for Direct Observation of Structure and Imperfections, vol. 1, pp. 177–219. Interscience Publishers, New York (1968)
158. Grund, D., Rouze, S.R.: Thermionic-emission microscopy. In: Weinberg, F. (ed.) Tools and Techniques in Physical Metallurgy, vol. 2, pp. 191–203. Marcel Dekker Inc., New York (1970)
159. Pfefferkorn, G., Schur, K.: Bibliography on emission electron microscopy (EEM), especially on Photo EEM. In: Pfefferkorn, G., Schur, K. (eds.) Proceedings of the First International Conference on Emission Electron Microscopy: Beiträge zur elektronenmikroskopischen Direktabbildung von Oberflächen, Tubingen/Germany 1979, pp. 205–233. Verlag R.A. Remy, Münster (1979)

Chapter 2
Basic Interactions

Abstract Understanding the interactions of photons and slow electrons with condensed matter is fundamental for the understanding of the imaging process in emission and reflection electron microscopy with slow electrons. While some of these interactions are common to both emission and reflection imaging, such as the propagation of the electrons in the sample to its surface, there are fundamental differences between the two imaging modes. Following the historical evolution of surface imaging with slow electrons, this chapter discusses first the processes involved in emission microscopy and then turns to those important in reflection microscopy with electrons entering the sample, i.e., to low energy electron microscopy (LEEM). In mirror electron microscopy the incident beam does not interact with the sample but only with the field modification caused by the sample, which is a problem of electron optics and image formation and will be discussed in Chap. 4.

2.1 Fundamental Theories of Electron Emission

In emission imaging the dominant mode has become photoelectron emission while secondary electron emission caused by electron, ion, or fast atom bombardment, which was the most important mode before the advent of ultrahigh vacuum, is not used anymore and will, therefore, be discussed only briefly. Similarly thermionic emission, the main method in the early days, has been relegated to studies of thermionic emitters while in many cases it is a disturbing factor, for example in LEEM at high temperatures. In all imaging methods the sample is in a high field so that in principle field emission can occur too, though only in special cases. For these reasons we will first discuss photoemission, thermionic emission, and field emission briefly jointly within the traditional formalism, then concentrate on the various modes of photoemission on a broader basis and finally consider secondary emission caused by incident particles in connection with photon-created secondary electrons.

E. Bauer, *Surface Microscopy with Low Energy Electrons*,
DOI 10.1007/978-1-4939-0935-3_2, © Springer Science+Business Media New York 2014

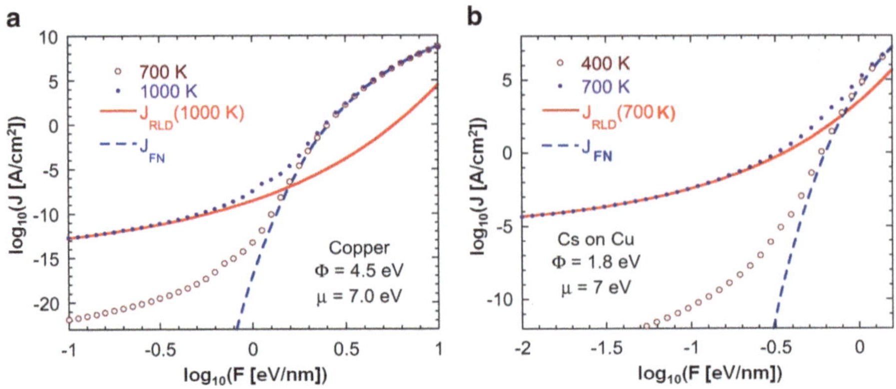

Fig. 2.1 Thermal field emission current density as a function of the applied field strength according to several theoretical models: Fowler-Nordheim (FN, *dashed line*), Richardson-Laue-Dushman (RLD, *full line*), and general field emission model (*dotted* and *open circle lines* for two different temperatures). μ is the Fermi energy, Φ the work function. (**a**) clean Cu, (**b**) cesiated Cu. Adapted from Ref. [2] with permission from the American Institute of Physics, © 2007

The fundamental theories of thermionic emission (TE), photoemission (PE), and field emission (FE) are the Richardson-Laue-Dushman (RLD) theory, the Fowler-DuBridge theory, and the Fowler-Nordheim (FN) theory, respectively. In emission microscopy the sample is in a high field, at nonzero temperature and in photoemission it is also illuminated so that in principle all emission modes can occur simultaneously. It is therefore useful to consider first generalized theories, which take account of all emission modes in order to see the possible influence of simultaneous TE, FE and/or PE. Purely analytical theories [1] do not allow general predictions because they require numerical evaluation of the resulting equations for specific values of the relevant quantities such as temperature, field, work function, etc. Figure 2.1 shows an example of the total current density for simultaneous TE and FE for two work function values and two temperatures at a chemical potential μ (= Fermi energy E_F at $T = 0$ K) of 7 eV [2]. The dashed curves are from the classical (temperature-independent) Fowler-Nordheim theory, the solid curves from the classical (field-independent) RLD-theory for the higher of the two temperatures shown. The open and solid circles are from the general thermal field theory developed in Ref. [2]. Figure 2.1a is typical for Cu, Fig. 2.1b for cesiated Cu. At the lower temperatures there is a clear influence of the field on the emission in both cases but the current densities at typical field strengths used in emission microscopy (0.1 V/nm) are far below those useful for imaging. Only when features with small radius of curvature, which cause strong local field enhancement, are present on the surface, the general thermal field theory becomes relevant. At temperatures at which TE sufficient for imaging occurs, the field influence on emission can be completely neglected. A qualitative way to look at this problem is to compare the field term with the work function Φ in the Schottky equation for the total emitted current

$$J = A_0 T^2 \exp\left[\left(\sqrt{e^3 F} - \Phi\right)/kT\right], \tag{2.1}$$

where $A_0 = 4\pi mk^2e/h^3 = 1.20 \times 10^6$ A/m^{-2} K^{-2} is the universal Richardson constant and F the field strength. For typical fields in emission microscopy $(e^3F)^{1/2} \leq 0.12$ eV, so that the field F has a significant influence only for small Φ. Having thus excluded the influence of the field on TE microscopy—except FE through an insulating barrier in MIM (metal–insulator–metal) or SIM (semiconductor–insulator–metal) cathodes—we restrict the further discussion to the generalized RLD equation for the total emitted current (Eq. (2.1) without the field term). At the same time we take two aspects into account in a phenomenological manner, which are not included in the RLD equation: the averaged quantum-mechanical reflection factor at the surface r and the influence of the band structure with a factor λ_B, by replacing A_0 with the generalized Richardson constant A_G [3]:

$$A_G = \lambda_B(1 - r)A_0. \tag{2.2}$$

Usually also the temperature dependence of $\Phi \approx \Phi_0 + cT$ is included in λ_B via the factor $\exp(c/k)$. Calculations of the influence of these parameters can be found in Ref. [3], experimental results for various tungsten surfaces in Ref. [4]. The total emission current determines the image brightness in emission microscopy and thus the contrast between regions with different Φ, r, and λ_B. Details can be found in the classic paper by Herring and Nichols [5] and in reviews of the subject [3, 6, 7]. For the image resolution the energy, energy distribution, and angular distribution of the emitted electrons are of importance. The energy distribution is well approximated by a Maxwell distribution

$$\mathrm{d}J/\mathrm{d}E \sim E\exp(-E/kT), \tag{2.3}$$

which has a maximum E_0 at kT and a full width at half maximum ΔE of $2.45\,kT = 21 \times 10^{-5}\,T$. TE microscopy uses in general temperatures below 2000 K so that $E_0 \leq 0.17$ eV and $\Delta E \leq 0.4$ eV. The angular distribution can be approximated by a $\cos^2\theta$ function because only the normal component of the energy is important for escape into the vacuum. The consequences of angular and energy distribution will be discussed in the chapters on image formation and electron optics (Chaps. 3 and 4).

Next we consider the case of simultaneous TE and PE. Figure 2.2 [2] is an example, again for clean Cu and cesiated Cu parameters. Here the dashed and solid curves show the relative quantum yields $P(\hbar\omega)/P(\hbar\omega_0)$ ($\hbar\omega_0 = 6.2$ eV) for clean Cu and cesiated Cu, respectively, the solid and open circles the corresponding curves from the generalized thermal and photo emission theory developed in Ref. [2]. It is evident that at the higher work function values, which are encountered in general, the classical Fowler-DuBridge theory is completely sufficient and that at extremely low work functions only minor deviations occur from it at 600 K. It should be mentioned, however, that the inverse effect to thermal-enhanced

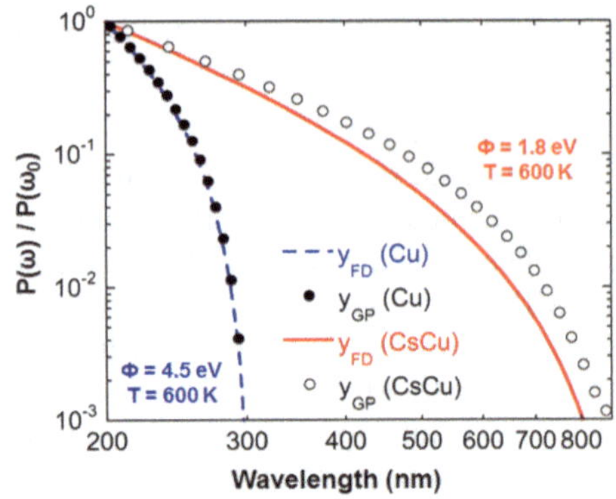

Fig. 2.2 Comparison between Fowler-Dubridge equation (*lines*) and general photoemission equation (*points*) for Cu and cesiated Cu. Adapted from Ref. [2] with permission from the American Institute of Physics, © 2007

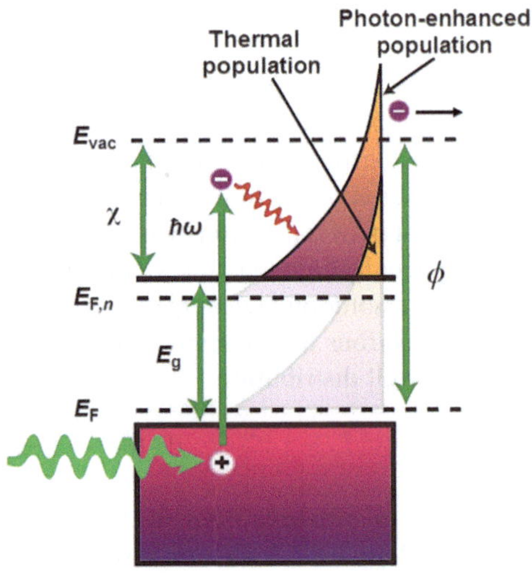

Fig. 2.3 Photon-enhanced thermionic emission from a p-type semiconductor. Photoabsorption shifts the thermal energy distribution of the electrons from the Fermi level E_F to the quasi-Fermi level $E_{F,n}$, thus allowing the high energy tail of the electrons to overcome the work function Φ, electron affinity χ. Adapted from Ref. [8] with permission from Nature Publishing Group (UK)

PE emission, photon-enhanced TE emission, also occurs and can be important in semiconductors [8]. Photoexcitation increases the electron concentration n in the conduction band, which shifts the Fermi level from E_F to $E_{F,n} = E_F + kT \ln(n/n_{eq})$ (n_{eq} equilibrium electron concentration in the conduction band) as indicated in Fig. 2.3. An example is GaN, which has a band gap of 3.4 eV. If its electron affinity is reduced to 0.3–0.4 eV by Cs adsorption then illumination with 3.5 eV photons more than doubles the quantum efficiency from about 320 to 500 K [8]. Of course, these are unusual conditions in emission microscopy. Therefore, we will not dwell any further on the combined effects of TE, FE, and PE, having seen that they are in general not significant in emission microscopy, and concentrate now on the various aspects of pure photoemission.

2.2 Photoemission

2.2.1 General Considerations

The modern theory of photo electron emission phenomena is based in the so-called one-step model [9, 10]. This model treats all aspects of the phenomena within the same framework, from the photo-excitation process via the electron transport to the surface and the transmission through the surface barrier to the detector. These three steps, which were originally treated separately in the properly named three-step model [11] are intimately connected, which makes the three-step model only a first but in many cases a useful approximation. Using it means that the relaxation of the initial N electron system into the final electron system, multiple scattering of the photoelectron and other processes involved in photoemission are not taken into account, a deficiency which is important in photoelectron spectroscopy and photoelectron diffraction but less critical in photoelectron microscopy where other factors such as surface topography and lens aberrations are more important. With this in mind, the discussion in this section will be mainly in the framework of the three-step model. For detailed descriptions of the one-step model and its relation to the three-step model the reader is referred to advanced books which emphasize spectroscopy, diffraction, and other phenomena [12–17].

In the following we outline the general theory of photoemission without going into detail because this is beyond the scope of this book. For the details the reader is referred to the literature on photoelectron spectroscopy cited above, in particular to Ref. [15]. The starting point of the theory of photoemission excited by a photon with energy $\hbar\omega$ is in general Fermi's Golden Rule

$$T_{if} = \frac{2\pi}{\hbar}\left|\langle f|H_{\text{int}}|i\rangle + \sum_n \frac{\langle f|H_{\text{int}}|n\rangle\langle n|H_{\text{int}}|i\rangle}{\varepsilon_i - \varepsilon_n}\right|^2 \delta\left(\varepsilon_f - \varepsilon_i - \hbar\omega\right), \tag{2.4}$$

which gives the transition probability per unit time up to second order from the initial state i (ψ_i) of the system with energy ε_i to the final state f (ψ_f) with energy ε_f either directly or via virtual intermediate states n (ψ_n) in response to the Hamiltonian H_{int}. The task is now to choose the wave functions ψ_i, ψ_f, ψ_n and the Hamiltonian H_{int}, which describe the states of the system and its interaction with the electromagnetic wave best. In principle the interaction of an electromagnetic (EM) wave with matter is a problem of quantum electrodynamics, but in practice such a treatment is too complex in general. In view of the fact that the energies involved in photo electron emission microscopy (PEEM) are usually less than 1 keV and only in hard X-ray PEEM (XPEEM) up to several keV, energies which are small compared to the rest mass energy of the electron (511 keV), nonrelativistic approximations of the relativistic equations are sufficient for the interpretation of most PEEM results.

A useful starting point is the nonrelativistic limit of the Dirac equation of a single electron in an electromagnetic field [18], the Pauli-Schrödinger equation. If we neglect relativistic correction terms, which are not important in PEEM, then its Hamiltonian is

$$H_{\text{int}} = \frac{1}{2m}\left(\boldsymbol{p} - \frac{e}{c}\boldsymbol{A}\right)^2 + e\Phi - \frac{e\hbar^2}{2me}\boldsymbol{\sigma}\cdot\boldsymbol{B} - \frac{e\hbar^2}{4mc^2}\boldsymbol{\sigma}\cdot(\boldsymbol{E}\times\boldsymbol{p}). \tag{2.5}$$

$\boldsymbol{p}$ $(= -i\hbar\nabla)$ is the momentum (operator), $\boldsymbol{A}$ the vector potential of the electromagnetic field, Φ the scalar potential, $\boldsymbol{\sigma}$ the spin (operator), $\boldsymbol{B}$ the magnetic field, and $\boldsymbol{E}$ the electric field. The last term is the only relativistic correction and represents the spin–orbit interaction H_{so}, the term preceding it is the Zeeman term H_{B}. Both of them cause spin-dependent energy level splitting, which plays an important role in several PEEM subfields. The wave functions Ψ are in the nonrelativistic approximation two-component spinors with the components $|\uparrow\rangle$, $|\downarrow\rangle$, and the corresponding space-dependent wave functions $\varphi_\uparrow(\boldsymbol{r})$, $\varphi_\downarrow(\boldsymbol{r})$. Usually the wave function Ψ is a linear combination of the two spinor functions $\varphi_\uparrow(\boldsymbol{r})\,|\uparrow\rangle$ and $\varphi_\downarrow(\boldsymbol{r})\,|\downarrow\rangle$, in the following abbreviated with $\psi_\uparrow(\boldsymbol{r})$ and $\psi_\downarrow(\boldsymbol{r})$ or briefly $\psi(\boldsymbol{r})$.

Equation (2.5) is a one-particle Hamiltonian. The specimens in PEEM are many-particle systems with strong electron–electron interactions, usually in a periodic potential. Electron–electron interactions are generally described by a spin-dependent exchange-correlation potential $V_{\text{xc}}^{\uparrow\downarrow}(\boldsymbol{r})$. This adds an additional term $H_{\text{xc}} = eV_{\text{xc}}^{\uparrow\downarrow}(\boldsymbol{r})$ to Eq. (2.5), which can then be written in the form

$$H_{\text{int}} = H_0 + H_{\text{rad}} + H_{\text{xc}} + H_{\text{B}} + H_{\text{so}}, \tag{2.6}$$

with

$$H_0 = -\frac{\hbar^2}{2m}\Delta + eV(\boldsymbol{r}) \tag{2.7}$$

and

$$H_{\text{rad}} = -\frac{e}{2mc}(\boldsymbol{A}\cdot\boldsymbol{p} + \boldsymbol{p}\cdot\boldsymbol{A}) + \frac{e^2}{2mc^2}A^2. \tag{2.8}$$

H_0 describes the electron in the potential field $V(\boldsymbol{r})$ without electron–electron interactions, H_{rad} the interaction of the electron with the EM field. Rewriting $\boldsymbol{A}\cdot\boldsymbol{p} + \boldsymbol{p}\cdot\boldsymbol{A} = 2\boldsymbol{A}\cdot\boldsymbol{p} - \boldsymbol{A}\cdot\boldsymbol{p} + \boldsymbol{p}\cdot\boldsymbol{A}$, using the commutation rule for $\boldsymbol{A}$ and $\boldsymbol{p}$, $\boldsymbol{p}\cdot\boldsymbol{A} - \boldsymbol{A}\cdot\boldsymbol{p} = -\text{i}\hbar\nabla\cdot\boldsymbol{A}$, and expressing $\boldsymbol{p}$ by $-\text{i}\hbar\nabla$ we finally arrive at

$$H_{\text{int}} = -\frac{\hbar}{2m}\Delta + i\frac{e\hbar}{mc}\boldsymbol{A}\cdot\nabla + i\frac{e\hbar}{2mc}\nabla\cdot\boldsymbol{A} + \frac{e^2}{2mc^2}A^2 + V(\boldsymbol{r}) + H_{\text{xc}} + H_{\text{B}} + H_{\text{so}}. \tag{2.9}$$

From this starting point we can now consider photoemission at different levels of simplification without going into the details of the calculations. The $\boldsymbol{A}$-containing terms determine H_{int}, the remaining terms the initial and final wave functions.

2.2.2 The Free Electron Gas Approximation

The simplest model of photoemission assumes that the electrons are free and move in a positive background produced by the homogeneously distributed charge of the ion cores. This produces a potential well with constant potential V_0, which is filled with electrons up the Fermi level E_F. The Fermi level and the top of the potential well (the vacuum level) differ by the work function Φ. Furthermore it is assumed that there is no magnetic field B. Together with the absence of electron–electron interactions these assumptions reduce the Hamiltonian (2.6) to the first two terms (2.7) + (2.8) with $V(\boldsymbol{r}) = V_0$. Furthermore it is assumed that the boundary between the semi-infinite potential well and vacuum at $z = 0$ is abrupt and that A is constant so that $\nabla \bullet \boldsymbol{A} = 0$. In classical ultrahigh vacuum PEEM (UVPEEM) with Hg arc, D_2, or He lamps the electromagnetic wave is weak enough so that the A^2 term can be neglected but when high-power lasers are used in UVPEEM it becomes important. The spin is taken into account simply by double occupancy of each state in the free electron system (↑,↓). With these assumptions the Pauli-Schrödinger equation becomes the Schrödinger equation with the Hamiltonian

$$H_{\text{Sch}} = -\frac{\hbar^2}{2m}\Delta + eV_0 + i\frac{e\hbar}{mc}\boldsymbol{A}\bullet\nabla + \frac{e^2}{2mc^2}A^2. \tag{2.10}$$

Kantorovich [19] has solved with this Hamiltonian the Schrödinger equation for a plane incident wave $\boldsymbol{A} = \boldsymbol{A}_0 \cos \omega t$ with the Ansatz for the wave function

$$\psi = \sum_{n=0}^{\infty} \psi_n \exp\left[-i\left(n\omega - \frac{p^2}{2m\hbar}\right)t\right], \tag{2.11}$$

With $|\psi_n| \sim |\boldsymbol{A}_0|^n \sim (E_0 c/\omega)^n$, where E_0 is the amplitude of the electric field of the wave and $p^2/2\,m$ is the kinetic energy of the free electron in the absence of the EM wave. The results of his very condensed paper are shown in Refs. [20, 21]. In brief, the current density of electrons which have absorbed n photons, J_n is proportional to the 2nth power of the E_0, i.e., proportional to nth power of the intensity. For $n = 0$ his results reduces to the RLD (thermionic emission) result, for $n = 1$ to that of the Fowler-duBridge theory [22, 23] (photoemission, which will discussed next) because the fundamental assumptions are identical in these studies.

These fundamental assumptions have some fundamental flaws, the most important one being the assumption of free electrons. Free electrons have a parabolic band structure $E(\boldsymbol{k}) \sim k^2$. A transition from a lower (at $\boldsymbol{k}_1$) to a higher (at $\boldsymbol{k}_2$) energy level is therefore possible only if simultaneously a momentum

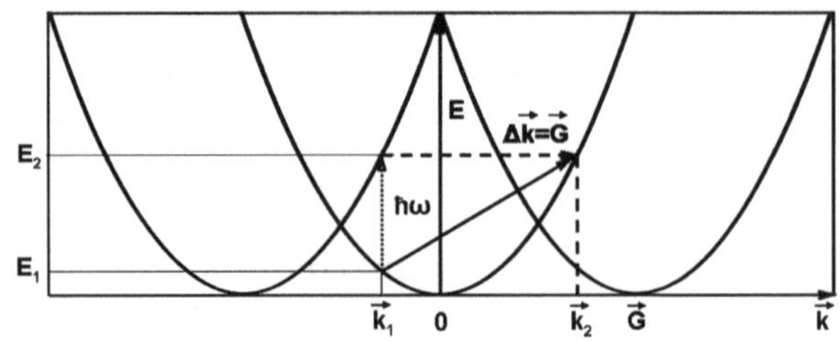

Fig. 2.4 Extended zone scheme of a quasi-free electron metal with the reciprocal lattice periodicity $\boldsymbol{G}$. The momentum transfer $\boldsymbol{G} = \boldsymbol{k}_2 - \boldsymbol{k}_1$ allows the vertical transition to the higher energy state $E_2 = E_1 + \hbar\omega$

transfer $\Delta\boldsymbol{k} = \boldsymbol{k}_1 - \boldsymbol{k}_2$ occurs, which is much larger than what the photon can supply ($k_{\mathrm{Ph}} = h/\lambda$). This problem was "solved" by assuming that the surface was providing the necessary momentum, therefore the old name surface photoelectric effect. This should not be confused with today's surface photoemission from surface states and surface resonances, which are true surface effects. The assumption that the EM wave is constant in the semi-infinite potential well ($A = \mathrm{const}$) is incompatible with the very process of electron excitation, which attenuates the EM field. Likewise, the assumption that all excited electrons reach the interface is also unrealistic, mainly because they lose energy by inelastic scattering on their way to the surface and cannot escape any longer. Mahan [24] has taken these two effects into account. The momentum transfer problem was solved by replacing the homogeneous background charge distribution by the periodic arrangement of the ion cores, which in $\boldsymbol{k}$ space leads to a periodic arrangement of $E(\boldsymbol{k})$ parabolas instead of a single parabola. Now the interaction of the electron with the lattice provides the necessary momentum, while the photon provides the energy as shown in Fig. 2.4. The surface is now involved only in as much as it imposes the conservation of the parallel momentum of the electron during emission. Mahan calculated also the angular distribution and finally took the band splitting at the reciprocal lattice zone boundaries into account.

Although the free electron model has many inconsistencies as reviewed, for example, by Dobretsov [6], its predictions are in surprisingly good quantitative agreement with many experimental observations, not only from free electron-like metals but also from semiconducting and insulating materials. Therefore, it is appropriate to examine to what extent it is useful for UVPEEM. Obviously it is not applicable for studies, in which the excluded Hamiltonian components play a role, such as in magnetism, spin–orbit splitting, or strong correlations. However, for questions such as the dependence of the photoemission yield upon photon energy or work function the Fowler-DuBridge theory seems to be sufficient. An example is the yield dependence upon photon energy. The theory predicts for the current density

$$J = \alpha A_0 T^2 f\left(\frac{h\nu - \Phi}{k_{\mathrm{B}}T}\right) \tag{2.12}$$

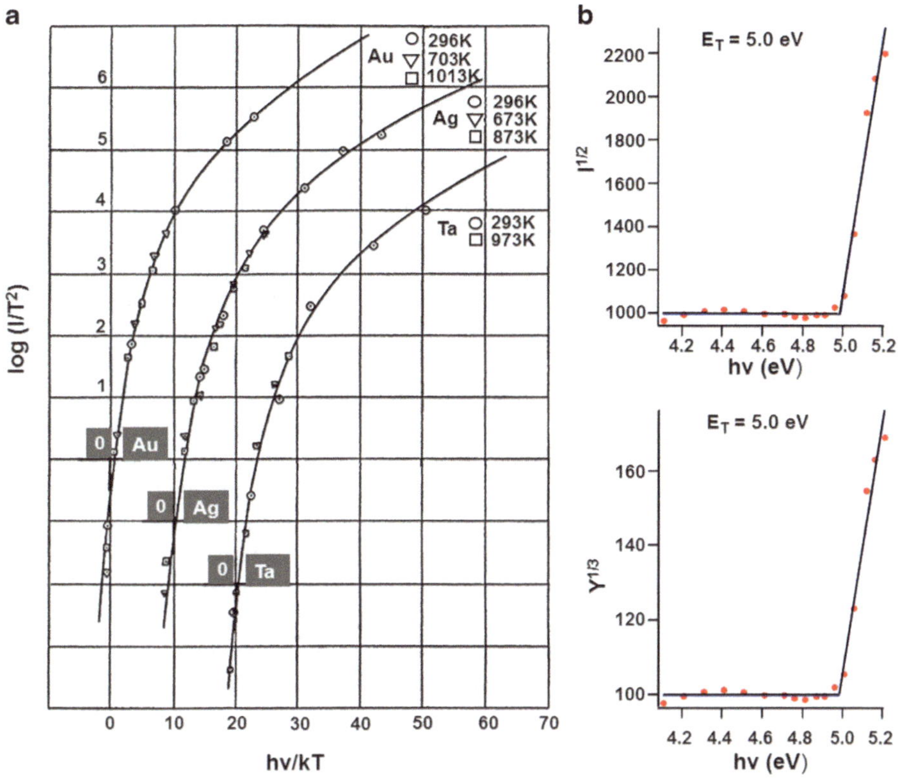

Fig. 2.5 (**a**) Fowler function of three metals at several temperatures. The zeros are shifted for clarity as indicated. Adapted from Ref. [22]. (**b**) Photoelectron yield at constant temperature from a single fibrogen molecule as a function of photon energy fitted with different exponents m according to the semiempirical law $Y \sim (h\nu - E_{\mathrm{th}})^m$, where E_{th} is a threshold energy. (**a**) Adapted with permission from Ref. [22]. Copyright 1931 by the American Physical Society. (**b**) Reproduced from Ref. [25] with permission from the American Institute of Physics, © 2008

where α is a proportionality constant, A_0 the Richardson constant mentioned earlier and

$$\begin{aligned} f\left(\frac{h\nu - \Phi}{k_{\mathrm{B}}T}\right) &\equiv f(x) = \int_{-\infty}^{x} \ln\left(1 + e^t\right)\mathrm{d}t \\ &= \frac{x^2}{2} + \frac{\pi^2}{6} - e^{-x} + \frac{e^{-2x}}{2^2} - \frac{e^{-3x}}{3^2} + \ldots \quad \text{for} \, x \geq 0, \qquad (2.13) \\ &= e^x - \frac{e^{2x}}{2^2} + \frac{e^{3x}}{3^2} - \ldots \quad \text{for} \, x \leq 0 \end{aligned}$$

is the Fowler function. In Fig. 2.5a [22] this function is compared with experimental data for several nonfree electron-like metals, illustrating excellent agreement between theory and experiment.

In UVPEEM a similar agreement should be expected because of the low electron energies all electrons are transmitted through the objective lens so that the total current determines the intensity. In most UVPEEM studies the light source is not monochromatic so that a direct comparison with the Fowler function is not possible. Free electron laser (FEL) light, however, is tunable over a sufficient energy range to allow fitting the emission current with the Fowler formula. Because of the small accessible energy range only the leading term in $f(x)$, $x^2/2$, is taken into account, so that a plot of $J^{1/2}(h\nu)$ should be linear as evidenced in Fig. 2.5b [25] for a biological molecule (fibrinogen). A fit with $J^{1/3}(h\nu)$ is, however, equally good as shown in the lower part of Fig. 2.5b. This power law is one of several, which Kane [26] has suggested for different emission processes from semiconductors. This example illustrates that little can be deduced about the emission process from fits over a small energy range, except an approximate determination of the work function. Data obtained over a wider photon energy range, which can be obtained with synchrotron radiation and are shown later in Fig. 3.48, are more illustrative. They show considerable structure, saturation, or even a decrease of the emission current at high energies, features which are due to changes in the reflectance of the surface with energy, band structure effects, variations of the excitation cross section etc, phenomena not taken into account in the simple free electron-based photoemission theory. At best, we can use the emission current to determine the work function and make use of varying work function across the specimen for work function contrast, but also this can be used only with limitations, which will be discussed later.

2.2.3 Band Structure UV Photoemission

The next aspect, which we have to take into account in photoemission, is the band structure, which has been and is studied extensively with angle-resolved photoelectron spectroscopy (ARUPS, when a He lamp is used or ARPES, when synchrotron radiation is used) [13–17]. UVPEEM without contrast aperture is angle-integrating at very low photon energy—and also energy-integrating without an energy filter—so that the contrast in the real image is determined by the total current. Even with a contrast aperture a wide angular range contributes to the image when a Hg arc lamp is used. All PEEM instruments allow also imaging of the reciprocal space. If equipped with an energy filter, photoelectron diffraction (PED) patterns, and in particular angle-resolved spectra can be obtained. PEEM is increasingly being used together with PED and ARPES, so that we have to discuss here these fields too.

The ARPES spectrum is determined by the band structure and the transition selection rules, which are governed by the crystal symmetry. Although PEEM instruments with energy filters allow fast acquisition of ARPES spectra, their energy and angular resolution if far inferior to that of instruments dedicated solely for this purpose. Thus fine details of these spectra cannot be recorded with PEEM instruments. In addition, the polar and generally also the azimuthal angle of incidence are fixed in PEEM. Furthermore, the accessible emission angle is limited to a cone normal to the surface, whose cone angle decreases rapidly with increasing electron energy as will be discussed in Chap. 4. Therefore ARPES studies in PEEM are limited in several respects. However, PEEM instruments have the important advantage that the spectra can be taken from very small regions of the specimen down to less than 1 μm in diameter and can be correlated with the microstructure of the specimen. This advantage is enhanced even more in combined spectroscopic photoemission and low energy electron microscopes (SPELEEMs), which allow correlation with the local crystal structure via LEEM and low energy electron diffraction (LEED). (This is important because many specimens are intentionally or unintentionally heterogeneous. High energy and angle resolution instruments generally average over different regions in such specimens.) For this reason we will discuss here, though in a very condensed and somewhat superficial manner, the influence of the crystal structure on the energy and angular distribution of photoelectrons. Detailed discussions of these aspects can be found in Chaps. 6 and 7 of Ref. [15] and in Ref. [27].

In a crystal the band crossings at the periodicity boundaries (Brillouin zones) open up into band gaps and the different periodicities in different directions produce direction-dependent band structures. Simultaneously the localization of the positive charge of the ion cores produces localized fields, which makes the band structure spin-dependent, according to the relativistic correction term in the Hamiltonian (2.5) in the spin–orbit coupling. This is not only true for crystals containing atoms with high nuclear charge but has also to be taken into account in crystals with lighter atoms such as in GaAs for which fully relativistic calculations have been made early in connection with photoemission [28]. In addition, the electron–electron interactions have to be taken into account with the H_{xc} term in the Hamiltonian. Band structure calculations with all these ingredients give the band structure of the ground state of the system with the wave functions ψ_i of the initial state of the photoemission process. The result is usually not presented in the extended zone scheme shown in Fig. 2.4 but in the reduced zone scheme, in which the bands, displaced by the reciprocal lattice vectors $\boldsymbol{G}$, are folded back on the zone centered at $\boldsymbol{G} = 0$. Simultaneously, the reduced zone band structures in the mayor crystal directions are shown side by side. In three dimensions the zone boundaries of the first Brillouin zone are shown in Fig. 2.6a for a fcc lattice with the notation of symmetry points and symmetry directions. Also shown in this figure (Fig. 2.6b, c) [29] is the band structure of Cu, one of the metals whose $J(h\nu)$ dependence fits so well the free electron Fowler function (Fig. 2.5a), despite the strong deviations from the free electron band structure [29].

The interaction Hamiltonian with the electromagnetic wave, which describes the excitation process into the final state ψ_f is usually considerably simplified. We have already discarded the A^2 term and assumed $\nabla \bullet \boldsymbol{A} = 0 = 0$ for photoemission from the bulk. We also restrict ourselves to the first term in Fermi's golden rule. As a consequence the Hamiltonian H_{rad} in the matrix element M_{if}, which determines the probability for a transition from an initial state i to a final state f, caused by the interaction with the EM wave, becomes

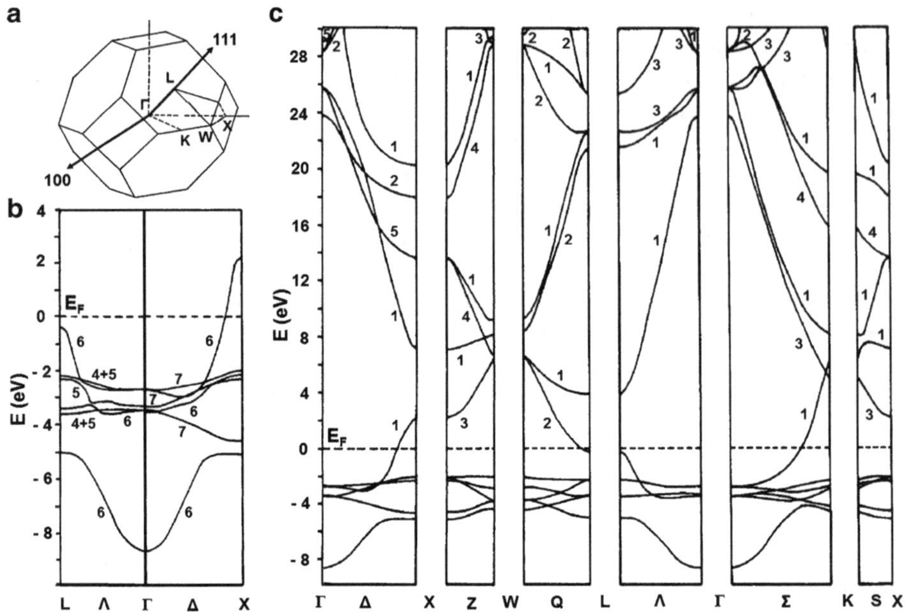

Fig. 2.6 Electrons in an fcc crystal. (**a**) First Brillouin zone. (**b**) Relativistic band structure of Cu along the $\langle 111 \rangle$ and $\langle 100 \rangle$ direction. (**c**) Skalar relativistic band structure of Cu along several high symmetry directions. For explanation see text. (**b**, **c**) Reproduced with permission from Ref. [29]. © IOP Publishing 1984. All rights reserved

$$M_{if} = i\frac{e\hbar}{mc}\psi_f|\boldsymbol{A}\bullet\nabla|\psi_i. \tag{2.14}$$

This matrix element can be expressed in various ways [13–17], usually with the assumption that the wavelength of the EM wave is large compared to the interatomic distances—called dipole approximation—which is well-fulfilled in the wavelength range of interest in band structure UVPEEM. Taking conservation of energy and of momentum into account we can finally write the transition probability as

$$T_{if} = \frac{2\pi}{\hbar}\left|\psi_f|\boldsymbol{A}\bullet\nabla|\psi_i\right|^2\delta\left(\varepsilon_f - \varepsilon_i - \hbar\omega\right)\delta\left(\boldsymbol{k}_f - \boldsymbol{k}_i - \boldsymbol{G}\right). \tag{2.15}$$

The wave function ψ_i is obtained by solving the Pauli-Schrödinger equation with the Hamiltonian without H_{rad}. This is in general justified because the incident wave is only a weak disturbance of the system. In the three-step model ψ_f is obtained in the same manner and the propagation to and through the surface is taken into account separately. The one-step model takes into account that the final state consists of the photoelectron and a screened, decaying hole in the occupied states. The photoelectron not only experiences the exchange-correlation interaction ("self-energy" Σ) and attenuation (imaginary potential V_i), both energy-dependent, but also multiple scattering in the crystal lattice ("inverse LEED state"). The formalisms used in the one-step model can be found in Refs. [13–17]. Even the one-step model calculations do not give quantitative agreement with experiment without fitting Σ and V_i to the experiment as illustrated, for example, in Ref. [30]. Therefore we restrict ourselves to the three-step model, which gives at least semi-quantitative results which are sufficient in general for the interpretation of the ARPES results obtained in PEEM instruments in terms of the *electronic* structure. The extraction of the *atomic* structure from photoelectron diffraction patterns (PED) in the extended UV range up to about 100 eV is difficult because of strong multiple scattering and useful only when no other methods are available such as PED with high photon energies, where single forward scattering is a good approximation in many cases, or LEED when the specimen has sufficient long range order to allow a quantitative LEED analysis.

The knowledge of ψ_i and ψ_f gives also the band structure $E(\boldsymbol{k})$ which is illustrated in Fig. 2.6 for Cu, together with the Brillouin zone and the symmetry points and symmetry directions (a). Figure 2.6b shows the relativistic band structure (with spin–orbit coupling) in two main symmetry directions ($\langle 111\rangle, \langle 100\rangle$), Fig. 2.6c the scalar-relativistic band structure (without spin–orbit coupling) for more high symmetry directions up to 30 eV above the Fermi level [29]. The numbers indicate the symmetry properties of the bands, which determine the allowed (vertical) interband transitions between occupied and unoccupied bands. The relevant selection rules, which depend upon the polarization direction of the vector potential (see Eq. (2.15)), for normal incidence and some high symmetry directions can be found, for example, in Ref. [15]. The good agreement between ARPES measurements and theory is also illustrated in the same reference.

In UVPEEM the angle of incidence, measured from the normal is in general between 75º and 45º, the azimuth not in a high symmetry direction and the light is not polarized, laser-UVPEEM experiments excepted. The emission angle range of the photoelectrons around the surface normal can be limited at higher energies with the contrast aperture to a small cone so that some emission angle sensitivity can be achieved in imaging, for example along the $\langle 111\rangle$ (Λ) direction of a (111)-oriented crystal. Typical cone half-angles range from 2º at 22 eV to 1º at 90 eV electrons with an aperture used for high-resolution imaging. However near threshold the angle is quite large, for example nearly 10° at 1 eV. Without aperture all electrons are transmitted through the objective lens at this energy so that the image brightness is determined by the angle-integrated intensity as already mentioned earlier. This can be calculated from the band structure via the joint density of states defined

$$D(\hbar\omega) = \frac{1}{2\pi^3}\sum_{i,f}\int \frac{d\Omega_{f,i}(\boldsymbol{k})}{\left|\nabla_k E_f(\boldsymbol{k}) - \nabla_k E_i(\boldsymbol{k})\right|}, \tag{2.16}$$

where $d\Omega_{f,i}(\boldsymbol{k})$ is an element of a surface of constant energy difference between initial and final bands and $\nabla_k E(\boldsymbol{k})$ is the slope of the energy surface $E(\boldsymbol{k})$.

More conveniently this information is obtained by measurements from a flat polycrystalline surface with completely random orientation of the crystals. In any case the brightness of an UVPEEM image is not simply determined by the work function according to the Fowler-DuBridge theory (Eqs. (2.13) and (2.14)) but depends in general also upon the band structure, the light polarization, the angles of incidence, and the emission angle range used for imaging.

Up to now we have assumed that the amplitude A_0 of the vector potential is constant at the atomic level with the argument that the wavelength in the UVPEEM range is much larger than the interatomic distance. However, at the surface the dielectric constant changes within a short distance and, therefore, also $\boldsymbol{A}$, invalidating the assumption $\nabla \bullet \boldsymbol{A} = 0$ made up to now. This case has been discussed in a profound manner by Feibelman [31] and studied extensively in the literature. Strongly simplified, the sudden termination of the crystal lattice produces bonding states (surface states or surface resonances) of the electrons different from those in the bulk and the weaker bonding causes a stronger interaction with the incident wave, modifying A_0 within a short distance. This intimate coupling between EM wave and local electronic properties is, for example, clearly evident in a surface excitation, which has been and is being studied extensively with PEEM, the surface plasmon-polariton. We will come back to this later.

Quite general, surface excitations are important in PEEM because the polar angle of incidence θ is usually large, for example 74° in one of the commercial instruments. Therefore, the normal component $A_\perp$ of $\boldsymbol{A}$ and consequently also of the electric field, $E_\perp$, which determines the surface excitation probability, is large. This is evident in many experimental results, for example in the early experiments by Gartland et al. [32, 33] on Cu(111), who observed a maximum in the total photoemission intensity at an angle of incidence of 71°. More recent results for this surface (Fig. 2.7 [34]) show in (a) the quantum efficiency—the angle-integrated number of photoelectrons emitted per incident photon—as a function of θ for three different photon energies. At normal incidence ($A_\perp = 0$) there is only emission from the bulk, which increases with increasing number of accessible bulk states (increasing photon energy), but with increasing θ ($A_\perp$) emission from the surface state dominates at all energies. The transitions from bulk states, which are possible at the three energies, are indicated by different shading in the surface-projected band structure (b).

This example illustrates that when a surface state exists, it will determine the PEEM image intensity near threshold because of the large angle of incidence and the normal emission. At higher energies the relative contribution of surface states to bulk states decreases strongly because the number of contributing bulk states increases considerably. When a small contrast aperture is used in PEEM, only electrons close to the surface normal contribute to the image. This limits the number of accessible final states and produces intensity variations with energy corresponding to the band structure of the unoccupied states, as illustrated by experiments with Cu(111) and Ag(111) surface states [35]. The band structure causes also a strong nonmonotonic variation of the inelastic mean free path. In view of the strong influence of surface states a brief look at their spatial extension

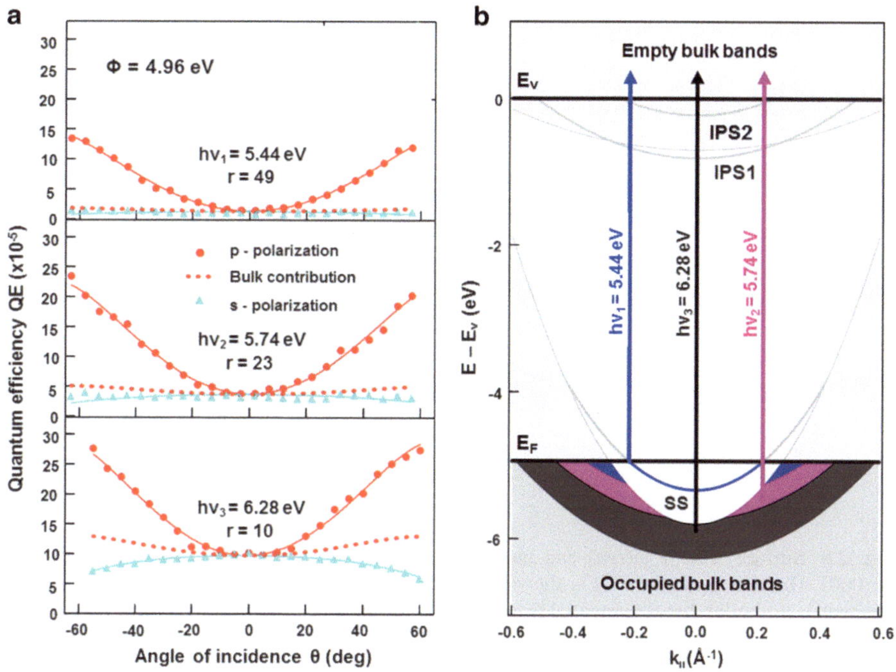

Fig. 2.7 Vectorial photoeffect in photoemission with polarized light. (**a**) Quantum efficiency from a Cu(111) surface as a function of the incident light angle θ and polarization for the three different photon energies. r is a measure for the relative absorption efficiency of the $\boldsymbol{E}_{\text{field}}$ components (or vector potential $\boldsymbol{A}$) perpendicular and parallel to the surface. (**b**) Surface-projected band structure of the Cu(111) surface showing the initial states from which electrons can be excited above the vacuum level. Low energy light (*blue*) can cause emission nearly only from surface state (SS) but with increasing energy (*pink* and *black*) also bulk states are increasingly accessible. What is important for PEEM is the strong increase with increasing θ (increasing $E_{\perp}$). Adapted from Ref. [34] with permission from the American Institute of Physics, © 2008

normal to the surface is appropriate. Figure 2.8 [36] shows the wave functions of the surface state and of a bulk state with comparable energy, together with the potential $V(z)$ as a function of distance from the surface for Al(111) and Be(0001). It is evident that strong potential variations cause a strong localization and high amplitude of the surface state while weak potential variations cause significant overlap of surface and bulk state.

The results shown in Fig. 2.7 were obtained with single photon photoexcitation. Multiphoton excitation opens up additional channels for photoemission. This is illustrated in Fig. 2.9 for multiphoton photoemission (nPPE) normal to the Cu(100) surface [37]. This surface has no surface state but a surface resonance (SR) and similar to Cu(111) two image potential states (IP). One of them is indicated in the right hand side of the figure, which shows the band structure along the surface normal (in the lower half) and the nPPE transitions induced by multiple absorption of 3.07 eV photons. The resulting photoelectron yields are shown on the left side of the figure. It is apparent that the high intensity of 3PPE and 4PPE is mediated

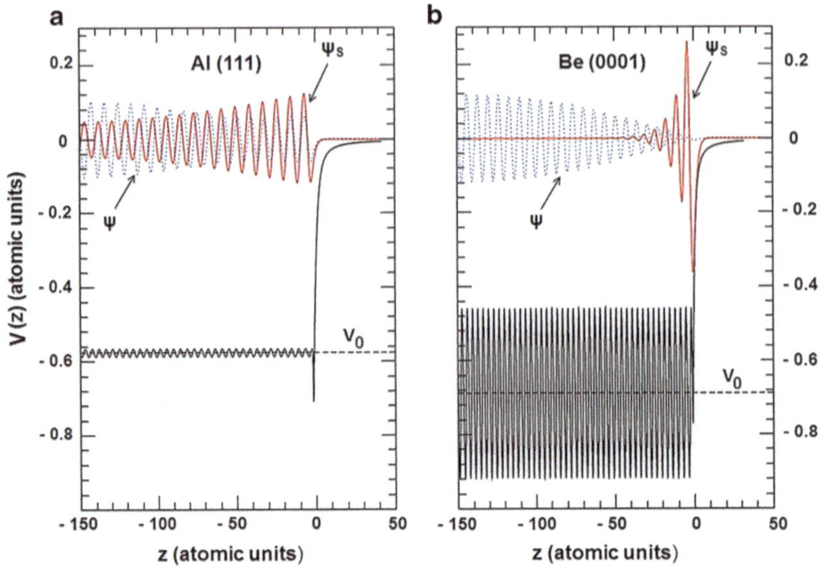

Fig. 2.8 Surface state Ψ_S (*red*) and bulk state Ψ wave functions (*gray*) of Al(111) (**a**) and Be(0001) (**b**). The potential $V(z)$ is shown in *black*. Although the mean inner potential V_0 of the two metals is similar, the Be potential is much more corrugated, which is responsible for the rapid decay of the surface state into the bulk compared to Al. Reprinted with permission from Ref. [36]. Copyright 2012 by the American Physical Society

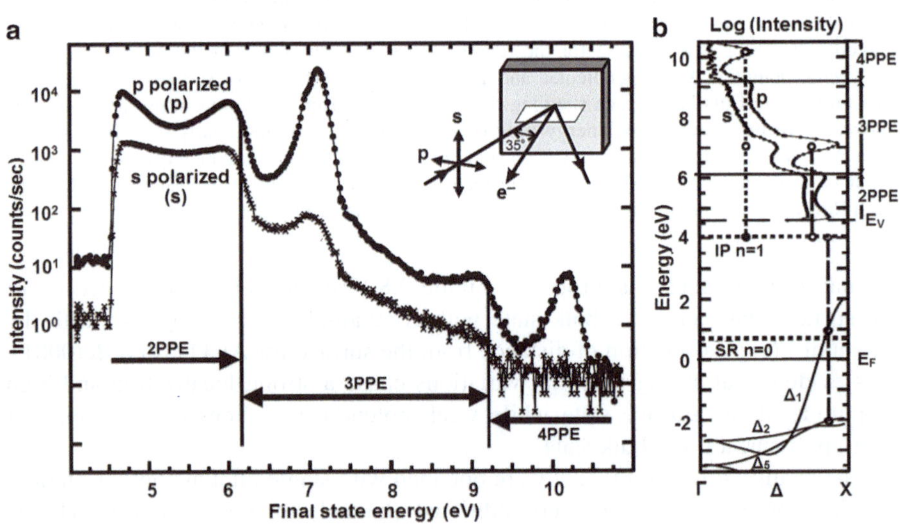

Fig. 2.9 Above threshold photoemission mediated by n-photon absorption from Cu(100) (work function $\Phi = 4.63$ eV). (**a**) Normal emission multiphoton photoelectron yield excited by p- and s-polarized frequency-doubled light from a Ti:sapphire laser ($\hbar\omega = 3.07$ eV) in logarithmic scale; $\theta = 35°$. The *arrows* show the contributions of the various nPPE processes, the inset the experimental geometry. (**b**) Illustration of the 3PPE and 4PPE transitions via an image potential state IP leading to the $I(E)$ dependence shown in (**a**). Adapted with permission from Ref. [37]. Copyright 2006 by the American Physical Society

by transition to the IP 1 state followed by absorption of one or two more photons. A full understanding of the spectrum, in particular of the lower efficiency of 2PPE compared to 3PPE, requires a detailed discussion of the band structure, which can be found in the original publication [37]. Much of the information from photoelectron spectroscopy discussed here has not been used yet in PEEM but should become important with increasing use of spectroscopic instruments. The *n*PPEEM studies made to date usually show many spots with much higher intensity ("hot spots"), which very likely are also present in the laterally averaging spectroscopy studies and may distort the spectra unless they have the same spectrum as the rest of the surface. An interesting phenomenon involving *n*PPE is the emission from negative electron affinity diamondoid (tetramantane-6-thiol) monolayers on Ag(111) and Au(111) surfaces. With 400 nm light $n = 2$ was observed, with 800 nm light $n = 3.7$ [38]. The narrow energy distribution and high yield makes this phenomenon promising for high brilliance pulsed electron sources.

2.2.4 Spin Effects in UV Photoemission

Up to now we have not taken the spin of the electron into account although spin–orbit coupling makes the band structure and consequently also interband transitions spin-dependent, irrespective of whether or not the material is ferromagnetic or antiferromagnetic or not. Spin-polarized photoemission has been studied for a long time, both with conventional light sources such as HeI light (21.2 eV) and with synchrotron radiation [39–42]. Most work was done with circular-polarized light but for special geometries and symmetries linear-polarized light also allows spin-discrimination. Here we show only a recent example of spin selection in a nonferromagnetic material, Cu(100), using 3PPE, circular polarization, and normal emission. Figure 2.10 [43] illustrates the principle and the result: (a) the 3PPE transition from the spin–orbit split bands of the relativistic band structure shown already in Fig. 2.6 near the zone boundary via the unoccupied sp band and the image potential state to final unoccupied states; (b) the *k*-region, in which the transitions occur is re-plotted strongly distorted along the energy scale together with the transitions A and B with opposite spin orientation (top). The non-spin-resolved spectrum obtained with a fundamental energy of 3.05 eV, displayed in (b) (bottom), shows that the two spin signals differ by more than a factor of two. This bodes well for energy-resolved magnetic imaging of ferromagnetic materials without spin detection. Its usefulness for PEEM of nonmagnetic materials is not obvious, although strain, step bunches, adsorbate islands can modify the spin–orbit splitting so that they could be imaged this way in principle. However, the work function changes caused by these surface modifications provide a much direct and easier way to image them.

In ferromagnetic materials the situation is quite different. In addition to the spin-splitting of the energy bands caused by the spin–orbit interaction we have the spin-splitting caused by the exchange interaction. This opens up additional routes for spin-dependent photoelectron emission as illustrated in Fig. 2.11 for a

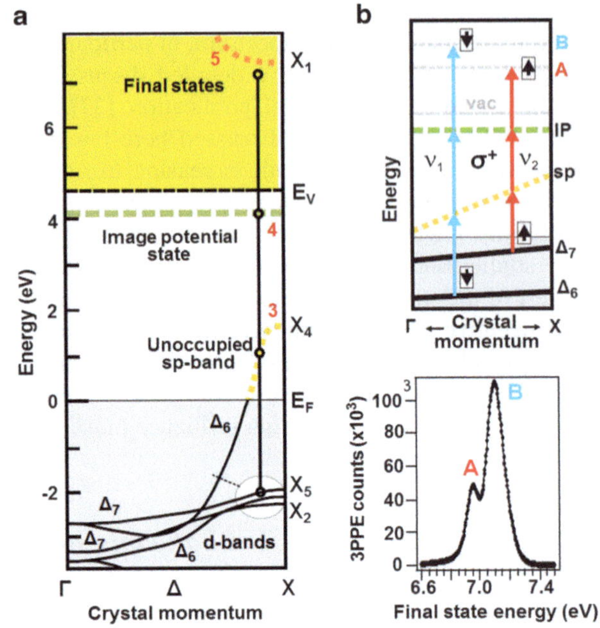

Fig. 2.10 Spin-selected photoemission via circular-polarized three photon photoemission (3PPE, $\hbar\omega = 3.07$ eV) from spin–orbit split states of Cu(100); $\theta = 42°$. For explanation see text. Adapted with permission from Ref. [43]. Copyright 2008 by the American Physical Society

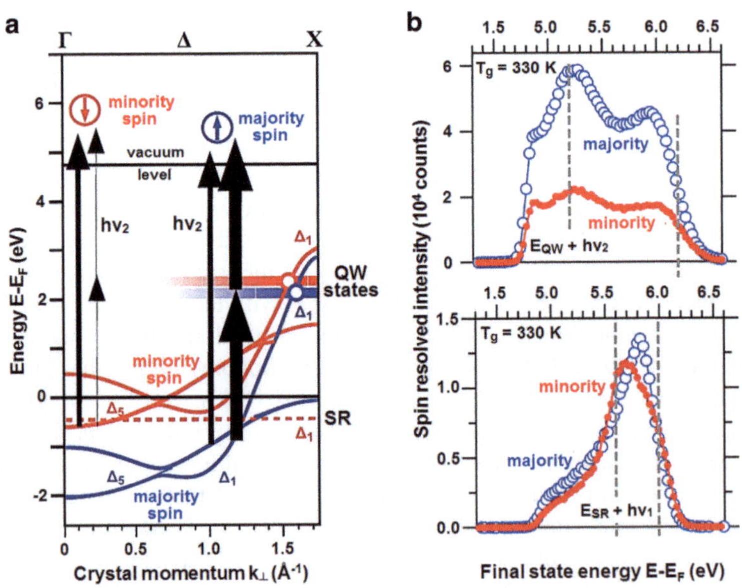

Fig. 2.11 Spin-selected photoemission via linear-polarized one- and two-photon photoemission ($\hbar\omega = 3.1$ eV) from exchange-split states in a fcc Co film on Cu(100); $\theta = 45°$. For explanation see text. Adapted with permission from Ref. [44]. Copyright 2012 by the American Physical Society

6-monolayer thick epitaxial fcc Co film on Cu(100) for emission normal to the surface upon excitation with p-polarized light incident at 45° [44]. Co(100) has a surface resonance and the film has a quantum well state near the zone boundary. In the band structure picture (a) the transitions possible from the exchange-split bands via one and two photon absorption with photon energies of $h\nu_1 = 6.0$ eV and $h\nu_2 = 3.1$ eV, respectively, are indicated by arrows whose width indicates their relative contribution to the intensity; (b) shows the spin-resolved intensities for 2PPE as a function of final state energy (top) and the corresponding results for 1PPE (bottom). It is obvious that the transition from the majority spin band via the quantum well state is more than a factor of 2 stronger than the transition from the minority spin band, while in the corresponding 1PPE transition the difference is much smaller. This example points out the importance of multiphoton photoemission and of sufficiently long-lived intermediate states for spin-resolved measurements such as for spin-resolved PEEM. It is interesting to note that the energy dependence of the final state intensity in the 2PPE transition in the Co film is quite sensitive to the film structure but not the spin polarization of the emitted electrons.

Spin-resolved PEEM becomes important whenever the direction of the spin alignment is not homogeneous. In the bulk the alignment is determined by the magnetocrystalline anisotropy, which is caused by the spin–orbit coupling but in thin films or small particles other anisotropies dominate the alignment: surface and interface anisotropy, magnetoelastic anisotropy caused by misfit strain, and shape anisotropy, a consequence of the shape- and size-dependent strayfield energy. The competition between all these anisotropies breaks up the homogeneous spin alignment into domains with different spin-orientations, separated by domain walls. Spin-resolved PEEM allows studying these domains and domain walls as a function of applied field, temperature, and other influences such as current through thin magnetic wires (spin current). An example of the application of the 2PPE process illustrated in Fig. 2.11 is the imaging of the domain structure of a 8-monolayer thick Co film on Cu(100) with an imaging spin-filter [45], using p-polarized light, which will be described in Chap. 3.

It should be mentioned that magnetic domain images excited by p-polarized light can also be obtained in near threshold emission without spin filter. Actually the first UVPEEM domain images were from a 100 nm thick polycrystalline Fe film, excited with light from a Hg high-pressure arc lamp [46]. Without spin selection magnetic contrast is obtained by changing the magnetization direction, subtracting the resulting images, and dividing by their sum pixel by pixel. This gives the magnetic linear dichroism (MLD) asymmetry image $A_{MLD} = [I(M_+) - I(M_-)]/[I(M_+) + I(M_-)]$. The magnetic contrast in this film cannot be explained in terms of a picture similar to that shown in Fig. 2.11 because the film is polycrystalline, has no quantum well state and the photon energy is too low for excitation into one of the known unoccupied states. Therefore, a phenomenological explanation in terms of a photoemission counterpart to the magneto-optical Kerr effect was given.

Magnetic *linear* dichroism is restricted to certain measurement geometries and crystal symmetries [27, 39–42]. Therefore, most of the magnetic

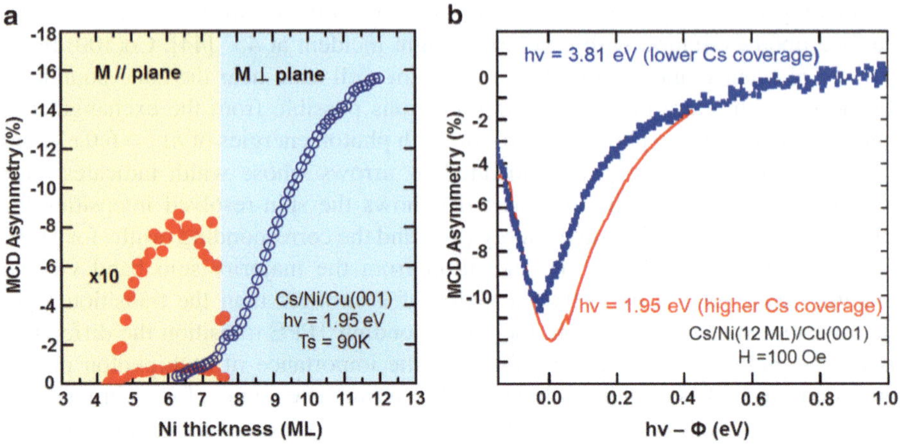

Fig. 2.12 Threshold magnetic circular dichroism (MCD) in photoemission. (**a**) Thickness dependence of the MCD of a cesiated Ni film on Cu(100) surface with work function $\Phi = 1.8$ eV, measured via the drain current from the sample. (**b**) Dependence of the MCD of a cesiated 12 monolayer thick Ni film on Cu(001) upon the difference between photon energy and work function. Adapted with permission from Ref. [47]. Copyright 2006 by the American Physical Society

dichroismspectroscopy studies used *circular* polarized light, which has fewer restrictions. In magnetic circular dichroism (MCD) measurements usually the helicity of the circular polarized light is changed instead of the magnetization direction and the MCD asymmetry A_{MCD} is defined by $(I_{\mathrm{r}} - I_{\mathrm{l}})/(I_{\mathrm{r}} + I_{\mathrm{l}})$, where r and l stand for right-handed and left-handed helicity. As an introduction to threshold MCDPEEM Fig. 2.12 [47] shows some results of the MCD calculated from total emission current measurements, that is measurements without angle and energy selection, from epitaxial Ni films on Cu(100). Illumination with 1.95 and 3.81 eV light was used. Up to a thickness of 7.5 monolayers the magnetization in these films is in the film plane, above this thickness perpendicular to the film plane as indicated in Fig. 2.12a. The work function of the surface was adjusted by Cs deposition so that with 1.95 eV photons threshold emission occurred. At threshold the MCD is largest as seen in Fig. 2.12b, which shows the dependence of the MCD upon the difference between photon energy and work function for two different photon energies. In contrast, the total emission current increases strongly. This difference is due to the fact that additional transitions with opposite sign of the MCD or zero MCD from different energy bands near the Fermi level become possible. This indicates the unique possibilities of threshold MCD for magnetic imaging, which were demonstrated soon thereafter [48] with Cs-adjusted threshold PEEM with 3.05 eV cw laser and a 3.06 eV pulsed laserlight incident at 55°, using the same 12 nm thick Ni film as in Fig. 2.12b. Thus threshold MCDPEEM promises to become an important magnetic imaging method, though somewhat limited by the need of adjusting the work function to the photon energy. This adjustment is limited by the availability of tunable bright light sources in the energy range of most work functions.

Before leaving the subject of spin-related effects another relativistic effect should be mentioned yet, which occurs on surfaces with a large potential gradient, equivalent to a strong electric field, the Rashba effect [49]. It causes an energy splitting which is proportional to the momentum k of the electron. In the nearly free electron picture the Rashba effect is at constant energy apparent in two concentric rings instead of one ring in k space, with the spin rotating in opposite direction in the two rings. The study of this phenomenon has been the domain of high-resolution ARPES but in cases, in which the energy splitting is large it is also accessible to spectroscopic PEEM in the reciprocal space imaging (LEED) mode. Using the field-limiting aperture allows studying variations of the Rashba effect across the surface in the submicron range.

2.2.5 Surface Plasmon Photoemission

The pulsed laser multiphoton photoemission experiments described in the previous paragraphs were made on well-prepared smooth surfaces and could be interpreted in terms of the band structure including quantum well and image potential states. It had been observed already around 1990 that pulsed lasers with photon energies below the work function could produce emission of energetic electrons. Aeschlimann et al. [50] studied this phenomenon on Cu(100) and Cu(110) single crystal surfaces. They noticed that the electron yield depended strongly on the surface roughness and attributed the strong emission to surface plasmon excitation, based on Raether's work [51]. Raether's group had studied surface plasmons in depth with electron energy loss spectroscopy and optical methods and had found that their oscillations can enhance the local field strength of the incident wave by orders of magnitude. The physics of localized surface plasmons (LSPs) and of surface plasmon polaritons (SPPs, evanescent surface electromagnetic waves) [52–56] connected with propagating surface plasmons (charge density wave) has developed into an important field, mainly because of its possible practical application in many fields, in particular in nano-optics ("plasmonics") [57–62].

Here we briefly summarize only the basics needed later in the discussion of plasmon-PEEM (Chap. 6). Plasmons can be excited by light in materials, whose real part ε_1 of the complex dielectric function $\varepsilon = \varepsilon_1 + i\varepsilon_2$ is negative and its imaginary part $\varepsilon_2 \ll \varepsilon_1$. Ag is the best material from this point of view, followed by Au, with Cu in some distance. In these materials plasmons can be excited in many forms of the specimen. We discuss here only the limiting cases, the perfectly smooth planar surface and the small spherical or ellipsoidal particle. The charge distribution and associated electric field of a SPP on a planar surface is illustrated in Fig. 2.13a [63]. The energy of a SPP depends upon its wavelength as shown by the dispersion relation marked SPP in Fig. 2.13b [63]. Its energy comes from absorption of light, whose dispersion relation is shown by the light line. For absorption to occur energy and momentum have to be conserved. This can be achieved (1) by changing the slope of the light line so that it intersect the SPP curve or (2) by shifting the light line to

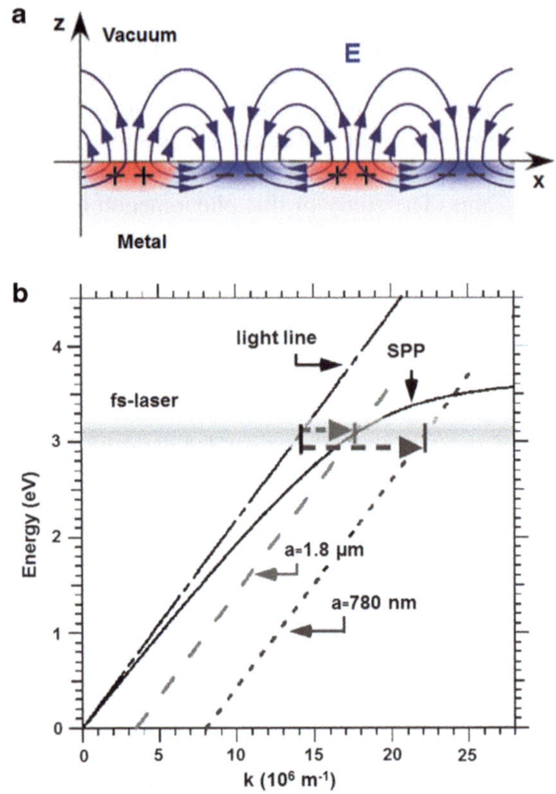

Fig. 2.13 Surface plasmons. (**a**) Charge distribution and associated electric field of a surface plasmon. (**b**) Dispersion relation of surface plasmons and coupling with light. For explanation see text. From Ref. [63] with permission. Copyright 2008 The Vacuum Society of Japan

the right as indicated in Fig. 2.13b. Method (1) is the classical Kretschmann method of illuminating a very thin film on a prism with refractive index *n* from the backside, which reduces the slope of the light line by a factor 1/*n* so that it intersects the SPP curve. This is impractical for PEEM, though possible in principle, while solution (2) is simpler to realize because it can be done by illumination from the front. The shift implies a momentum transfer. This is achieved either by diffraction from a surface grating or by scattering from small surface irregularities, which can be produced in nearly any size, shape, and distribution. In the first method a true SPP is produced, also called surface plasmon wave (SPW) [64]. In the second case localized surface plasmons (LSPs) are excited, which can also support a SPW via coupling between the LSPs. In both cases the radiation fields of the oscillating charges cause a strong enhancement of the incident wave field, producing the high fields mentioned above. For example, for plasmon excitation in a flat Ag film in the Kretschmann set-up the plasmon field is under optimum conditions (zero optical reflectivity) up to 250 stronger than the incident field at a wavelength of 600 nm. The same field *intensity* enhancement $E^2_{\mathrm{Plas}}/E^2_{\mathrm{inc}}$ can also be achieved with grating excitation [51]. However, with increasing deviation from the optimum conditions the field enhancement decreases strongly, for example to 80 at 420 nm. The decay length ($1/e$ distance) of the plasmon field into the vacuum is in the 100 nm range, for example 390 nm in the case of Ag under the optimum excitation conditions mentioned, *into* the metal it is the skin depth, 24 nm and the decay length of the SPP *along* the surface is 36 μm. Taking the group velocity $d\omega/dk$—the slope of the SPP curve in Fig. 2.13b, divided by $\hbar$—into account, a lifetime of the SPP in the low 10 fs range follows. The short decay length and life time of these evanescent waves is due to energy transfer from the collective charge fluctuations of the plasmon to phonons (heating), to photons (radiation), and to electrons (single electron excitations), the last transfer being the process of interest in PEEM.

The second limiting case, the small particle is illustrated in Fig. 2.14a [65] for a spherical particle. Under the influence of the oscillating field the electrons oscillate collectively, forming localized charge oscillations in the field direction. The field *amplitude* enhancement factor $FEF = |F_n|/|F_{\text{inc}}|$($F_n$ normal component of oscillating field) caused by these oscillations are shown for spherical and ellipsoidal particles with an axial ratio $a/b = 3$ along the long axis in Fig. 2.14b, c [66] for different particle sizes and shapes. For $a/b = 5$ and $a = 25$ nm $FEF = 250$ at the optimum wavelength, similar to the maximum values for the flat film. This change in the interaction with aspect ratio is illustrated by extinction measurements from well-defined particles prepared by electron beam lithography in Fig. 2.14d, e [67] for polarization parallel and perpendicular to the long axis, respectively for $b \approx 40$ nm. The large difference between the two polarization directions due to the different polarizabilities and the log scale should be noted. A detailed account of this limiting case can be found in Ref. [68].

The two limiting cases show that large field enhancements can occur in a wide range of specimen configurations and excitation modes. The question is now how large are the plasmonic fields. The average intensity of the incident wave is connected with the amplitude of the electric field by $I = 0.5c\varepsilon_0 E_0^2 = 26.544 \times 10^{-4} E_0^2$, where E_0 is measured in V/m and I in W/m^2. For an average intensity of 10^{12} W/m^2, which is in the range used in many experiments, the peak field strength E_0 is $\approx 0.2 \times 10^8$ V/m. At this field strength no field emission can occur. The onset of field emission is usually characterized by the Keldysh parameter $\gamma = \omega(2m\Phi)^{1/2}/eE$, which depends upon the frequency ω of the light wave, the work function Φ, and the field E. For $\gamma \ll 1$ electrons tunnel through the barrier, for $\gamma \gg 1$ they are excited over the surface barrier via multiphoton photoemission. At the frequently used wavelength $\lambda = 800$ nm, corresponding to $\omega = 2.35 \times 10^{15}$ Hz, $\gamma = 1$ at a field strength of 2.26×10^{10} V/m for Ag whose work function is 4.26 eV. We will discuss here only the case $\gamma > 1$, i.e., multiphoton photoemission (nPPE), which is experimentally verified by the dependence of the emission current upon light intensity, $I_n \sim E^{2n}$. At low intensities, e.g., in the 10^{11} and low 10^{12} W/m^2 range 2PPE is observed and the energy distribution of the emitted electrons reflects the density of states of the material [50, 69–71]. With increasing power 3PPE and even 4PPE occurs and the energy distribution changes dramatically, with energies far above those expected from the absorbed photon energy. For example, Irvine et al. [72] observed at an intensity of

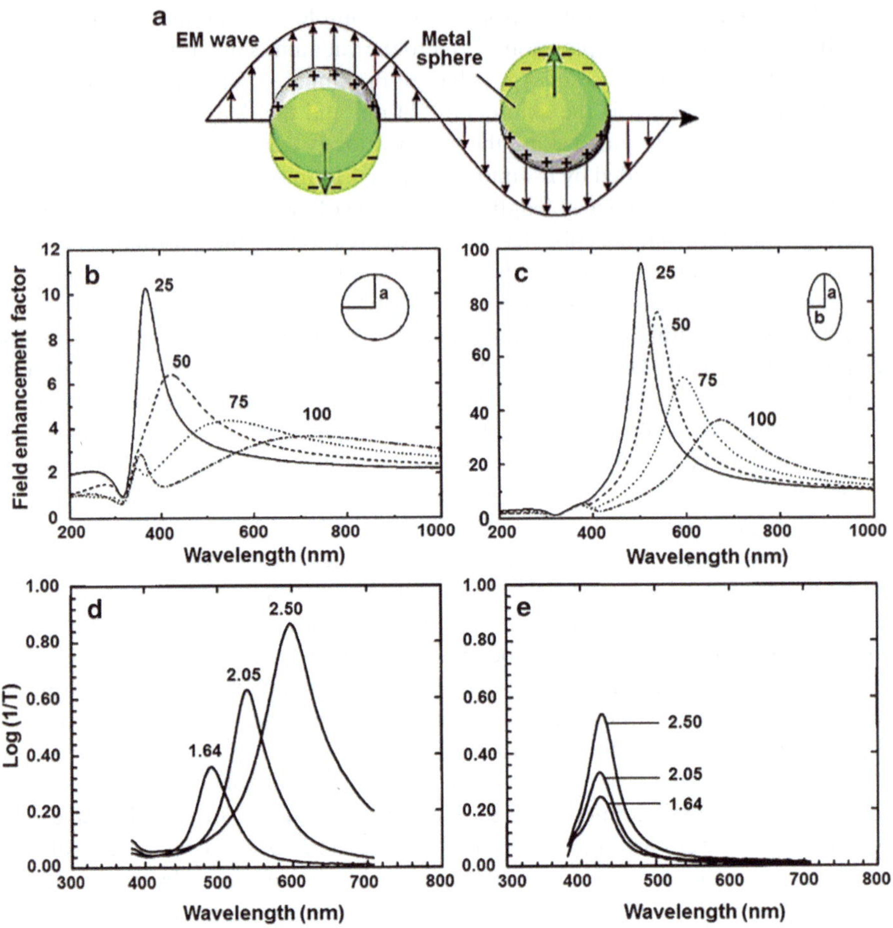

Fig. 2.14 Plasmons in small particles. (**a**) Schematic of the excitation of synchronous charge fluctuations in the particles by an incident electromagnetic wave, which produces plasma radiation, leading to a strong enhancement of the electromagnetic field. (**b**, **c**) Dependence of the field enhancement on the size of (**b**) spherical particles and (**c**) ellipsoidal particles with $a/b = 3$ as function of wavelength; a in nm, dielectric constants of Ag, wave incident along horizontal direction. (**d**, **e**) Extinction (1/Transmission) of light incident on 30 nm thick elliptically-shaped Ag particles for various a/b ratios with $b \approx 40$ nm as a function of wavelength incident (**d**) parallel and (**e**) perpendicular to the long axis. (**a**) Adapted with permission from Ref. [65]. Copyright 2003 American Chemical Society. (**b**, **c**) Adapted with permission from Ref. [66]. Copyright 2004 by the American Physical Society. (**d**, **e**) From Ref. [67] with permission from Optical Society of America

1.8×10^{13} W/m^2 3PPE and a maximum of the energy distribution at 315 eV with a full width at half maximum (FWHM) of 83 eV. They attributed the high energy and large energy width to acceleration of the electrons by the ponderomotive force of the oscillating surface field. The energy of an emitted electron is determined by the phase of the field at which it is ejected from the field. In order to achieve agreement between calculation and experiment a field enhancement factor of about 10^3 was necessary, higher than what has been mentioned above. At somewhat lower intensities ((0.17–0.89) $\times$ 10^{13} W/m^2) Aeschlimann et al. [50] observed an energy-dependent power dependence of the emission ($n = 2.8$ at 0.8 eV and 3.2 at 2.5 eV) and an exponential decrease of the intensity with energy up to 6 keV at the highest power, which was attributed to field enhancement at hot spots. The wide energy spread reduces the resolution in PEEM so that most PEEM experiments (Chap. 6) use lower laser powers to which we restrict the following discussion of the recent theoretical results.

Most calculations assume a free electron metal. The most detailed treatment of interest for *n*PPE has been given by Yalinin et al. [73]. They calculated the total emission current and its angular distribution for 800 nm light ($\omega = 2.35 \times 10^{15}$ Hz), work function $\Phi = 5.5$ eV and initial-state energy (Fermi energy) of 4.5 eV as a function of the amplitude E_0 of the incident light for pulse durations between 2.7 and 27 fs. They found good agreement between three different theoretical approaches and approximations, including the solution of the one-dimensional Schrödinger equation. In the range of interest for PEEM—their weak-field regime—they obtained a power dependence of the emission probability upon E_0 with $n = 4$ for $\Phi = 5.5$ eV and with $n = 5$ for a fictive $\Phi = 7.1$ eV. However with increasing E_0 increasing deviations from the integer values occurred (Fig. 2.15a) [73], which they attributed to decreasing reflection of the oscillating electron in the oscillating ponderomotive field at the surface. This could, for example, explain the deviations mentioned above. Details of the angular distribution, of the time-dependent field, etc., can be found in their paper. Here it should only be mentioned that it does not include the image potential and the field penetration but nevertheless accounts for many experimental observations. The field penetration (or induced potential) plays an important role at and near the surface plasmon resonance. This is illustrated in Fig. 2.15b [74], which shows the energy distribution of the electrons emitted normal to the surface upon excitation by very short pulses with the surface plasmon frequency ω_s. Figure 2.15c [74] gives a visual impression of the induced potential. In an earlier study of the same system (Al(111)) [75], which was limited to $\omega > \omega_s$, the field penetration was not taken into account. The results showed features observed in experiments, for example the exponential decrease of the intensity with energy mentioned above. Details can be found in this paper. Band structure effects, including surface states seem to change the energy distributions calculated for the free electron model little for Al but have a strong influence in Be. This has been attributed to the much stronger localization of the surface state in Be than in Al, illustrated in Fig. 2.8, with a correspondingly much higher density of states in it in Be than in Al [36]. Thus, while it appears that the free electron model can explain many experimental results, there are also many details, which are material-specific. The details of plasmon-mediated photoemission depend on many factors: (1) the properties of the plasmon-exciting wave, such as pulse frequency, length, shape, intensity, polarization, (2) the properties of the specimen such as period, profile, and length of the grating, size, shape, and distribution of particles, (3) the relation between incident wave and specimen,

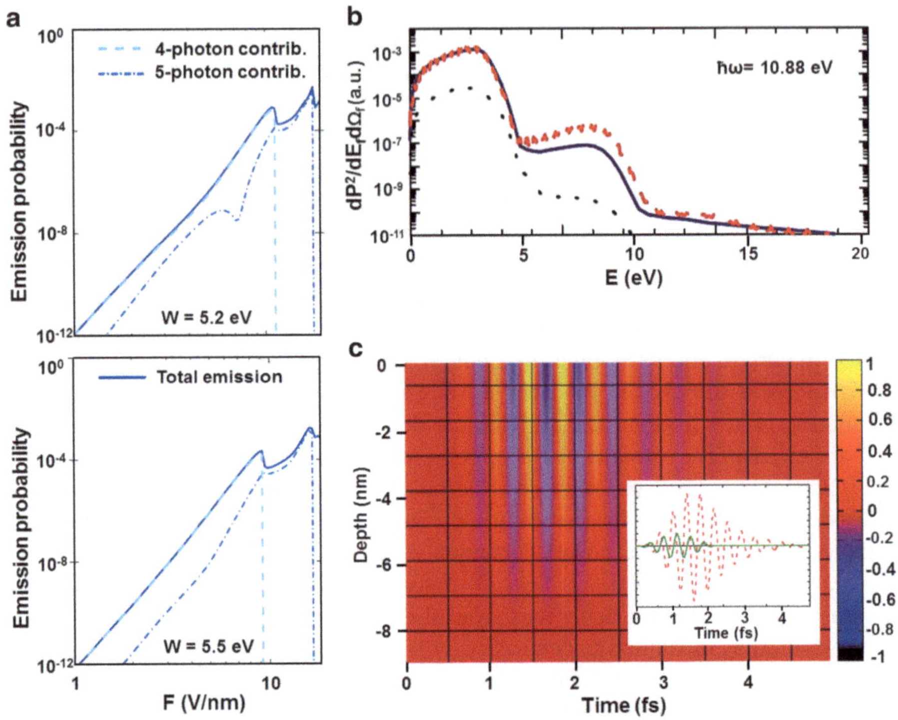

Fig. 2.15 Influence of field enhancement on photoemission due to plasmon excitation by high power fs pulses. (**a**) Contributions of 4PPE and 5PPE processes to the total (full line) emission probability for a work functions of 5.2 eV (*top*) and 5.5 eV (*bottom*) and $\hbar\omega = 1.55$ eV (Ti:sapphire laser) as function of field strength. The 4PPE is the threshold multiphoton process and dominates the total emission. Above 18 V/nm tunneling occurs. (**b**) Differential electron emission probability as a function of the electron energy for normal emission, excited at the surface plasmon frequency. The *dashed* and *solid curves* take the induced potential into account, the *dotted curve* not, indicating the strong influence of the induced potential on the emission probability at all energies. (**c**) Induced potential as a function of time and distance z perpendicular to the surface for excitation at the plasma frequency. The *inset* shows the laser pulse (*solid line*) and plasmon pulse (*dashed line*) at a depth of 2.5 nm. (**a**) Reprinted with permission from Ref. [73]. Copyright 2011 by the American Physical Society. (**b, c**) From Ref. [74] with permission from Optical Society of America

such as angle of incidence—also azimuthal angle in case of non-isometric particles-, match between frequency of incident wave and particle size- and shape-dependent plasmon frequency, and other parameters. Therefore a comparison between theory and experiment and even between different experiments is difficult.

For applications in plasmon nanooptics the decay length along plasmon stripes or wires, which is determined by the plasmon life time, and the coupling between wires and particles via the plasmonic radiation field, which is determined by the decay length is important. In the past optical and laterally integrating photoemission measurements were used to characterize the radiation field and scanning electron microscopy (SEM) to characterize the structure of the patterns. PEEM can accomplish both in the 10–100 nm range and has been used very successfully for this purpose as we will see in Chap. 6. In both cases the same light sources and plasmon materials are used: predominantly Ti:sapphire laser-based systems with repetition rates between 75 and 95 MHz, pulse widths of several 10 fs, pulse energies of the order 1 nJ, power densities between 10^2 and 10^3 MW/cm^2 resulting in field strengths in the 10^6 V/cm range. These systems are tunable over a narrow wavelength range (typically from 770 to 830 nm, corresponding to 1.61–1.49 eV), which gives some flexibility in the adaption to the plasmon material. In order to overcome the work function of the material the pulses are fist frequency-doubled so that emission via 2PPE becomes possible or high radiation power is used to excite nPPE with $n > 2$. Free-electron-like metals, for which most of the theoretical work has been done, have either very high surface plasmon energies, such as Al, or are very reactive such as the alkali metals, or have no well-defined plasma resonances. For these reasons Ag, Au, and to a certain extent also Cu, are the materials of choice. Furthermore their surface plasmon energies are in the range accessible with frequency-doubled light from the Ti:sapphire laser. As will become evident in Chap. 6 there are additional reasons for the choice of this laser in PEEM.

2.2.6 X-Ray Photoemission

From the point of view of PEEM X-ray photoemission may be divided into three groups: (1) slow secondary electron emission, (2) Auger electron emission, and (3) true photo electron emission. Auger electron emission is rarely used in PEEM because the strong peaks involve two electrons in the valence band. As a consequence the peaks are broad and much less suited for imaging than the true photoelectron peaks. Therefore, they will not be discussed here. For the first group a more appropriate name would be XSEEM but the acronym XPEEM has been generally accepted also for secondary electrons, not only for true photoelectrons. The argument for this usage is that the intensity of the secondary electrons is determined by the ionization cross section of inner shells just like in true photoemission. We start, therefore, with the third group although it is less widely used for two reasons: (1) it requires an energy filter, with which in many instruments are not equipped and (2) the intensity of secondary electrons is usually much higher than that of the true photoelectrons, which allows faster image acquisition. We call in the following the true photoelectrons simply photoelectrons and the electrons in the first group secondary electrons.

2.2.6.1 Photoelectron Emission

The characteristic parameters in this emission process are photoelectron energy, intensity, and angular distribution. The first two are determined by the binding energy E_B of the core level, by the polarization-dependent selection rules and by the photon energy $\hbar\omega$ used for imaging, the third one by the structure of the specimen. With energies referred to the Fermi energy the kinetic energy of the photoelectron is then with the work function Φ

$$E_{\text{kin}} = \hbar\omega - \Phi - E_{\text{B}}. \tag{2.17}$$

The binding or ionization energy is defined by the energy difference of the *total* N-electron system before and after ionization of the selected core level:

$$E_{\text{B}} = E_{\text{tot}}(N-1) - E_{\text{tot}}(N). \tag{2.18}$$

Core level ionization is a localized process in an extended, usually crystalline environment. The calculation of $E_{\text{tot}}(N-1)$ and $E_{\text{tot}}(N)$ is usually done by imbedding the atom/ion in which the hole is created and its immediate neighbors (ligands) into the environment represented by an effective crystal field simulated by point charges, as illustrated by calculations of the p-shell ionization of MnO [76]. The other approach is to start from the crystalline environment by calculating the crystal potential of an infinite crystal with one of the state of the art computational methods such as the density functional theory and deriving from it a set of Wannier functions localized at the central atom and its ligands. This method has recently also been applied to MnO and other oxides [77]. Both methods are called multiplet ligand-field theories but differ in the representation of the localized state and its environment. The calculations are quite elaborate and have not only to take exchange, correlation, and spin–orbit coupling of the two states into account but also the relaxation processes following the hole creation in the core level, which may be divided roughly into local (intra- and interatomic) and collective processes or—in XPS jargon—into shake-up and shake-down processes. The local relaxation involves cascading processes, in which an electron with lower binding energy fills the hole, thus creating another hole to be filled again. These electrons do not have to come from the atom with the original hole but also from a neighbor atom. As a result a number of electron configurations with different total energies are formed in the final state, giving rise to the multiplet structure in the photoelectron spectrum, which is usually described by atomic quantum numbers. In as much as in this relaxation process also the most weakly bound valence electrons are involved, there is also a collective response to the hole creation, which is particularly evident in 3d transition metals and their compounds. Interestingly the relaxation processes can not only involve transitions to more strongly bound states but also to unoccupied higher energy level states, with the energy increase compensated by parallel transitions to more strongly bound states. An example is the so-called frustrated Auger configuration process [78], in which an electron is promoted into an unoccupied 4f state.

Although serious efforts started already 40 years ago [79], the calculation of the multiplet structure is still an active field of research and the advantages and accuracy of different approaches are still a matter of discussion [80, 81]. Relative intensities and electron configuration assignments may well change with further improvements of the models and methods used in these calculations so that we will not dwell further on this subject but refer to reviews of this subject [82, 83]. The most important aspect of these calculations for spectroscopy in PEEM is the chemical shift caused by the bonding to the ligands and the relaxations in the emission process, which is used for identifying the bonding state, usually the oxidation state. Due to the complexity of the calculations reliable theoretical data are available only for a limited number of systems so that the experimentalist has to rely in his interpretation on simpler calculations (see e.g., [84]) and on the results of spectroscopic studies of well-defined systems if available.

While multiplet calculations are difficult, experiment has also its problems, which we illustrate now with a few examples of relative simple systems. These problems range from nonconductive specimens to background subtraction and to peak fitting. Charging not only causes peak shifts, which usually can be accounted for by comparison with peaks with known binding energy, but can also cause peak broadening, making correct peak fitting questionable. For background subtraction and peak fitting a number of proven methods are available [85] from which the experimentalist can choose, which puts some uncertainty into the final result [86]. Furthermore experimental conditions such as energy, energy width, polarization, angle of incidence of the X-rays, emission direction of the electrons ("take-off angle"), energy resolution of the analyzer and other factors have a strong influence. Figure 2.16a–f [87–91] illustrates some of these problems. Except for (c) and (d) the spectra are taken with Al Kα radiation with high-resolution spectrometers. They show the S 2p peak in different environments taken with different experimental conditions. In all cases the dominating features can be fitted with the spin–orbit split S $p_{3/2}$ and $p_{1/2}$ doublet with 1.20 eV spacing and 2:1 intensity ratio as expected from theory. However, there are significant differences in the peak shapes between (a) and (b) which are from UHV-cleaved sphalerite ($Zn_{1-x}Fe_xS$) surfaces, and between (d), (e) and (f) which are from UHV-cleaved pyrrhotite (Fe_7S_8, $Fe_{10}S_{11}$) surfaces. The difference between spectrum (a) with $x=0$ [87] and (b) with $x=0.25$ [88] cannot be explained by differences in take-off angle ($\theta=45°$ in a), 90° (open symbols), and 20° (full symbols) in (b) nor by different instrument resolution. The difference in x should rather have the opposite effect because of the more complex environment of S in (b). A possible cause for the broader peak in (a) is charging [88]. The difference between 90° and 20° take-off angle in (b) is attributed to the larger contribution of S surface atoms with slightly different bonding. Charging-induced peak broadening may also be responsible for the difference between the spectra (e) [90] and (f) [91], taken under very similar conditions (electron kinetic energy 1325 eV), but the large width of spectrum (d) [90] clearly has a different reason. It was taken with 210 eV synchrotron radiation light so that the spectrum is very surface-sensitive because of the low energy (48 eV) of the photoelectrons. As a consequence, a strong surface peak doublet with lower binding energy is needed for fitting the spectrum. A high binding energy peak doublet is needed in the spectra (d), (e), and (f), connected with different bonding sites in the bulk of these materials. The complication of the separation of volume and surface contributions does not exist of course in adsorbed

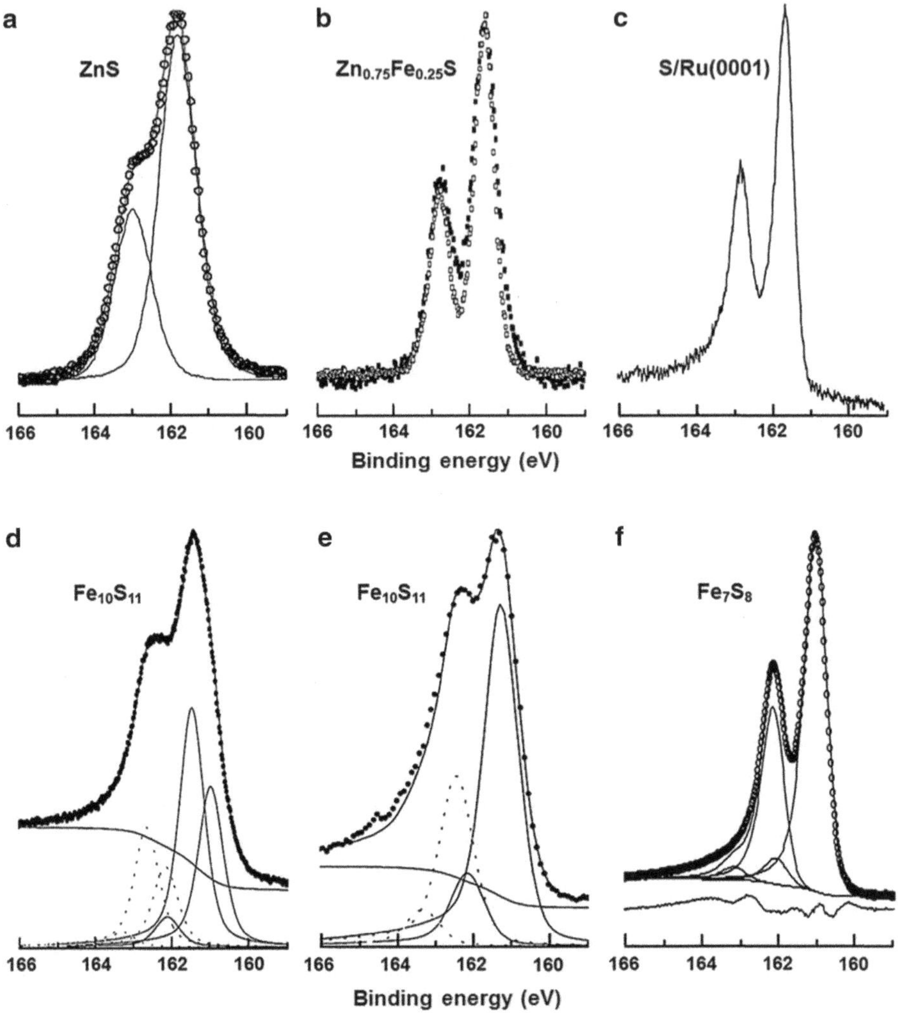

Fig. 2.16 Sulfur 2p X-ray photoelectron spectra in different environment and taken under different experimental conditions illustrating the influence of type of X-ray source, take-off angle, surface condition, and possibly charging. For explanation see text. (**a**) Reproduced from Ref. [87] by permission of The Electrochemical Society. (**b**) Reproduced with permission from Ref. [88]. Copyright 2007 Elsevier. (**c**) Reproduced with permission from Ref. [89]. Copyright 1998 Elsevier. (**d, e**) Reproduced from Ref. [90] with permission from Mineralogical Society of America. (**f**) Reproduced with permission from Ref. [91]. Copyright 2004 Elsevier

layers as indicated in spectrum (c) from 0.2 monolayers of S on Ru(0001) obtained with 260 eV photons [89].

The XPS spectra of monatomic materials seem at first sight simpler to analyze. That this is not so in general is illustrated with Fig. 2.17a–c [92–94] for the Fe 3p spectrum. Here we have in addition to the spin–orbit coupling and the relaxation

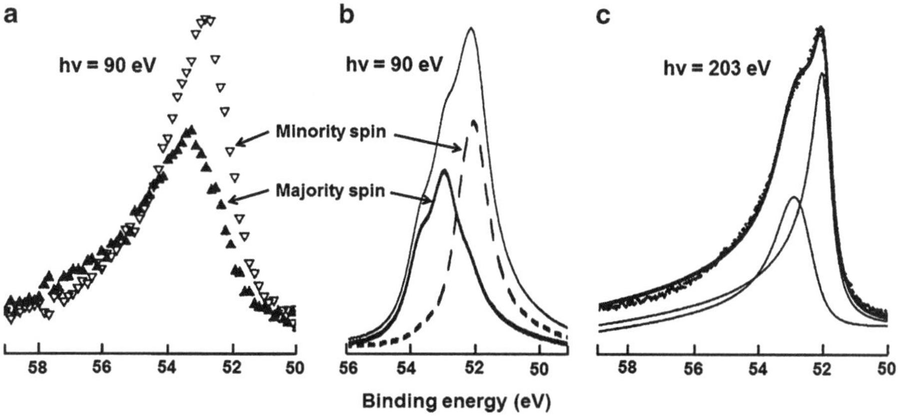

Fig. 2.17 Fe 3p XPS spectra. (**a**) Experimental spin-resolved spectrum [92], (**b**) Theoretical spin-resolved spectrum [93], (**c**) spin-averaged spectrum [94]. For explanation see text. (**a**) Reprinted with permission from Ref. [92]. Copyright 1990 by the American Physical Society. (**b**) Reprinted with permission from Ref. [93]. Copyright 1994 by the American Physical Society. (**c**) Reproduced from Ref. [94] with permission from The American Chemical Society, © 2008

effects also exchange effects. Spin-resolved XPS [92, 95, 96] has shown that the spin-up and spin-down peaks are shifted by about 0.5 eV (Fig. 2.17a [92]). The relative height of the peaks was found to depend upon the direction and polarization of the incident light with respect to the magnetization direction (magnetic dichroism) [97, 98] in agreement with theory [93]. Additional studies as a function of the magnetic field direction [99–102] revealed the complexity of the nature of the Fe 3p peak, for example its dependence upon energy, type of polarization (linear, circular) of the photons [102] and allowed to separate spin–orbit and exchange splitting contributions to the peak shape (0.67 and 0.41 eV). The spin-averaged spectrum (Fig. 2.17c) [94] is in approximate agreement with that shown in Fig. 2.17b [93], calculated for a specific combination of direction, polarization, and magnetization [93] but other peak shapes have been reported too. The 3p spectra of Co [99] and Ni [103] are of similar complexity. In materials with larger spin–orbit splitting, in which the peaks of the various multiplet configurations are separated better and which have been studied in far more detail, such as oxides and halides with well-defined oxidation states, multiplet analysis allows to determine the composition. However, as soon several oxidation states are present simultaneously or resolution and signal to noise ratio is limited, for example by the size of the features of interest, compositional analysis becomes elusive, with few exceptions such as the well-studied Si-O system. We will discuss such a difficult system in Chap. 6.5, where the nature of the features of interest, submicron crystals formed by reaction with the substrate, finally had to be determined by μ-LEED.

Despite of these limitations XPS is a valuable complement to the various imaging methods discussed in this book because it gives information on the atomic species present. XPEEM generally works with synchrotron radiation sources and is rarely used with the traditional Al or Mg Kα radiation, for two reasons: the photoelectrons produced by the traditional sources probe mainly the bulk and the intensity of even the most powerful X-ray sources is too low for imaging at magnifications of interest. Therefore, the quantification methods developed in the past for relative atomic composition determination are of limited value, in particular for small three-dimensional features. However, at least qualitative information can be obtained from binding energy and peak height in the spectrum. There are many tables of binding energies. The one generally used in XPEEM can be found in the X-ray data booklet of the Advanced Light Source [104, 105]. In contrast to definition (Eq. (2.17)) the binding energies of semiconducting elements are defined with respect to the top of the valence band, those of elements, which are gaseous at room temperature, with respect to the vacuum level. Of course, in compounds the chemical shifts, discussed above have to be taken into account, which rarely poses a problem because the binding energies of the elements differ sufficiently.

The XPS peak intensities characterized either by peak height or by area below the peak, are determined by a number of factors. First of all they depend on the initial ionization process and the process, which fills the hole created in the ionization. The first process is characterized by the ionization cross section, the second by the Auger yield. The ionization cross section depends upon the photon energy as illustrated for the core levels of Au in the upper part of Fig. 2.18 [106]. The Auger yield is the fraction of the ionization events, in which the system does not relax via fluorescence radiation but by ejecting a photoelectron and filling the hole via Auger electron processes. It depends upon nuclear charge as shown in the lower part of Fig. 2.18 [107] for different electron shells. In the energy range of interest in XPEEM there are always shells available for which the Auger yield is practically 1, so that all ionization processes lead to photoelectron emission. From the upper part of Fig. 2.18 the great advantage of synchrotron radiation becomes evident: by choosing the photon energy the ionization cross section can be maximized or the photoelectron energy can be selected for optimum surface sensitivity via the energy-dependent attenuation of the photoelectron current during its way to the surface.

The attenuation is caused foremost by inelastic scattering but elastic scattering also contribute to it. It is characterized by a number of quantities with the acronyms IMFP, TMFP, EAL, MED, and ID, which stand for inelastic mean free path, transport mean free path, effective attenuation length, mean escape depth, and information depth [108]. Other names such as electron sampling depth or probing depth are used too. All of them describe in one way or other the information depth of surface analytical methods and are frequently used interchangeably. A rough value for the IMFP can be extracted from the frequently used universal curve but there are many more accurate calculations and compilations of which only the most recent ones are cited here [108–113]. In particular Tanuma et al. [112] give an extensive list of IMFP values for many materials and electron energies from about 50 to 30000 eV. However care has to be taken when using them in XPS data interpretation. For example, Nakajima et al. [114] have measured the increase of the L edge signal of Fe, Co, and Ni as a function of thickness and deduced from it

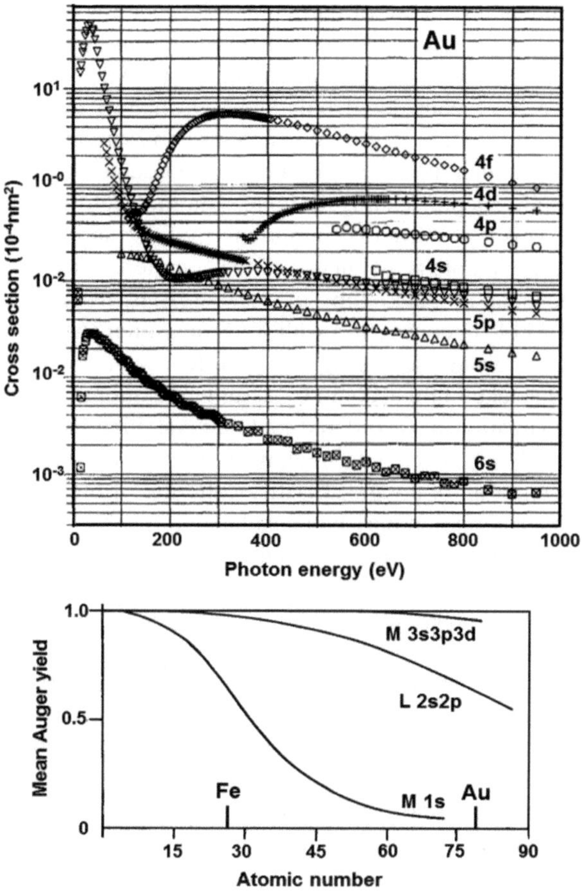

Fig. 2.18 *Top*: Atomic subshell photoionization cross sections of Au as a function of photon energy [106]. *Bottom*: Mean Auger electron yield Y_A for some subshells as a function of atomic number [107]. Y_A is a measure for the filling of the holes created by photoionization by electron emission instead of X-ray radiation. For example $Y_A \approx 1$ for the Au 4f and the Fe 3p shells, which are frequently used in PEEM, indicating that in the energy range of interest in PEEM nearly all holes in free atoms are filled by electron emission. *Top*: Reproduced from Ref. [106] with permission. Copyright 1985 Elsevier. *Bottom*: Reproduced from Ref. [107] by permission of Taylor and Francis Group, LLC

electron escape depths of 17 ± 2, 25 ± 3 and 25 ± 3 Å at the edge energies 707, 778, and 853 eV, respectively. In contrast, the most recent tabulated IMFP values for the same energies in these metals are 13.0, 12.1, and 13.3 Å, respectively [112]. Similar differences between IMFP values and mean probing depth values λ_p ($1/e$-values) were found in studies of the attenuation of photoelectrons of various energies E in Cr films as a function of thickness [115]. For $E = 101$, 464, 707 and 929 eV $\lambda_p = 9.8$, 15.6, 25.5, and 47 Å, while the corresponding IMFP values are 4.4, 9.5, 12.9, and 15.4 Å [112]. Thus, the values differ by as much as a factor of 3.

These last experimental data are of particular relevance for XPEEM because they were measured in a PEEM instrument with an angle-limiting aperture as used for imaging and show that caution is in place when using IMFP data in XPEEM.

Up to now we have considered only the influence of the attenuation of the *electrons* on the information depth. When X-ray absorption is strong as it is at the absorption edges then also the finite penetration depth of the *X-ray* beam can influence the sampling depth, leading to signal saturation before electron attenuation limits the signal. This is of particular importance at large angles of incidence of the photon beam as for example at 74° in one of the commercial instruments and has to be taken into account in thick films or bulk crystals. This saturation effect is discussed in Ref. [114] in connection with XAS measurements.

The influence of elastic scattering on the effective attenuation length (EAL) mentioned above is taken into account by assuming isotropic scattering. This is an acceptable approximation in disordered or fine-crystalline materials with random crystal orientation. Many specimens are, however, single crystals or epitaxial layers and then this assumption is invalid. Elastic scattering of the photoelectron wave, which to a first approximation may be considered as a spherical wave at its origin, produces then a photoelectron diffraction pattern similar to what has been mentioned already in Sect. 2.3.3. The amplitude of a spherical wave decreases like $1/r$ with distance r so that the immediate neighbors of the source atom scatter back most strongly, creating an interference pattern characteristic for the local atomic configuration. Examples of such X-ray photoelectron diffraction (XPD) patterns are shown in Fig. 2.19 together with the XPS spectrum of an oxidized Si(111) surface from which they were taken [116]. The spectrum clearly shows the chemical shift between the different oxidation states. The XPD patterns taken at their peak energies allow by comparison with multiple scattering calculations for model SiO_x clusters determination of the bonding distances and angles in them. For a recent review of XPD experiments see Ref. [117], for a review of the theory of XPD and other diffraction processes see Ref. [118], for software packages for the analysis of X-ray multiple scattering patterns see Ref. [119]. With increasing energy the scattering occurs increasingly in the forward direction, which allows considerable simplifications in the analysis [120]. In PEEM, high-energy XPD is limited by the limited angular transmission of the objective lens. However at lower energies it can be quite useful. For example much of the XPD pattern shown in Fig. 2.19 can be observed in a PEEM instrument in the diffractive plane much faster than with most spectrometers. Alternately local variations of the oxidation states can be imaged by inserting the contrast aperture in regions, in which the intensity differs strongly between different oxidation states.

2.2.6.2 Secondary Electron Emission

The intensity of slow secondary electrons with kinetic energies from zero to several eV is orders of magnitude higher than that of most photoelectrons and, therefore, particularly well-suited for imaging, except for its wide energy distribution, which

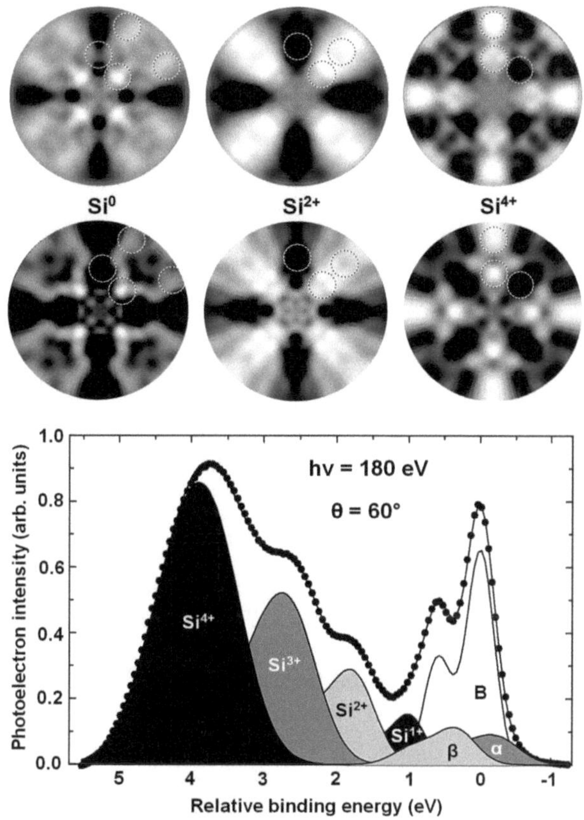

Fig. 2.19 X-Ray photoelectron diffraction patterns (XPD, *top*) and XPS spectrum (*bottom*) of an oxidized Si(100) surface. The *top row* shows the experimental XPD patterns taken at the energies of the maxima of the Si^0 (B), Si^{2+}, and Si^{4+} states of Si shown in the spectrum, the *second row* the XPD patterns simulated with multiple-scattering calculations. Adapted with permission from Ref. [116]. Copyright 2004 by the American Physical Society

is detrimental to resolution, even with energy filtering because of the large $\Delta E/E$ ratio. Nevertheless imaging with secondary electrons is the most widely used imaging mode because of its wide application range. In contrast to the electron *emission* process discussed in the previous section, this mode makes use of the secondary products of the X-ray *absorption* process. While previously the photon energy was fixed and the energy of the photoelectrons was scanned, now the photon energy is scanned and the electron energy is fixed. The various imaging modes are described by the prefixes XANES or NEXAFS, XMCD, XMLD and XNLD added to PEEM. XANES stands for X-ray absorption near edge structure, NEXAFS for near edge X-ray absorption fine structure, XMCD for X-ray magnetic circular dichroism, XMLD for X-ray magnetic linear dichroism and XNLD for X-ray natural magnetic dichroism. It is usual not to take a series of images as a function of photon energy but first to take a spectrum—or two in the case of XMCD and XMLD—and select the energy of interest for imaging. The spectra are the subject of this section.

Because if its importance for the spatial resolution in XPEEM we have to take a look at the secondary electron distribution produced by X-rays. Henke et al. have

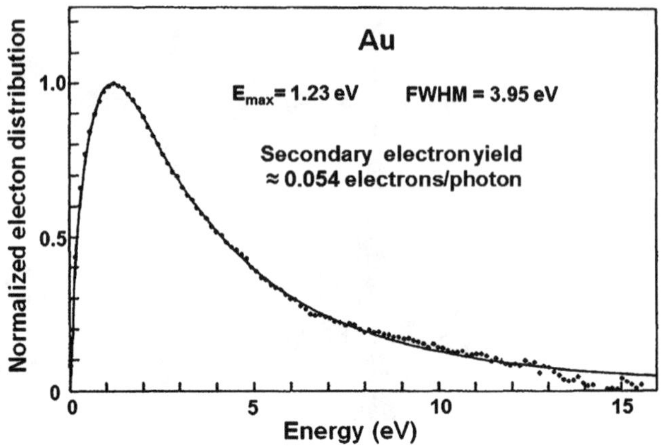

Fig. 2.20 Energy distribution of secondary electron from an evaporated Au film excited by Al Kα radiation (1487 eV) at normal incidence. Reprinted with permission from Ref. [122]. Copyright 1979 by the American Physical Society

studied the low energy region of the electron distribution theoretically and experimentally for polycrystalline Au and Al [121] and for a number of metals, semiconductors, and compounds [122] using Al Kα radiation. Figure 2.20 shows the secondary electron distribution of Au [122]. For metals and narrow band gap materials good approximations are

$$I_s(E) \sim \frac{E}{(E+\Phi)^n}, \tag{2.19}$$

with $n = 4$ for metals [121] and $n = 3$ for narrow band gap materials [122]. Here Φ is a fit parameter, which can be approximated by the work function Φ and the electron affinity χ, respectively. The distributions have maxima at $\Phi/3$ and $\chi/2$ and full widths at half maximum of 1.10 Φ and 1.90 χ, respectively. Unfortunately the secondary signal is still high up to several ten eV. This is an impediment to spectroscopy and imaging with low energy photoelectrons because they are superimposed on the strong secondary electron background from which they are increasingly more difficult to separate with decreasing energy.

With this background in mind we can now address the first XAS mode, NEXAFS, or XANES. These two acronyms stand for the same type of spectroscopy, the determination of chemical bonding via the bonding-specific ionization energies of different species, using the secondary electrons produced by the photoelectrons and Auger electron cascade in the filling of the hole created in the ionization process. With linear and circular polarized light in addition the orientation and chirality of adsorbed species can be obtained. The different bonding states produce partially overlapping absorption peaks, from which bonding distances and bond angles can be extracted by comparison with theoretical models. There is an excellent book on this subject [123] so that we will concentrate here only on the

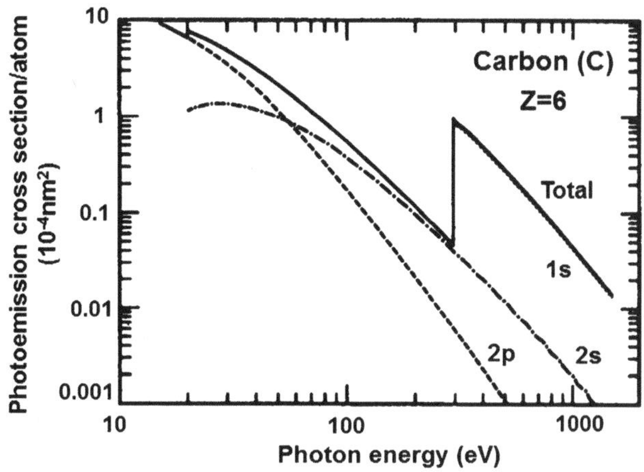

Fig. 2.21 Atomic photoionization cross section of carbon, showing the K absorption edge. Reproduced from Ref. [104] with permission

aspects important for PEEM. An example of a photoionization cross section is shown in Fig. 2.21 for atomic C [104, 105]. At the K absorption edge (277 eV), where most of the XANES imaging of C-containing materials is done, the cross section is two orders of magnitude higher than at the Al Kα energy, another advantage of synchrotron radiation compared to conventional X-ray sources. Furthermore, the cross section jumps by a factor of 20 at the K edge so that chemical shifts can be measured with high signal/noise ratio. One problem frequently encountered is C-contamination of the mirrors in the synchrotron radiation beamline, which can produce a significant background in the XAS spectrum. In imaging this is less critical then in spectroscopy, because it produces only a location-independent background. However it reduces the contrast from local variations of the bonding such as in copolymers (see Chap. 6.3). Keeping the decrease of the ionization cross section with photon energy in mind, useful absorption edges for imaging are the K edges up to F and the L edges up to Ni. As an example, the sulfur L edge XAS spectra in Fig. 2.22 illustrate the sensitivity of this method to the chemical bonding, which is discussed in Ref. [124]. It should be noted that the ionization cross sections in Refs. [104, 105] and plotted in Fig. 2.18 are for free atoms with energies referred to the vacuum level. With correction for the binding energy difference between free atoms and atoms in condensed matter these data are, however an excellent criterion for the selection of the core level and photon energy. Another useful criterion for imaging with secondary electrons is the mass absorption coefficients for X-rays, which takes the atomic density in condensed matter into account [125–127].

It is useful to compare XANES with XPS. XANES measures the intensity of all slow secondary electrons, not only those produced at the selected absorption edge but also by ionization of states with lower binding energy. On the other hand the energy filter, which selects the desired photoelectrons transmits only a small fraction of the always present secondary background, resulting in a much better atom-specific signal-to-noise background ratio. The IMFP of the secondaries is

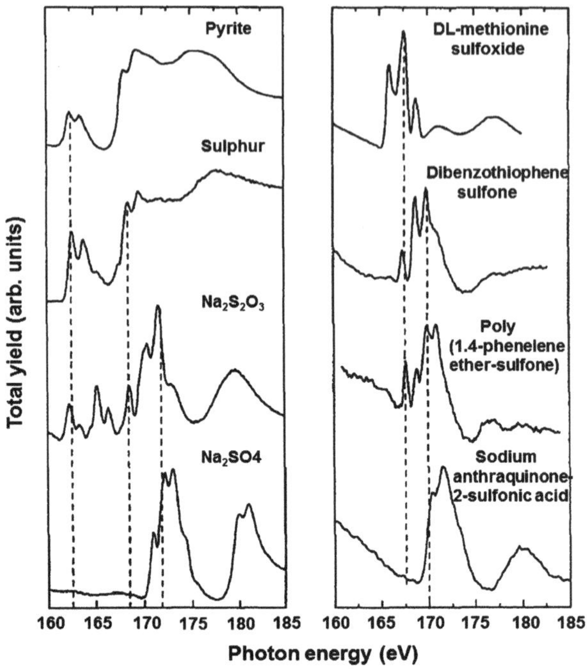

Fig. 2.22 X-Ray absorption near edge structure (XANES) at the L edge of elemental sulfur and sulfur-containing compounds, illustrating the high sensitivity of XANES to the chemical environment of the selected atom. Reproduced with permission from Ref. [124]. Copyright 1996 Elsevier

generally large compared to that of photoelectrons, in particular if the photon energy is chosen in such a manner that the photoelectrons have an energy, at which their IMFP is small. This makes photoelectron imaging more surface sensitive than XAS imaging. Of course there are also other considerations, which determine the selection of the method such as the much higher intensity of secondary electrons and the absence of the energy filter needed for XPS in many PEEM instruments, the latter leaving no choice.

While XANES is the method for chemical surface characterization in PEEM instruments without energy filter, XMCD it the workhorse for atom-specific magnetic characterization of ferromagnetic materials. Again, there is an excellent book covering this subject [128] so that we will limit the discussion here only to a brief discussion of the basic aspects of XMCD and to aspects important for XMCD imaging. Circular polarized light has angular momentum as demonstrated experimentally already 1936 [129]. In the wave picture (Maxwell's equations) this is surprising but can be derived with some effort [130, 131]. In brief, an electromagnetic wave with the energy density $\varepsilon_0 \boldsymbol{E}^2$ has an angular momentum $\boldsymbol{L} = \varepsilon_0 \boldsymbol{E}^2 \; (|\psi_R|^2 - |\psi_L|^2)/\omega$ in direction of the propagation vector of the wave, where ψ_R and ψ_L are the amplitudes of right hand and left hand polarized waves and ω is the angular frequency. According to the correspondence principle the classical energy density $\varepsilon_0 \boldsymbol{E}^2$ of the wave corresponds in the particle picture the energy density $n\hbar\omega$ of n photons with energy $\hbar\omega$. Comparison with the classical expression for $\boldsymbol{L}$ shows immediately that the angular momentum of the photon

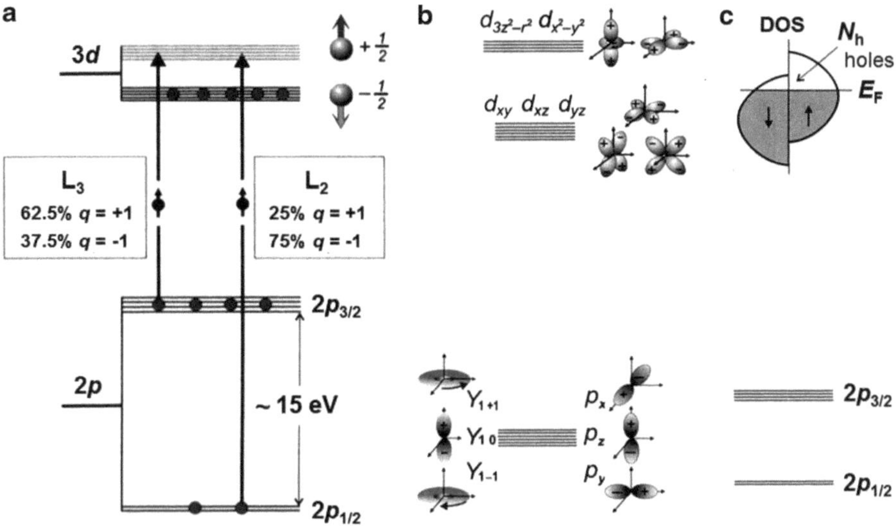

Fig. 2.23 Model of the transition processes leading to X-ray magnetic circular dichroism. (**a**) atomic model, (**b**) ligand field model, (**c**) band structure model. Not to scale. The energy distances between core and valence states are shown much smaller than the splitting of the 2p and 3d states. The *insets* in (**a**) are the relative transition probabilities as calculated in the ligand field model. Adapted from Ref. [128] with permission from Springer Science + Business Media

is $\boldsymbol{l} = \hbar\,(|\psi_R|^2 - |\psi_L|^2)$. For right (left) hand polarized light the second (first) term is zero and the other one 1 so that $|\boldsymbol{l}| = l_z = \pm\hbar$ or in the usual atomic angular momentum units ($\hbar$) $l_z = \pm 1$. This angular momentum is a spin angular momentum and we will call it from now on therefore s_P. From the particle point of view this is immediately obvious: photons are bosons and bosons have angular momentum of magnitude $\hbar$. This quantum number has to be taken into account in the absorption process, i.e., in the transition from the initial core state to the final unoccupied state.

XMCD is studied in PEEM by imaging with secondary electrons, which are created during the relaxation cascade from the nonequilibrium state produced in the excitation of core electrons by circular polarized light. The nonequilibrium states are the spin-split 3d and 4f states of ferromagnetic and antiferromagnetic materials just above the Fermi level, the corresponding core electrons are the 2p and 3d electrons ($L_{2,3}$ and $M_{4,5}$ states). As an example we consider a cubic 3d metal, first an isolated atom, then an atom surrounded by six atoms as in a cubic crystal (octahedral site) and finally an "atom" in a periodic crystal. Figure 2.23a [128] shows schematically the energy level structure of an isolated 3d metal atom with five electrons in the d band. The 2p levels are split by spin–orbit interaction into $2p_{1/2}$ (L_2) and $2p_{3/2}$ (L_3) levels, the 3d levels by exchange interaction into spin-down and spin-up level groups, each of which contains 5d orbitals. Only the spin-down levels are occupied according to Hund's rule. When the atom interacts with neighboring atoms, which create a so-called ligand field ("crystal field"), the sequence of the energy levels is rearranged into two level groups, e_g and t_{2g}, which contain the d

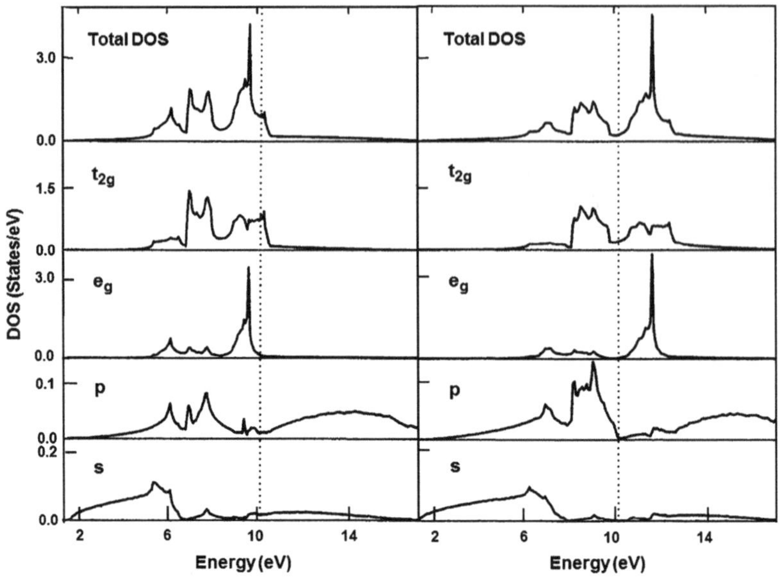

Fig. 2.24 Angular momentum-resolved and total density of states of Fe. *Left*: spin down states, *right*: spin up states. There is also a significant spin polarization in the p states although it contributes little to the total spin imbalance, which is dominated by the e_g and t_{2g} states. From Ref. [132] with permission from Springer Science + Business Media

orbitals shown in Fig. 2.23b [128]. A further redistribution occurs when the electrons become delocalized in the periodic lattice of the crystal and form majority and minority spin bands as indicated by the schematic density of states (DOS) in Fig. 2.23c [128]. The true density of states as obtained from band structure calculations for Fe, for example, shows that e_g and t_{2g} levels are no longer separated (Fig. 2.24 [132]). They both have regions above the Fermi level into which electrons from 2p levels can be excited with near-threshold radiation. The relative transition probabilities into the unoccupied d states upon right hand ($q = +1$) and left hand ($q = -1$) radiation excitation determines the probability of secondary electron emission. They can be calculated to a first approximation in the ligand field model, which gives the results shown in Fig. 2.23a by their percentages. The XMCD signal is the difference between the secondary electron emission current produced by right and left circular polarized light, which is 50 % for excitation from the L_2 state and 25 % from the L_3 state. However the L_3 state has twice the number of electrons than the L_2 state so that the XMCD signal resulting from the excitation of the two states has the same magnitude but opposite sign. Electron delocalization in the crystal modifies the relative magnitude of the two signals but the ligand field theory model gives a first good approximation to the mechanism of XMCD. It explains qualitatively the experimental absorption and XMCD spectra shown in Fig. 2.25a [128], whose quantitative description requires more sophisticated calculations as described in Ref. [128].

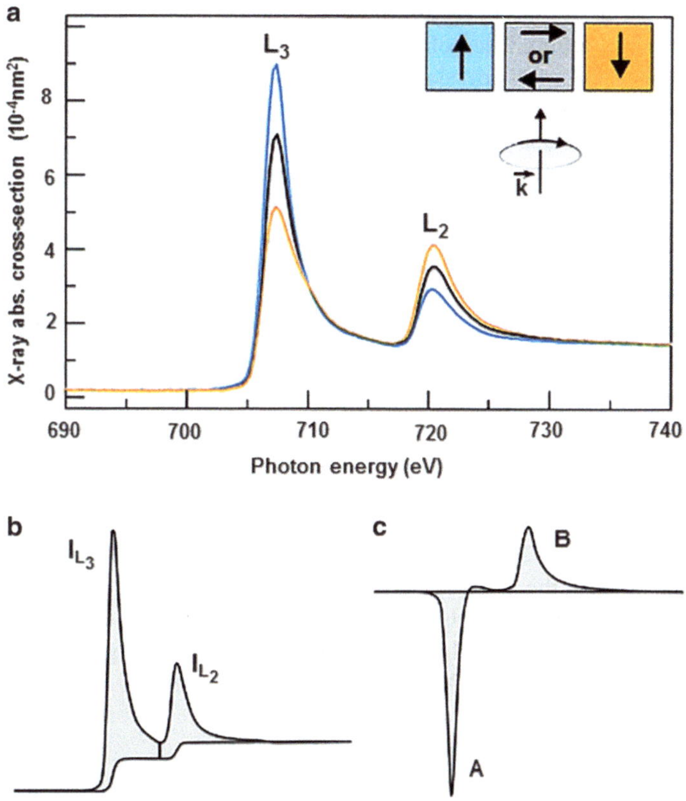

Fig. 2.25 X-Ray absorption spectrum (XAS) at the Fe $L_{2,3}$ edge (secondary electron yield) of circular polarized light (**a**). The directions of the angular momentum $\boldsymbol{L}$ of the photon ($\boldsymbol{L} \parallel \pm \boldsymbol{k}$) and of the magnetization $\boldsymbol{M}$ are indicated by *arrows*. As seen in Fig. 2.23 the transition probability from the L_3 level to the levels above the Fermi level is higher when $\boldsymbol{L} \parallel \boldsymbol{M}$ than when $\boldsymbol{L} \parallel -\boldsymbol{M}$. while for the L_2 level the opposite is the case (not shown). The difference between the two spectra are the XMCD signals (**b**) and (**c**). Their magnitude is proportional to the magnetic moment of the absorbing atom in its environment and thus a measure for the magnetization. As described in the text from the areas below the peaks the number of empty d states and the orbital and spin magnetic moments can be extracted after proper background correction, which takes into account energy losses (smoothed step function seen in (**b**)). Adapted from Ref. [128] with permission from Springer Science + Business Media

Figure 2.25a shows that the photon spin direction has to be (anti)parallel with the spin direction of the electron for the level selection of the final state. The magnetic dichroism is defined by convention as the intensity difference between the two spectra $\Delta I = I_{\uparrow\downarrow} - I_{\uparrow\uparrow} \equiv I_{\mathrm{XMCD}}$ and is directly proportional to the atomic magnetic moment. Frequently the angular momentum of the photon $\boldsymbol{s}_{\mathrm{P}}$ is not parallel to that of the magnetic moment $\boldsymbol{m}$ connected with the angular momentum of the 3d shell and the circular polarization P of the photon beam is not 100 %. Then

$$I_{\mathrm{XMCD}} = P\boldsymbol{s}_{\mathrm{P}} \boldsymbol{\cdot} \boldsymbol{m} = \boldsymbol{L}_{\mathrm{P}} \boldsymbol{\cdot} \boldsymbol{m} = L_{\mathrm{P}} m \cos\theta, \tag{2.20}$$

where θ is the angle between the direction of the incident beam and the local magnetization direction. The polar angle of incidence is fixed in all PEEM instruments and in most of them also the azimuthal angle unless the instrument is equipped with a rotatable specimen manipulator. Therefore in XMCDPEEM usually only the component of $\boldsymbol{m}$ in the direction of the incident light is measured.

The spectra in Fig. 2.25 a contain not only information about the atomic moment but also about the total number of unoccupied states N_h in the 3d band and allows separation of the orbital magnetic moment m_O from the spin magnetic moment m_S. For this it is in principle necessary to saturate the specimen magnetically in the direction of the incident beam and to average over all directions, that is to make three measurements in three orthogonal directions. In XMCDPEEM these conditions are not fulfilled so that only approximate values can be obtained. These are nevertheless frequently close to the values determined with averaging measurements, with which they agree when the specimen is polycrystalline and magnetically not aligned because the averaging is done by the specimen. The information about N_h, m_S, and m_O is obtained from the shaded areas in Fig. 2.25b, c [128]. The first one shows the sum $I_{\uparrow\downarrow}+I_{\uparrow\uparrow}$ of the spectra taken with opposite helicity light, from which the transitions from the 2p level to the unoccupied 4 s states have been subtracted so that only the holes in the 3d band contribute to the signal. The 3d contribution to the L_3 peak is twice as large as that to the L_2 peak because of the higher transition probability from the $2p_{3/2}$ level compared to the $2p_{1/2}$ level. Figure 2.25c shows the difference $I_{\uparrow\downarrow}-I_{\uparrow\uparrow}$, that is the XMCD spectrum. The symbols I_{L3}, I_{L2}, A, and B are the areas under the peaks. The sum rules then state that

$$N_h = I_{L3} + I_{L2}/C \; m_S = \mu_B - A + 2B/C \; m_O = -2\mu_B A + B/3C \qquad (2.21)$$

where the brackets indicate the average mentioned and C is about 10 for Fe [128]. Frequently only the ratio m_O/m_S is determined, thus eliminating C. For a deeper insight in the use of XMCD for the determination of m_S, m_O, and the magnetic anisotropy energy from angular-dependent measurements of m_O see Ref. [133], for the influence of saturation effects see Ref. [134].

Two notes in passing: (1) in the literature the arrow in the majority band frequently is shown pointing upward and the band is correspondingly called spin-upband in contrast to the direction shown here, which follows from the atomic model. What is really meant with the direction of the arrow in this other description is the magnetic moment $\boldsymbol{m}$ whose sign is opposite to the spin. (2) The high peaks at the absorption edges, in particular the L_3 peak, which are a consequence of the high density of 3d states at the Fermi edge, are frequently called white lines for historical reasons when spectra were recorded on photo plates.

Summarizing these qualitative considerations, XMCD gives atom-specific information on the magnitude and direction of the average magnetic moment and allows with some limitations separation into orbital and spin magnetic moment. Details can be found in Refs. [83, 128, 133] and in many other books and reviews, too numerous to be listed here.

X-Ray linear dichroism (XLD) is not only important for magnetic studies as X-ray magnetic linear dichroism (XMLD) but also for nonmagnetic studies as X-ray natural magnetic dichroism (XNLD). XLD occurs whenever the charge distribution around the absorbing atom is nonspherical so that the absorption depends upon the angle of the ***E*** vector of the *linear* polarized wave with respect to the direction of the charge distortion. Thus whenever the specimen has directional bonds, which are well aligned over the region measured, XNLD is present, independent whether the specimen is nonmagnetic, ferromagnetic, or antiferromagnetic. In the second two cases it may be induced by symmetry reduction of a spherical charge distribution by magnetization along a certain axis. This effect is most pronounced when the 3d electrons are strongly correlated and the 3d levels are split by crystal field and spin–orbit interactions. Many antiferromagnetic oxides containing 3d transition metal ions with nonsymmetric charge distribution are systems of this type such as NiO, which has been the subject of many studies and discussion. Cubic crystals have several easy magnetic axes, which leads to multidomain formation unless some anisotropy is induced, for example by strain. Here XMLDPEEM is very useful for the understanding of the antiferromagnetic structure [135]. The XMLD can be separated from the XNLD either by measurements at the $L_{2,3}$ edge above the Néel temperature or via the O 2p orbital anisotropy at the O K edge. The calculations needed for the interpretation of the spectra are equally as or more demanding than those for XMCD spectra. Their principles and references to them can be found in Ref. [128]. Here we consider only briefly the essentials needed for XLDPEEM.

As already mentioned at the beginning, for XLD to occur ***E*** must be parallel to direction of the oriented bonds in molecules, to the distortion in nonmagnetic crystals or to an easy magnetic axis in a ferromagnet or antiferromagnet, which we summarize under the term symmetry axis. The transition probability is proportional to the number of unoccupied bonding states in the direction of ***E*** ("search light" effect). The XLD signal is then obtained from two measurements one parallel to the symmetry axis, the other perpendicular to it: $I_{\mathrm{XLD}} = I_{\|} - I_{\perp}$. When the symmetry axis has more than twofold rotational symmetry then the angular dependence of the intensity can be written as

$$I(\theta) = I_{\|} \cos^2\theta + I_{\perp} . \sin^2\theta \ \text{ or } \ I(\theta) = I_0 - \left(1/\sqrt{2}\right)\left(3\cos^2\theta - 1\right) I_2. \quad (2.22)$$

Here θ is the angle with the symmetry axis, I_0 and I_2 are the isotropic and the angular part of the total absorption cross section, respectively. In the case of a ferromagnetic specimen the XMLD intensity can be expressed as

$$I_{\mathrm{XMLD}} \sim P\left|\boldsymbol{m} \cdot \boldsymbol{E}\right|^2 \sim P\, m^2 \cos^2\theta, \quad (2.23)$$

where P is the degree of linear polarization and two measurements with ***E*** parallel and perpendicular to ***m*** are taken. Note the difference between Eqs. (2.20) and (2.23). The choice of the direction of ***E*** is usually limited so that only components of ***m*** can be measured.

For XMLD and XNLD there are sum rules similar to those for XMCD, which are discussed briefly in Ref. [128], where also references to their derivation can be found. The main application field of XNLD is the determination of the orientation of adsorbed molecules, that of XMLD is the study of antiferromagnetic materials and their interfaces with ferromagnetic materials. In ferromagnetic materials the XMLD effect is much weaker than the XMCD effect and, therefore, used much less. For specific studies such as of the magnetocrystalline anisotropy it may be very useful however [136]. Another important application field of XMLD together with XMCD is in multiferroics [137].

2.3 Electron Reflection

2.3.1 General Considerations

In this section we will discuss the interaction processes between the incident electron beam and the specimen which are important for image formation in LEEM and SPLEEM. As already mentioned earlier, these processes are not relevant for mirror electron microscopy because in this case the electrons are reflected in front of the surface. The processes involved are elastic, inelastic and thermal diffuse scattering and foremost diffraction, which is dominated by the periodicity of the specimen but also depends on the various scattering processes. Inelastic scattering is not only of interest for energy electron loss spectroscopy (EELS) but also for energy electron loss microscopy. Finally, we also have to consider briefly Auger electron emission, which can be used in the laboratory for spectroscopic Auger electron emission microscopy (AEM), provided that the microscope is equipped not only for spectroscopy but also allows to use different energies in illuminating and imaging beam.

2.3.2 Elastic Scattering

In the past LEEM made use of the fact that in crystalline materials the backscattered intensity is concentrated in a few diffracted beams so that images could be acquired with short exposure times. With increasing availability of aberration-corrected instruments, which allow much larger contrast apertures, imaging of materials without long-range order, such as liquids, polymers, or biological materials will become possible. Therefore we consider first backscattering in the absence of long range order. This is also relevant for the understanding of the influence of disordered adsorption layers on crystal surfaces such as two-dimensional gases. As a starting point we consider the scattering of slow electrons by an atom in a completely disordered monatomic material but with the density of its crystalline phase

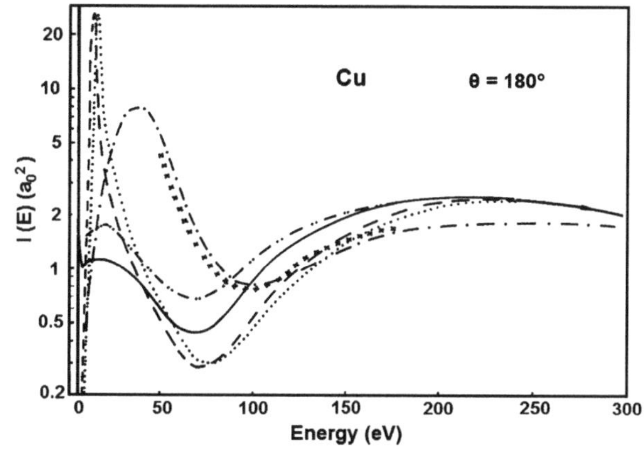

Fig. 2.26 Elastic 180° scattering of Cu atoms as a function of energy calculated for different potentials. The *cross* and *dash-dotted curves* are from free atom potentials, the others from solid state potentials, the best ones being the *dotted* and *dashed curves*, which differ only slightly in the exchange potential used in the atomic potential calculations. The curves from solid state potentials have to be shifted by 13.1 eV to lower energies because their zero is the muffin tin potential. Adapted from Ref. [141] with permission from Springer Science + Business Media

("randium"), based on calculations which Browne and Bauer made in the early years of the development of LEEM (1964–1967). Their purpose was to ensure that there was enough backscattering at low energies for imaging and to determine the optimum energies for various materials. Except for the initial calculations for free atoms [138] only some of the results were published later [139–146].

The scattering potential is not that of a free atom as used in the calculations of the scattering of fast electrons [147] and of slow electrons sometimes down to 50 eV [148–151]. The reason is simple: the outer part of the potential, which is missing in condensed matter because of the overlap with that of the neighbors, becomes increasingly important with decreasing energy, in particular below 100 eV. This is illustrated in Fig. 2.26 [141] which compares the 180° scattering cross section of Cu atoms for different scattering potentials [140, 141, 146]. The cross sections from the free atom potentials [152, 153] differ significantly from those of the solid state atoms [154], whose overlapping potentials are cut off at some radius forming muffin tin potentials with constant potential between them, called muffin tin zero. The influence of this cut-off is even more pronounced than shown in the figure because the muffin tin zero is 13 eV below the vacuum level, which is used as zero in the figure. Taking this into account shifts the sharp maximum, which is due to a maximum in the $l = 2$ partial wave cross section, to zero. Based on this result and similar ones for other atoms we will consider in the following only scattering from muffin tin potentials used in band structure calculations.

Figure 2.27 [141] compares the backscattering of several atoms into a 30° wide cone around the backward direction [141, 142, 144–146]. Again the curves have to be shifted by the muffin tin zero, which ranges from 12 to 17 eV for these metals, to the left. It is seen that the backscattering of Cu (W) is much weaker than that of

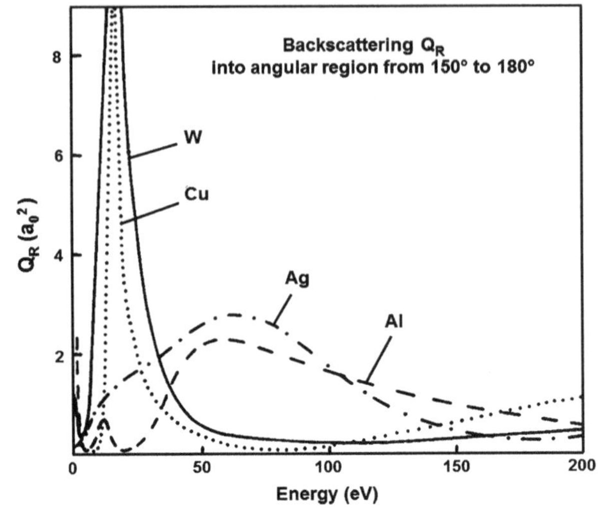

Fig. 2.27 Elastic scattering cross sections of W, Cu, Ag, and Al atoms integrated from 180° to 150°, based on potentials from band structure calculations. The muffin tin zero has to be subtracted in order to reference the energy to the vacuum level. Adapted from Ref. [141] with permission from Springer Science + Business Media

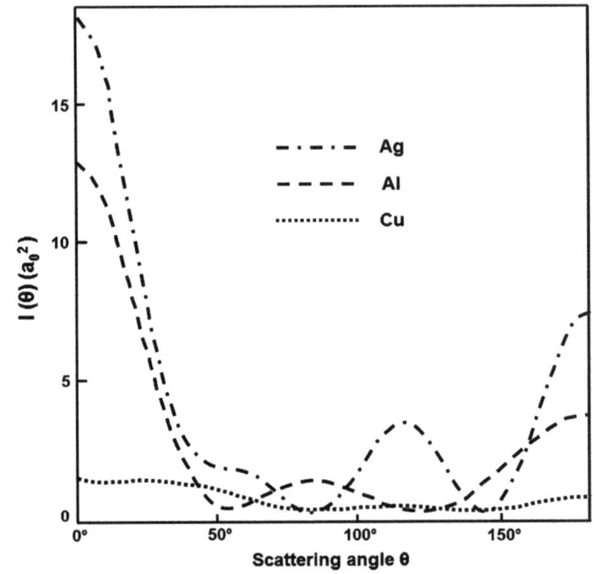

Fig. 2.28 Angular distribution in low energy electron scattering from realistic solid state atomic potentials of Al, Cu, and Ag at 50 eV. Reproduced from Ref. [139] with permission from the American Institute of Physics, © 1970

the lighter atoms Al (Ag) over a wide energy range but that at the lowest energies Cu and W scatter much more than Al and Ag. This example shows that the scattering cross section changes nonmonotonically with nuclear charge Z and energy in contrast to high energy electron scattering so that in LEEM no Z-contrast imaging like in transmission electron microscopy is possible. That the weak scattering of Cu around 50 eV is not limited to backscattering [139, 141, 142, 144] is seen in the angular distribution of the scattering from Cu, Al, and Ag muffin tin potentials in Fig. 2.28 [139]. This is also confirmed by angle-integrated measurements of several polycrystalline metals, which extend from 2.5 keV down to 60 eV and illustrate that the nonmonotonic behavior extends up to about 1.5 keV [155]. Otherwise there are very few experimental data below 100 eV, which could be compared with theory as the most recent review of elastic backscattering from disordered and polycrystalline surfaces shows [156]. The angular distribution of electrons scattered from liquid Hg was measured only down to 100 eV [157]. Elastic backscattering cross sections of amorphous Si integrated over the angular range of a LEED optics agreed well with calculations taking double scattering and the muffin tin zero shift into account [158].

2.3.3 Inelastic Scattering

For a complete description of electron reflection without energy loss, elastic multiple scattering, and losses due to inelastic scattering have to be taken into account. Inasmuch as one inelastic scattering event is sufficient to remove an electron from the elastic channel, inelastic scattering can be taken into account via an inelastic mean free path (IMFP) λ_i in the elastic multiple scattering calculation. In contrast to the scattering in the much-studied higher energy range ($E > 100$ eV), where frequently strong simplifications of the angular distributions are made, in the energy range of interest in LEEM the angular distribution of the scattering by the individual atoms has to be fully taken into account. To a first approximation this can be done by considering only double scattering. This is justified by the fact that λ_i is frequently shorter than the elastic mean free path λ_e so that inelastic scattering has already caused considerable attenuation before the second elastic scattering process occurs. In the limit of very strong inelastic scattering single elastic scattering, can be used to estimate penetration depth and backscattered intensity [139–141]. In general, however, a more rigorous approach such as Monte Carlo simulation or Boltzmann transport equation methods has to be used. These methods have been developed well for higher energies as described repeatedly in reviews, for example in Refs. [159–162], but have been applied with realistic potentials to energies below 100 eV only rarely [163]. Reference [160] gives some results for the elastic MFP down to 10 eV but calculated with free atom potentials, which are inadequate at these low energies.

The theory of inelastic scattering of slow electrons is on much better footing. The original approach in connection with LEEM [139–141] assumed that the incident electron interacts with a degenerate electron gas (jellium approximation, Lindhard dielectric function), taking into account exchange and correlation, which were not included in the calculations of elastic scattering at low energies. Now calculations with more realistic descriptions of the interactions are available. In them material-specific dielectric functions are used [163, 164] or the Lindhard dielectric function is effectively replaced by a dielectric function obtained from experimental optical data (Penn algorithm [165]). In one case [163, 164] IMFP calculations were performed for SiO_2 down to below 0.1 eV, taking optical phonon excitation into account, and for C, Al and Ag down to the 1 eV range. IMFP results

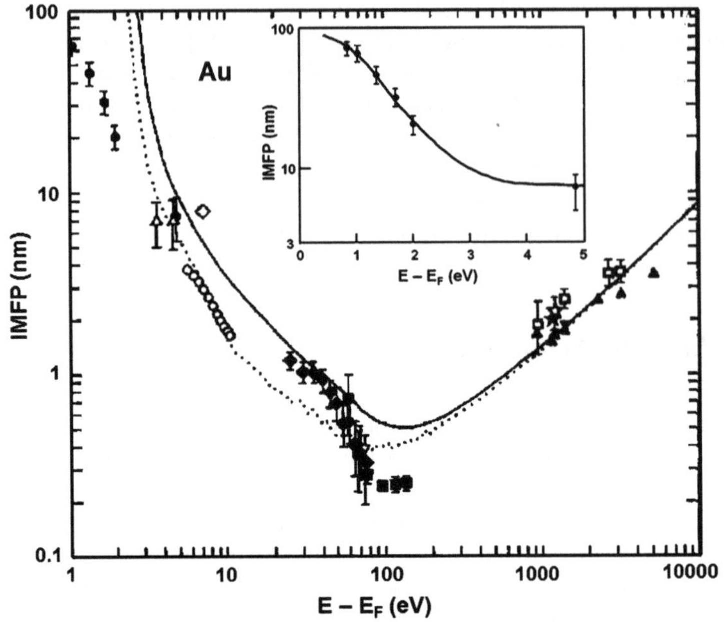

Fig. 2.29 Inelastic mean free path of electrons in Au as a function of energy. *Symbols* are from different measurements, lines from calculations using different plasmon dispersion relationships, the *dotted one* being more accurate. Adapted with permission from Ref. [166]. Copyright 1989 Elsevier

for 41 elemental solids for energies down to 50 eV calculated in this manner have been published by Tanuma et al. [112]. Below 100 eV they show a clear difference between the IMFPs of diamond on the one hand and graphite and glassy carbon on the other, indicating the strong influence of the electronic structure at low energies. For several elements the calculations down to 1 eV have been compared with experimental results [166]. Figure 2.29 shows as an example Au [166]. A difference in the dispersion relation assumed for the plasmon is responsible for the difference between the solid and the dashed curve, which gives a somewhat better fit to the experimental data. However, both curves deviate considerably from experiment, in particular at the lowest energies shown in more detail in the inset, which are, however, below the vacuum level and therefore not important for LEEM. The strong decrease above about 60 eV can be attributed to 5p excitations, which are apparently not taken into account in the dielectric function. Similar deviations below 100 eV have been found in a comparison of the calculations with elastic-peak electron spectroscopy measurements from many elemental polycrystalline solids, while above 200 eV theory and experiment agree very well [167] and follow the universal IMFP curve.

However at very low energy, let's say below 10 eV, the details of the electronic structure cause very large deviations from the "universal" IMFP, which is based on measurements from materials whose energy bands in this energy regions are free electron-like (parabolic), such as Au, Ag, and Au. The density of states $D(E) \sim \int dS_E/\nabla_k E(\boldsymbol{k})$, ($dS_E$ surface with constant energy) into which electrons can be excited in these materials, is small. The situation is quite different in materials, which have d or f bands above the Fermi level. These are rather flat ($\nabla_k E(\boldsymbol{k}) \to 0$) so that $D(E)$ is very high and the probability of low energy electron excitations into these states is high, in particular if there is a high density of occupied states close to the Fermi level. For example in Fe $\lambda_i \approx 0.5$ nm below 25 eV in contrast to Au, in which λ_i rises from 1 nm into the 10 nm range. We will return to the short IMFPs in the 3d transition metals later in connection with spin-polarized low energy electron interactions. Thus, in the energy range of interest for LEEM the energy dependence of the IMFP is still poorly described by theory and—as we will see later in the discussion of diffraction by crystalline materials—is used as an adjustable parameter to fit experiment. The old calculations for low energies based on the jellium model [139–141, 168, 169] are of limited value because only Al and a few other metals are free electron-like. In the absence of reliable data for many materials an interpolation formula between a low energy expression, which had been proposed already by McRae [169], and one of the various high energy expressions of the IMFP gives probably useful IMFP values, although it contains parameters which have to be determined from experiment [170].

2.3.4 Surface Effects

The third important interaction affecting the reflection of slow electrons from surfaces in addition to the elastic and inelastic scattering *in* the specimen is scattering *from* the surface. This subject has a long history and will certainly also be studied in the future. It is immediately obvious that the reflection of a slow electron is strongly influenced by the boundary between vacuum and condensed matter in which the mean potential differs from that outside typically by 10–20 V. The relative magnitude of this influence compared to the processes which occur inside the specimen depends strongly on the shape of the boundary. The simplest model is that of a sharp planar interface between vacuum and a polarizable medium with the dielectric constant ε. The incident electron then produces an image charge in the medium. This in turn produces the classical image potential $V(z)$ acting on the electron

$$V(z) = -\frac{\varepsilon - 1}{\varepsilon + 1}\frac{1}{4z}, \tag{2.24}$$

which for ideal metals ($\varepsilon = -\infty$) assumes the familiar form $V(z) = -1/4z$ (in atomic (Hartree) energy units) or $-1/2z$ (in Rydberg units). For elemental semiconductors such as Si ($\varepsilon = 11.9$) and organic semiconductors such as pentacene ($\varepsilon = 5.3$) $V(z)$ is only slightly smaller, 0.84 times for Si and 0.68 for pentacene. There have also been numerous quantum mechanical calculations for the approximation of the metal by a free electron gas (jellium), which is inhomogeneous near the surface. They give

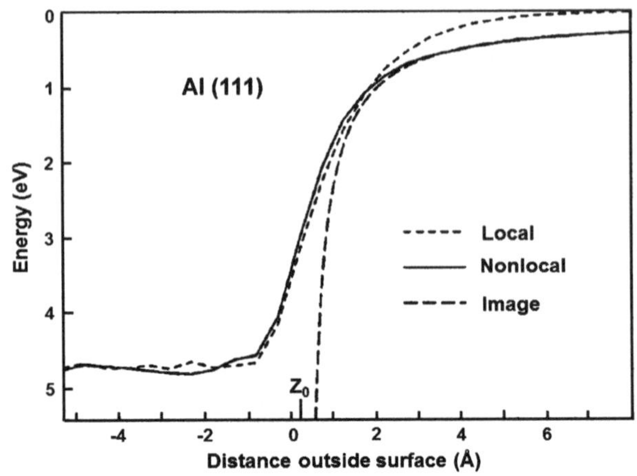

Fig. 2.30 Surface-averaged effective local potential experienced by electrons in the near-surface region of the Al(111) surface. *Long-dash line*: classical image potential, *short-dash line*: local density approximation, *continuous line*: *GW* approximation for the self-energy. z_0 is the image plane position obtained from the *GW* calculation. (*G* single-particle Green's function, *W* screened Coulomb interaction). Adapted with permission from Ref. [177]. Copyright 1998 by the American Physical Society

essentially the same $1/z$ dependence for the asymptotic form ($z \rightarrow \infty$) of $V(z)$. It has been suggested to add a correction term α_{KSx}/z to account for the exchange potential at the surface, which depends upon the potential step at the surface and the Fermi energy in the bulk: $V(z) = -(\alpha_{\mathrm{KSx}} + 1/4)/z$ [171, 172]. α_{KSx} is approximately ¼ for most metals, ranging from about 0.22 for Fe to 0.26 for Cs.

While there is general agreement on the asymptotic $1/z$ form of the image potential the region close to the surface poses a problem because the potential goes to $-\infty$. The simplest and frequently used solution of this problem is to cut the potential at a distance z_0 and fit it in some manner to the potential inside the medium so that $V(z) = -1/4(z - z_0)$. In principle z_0 is the centroid of the image charge distribution but in general it is used as an adjustable parameter. Most early surface barrier models which simulate the transition to the asymptotic form have been reviewed repeatedly [173, 174] but others have been proposed too [175]. The most advanced calculations start from the electronic structure of crystal, taking the *nonlocality* of the exchange-correlation potential V_{xc} near the surface into account, similar to what has been done for jellium (see above), and fit $V(z) = -1/4(z - z_0)$ to the numerical results [176–178]. Referred to the geometric edge of the crystal, assumed to be half a layer spacing outside the outermost crystal plane, the z_0 value for Al (111) is 0.20 Å [177, 178], for Al(100) 0.29 Å [178] and for the jellium approximation of Al 0.38 Å [178]. Figure 2.30 [177] show the potential near the Al(111) surface, averaged over planes parallel to the surface. The shape of the potential in the *local* density approximation (LDA) does not fit the image potential at large distances. The potential modulation parallel to the surface outside the surface is small for these densely packed surfaces so that the lateral averaging is good approximation. These calculations

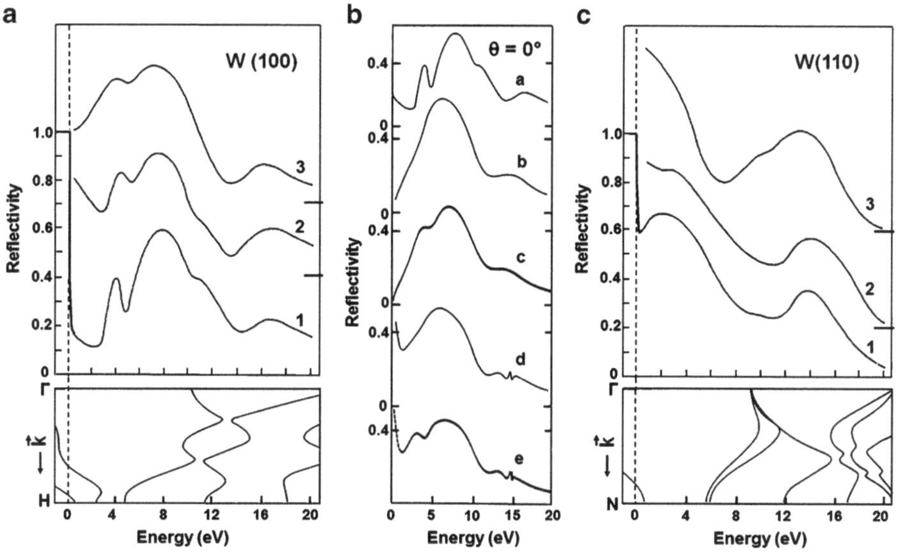

Fig. 2.31 Normal incidence specular electron reflectivity at very low energies. Comparison between experiment and theory for W(100) and W(110). The *curves 1* and *a* are experimental curves, the *curves 2*, *3*, and *b–e* are calculated for various models of the surface. *d* and *e* are from spin-dependent calculations. For explanation see text. (**a**, **c**) Adapted from Refs. [183, 184] with permission. Copyright 1981 Elsevier. (**b**) Reproduced from Ref. [185] with permission. Copyright 1983 by the American Physical Society

do not require any input from other calculations or from experiment but are very demanding because of the inclusion of the nonlocal V_{xc}.

Neglecting the nonlocality allows much less computer-intensive first-principle pseudopotential calculations but requires methods to bridge the gap between the self-consistent crystal potential and the correct image potential. This was achieved by several functions of z, which fit to each other also in the first derivatives and contain four parameters that are adjusted to reproduce the experimental binding energies of image states [179]. Calculations of this type for several densely packed free electron-like metals allowed construction of model potentials for 14 simple and noble metal surfaces [179]. This method was also extended to fcc(110) surfaces [180]. Still another approach to bridge the gap between crystal and asymptotic image potential is to mix the two potential $V_{mix}(z) = V_{DFT}(z)f(z) + V_{im}(z)\,(1 - f(z))$, with $f(z)$ changing from 0 to 1 as $\cos(z)$ in a chosen mixing region up to a chosen z_{mix} [181]. Thus theory has provided the experimentalist with a variety of surface barrier potentials which could be used in the interpretation of the reflection of the very slow electrons generally used in LEEM.

On the experimental side the surface barrier was introduced originally into LEED structure calculations more than 40 years ago to explain the resonances observed previously in nonspecular diffraction [182]. Subsequently the few specular reflectivity data at normal incidence, available before LEEM was available, were used to test various surface potential barriers. Figure 2.31 shows the fits to the best experimental results, which were obtained in a trochoidal electron spectrometer with an energy resolution of 20 meV [183]. The fits on the left and right hand side are from Ref. [184], those in the center from Ref. [185]. Curves (1) (left and right) and (a) (in center) are the experimental data. Curves (3) (left and right) are fits with a nonreflecting barrier as used in standard LEED calculations above about 50 eV and curves (2) (left and right) are obtained by fitting the bulk potential to the asymptotic image potential with an exponential function. In the center panel (b) and (c) are for nonreflecting barrier fits without and with spin taken into account, while (d) and (e) are the corresponding fits with a bridging function to the asymptotic image potential.

It should be noted that some of the features such as the high reflectivity between 0 and 5 eV on the (110) surface and the reflectivity peak at 4 eV on the (100) surface can already be explained with the energy gaps in the bulk band structure. Most other features and the detailed energy dependence of the reflectivity however require a surface potential barrier. A band gap as the cause of a high reflectivity was already invoked by Andersson [186] to explain his results for the Cu(100) surface. That the reflectivity does not reach the value 1 is in part due to inelastic scattering of the evanescent wave decaying into the crystal in the energy range of the gap and in part due to the influence of the surface barrier. A comparison of the fits in Fig. 2.31 shows that none of them can reproduce the experiment. It remains to be seen if the methods for the calculation of the surface potential barrier, which have been developed more recently as described above, will produce quantitative agreement with experiment. Several more factors in addition to the proper magnitude and energy dependence of the imaginary potential, which is a measure for the inelastic mean free path, and of the real part of the self-energy, which will be discussed below, have to be taken onto account: surface relaxation, enhanced thermal vibrations of the surface atoms and the deviation of the scattering potential of the atoms in the topmost layers from that in the bulk. In addition also anisotropic damping can occur at energies at which the first diffracted beam emerges from the surface [187] as shown for the Cu(111) surface [188].

2.3.5 VLEED, LEETS, TCS

One of the main driving forces for very low energy electron diffraction (VLEED) was and is to obtain information on the electronic band structure. This can be obtained also by the complementary method of measuring the current to the crystal. Early efforts in this direction started already more than 50 years ago but did not became more broadly applied until the late 1970s, mainly thanks to the efforts of Sanche [189] and Komolov et al. [190]. Sanche called his method low energy electron transmission spectroscopy (LEETS) because he was interested in the interaction of slow electrons with thin crystalline organic and noble gas films on metals, while Komolov, whose interest was the electronic structure of crystals, called it total current spectroscopy (TCS), sometimes also called absorbed

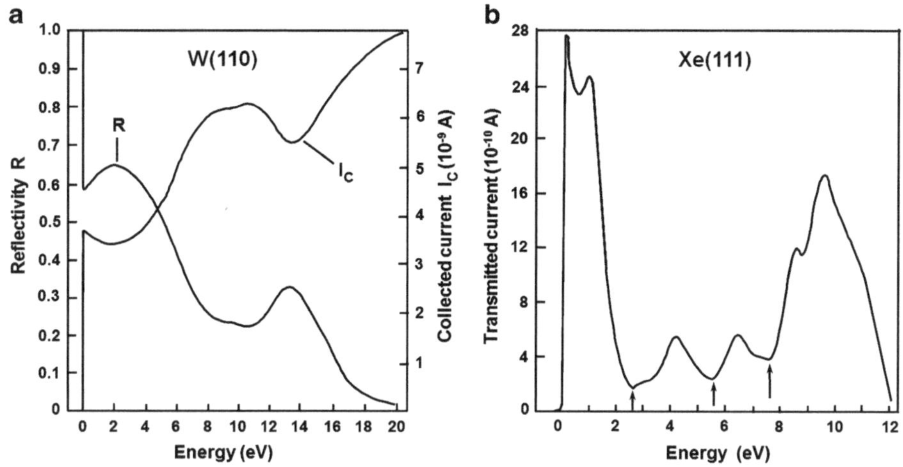

Fig. 2.32 (**a**) Complementarity of specular reflected current at normal incidence and collected current ("drain current") I_c at low energies, at which inelastic scattering is small and no nonspecular scattering occurs. Example W(110). (**b**) Transmission of low energy electrons through a 100 monolayer thick (111)-oriented Xe film on (111)-oriented polycrystalline Pt (Pt drain current, "LEETS"). Normal incidence. The energy dependence is interpreted in terms of the density of unoccupied states and the energy-dependent inelastic mean free path. (**a**) From Ref. [183]. (**b**) Adapted with permission from Ref. [191]. Copyright 1985 by the American Physical Society

current spectroscopy. The complementarity with reflectivity is illustrated by the reflectivity $R(E)$ and $T(E)$ curves of W(110) in Fig. 2.32a [183], taken with the trochoidal electron spectrometer mentioned earlier. By proper choice of the potential of the electrodes in this system it is also possible to separate elastic from inelastic scattering [183]. Figure 2.32b [191] shows a LEETS spectrum of a (111)-orientated Xe layer on Pt from Sanche's group [192]. The lowest energy loss that can occur in Xe is a transition into an exciton state at 8.45 eV above the conduction band. Therefore the structure seen in this figure is solely due to band structure effects and could be correlated with the electronic density of states: minima (maxima) correspond to energies with low (high) density of states. In particular, the strong decrease in transmission above 1.5 eV is due to a band gap and the subsequent oscillations are connected with variations in the band structure [193]. Above the first excitation threshold inelastic processes are superimposed on the loss-free features and give information on electronic transition.

A detailed analysis of the unoccupied density of states and of the *elastic* mean free path was made for solid Ar [194]. LEETS has been extended from simple atomic and molecular crystal layers to large biomolecules, which are difficult to crystallize. Although self-assembled monolayers can be prepared of them, for example of DNA [195], the complexity of the molecules requires an extension of the theory for the description of the scattering processes. A multiple scattering theory has been developed over the years by Caron and Sanche [196, 197] but another approach has been taken too [198]. Several reviews [199–202] give a good impression of the possibilities and limitations of LEETS and also useful information on the interaction of slow electrons with insulating materials in general and biomolecules in particular. LEETS can become a useful complementary method in LEEM instruments without energy filter or with small beam separator angles (small energy dispersion), for example in microspectroscopy of organic materials or energy loss spectroscopy similar to Andersson's early work [203].

Total current spectroscopy (TCS) also would be a useful complement to LEEM for the same reasons, giving information on the electronic structure and excitations of the specimen. Early work has been reviewed by Komolov [204] and Strocov [205]. Close cooperation between experiment and theory has given detailed insight in the electronic structure of crystals such as $NbSe_2$ [206], graphite [207], $TiTe_2$ [208] or TiS_2 [209]. These results have been obtained in display type LEED systems with modified potentials by measuring the crystal target current $T(E)$, but have been presented as VLEED results, based on the relation $R(E) = 1 - T(E)$. While the experimental set-up of these measurements is simple, doing them in a LEEM system offers several advantages: (1) LEEM imaging allows first to select a region with well-defined microstructure, for example a step-free region, from which micro-TCS measurements can be made using a small illumination aperture; (2) the crystal surface can be aligned very precisely normal to the beam; (3) the coherence length is large; (4) in a system with an energy filter all electrons, which have lost more than 0.2–0.3 eV energy can be eliminated. Thus the reverse to what was said at the beginning of this paragraph is true too: LEEM is a good complement to TCS.

An important point, which has been mentioned several times, still has to be emphasized: the influence of the mean inner potential or muffin tin potential V_0 on the analysis of $R(E)$ and $T(E)$ measurements and of the LEED $I(V)$ analysis, which will be discussed briefly below. The value of V_0 depends upon the radius at which the periodic potential of the ion cores is cut off. It can be optimized for specific ion core combinations and can take into account surface core level shifts [210]. An additional electron injected into the N-electron system, for which the band structure was calculated, has exchange and correlation interactions with this N-electron system. Also, its energy is far above the Fermi energy and as a consequence is loses energy, which amounts to a loss of electrons in the incident electron beam (absorption). These two processes, which depend on the energy of the injected electron, are taken into account by a real part V_{or} in addition to V_0, frequently called ΔE, and an imaginary part V_{0i} in the interaction potential ("optical potential", "self energy"). Furthermore, as already pointed out by Feibelman [9], the additional electron can also penetrate as an evanescent wave into the crystal in energy ranges, which are forbidden in the N-electron system, and lose energy. Thus the complex band structure must be used in the comparison between theory and experiment. This method is now well-developed (see for example Refs. [211, 212]). While V_{or} and V_{oi} could be calculated in principle, they are usually determined by comparison with experiment, V_{or} by shifting characteristic features in theory and experiment into coincidence, V_{oi} by fitting the widths of the peaks in the spectrum, which is determined by absorption. More details can be found in the original papers.

2.3.6 Quantum Well Effects

In an energy region in which absorption is weak, so that the IMFP is long, that is at low energies, information on the unoccupied band structure above the vacuum level normal to the surface can be obtained from interference effects in thin films with parallel boundaries, called quantum size effects (QSE). Interference effects in thin films had already been observed in specimen current (LEETS) measurements of (111)-oriented Au films on Ir by Thomas [213] and by Jonker et al. [214] in (111)-oriented Au and Ag films on W(110). While a rigorous interpretation of the interference phenomena requires a quantum mechanical treatment, to a first approximation the film may be considered as the electron wave optical analogue to the light-optical Fabry-Perot etalon. A more precise analysis takes absorption into account in a manner as described in Ref. [215] for an absorbing film on a transparent substrate, but generalized to an absorbing substrate. Extraction of the band structure does not have to go into this detail. It proceeds along the following line. For a wave with wavelength λ the condition for constructive interference in a film with thickness t is given by

$$n\lambda/2 + \varphi_1(E) + \varphi_2(E) \quad = t \quad \text{or with } \lambda = 2\pi/k \quad kt - \Phi(E) = n\pi, \tag{2.25a}$$

where $\Phi(E) = k(E) \bullet (\varphi_1(E) + \varphi_2(E))$ and $\varphi_1(E)$, $\varphi_2(E)$ are the phase changes upon reflection at the two film boundaries. The unknown function $\Phi(E)$ can be eliminated using pairs n, t with the same energy, from which one can obtain

$$k(E) = \pi(n_2 - n_1)/(t_2 - t_1) \tag{2.25b}$$

and by inversion $E(k)$ in the direction $\boldsymbol{k}$ normal to the film surface. The first band structure determinations from thin film interferences were done in a slightly different manner with LEETS using a trochoidal spectrometer by Perluzzo et al. [216]. Figure 2.33a shows their results for (111)-oriented Ar films on (111)-oriented Pt. The interference maxima in the film transmission are superimposed on the general energy dependence which is low at intermediate energies due to a band gap in Ar. The band structure deduced from these maxima is shown in Fig. 2.33b, together with the calculated band structure. More recent band structure determinations with this method have been done in the reflection mode in LEEM and will be discussed there together with other QSE effects in thin films in Chap. 5. QSE intensity oscillations in thin films occur also in reflection high energy electron diffraction as a function of angle of incidence at fixed thickness [217]. In the reflection of slow electrons at oblique incidence they have been observed as a function of film thickness [218]. In general QSE effects are unavoidable in thin films with parallel boundaries and provide many possibilities to extract information on the specimen, not only in reflection and transmission of electrons but also in photoemission and inverse photoemission, two methods, with which quantum size effects, also called quantum well effects, have been studied extensively [219].

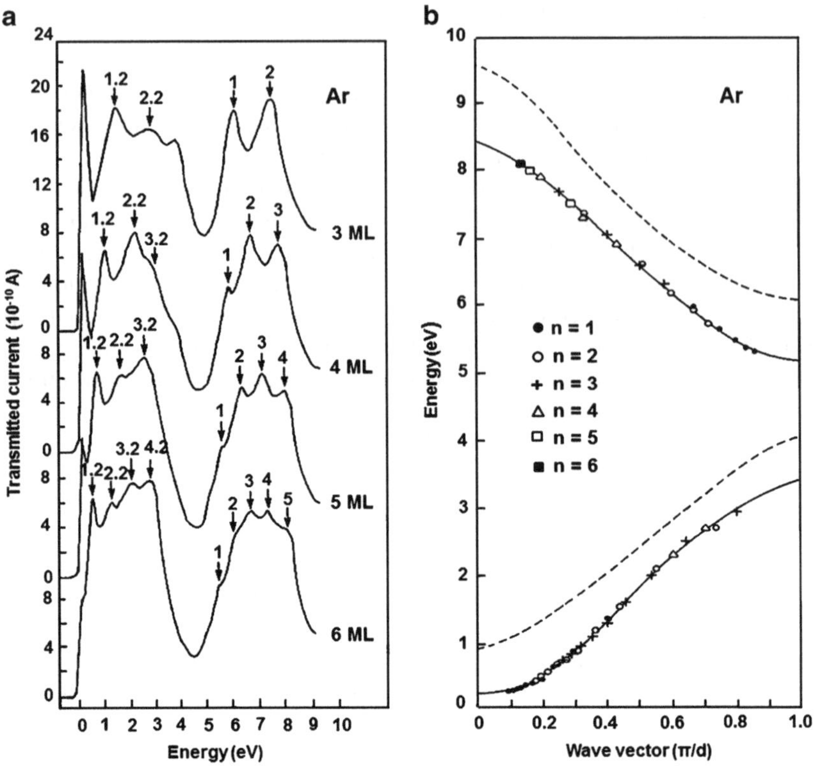

Fig. 2.33 (**a**) Quantum size oscillations in the transmitted current through (111)-oriented 3–6 monolayer thick Ar films on (111)-oriented polycrystalline Pt (normal incidence). The *numbers* at the maxima indicate the interference order including a phase shift factor characteristic for a given band. (**b**) Band structure $E(\boldsymbol{k})$ of Ar along the Γ–L direction derived from the data shown in (**a**). Adapted with permission from Ref. [216]. Copyright 1985 by the American Physical Society

2.3.7 Other Aspects

Concluding the discussion of the interaction of very slow electrons with surfaces, an important interaction still has to be mentioned: the specular reflection of spin-polarized electrons from ferromagnetic materials. At normal incidence and specular reflection as used in cathode lenses the reflected intensity is not influenced by spin–orbit coupling in contrast to spin-polarized LEED (SPLEED), which works with oblique angles of incidence. Therefore the only magnetic interaction, which the spin-polarized electron experiences is the exchange interaction. As a consequence the reflectivity R depends upon the relative orientation of the spin of the incident electron and of the electrons in the ferromagnetic material: $R \sim \boldsymbol{P} \bullet \boldsymbol{M}$, where $\boldsymbol{P}$ is the spin polarization of the incident beam and $\boldsymbol{M}$ the magnetization. All normal incidence specular reflection experiments made up to now, including the first one in the diffraction mode [220], were made in LEEM instruments. Therefore the subject will be discussed in more detail in connection with magnetic imaging in Chap. 7. It should only be mentioned that very low energy spin-polarized electron reflectivity measurements can also be done at oblique incidence [218, 221] but the analysis is then more complicated because of the influence of the spin–orbit interaction.

Up to now we have been concerned with interactions in the energy range most frequently used in LEEM, i.e., up to about 20–30 eV. Higher energies are of relevance in combined LEEM-LEED studies using conventional LEED with energies between about 50 eV and several 100 eV. LEED studies in this energy range complement LEEM by giving information (1) on the lateral periodicity of the surface and on the surface defect structure below the resolution limit of LEEM and (2) on the atomic structure of the surface. In the first case the information is contained mainly in the intensity *distribution* in the LEED pattern, i.e., in the spot positions and shapes, which is described by the lattice factor, and its analysis requires only the kinematic theory of diffraction. In the second case the *relative intensities* of the diffraction spots allow extraction of the atomic positions within the lateral periodicity units (unit meshes) including their thermal vibrations. For this purpose the dynamical theory of diffraction which takes account of multiple scattering and absorption is used. In particular this second aspect of LEED has been covered in several books and review papers [222–229], to which the reader is referred. Reviews on spin-polarized LEED can be found in Refs. [40, 230, 231]. A number of software packages are available for structure analysis but only a few have been used up to now for LEED structure analysis in LEEM instruments. This is surprising in view of the significant advantages of LEED in a LEEM instrument compared to a standard LEED system: the (00) beam, which contains most of the information on the atomic arrangement normal to the surface, can be measured, the spots do not move with energy, there are no grids, which distort the diffraction spots and with an energy filter electrons, which suffered even only very small losses, are filtered out. Furthermore, the microstructure of surface region from which the LEED pattern is taken can be inspected. The disadvantages of slight distortions due to lens aberrations and misalignment or the limitation of the number of diffraction spots by the angular transmission of the objective lens are in general insignificant except perhaps when the lattice constants are very small.

The other application of standard LEED, determination of the lateral periodicity and of surface defects [232–235], is much more straightforward and has, therefore, found already considerable application in LEEM. Examples of such surface defects are two-dimensional island, antiphase domains, steps, in particular on vicinal planes, facet orientations, to name a few. To extract quantitative information beyond the lateral periodicity, however, requires delving somewhat into the instrument properties, which are characterized by the instrument response function. We will encounter it again in connection with the theory of image formation in Chap. 4. For LEED this function has been introduced by Park et al. [236] and subsequently discussed in more detail by others [237, 238]. The instrument response function

$T(\boldsymbol{k})$ describes how the instrument modifies the intensity $I_S(\boldsymbol{k})$ scattered from the specimen into direction $\boldsymbol{k}$ as seen on the detector ($I_D(\boldsymbol{k})$):

$$I_D(\boldsymbol{k}) = I_S(\boldsymbol{k}) * T(\boldsymbol{k}), \tag{2.26}$$

where the sign $*$ means convolution. $T(\boldsymbol{k})$ includes the broadening of $I_S(\boldsymbol{k})$ caused by different angles of incidence resulting from the finite source size, the energy spread of the beam, the finite beam width, the resolution of the detector and other instrument properties. If $T(\boldsymbol{k})$ is known then $I_S(\boldsymbol{k})$ can be obtained by deconvolution and from it the real space information via Fourier transformation. Determination of $T(\boldsymbol{k})$ requires an ideal, defect-free specimen so that $I_S(\boldsymbol{k})$ is basically a δ function. In a conventional LEED system it is difficult to identify such specimen regions but in a LEEM system imaging allows to search for step-free regions and proper surface treatment can also minimize the number of adatoms and vacancies. Some refractory surfaces such as the W(110) surface are very well suited for this purpose. $T(\boldsymbol{k})$ is usually a Gaussian with Lorentzian tails. Of course, $T(\boldsymbol{k})$ depends strongly on the instrument settings such as cathode temperature, Wehnelt voltage, illumination focusing, and aperture and has to be determined for the specific settings used in the LEED experiment. Limitations of the fitting process in LEED have been pointed out in Ref. [239].

References

1. Brodskii, A.M., Gurevich, Y.Y., Levich, V.G.: General threshold theory of electron emission from the surface of a metal. Phys Status Solidi **40**, 139–151 (1970)
2. Jensen, K.L., Moody, N.A., Feldman, D.W., Montgomery, E.J., O'Shea, P.G.: Photoemission from metals and cesiated surfaces. J. Appl. Phys. **102**, 024911, 11 pages (2007)
3. Modinos, A.: Field, Thermionic, and Secondary Electron Emission Spectroscopy. Plenum, New York (1984)
4. Hutson, A.R.: Velocity analysis of thermionic emission from single-crystal tungsten. Phys. Rev. **98**, 889–901 (1955)
5. Herring, C., Nichols, M.H.: Thermionic emission. Rev. Mod. Phys. **21**, 185–270 (1949)
6. Dobretsov, L.N.: Elektronen- und Ionenemission. NASA Techn. Transl. TT F-73, NASA, Washington 1963. VEB Verlag Technik, Berlin (1954)
7. Nottingham, W.B.: Thermionic Emission. In: Flügge, S. (ed.) Encyclopedia of Physics, vol. 21, pp. 1–175. Springer, Berlin (1956)
8. Schwede, J.W., Bargatin, I., Riley, D.C., Hardin, B.E., Rosenthal, S.J., Sun, Y., Schmitt, F., Pianetta, P., Howe, R.T., Shen, Z.X., Melosh, N.A.: Photo-enhanced thermionic emission for solar concentrator systems. Nat. Mater. **9**, 762–767 (2010)
9. Feibelman, P., Eastman, D.E.: Photoemission spectroscopy: correspondence between quantum theory and experimental phenomenology. Phys. Rev. B **10**, 4932–4947 (1974)
10. Pendry, J.B.: Theory of photoemission. Surf. Sci. **57**, 679–705 (1976)
11. Berglund, C.N., Spicer, W.E.: Photoemission studies of silver and copper: theory. Phys. Rev. **136**, A1030–A1044 (1964)
12. Feuerbacher, B., Fitton, B., Willis, R.F. (eds.): Photoemission and the Electronic Properties of Surfaces. John Wiley & Sons, Chichester (1978)
13. Kevan, S.D. (ed.): Angle-Resolved Photoemission. Elsevier, Amsterdam (1992)

14. Hedin, L.: On correlation effects in electron spectroscopies and the GW approximation. J. Phys. Condens. Matter **11**, R489–R528 (1999)
15. Hüfner, S.: Photoelectron Spectroscopy Principles and Applications, 3rd edn. Springer, Berlin (2003)
16. Schattke, W., van Hove, M.A. (eds.): Solid-State Photoemission and Related Methods. Wiley-VCH, Weinheim (2003)
17. Hüfner, S. (ed.): Very High Resolution Photoelectron Spectroscopy. Springer, Berlin (2007)
18. Schwabl, F.: Advanced Quantum Mechanics, p. 187. Springer, Berlin (1999)
19. Kantorovich, I.I.: Nonlinear surface photoelectric effect in metals subjected to intense light. Sov. Phys. Tech. Phys. **22**, 397–399 (1977)
20. Anisimov, S.I., Benderskiĭ, V.A., Farkas, G.: Nonlinear photoelectric emission from metals induced by laser irradiation. Sov. Phys. Usp. **20**, 467–488 (1977)
21. Ferrini, G., Banfi, F., Giannetti, C., Parmigiani, F.: Non-linear electron photoemission from metals with ultrashort laser pulses. Nucl. Instrum. Meth. Phys. Res. A **601**, 123–131 (2009)
22. Fowler, R.H.: The analysis of photoelectric sensitivity curves for clean metals at various temperatures. Phys. Rev. **38**, 45–56 (1931)
23. DuBridge, L.A.: Theory of the energy distribution of photoelectrons. Phys. Rev. **43**, 727–741 (1933)
24. Mahan, G.D.: Theory of photoemission in simple metals. Phys. Rev. B **2**, 4334–4350 (1970)
25. Kong, X., Rowe, J.E., Nemanich, R.: Single molecule measurements with photoelectron emission microscopy. J. Vac. Sci. Technol. B **26**, 1461–1465 (2008)
26. Kane, E.O.: Theory of photoelectric emission from semiconductors. Phys. Rev. **127**, 131–141 (1962)
27. Kuch, W., Schneider, C.M.: Magnetic dichroism in valence band photoemission. Rep. Prog. Phys. **64**, 147–204 (2001)
28. Braun, J., Borstel, G.: Relativistic photoemission theory applied to GaAs(110). Phys. Rev. B **48**, 14373–14380 (1993)
29. Eckardt, H., Fritsche, L., Noffke, J.: Self-consistent relativistic band structure of the noble metals. J. Phys. F Met. Phys. **14**, 97–112 (1984)
30. Krasovskii, E.E., Schattke, W., Jiříček, P., Vondráček, M., Krasovska, O.V., Antonov, N.V., Shpak, A.P., Bartoš, I.: Photoemission from Al(100) and (111): experiment and ab initio theory. Phys. Rev. B **78**, 165406, 9 pages (2008)
31. Feibelman, P.: Surface electromagnetic fields. Progr. Surf. Sci. **12**, 284–408 (1982)
32. Gartland, P.O., Berge, S., Slagsvold, B.J.: Surface-effect characteristics of photoemission from clean copper-crystal surface. Phys. Rev. Lett. **30**, 916–919 (1973)
33. Gartland, P.O., Slagsvold, B.J.: Transitions conserving parallel momentum in photo emission from the (111) face of copper. Phys. Rev. B **12**, 4047–4058 (1975)
34. Pedersoli, E., Greaves, C.M.R., Wan, W., Coleman-Smith, C., Padmore, H.A., Pagliara, S., Cartella, A., Lamarca, F., Ferrini, G., Galimberti, G., Montagnese, M.: dal Conte, S., Parmigiani, F.: Surface and bulk contribution to Cu(111) quantum efficiency. Appl. Phys. Lett. **93**, 183505, 3 pages (2008)
35. Lobo-Checa, J., Ortega, J.E., Mascaraque, A., Michel, E.G., Krasovskii, E.E.: Effect of photoelectron mean free path on the photoemission cross-section of Cu(111) and Ag(111) Shockley states. Phys. Rev. B **84**, 245419, 6 pages (2011)
36. Rios Rubiano, C.A., Gravielle, M.S., Mitnik, D.M., Silkin, V.M.: Band-structure effects in photoelectron-emission spectra from metal surfaces. Phys. Rev. A **85**, 043422, 8 pages (2012)
37. Bisio, F., Nývlt, M., Franta, J., Petek, H., Kirschner, J.: Mechanisms of high-order perturbative photoemission from Cu(001). Phys. Rev. Lett. **96**, 087601, 4 pages (2006)
38. Roth, S., Leuenberger, D., Osterwalder, J., Dahl, J.E., Carlson, R.M.K., Tkachenko, B.A., Fokin, A.A., Schreiner, P.R., Hengsberger, M.: Negative-electron-affinity diamondoid monolayers as high-brilliance source for ultrashort electron pulses. Chem. Phys. Lett. **495**, 102–108 (2010)

39. Feder, R. (ed.): Polarized Electrons in Surface Physics. World Scientific Publications, Singapore (1985)
40. Kirschner, J.: Polarized Electrons at Surfaces (Springer Tracts in Modern Physics), vol. 106. Springer, Berlin (1985)
41. Heinzmann, U.: Angle-, energy- and spin-resolved photoelectron emission using circularly polarized synchrotron radiation. Phys. Scr. **T17**, 77–88 (1987)
42. Johnson, P.D.: Spin-polarized photoemission. Rep. Prog. Phys. **60**, 1217–1304 (1997)
43. Winkelmann, A., Lin, W.-C., Bisio, F., Petek, H., Kirschner, J.: Interferometric control of spin-polarized electron populations at a metal surface observed by multiphoton photoemission. Phys. Rev. Lett. **100**, 206601, 4 pages (2008)
44. Chiang, C.-T., Winkelmann, A., Henk, J., Bisio, F., Kirschner, J.: Spin-selective pathways in linear and nonlinear photoemission from ferromagnets. Phys. Rev. B **85**, 165137, 4 pages (2012)
45. Tusche, C., Ellguth, M., Ünal, A.A., Chiang, C.-T., Winkelmann, A., Krasyuk, A., Hahn, M., Schönhense, G., Kirschner, J.: Spin resolved photoelectron microscopy using a two-dimensional spin-polarizing electron mirror. Appl. Phys. Lett. **99**, 032505, 3 pages (2011)
46. Marx, G.K.L., Elmers, H.J., Schönhense, G.: Magneto-optical linear dichroism in threshold photoemission electron microscopy of polycrystalline Fe films. Phys. Rev. Lett. **84**, 5888–5891 (2000)
47. Nakagawa, T., Yokoyama, T.: Magnetic circular dichroism near the fermi level. Phys. Rev. Lett. **96**, 237402, 4 pages (2006)
48. Nakagawa, T., Yokoyama, T., Hosaka, M., Katoh, M.: Measurements of threshold photoemission magnetic dichroism using ultraviolet lasers and a photoelastic modulator. Rev. Sci. Instrum. **78**, 023907, 5 pages (2007)
49. Dil, J.H.: Spin and angle resolved photoemission on non-magnetic low-dimensional systems. J. Phys. Condens. Matter **21**, 403001, 22 pages (2009)
50. Aeschlimann, M., Schmuttenmaer, C.A., Elsayed-Ali, H.E., Miller, R.J.D., Cao, J., Gao, Y., Mantell, D.A.: Observation of surface enhanced multiphoton photoemission from metal surfaces in the short pulse limit. J. Chem. Phys. **102**, 8606–8613 (1995)
51. Raether, H.: Surface Plasmons on Smooth and Rough Surfaces and on Gratings. Springer Tracts in Modern Physics, vol. 111. Springer, New York (1988)
52. Liebsch, A.: Electronic Excitations at Metal Surfaces. Plenum, New York (1997)
53. Zayats, A.V., Smolyaninov, I.I., Maradudin, A.A.: Nano-optics of surface plasmon polaritons. Phys. Rep. **408**, 131–314 (2005)
54. Pitarke, J.M., Silkin, V.M., Chulkov, E.V., Echenique, P.M.: Theory of surface plasmons and surface-plasmon polaritons. Rep. Prog. Phys. **70**, 1–87 (2007)
55. Berini, P.: Long-range surface plasmon-polariton waveguides in silica. J. Appl. Phys. **102**, 053105, 9 pages (2007)
56. Maier, S.A.: Plasmonics: Fundamentals and Applications. Springer, New York (2007)
57. Kawata, S. (ed.): Near-Field Optics and Surface Plasmon Polaritons. Springer, Berlin (2001)
58. Homola, J.: Surface Plasmon Resonance Based Sensors. Springer, Berlin (2006)
59. Ozbay, E.: Plasmonics: merging plasmonics and electronics at nanoscale dimensions. Science **311**, 189–193 (2006)
60. Shalaev, V.M., Kawata, S. (eds.): Nanophotonics with Surface Plasmons. Elsevier, Amsterdam (2007)
61. Halas, N.J.: Plasmons in strongly coupled metallic nanostructures. Chem. Rev. Mod. Phys. **111**, 3913–3961 (2011)
62. Stockman, M.I.: Nanoplasmonics: past, present, and glimpse into future. Opt. Express **19**, 22029–22106 (2011)
63. Kubo, A., Petek, H.: Femtosecond time-resolved photoemission electron microscope study of surface plasmon dynamics. J. Vac. Soc. Jap. (Japanese) **51**(368–376) (2008)
64. Shalaev, V.M., Douketis, C., Haslett, T., Stuckless, T., Moskovits, M.: Two-photon electron emission from smooth and rough metal films in the threshold region. Phys. Rev. B **53**, 11193–11206 (1996)

65. Kelly, K.L., Coronado, E., Zhao, L.L., Schatz, G.C.: The optical properties of metal nanoparticles: the influence of size, shape, and dielectric environment. J. Phys. Chem. B **107**, 668–677 (2003)
66. Geshev, P., Klein, S., Witting, T., Dickmann, K., Hietschold, M.: Calculation of the electric-field enhancement at nanoparticles of arbitrary shape in close proximity to a metallic surface. Phys. Rev. B **70**, 075402, 16 pages (2004)
67. Gotschy, W., Vonmetz, K., Leitner, A., Aussenegg, F.R.: Optical dichroism of lithographically designed silver nanoparticle films. Optics Lett. **21**, 1099–1101 (1996)
68. Kreibig, U., Vollmer, W.: Optical Properties of Metal Clusters. Springer, Berlin (1995)
69. Lehmann, J., Merschdorf, M., Pfeiffer, W., Thon, A., Voll, S., Gerber, G.: Surface plasmon dynamics in silver nanoparticles studied by femtosecond time-resolved photoemission. Phys. Rev. Lett. **85**, 2921–2924 (2000)
70. Kennerknecht, C., Hovel, H., Merschdorf, M., Voll, S., Pfeiffer, W.: Surface plasmon assisted photoemission from Au nanoparticles on graphite. Appl. Phys. B **73**, 425–429 (2001)
71. Merschdorf, M., Kennerknecht, C., Pfeiffer, W.: Collective and single-particle dynamics in time-resolved two-photon photoemission. Phys. Rev. B **70**, 193401, 4 pages (2004)
72. Irvine, S.E., Dechant, A., Elezzabi, A.Y.: Generation of 0.4-keV femtosecond electron pulses using impulsively excited surface Plasmons. Phys. Rev. Lett. **93**, 184801, 4 pages (2004)
73. Yalunin, S.V., Gulde, M., Ropers, C.: Strong-field photoemission from surfaces: theoretical approaches. Phys. Rev. B **84**, 195426, 14 pages (2011)
74. Faraggi, M., Aldazabal, I., Gravielle, M.S., Arnau, A., Silkin, V.M.: Study of the induced potential produced by ultrashort pulses on metal surfaces. J. Opt. Soc. Am. B **26**, 2331–2336 (2009)
75. Faraggi, M.N., Gravielle, M.S., Mitnik, D.M.: Interaction of ultrashort laser pulses with metal surfaces: impulsive jellium-Volkov approximation versus the solution of the time-dependent Schrödinger equation. Phys. Rev. A **76**, 012903, 6 pages (2007)
76. Bagus, P.S., Ilton, E.S.: Effects of covalency on the p-shell photoemission of transition metals: MnO. Phys. Rev. B **73**, 155110, 14 pages (2006)
77. Haverkort, M.W., Zwierzycki, M., Andersen, O.K.: Multiplet ligand-field theory using Wannier orbitals. Phys. Rev. B **85**, 165113, 20 pages (2012)
78. Bagus, P.S., Broer, R., Ilton, E.S.: Atomic near-degeneracy for photoemission: generality of 4f excitations. J. Electron. Spectrosc. Relat. Phenom. **165**, 46–49 (2008)
79. Bagus, P.S., Freeman, A.J., Sasaki, F.: Prediction of new multiplet structure in photoemission experiments. Phys. Rev. Lett. **30**, 850–853 (1973)
80. Nelin, C.J., Bagus, P.S., Ilton, E.S., Chambers, S.A., Kuhlenbeck, H., Freund, H.-J.: Relationships between complex core level spectra and materials properties. Int. J. Quantum Chem. **110**, 2752–2764 (2010)
81. Uldry, A., Vernay, F., Delley, B.: Systematic computation of crystal-field multiplets for x-ray core spectroscopies. Phys. Rev. B **85**, 125133, 14 pages (2012)
82. Egelhoff Jr., W.F.: Core-level binding-energy shifts at surfaces and in solids. Surf. Sci. Rep. **6**, 253–415 (1987)
83. de Groot, F., Kotani, A.: Core Level Spectroscopy of Solids. CRC Press, Boca Raton (2008)
84. Gerson, A.R., Bredow, T.: Interpretation of sulphur 2p XPS spectra in sulfide minerals by means of ab initio calculations. Surf. Interface Anal. **29**, 145–150 (2000)
85. www.casaxps.comhelp_manualmanual_updatespeak_fitting_in_xps.pdf
86. Yamashita, T., Hayes, P.: Effect of curve fitting parameters on quantitative analysis of Fe0.94O and Fe_2O_3 using XPS. J. Electron. Spectrosc. Relat. Phenom. **152**, 6–11 (2006)
87. Pratt, A.: Fine structure in XPS and XANES spectra acquired from a series of sphalerite samples. ECS Trans. **28**, 95–103 (2010)
88. Harmer, S.L., Goncharova, L.V., Kolarova, R., Lennard, W.N., Munoz-Marquez, M.A., Mitchell, I.V., Nesbitt, H.W.: Surface structure of sphalerite studied by medium energy ion scattering and XPS. Surf. Sci. **601**, 352–361 (2007)

89. Jirsak, T., Rodriguez, J.A., Chaturvedi, S., Hrbek, J.: Chemistry of SO_2 on Ru(001): formation of SO_3 and SO_4. Surf. Sci. Rep. **418**, 8–21 (1998)
90. Nesbitt, H.W., Schaufuss, A.G., Scaini, M., Bancroft, G.M., Szargan, R.: XPS measurement of fivefold and sixfold coordinated sulfur in pyrrhotites and evidence for millerite and pyrrhotite surface species. Am. Mineral. **86**, 318–326 (2001)
91. Skinner, W.M., Nesbitt, H.W., Pratt, A.R.: XPS identification of bulk hole defects and itinerant Fe 3d electrons in natural troilite (FeS). Geochim. Cosmochim. Acta **68**, 2259–2263 (2004)
92. Sinkovic, B., Johnson, P.D., Brookes, N.B.: Magnetic structure of oxidized Fe(001). Phys. Rev. Lett. **65**, 1647–1650 (1990)
93. Tamura, E., Waddill, G.D., Tobin, J.G., Sterne, P.A.: Linear and circular dichroism in angle resolved Fe 3p photoemission. Phys. Rev. Lett. **73**, 1533–1536 (1994)
94. Tait, S.L., Wang, Y., Costantini, G., Lin, N., Baraldi, A., Esch, F., Petaccia, L., Lizzit, S., Kern, K.: Metal-organic coordination interactions in Fe-terephthalic acid networks on Cu (100). J. Am. Chem. Soc. **130**, 2108–2113 (2008)
95. Jungblut, R., Roth, Ch., Hillebrecht, F.U., Kisker, E.: Spin-polarized electron spectroscopy as a combined chemical and magnetic probe. Surf. Sci. **269/270**, 615–621 (1992)
96. Van Campen, D.G., Pouliot, R.J., Klebanoff, L.E.: Spin-resolved x-ray-photoelectron-spectroscopy study of ferromagnetic iron. Phys. Rev. B **48**, 17535, 5 pages (1993)
97. Roth, C., Hillebrecht, F.U., Rose, H.B., Kisker, E.: Linear magnetic dichroism in angular resolved Fe 3p core level photoemission. Phys. Rev. Lett. **70**, 3479–3482 (1993)
98. Roth, C., Rose, H.B., Hillebrecht, F.U., Kisker, E.: Magnetic linear dichroism in soft X-ray core level photoemission from iron. Solid State Commun. **86**, 647–650 (1993)
99. Rossi, G., Sirotti, F.: 3p Fine structure of ferromagnetic Fe and Co from photoemission with linearly polarized light. Solid State Commun. **90**, 557–562 (1994)
100. Tobin, J.G., Goodman, K.W., Schumann, F.O., Willis, R.F., Kortright, J.B., Denlinger, J.D., Rotenberg, E., Warwick, A., Smith, N.V.: Generalized description of magnetic x-ray circular dichroism in Fe 3p photoelectron emission. J. Vac. Sci. Technol. A **15**, 1766–1769 (1997)
101. Huang, D.-J., Riffe, D.M., Erskine, J.I.: Simultaneous determination of Fe 3p spin–orbit and exchange splittings in photoemission. Phys. Rev. B **51**, 15170–15179 (1995)
102. Bansmann, J., Lu, L., Meiwes-Broer, K.H., Schlatholter, T., Braun, J.: Relationship between magnetic circular and linear dichroism in photoemission from Fe 3p core level: an experimental and theoretical investigation. Phys. Rev. B **60**, 13860–13868 (1999)
103. Liu, Y., Xu, Z., Johnson, P.D., van der Laan, G.: Spin–orbit coupling, exchange interaction, and hybridization in the photoexcitation of the Ni 3p core level. Phys. Rev. B **52**, R8593–R8596 (1995)
104. Thompson, A., et al. (eds.): X-Ray Data Booklet. Lawrence Berkeley National Laboratory, Berkeley, CA (2009)
105. http://xdb.lbl.gov
106. Yeh, J.J., Lindau, I.: Atomic subshell photoionization cross sections and assymetry parameters: $1 < Z < 103$. At. Data. Nucl. Data Tables **32**, 1–155 (1985)
107. Ferguson, I.: Auger Microprobe Analysis. Adam Hilger, Bristol (1989)
108. Powell, C.J., Jablonski, A.: Surface sensitivity of X-ray photoelectron spectroscopy. Nucl. Instrum. Meth. Phys. Res. A **601**, 54–65 (2009)
109. Jablonski, A., Powell, C.J.: Practical expressions for the mean escape depth, the information depth, and the effective attenuation length in Auger-electron spectroscopy and x-ray photoelectron spectroscopy. J. Vac. Sci. Technol. A **27**, 253–261 (2009)
110. Jablonski, A., Powell, C.J.: Improved analytical formulae for correcting elastic-scattering effects in X-ray photoelectron spectroscopy. Surf. Sci. **604**, 327–336 (2010)
111. Powell, C.J., Jablonski, A.: Progress in quantitative surface analysis by X-ray photoelectron spectroscopy: current status and perspectives. J. Electron. Spectrosc. Relat. Phenom. **178–179**, 331–346 (2010)

112. Tanuma, S., Powell, C.J., Penn, D.R.: Calculations of electron inelasticmean free paths. IX. Data for 41 elemental solids over the 50 eV to 30 keV range. Surf. Interface Anal. **43**, 689–713 (2011)
113. Seah, M.P.: An accurate and simple universal curve for the energy-dependent electron inelastic mean free path. Surf. Interface Anal. **44**, 497–503 (2012)
114. Nakajima, R., Stöhr, J., Idzerda, Y.U.: Electron-yield saturation effects in L-edge x-ray magnetic circular dichroism spectra of Fe, Co, and Ni. Phys. Rev. B **59**, 6421–6429 (1999)
115. Frazer, B.H., Gilbert, B., Sonderegger, B.R., De Stasio, G.: The probing depth of total electron yield in the sub-keV range: TEY-XAS and X-PEEM. Surf. Sci. Rep. **537**, 161–167 (2003)
116. Dreiner, S., Schurmann, M., Westphal, C.: Structural analysis of the SiO_2/Si(100) interface by means of photoelectron diffraction. Phys. Rev. Lett. **93**, 126101, 4 pages (2004)
117. Woodruff, D.P.: Surface structural information from photoelectron diffraction. J. Electron. Spectrosc. Relat. Phenom. **178–179**, 186–194 (2010)
118. Sébilleau, D., Natoli, C.R.: Some insight into the convergence of the multiple scattering series expansion. In: 14th Int. Conf. on X-Ray Absorption Fine Structure (XAFS14) 2009, pp. 1–12. J. Phys. Conf. Ser. (2009)
119. Sébilleau, D., Natoli, C., Gavaza, G.M., Zhao, H., Da Pieve, F., Hatada, K.: MsSpec-1.0: a multiple scattering package for electron spectroscopies in material science. Comput. Phys. Commun. **182**, 2567–2579 (2011)
120. Fadley, C.S.: X-ray photoelectron spectroscopy: progress and perspectives. J. Electron. Spectrosc. Relat. Phenom. **178–179**, 2–32 (2010)
121. Henke, B.L., Smith, J.A., Attwood, D.T.: 0.1-10-keV x-ray-induced electron emissions from solids: models and secondary electron measurements. J. Appl. Phys. **48**, 1852–1866 (1977)
122. Henke, B.L., Liesegang, J., Smith, S.D.: Soft-x-ray-induced secondary-electron emission from semiconductors and insulators: models and measurements. Phys. Rev. B **19**, 3004–3021 (1979)
123. Stöhr, J.: NEXAFS Spectroscopy. Springer, Berlin (1992)
124. Kasrai, M., Brown, J.R., Bancroft, G.M., Yin, Z., Tan, K.H.: Sulphur characterization in coal from X-ray absorption near edge spectroscopy. Int. J. Coal Geol. **32**, 107–135 (1996)
125. Henke, B.L., Lee, P., Tanaka, T.J., Skimabukuro, R.L., Fujikawa, B.K.: Low-energy X-ray interaction coefficients: photoabsorption, scattering and reflection, E = 100-2000 eV, Z = 1–94. At. Data. Nucl. Data Tables **27**, 1–144 (1982)
126. Henke, B.L., Gullikson, E.M., Davis, J.C.: X-ray interactions: photoabsorption, scattering, transmission, and reflection at E = 50-30,000 eV, Z = 1–92. At. Data. Nucl. Data Tables **54**, 181–342 (1993)
127. www-cxro.lbl.gov/optical_constants/
128. Stöhr, J., Siegmann, H.C.: Magnetism. Springer, Berlin (2006)
129. Beth, R.A.: Mechanical detection and measurement of the angular momentum of light. Phys. Rev. **50**, 115–125 (1936)
130. Stewart, A.M.: Angular momentum of the electromagnetic field: the plane wave paradox resolved. Eur. J. Phys. **26**, 635–641 (2005)
131. Mansuripur, M.: Angular momentum of circularly polarized light in dielectric media. Opt. Express **13**, 5315–5324 (2005)
132. Papaconstantopoulos, D.A.: Handbook of the Band Structure of Elemental Solids. Plenum, New York (1986)
133. Stöhr, J.: Exploring the microscopic origin of magnetic anisotropies with X-ray magnetic circular dichroism (XMCD) spectroscopy. J. Magn. Magn. Mater. **200**, 470–497 (1999)
134. Wilhelm, F., Poulopoulos, P., Srivastava, P., Wende, H., Farle, M., Baberschke, K., Angelakeris, M., Flevaris, N.K., Grange, W., Kappler, J.-P., Ghiringhelli, G., Brookes, N. B.: Magnetic anisotropy energy and the anisotropy of the orbital moment of Ni in Ni/Pt multilayers. Phys. Rev. B **61**, 8647–8650 (2000)
135. Arai, K., Okuda, T., Tanaka, A., Kotsugi, M., Fukumoto, K., Oura, M., Senba, Y., Ohashi, H., Nakamura, T., Matsushita, T., Muro, T., Kakizaki, A., Kinoshita, T.: Complete assignment of

spin domains in antiferromagnetic NiO(100) by photoemission electron microscopy and cluster model calculation. J. Phys. Soc. Jpn. **79**, 013703, 4 pages (2010)
136. Kunes, J., Oppeneer, P.M.: Anisotropic x-ray magnetic linear dichroism at the L2,3 edges of cubic Fe, Co, and Ni:Ab initio calculations and model theory. Phys. Rev. B **67**, 024431, 9 pages (2003)
137. Spaldin, N.A., Cheong, S.-W., Ramesh, R.: Multiferroics: past, present, and future. Phys. Today **63**, 38–43 (2010)
138. Bauer, E., Browne, H.N.: Elastic Scattering of Electrons by the Many-Electron Atom. In: McDowell, M.R.C. (ed.) Proc. 3rd Int. Conf. on the Physics of Electronic and Atomic Collisions, London 1963. Atomic Collision Processes, pp. 16–27. North-Holland, Amsterdam (1964)
139. Bauer, E.: Interaction of slow electrons with surfaces. J. Vac. Sci. Technol. **7**, 3–12 (1970)
140. Bauer, E.: Interaction of slow electrons with randium-jellium. In: Les Interactions des Electrons avec la Matiere Condensee. Applications a I'Etude du Solide. pp. 42–96. AVCP (1972)
141. Bauer, E.: Low Energy Electron Diffraction (LEED) and Auger Methods. In: Gomer, R. (ed.) Interactions on Metal Surfaces, pp. 225–274. Springer, Berlin (1975)
142. Bauer, E., Telieps, W.: Low energy electron microscopy. In: Scanning Microscopy, vol. Suppl. 1. pp. 99–108. Scanning Microscopy Int., Chicago (1987)
143. Bauer, E.: Low energy electron microscopy and normal incidence VLEED. In: Koukal, J. (ed.) Physics of Solid Surfaces. Studies in Surface Science and Catalysis, vol. 40, pp. 26–36. Elsevier, Amsterdam (1988)
144. Bauer, E., Telieps, W.: Emission and low energy reflection electron microscopy. In: Howie, A., Valdre, U. (eds.) Surface and Interface Characterization by Electron Optical Methods, vol. 191 NATO ASI Series B: Physics, pp. 195–233. Plenum, New York (1988)
145. Bauer, E.: Low energy electron microscopy. Rep. Prog. Phys. **57**, 895–938 (1994)
146. Bauer, E.: LEEM basics. Surf. Rev. Lett. **5**, 1275–1286 (1998)
147. Berger, M.J., Seltzer, S.M., Wang, R., Schechter, A.: NISTIR Report No. 5188. In. Database ELAST (1993)
148. Jablonski, A., Salvat, F., Powell, C.J.: NIST Standard Reference Database 64 version 3.1. (2003)
149. www.nist.gov/srd/nist64.htm.
150. Elastic scattering of electrons and positrons, ICRU report no. 77. In. ICRU – International Commission on Radiation Units and Measurements. (2007)
151. Salvat, F., Jablonski, A., Powell, C.J.: ELSEPA-Dirac partial-wave calculation of elastic scattering of electrons and positrons by atoms, positive ions and molecules. Comput. Phys. Commun. **165**, 157–190 (2005)
152. Herman, F., Skillman, S.: Atomic Structure Calculations. Prentice Hall, Englewood-Cliffs, NJ (1963)
153. Fink, M., Martin, M.R., Somorjai, G.A.: Comparison of backscattering intensities for low energy electrons from various surface atoms (H, Li, Be, C, 0, Al, Si, S, V, Cr, Ni, Cu, Ag, Pt, Au). Surf. Sci. **29**, 303–308 (1972)
154. Snow, E.C.: Self-consistent energy bands of metallic copper by the augmented-plane-wave method. II. Phys. Rev. **171**, 785–789 (1968)
155. Schmid, R., Gaukler, K.H., Seiler, H.: Measurement of elastically reflected electrons ($E \leq 2.5$ keV) for imaging of surfaces in a simple ultra high vacuum scanning electron microscope. In: Scanning Electron Microscopy vol. II, pp. 501–509. Scanning Electron Microscopy, Inc., O'Hare, Chicago (1983)
156. Gergely, G.: Elastic backscattering of electrons: determination of physical parameters of electron transport processes by elastic peak electron spectroscopy. Prog. Surf. Sci. **71**, 31–88 (2002)
157. Schilling, J.S., Webb, M.B.: Low-energy electron diffraction from liquid Hg: multiple scattering, scattering factor, and attenuation. Phys. Rev. B **2**, 1665–1676 (1970)
158. Dietzel, W., Meister, G., Bauer, E.: Elastic and inelastic backscattering of slow electrons from silicon. Z. Phys. B **47**, 189–194 (1982)

159. Rösler, M., Brauer, W., Devooght, J., Dehaes, J.-C., Dubus, A., Cailler, M., Ganachaud, J.-P.: Particle Induced Electron Emission. Springer, Berlin (1991)
160. Ding, Z.-J., Shimizu, R.: A Monte Carlo modeling of electron interaction with solids including cascade secondary electron production. Scanning **18**, 92–113 (1996)
161. Dapor, M.: Electron-Beam Interactions with Solids: Application of the Monte Carlo Method to Electron Scattering. Springer, Berlin (2003)
162. Salvat, F., Fernandez-Varea, J.M.: Overview of physical interaction models for photon and electron transport used in Monte Carlo codes. IOP Publ. Metrol. **46**, S112–S138 (2009)
163. Fitting, H.-J., Schreiber, E., Kuhr, J.-C., von Czarnowski, A.: Attenuation and escape depths of low-energy electron emission. J. Electron. Spectrosc. Relat. Phenom. **119**, 35–47 (2001)
164. Kuhr, J.-C., Fitting, H.-J.: Monte Carlo simulation of electron emission from solids. J. Electron. Spectrosc. Relat. Phenom. **105**, 257–273 (1999)
165. Penn, D.R.: Electron mean-free-path calculations using a model dielectric function. Phys. Rev. B **35**, 482–486 (1987)
166. Ding, Z.-J., Shimizu, R.: Inelastic collisions of kV electrons in solids. Surf. Sci. **222**, 313–331 (1989)
167. Tanuma, S., Shiratori, T., Kimura, T., Goto, K., Ichimura, S., Powell, C.J.: Experimental determination of electron inelastic mean free paths in 13 elemental solids in the 50 to 5000 eV energy range by elastic-peak electron spectroscopy. Surf. Interface Anal. **37**, 833–845 (2005)
168. Penn, D.R.: Electron mean free paths for free-electron-like materials. Phys. Rev. B **13**, 5248–5254 (1976)
169. McRae, E.G.: Calculation of absorptive potentials for low energy electron scattering from optical data for solid Mg, Al, Cu, Ag, Au, Bi, C and Al_2O_3. Surf. Sci. Rep. **57**, 761–765 (1976)
170. Ziaja, B., London, R.A., Hajdu, J.: Ionization by impact electrons in solids: electron mean free path fitted over a wide energy range. J. Appl. Phys. **99**, 033514, 9 pages (2006)
171. Qian, Z., Sahni, V.: Quantum mechanical image potential theory. Phys. Rev. B **66**, 205103, 4 pages (2002)
172. Qian, Z., Sahni, V.: Exact electronic properties in the classically forbidden region of a metal surface. Int. J. Quantum Chem. **104**, 929–945 (2005)
173. Jones, R.O., Jennings, P.J.: LEED fine structure: origin and applications. Surf. Sci. Rep. **9**, 165–196 (1988)
174. Jennings, P.J., Jones, R.O., Weinert, M.: Surface barrier for electrons in metals. Phys. Rev. B **37**, 6113–6120 (1988)
175. Malmstrom, G., Rundgren, J.: A program for calculation of the reflection and transmission of electrons through a surface potential barrier. Comput. Phys. Commun. **19**, 263–270 (1980)
176. Manghi, F.: Nonlocal exchange and correlation in surface calculations: an application to GaAs(110). Phys. Rev. B **33**, 2554–2558 (1986)
177. White, D., Godby, R.W., Rieger, M.M., Needs, R.J.: Dynamic image potential at an Al(111) surface. Phys. Rev. Lett. **80**, 4265–4268 (1998)
178. Heinrichsmeier, M., Fleszar, A., Hanke, W., Eguiluz, A.G.: Nonlocal density-functional calculations of the surface electronic structure of metals: application to aluminum and palladium. Phys. Rev. B **57**, 14974–14982 (1998)
179. Chulkov, E.V., Silkin, V.M., Echenique, P.M.: Image potential states on metal surfaces: binding energies and wave functions. Surf. Sci. **437**, 330–352 (1999)
180. Tsirkin, S.S., Eremeev, S.V., Chulkov, E.V.: Model pseudopotential for the (110) surface of fcc noble metals. Surf. Sci. **604**, 804–810 (2010)
181. Hanuschkin, A., Wortmann, D., Blügel, S.: Image potential and field states at Ag(100) and Fe(110) surfaces. Phys. Rev. B **76**, 165417, 6 pages (2007)
182. Jennings, P.J.: Surface barrier effects in low-energy electron diffraction. Surf. Sci. **25**, 513–525 (1971)
183. Herlt, H.-J.: Elastische Rückstreuung sehr langsamer Elektronen an reinen und an gasbedeckten Wolfram-Einkritalloberflächen. Ph.D. thesis, TU Clausthal (1982)

184. Herlt, H.-J., Feder, R., Meister, G., Bauer, E.: Experiment and theory of the elastic electron reflection coefficient from tungsten. Solid State Commun. **38**, 973–976 (1981)
185. Jones, R.O., Jennings, P.J.: Fine-structure analysis of spin-polarized low-energy electron diffraction from W(001). Phys. Rev. B **27**, 4702–4711 (1983)
186. Andersson, S.: Low-energy electron diffraction intensities from the clean cooper (001) surface. Surf. Sci. **18**, 325–340 (1969)
187. Bartos, I., Koukal, J.: On electron damping in VLEED. Surf. Sci. **251/252**, 508–510 (1991)
188. Bartos, I., van Hove, M.A., Altman, M.S.: Cu(111) electron band structure and channeling by VLEED. Surf. Sci. **352–354**, 660–664 (1996)
189. Sanche, L.: Transmission of 0–15 eV monoenergetic electrons through thin-film molecular solids. J. Chem. Phys. **71**, 4860–4881 (1979)
190. Komolov, S.A., Chadderton, L.T.: Total current spectroscopy. Surf. Sci. **90**, 359–380 (1979)
191. Plenkiewicz, B., Plenkiewicz, P., Perluzzo, G., Jay-Gerin, J.-P.: Analysis of low-energy electron transmission experiments through thin solid xenon films in the elastic scattering region. Phys. Rev. B **32**, 1253–1256 (1985)
192. Bader, G., Perluzzo, G., Caron, L.G., Sanche, L.: Elastic and inelastic mean-free-path determination in solid xenon from electron transmission experiments. Phys. Rev. B **26**, 6019–6029 (1982)
193. Kessler, B., Eyers, A., Horn, K., Muller, N., Schmiedeskamp, B., Schonhense, G., Heinzmann, U.: Determination of xenon valence and conduction bands by spin-polarized photoemission. Phys. Rev. Lett. **59**, 331–334 (1987)
194. Goulet, T., Jung, J.-M., Michaud, M., Jay-Gerin, J.-P., Sanche, L.: Conduction-band density of states in solid argon revealed by low-energy-electron backscattering from thin films: role of the electron mean free path. Phys. Rev. B **50**, 5101–5109 (1994)
195. Pan, X., Sanche, L.: Mechanism and site of attack for direct damage to DNA by low-energy electrons. Phys. Rev. Lett. **94**, 198104, 4 pages (2005)
196. Caron, L.G., Sanche, L.: Low-energy electron diffraction and resonances in DNA and other helical macromolecules. Phys. Rev. Lett. **91**, 113201, 4 pages (2003)
197. Caron, L., Sanche, L., Tonzani, S., Greene, C.H.: Low-energy electron scattering from DNA including structural water and base-pair irregularities. Phys. Rev. A **80**, 012705, 6 pages (2009)
198. Orlando, T.M., Oh, D., Chen, Y., Aleksandrov, A.B.: Low-energy electron diffraction and induced damage in hydrated DNA. J. Chem. Phys. **128**, 195102, 7 pages (2008)
199. Bass, A.D., Sanche, L.: Absolute and effective cross-sections for low-energy electron-scattering processes within condensed matter. Radiat. Environ. Biophys. **37**, 243–257 (1998)
200. Sanche, L.: Primary interactions of low energy electrons in condensed matter. In: Ferradini, C., Jay-Gerin, J.-P. (eds.) Excess Electrons in Dielectric Media, pp. 1–42. CRC Press, Boca Raton (1991)
201. Bass, A., Sanche, L.: Interactions of Low-Energy Electrons with Atomic and Molecular Solids. In: Hatano, Y., Mozumder, A. (eds.) Charged Particle and Photon Interactions with Matter: Chemical, Physicochemical and Biological Consequences with Applications, pp. 207–257. Marcel Dekker, New York (2004)
202. Sanche, L.: Nanoscale dynamics of radiosensitivity: role of low energy electrons. In: Gómez-Tejedor, G., Fuss, M.C. (eds.) Radiation Damage in Biomolecular Systems, pp. 3–44. Springer, Dordrecht (2012)
203. Andersson, S.: Plasmon thresholds in the secondary electron yield – I experiment. Solid State Commun. **11**, 1401–1404 (1972)
204. Komolov, S.A.: Total Current Spectroscopy of Surfaces. Gordon and Breach, Philadephia, PA (1992)
205. Strocov, V.N.: Unoccupied band structure of layered materials by very-low-energy electron diffraction: implications in photoemission. In: Hughes, H.P., Starnberg, H.I. (eds.) Electron Spectroscopies Applied to Low-Dimensional Structures, pp. 161–208. Kluwer, New York (2000)
206. Krasovskii, E.E., Schattke, W., Strocov, V.N., Claessen, R.: Unoccupied band structure of NbSe2 by very low-energy electron diffraction: experiment and theory. Phys. Rev. B **66**, 235403, 11 pages (2002)

207. Barrett, N., Krasovskii, E.E., Themlin, J.-M., Strocov, V.N.: Elastic scattering effects in the electron mean free path in a graphite overlayer studied by photoelectron spectroscopy and LEED. Phys. Rev. B **71**, 035427, 9 pages (2005)
208. Strocov, V.N., Krasovskii, E.E., Schattke, W., Barrett, N., Berger, H., Schrupp, D., Claessen, R.: Three-dimensional band structure of layered TiTe2: photoemission final-state effects. Phys. Rev. Lett. B **74**, 195125, 14 pages (2006)
209. Krasovskii, E.E., Strocov, V.N.: Very-low-energy electron diffraction from TiS2: experiment and ab initio theory. J. Phys. Condens. Matter **21**, 314009, 7 pages (2009)
210. Rundgren, J.: Optimized surface-slab excited-state muffin-tin potential and surface core level shifts. Phys. Rev. B **68**, 125405, 9 pages (2003)
211. Krasovskii, E.E., Schattke, W.: Surface electronic structure with the linear methods of band theory. Phys. Rev. B **56**, 12874–12883 (1997)
212. Krasovskii, E.E.: Augmented-plane-wave approach to scattering of Bloch electrons by an interface. Phys. Rev. B **70**, 245322, 11 pages (2004)
213. Thomas, R.E.: Interference effects in the reflection of low-energy electrons from thin films of Au on Ir. J. Appl. Phys. **41**, 5330–5334 (1970)
214. Jonker, B.T., Bartelt, N.C., Park, R.L.: Quantum size effect in electron transmission through Cu and Ag films on W(110). Surf. Sci. **127**, 183–199 (1983)
215. Born, M., Wolf, E.: Principles of Optics, p. 628. Cambridge University Press, Cambridge (1980)
216. Perluzzo, G., Bader, G., Caron, L.G., Sanche, L.: Direct determination of electron band energies by transmission interference in thin films. Phys. Rev. Lett. **55**, 545–548 (1985)
217. Jalochowski, M., Bauer, E.: Quantum size and surface effects in the electrical resistivity and high-energy electron reflectivity of ultrathin lead films. Phys. Rev. B **38**, 5272–5280 (1988)
218. Kerkmann, D., Pescia, D., Krewer, J.W., Vescovo, E.: Low energy electron oscillations during epitaxial growth of thin films. Z. Phys. B **85**, 311–314 (1991)
219. Chiang, T.-C.: Photoemission studies of quantum well states in thin films. Surf. Sci. Rep. **39**, 181–235 (2000)
220. Wurm, K.: Spin-polarisierte LEEM-Untersuchungen an dunnen Kobalt-Epitaxieschichten auf W(110). M.S. thesis, Technische Universität Clausthal (1994)
221. Egger, S., Back, C.H., Krewer, J., Pescia, D.: A spin selective electron interferometer. Phys. Rev. Lett. **83**, 2833–2836 (1999)
222. Pendry, J.B.: Low Energy Electron Diffraction. Academic Press, London (1974)
223. Van Hove, M.A., Tong, S.Y.: Surface Crystallography by LEED. Springer, Berlin (1979)
224. Heinz, K., Müller, K.: LEED intensities – experimental progress and new possibilities of surface structure determination. In: Structural Studies of Surfaces (Springer Tracts in Modern Physics), vol. 91, pp. 1–53. Springer, Berlin (1982)
225. Marcus, P.M., Jona, F. (eds.): Determination of Surface Structure by LEED. Plenum, New York (1984)
226. Van Hove, M.A., Weinberg, W.H., Chen, C.-M.: Low-Energy Electron Diffraction. Springer, Berlin (1986)
227. Heinz, K.: LEED and DLEED as modern tools for quantitative surface structure determination. Rep. Prog. Phys. **58**, 637–704 (1995)
228. Garcıa de Abajo, F.J., Van Hove, M.A., Fadley, C.S.: Multiple scattering of electrons in solids and molecules: a cluster-model approach. Phys. Rev. B **63**, 075404, 16 pages (2001)
229. Soares, E.A., de Castilho, C.M.C., de Carvalho, V.E.: Advances on surface structural determination by LEED. J. Phys. Condens. Matter **23**, 303001, 19 pages (2011)
230. Feder, R.: Spin-polarised low-energy electron diffraction. J. Phys. C Solid State Phys. **14**, 2049–2091 (1981)
231. Elmers, H.-J.: Spin-polarized low-energy electron diffraction. In: Kronmüller, H., Parkin, S. (eds.) Handbook of Magnetism and Advanced Magnetic Materials. John Wiley & Sons, Chichester (2007)
232. Henzler, M.: Defects in surface structure: informations with LEED. In: Advances in Solid State Physics, vol. 19. pp. 193–205. Vieweg, Braunschweig (1979)

233. Henzler, M.: Measurments of surface defects. Appl. Phys. A **34**, 205–214 (1984)
234. Henzler, M.: Quantitative analysis of LEED spot profiles. In: Van Hove, M.H., Tong, S.Y. (eds.) The Structure of Surfaces (Springer Series in Surface Science), vol. 2, pp. 351–357. Springer, Berlin (1985)
235. Lagally, M.G.: Diffraction Techniques. In: Park, R.L., Lagally, M.G. (eds.) Solid State Physics: Surfaces, pp. 237–298. Academic Press, Orlando, FL (1985)
236. Park, R.L., Houston, J.E., Schreiner, D.G.: The LEED instrument response function. Rev. Sci. Instrum. **42**, 60–65 (1971)
237. Comsa, G.: Coherence length and/or transfer width? Surf. Sci. **81**, 57–68 (1979)
238. Lu, T.-M., Lagally, M.G.: The resolving power of a low-energy electron diffractometer and the analysis of surface defects. Surf. Sci. **99**, 695–713 (1980)
239. Vuorinen, J., Pussi, K., Diehl, R.D., Lindroos, M.: Correlation of electron self-energy with geometric structure in low-energy electron diffraction. J. Phys. Condens. Matter **24**, 015003, 7 pages (2012)

Chapter 3
Instrumentation

Abstract This chapter describes the instruments used in emission and reflection microscopy starting from the simplest configuration to the most sophisticated instruments. These include aberration-corrected microscopes and microscopes which allow spectroscopic and spin-resolved imaging; time-resolved imaging too will be discussed briefly. Next, the components used in them are discussed: objective lens and other axial symmetric lenses, beam separators, electron mirrors, aberration correction, energy filters, Wien filters, spin polarizers, stigmators/deflectors, apertures, photon sources, electron sources, image detection, and specimen manipulation. Other instrumentation aspects such as vacuum generation, airlock and preparation chamber, electronics, and computer control/operational procedures will only briefly be touched upon.

3.1 Instruments: From Simple to Complex

3.1.1 PEEM

We begin with the discussion of pure emission microscopes. The simplest microscope consists only of an electrostatic triode objective lens (two electrodes plus object) with contrast aperture, microchannel plate (MCP)-fluorescent screen detector and a simple specimen manipulator. Tonner and Harp used such a configuration in the first demonstration of imaging with synchrotron radiation-excited secondary electrons [1], called photoyield spectromicroscopy [2]. In an improved version of this set-up the triode objective was replaced by a tetrode (three electrodes plus object) [3]. Adding a projective lens allows changing the magnification but does not improve image quality. This can be achieved with improved objective lenses equipped with a stigmator for the correction of the axial astigmatism. An example is the instrument of Bethge and Klaua [4], which used a magnetic objective and a

E. Bauer, *Surface Microscopy with Low Energy Electrons*,
DOI 10.1007/978-1-4939-0935-3_3, © Springer Science+Business Media New York 2014

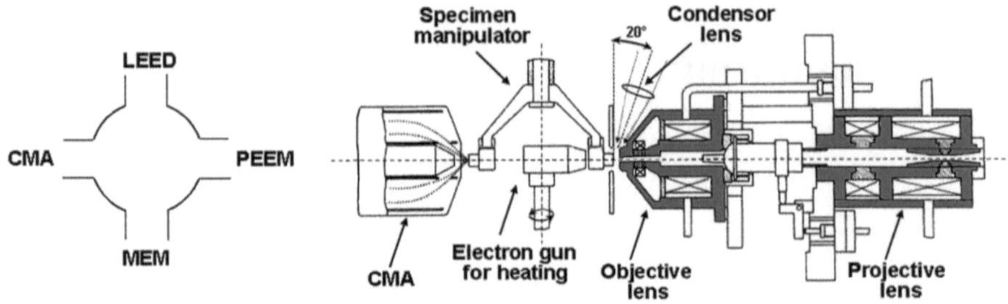

Fig. 3.1 Schematic of the PEEM and Auger electron spectrometer section through Bethge's UHV surface analysis system. The LEED optics and other components are in different sections. The objective lens was water-cooled and had an adjustable angle-limiting aperture in addition to the stigmator. Adapted with permission from Ref. [4]. Copyright 1983 Elsevier

double projective lens. It was mounted at a multimethod surface analysis system including LEED, AES, and a simple mirror electron microscope attachment. The first of the projective lenses can be used as an intermediate lens as described below or simply to vary the magnification between 60 and 10000 times. Figure 3.1 shows this arrangement without the detection system (fluorescent screen without or with MCP). The resolution in the system was limited mainly by the specimen manipulator, which had to be rotated into the various measurement positions at the expense of stability and could not be tilted to align the specimen exactly normal to the optical axis. The objective lens, however, could be aligned in situ precisely to the projective lens system.

The instruments described up to now allow only producing a real image of the specimen with a limited magnification range. Adding an intermediate lens between objective lens and projective lens not only increases this range but also provides imaging of the back focal plane of the objective lens into the object plane of the projective lens and imaging this plane onto the detector. As a result the angular distribution of the emitted electrons, which is imaged by the objective lens into its back focal plane, can be imaged. This is important at higher photon energies in instruments equipped with a band pass energy filter because it allows imaging of the photoelectron diffraction (PED) pattern and thus of the reciprocal space, which will be discussed later. All modern PEEM instruments have this facility. As an example Fig. 3.2 shows the schematic of the PEEM system of Engel et al. [5]. In contrast to the systems described up to now, in which the specimen is at high negative potential (up to −10 kV [1–3], −50 kV [4]), the specimen here is at ground potential and the optical column at high positive potential (+20 kV). The specimen can be tilted for exact alignment onto the optical axis of the objective lens, which is a tetrode without stigmator. A fixed aperture in the last electrode limits the emission angle for electrons with emission energies above 2 eV and allows at the same time a pressure difference of three orders of magnitude between the optics column and the specimen region. After the projective lens the electrons are decelerated in a two-element tube lens to energies below 2 keV in order to optimize the sensitivity of the MCP. With the intermediate lens switched off, a field of view of 400 μm can

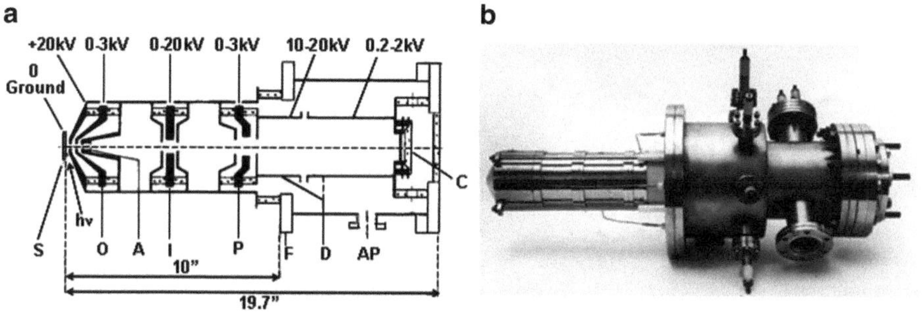

Fig. 3.2 Engel's flange-on PEEM. (**a**) Schematic. S: specimen, O: objective lens, A: angle-limiting aperture (fixed), I: intermediate lens, P: projective lens, D: decelerating tube lens, C: channel plate electron multiplier, F: mounting flange and AP: auxiliary pump connection. (**b**) Actual instrument. Adapted with permission from Ref. [5]. Copyright 1991 Elsevier

be imaged, while with maximum magnification a resolution of 100 nm is possible. This instrument was the predecessor of the commercial instrument produced by Staib Instruments GmbH [6], which will be discussed below.

Another electrostatic three-lens system was developed by Grzelakowski, based on his experience in the development of an electrostatic LEEM instrument [7], and further improved as IS-PEEM by Focus GmbH [8, 9]. In this instrument the specimen holder is integrated with the tetrode cathode lens ("IS"), which has exchangeable apertures and an octopole stigmator/deflector in the back focal plane. In the image plane of the objective lens a continuously adjustable iris aperture allows to select small specimen regions ("area selection aperture"). Image detection is via a MCP and a YAG scintillation crystal. The specimen is at negative potential up to -15 kV, the magnification ranges from 20 to 7200 times and the field of view can be changed from 2.5 to 1800 μm diameter with the help of the intermediate lens. A lateral resolution of 40 nm is specified. A simpler version of a three lens flange-on system with electrostatic four-electrode objective is produced by Elmitec GmbH [10].

The flange-on instruments described up are intended for mounting on UHV systems with suitable flanges and have found widespread application in surface science instruments. We turn now to systems, which can be used as stand-alone microscopes. The first one is that of Griffith et al. [11, 12], which was an improved version of an earlier two-lens system [13]. It is an electrostatic three-lens system with a tetrode objective lens, which is shaped for efficient illumination with reflected light as introduced by Engel [14] and used in the Balzers K3 PEEM [15]. A cut-away schematic of Griffith's instrument is shown in Fig. 3.3 and a very detailed and instructive description of it can be found in Ref. [12]. An electrostatic hexapole stigmator corrects the astigmatism of the objective lens and a 50 μm diameter contrast aperture limits the angular distribution. The specimen is at 30 kV. By varying the potential of the three lenses magnifications between 25000 and 75 times can be selected. The incorporation of an internal film camera limits the

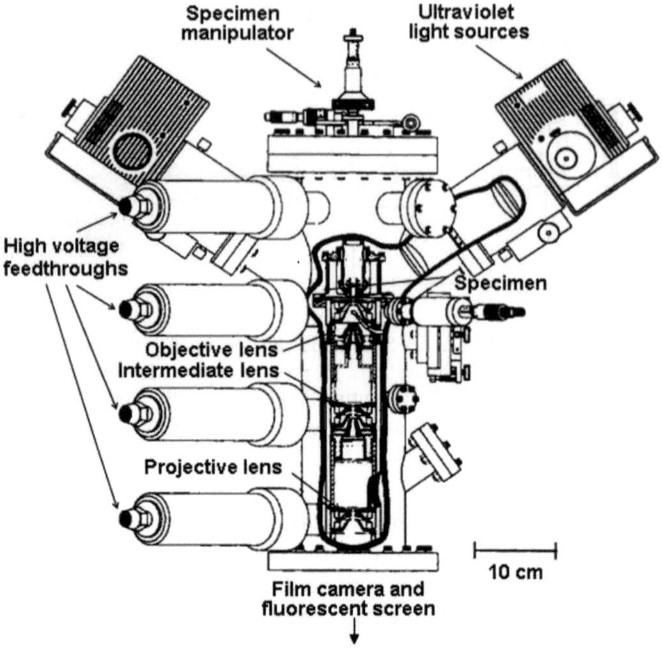

Fig. 3.3 Cutaway schematic of Griffith-Rempfer PEEM. For explanation see text. Reproduced with permission from Ref. [12]. Copyright 1991 Elsevier

pressure to about 10^{-7} mbar compared to the 10^{-9} to 10^{-10} mbar range in the earlier version [11], which used a MCP-fluorescent screen image detector. As this instrument was pumped with oil-free pumps and used for the study of biological specimens the lower pressure was not needed. An extension of Tonner's electrostatic two-lens system to three lenses with an electrostatic tetrode objective was the heart of the MEPHISTO PEEM at the Wisconsin synchrotron radiation center [16]. In this case UHV is a must. The instrument operated at 20 kV. The objective lens is shaped so that illumination is possible at a grazing angle of incidence of 30°. Interchangeable apertures (150, 50, and 20 μm in diameter) in the back focal plane are used mainly to limit the wide energy spread of secondary electrons created by the synchrotron radiation, making use of the chromatic aberration of the objective lens. Another electrostatic three-lens system with a tetrode objective lens, called transmission X-ray secondary-electron conversion microscope [17], was used for transmission X-ray microscopy using secondary electrons for imaging similar to Huang's work [18]. This was a significantly advanced design, incorporating an octopole stigmator and a hexapole deflector behind the angle-limiting aperture in the objective lens and a second deflector near the back-focal plane of the intermediate lens.

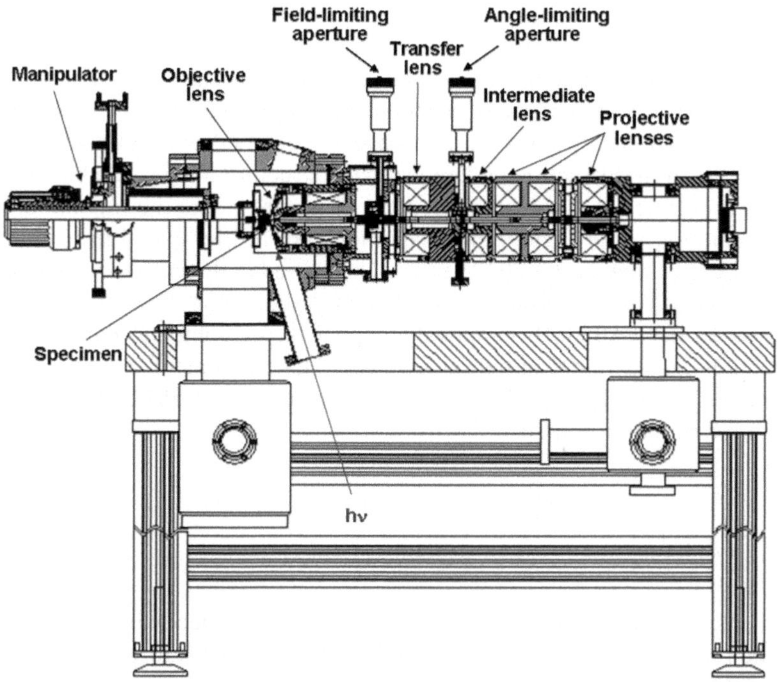

Fig. 3.4 Schematic of a magnetic lens PEEM. The photon beam enters the specimen chamber (*left side*) at a glancing angle of 16° with respect to the specimen surface. A sturdy table and active vibration damping provide the stability necessary for high-resolution imaging. See also text. Reproduced with permission from Ref. [21]. Copyright 2001 Elsevier

The next step in instrument design was realized in the PEEM2 instrument at the synchrotron radiation facility Advanced Light Source (ALS) in Berkeley [19]. The electron optics was similar to that in the instrument just described but a transfer lens was inserted between objective lens and intermediate lens. This allowed placing the angle-limiting aperture into the image of the back focal plane of the objective, which is produced by the transfer lens. In this way the space problem in/near the objective lens caused by the stigmator/deflector is alleviated and more space for interchangeable apertures (2 mm, 50 μm, 20 μm, and 12 μm diameter) for reduction of the energy spread of the beam is provided. Finally a PEEM instrument with magnetic lenses should be mentioned. It is basically the imaging column of a LEEM instrument based on the original design of Veneklasen [20]. Figure 3.4 [21] shows a schematic cross section of it. The objective is a magnetic triode or diode lens with integrated electromagnetic stigmator/deflector, followed by a transfer lens, an intermediate lens, and three projective lenses, all of which are equipped with electromagnetic deflectors. The main purpose of the transfer lens is to image the angular distribution in the back focal plane of the objective lens in front of the intermediate lens which, together with the projective lenses, can either image the specimen or the angular distribution of the emitted electrons, onto the MCP-fluorescent screen image detector. Like in the ALS PEEM2, the angle-limiting aperture is therefore placed after the transfer lens and a field-limiting aperture is placed in the image plane of the objective lens. The field-limiting aperture is important when the microscope is equipped with an energy filter so that photoelectron diffraction pattern and band structure images can be taken from small areas selected by the aperture. It plays the same role for the angular resolution in these patterns as the angle-limiting aperture does for the spatial resolution in the image. The specimen is at −20 kV and can be moved in three orthogonal directions and tilted about two orthogonal axes for precise alignment onto the optical axis of the objective lens. All electron-optical components are computer-controlled. More information can be found in the review [22]. This instrument is used either free-standing as shown in Fig. 3.4 in UVPEEM and laser-PEEM or connected to synchrotron radiation sources.

3.1.2 LEEM

LEEM instruments are more complex because of the necessity to separate the incident illuminating beam from the reflected imaging beam and the need of a high-brightness electron source. The first glass LEEM system was presented already in 1962 by Bauer [23]. By 1965 the basic metal system with a 60° beam separator and an Orbitron pump was finished (Turner and Bauer [24]). Figure 3.5 [25] shows the metal system with the reassembled beam separator setup in 1969 before it was transferred from the Michelson Laboratory (California) to the Technical University Clausthal (Germany). It took a long further development before this instrument became fully operational. In 1983 Telieps in his Ph.D. thesis [26] could finally demonstrate all the imaging methods and LEED. The first good LEEM images with a resolution of 20 nm published by Telieps and Bauer in 1985 [27] showed that LEEM is capable of studying the dynamics of surface processes such as phase transitions. (For the history of LEEM see Refs. [22, 25]). This instrument had a 60° beam separator, an electrostatic triode objective lens with a magnetic stigmator/deflector and an exchangeable angle-limiting aperture in its back focal plane, a field emission electron source, magnetic intermediate and projective lenses, an electrostatic filter lens, and other components shown in Fig. 3.6 [27]. Deflection coils, translation and tilting of the magnetic lenses allowed instrument alignment. Because it had no magnetic shielding the magnetic earth field and AC fields were compensated by large active Helmholtz coils. Despite of the mostly 20 year old technology of the instrument Telieps [27] was able to achieve a resolution of 20 nm in LEEM and 50 nm in UVPEEM. A significantly improved 60° beam separator LEEM, built with state of the art technology, was developed next by Tromp and Reuter [28]. It used a magnetic diode lens with magnetic stigmator and a commercial field emission gun followed by two magnetic condenser lenses. The three magnetic lenses in the imaging column allowed a

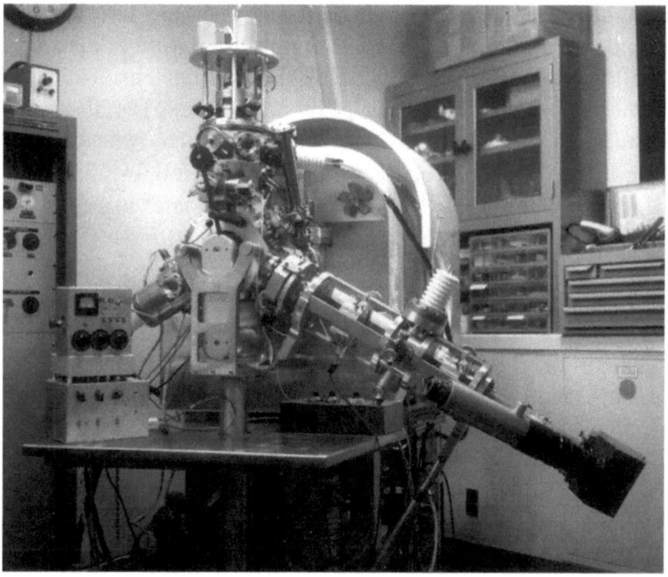

Fig. 3.5 The original LEEM instrument in the 1960s. The illumination column (*left*) consisted only of the field emission gun and two sets of quadrupoles, which were used for focusing, deflection, and correction of the astigmatism of the 60° beam separator. The objective was an electrostatic triode, the earth magnetic field in the specimen chamber was compensated by coils surrounding it and a motorized specimen manipulator allowed three translations, eucentric tilting, and a large tilt towards a window. For more details see the schematic in Fig. 3.6. Reproduced with permission from Ref. [25]. Copyright 2012 Elsevier

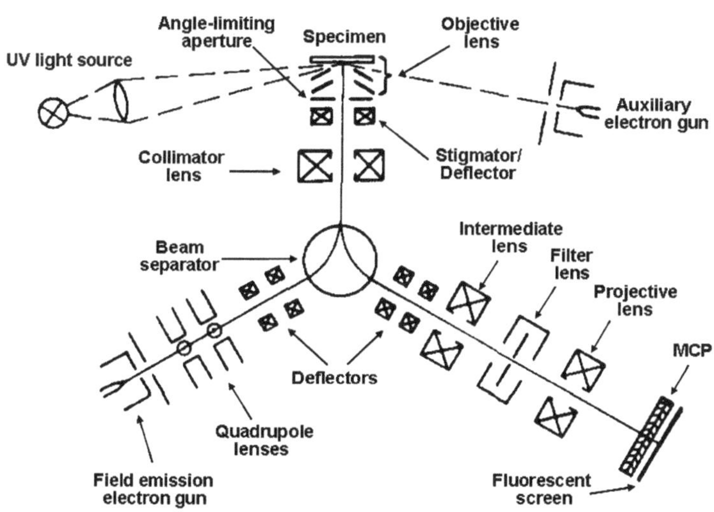

Fig. 3.6 Schematic of the first LEEM system. For explanation see text. Reproduced with permission from Ref. [27]. Copyright 1985 Elsevier

maximum magnification of 10000 times. The lenses are fixed and alignment is achieved with deflection coils. The angle-limiting aperture is placed in the image plane of the diffraction pattern produced by the transfer lens. The specimen is not moved manually but electromechanically by inchworms. The absence of magnetic shielding required magnetic field compensation, which was done in a similar manner as described above. The theoretical resolution in LEEM was at least 50 nm but a better resolution ($\approx$10 nm) was achieved later in experiment. Another interesting LEEM design was reported by Liebl and Senftinger [29]. It used a novel beam separator and an electrostatic objective lens combined with a Schwarzschild objective for optimized illumination with UV light, which will be discussed later. Unfortunately this instrument was never put into operation. Another small instrument also seems to not have become fully operational [30].

The problems encountered with magnetically unshielded instruments were largely eliminated in the instrument designed by Veneklasen [20] in Bauer's group in the Technical University Clausthal. A prototype was exhibited in 1989 at the 11th International Vacuum Congress in Cologne, Germany. In this instrument all magnetic lenses are close-packed and only the regions near the specimen, at the beam separator, and beyond the last projective lens require local shielding. Initially it was still vertical (Fig. 1 in [20]) like the previous ones, but soon changed to horizontal mounting on a table for better access to all components. The system was designed from the beginning for complementary spectroscopic imaging with synchrotron radiation-created photoelectrons. In the pure LEEM set-up the refocusing mirror for the synchrotron radiation and the energy analyzer with its associated energy selection slit manipulator shown in Fig. 3.7 [31] are missing. This instrument has become the basic building block of a number of more complex systems, which will be discussed below. It operates at 20 kV, uses a LaB_6 emitter in the electron gun, three condenser lenses in the illumination column, a close-packed prism array beam separator and in the imaging column a transfer lens, a field lens, an intermediate lens, and two projective lenses. All lenses are equipped with magnetic deflection coils. For the objective lens an electrostatic tetrode or a magnetic triode has been used. Now most instruments are equipped with a magnetic diode with magnetic stigmator/deflectors. The objective lens forms an image close to the center of the beam separator, where the area-selecting (field-limiting FA) aperture is located while the back focal plane of the objective lens is imaged by a transfer lens after the beam separator into a field lens, where the angle-limiting (contrast CA) aperture is inserted. An intermediate lens after the transfer lens, together with the projective lenses allows fast switching between diffraction and imaging over wide magnification range. The broad energy distribution of the LaB_6 cathode limits the resolution to slightly better than 10 nm. With a Schottky field emission gun the limit is about 5 nm. This instrument is commercially available [10]. Another well-designed LEEM with the same magnetic shielding principle via close-packed magnetic lenses, developed by Tromp et al. [32], is also commercially available [33]. It operates at 15 kV, has an improved 90° beam separator and a cold field emission gun whose narrow energy distribution limits the resolution to slightly below 5 nm. Other significant differences from the system shown in Fig. 3.7 are the

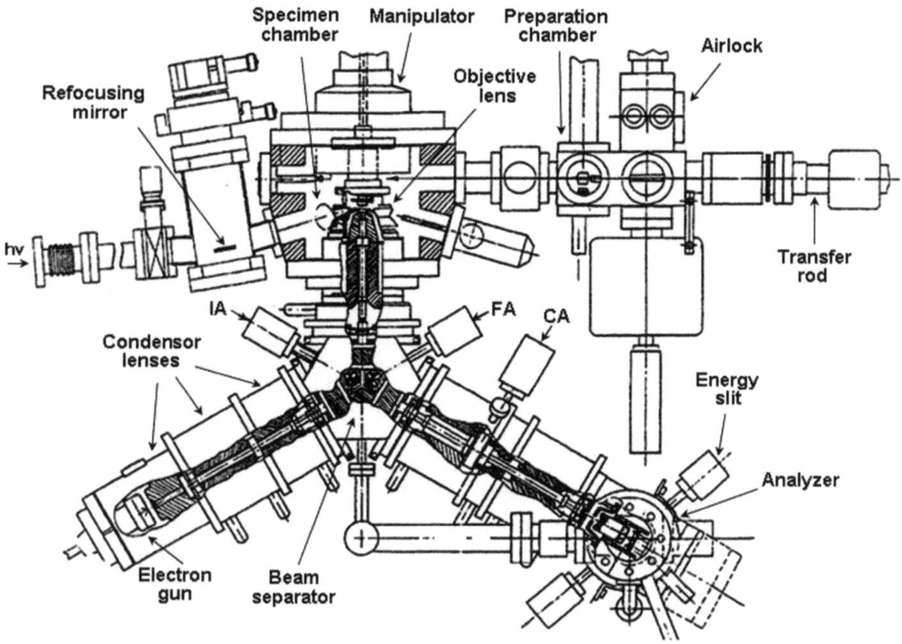

Fig. 3.7 Schematic of the first magnetic LEEM system. For explanation see text. Reproduced with permission from Ref. [31]. Copyright 1997 Elsevier

rotation-free double objective lens, the in-vacuum specimen manipulator, and the vertical mounting like in conventional electron microscopy.

A quite different LEEM system, also capable of spectroscopic imaging, has been developed by JEOL [34–36]. In this instrument the beam separator is a Wien filter with the reflected beam along its axis and the incident beam entering at an angle of $\pi/5$ relative to its axis (Fig. 3.8). It operates at 10 kV, uses a LaB_6 electron gun, an electrostatic tetrode, magnetic lenses for the incident and imaging beams with stigmators and deflectors and a second Wien filter as energy filter. A screen before the Wien filter allows the observation of unfiltered low-magnification images and the LEED pattern produced by the first two lenses. Due to the difficulty of aligning incident and reflected beam properly with the Wien filter, the instrument was used mainly for MEM, metastable impact electron emission microscopy (MIEEM) and PEEM.

In a first attempt to use LEEM for imaging of biological specimens Griffith et al. [37] converted Rempfer's electron-optical bench [38, 39] into a simply LEEM set-up with electrostatic lenses, magnetic deflectors, no apertures and stigmator and poor vacuum. It was not useful for surface studies but stimulated an UHV flange-on system of similar design, which is shown in Fig. 3.9 [7]. It uses two condenser lenses, an electrostatic tetrode objective lens, an intermediate lens, two projective lenses, and three magnetic deflectors, schematically indicated by circles.

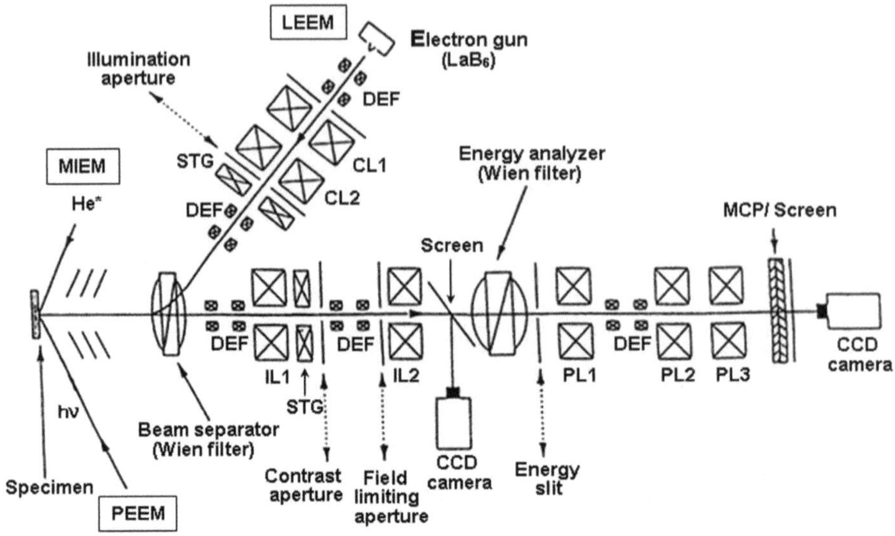

Fig. 3.8 Schematic of the JEOL LEEM system. For explanation see text. Reproduced with permission from Ref. [35]. Copyright 1995 Elsevier

The deflectors are surrounded by einzel slit lenses, whose electrode next to the deflector is magnetic, acting as a magnetic shield. This set-up allows parallel illuminating and imaging beams, mounting on a 10 in. base flange and connection to an UHV system via 6 in. flange. Electrostatic deflectors and objective stigmator as well as the illumination, angle-limiting, and field-limiting apertures, which can be manipulated with inchworms, are not shown. The specimen, also manipulated by inchworms, is at a potential, which differs from that of the grounded filament of the electron gun by the start voltage. The rest of the system is at 5 kV, except the first objective electrode, which is at 15 kV for a high field at the specimen. The valve at the bottom of the electron gun allows easy exchange between the LaB_6 gun and a spin-polarized electron gun. The system is shielded by μ-metal (removed in (b)), and has a resolution of about 20 nm. A further flange-on instrument simplification is realized in the LEEM system shown in Fig. 3.10 [40, 41]. By using small deflection angles in the beam separator the deflection magnets in Fig. 3.9 can be avoided and the complete system mounted on an 8 in. base flange. The potential distribution is the same as in the preceding instrument. The theoretical resolution of the electrostatic tetrode at 10 eV initial energy is 10 nm. The beam deflector has small aberrations and these, together with the astigmatism of the objective lens, are corrected by three deflector/stigmators DS, which at the same time are the only alignment tools. One transfer lens TL focuses the illuminating beam into the back focal plane of the objective lens, another one the image, which is produced by the objective lens in the center of the beam separator or the diffraction pattern to the front of the projective lenses P_1, P_2, and P_3. These allow a magnification range from 25 to 4000. Apertures are inserted

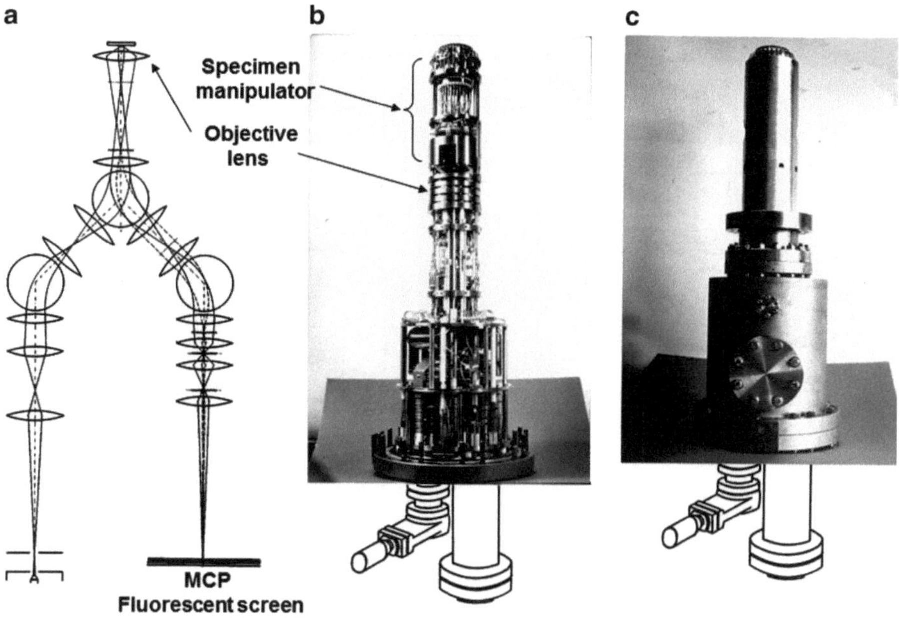

Fig. 3.9 Flange-on LEEM system with parallel illumination and imaging column. (**a**) Schematic (not to scale), (**b**) and (**c**) actual configuration, valved-off electron gun, and channel plate/fluorescent detector indicated. The beam separation system is in the lower, wider part of the system. See also text. Adapted from Ref. [7] with permission from the American Institute of Physics, © 1996

normal the plane of the cross-sectional drawings, mounted on ceramic rod-insulated drives between the TLs and DSs. With the μ-metal shielding indicated in (c) a resolution better than 20 nm has been demonstrated. This instrument has also been developed into a commercial product [10].

Finally a LEEM system specifically designed as electron beam inspection system of semiconductor devices should be mentioned [42–48]. This application requires fast distortion-free imaging of large areas of insulating surfaces in the MEM mode, which causes problems not encountered in the LEEM systems discussed up to now. They were overcome with a number of novel electron-optical components. Only one should be mentioned here: the elimination of surface charging of the specimen whose work function varies on the scale of the features, which have to be imaged. For good resolution of these features the incident beam has to be reflected as close as possible from the surface. As the beam has a finite energy width, part of it hits insulating regions leading to charging. One of the solutions of this problem was illumination with a second electron beam with higher energy (several hundred eV), which compensates the charge via secondary electron emission. This beam is introduced via a second beam separator ("combiner") as shown on Fig. 3.11 [43]. Many other details useful for standard LEEM can be found in the descriptions of this instrument, which is being further developed, in particular

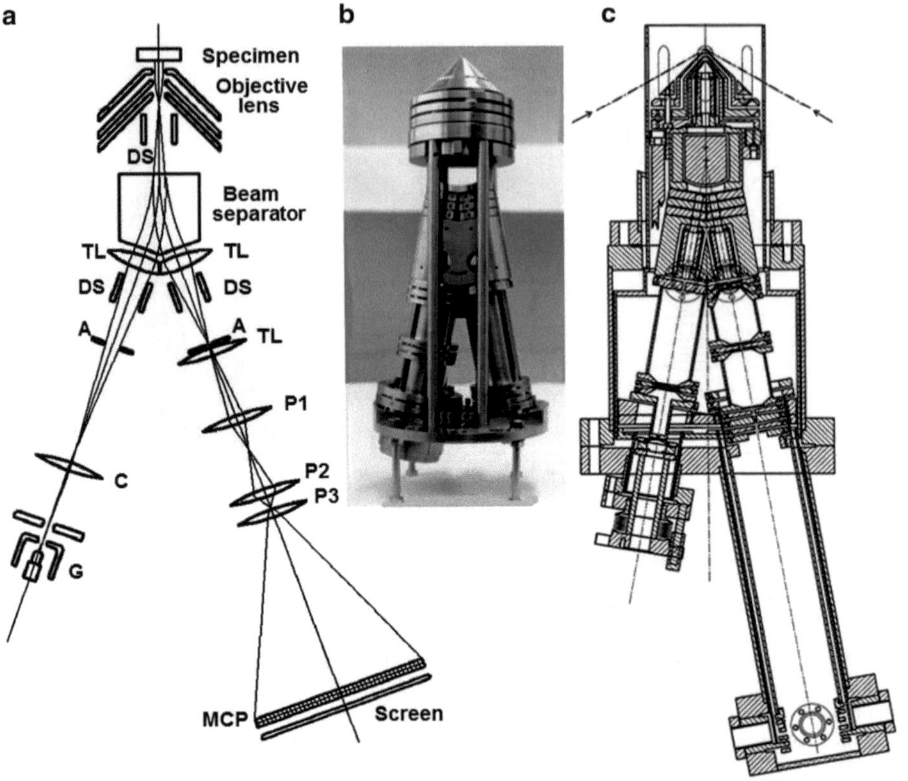

Fig. 3.10 Flange-on LEEM system with minimum deflection angle. (**a**) Schematic (not to scale), (**b**) actual configuration, (**c**) cross-sectional drawing. The location of the apertures are indicated by the *circles* in (**c**). For further explanation see text. (**a** and **b**) Adapted from Refs. [40, 41] with permission from the American Institute of Physics, © 1998. (**c**) Reproduced from Ref. [41] with permission from P. Adamec

for imaging of biological molecules. In addition to the dual beam set-up this instrument has a monochromator and aberration correction ("MAD-LEEM"), which should allow high resolution even at very low energies [48]. Another much simpler and less versatile MEM electron beam inspection system was developed at Hitachi [49]. It uses a Wien filter as beam separator and an electrostatic objective lens, two magnetic condenser lenses, and three magnetic lenses in the imaging column. Although cathode and specimen are at or near only 3 kV, a high acceleration potential at the electrode closest to the specimen giving a field of almost 6 kV/mm, allows good resolution. To end this section a pulsed beam MEM for time-resolved measurements should be briefly mentioned yet [50]. A thermionic emitter heated by 5 ns laser pulses focused into a 15 μm spot was used as electron gun, a 90° magnetic prism as beam separator and a two-electrode mirror as objective lens. A second pulsed laser is focused onto the specimen. The changes, which it causes, are imaged by mirror-imaging with the electron bunches created by

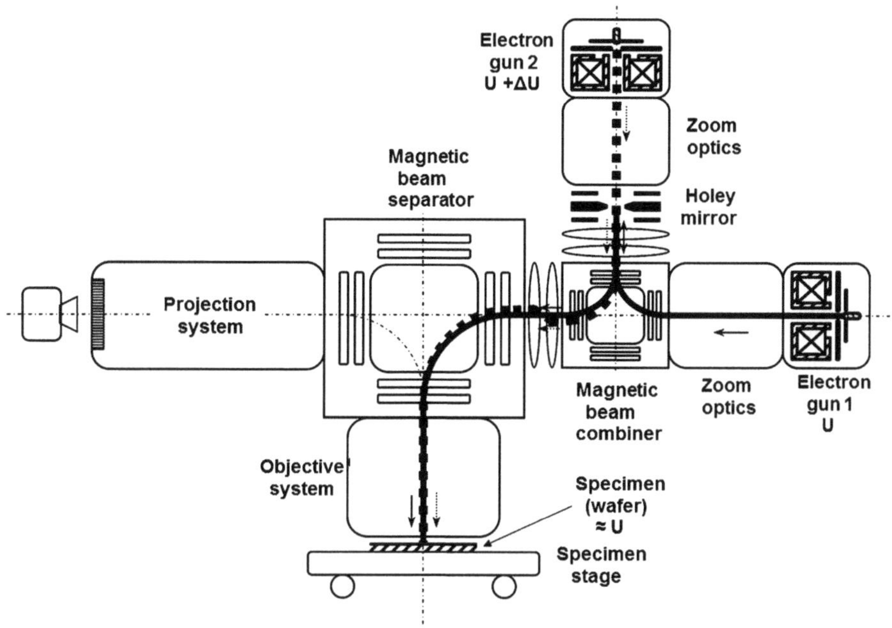

Fig. 3.11 Schematic of the dual beam LEEM system of Mankos et al. The system uses two 90° beam deflectors, one for combining illumination and discharging beams, the other for separating illumination and imaging beams. For details see one of the original publications. Reprinted from Ref. [43] with the permission of Cambridge University Press

the first laser with selected time delay. This "pump-probe" type time-resolved imaging will be discussed later in more detail.

3.1.3 Aberration-Corrected Instruments

The resolution of the instruments described up to now is ultimately limited by the spherical and chromatic aberrations of the objective lens. As Scherzer has shown already in 1936 [51] all electron lenses with rotational symmetry have these aberrations, in contrast to light optical lenses. In 1947 he presented several proposals on how they could be corrected [52]. The story of aberration correction is long and well told in two publications [53, 54]. Here only the correction with an electron mirror, which has been implemented in PEEM and LEEM instruments, will be discussed. Other methods will be mentioned briefly in Sect. 3.2.4, together with a detailed description of mirror correctors. Three types of aberration-corrected instruments have been reported, differing in the type of mirror and/or the beam separator which is needed to separate the beam incident on the mirror from the beam reflected from it. The first type is a pure PEEM instrument with a hyperbolic mirror and three deflection magnets (Fig. 3.12 [55, 56]) similar to

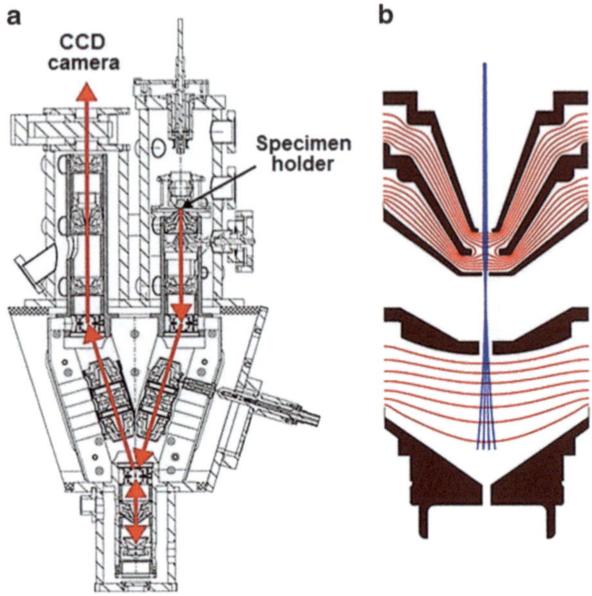

Fig. 3.12 Schematic of Könenkamp's aberration-corrected electrostatic PEEM based on Rempfer's beam separator and hyperbolic electron mirror. (**a**) Overall view, (**b**) hyperbolical mirror and pre-lens detail with potential distribution. For details see the original publication. Reproduced with permission from Ref. [56]. Copyright 2010 Elsevier

the one shown in Fig. 3.9. The other instruments are LEEM systems, which use a four electrode mirror whose electrodes are shaped for optimum compensation of the spherical and chromatic aberrations of the respective objective lenses. They differ in the method of beam separation. In the first one built and installed in the synchrotron radiation facility BESSY II in Berlin, called SMART (spectro-microscope for all relevant techniques) one complex beam separator fulfills two functions: separation of the incident and reflected beams at the specimen and at the mirror. In addition it has an energy filter. The system, which uses a magnetic objective lens, is shown schematically in Fig. 3.13 and has been described repeatedly [57–59], in most detail in [60]. The mirror corrector/beam separator produces an aberration-free image in the entrance plane of the transfer optics, which produces an image of either the specimen or of its diffraction pattern on the entrance plane of the magnetic energy filter. The filtered image after the energy selection slit is imaged with three projective lenses on the image detector. This instrument had achieved in 2010 a line resolution of 2.6 nm in LEEM with 15 eV electrons [61]. An improved version of this system with an electrostatic energy filter is in construction (T. Schmidt, private communication). A similar instrument with an electrostatic tetrode lens, the PEEM3, is being built at the ALS, though initially for PEEM without energy analyzer [62–65].

The third mode of aberration-correction grew from uncorrected LEEM systems and the difficulties encountered in the initial adjustment of the double deflection beam separator in the SMART instrument. In this mode the beam separation at the specimen and at the mirror are separated which makes alignment much easier. A photo of the first instrument of this type (Fig. 3.14) and first test images were

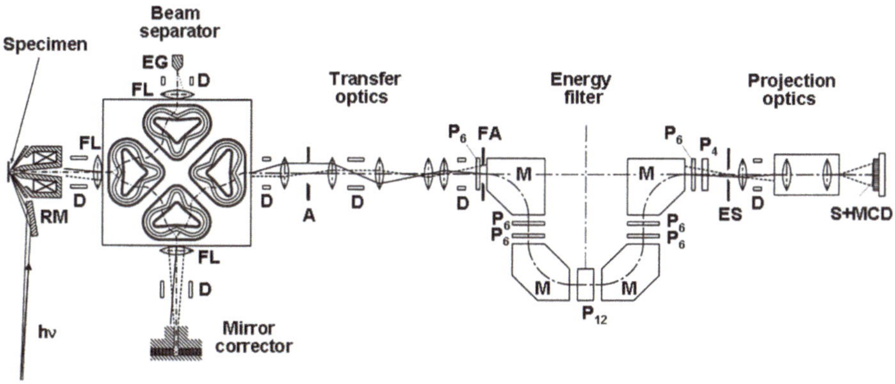

Fig. 3.13 Schematic of aberration-corrected LEEM system SMART based on the beam separator and mirror design by Rose et al. The system is equipped with an omega energy filter for spectroscopic imaging. D deflectors, FL field lenses, A, FA apertures, P_4, P_6, P_{12} n-pole elements, M deflection magnets, ES energy selection slit, EG electron gun, RM refocusing mirror for synchrotron radiation photon beam, S + MCD fluorescent screen and multichannel plate detector. Reproduced with permission from Ref. [58]. Copyright 1997 Elsevier

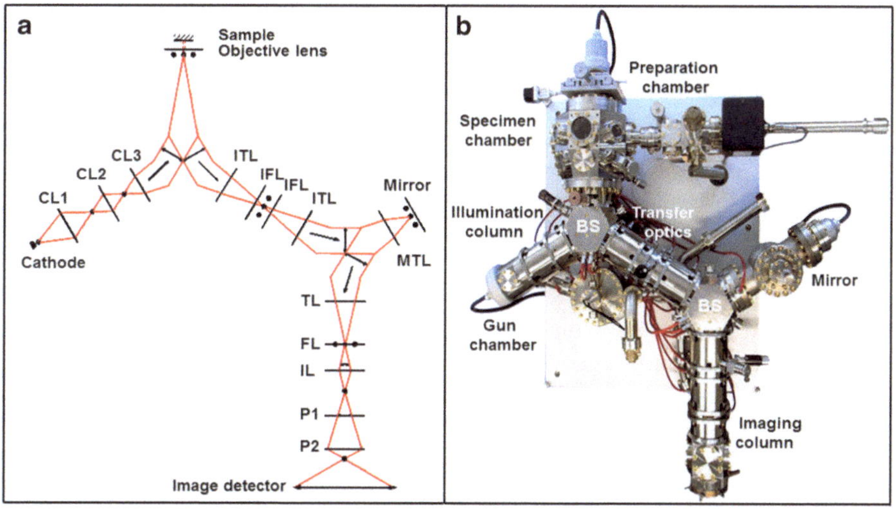

Fig. 3.14 Schematic (**a**) and actual configuration (**b**) of the aberration-corrected Elmitec LEEM. In (**a**) CL1–CL3 represent the illumination column of the standard LEEM, ITL intermediate transfer lenses, IFL a double field lens, MTL the mirror transfer lens and TL–P_2 the imaging column of the standard LEEM. In (**b**) BS stands for beam separator. The system is horizontal. (**a**) Courtesy of T. Schmidt; (**b**) Adapted with permission from Ref. [25]. Copyright 2012 Elsevier

presented at the LEEM/PEEM5 conference in 2006 but nothing was published about it. The schematic in Fig. 3.14 shows its basic set-up without energy analyzer. A comparison with Fig. 3.7 shows how it evolved from the uncorrected system. It is commercially available [10]. A 16 %/84 % line resolution better than 2 nm has

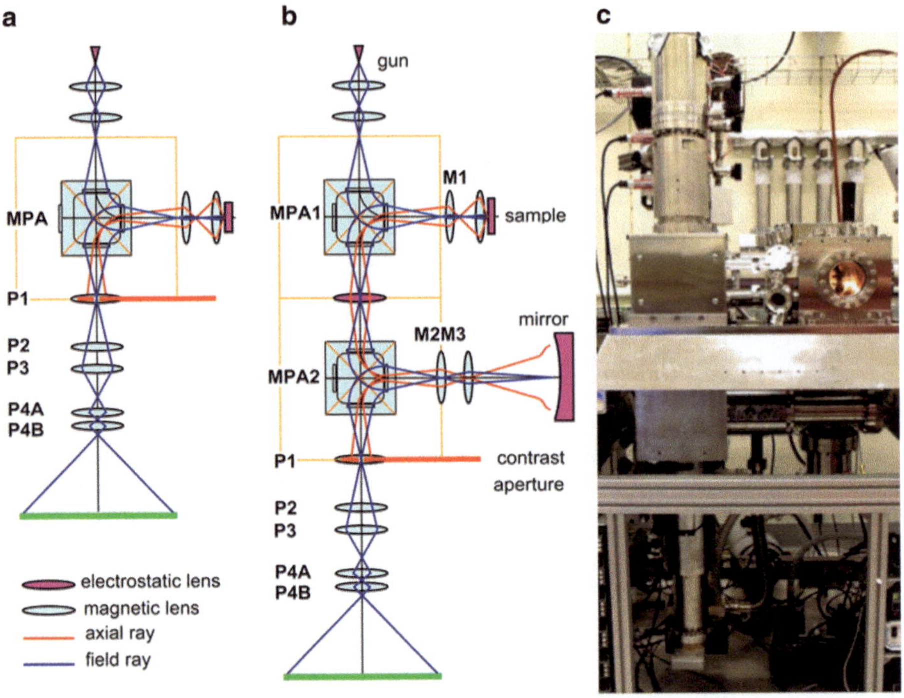

Fig. 3.15 Schematic (**b**) and actual configuration (**c**) of the Tromp/SPECS aberration-corrected LEEM. For comparison the schematic of the standard LEEM is shown in (**a**). In (**a**) and (**b**) the M's denote transfer lenses, the MPAs magnetic prism arrays, P_1 the transfer lens between them and the other P's represent the lenses in the imaging column. In (**c**) MPA1 and the specimen chamber with objective lens are above the table, MPA2, the mirror and the imaging column below the table. (**a** and **b**) Reproduced with permission from Ref. [66]. Copyright 2010 Elsevier; (**c**) Courtesy of R. Tromp

been reported for it. The second instrument of this type was developed by Tromp et al., who gave not only a good description of the design but also of the principles [66]. This instrument is also commercially available [33]. Tromp has achieved a resolution of 1.4 nm (Tromp, private communication). Its evolution from the uncorrected LEEM is illustrated in the schematics on the left side, its physical appearance on the right side of Fig. 3.15 [33, 66]. The instruments shown in Figs. 3.14 and 3.15 differ not only in their overall geometry but also somewhat in the optical system. The first one has no transfer lens between objective and first beam separator but transfer lenses between them and after the second beam separator. These are not needed in the second instrument because of the different properties of the beam separators. Irrespective of that they have comparable resolution as illustrated in the calculated resolution curves with and without aberration correction (Fig. 3.16) (Schmidt, private communication). The experimentally achievable resolution will depend on the technical perfection of the instrument and the skill of its operator. Another more complex design for a

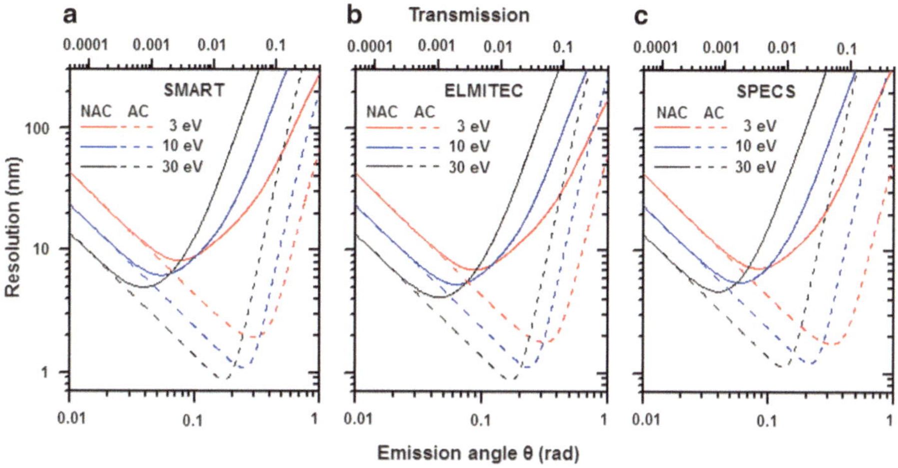

Fig. 3.16 Comparison of the resolution of the three uncorrected and aberration-corrected magnetic LEEM systems based on geometric optics calculations using calculated aberration constants of their objective lenses. (**a**) SMART, (**b**) Elmitec and (**c**) Tromp/SPECS instruments. Courtesy of T. Schmidt

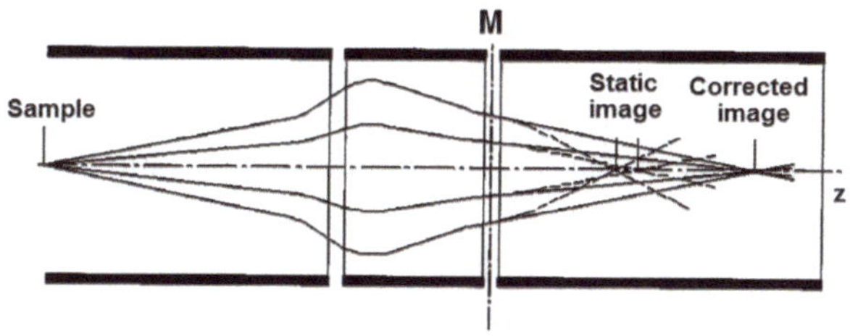

Fig. 3.17 Schematic of the spherical aberration-correction using pulsed beams. The first two electrodes of the three-electrode lens (einzel lens) act as a retarding lens for the electron bunch. When the bunch reaches plane *M*, the potential of the center electrode is suddenly decreased so that the second two electrodes act as an accelerating lens. This has diverging lens action with spherical aberration opposite to that of real image-forming lenses. The *full* (*dashed*) *lines* in the third electrode illustrate the corrected (uncorrected) trajectories. Reproduced from Ref. [68] with permission from the American Institute of Physics, © 2002

LEEM aberration-corrected with a mirror has been proposed by Tsuno et al. [67] but has not been realized. A pentode mirror-corrected LEEM system is in development [47, 48].

Finally two other proposals for the correction of the aberrations of cathode objective lenses that do not use an electron mirror should be mentioned. One is applicable only to systems with pulsed beams [68, 69]. The chromatic aberration is achieved by accepting only electrons with velocities within a narrow flight time window on the detector. Correcting the spherical aberration requires fast switching of the lens focus as illustrated in Fig. 3.17 [68] for an einzel lens (three-electrode lens) so that the electron bunch feels a weaker focusing action when it passes the plane *M*. This type of dynamic aberration correction puts high demands on fast electronics, in particular on the image detector, and has not been implemented. The second proposal is based on the fact that the images produced by electrons with different energies or emission angles are located in different planes along the optical axis. Thus by taking a stack of images at different focus and summing them with appropriate weights the aberrations can be reduced significantly. This method has been used previously in aberration-free optical microscopy for producing three-dimensional images and also successfully in transmission electron microscopy for aberration correction [70]. Koshikawa et al. [71] have applied this method, which does not require instrument modification but shifts the correction to image acquisition and analysis, and obtained noticeable resolution improvement. Aberration-corrected instruments have, however, displaced this approach more or less.

3.1.4 Spectroscopic Imaging Instruments

Spectroscopic imaging requires the addition of an energy filter to the instruments described up to now. Low pass, high pass, and band pass energy filters have been and are used. Low- and high-pass filters can also be used for imaging with nearly monochromatic electrons by taking a pair of images with energy E and $E + \Delta E$ and subtracting them so that only electrons with energies between these two energies contribute to the signal S_Δ. The signal-to-noise ratio S_Δ/N in such images is, however, poor because the noise N is proportional to $\sqrt{S}$, where S is from all electrons in the complete energy range of the two images. Therefore this method is hardly used. A low-pass filter for filtering out the high energy part of the secondary electron distribution in electron or ion bombardment-induced secondary electron imaging has been built by Möllenstedt et al. [72, 73] using a Castaing-Henry type magnetic prism–electron mirror combination (see Sect. 3.2.5). They could reduce the width of the energy distribution by a factor of 5 and demonstrate some resolution and contrast improvement. High-pass filters in form of retarding-accelerating grid structures are used in image energy filtering versions of some commercial PEEM instruments [6, 9, 74]. An energy resolution of 70 meV has been reached in UVPEEM with a drift energy of 50 eV [75]. They can be used for the elimination of the secondary electrons in imaging with faster electrons such as Auger electrons or photoelectrons with characteristic energies, which, however, requires in practice the image subtraction mentioned already above. Another high-pass filter, the electrostatic filter lens, has been used in the first LEEM instrument in order to separate the secondary electrons from the reflected electrons [27]. However, it made instrument alignment much more difficult and was therefore eliminated, in particular because it was found to be unnecessary. At the low energies used most of the time in LEEM the number of reflected electrons is large compared to that of the secondary electrons with the same energy passing through the angle-limiting aperture. Therefore the secondary electrons cause only a weak background in the image.

Band pass energy filters do not suffer from this deficiency because they use only electrons within a selected energy window for imaging. The energy selection can be achieved in two ways: either by displacement in space by a dispersive deflecting system combined with a spatial window or by displacement in time (without deflection) combined with a temporal window (time-of-flight PEEM). In this "TOF-PEEM" a drift tube is inserted between the last projective lens and the image detector. The electrons are decelerated to lower energies upon entering the drift tube for better time resolution. Flight time discrimination—and therefore energy discrimination of the electrons in the pulsed beam—is achieved with a fast gated intensified CCD camera or with a two-dimensional delay line detector behind the MCPs [76–81]. A spatial resolution of about 25 nm and an energy resolution of about 50 meV has been reported for this type of spectroscopic imaging [82].

The most widely used band pass filters use dispersive deflection systems. These systems can either be magnetic or electrostatic. Magnetic deflection systems are the Castaing-Henry filter [83, 84] and the Omega filter [85, 86] which were originally developed for transmission electron microscopy. The first method is illustrated in Fig. 3.18 [84]. Möllenstedt et al. used a simpler version without energy selection slit so that they had only a low-pass filter. Cazaux suggested using this filtering method for XPEEM already in 1973 [87], long before the necessary light sources were available. While this type of band pass filter instrument was never built for surface imaging with slow electrons, the Omega filter is incorporated at least in one instrument, the SMART. It is indicated in Fig. 3.13 and consists of the four magnetic 90° deflectors M and the eight multipole correctors P_n. It has an energy dispersion of 35 μm/V at 15 keV electron energy which allows an energy resolution of 0.18 eV in a 17 eV wide window on the detector.

Most spectroscopic instruments today use electrostatic concentric hemispherical deflection analyzers (CHA), although smaller instruments with smaller deflection angles such as 70° have been proposed [29]. In the first version of the "Spectroscopic Photo Emission and Low Energy Electrons Microscope" (SPELEEM) a 90° deflection was used temporarily [20, 31, 88, 89] (Fig. 3.7 [31]) before it was replaced by a 180° deflector [90]. It was developed by Bauer et al. in the 1990s. A pure PEEM with a hemispherical analyzer was built by Tonner et al. [91] but apparently used very little, while the PEEM shown in Fig. 3.4, equipped with such an analyzer, is used now in several synchrotron radiation facilities and other laboratories. The second order aberrations of the 180° spherical deflector can be eliminated by using a toroidal shape of the plates [92]. A more elegant solution is to use a second hemispherical field with a transfer lens in between, as in [93] and in the "imaging double energy analyzer" (IDEA) [94, 95] (Fig. 3.19), which is used in the NanoESCA [96] with the input optics of the integrated sample stage PEEM (IS-PEEM) [8, 9]. Depending upon the voltages applied to the hemispheres, either unfiltered PEEM (no voltage applied), selected area electron spectroscopy (no voltage on second hemisphere) or energy filtered PEEM (voltage on both hemispheres) is available.

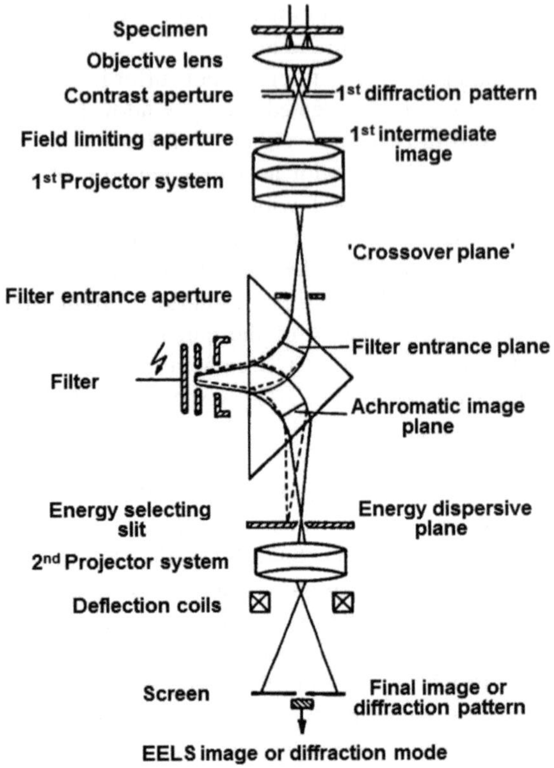

Fig. 3.18 Schematic of the Castaign-Henry energy-filtered microscope, originally developed for transmission electron microscopy but with a cathode objective lens also suitable also for PEEM. Reproduced with permission from Ref. [84]. Copyright 1988 Elsevier

The lateral resolution in XPEEM with secondary electrons is 80 nm, the energy resolution 170 meV. Another energy-filtering instrument, the "Dual Emission Electron Microscope" (DEEM) will be described briefly in Sect. 3.2.5.

An elegant energy filter, which does not require deflectors, is the Wien filter. It was developed already in the 1960s for electron spectroscopy but its use for imaging was suggested much later [97, 98]. A PEEM instrument with a Wien filter as energy filter has been built by Marx et al. [99] but no results of energy-filtered images have been published. A Wien filter is also used for energy filtering in the JEOL LEEM instrument [34–36] but only low-magnification, low-resolution, and low-contrast images have been published. We will discuss the possible reasons for the deviations from the theoretical expectations in Sect. 3.2.6.

Finally, an ingenious simple energy filter should be mentioned, which makes use of the chromatic dispersion of the beam separator in LEEM. Tromp et al. [100] have inserted a movable slit into the diffraction plane before the beam separator and a slit or aperture into the diffraction plane after the beam separator, in which electrons with different energy are dispersed because of the chromatic aberration of the sector field. By positioning the aperture in the spectrum, energy-filtered images can be

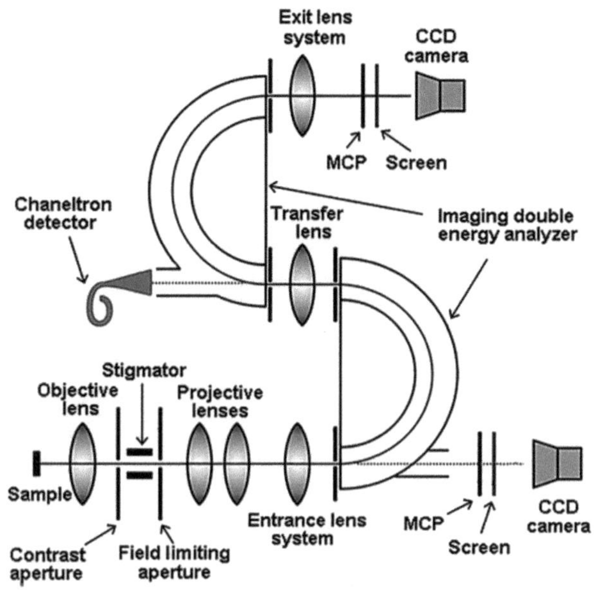

Fig. 3.19 Schematic of the nano-ESCA PEEM, which uses a standard electrostatic PEEM before the imaging double energy analyzer (IDEA) consisting of two concentric hemispherical analyzers (CHA) connected by a transfer lens. Depending upon which CHAs are activated unfiltered PEEM, selected area spectroscopy and energy-filtered PEEM or angle-resolved spectroscopy is possible. Reproduced with permission from Ref. [94]. © IOP Publishing 2005. All rights reserved

taken at various energies, for example with plasmon losses. By displacing the entrance slit parallel to the exit slit and taking many spectra not only the spectrum for one direction in the diffraction pattern can be taken but also for the other direction and the complete diffraction pattern as a function of energy can be reconstructed. For valence band electrons this means that the valence band structure can be determined. Figure 3.20 [100] shows a schematic of this energy filter setup.

3.1.5 Spin-Resolved Imaging Systems

Spin separation can be achieved either before or after interaction with the specimen. In the first case a spin-polarized electron source and a spin manipulator are needed, in the second case a spin selector. A variety of spin-polarized electron sources has been around for a long time (see Sect. 3.2.8) but in spin-polarized LEEM (SPLEEM) instruments only III–V semiconductor photocathodes, activated to negative electron affinity, are used. GaAs(100) is the material used in the first two generations of

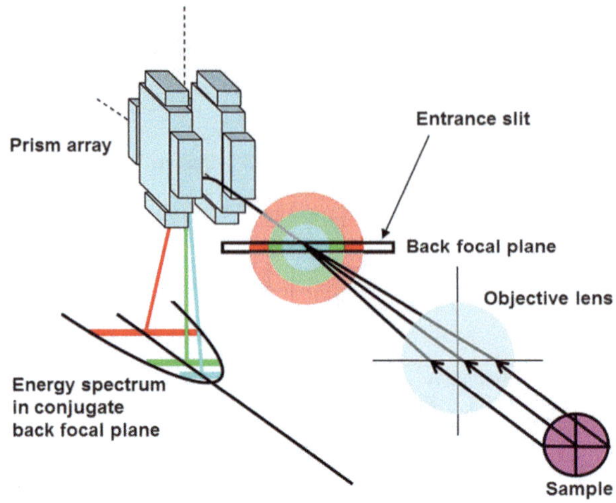

Fig. 3.20 Schematic of energy filtering in LEEM using the chromatic dispersion of the beam deflector and an angle-limiting slit in the diffraction pattern transferred from the objective lens to the entrance plane of the beam deflector. The conjugate plane after the separator displays the angle-resolved energy spectra from which the desired energy for imaging can be selected with a slit. Reproduced with permission from Ref. [100]. © IOP Publishing 2009. All rights reserved

SPLEEM instruments. In a more recent instrument in Koshikawa's group in Osaka it was replaced by GaAs–GaAsP superlattice cathodes grown on GaP(100) [101] (see Sect. 3.2.8). The first SPLEEM instrument was developed in Bauer's group in the TU Clausthal in the early 1990s. It resulted from the first LEEM instrument schematically shown in Fig. 3.6 [27] by replacing the illumination system with that shown in Fig. 3.21a [102] with a spin-polarized electron gun and spin manipulator [103]. The spin manipulator consists of a combined electrostatic-magnetic 90° sector field and a magnetic rotator lens as indicated in Fig. 3.21b [102]. A photo of the first SPLEEM instrument is shown in Fig. 3.22. A second SPLEEM instrument was developed also in the Arbeitsgruppe Bauer in collaboration with Helmut Poppa for IBM California. Later this instrument was transferred to Berkeley where it is used up to now. In this new instrument [104] the spin-polarized gun (Fig. 3.21b) was mounted to the flange-on LEEM system shown in Fig. 3.9 [7]. More information about these instruments can be found in [105, 106]. In both instruments the cathode is illuminated by a laser from the front through a hole in the beam separator, resulting in a large illuminated region (about 1 mm in diameter) on the surface from which the electrons are extracted with a relatively low voltage (1 kV). The same is true for the commercial instrument [10], which is a version of the standard LEEM instrument, modified by replacing the illumination system with rotation-free lenses. In this case only the cathode and the sector field are at high potential.

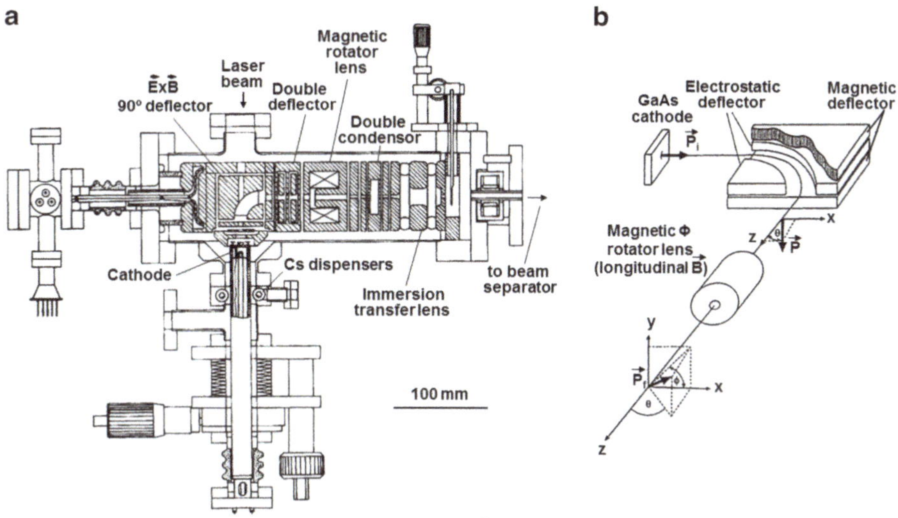

Fig. 3.21 Illumination system used for the conversion of the first LEEM system for SPLEEM. (**a**) Cross-sectional drawing of the complete illumination system. All components are at high potential terminating in the immersion transfer lens. All lenses except the rotator lens are electrostatic. (**b**) Schematic of the spin gun and spin rotator, showing the magnetic-electrostatic deflector for in-plane spin rotation and the magnetic rotator lens further downstream. Reproduced from Ref. [102] with permission from Wiley-VCH Verlag GmbH & Co. KGaA, Weinheim

Fig. 3.22 First SPLEEM system converted from strongly modified first LEEM system

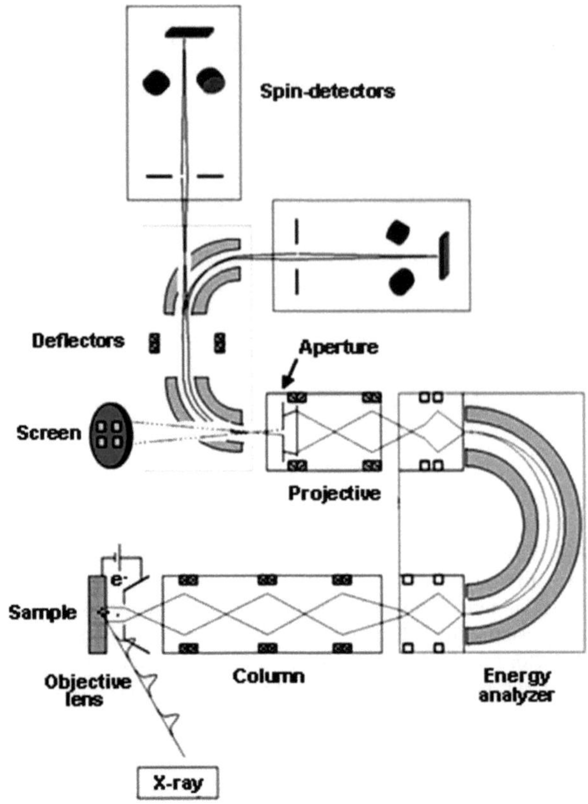

Fig. 3.23 Schematic of the spin-resolved PEEM (S-PEEM) based on a Elmitec PEEM system with energy analyzer. The spin selection is achieved with two Mott detectors. For more information see text. Reproduced from Ref. [110] with permission from John Wiley & Sons

While these instruments have produced many results in thin film magnetism they suffer from low brightness and short cathode lifetime of the order of days. These problems have been overcome by replacing the GaAs cathode with the superlattice cathode mentioned above, by increasing the extraction voltage to 20 kV and by improving the vacuum in the cathode region. Illumination of the cathode from the backside allowed focusing the laser beam to nearly 1 μm diameter, which together with the high extraction voltage resulted in a three orders of magnitude higher brightness compared to the older spin guns and higher brightness than a LaB_6 electron gun. Spin rotation at the high electron energy is achieved with a Wien filter [107–109].

Spin separation after interaction with the specimen has long been done in spin-resolved spectroscopy and spin-polarized scanning electron microscopy (spin-SEM or SEMPA) with Mott detectors. A spin-resolved PEEM instrument using mini-Mott detectors has been developed by Kronast et al., called S-PEEM [110]. Its principle is illustrated in the schematic of Fig. 3.23 [110]. The spin-analysis part of the instrument is attached to the end of a commercial spectroscopic PEEM [10]. It consists of two electrostatic 90° deflectors, which can switch the electron beam

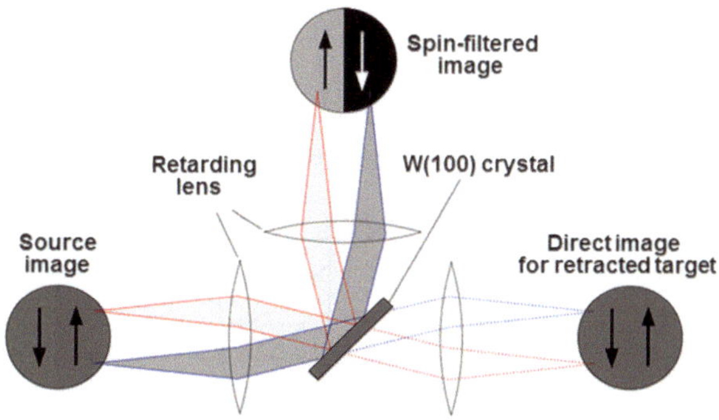

Fig. 3.24 Schematic of the spin-selection system of the spin-resolved PEEM based on a nanoESCA system. The spin selection is achieved by the spin–orbit coupling-caused spin-dependent specular reflection on a W(100) crystal at energies in the 10 eV range. For more information see text. Reproduced from Ref. [111] with permission from the American Institute of Physics, © 2011

between two mini-Mott detectors and detect all three components of the spin. A small aperture near the end of the energy filter allows selecting regions as small as 200 nm in diameter for spin analysis and by scanning the image across the aperture spin-resolved images are obtained. With the deflectors switched off and the aperture removed, the energy-filtered spin-averaged image can be obtained. If circular polarized synchrotron radiation or laser light is used for illumination, also magnetic moment-selected imaging (MCDPEEM) is possible for comparison with the spin-selected images. Although MCDPEEM imaging is much faster than the (scanning) spin imaging because of the parallel image acquisition and the low efficiency of the spin detectors, S-PEEM imaging is expected to provide new insight into the spin structure and dynamics of surfaces.

A second spin-resolved PEEM instrument [111] avoids the slow image acquisition caused by scanning and by the low efficiency of Mott detectors. Instead of Mott scattering for spin analysis it relies on the strong spin dependence of the diffraction of slow electrons from W single crystals. Similar to the S-PEEM it starts from a commercial spectroscopic PEEM, in this case the nanoESCA [94–96]. The spin filter, schematically indicated in Fig. 3.24 [111], is mounted at the end of the double-CHA energy analyzer. The center piece is a W(100) crystal set at an angle of 45° with respect to the exit direction of the electrons. Between the spin-averaged image at the exit of the energy analyzer and the crystal an electrostatic retarding lens reduces the energy of the electrons to the optimum energy for spin-selecting specular diffraction from the crystal and produces the diffraction pattern of the selected specimen region on the crystal. An accelerating lens mounted at right angles to the incident beam converts the diffraction pattern into the spin-selected image. The efficiency of this imaging mode is orders of magnitude higher than that of the S-PEEM. By retracting the W crystal from the beam, spin-averaged images can be recorded similar to the S-PEEM.

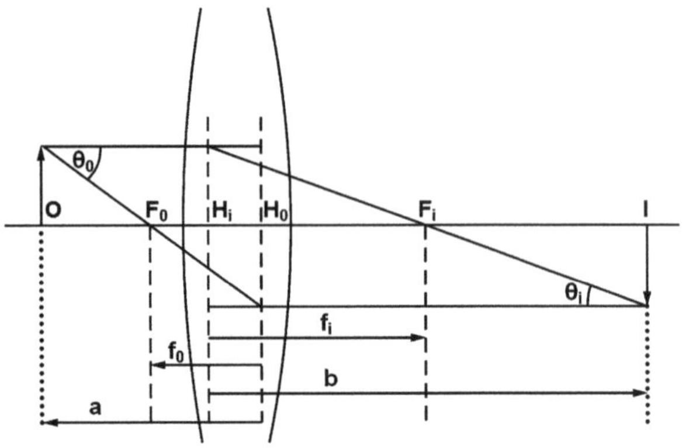

Fig. 3.25 Optical parameters of a lens. Reproduced from Ref. [112] with permission from Taylor & Francis

3.2 Components

3.2.1 Objective Lens and Other Axial-Symmetric Lenses

The objective lens is the heart of all instruments discussed in this book. It determines the resolution and in the case of a wide angular distribution of the emitted electrons also the transmission of the instrument unless it is limited by other parts of the system such as apertures. It is frequently called a cathode lens because the specimen is the negative electrode of the lens and because emission microscopy started with the study of thermionic cathodes. It is also called an immersion lens because the electron velocity on the exit side of the lens is different from that on the specimen side. Like all lenses the imaging properties of a cathode lens are characterized by its focal properties and by its aberrations. It is a thick lens (Fig. 3.25) [112], so that it has two principal planes H_0, $H_{i,}$ and two corresponding focal lengths f_0, f_i. f_0, and f_i are interrelated by the refractive indices n_0, n_i, which are determined by the initial and final velocities v_0, v_i and energies E_0, E_0 of the electrons, via

$$\frac{f_0}{f_i} = \frac{n_0}{n_i} = \frac{v_0}{v_i} = \left(\frac{E_0}{E_i}\right)^{1/2}. \tag{3.1}$$

Image and object positions are related through the lens equation

$$\frac{f_i}{b} + \frac{f_0}{a} = 1, \tag{3.2}$$

where a, b are object and image distance, respectively, with all distances measured from the corresponding principle planes. Inasmuch as the specimen position a is part of the lens and the ratio n_0/n_i is determined by initial and final energies only one principle plane and one focal length is sufficient for characterizing the lens.

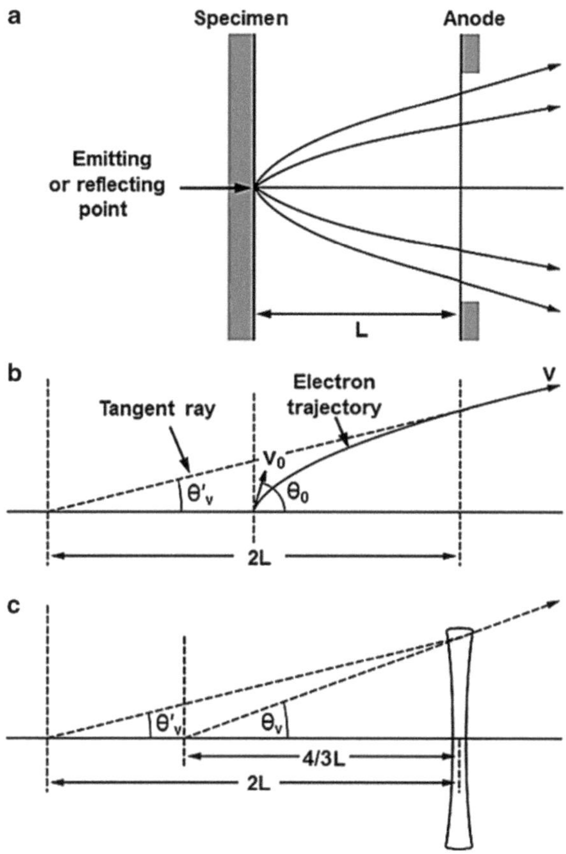

Fig. 3.26 Properties of the acceleration part of a cathode lens. (**a**) Schematic of the trajectories, with lateral dimensions strongly enhanced. (**b**) Transition from real object to virtual object plane, neglecting the influence of the aperture. (**c**) Influence of the aperture acting as a divergent lens. Adapted with permission from Ref. [113]. Copyright 1980 Elsevier

To a first approximation a cathode lens may be considered to consist of two parts: a homogeneous field, in which the electrons are accelerated to their final energy, followed by a lens. The homogeneous field forms a virtual image of the specimen behind it, of which the following lens forms a real image (Fig. 3.26 [113]). Elementary calculations show that the resolution of the total lens is determined in general by the homogeneous field. Early calculations were made for emission electron microscopy, assuming that all emitted electrons (emission angle θ up to 90°) contribute to the image, that their emission energy is small compared to their final energy E and that their energy distribution ΔE can be replaced by the mean emission energy E_0. As already mentioned in Chap. 1, Recknagel [114] found for the resolution

$$\delta = 4E_0/eF, \tag{3.3}$$

where $F = E/eL$ is the field strength and L the length of the field. Later other authors derived under somewhat different assumptions similar expressions where the factor 4 was replaced by values as low as 0.6, but the Recknagel formula $\delta \approx E_0/eF$ has long been accepted as an approximate value of the resolution limit. This impeded the evolution of LEEM because LEEM uses a larger E_0 so that a poorer resolution is expected. What was not taken into account in this reasoning is that in LEEM the reflected electrons are confined into diffraction beams, which have a narrow angular distribution in contrast to emission microscopy where $\theta_{max} = 90°$. When the angle of the electron bundle used for imaging is limited then Eq. (3.1) has to be multiplied by

$$\sin\theta(1 - \cos\theta) \tag{3.4}$$

as shown already by Recknagel in 1942 [115]. Boersch [116] has suggested in the same year to achieve this angle limitation by an aperture in the back focal plane and used it to improve the resolution. For small angles Eq. (3.4) simplifies to the standard form of the angle dependence of the spherical aberration, θ^3 with the prefactor of Eq. (3.3) except for the factor 4.

In addition to the difference of the angular distribution between LEEM and emission microscopy also the energy distribution differs. In LEEM usually the energy width $\Delta E \ll E_0$, for example for a field emission gun 0.25 eV vs. 10 eV for typical values of E_0, while in thermionic and UV photo emission microscopy $\Delta E \approx E_0$. For these reasons Bauer re-derived in the early phase of the LEEM development the resolution of the homogeneous field in general terms (finite θ, $\Delta E \neq E_0$) and reported it in 1964 [117] but published it much later [118] in order to convince the scientific community of the possibilities of LEEM. Also in 1964 Cruise and Bauer [119] confirmed these possibilities with numerical calculations for the lens used in the first LEEM instrument, a triode objective. For small angles the resolution is proportional to $(\Delta E/E)\ \theta$. This is the chromatic aberration. The spherical and chromatic aberration-limited resolution is usually written as $\delta_s = C_s\theta^3$ and $\delta_c = C_c\rho\theta$, where $C_c = -C_s = -L(E_0/E)$ are the lowest order chromatic and spherical aberration constant and $\rho = \Delta E/E$.

The assumption of a homogeneous field is of course only an approximation because the anode has a hole and acts as a lens, called aperture lens, as already noticed by Davisson and Calbick [120] in the early 1930s and mentioned in Chap. 1. This lens forms from the virtual image, produced by the homogeneous field at the distance $2L$ from the aperture, a virtual image with magnification $M = 2/3$ at a distance of $4/3L$ from the aperture, i.e. at $L/3$ behind the specimen (Fig. 3.26). This is then the object for the lens after the aperture lens. Lenc and Müllerova [121] extended the calculations for the homogeneous field to include the effect of the aperture lens and provided useful relationships for the dependence of the aberrations on the ratio of the final energy to the starting energy of the electron, the "immersion ratio" $k \doteq (E - E_0)/E_0$. In a further step they added an imaging lens, characterized by the two principal planes and respective foci and calculated C_c and C_s as a function of k for the complete objective lens [122]. For k values of interest in LEEM and PEEM they showed that the influence of the aperture lens on these constants is very small, a result which was confirmed in a different context by Wang [123]. Later Tromp et al. [124] calculated analytically the aberration constants up to fifth order for energies from 1 to 100 eV and obtained simple analytical expressions which showed little influence of the aperture on the aberration constants. They also calculated the aberrations of their complete magnetic cathode lens by raytracing and found that they agreed very well with those of the analytical theory when the additional aberrations of the magnetic imaging field were taken into account.

Analytical calculations, including calculations for arbitrary rotation-symmetric lens potentials, which are quite demanding, are useful for general considerations. However, quantitative values of the focal properties and aberration coefficients have to be determined for each specific lens configuration. This has been done already before fast computers were available with either analog methods or by systematic measurements. Analog methods simulated the fields produced by the lens electrodes by electrodes in an electrolytic tank [125, 126] or by resistance networks [127]. These methods were developed to a high level of accuracy and even allowed automatic determination of the electron trajectories [125, 126]. An example of the quality of the objective lenses achieved in this manner is Engel's magnetic lens, with which he achieved a resolution of 12 nm in the PEEM mode [14]. Figure 3.27 [14] shows the near-axis region of this lens and the field distribution along the optical axis determined with the resistance network and by direct measurements, compared to Glaser's bell-shaped model field. Many other good lenses were developed in a similar manner. The other approach was to measure the lens properties directly in electron-optical benches. One example is Rempfer's study in the 1950s of electrostatic unipotential einzel lenses, which she published much later [128]. Her results have been used by many instrument designers. Comparison with computer calculations confirmed her measurements. Another example is Soa's study of the influence of the electrode configuration of electrostatic triodes on their focal properties [129].

With the emergence of fast computers with large memory the previous methods became obsolete. One of the first precise computational methods, the so-called charge density or boundary element method, was developed by Cruise already in 1963 [130] for the calculation of the first electrostatic objective lens used in LEEM, a triode lens [119]. The high accuracy is necessary because of the low velocity and the strong potential field gradient near the specimen. A number of other computational methods have been developed since, which the interested reader can find in reviews on electron optics such as [131] or [132]. Also excellent software is available for all kinds of electron-optical calculations, such as [133–135]. Electron optics in general is treated by Liebl [136] and Rose [137] on an introductory and advanced level, respectively, and electron optics of electrostatic lenses by Heddle [138]. Here we discuss only a few examples of objective lenses calculated with the charge density method or with one of the commercial computer codes. Beforehand,

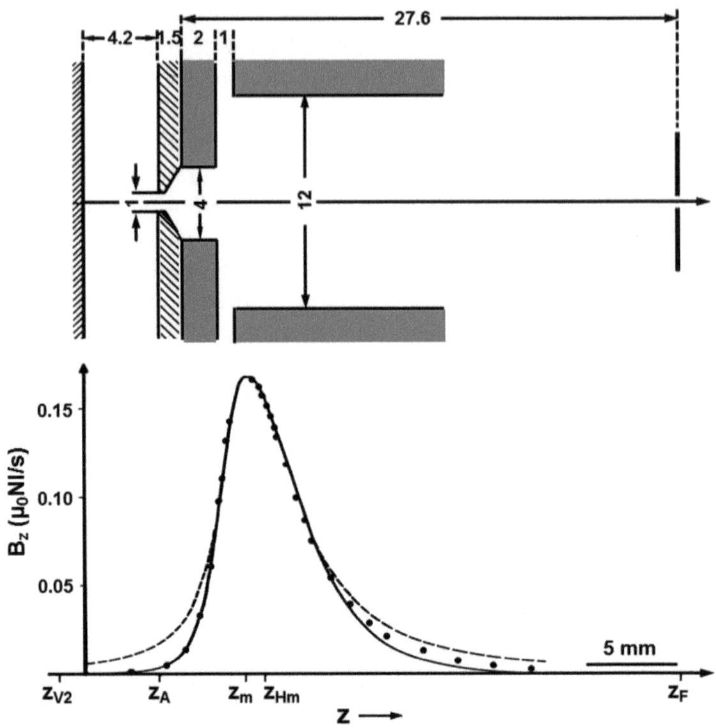

Fig. 3.27 Near axis region of a magnetic immersion lens from specimen to focal plane (*top*) and axial magnetic field distribution (*bottom*). *Full line*: from resistance network, *dashed line*: Glaser's bell-shaped curve, *dots*: from direct field measurements. The first magnetic pole piece is covered by highly polished Al anode acting as reflector of the UV light onto the specimen. Adapted from Ref. [14]

however, a few comments on the units of the aberration constants and on the definition of the resolution are necessary.

The aberration constants can be defined in three different ways: with respect to the specimen plane, to the virtual image plane at distance L, where the magnification is 1, or to the real image plane with magnification M. As seen in Fig. 3.26 with the second choice the imaging process can be treated as if the electrons would be starting in this plane with their final energy if one neglects the shift of the virtual image plane to $-L/3$ and demagnification to 2/3 caused by the lens effect of the anode aperture. This is permissible because of small influence of the aperture effect [121, 124]. With this choice of the reference plane the aberration constants refer to the angle θ and to the $\Delta E/E$ ratio ρ at the final energy E. They are obtained from those referred to the emission angle θ_0 and start energy E_0 at the specimen by multiplying the C values with factors, which take into account the transition from initial to final energy. For each power of θ the factor is the refractive index ratio $n_i/n_0 = v_i/v_0 = (E/E_0)^{1/2}$ to account for the change of angle from θ_0 to θ due to refraction in the lens field, for each power of ρ this factor is E/E_0 to account for the transition from E_0 to E. The transition from virtual image ($M=1$) to real image in the Gaussian image plane—the plane at which electrons leaving the object at very small angles intersect the optical axis—is accomplished by multiplying the C values with the real magnification M. The resolution itself is obtained to a first approximation by assuming that the contributions of the various aberrations are independent. Taking all spherical and chromatic aberrations into account, which occur in an axially symmetric field up to fifth order and also the influence of diffraction at the angle-limiting aperture after the objective lens, the resolution δ is given by

$$\delta = \left[(0.6\lambda/\theta)^2 + (C_c\rho\theta)^2 + \left(C_3\theta^3\right)^2 + \left(C_{cc}\rho^2\theta\right)^2 + \left(C_{3c}\rho\theta^3\right)^2 + \left(C_5\theta^5\right)^2 + \cdots\right]^{1/2}. \tag{3.5}$$

Before aberration correction ($C_c = C_3 \equiv C_s = 0$) was introduced into cathode lens microscopy only C_c and C_3 were calculated. An example are the calculations of Chmelik et al. [139] with the charge density method and integration of the paraxial ray equations. For a field strength $F = 10$ kV/mm they compared the optical properties of the electrostatic triode, electrostatic tetrode, and magnetic triode shown in Fig. 3.28a, b, d [140] with that of the homogeneous field. The magnetic triode has a gap in the magnetic circuit so that one pole piece can be put on a potential different from that of the other, giving an additional degree of freedom. In the calculations the pole pieces were at the same potentials. The resolution δ in the Gaussian image plane was obtained by the usual assumption that chromatic aberration, spherical aberration, and diffraction at the angle-limiting aperture contribute independently to it, so that

$$\delta = \left[(C_c\rho\theta_0)^2 + \left(C_s\theta_0^3\right)^2 + (0.6\lambda_0/\theta_0)^2\right]^{1/2}, \tag{3.6}$$

where θ_0 is the optimum aperture at which $\delta(\theta_0)$ is minimum and λ_0 the wavelength corresponding to the emission energy in object space. The optical properties of the electrostatic triode (a), the workhorse of the past, are inferior to the other two lens types because of the much lower field at the specimen as evident in Fig. 3.28a and will not be discussed further. Figure 3.28e, f [141] shows the optical properties of the electrostatic tetrode (b) and the magnetic triode (d) as a function of initial energy E_0 for a field strength F at the object of 10 kV/mm, an energy width ΔE of 0.5 eV and a final energy E of 20 keV. The properties of the magnetic diode (c), which is now used in general, are similar to that of the triode but allows larger distances between object and the lens for the same field strength. The lenses (b)–(d), in particular the magnetic lenses, approach with decreasing energy the resolution of the homogeneous field but at high energy the aberrations of the imaging part of the lens become dominant. The various contributions to the resolution show an important difference between the electrostatic and the magnetic lens: at high energies the

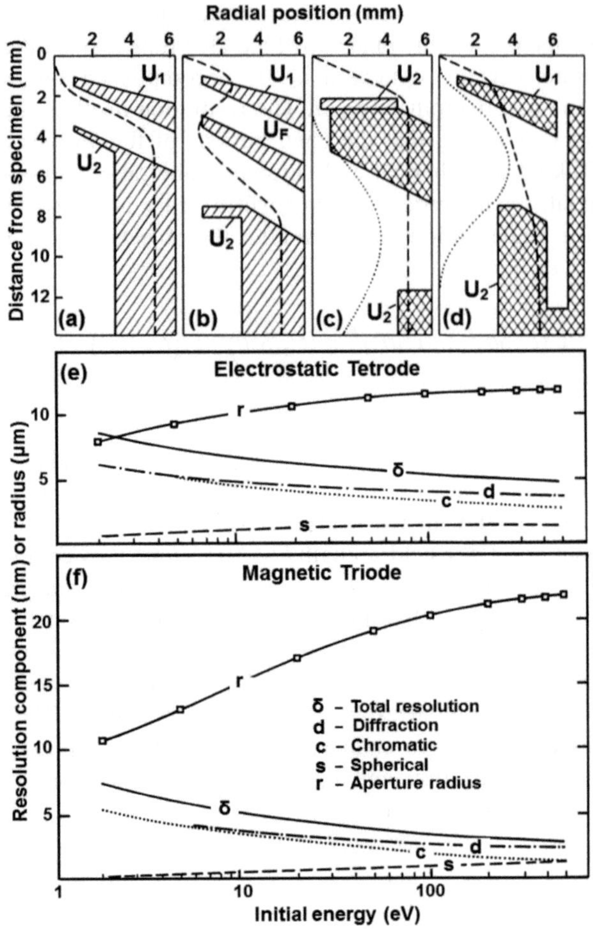

Fig. 3.28 Comparison of cathode lens configurations. (**a**) Electrostatic triode, (**b**) electrostatic tetrode, (**c**) magnetic diode, (**d**) magnetic triode. The *dashed lines* show the axial electric field, the *dotted lines* the axial magnetic field. (**e** and **f**) Contributions of the various resolution-limiting factors to the total resolution δ and aperture radius r for optimum resolution as function of initial energy. Field strength at object 10 kV/mm, energy width 0.5 eV. (**a–d**) Reproduced with permission from Ref. [140]. © IOP Publishing 1994. All rights reserved. (**e** and **f**) Adapted from Ref. [141] with permission from Springer Science + Business Media

optimum aperture of the magnetic lens is much larger. This is important for the transmission in PEEM with fast photoelectrons. The lenses shown in Fig. 3.28a, b, d were not optimized. Chmelik (private communication) compared also the properties of several optimized lenses whose C_c and C_3 values are shown in Table 3.1.

In agreement with the δ-curves in Fig. 3.28e, f the magnetic lens has the lowest aberrations, approaching those of the acceleration field at low energy. Amongst the four electrostatic tetrode lenses the Orloff-Swanson lens [142] has the lowest

Table 3.1 Aberration coefficients of several immersion objective lenses, referred to the object plane. All values in μm

Aberration constant	$C_s \equiv C_3$			C_c		
Start energy (eV)	2.0	20	200	2.0	20	200
Homogeneous field	0.22	2.2	22	0.22	2.2	22
Magnetic objective[a]	0.35	6.2	130	0.23	2.6	33
Short electrostatic tetrode[a]	0.96	25	620	0.28	4.2	80
Long electrostatic tetrode[a]	0.61	14	330	0.30	4.6	92
Orloff-Swanson lens [142]	0.56	13	300	0.28	4.4	89
Liebl-Senftinger lens [29]	0.96	24	570	0.40	7.1	160

[a]J. Chmelik (private communication)

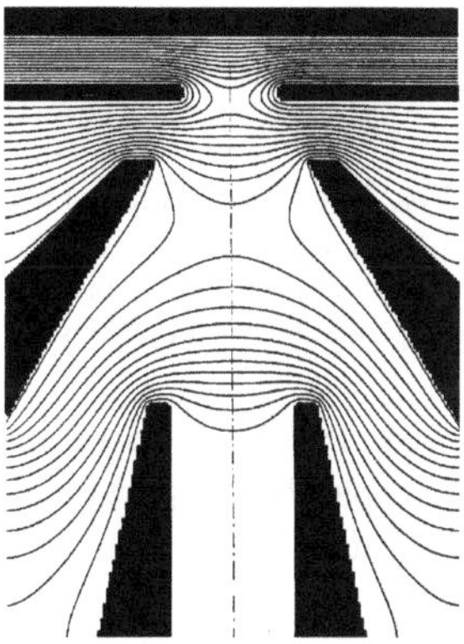

Fig. 3.29 Electrode configuration and potential distribution in a tetrode lens optimized for simultaneous light-optical imaging. Reproduced with permission from Ref. [29]. Copyright 1991 Elsevier

overall aberrations, the Liebl-Senftinger lens (Fig. 3.29) [29] the highest. The larger aberrations in the last lens are a consequence of the compromise between electron optics and light illumination and observation via the Schwarzschild optic, requiring electrode configurations, which reduce the field strength at the specimen. However, the resolutions do not differ strongly. For example a lens similar to the Orloff-Swanson lens has $\delta = 7$ nm at $E_0 = 10$ eV, $E = 20$ keV and $\Delta E = 0.5$ eV (J. Chmelik, private communication) while for the Liebl-Senftinger lens $\delta = 10$ nm at $E_0 = 5$ eV and $\Delta E = 0.6$ eV (from Fig. 10 in Ref. [29]), both at $E = 15$ keV. The lower energy and the larger energy width in the second case are largely responsible for the difference in δ. Table 3.1 shows also another important aspect: the aberration "constants" are energy-dependent. When referred to the specimen this dependence is very strong but when referred to the virtual image it is less pronounced due to the transition from low energy to high energy space. This is illustrated in Fig. 3.30 [143], which compares the exact C values from raytracing with the values obtained analytically from the homogeneous field without and with taking the aperture effect into account for small ρ_0. The values are for the real image plane with $M = -4.27$ [143]. It is evident that the C_c is determined mainly by the acceleration field in the low energy range, which is important in LEEM and in particular in PEEM with low energy electrons.

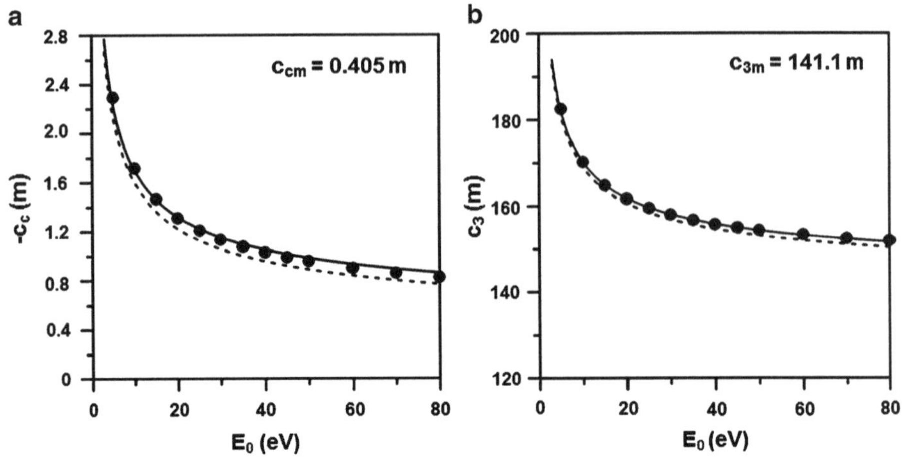

Fig. 3.30 Chromatic (**a**) and spherical (**b**) aberration constants of the magnetic lens in Tromp's LEEM without (*solid line*) and with (*dashed line*) aperture contribution included as function of initial energy. The contribution of the magnetic imaging part is shown in the *upper right*. Reproduced with permission from Ref. [143]. Copyright 2011 Elsevier

In PEEM, in particular in XPEEM, transmission, and resolution of the lens are of comparable importance. This forces the use of apertures larger than the optimum aperture which causes a concomitant deterioration of the resolution. Depending upon the relative importance of these two factors, several compromises have been considered [144]. Feng et al. made detailed calculations for the objective lens used in the ALS PEEM3 shown in Fig. 3.31a [145] and determined the resolution and transmission in secondary electron imaging as a function of the size of the angle-limiting aperture for realistic angular and energy distributions (Fig. 3.31b, c) [145]. The strong interrelation between transmission and resolution is immediately evident. The aperture eliminates electrons emitted with high energy at large angles but not at small angles so that only limited "monochromatization" with apertures is possible. While with small apertures the chromatic aberration of the acceleration field dominates the resolution, at large apertures it is the spherical aberration. For the smallest aperture for practical applications (12 μm diameter) the transmission is about 3 %.

When comparing the aberration constants and resolution reported in different publications it should be kept in mind that different definitions exist not only for the aberration constants but also for the resolution: it can be defined for the Gaussian image plane, for the plane of least confusion or wave-optically or in some other way. We will discuss these questions in Chap. 4. For example, there are publications, based on trajectory calculations with the Simion 7 program for the ALS PEEM3 objective, which claim that in the plane of least confusion the diameter of the disc of confusion—which is about half as large as in the Gaussian image plane (Eq. 3.1)—should allow a resolution of 2 nm for $\Delta E = 4$ eV [146] or even less than 2 nm for $\Delta E = 10$ eV [147], both for a 30 μm diameter aperture. This is obviously

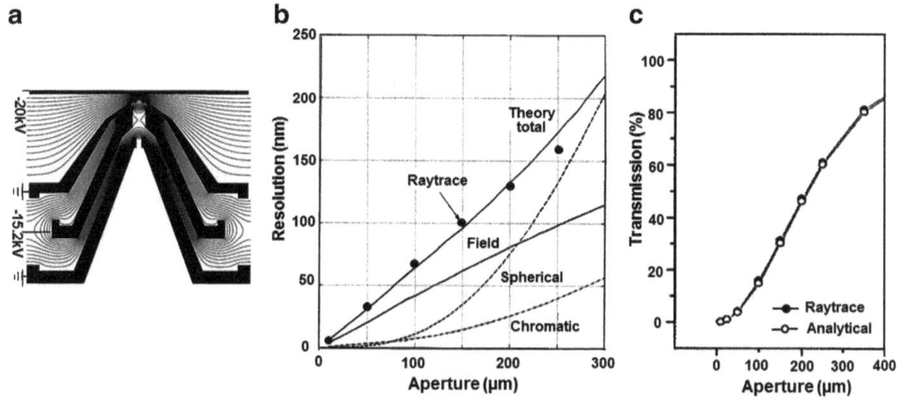

Fig. 3.31 Electrostatic tetrode lens of the ALS PEEM3 system. (**a**) Electrode configuration and potential distribution. (**b**) Contributions of the acceleration field, and of the chromatic and spherical aberrations of the imaging part of the lens to the total resolution for imaging with secondary electrons as a function of aperture diameter. (**c**) Transmission as function of aperture diameter. Adapted from Ref. [145] with permission from the American Institute of Physics, © 2002

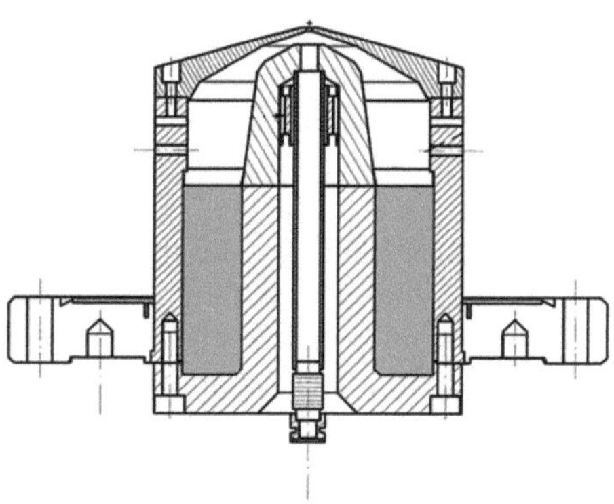

Fig. 3.32 A typical modern magnetic cathode lens similar to that used in a commercial system. The coil is sealed in the body of the lens which ends in the lower pole piece. The upper pole piece is made of a different material with a high saturation magnetization. The beam tube with stigmator and alignment coils on top is inserted into the lens body

incompatible with the results of Fig. 3.31 for the same objective and with all other theoretical and experimental results.

Instead of defining the resolution in ray-optical terms, it can also be defined by wave optics. This subject will be discussed in the Chap. 4 but here we give only one number for comparison: the 84 %/16 % resolution at a sharp edge, at which the reflectivity abruptly changes. It has been calculated wave-optically by Pang et al. [148] for the magnetic lens shown in Fig. 3.32, whose C_s and C_c have been calculated by Chmelik (private communication). For $E_0 = 10$ eV, $E = 20$ keV and

$\Delta E = 0.5$ eV they obtained with these constants a resolution of 4.1 nm, while the Raleigh criterion gives 5.4 nm, which is in good agreement with the ray-optical calculations.

Summarizing the theoretical work, in LEEM the theoretical resolution limit of magnetic lenses is between 4 and 8 nm, depending upon the value of ΔE, E_0, and E, while for electrostatic lenses a realistic limit is about 10 nm. While these limits have been reached in special cases, in practice the resolution δ is worse because the experimentalist would need a continuously variable aperture in order to adjust it to its optimum value for each E_0 and ΔE. In addition high voltage and lens current fluctuations of the order $\Delta U/U$, $\Delta I/I \approx 10^{-5}$ can already increase δ value by 1 nm. Additional experimental limitations are of course vibrations and specimen drift.

Comparing magnetic and electrostatic objectives, both have their advantages and disadvantages. Magnetic objectives are superior in resolution and transmission but have some stray field at the specimen position, for example about 10 G in the lens shown in Fig. 3.32. They are also more expensive and require that the specimen is near ground potential because it is difficult to have magnetic lenses floating at high potential. Furthermore they cause image rotation, which is inconvenient when changing magnification or switching from imaging to diffraction. However, this problem can be overcome by double lenses with opposite current directions. Electrostatic lenses have no stray field, can be floating at high potential, so that the specimen can be near ground potential, and are cheaper. On the other hand they pose high voltage insulation problems between the electrodes, which have been solved in a variety of ways. Many systems use precision ceramic cylinder insulators, which have been used already in the electrostatic transmission electron microscopes in the 1960s and are used in the Rempfer design (Fig. 6 in [12]). In other systems precision ceramic balls provide the necessary insulation as illustrated in Fig. 3.38. In all cases the proper design of the transition between the negative electrode and the insulator is important in order to avoid field emission and leakage current along the insulator, a fact which has been known for a long time [149, 150]. The electrodes in electrostatic lenses are usually made of Al or Ti, sometimes of silicon bronze, the magnetic parts of magnetic lenses of forged alloys with high saturation magnetization and permeability. This allows reaching the high axial fields needed for imaging with relative low currents, for example with less than 1000 Ampere-turns in the lens shown in Fig. 3.32. As a consequence the heat generated by the current is easily conducted away by the mounting flange and the bottom part of the flange, so that air cooling is sufficient.

The other lenses used in low energy surface electron microscopes are essentially the same as those, which have been used for many years in transmission and scanning electron microscopes, except they have to be bakeable. References to them can be found in the comprehensive books mentioned earlier [131, 132]. Here only Rempfer's frequently used experimental data for electrostatic lenses [128] and the calculations for asymmetric lenses, which have lower aberrations [151], should be mentioned. As far as magnetic lenses are concerned, the lenses of the first LEEM were designed based on the experimental data of Liebmann et al. [152–154]. Magnetic lenses are reviewed in a book edited by Hawkes [155].

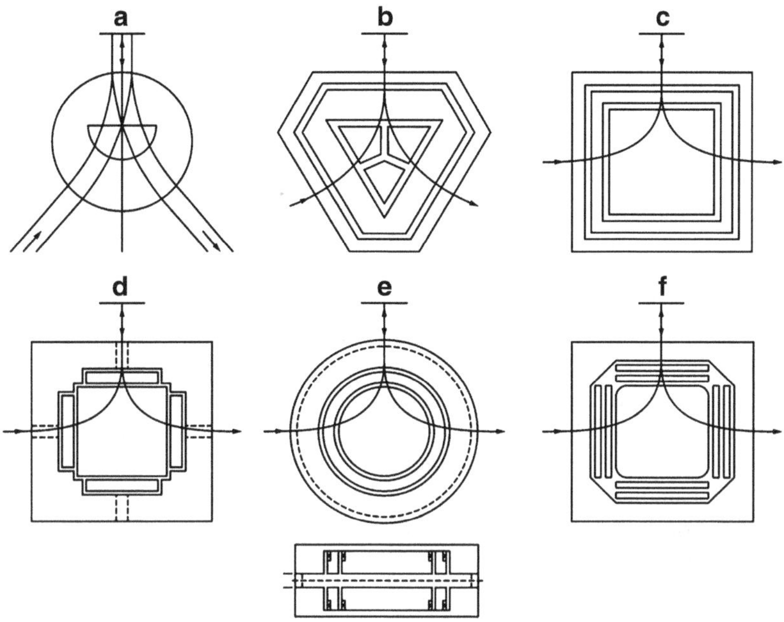

Fig. 3.33 Beam separator types. (**a**) Archard-Mulvey type, (**b**) close-packed prism 60° separator with different deflections of the two beams, (**c**) close-packed prism 90° separator, (**d**) close-packed 90° separator with independent segments, (**e**) circular separator with cross section showing the coil locations, (**f**) multisegment separator. The cross section of (**e**) is shown at the *bottom* with the excitation coils indicated

3.2.2 Magnetic Deflectors

The physics and technology of magnetic deflectors is an old field [156] but plays an important role in the instruments discussed in Sect. 3.1. Magnetic deflectors are used in aberration corrected instruments and are needed for imaging with reflected electrons (LEEM and MEM). As mentioned in Chap. 1 simple magnetic deflectors have been used already in the 1930s in MEM systems but here only more recent deflectors will be discussed. A general description of magnetic deflectors is given in Ref. [136]. Magnetic deflectors have different focusing properties in the plane of deflection and normal to it so that they are inherently astigmatic; they are energy-dispersive, i.e. the deflection is energy dependent, and they cause also distortions in large field of views. In the original LEEM (Fig. 3.6), which had round pole pieces, the astigmatism of the collimator lens limited the resolution to 50 nm, while the astigmatism caused by the sector field was calculated to be in the low nanometer range when the image was located in the center of the deflector [26]. Nevertheless, the original pole piece was replaced by the astigmatism-free Archard-Mulvey pole piece with a semicircular cut-out [157] (Fig. 3.33a). The same method was used by Tromp and Reuter [28].

The next step of deflector development was motivated by the goal to combine LEEM with Auger electron microscopy (AEM), which requires different energies of the incident and reflected beam and therefore different magnetic field strengths in the two sides for symmetric optical path. This was achieved by dividing the pole pieces into an outer annular sector and several inner sectors, which could be excited independently as illustrated in Fig. 3.33b [20]. The first order properties of these so-called close-packed prism arrays [158, 159] were calculated for several prism configurations, in particular also for 90° deflection. Unlike in previous calculations, which assumed a sharp decrease of the field at the magnet boundaries (sharp cut-off approximation (SCOFF)), the fringing fields at the boundaries were taken into account with effective edges. Later calculations [160] for one of these prism configurations estimated its second order aberrations and noted that the SCOFF approximation failed to give the correct angles in the prisms for stigmatic focusing. The failure of the SCOFF approximation had already been noted before by Degenhardt [161], who explicitly introduced symmetry considerations into the design. He calculated iteratively the exact field shape at the boundaries and achieved not only stigmatic but also distortion-free imaging with the configuration shown in Fig. 3.33c [161]. Tromp et al. [32] modified Degenhardt's design by dividing the ring part in four segments, which can be excited independently (Fig. 3.33d [32]) for more flexibility. Kan et al. [162] used a different accurate field calculation method and applied it to round deflectors consisting of an annular outer and an inner pole piece. The calculated results showed very good agreement with experiment, so that they could apply their calculation method for a comparison of Degenhardt's square design with its round analog (Fig. 3.33e [163]), with the result that the square deflector has larger distortions than the round deflector [163]. A square deflector with more segments (Fig. 3.33f [42]) was optimized by Mankos et al. [42, 44, 45]. It images the image produced by the objective lens stigmatically and distortion-free over a large field of view.

In all deflectors discussed up to now electrons with different velocity experience different deflections so that the deflectors are dispersive. The deflector needed for aberration correction in LEEM puts additional constraints in the design of the beam separator: it must not only be stigmatic and distortion-free but also dispersion-free. These conditions are fulfilled in the four quadrant beam separator developed by Rose's group in Darmstadt [164–166] (Fig. 3.13) for the SMART instrument [60, 61] and tried later in the first version of the beam separator of the ALS PEEM3 [167, 168]. This beam separator has in addition to the mid section symmetry double mirror symmetry in the quadrants, which together cancel all second-order aberrations and dispersions of first and second degree (Fig. 3.34a [60]). In contrast to the simpler beam separators discussed above, in which the image formed by the objective lens is placed into the midsection plane, the image is here in front of the beam separator. It is imaged by the first quadrant into the midsection plane and by the second quadrant with unit magnification to its exit plane without dispersion. The symmetric arrangement about the symmetry planes S_1 and S_2 cancels the geometric second order aberrations (Fig. 3.34b [60]). Field lenses before the entrance and after the exit of the beam from the separator transfer the

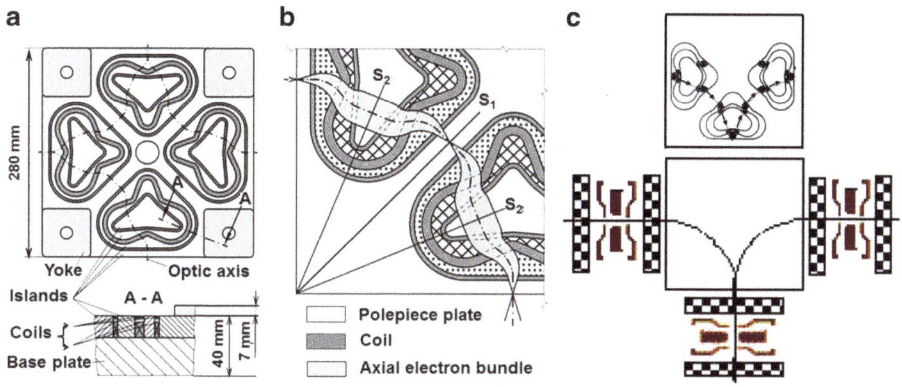

Fig. 3.34 Beam separators for aberration-corrected systems. (**a** and **b**) SMART beam separator, (**c**) ALS PEEM3 beam separator. For explanation see text. Adapted from Ref. [60] (**a** and **b**) and [65] (**c**). The *top* part of (**c**) shows schematically the attempted three-deflector analog of the four-deflector separator (**a**). (**a** and **b**) Adapted with permission from Ref. [60]. Copyright 2002 Elsevier. (**c**) Adapted from Ref. [65] with permission from the American Institute of Physics, © 2007

diffraction plane accordingly. In order to adjust the coil currents exciting the various magnet sections in the four quadrants tuning coils are added to the main coils as indicated in Fig. 3.34a. In addition the mechanical tolerances were very demanding. For these reasons a new beam separator was developed for the ALS PEEM3, consisting of a simple square deflector in the center, surrounded by round electrostatic lenses and quadrupoles [65, 169]. It is shown schematically in Fig. 3.34c [65]. In this design the magnet is used only for the 90° deflection. The quadrupoles correct the astigmatism of the deflector after each 90°deflection. The round lenses focus the image before the beam separator to its exit and the mirror symmetry about the center plane of the magnet ensures that this image is achromatic. The calculated resolution and transmission are equal or better than that of the original beam separator, in particular for off-axis points on the specimen. However, unlike in the previous beam separators, the image after the first 90° pass is not achromatic, a deficiency, which is problematic for correction with the mirror.

The initial difficulties encountered with the operation of the SMART beam separator/mirror combination stimulated the search for still other solutions. These were suggested already many years ago for combinations of deflectors, making use of symmetry [170]. The double deflector aberration correctors of Elmitec [10] illustrated in Fig. 3.14 and SPECS [33] shown in Fig. 3.15 are based on them. The basic idea is illustrated in Fig. 3.35: (a) shows the 120° Elmitec deflector, which was stimulated by Rose and Plies' imaging 90° deflector without dispersion and second-order aberrations [170], (b) the 180° SPECS deflector. While nothing was published about the Elmitec system, Tromp et al., who developed the SPECS system, gave a detailed description of the properties of their double deflector setup [66]. As seen in

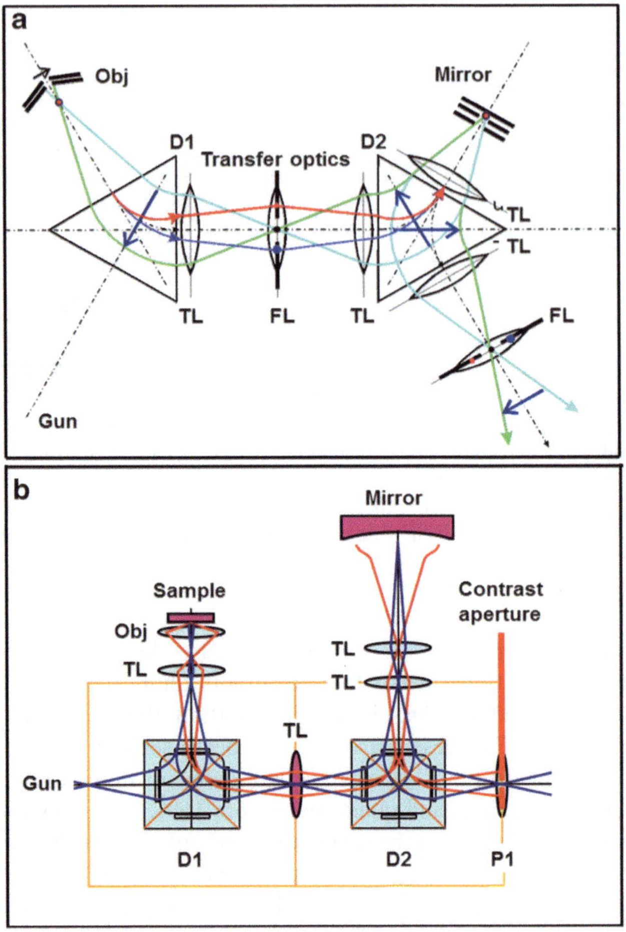

Fig. 3.35 Two-separator arrangement for aberration-corrected systems. (**a**) Elmitec system, (**b**) Tromp/SPECS system. In (**b**) the coupling between the two deflectors is achieved with an electrostatic field lens (TL), in (**a**) with a magnetic field lens pair (FL) and two magnetic transfer lenses, which are needed to transfer the diffraction pattern into the field lens. (**a**) Based on discussion between E. Bauer, C. Koziol and T. Schmidt. (**b**) Reproduced with permission from Ref. [66] Copyright 2010 Elsevier

the trajectory plots the image plane is located in the diagonal plane of the first deflector, which transfers the diffraction plane before the deflector into the center between the two deflectors. An electrostatic field lens in this plane in the SPECS system—or in the Elmitec system a rotation-free magnetic lens pair—images the diagonal plane of the first deflector into the diagonal plane of the second deflector. Due to symmetry the dispersion caused by the first deflector in the plane of the diffraction pattern is removed by the second deflector. The resulting symmetry cancels out the major aberrations of the deflectors. The image in the diagonal plane

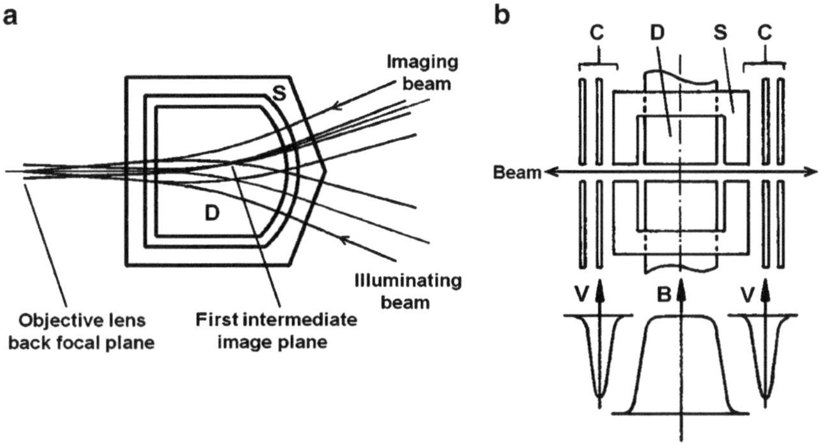

Fig. 3.36 Beam separators in the electrostatic flange-on LEEM systems. (**a**) 10° Deflector with curved and inclined entrance and exit boundaries with minimum aberrations. (**b**) Magnetic-electrostatic compound deflector. It consists of the central magnetic deflector (D) surrounded by the shunt (S), The shunt is at the same time one of the electrodes of the electrostatic slit lenses (C) on both sides of the deflector, which compensate the astigmatism of the deflection field. The field distributions are shown *below*. (**a**) Adapted from Ref. [40] with permission from the American Institute of Physics, © 1998. (**b**) Adapted from Ref. [7] with permission from the American Institute of Physics, © 1996

of the second deflector is finally imaged into the entrance plane of the mirror, which will be discussed below.

Another double deflector aberration corrector with 45° deflection has been proposed by Tsuno et al. [67] but no details are given. Finally a deflector–mirror combination should be mentioned, the Castaing-Henry filter [83], the first combination of this type. Equipped with an energy selecting slit in the energy dispersive plane it was used for spectroscopic imaging as already mentioned earlier [72, 73]. In another application of the Castaing-Henry filter Lichte and Möllenstedt [171] inserted electron-optical biprisms before and after the deflector and measured the smoothness of the mirror with electron mirror interference microscopy with sub-nanometer height resolution.

Compact LEEM and mirror-corrected PEEM systems require simpler deflectors. In the LEEM system of Adamec et al. (Fig. 3.10) [40] illumination and imaging beam are tilted as little as possible against the optical axis of the objective lens, forming an angle of only 10° with it. This allows a very small pole piece with about 40 mm field length (Fig. 3.36a [40]). By tilting the exit plane a few degrees with respect to the plane normal to the beam and placing the image produced by the objective lens into the center of the deflector all low-order aberrations are eliminated. In the LEEM system of Grzelakowski and Bauer (Fig. 3.9) [7] double focusing of the three deflectors is not achieved by tilting the pole boundaries as in the previous system but by adding electrostatic cylinder lenses as indicated in Fig. 3.36b [7]. These lenses are einzel lenses, whose innermost electrode is the

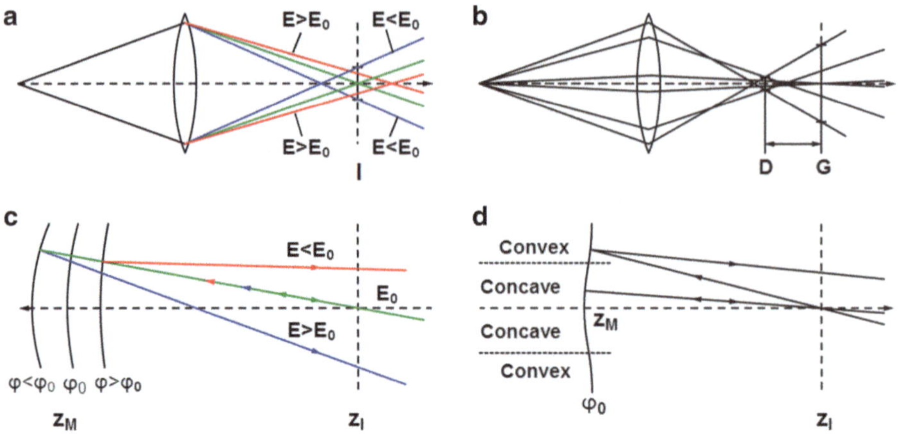

Fig. 3.37 Schematic of the aberration correction with an electron mirror. (**a**) and (**b**) Chromatic and spherical aberration in a focusing electron lens, (**c**) and (**d**) chromatic and spherical aberration in an electron mirror. For explanation see text. *G* Gaussian image plane at z_I, *D* disc of least confusion

magnetic shunt limiting the magnetic field of the deflector. In both instruments the symmetry reduces the aberrations of the deflectors to such a degree that the diameters of the discs of confusion caused by them are significantly below the resolution limit of the objective lens. Rempfer's deflection system for her mirror aberration corrector [37, 38, 172–174], also uses three deflectors but they are round without fringing field termination. However, the intermediate images are transferred by relay lenses into the center of the deflectors, which reduces aberrations, while in the system of Fig. 3.9 the images are before and after the deflectors, but apparently without noticeable influence on resolution. In a more recent version of Rempfer's corrector [56] the deflectors have been improved by adding electrostatic correction plates to the magnets.

Summarizing this section, there are many magnetic deflector designs for LEEM without and with aberration correction and for aberration-corrected PEEM. All of them make use of symmetry considerations already proposed in 1973 [170] for minimizing the influence of aberrations, resulting in deflectors, which do not deteriorate the resolution of the overall system.

3.2.3 Electron Mirrors

Already in the 1930s it was discovered that mirrors (reflecting electron lenses) have properties opposite to those of (transmitting) electron lenses. This is illustrated in Fig. 3.37 for the chromatic and spherical aberrations. In transmission slower electrons are more deflected than faster electrons, in reflection the reverse is true because slower electrons do not penetrate as deeply into the lens field and therefore experience less force. Thus the chromatic aberrations have opposite signs.

The same is true for the spherical aberrations. In transmission electrons further away from the optical axis pass through a region with larger electric field strength and are deflected more than electrons closer to the axis. In the mirror the curvature of the potential surfaces decreases with distance from the axis so that electrons further away from the axis are reflected with less than the 180° reflection of the beam along the axis. Zworykin et al. apparently were the first to propose in 1945 using these opposing actions of mirrors to correct the aberration of electron lenses in their classical book on electron optics and electron microscopy [175]. Scherzer suggested it in his much-cited work in 1947 [52]. However, the first calculations of the aberrations of mirrors came to the conclusion that this method was impractical [176]. Although there was some interest in mirror correctors thereafter little effort was invested until Rempfer and Mauck proposed a rotationally symmetric hyperbolic mirror corrector, whose potential distribution can be expressed analytically [177]. This proposal was studied further [38, 39, 172, 173, 178] finally leading to a convincing demonstration of partial correction [174]. The configuration of this mirror is shown in Fig. 3.12b [55, 56], together with the preceding adaption lens to the rest of the system in Fig. 3.12a. With only two electrodes the mirror potential is the only adjustable parameter. As a consequence simultaneous spherical and chromatic aberration compensation is possible for only one magnification and one energy. This restriction is eliminated by adding a third electrode [179].

Even more flexibility is obtained with an additional electrode, i.e. with a tetrode mirror [180]. This type of mirror has been studied numerically thoroughly and is used in most aberration-corrected instruments [60, 62, 63, 66, 67, 164, 178, 181–184]. Figure 3.38 [60] shows as an example the tetrode mirror used in the SMART system [60]. With the specimen at high voltage the top and bottom electrodes are at ground potential. The mirror electrode with its approximately spherical front face is at a somewhat higher potential U_1 than the specimen to provide the reflecting potential shape and U_2 and U_3 are used for focusing. For fixed electrode configuration these three potentials determine the focal length f and the parameter space of the C_c and C_s values of the mirror available for compensation of the C_c and C_s values of the objective lens. This will be discussed in the next section.

3.2.4 Aberration Correctors

The goal of aberration correction is to minimize δ in Eqs. (3.6) and (3.7). Inasmuch as ρ and θ are determined by the experimental conditions, C_c and C_s are the goals for minimization. The importance of the other aberration coefficients is to a certain degree determined by the relative energy width ρ and the desired transmission, which depends upon θ. For example, when ρ is large—as in case of in electron, ion, or fast atom bombardment-induced secondary electron imaging or in XPEEM with slow electrons without energy filtering—optimization of C_{3c} may become as important as optimizing for C_s. Nevertheless, even C_c and C_s optimization brings significant improvements as illustrated in Fig. 3.39 for PEEM3 [63]: compared to

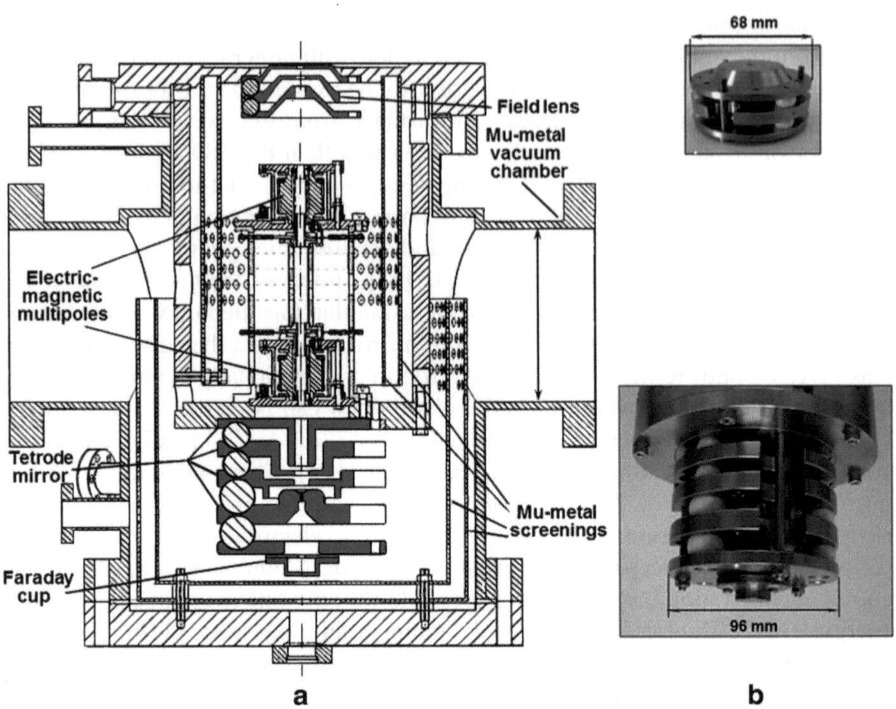

Fig. 3.38 Mirror section of the SMART system. *Left*: cross section showing mirror at *bottom*, electric-magnetic multipole correctors in the *middle* and transfer lens on *top*. *Right*: photographs of the field lens and of the mirror. Adapted with permission from Ref. [60]. Copyright 2002 Elsevier

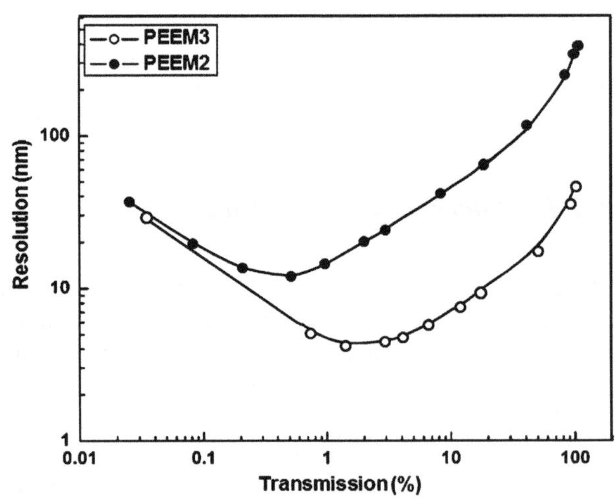

Fig. 3.39 Calculated resolution and transmission improvement by aberration correction of the electrostatic ALS PEEM system. Field strength 10 kV/mm. Reproduced from Ref. [63] with permission from the American Institute of Physics, © 2004

the uncorrected system the resolution and transmission in secondary electron emission imaging at the optimum aperture is improved from 11 nm to 4 nm and from 0.5 % to 2 %, respectively, taking electrons with energies up to 20 eV and emissions angles up to 90° into account. At fixed resolution the transmission increases by a factor of 50. Aberration correction involves the objective lens, the mirror, and the optics in between them. The optical path in this combination has already been discussed briefly in Sect. 3.1.3 and illustrated in Figs. 3.12, 3.13, 3.14, and 3.15, while the three components have been described in Sects. 3.2.1–3.2.2. Here we discuss the matching conditions, which must be fulfilled for optimum aberration correction. Objective lenses are traditionally designed for minimum C_c and C_s without regard for higher order aberrations and the aberrations of the mirror. Future optimization methods taking more aberrations into account simultaneously might lead to better correction solutions but at the present state of the art the mirror has to be operated with C_c and C_s values adapted to those of the objective lens. Calculations, which take higher order aberration of the objective lens into account have already shown that this approach poses problems for aberration correction [185], making it extremely sensitive to small variations of the adjustable parameters of the mirror which determine C_c and C_s.

We consider here two examples, systems with one beam separator (SMART and PEEM3) and systems with two beam separators (Tromp/SPECS and Elmitec), assuming that the symmetry of the separator(s) and the properties of the other intermediate components do not deteriorate the correction noticeably. All these systems use a four electrode mirror. Subsequently correction with a hyperbolic two and three electrode mirror and other correction possibilities will be discussed.

In the four-electrode mirror with a given geometry, three parameters, the three potentials V_1–V_3, are available for controlling its focal length—that is the position of the input image = output image and the aberration constants. All these quantities change simultaneously with the values of these potentials. The goal is not only to compensate or minimize the aberrations of the objective lens but also to control the focal length, so that the image position is in the place determined by the preceding electron optics, mainly by the beam separator configuration. Adaptation lenses between the separator and the mirror provide some flexibility. In Tromp's system this is achieved by the lenses M_2 and M_3 in Fig. 3.15. M_2 is located in the diffraction plane and images the center of the second deflector into the center of M_3, into which the mirror has to focus the reflected image. This is possible only within a limited magnification M range, indicated in Fig. 3.40 [66] for the mirror in Tromp's instrument [66]. In order to match the mirror-dictated magnification, the magnification produced by the preceding optics has to be adapted to this value. The parabola in Fig. 3.40 passing from $C_c = 10$ to $C_s = C_c = 0$ represents the C_s, C_c values of the objective, taking into account the appropriate powers of M (see Sect. 3.2.1). The dashed region in the C_s–C_c space, which is accessible within the range of the potential values shown by solid and dashed lines, indicates operating conditions with practical potential values, requiring $M \approx 8.5$. It should be kept in mind that aberration coefficients depend upon energy (see for example Table 3.1 or Fig. 3.30 [143]) and upon specimen-object distance, as does—though to a much

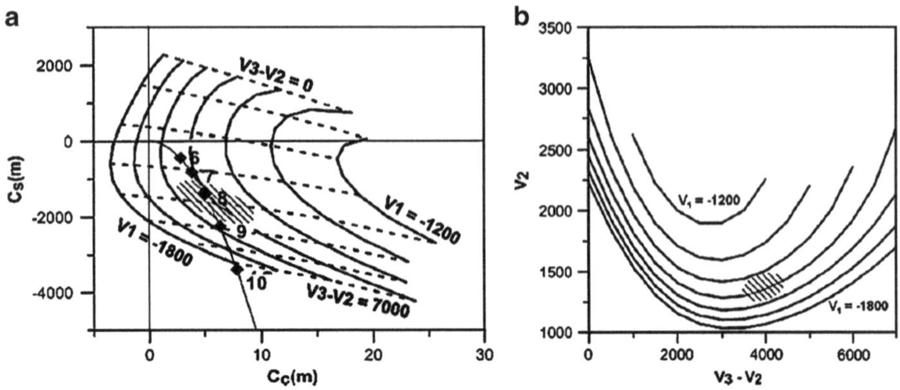

Fig. 3.40 Operational parameter of the four-electrode mirror of the Tromp/SPECS system. V_1 is the mirror electrode potential, V_2 and V_3 the potential of the following electrodes, all referred to the column potential (−15010 V). The *solid curves* in (**a**) show the C_3–C_c relationship of the mirror for various V_1 values, the *dashed lines* the corresponding relationship for various $V_3 - V_2$ values, the upward parabola the C_3–C_c relationship of the objective lens for magnifications from 6 to 10 in the image plane in M_3 (Fig. 3.15). (**a**) Shows that only a narrow range around $V_3 - V_2 = 4000$ V and $V_1 = 1500$ V allow correction, (**b**) shows that for this combination $V_2 \approx 1300$ V. For the calculation method see the original publication. No corresponding data have been published for the magnetic aberration-corrected Elmitec microscope. Reproduced with permission from Ref. [66]. Copyright 2010 Elsevier

lesser degree—the focus and magnification of the objective lens. As a consequence, whenever one of these parameters is changed, the settings of the mirror have to readjusted, a procedure, which is possible only via determining and storing a large data set and computer control.

Turning now to aberration correction with the hyperbolic mirror pioneered by Rempfer it should be remarked first that all the considerations basic to aberration correction such as the symmetry of the system, the location of the intermediate image planes or the need for adaption of the magnification to the mirror via an adaptation lens were already presented in her fundamental work on aberration correction in electron optics in 1990 [172]. Experimentally correction of both spherical and chromatic aberration was demonstrated qualitatively in 1997 in an electron-optical bench [174] and semi-quantitatively in 2010 an aberration-corrected PEEM instrument, which improved the resolution from about 50 nm before correction [55] to about 5 nm after correction [56]. This instrument is illustrated in Fig. 3.12a and gives a total magnification of 3×10^5. The transfer lens system below the three deflection magnets, in which the intermediate images are located, provides the necessary magnification adaption to the mirror so that C_s and C_c could be reduced to about 0.01 m. With only one adjustable parameter in this diode mirror its correction possibilities are very limited. The route to improve them is obviously to go to more electrodes. Fitzgerald et al. [179, 186] have examined the limitations of the diode mirror in detail and explored the possibilities of the triode mirror. Figure 3.41a [179] illustrates how the ratio for C_c/C_s varies with the ratio of

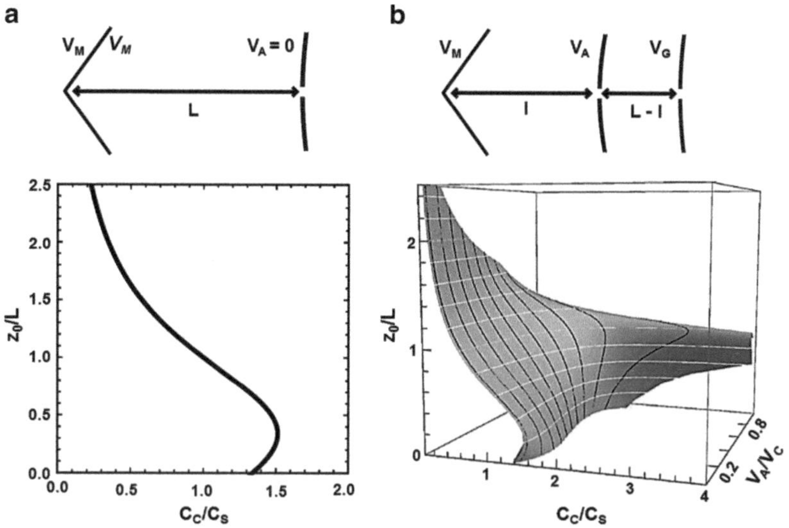

Fig. 3.41 Aberration coefficients of the hyperbolic mirror. (**a**) Diode mirror. For a given object-image distance z_0 and beam energy the ratio C_c/C_3 is determined by z_0/l. The triode mirror (**b**) allows a wider range of the C_c/C_3 ratio by adjusting l, L and/or V_A as shown for the dependence upon V_A/V_C and z_0/L. Adapted with permission from Ref. [179]. Copyright 2011 Elsevier

image position z_0 to mirror length in the diode mirror, Fig. 3.41b [179] how it changes with the ratio of center electrode potential V_A to the potential V_C at which the electrons are reflected in the triode mirror. In the first case the mirror potential V_M is the only adjustable parameter for a given mirror length. As a consequence for a given C_c/C_s of the mirror—which has to match the corresponding ratio of the objective lens, taking the magnification into account—the image position is fixed. Matching of the absolute values of the C values of mirror and objective is possible only via the mirror length. In the triode mirror the additional parameter V_A gives a wider range of C_c/C_s for correction for a given geometry L, l (here $l/L = 0.66$) of the mirror. Data for other l/L ratios (0.3, 0.8) show that the mirror is not extremely sensitive to this ratio. Now one of the C values can be chosen freely within a limited range to fit the corresponding value of the objective lens. As seen in Table 3.1, at 2–3 eV emission energy the C_c/C_s ratio of the objective lenses is about 0.5 referred to the specimen but the magnification M changes it to $0.5\,M^2$ before the mirror. Fitting to the ratio range shown in Fig. 3.40 requires therefore demagnification. Adding a suitable combination of lenses before the mirror not only accomplishes this but also provides some flexibility in the image position z_0 [186]. Demagnification is also used in the tetrode mirror so that the tetrode mirror with its additional control parameter is superior to the triode mirror, in particular in LEEM and XPEEM with photoelectrons, where the aberration constants vary significantly due to the wide energy range used for imaging. In UVPEEM the energy range is more or less fixed and thus also the aberration constants, so that a triode mirror is probably sufficient.

While in LEEM the mirror corrector is the natural choice for aberration correction because a beam separator is needed anyway, this is not true for PEEM with its straight beam path. In principle C_c- and C_s-correcting aberration correctors used in or proposed for transmission electron microscopy (TEM) [54, 187] appear suitable for PEEM too. These include multipole systems [188–190] and retarding Wien filter correctors [191, 192]. Compared to mirror correctors they are too complex to compete with them. For example, the C_c/C_s corrector developed by Haider et al. [189, 190] consists of ten quadrupole elements, octupole elements, several deflectors, and round lenses, needs 151 ultra-stable current supplies and four voltage supplies for operation and is 83 cm long. The control and the cost of such a corrector put it out of range for PEEM. Part III of Ref. [193] gives an introduction and overview of these correctors. A more appropriate route to image improvement, though not to complete aberration correction, in systems without corrector is probably the moving focus method mentioned in Sect. 3.1.3 [71]. It requires, however, very good beam and specimen alignment and precise focus control to be viable.

3.2.5 Energy Filters

In this section we discuss band pass energy filters. Low- and high-pass energy filtered and other energy selection methods have been mentioned briefly already in Sect. 3.1.4. Band pass energy filters are used to cut out a narrow energy window from the energy distribution of the emitted or reflected electrons for imaging the specimen in real space or in reciprocal space. Several devices are used or have been used for this purpose: magnetic sectors or sector combinations, Wien filters and electrostatic sectors or sector combinations. The magnetic beam separator in LEEM is the simplest magnetic sector filter for reciprocal space imaging, which Tromp has artfully made use of as described in Sect. 3.1.4. Its dispersion is limited because the electrons pass with full energy the dispersive field of the magnet. Despite of its low value, 6 μm/eV, its usefulness has been clearly demonstrated [100]. The dispersion is significantly improved to 35 μm/eV in the magnetic multisector Omega filter (Fig. 3.42), which was originally developed for TEM [97, 98] and is used in the SMART instrument. The Omega filter consists of four 90° deflectors, four sextupoles symmetrically arranged about the central symmetry plane and four 12-poles at the ends and in the center of the filter, which allow producing quadrupole fields in addition to the sextupole fields. The symmetry of the arrangement eliminates in principle second-order axial aberrations and distortions, the sextupoles any residual second-order aberrations and the quadruple fields any residual astigmatism, so that both image plane and diffraction plane are imaged stigmatically. While in TEM the filter has to be magnetic because of the high electron energy in order to achieve sufficient energy dispersion, in LEEM and PEEM electrostatic deflectors can be used, as planned for the second version of the SMART instrument.

Fig. 3.42 Omega filter in the SMART system. The electron path is indicated by the *red arrows*. Base dimensions 30 cm × 30 cm. For description see text. Courtesy of T. Schmidt

Wien filters seem to be ideally suited for energy filtering on first sight because of their straight optical axis and their potential high energy resolution at low electron energy. Under the action of the Lorentz force $\boldsymbol{F} = e(\boldsymbol{E} + \boldsymbol{v} \times \boldsymbol{B})$ only electrons with a velocity $\boldsymbol{v}$ for which $\boldsymbol{F} = 0$ are not deflected, while electrons with different velocity are deflected and can be removed from the image by an energy-selecting slit. Because of this apparently simple phenomenon imaging Wien energy filters have been proposed repeatedly [97, 98, 194–199]. A Wien filter has been incorporated in a commercial LEEM/PEEM instrument [35, 36] but not used or used with limited success [200–202] and was replaced later by a hemispherical analyzer [203]. No energy-filtered images were published from another PEEM instrument with a Wien filter [99] and the results from a more recent Wien energy filter show very limited resolution [198]. For these reasons we will postpone the discussion of the use of Wien filters to the discussion of their general features in the following section.

The workhorse for energy filtering in microscopy with slow electrons is the concentric hemispherical electrostatic energy analyzer (CHA) [204] and its variations. Its properties have been studied extensively and reviewed recently by Zouros et al. [205–210]. Other electrostatic deflection systems have also been used as described in Ref. [136]. For example, the first spectroscopic LEEM/PEEM instrument [20] used a 90° sector with which the first Auger electron and photoelectron images were taken [31, 88, 89]. This energy filter has several weak spots: the dispersion is limited, it is not isochromatic and the energy-dispersive plane is strongly tilted. Therefore it was replaced by a CHA, which produced soon photoelectron images with about 20 nm resolution and selected area XPS spectra with 0.2 eV resolution [90]. The experimental set-up is shown schematically in Fig. 3.43 [211]. For spectroscopic imaging the image is transferred by the preceding optics into the center of the hemispheres, for selected area diffraction the

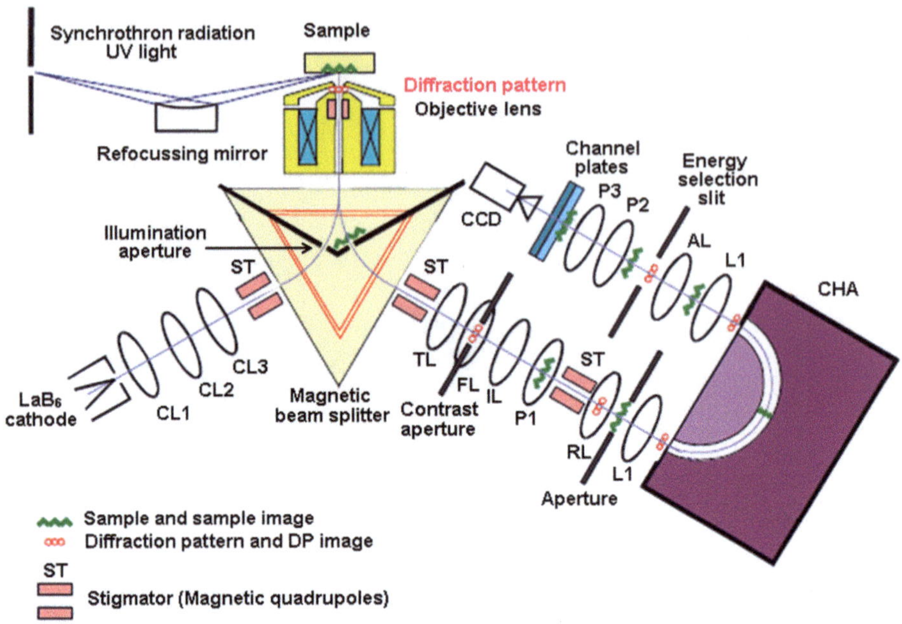

Fig. 3.43 Schematic of the electron optics of the first spectroscopic LEEM/PEEM system. The retarding lens (RL) before the concentric hemispherical analyzer (CHA) reduces the energy, the transfer lens L1 transfers the image into the center of the CAH, in which the electrons have a typical pass energy of about 1.8 kV. The transfer lens L1 and the accelerating lens (AC) after the CHA bring the electrons to their original energy and focus the image in front of the projective lens. The diffraction pattern is located at the entrance and exit of the CHA and is imaged by L1 after the exit on the energy selection slit. Energy-filtered diffraction and spectroscopy is achieved by changing the excitation of the intermediate lens IL so that image and diffraction planes are interchanged. The calculated dispersion and resolution were 54.6 μm/eV and 400 mV, respectively. Reproduced from Ref. [211] with permission from Czechoslovak Microscopy Society

diffraction pattern. Switching between these two modes is accomplished with the intermediate lens IL In the filter the electron energy is reduced from 20 to 1.8 kV, which is accomplished with the retarding and accelerating lenses RL and AL. The dispersion in the dispersive plane DP is 42 μm/eV. By removing the energy selection slit in the dispersive plane and imaging it a 10 eV wide spectrum can be recorded. A filter of this type is now part of a commercial instrument [10]. The properties of the CHA for imaging applications have been analyzed by Tonner in detail [212].

The large dispersion of the 180° filter which is useful for spectroscopic imaging is a disadvantage when the spectrum is wide and the intensity low because only a small fraction of the intensity is transmitted through the energy slit. This problem is eliminated by adding a second 180° filter, which not only eliminates the dispersion but also corrects the second order spherical aberration of the 180° filter. Two solutions for this arrangement exist: (1) by rotating the second 180° filter 180°

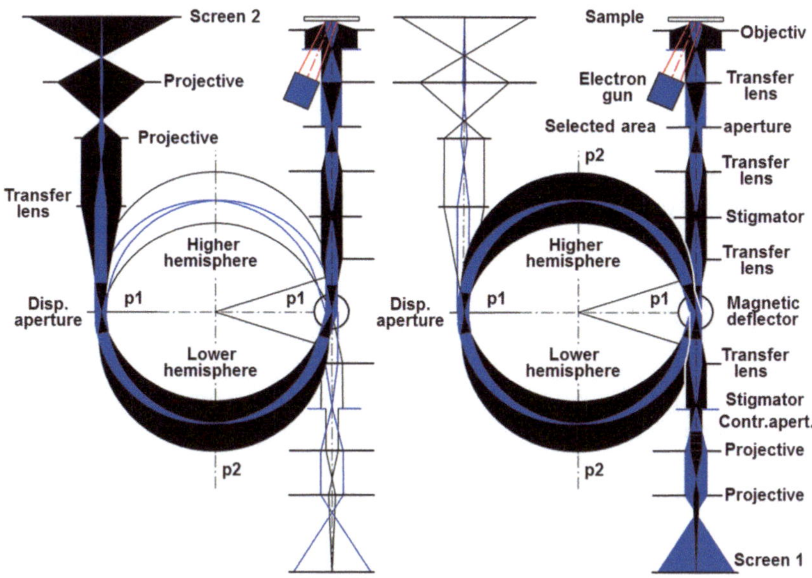

Fig. 3.44 Operation modes of the α-SDA (spherical deflector analyzer) in the flange-on dual emission electron microscope (DEEM). In contrast to the IDEA analyzer the α-SDA makes optimum use of the high symmetry of the central field of a 360° system. The electrons are fed tangentially into one hemisphere with a magnetic deflector and are energy-selected at its end with the dispersion aperture. An energy filtered reciprocal space (***k***-space) image is obtained when the second hemisphere is turned off, an energy-filtered real space image when the second hemisphere is turned on. Calculations show that the disturbance of the full symmetry by the feed-in cut-out, magnetic deflector and dispersion aperture cut-out do not reduce the resolution by more than 1 nm. Except for the magnetic deflector the system is purely electrostatic and allows fast switching between real and reciprocal image. Adapted with permission from Ref. [214]. Copyright 2013 Elsevier

with respect to the first one as indicated in Fig. 3.19 and inserting a transfer lens between the two filters [94, 96], (2) by not rotating the second 180° filter, cutting out a wedge from the resulting 360° filter and using magnetic deflection to feed the beam in and out of the analyzer [213, 214] as indicated in Fig. 3.44 [214]. The first design, called IDEA analyzer (Imaging Double Energy Analyzer) is used in the nanoESCA PEEM (Fig. 3.19) [96] with great success, the second one, called α-SDA (α-Spherical Deflection Analyzer) is still in the test phase [215]. The energy-selecting slit between the two hemispheres allows spectroscopic imaging, when it is removed then achromatic images with all transmitted electrons can be taken after the second hemisphere using the subsequent projective lens system. When in the IDEA analyzer the positions of image and diffraction planes in the analyzer are interchanged with the lenses before the analyzer and the energy slit is inserted, energy-selected reciprocal space images (LEED, PED, band structure patterns) can be acquired. In the α-SDA this is achieved by turning off the voltage on the second 180° filter, which allows very fast switching between real and reciprocal space imaging. In both instruments the central symmetry of the 360° system is reduced by the transfer lens and the cutout and magnetic deflector, respectively, so that the second-order aberrations are not completely eliminated. Nevertheless, the resolution deterioration caused by these effects is predicted to be only about 1 nm in the α-SDA. In the IDEA analyzer the aberrations of the transfer lens limit the resolution to about 40 nm [94, 96].

3.2.6 Wien Filters

Although Wien filters are not used extensively in microscopy with slow electrons their broad potential application range justifies a brief separate treatment. They have been suggested for beam separators, for aberration correctors, for energy filters and for spin manipulators and have actually been incorporated and used in some instruments as mentioned earlier. The most obvious use of a Wien filter is as a monochromator or an energy analyzer because only electrons with one selected velocity $\boldsymbol{v}$ are transmitted through it under the influence of the Lorentz force. Legler [216] and Boersch et al. [217] were the first to use it for this purpose and modern TEM and scanning TEM (STEM) instruments are equipped with a Wien monochromator in the illumination system to reduce chromatic aberration. In LEEM it would be useful too for the same purpose. Since Boersch's early work Wien filters have been studied extensively, in particular as energy filters. For a review see Ref. [218]. Here we discuss only the work relevant to its use for energy filtering of images. In the first studies of this kind octopole Wien filters were proposed [97, 98, 195]. These were found to have sixfold field components in the fringing region at the end of the filter. In 12-pole filters [191] the sixfold field components nearly vanish. In general the potentials generating the electric field components are applied to the magnetic poles, for example also in the 12-pole filter in Ref. [99]. Separating electric and magnetic poles, for example by using air-core coils outside the vacuum, eliminates insulation, cooling, and hysteresis problems and allows larger inside diameters. Also magnetic and electric fields can be shaped independently. N-pole Wien filters of this type have been calculated taking fringing fields into account and found to have minimum aberrations at $N = 12$ [196, 197]. An experimental realization of such a 12-pole filter with well-matched $\boldsymbol{E}$ and $\boldsymbol{B}$ fields has produced a resolution of about 1 μm in secondary emission with a 1.3 eV wide slit, similar to what was achieved with the earlier simple Wien filter [36], though with a much larger field of view (300 μm) [198]. The earlier filter had only two magnetic poles with superimposed electric poles and two electric poles as shown in Fig. 3.45a, b with an inside diameter of 10 mm and 40 mm length (a). The calculated electric and magnetic potential distribution is very symmetric over a large part of the cross section (b). Stigmatic imaging is achieved by superimposing on the magnetic poles potentials with the same magnitude but sign opposite to that on the electric poles. In the filter the energy of the electrons is reduced

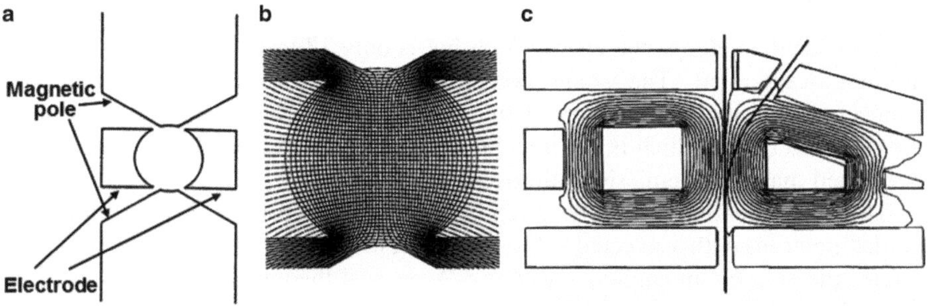

Fig. 3.45 Wien filter as energy filter (**a** and **b**) and beam separator (**c**) of the JEOL LEEM system [35]. (**a**) Cross section through the electrodes and poles, which form a cylinder with 10 mm diameter and 40 mm length. (**b**) Electric and magnetic potential distribution in the filter from simulations. (**c**) View along the optical axis. The outside electrodes are at ground potential, the inside at high voltage, which produces the electric potential contours shown. The magnetic poles are *above* and *below* the plane of the drawing. The cross section *perpendicular* to the plane of the drawing is similar to that shown in (**a**). The incident beam comes from the *upper right*. (**a** and **b**) Adapted from Ref. [36] with permission from World Scientific; (**c**) Reproduced with permission from Ref. [34]. Copyright 1994 Elsevier

from 10 keV to 200 eV, which results in a calculated resolution of 0.5 eV at a maximum angular aperture of 8 mrad [36]. However, the spectroscopic images and spectra produced with this instrument [200, 202, 203] were far inferior to those obtained 10 years earlier with a CHA [31, 88–90] Thus, much more development is necessary before the Wien filter can compete in spatial and energy resolution with the CHA. As far as energy resolution as function of dimensions and energy is concerned the two filters are theoretically comparable: $\Delta E = (w/2R)E_0$ for the CHA, $\Delta E = (\pi w/\sqrt{2}L)E_0$, for the CHA and the Wien filter, respectively, neglecting aperture aberrations; w is the energy selection width, R the radius of the center of the CHA, L the length of the Wien filter and E_0 the pass energy. A considerable amount of work has been done with the goal to reduce aberrations by combining two or even four [194, 218] Wien filters, which provide additional parameters and symmetries (for references see Ref. [218]), but no experimental data are available for them.

As far as the use of Wien filters for beam separation and correction of the aberrations of the objective lens is concerned, much less work has been done. Liu et al. [192] have calculated a double 12-pole Wien filter aberration corrector capable of correcting both C_c and C_s of a round lens as a demonstration of a new MEBS particle optics software package. Whether or not the C_c and C_s values needed for correction of cathode lenses can be achieved without introducing detrimental aberrations in the Wien filter remains to be seen. Beam separators have not only been calculated but also been built. They have been used to separate secondary electrons from primary electrons in SEM systems (for references see Ref. [218]). The JEOL LEEM/PEEM system uses a Wien filter for separating incident and reflected beam [34, 36]. Figure 3.45c shows a cross section of the beam separator of this instrument with the electric potential distribution, which is limited by field plates around the Wien filter in the center [34]. Stigmatic imaging is achieved in the same manner as in the energy filter of Fig. 3.45a, b. The filter has the same inside diameter as the energy filter but is only 18 mm long and the electrons are not decelerated. The slight potential asymmetry at the upper right side connected with the introduction of the incident beam causes some deflection of the reflected beam, which is detrimental for LEEM. Therefore, this instrument has been used mainly for emission microscopy.

The most effective use of the Wien filter is as a spin rotator. The Lorentz force guides electrons with a selected velocity $\boldsymbol{v}$ along its axis but its magnetic component rotates the spin by an angle $\alpha = eBL/mv$. A general theoretical treatment of Wien filters including spin-precession has been given by Scheinfein [107]. Here we discuss only some experimental realizations of such rotators. Kisker et al. [219] used the 2-pole Wien filter shown in Fig. 3.46a to rotate the spin 90° from the beam

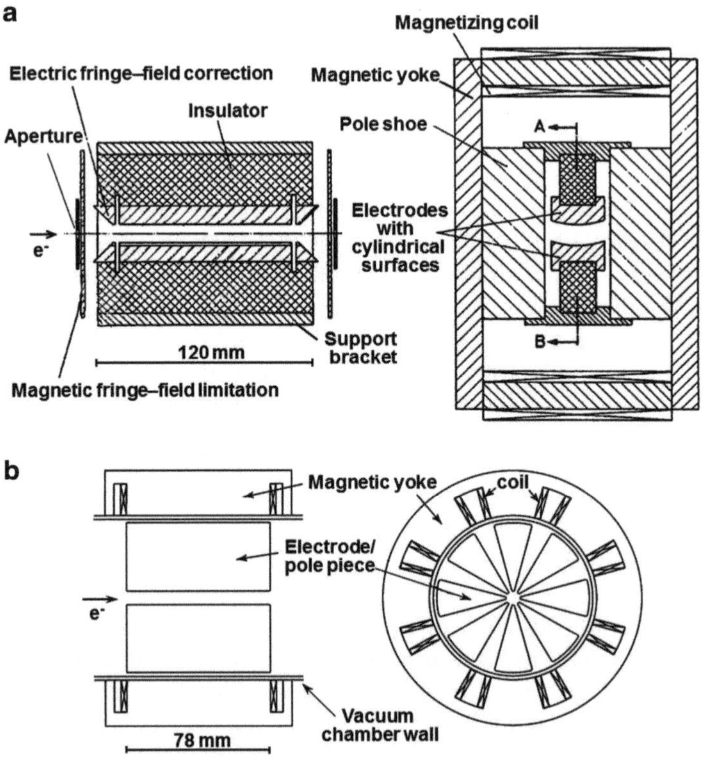

Fig. 3.46 Cross sections through spin rotation Wien filters parallel (*left* and perpendicular (*right*) to the beam direction. (**a**) Kisker's spin rotator for rotation of the spin from parallel to perpendicular to the optical axis. (**b**) 8-pole spin rotator of the spin-polarized LEEM system of Koshikawa et al. The magnetic poles inside the vacuum are at the same time electrodes, the exciting coils are outside the vacuum. The eight poles allow rotation of the spin of 20 keV electrons in any direction within the plane perpendicular to the initial axial spin direction by applying voltages $u = U\ \cos[(1+2n)\pi/8]$ and currents $i = I\ \sin[(1+2n)\pi/8]$ with $U = 3400$ V and $I = 260$ mA. (**a**) Adapted with permission from Ref. [219]. Copyright 1978 by the American Physical Society. (**b**) Courtesy of T. Koshikawa and T. Yasue

direction. Focusing in the plane perpendicular to the electric field was achieved not as in the energy filter and deflector shown in Fig. 3.44 but by curved pole faces of the electric poles as in Legler's energy filter [216]. A similar, more sophisticated 2-pole 90° spin rotator with a much larger diameter, based on three-dimensional computer simulation of the electric and magnetic fields, electron trajectories, and spin rotation, was built by Kohashi et al. for a spin-polarized SEM [108, 109]. A 2-pole 90° spin rotator was also used in a SPLEEM instrument, in which the spin gun of a commercial instrument was replaced by a spin gun with the high spin polarization and brightness mentioned in Sect. 3.1.5. In this 80 mm long rotator, which is briefly described in Ref. [67], the spin of 20 keV electrons is rotated from the propagation direction into the plane normal to it. Azimuthal rotation in this plane is subsequently done with the longitudinal field of the magnetic lenses of the commercial instrument. The magnetic field is created by a coil pair outside the vacuum. With a 4-pole Wien filter the spin can be rotated in only one direction. A 12-pole filter or octopole filter (Fig. 3.46b, T. Koshikawa, private communication) allows rotation into any direction by varying the excitation of the corresponding magnetic pole pairs, thus avoiding the necessity of a lens rotator. Such a spin rotator is already used in a SPLEEM instrument [220].

3.2.7 Photon Sources

In PEEM a wide range of photon energies is used, ranging from infrared to the hard X-ray range. In single-photon PEEM the photon energy has to be higher than the work function in the case of metals. In semiconductors and insulators it has to be higher than the electron affinity unless bulk doping or band bending, induced for example by $Cs + O_2$ adsorption, reduces the emission threshold. Except for the alkali, alkaline earth, and rare earth metals, at least 4 eV photons, that is UV light, is required for threshold photoemission. The Osram HBO 100 Hg Short-Arc Lamp, whose spectrum is shown in Fig. 3.47a [221] has been the workhorse in UV-PEEM for many decades. Its short small luminous area (0.25×0.25 mm^2) allows efficient specimen illumination using quartz lenses. The photon energy has been varied by (1) changing the focus [14] making use of the chromatic aberration of the quartz lens, whose refractive index varies from 1.550 at 200 nm to 1.488 at 300 nm, (2) with a monochromator (T. Koshikawa, private communication) or (3) with band pass filters at the expense of intensity. Although the most intense Hg lines in the UV are at 184.9 nm (6.7 eV) and 253.6 nm (4.9 eV), they are strongly suppressed by self-absorption so that 4.9 eV is in practice the maximum photon energy available. Higher energies can be achieved with deuterium lamps, which provide photons up to 10 eV in the MgF_2 window version of the lamp [222], though with low intensity.

Laser-driven discharge light sources such as the Energetiq LDLS EQ-99 lamp [223] produce a much more intense quasi-continuous spectrum in the energy range of interest for PEEM (Fig. 3.47b) [223]. In this source a laser beam is focused into

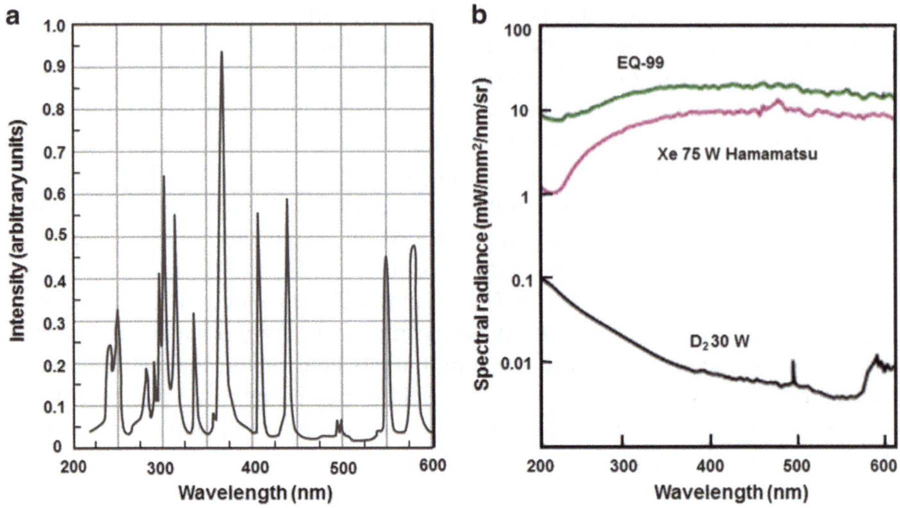

Fig. 3.47 UV light source characteristics. (**a**) Spectrum of the Osram 100B high-pressure Hg arc lamp. (**b**) Comparison of several other UV light sources. For explanation see text. Modified from Ref. [221] and with permission from Ref. [223]

the center of a fused silica bulb filled with high-pressure Xe gas, producing a plasma in a small region. The plasma is much hotter than in the short arc lamps and also much smaller (about 0.1 mm FWHM in diameter). This produces a much higher spectral radiance, reducing acquisition times of PEEM images from high electron affinity surfaces typically by a factor of 10 or allowing larger magnifications. In order to avoid absorption in air, the space between lamp and window has to be pumped or filled with a gas, which does not absorb light in this energy range and does not produce ozone. Furthermore it has to be kept in mind that the photoelectron spectrum will be quite broad when the work function or electron affinity is not close to the maximum photon energy, which will reduce the lateral resolution. However, if high resolution is not needed, photon energies a few eV above the work function/electron affinity are useful, because of the rapid increase of the photo yield above the threshold as illustrated in Fig. 3.48 [224]. The next step in photon energy is provided by the He gas discharge lamp which, however, has to be incorporated into the microscope with differential pumping. It has been developed over many years for electron spectroscopy and can be operated either in the HeI (21.6 eV) emission mode or also optimized for the much weaker HeII (40.6 eV) emission mode. It is commercially available from various sources, for example from Ref. [33]. Despite of its high spectral radiance it is used very little in PEEM, mainly because it is useful only in connection with an energy filter due to the large energy width of the emitted electrons.

The light sources discussed up to now, which make use of a gas discharge, provide continuous emission. Other sources for continuous wave emission are cw lasers.

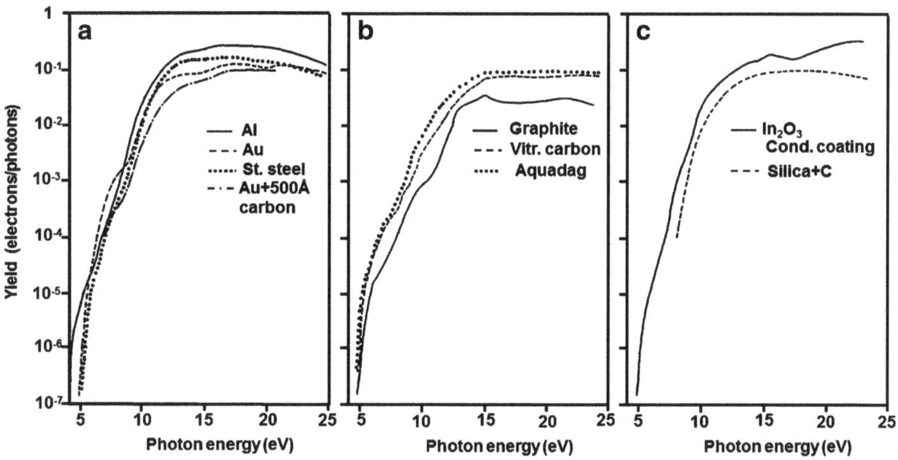

Fig. 3.48 Photoemission yield for various polycrystalline (**a**, **b**) and amorphous (**c**) materials exposed to atmosphere. Adapted from Ref. [224] with permission from the American Institute of Physics, © 1972

Most cw lasers have wavelengths in the infrared, visible, and near UV range but frequency-doubled Ar ion lasers provide photons with energies ranging from 4.66 to 5.08 eV. A frequency-doubled Ar ion laser with 257 nm wavelength (4.82 eV) was used successfully already in 1971 by Wegmann et al. [225]. Lasers with lower photon energies can be used too when the work function or affinity is reduced by suitable adsorbates such as Cs. In this manner photoelectrons can be excited from most specimens. Because of the continuous emission and the small angular aperture of the beam only a few mW are sufficient. An example of the application of such lasers is the UV-MCDPEEM study of Nakagawa et al. [226] with a 405 nm 1 mW laser. Suitable lasers are available from many companies.

Another group of UV light sources are pulsed lasers. They are used mainly in time-resolved studies in the pump-probe mode and provide flexibility in the choice of the photon energy via higher harmonic generation. Pulsed laser light sources were introduced in PEEM around 1980 by Jones et al. [227, 228]. They used a Q-switched Nd:YAG laser (1064 nm) and produced from it with optical nonlinear crystals ($LiIO_3$ and KD_2PO_4) second and fourth harmonic radiation corresponding to 2.33 and 4.66 eV. PEEM images were obtained not only with the 4.66 eV radiation but also with 2.33 eV, the latter case being the first demonstration of nonlinear or two-photon photoelectron PEEM (2PPEEM). Pulse length and repetition rate in these experiments were 100 ns and 1.5–4 kHz, respectively. A detailed study of the influence of the laser power on image resolution soon demonstrated the limitations of pulsed lasers in PEEM: space charge and other unexplained effects caused a dramatic decrease of resolution with increasing laser power [229]. This discouraged the use of these more expensive light sources compared to the various discharge lamps until strong interest in time-resolved PEEM and in the physics of multiphoton photoelectron PEEM (nPPEEM) developed around 2000.

The first attempt at time-resolved PEEM (TR-PEEM) with the pump-probe method used also a Q-switched, frequency-doubled Nd:YAG laser with a wavelength of 532 nm and 15 ns pulse width as a pump and a 248 nm KrF excimer laser with 4 ns pulse length as a probe to study the time dependence of melting, crystallization, and alloying of an Al film [230]. Space charge effects caused by the high power of the pulses reduced the lateral resolution so much that it discouraged this method. Interest in TR-PEEM was stimulated again by laterally-averaging time-resolved two photon photoemission (TR-2PPE) spectroscopy. These studies had shown phenomena, which were attributed to lateral variations of the emission process [231]. They used a Ti:Sapphire laser with 82 MHz repetition rate, 70–90 fs pulse width and 10 nJ pulse energy. It is tunable from 750 to 850 nm. Shorter wavelengths were obtained by frequency doubling and tripling. Due to the low pulse power, space charge processes are minimized and the high repetition rate provided enough average power for imaging.

Next, TR-PEEM imaging was attempted with a similar set-up: 40 fs pulses from a Ti:Sapphire laser served as pump pulses and pulses, frequency-doubled in a β-BaB_2O_4 (BBO) crystal to 3–3.4 eV, were used as probe pulses for imaging the decay of electron excitations in GaAs covered with Ag patterns [232]. The Ag pattern could not only be imaged with the frequency-doubled light via 2PPE but also with 3PPE from the fundamental frequency light (1.55 eV). The "hot spots" seen in the 3PPE image stimulated many subsequent studies of surface plasmon and polariton phenomena in Ag and Au nanostructures, which will be discussed in Sect. 6.2.

The Ti:Sapphire laser has become the most used pulsed laser. Its high repetition rate and low energy per pulse minimize space charge effects. It can be pumped and frequency-multiplied in a variety of ways. Usually it is operated between 75 and 90 MHz, mostly at 80 MHz, but much lower repetition rates such as 250 kHz [233] or a few Hz [82] have also been used. Frequency multiplication is usually achieved with nonlinear optical crystals such as BBO, sometimes in high-pressure noble gas cells (Xe [233]) or noble gas jets (Ne [82], Ar [234]). Pulse width and/or repetition rate are frequently changed with various methods; time-resolution is accomplished by beam splitters and adjustable optical path differences between the two resulting beams. A description of these methods is beyond the scope of this book and can be found in many elementary and advanced textbooks on laser physics or photonics. An example of such a laser source set-up for 2P-PEEM and TR-PEEM is shown in Fig. 3.49 [235] for illustration. In addition to the Ti:Sapphire laser several other lasers have been used, for example Nd:YAG lasers and a more complex laser set-up which provides a frequency-doubled and a frequency-tripled beam, which can be further doubled to reach photon energies as high as 5.6 eV with adjustable time difference for TR-PEEM [236].

The various laser sources cover a wide range of repetition rates up to nearly 100 MHz, photon energies $h\nu$ up to the 10 eV range via higher harmonic generation in gas jets, pulse energies up to about 100 nJ and pulse durations τ down to about 100 attoseconds. These four features can, however, combined only in a limited parameter range, because the time-averaged photon flux density on the specimen

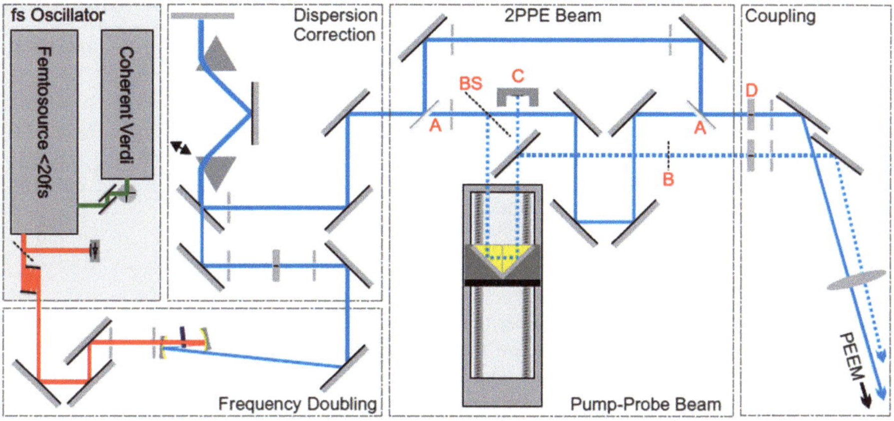

Fig. 3.49 Example of a laser source for time-resolved and multiphoton PEEM. From *left* to *right*: a Ti:sapphire laser pumped by a 532 nm cw laser produces 20 fs linearly polarized 800 nm pulses, which are frequency-doubled in a nonlinear optical crystal. The many optical components in the system cause a broadening of the 20 fs pulses, which is pre-compensated with prisms. The resulting beam is then either used directly for PEEM (2PPE Beam) or for pump-probe experiments. In the second case a beam splitter BS divides the beam into the pump pulse (*solid line*) and via a linear positioning stage, whose position determines the delay time, into the probe pulse (*dashed line*). Two $\lambda/2$ plates D are used to adjust the polarization of the two beams. A CCD camera (C) monitors the beam stability during changes of the delay time. Reproduced with permission from Ref. [235]. Copyright 2007 Elsevier

must be high enough for focusing. At the same time the flux density in the pulse must be low enough so that space charge effects do not limit the desired resolution. Thus the repetition rate should be high and the power per pulse low. This is the case with the 80 MHz laser set-ups with pulse durations between about 10 and 100 fs. Problems still can arise, because of the high intensity of the pulses which results from the fact that frequency multiplication in the dispersive nonlinear media requires high electric field amplitudes, that is high intensity of the input power. This is already evident at 80 MHz where an increase of the irradiation power density at the specimen from 0.6 to 2 W/cm^2 increased a magnetic domain wall width observed in a simple PEEM instrument (Elmitec PEEM Spector) from 0.5 to 1.5 μm, followed by a linear increase of nearly 0.3 μm/W cm^2 (Fig. 3c in [226]). Another illustration of the influence of the photon flux density on energy resolution can be found in Ref. [237].

The influence of space charge effects becomes even more evident at lower repetition rates, for example at 250 kHz were serious image deterioration was observed already at about 4×10^{-2} W/cm^2 in images taken with 140 fs, 200 nm pulses (Fig. 4f in [238]). In imaging at 1 kHz with pulse bunches consisting of 200 as pulses an even lower limit for the onset of space charge effects was found: a pulse charge corresponding to about 2 e/μm^2 [234]. Below this limit the resolution was still by a factor of 4–6 worse than with imaging with a Hg lamp and the image acquisition time had to be increased by a factor of 2–4. The resolution decrease

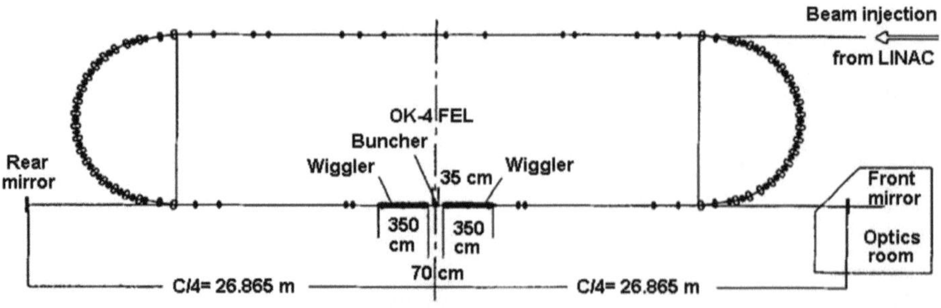

Fig. 3.50 Example of a free electron laser (FEL) source (DUKE University FEL). The electrons are injected from a linear accelerator into the storage ring. The wigglers produce spontaneous radiation along the straight section of the ring, which is also part of the laser cavity with multilayer mirrors. The spontaneous radiation can be varied from 0.6 eV to 1 keV by changing the energy of the electron beam from 0.25 to 1 GeV and/or the field strength of the wigglers. The PEEM system is located at the end of the cavity in the optics room but was used only in the spontaneous radiation mode up to 6 eV with a line width $\Delta E/E$ of about 3 %. Reproduced with permission from Ref. [239]. Copyright 1996 Elsevier

below the space charge limit is a consequence of the uncertainty principle $\Delta\nu\Delta t \approx 1/2\pi$, which leads to an energy width of the pulses in the 10 eV range. The chromatic aberration of the objective lens causes then a corresponding loss of resolution. Summarizing the experience in PEEM with pulsed lasers, their use seems appropriate only for time-resolved studies with pump-probe techniques. When time resolution is not needed and the work function or electron affinity is low enough, which can be achieved with Cs deposition, then CW lasers are sufficient. When no high magnification, high resolution, or fast image acquisition is necessary, the cheaper conventional light sources suffice.

The next step towards higher photon energies in the vacuum ultraviolet (VUV) is enabled by the free electron laser (FEL). FELs can be tuned over a wide energy range, which is determined by their design, and produce highly monochromatic light in the laser mode of operation. FELs are devices in which high energy electron pulses are injected into a device, which causes oscillations of the pulses (wiggler or undulator). This produces radiation, which is then either amplified in the device by self-amplification of the electron-beam spontaneous emission (SASE) or by injection of light from a conventional laser ("seeding"), which is frequency-multiplied by interaction with the oscillator light. We consider here briefly two FEL sources, which have been used in PEEM, a SASE source [239, 240] and a seeded source [241, 242]. The first source, which is shown in Fig. 3.50 [239], is designed for a photon energy range from 3.1 to 24.8 eV, pulse lengths from 3 to 30 ps and a pulse repetition rate of 2.8 MHz. In the PEEM studies the FEL was used below the lasing threshold in the spontaneous mode—because of the space charge effects caused by the high intensity in the lasing mode—with photon energies between 4 and 7 eV and a line width of about 0.13 eV. In the second source, which is connected to a synchrotron radiation source, the photon energy, pulse repetition rate, and length are determined by the seeding laser, whose frequency determines the harmonics produced in the FEL in a process described in Ref. [242]. In the test experiments light from a Ti:sapphire laser (794 nm, 1 kHz repetition rate) was frequency-doubled in a nonlinear crystal and injected into the modulator, which produced 100 fs third harmonics (132 nm) of these pulses. The problems of this light source for PEEM are obvious: the repetition rate is low, the FEL light pulses are superimposed on the synchrotron radiation pulses so that the same problem arises, which was already discussed in connection with pulsed laser illumination. Indeed, space charge effects caused a strong smearing of the energy spectrum [243] and image quality was far below that achieved with conventional and synchrotron radiation illumination (Locatelli, private communication). With the SASE-FEL in the spontaneous emission mode the intensity in the VUV range still was intense enough for imaging with a field of view of 1.5 μm with acceptable image acquisition times [244]. This allowed imaging and spectroscopy of larger molecules such as fibrinogen via changing the photon energy. Thus, when the lasing action with the concomitant high space charge-creating intensity is not used, FEL lasers are useful for imaging and spectroscopy in the VUV.

At photon energies above about 10 eV synchrotron radiation is the standard light source for PEEM, He lamps excepted. In the early years of synchrotron radiation-excited PEEM (XPEEM) bending magnets or wigglers produced the light but most PEEM instruments are now connected to undulators in third generation synchrotron light sources. The physics and applications of synchrotron light are described in many books, for example in Refs. [245, 246]. Here we consider only briefly undulators, which are straight insertion devices in synchrotron radiation rings. They consist of a periodic array of dipole magnets with alternating field direction normal to the plane of the ring [247]. These superimpose oscillations on the linear motion of the electron beam and thus radiation in a narrow cone in the forward direction because of the relativistic speed v of the electrons. The white radiation emitted from equivalent points in the periodic array interferes with each other like in a diffraction grating, which produces periodic intensity maxima at wavelengths given by

$$\lambda = \frac{\lambda_u}{2n\gamma^2}\left(1 + \frac{K^2}{2} + \theta^2\gamma^2\right) \tag{3.7}$$

where λ_u is the period of the magnet array, n the order of the interference (harmonic), $\gamma = [1 - (v/c)^2]^{1/2}$ the relativistic Lorentz factor, K a characteristic undulator parameter, which is proportional to the maximum field strength B_0 on the axis and to λ_u, and θ is the emission angle. The intensity distribution around the intensity maxima is given by the usual interference expression for a grating with N periods

$$I(N, \nu, \nu_1) = \frac{\sin^2\left(\frac{N\pi\nu}{\nu_1}\right)}{\sin^2\left(\frac{\pi\nu}{\nu_1}\right)}. \tag{3.8}$$

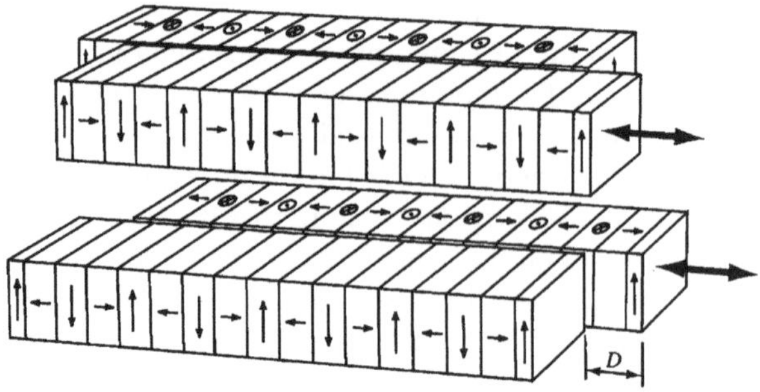

Fig. 3.51 Dipole magnet arrangement in an APPLE type undulator. The *arrows* indicate the direction of the magnetic field. Reproduced from Ref. [247] with permission of Oxford University Press

This function has the maximum value N^2 when $\nu = n\nu_1$, ν_1 being the fundamental frequency (first harmonic). The FWHM of the Gaussian-like peaks centered at these nth harmonics is $\Delta\nu_n = \nu_n/nN$.

Several undulator designs are used in the various synchrotron light source facilities, providing high-intensity light from about 10 eV to several keV, which is used for XPEEM. Here we use as example the undulator and light-processing beamline of the Elettra synchrotron radiation facility, which is the first one used with a SPELEEM. The light-producing insertion device in the synchrotron ring consists of two so-called Apple II undulators (Fig. 3.51 [247]), each consisting of $N = 20$ dipole magnet pairs with a period $\lambda_u = 10$ cm. The wavelength of the emitted light is controlled by the distance of the magnet poles, the polarization by the relative position of the magnet pairs as indicated in Fig. 3.50 by the arrows. A phase modulation electromagnet between the two undulators allows fast shifting of the phase of the light from the first undulator relative to that of the second at fixed undulator setting for changing the polarization, for example from linear to circular polarization. With 2 GeV electrons in the ring, $\lambda_u = 10$ cm and fields strengths from 0.1 to 1.02 T the wavelength of the first harmonic λ_1 can be varied from 151.33 to 4.68 nm, corresponding to energies $h\nu$ between 8.2 and 264 eV according to Eq. (3.7).

The SPELEEM instrument in Elettra is connected to the undulator with the beamline shown in Fig. 3.52 ([248] and Courtesy of A. Locatelli). Ten meter from the undulator a pin hole PH selects from the highly directional beam a small coherent fraction, which is focused by a toroidal mirror TM onto to the horizontal and vertical entrance slits EnS. The beam emanating from the entrance slit, which is now the source for the rest of the beamline, is reflected by the mirror MP onto the plane variable spacing line (VSL) gratings of the monochromator MG 21 m away from the source. Beyond the exit slit ExS two bendable elliptical mirrors VR and HR in the so-called Kirkpatrick-Baez configuration at 31.3 and 31.75 m from the source finally focus the beam vertically and horizontally onto the specimen at about 33 m from the source. With a 10 μm wide exit slit the illuminated region of about 20 μm × 2 μm in the SPELEEM is elongated because of the large angle of incidence (16° glancing angle). A 400 line/mm grating (VLS400) allows selection of photon energies from 100 to 1000 eV with a resolving power $E/\Delta E$ of 4000 at 400 eV,

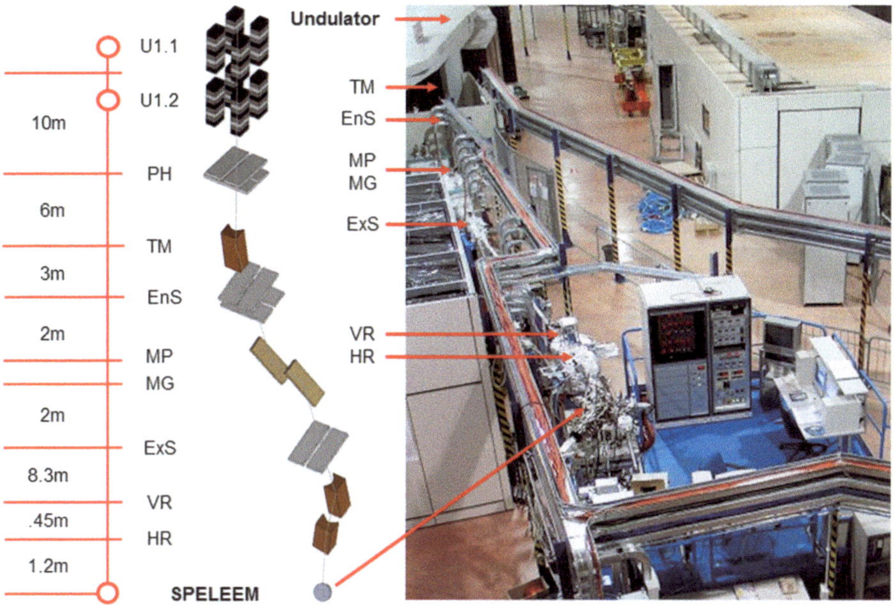

Fig. 3.52 Example of a synchrotron radiation beamline connected to a spectroscopic photoemission and low energy electron microscopy (SPELEEM) end station (Elettra Nanospectroscopy beamline). For explanation see text. Courtesy A. Locatelli and from Ref. [248] with permission

a 200 line/mm grating (VLS200) photon energies from 20 to 250 eV with a resolving power of 4000 at 200 eV, an additional seldom used spherical grating photon energies from 5 to 80 eV. Figure 3.52 shows also a bird's eye view of the beamline and the SPELEEM. For more detailed information see Refs. [249, 250]. The intensity of the light depends on the selected grating and harmonic as well as on the polarizations as shown for typical settings in Fig. 3.53 [250]. The indicated intensities result from the time-averaged radiation from 432 electron bunches 42 ps long, which circulate in the storage ring at a frequency of 1.157 MHz. A photon flux of about 10^{13} photons/s in Fig. 3.53 is thus concentrated in 5×10^8 pulses so that each 42 ps pulse contains 20000 photons. This leads to the similar space charge problems as in pulsed lasers [251], which, however, can be avoided by reducing the exit slit width and/or detuning the undulator. Of course, the reduced intensity requires then correspondingly longer image acquisition times, which depend on magnification, on the photo ionization cross section of the material studied and whether photoelectrons or secondary electrons are used for imaging. Fractions of a second to tens of seconds are common for materials with high ionization cross sections but significantly longer acquisition times have to be accepted when these cross sections are small. The high intensity, the ability to change the photon energy over a wide energy range with highly monochromatic light and the availability of any desired polarization state have made undulators the most important light sources for PEEM.

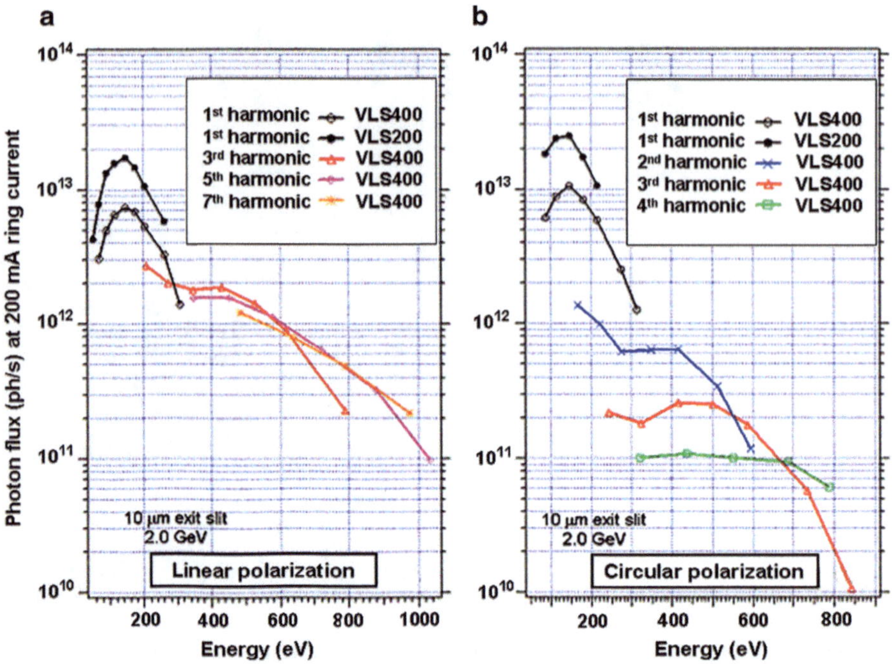

Fig. 3.53 Example of the photon flux in a synchrotron radiation beamline (Elettra Nanospectroscopy beamline) as a function of photon energy for two monochromator gratings and several harmonics in linear (**a**) and circular (**b**) polarization. The flux is measured after the exit slit ExS in Fig. 3.52, at the sample it is lower by a factor of 0.7. From Ref. [250]

While photon sources are by far the most important excitation sources for emission microscopy a few other sources should be mentioned briefly. Ion, fast atom, and fast electron sources have been used in the early years of surface microscopy with slow electrons as mentioned in Chap. 1. They are not used any longer because the wide energy distribution of the secondary electrons, which they produce, leads to poor spatial resolution unless a band pass filter is used. Also the UHV in today's instruments obviates the need of keeping the surface clean by ion or fast atom bombardment, which was the original motivation. One more recent method worthwhile mentioning is metastable He impact deexcitation electron emission microscopy, MIEEM, also abbreviated in the literature by the acronyms MIEM, MEEM. MIEEM is based on the fact that the potential energy of the 1s2s 2^3S He state He*—in which the two electrons have parallel spin—is 19.82 eV above the ground state and transferred to the specimen upon impact. In the deexcitation process an electron is emitted by one of two processes, the resonance ionization-Auger neutralization (RI-AN) process, which involves two electrons of the specimen and is dominant in metals, and the Auger neutralization (AN) or Penning ionization (PI) process, which involves only one electron of the specimen and is dominant in insulators.

The energy distribution of the emitted electrons reflects in the first case to a first approximation the self-convolution of the density of states in the object, in the second case it gives information similar to photoelectron spectroscopy [252, 253]. In both cases the information provided by the spectrum can be obtained easier otherwise. What makes this method useful is the large extension of the 2s wave function of the 2^3S state so that the interaction occurs with the tails of the wave function above the surface, in particular with the wave functions of the topmost atoms of molecules. Because of the high energy available for emission the spectrum is broad. Therefore imaging without energy selection gives poor spatial resolution and—most importantly—loses the information unique to MIEEM about the outermost atoms. Comparisons between unfiltered MIEEM, MEM, LEEM, and PEEM [254–256] show that the intensity contrast seen in MIEEM does not give information which cannot be obtained with the other methods. This is even true for energy-filtered images and local spectra of surface features, which do not require the high surface sensitivity of MIEEM [257–260].

However, for molecular systems, which are damaged by other surface probes, the situation is different. Examples are organometallic molecules on oxides used in solar cells, surfactants, organic electronics, Langmuir-Blodgett films, to name a few systems. Here spectroscopic MIEEM can make valuable contributions with high-intensity He* sources (G. Lilienkamp, private communication), which will allow image and local spectra acquisition in acceptable times. With further source improvement MIEEM could become a very useful surface imaging method.

3.2.8 Electron Sources

The electron sources used in microscopy with slow reflected electrons (LEEM, MEM) are based on a variety of emission modes: thermionic emission (TE), field emission (FE), photoemission (PE), and combinations of them such as thermal field emission or photo field emission. The electron source, together with the electron optics of the illumination system, must produce in the back focal plane of the objective lens a sharp focus with high current density to obtain parallel illumination of the specimen. This requires a high current per unit area and angular range, that is a high brightness

$$B = \frac{\mathrm{d}I}{\mathrm{d}\Omega \mathrm{d}A} = \frac{I}{\pi\theta_0^2 r_0^2 \pi}\left[A/\mathrm{srd}\cdot\mathrm{m}^2\right], \tag{3.9}$$

where r_0 is the radius of the minimum beam cross section in the gun and θ_0 the corresponding half-angle of the emission cone. The source should also have a narrow energy width to minimize the influence of the spherical aberration of the objective lens. Cold tungsten field emitters (CFE) have been for a long time the sources with the highest B values. In the early years of the development of LEEM starting in the 1960s the author believed that only these sources could provide the

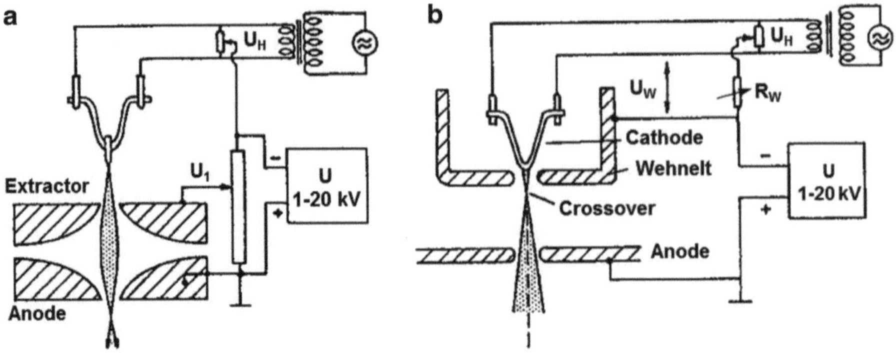

Fig. 3.54 Electron guns used in LEEM and MEM systems. (**a**) Field emission gun, (**b**) thermionic emission gun. For explanation see text. Adapted from Ref. [261] with permission from Springer Science + Business Media

necessary illumination conditions. At that time metal UHV technology was still not at a stage to produce sufficiently low pressures for reliable operation of field emitters. As a consequence their lifetime was extremely short due to bombardment of ions generated in the residual gas and by electron-stimulated desorption from the electrodes near the emitter tips. Lifetimes long enough to allow imaging for a limited time were not reached until the early 1980s, leading to the first LEEM images. Since then CFE gun development has reached a high level of sophistication, mainly driven by the needs of high resolution TEM and SEM instruments, most of which are equipped now with these guns, as is also one of the LEEM instruments mentioned earlier [32, 33, 66]. In CFE guns the emitter is a thin (310) oriented W single crystal wire, which has been electrochemically etched to form a sharp tip with a radius of curvature r of about 100 nm. It is spot welded to a W hair pin and the tip of the wire is a few mm behind the extractor electrode, which is at a lower potential than the anode (Fig. 3.54a [261]). The high field at the tip causes field emission, which seems to come from a virtual source behind the tip similar to the situation in LEEM, only with a much smaller size r_v than in LEEM, because of the large curvature of the tip. Typical emission properties of the CFE source are shown in Table 3.2 together with those of other sources used in LEEM instruments.

The normalized or reduced brightness $B_r = B/V$ (V = extraction voltage) is a measure of the quality of the source for imaging but the numbers have to be taken with caution because they depend strongly on the operation conditions as we will see below. When energy width ΔE is of importance—as it is for high resolution in microscopes without aberration correction and illumination monochromator—a better quality factor is the specific brightness $B_s = B/\Delta E$. Judged by the B_s value, the CFE source is far superior to the other sources at the low currents needed in LEEM and MEM. An alternative to the W(310) CFE source is the HfC (310) CFE source with similar tip radius. It is clamped between small pyrolytic blocks, which are pressed against the HfC single crystal wire by stiff refractory wires ("Vogel mount", Fig. 3.55a [263, 264]). The (310) surface of HfC has a

Table 3.2 Typical electron source performance comparison

Emitter type	Thermionic	Schottky FE	Cold FE
Cathode material	LaB_6	ZrO/W(100)	W(310)
Operating temperature (K)	1900	1800	300
Cathode radius r (nm)	10000	<1000	<100
Virtual source radius r_v (nm)	5000	15	2.5
Emission current density (A/cm^2)	30	5300	17000
Total emission current (μA)	80	200	5
Reduced brightness (A/cm^2 sr kV)	1×10^5	1×10^7	2×10^7
Maximum probe current (nA)	1000	10–100	0.2
Energy spread @ cathode (eV)	0.40	0.31	0.26
Energy spread @ gun exit (eV)	1.3–2.5	0.35–0.7	0.3–0.7
Beam noise (%)	1	1	5–10
Emission current drift (%/h)	0.2	<0.5	5
Operating vacuum (hPa/mbar)	$<1 \times 10^{-6}$	$<1 \times 10^{-9}$	$<1 \times 10^{-10}$
Typical Cathode life (h)	>1000	>5000	>2000
Cathode regeneration	Not required	Not required	6–8 h
Sensitivity to external influence	Minimal	Low	High

From Ref. [262] with permission (http://www.tedpella.com)

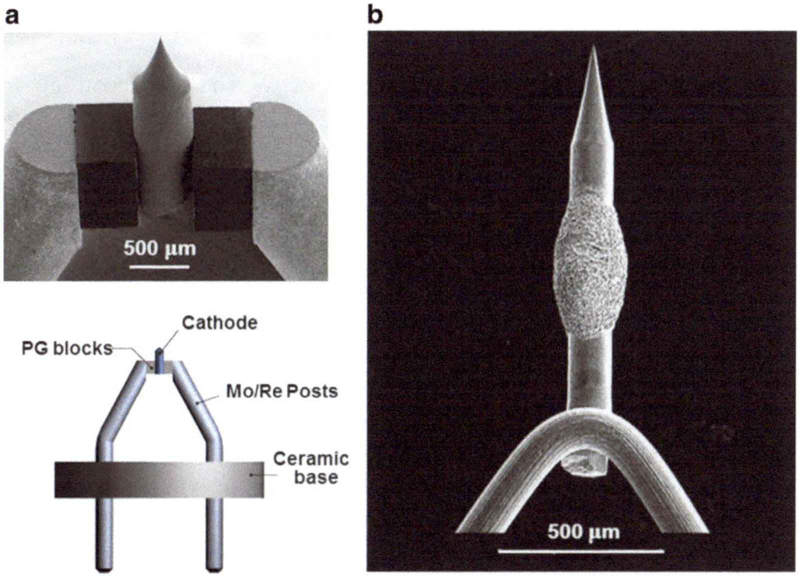

Fig. 3.55 Electron emitters. (**a**) Vogel-mounted cathodes. *Top*: HfC field emitter, *bottom*: LaB_6 thermionic emitter. (**b**) Zirconia-coated W(100) Schottky cathode. The bulge on the wire is the zirconia reservoir. (**a**) From Ref. [263] with permission from AP Tech and from Ref. [264] with permission from the American Institute of Physics, © 2011. (**b**) Reproduced from Ref. [269] by permission of Taylor and Francis Group, LLC

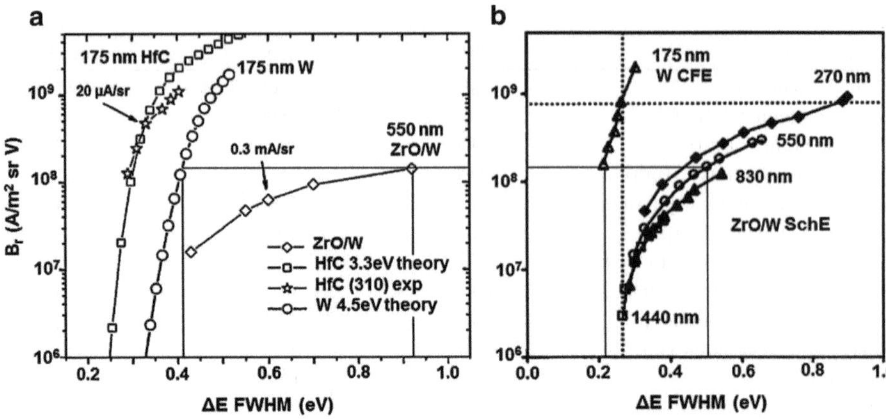

Fig. 3.56 Relationship between reduced brightness B_r and full width at half maximum ΔE of the energy distribution of several electron sources. (**a**) Comparison of HfC and W CFE sources with a ZrO/W Schottky emission (SchE) source. (**b**) Comparison of W CFE and SchE sources with different radii. The *horizontal* and *vertical solid lines* are drawn for easy comparison of the two data sets, showing differences of nearly a factor of 2 in ΔE between them at the same value of B_r. The *dotted lines* in (**b**) indicate the large difference between CFE and SchE source for constant B_r and ΔE. (**a**) Adapted from Ref. [265] with permission from the American Institute of Physics, © 2008. (**b**) Adapted from Ref. [270] with permission from the American Institute of Physics, © 2006

somewhat lower work function Φ than that of W so that according to the Fowler-Nordheim equation

$$J = \frac{e^3}{8\pi h}\frac{F^2}{\Phi}\exp\left[-\frac{8\pi}{3eh}\frac{\left(2m\Phi^3\right)^{1/2}}{F}\right] \tag{3.10}$$

the emission current J is higher under otherwise identical operation conditions as seen in Fig. 3.56a [265]. It is more robust than the W(310) emitter but nevertheless extremely good vacuum is necessary for stable operation and periodic refreshing by flashing as in the case of the W(310) emitter. It should be noted that the form (Eq. 3.10) of the FN equation is highly simplified [266] and that it becomes invalid at high current densities due to space charge effects [267, 268].

The next important source is the Schottky emitter, in which the applied field effectively reduces the work function according to the Schottky equation (Eq. (2.1)) for the emission current J for low field strength. With increasing field strength tunneling becomes more important and considerable deviation from this equation occur (extended Schottky emission) [269].

A more frequently used source is a (100)-oriented W wire with a sharp tip (radius r several 100 to 1000 nm) which is covered with a refractory oxide of an electropositive metal, usually Zr but also Y and Sc. The oxide is applied on the shank of the wire which is welded on a W hair pin (Fig. 3.55b [269]). The emitter is operated at a temperature, at which the oxide reservoir on the shank continuously replenishes the oxide coverage on the tip by surface diffusion. The oxide reduces the work function of the (100) surface to about 3 eV for Zr and to somewhat less for Sc. The tip is slightly behind the negatively biased electrode ("Wehnelt"), which controls the field at the tip. Typical operation data are shown in Table 3.2. The ZrO/W(100) SE source is discussed in detail in Ref. [269] and compared in Fig. 3.56b with the W(310) CFE source [270]. It is evident that its reduced brightness for a given energy width is much lower than that of the W CFE source and that at the same brightness the energy width is much larger than that of the W CFE source. A comparison of Fig. 3.56b with Fig. 3.56a illustrates why the published data have to be taken with caution because of different operation conditions: at a reduced brightness of 1.5×10^8 A/cm^2 sr V and a tip radius of 550 nm, the energy widths differ by a factor of nearly 2.

The most frequently used sources are LaB_6 or CeB_6 thermionic emitters whose emission current density is given by Eq. (3.12) for $F = 0$ (Richardson equation). These emitters are [100] orientated single crystal rods with a flat top, typically 10–20 μm wide on the conical end. The low work function of the (100) surface (2.70 eV for LaB_6, 2.65 eV for CeB_6) allows operation already at 1800 K and even lower temperatures. In order to suppress emission from the side faces of the cone the tip is located somewhat behind a control electrode (Wehnelt cylinder) (Fig. 3.54b [261]), which is somewhat negative with respect to the cathode, either biased via the emission current or separately controlled. Very high currents can be drawn, limited by the accompanying strong space charge, which cause the large energy width shown in Table 3.2. However, by operating at lower temperatures and emission currents, energy widths as low as 0.5 eV can be achieved. The large emission area requires demagnification of the source to be useful in LEEM. As a consequence the current on the specimen is significantly lower than the emission current, typically only 10 % or less of it. Because of the large volume of this cathode, its lifetime is nearly infinite, when used properly. This was the reason that Veneklasen used it—based on his long experience with SEM instrumentation—in the design of the SPELEEM to replace the CFE and Schottky emission (SchE) cathodes, which had caused so much trouble in the first LEEM instrument. For multiuser-dedicated instruments, for example in synchrotron radiation facilities, this is probably the best cathode, despite of its larger energy width, because it can tolerate relatively high pressures. More information on these cathodes can be found in Ref. [264] and the data sheets of the producers and distributors.

In addition to these proven and commercially available sources there are other promising emitters, once they can be produced reproducibly in an affordable manner: carbon nanotube and W nanofacet emitters. Both of them are field emission sources and have been around for some time but are still in the development phase. The carbon nanotube emitter is a multiwall nanotube with closed endcap, mounted on a W support tip similar to the W(310) CFE emitter. The effective emission area is an order of magnitude smaller than the diameter of the nanotube (5–30 nm). This leads to an order of magnitude higher reduced brightness compared to the W(310) CFE, with only slightly larger energy width, and to a much smaller

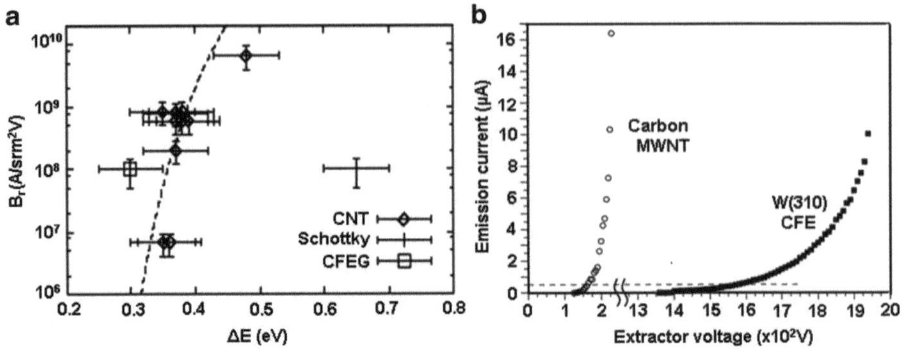

Fig. 3.57 Comparison of a carbon nanotube (CNT) emitter with a W CFE emitter. (**a**) B_r–ΔE relationship. The CNT and CFE data are comparable, the SchE data as seen already in Fig. 3.56 are shifted to much larger ΔE. (**b**) Emission current vs. extraction voltage. The *dashed line* defines an emission onset value, which is about a factor 10 times higher for the CFE source compared to the CNT source. (**a**) Reproduced from Ref. [271] with permission from Wiley-VCH Verlag GmbH & Co. KGaA, Weinheim. (**b**) Reproduced with permission from Ref. [272]. Copyright 2009 Elsevier

energy width than that of a Schottky emitter for comparable brightness (Fig. 3.57a [271]). Also the emission onset voltage is much smaller than that of a W(310) CFE emitter (Fig. 3.57b [272]). Furthermore, the stability is much better than that of the W(310) CFE, with a current noise of only about 0.1 % or less, independent of current from 100 to 1000 nA and of pressure up to 10^{-8} mbar [273]. Thus multiwall carbon nanotube emitters appear as ideal sources once the fabrication problems are solved. A review of these sources can be found on Ref. [274].

A second promising FE source is the nanopyramid W(111) emitter. It makes use of the fact that when the W(111) surface, covered with a monolayer of Au, Pd, Pt, Ir, Rh, or similar metals, is annealed at high temperature it facets into three {211} faces. Depending upon tip-forming and operation conditions, emission from the three {211} facets or from the top atom of the pyramid is dominating. The emission is concentrated within a cone of a few degrees around the tip axis. The resulting reduced brightness is two to three orders of magnitude higher than that of the CFE and SchE emitters discussed above [275, 276]. B_r values of 10^9 to 10^{11} A/m^2 sr V have been reported [277]. The vacuum requirements are the same as for the CFE source, the stability is superior to it, the metal overlayer provides some inertness to residual gases and the source is self-repairing as long as there is enough metal coverage. Lost metal can be replaced by surface diffusion by coating the shank with the metal that causes the faceting, similar to the method used for the ZrO/W(100) SE sources, or possibly by using alloy wires. Simple fabrication methods have been developed. One of them is illustrated in Fig. 3.58a [278]. The resulting tip shape is schematically shown in Fig. 3.58b. Other methods are described in Refs. [279, 280] where also more references can be found. Considering all these aspects, the nanopyramid W(111) FE source promises to become the preferred electron source

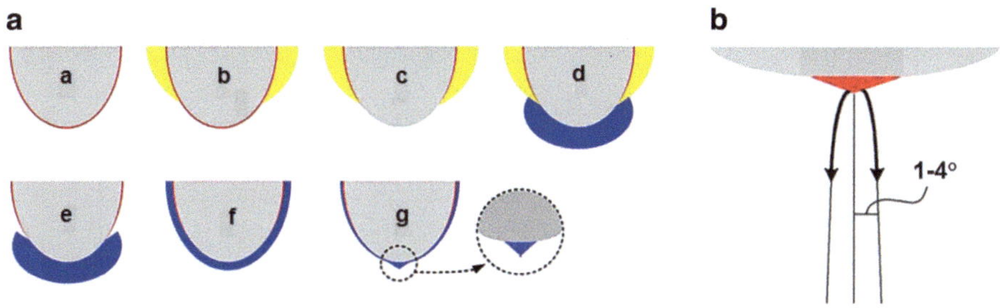

Fig. 3.58 The metal-covered W(111) single atom electron emitter. (**a**) Schematic of the various steps of one the preparation procedures. (*a*) Etch tip, (*b*) cover the tip with nail polish except for the apex, (*c*) reduce native oxide by cathodic polarization, (*d*) electroplate a noble-metal film on the tip under the condition of cathodic polarization, (*e*) remove the nail polish and transfer the tip into the vacuum, (*f*) anneal the tip to diffuse noble-metal, (*g*) field build-up the pyramidal metal emitter. (**b**) More realistic final emitter configuration and angular distribution of the emission. Reproduced from Ref. [278] by permission of The Electrochemical Society

in electron microscopy, at least in instruments with extreme high vacuum (XHV) electron gun chambers.

Before turning to photoemission electron sources a special W FE source should be mentioned yet, the Fe- and Co-coated W(100) and (111) emitter of spin-polarized electrons [281, 282]. In these emitters the tip diameter has to be large enough, so that layer is ferromagnetic at 300 K, the desired operation temperature, which requires a diameter in the 10 nm range. Six to ten pseudomorphic monolayers produce a polarization of the emitted electrons between 10 and 45 %. The polarization direction is parallel to the surface and stable at room temperature, pointing in one of the W⟨001⟩ directions in the case of Fe layers, in the ⟨011⟩ directions in the case of Co layers on the W(100) surface. The direction can be switched by brief heating above the Curie temperature, which in this thickness range is a few hundred degrees. This source like the first spin-polarized FE source, the EuS-W emitter [219], whose Curie temperature is 17 K, could be used in principle in SPLEEM, but it is not competitive with highly Zn-doped p type III–V semiconductor photocathodes. These were developed already in the middle 1970s, a development which has been well chronicled by one of its pioneers, Pierce [283], who gives also a detailed description of the physics, preparation, and application.

Photoemission sources of spin-polarized electrons are based on transfer of photon angular momentum of circular polarized light to electron angular momentum (spin, with opposite sign). Here we discuss only briefly the very basics and then the present state of the art. III–V semiconductors such as GaAs, InAs, GaP, and their alloys have a direct band gap between the $^2S_{1/2}$ conduction band and the $^2P_{3/2}$ valence band. In GaAs, for example [283], the conduction band is made up of Ga 4s electrons, the valence band of As 4p electrons, whose levels are split by spin–orbit interaction into $4p_{3/2}$ level and 0.34 eV lower lying $4p_{1/2}$ levels (Fig. 3.59a). Light with the energy of the band gap $E_g = 1.52$ eV can only excite transitions from the fourfold degenerate $4p_{3/2}$ state ($m_j = -3/2, -1/2, +1/2, +3/2$) to the $4s_{1/2}$ states

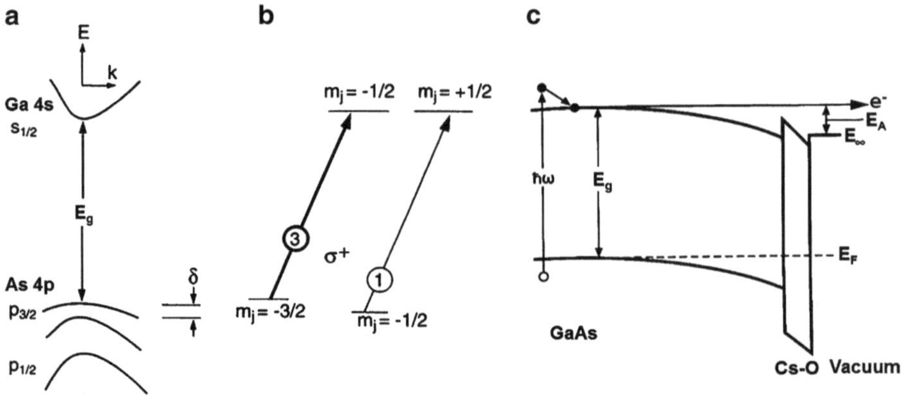

Fig. 3.59 Principle of spin-polarized photoelectron emission from GaAs and other III–V compounds. (**a**) Band structure of GaAs(100) around the Γ point for the case that the degeneracy of the four As $4p_{3/2}$ levels is lifted. When $\delta = 0$ transitions from all four sublevels are possible as indicated in (**b**) for right-circular polarized light; for left-circular polarized light the scheme in (**b**) is mirror-like. When $\delta \neq 0$ then the $m_j = -\frac{1}{2}$ level is about 0.1 eV below the $m_j = -3/2$ level so band gap light cannot excite transitions from it. (**c**) Schematic of the excitation into the conduction band and of the emission process enabled by the reduction of the electron affinity caused by a Cs-O overlayer. Adapted with permission from Ref. [283]. Copyright 1995 Elsevier

($m_j = -1/2$, $+1/2$). Right/left circular polarized light (angular momentum $+h/-h$) can excite only transitions with $\Delta m_j = +1/-1$ (Fig. 3.59b) creating electrons with the resulting angular momentum (spin up/spin down) in the conduction band. The transition probability for the $m_j = 3/2$ levels is three times larger than for the $m_j = \frac{1}{2}$ levels. As a result the ratio of the number of spin-up to spin-down electrons—and vice versa for the opposite helicity of the light—is 3:1, i.e. the polarization is $(3-1)/(3+1) = 0.5$. When the cubic symmetry of the crystal is destroyed by strain, the degeneracy of the $4p_{3/2}$ levels is lifted so that the $m_j = \pm\frac{1}{2}$ levels (heavy hole band) are about 0.1 eV below the $m_j = \pm 3/2$ levels (light hole band) (Fig. 3.59a) and band gap light can excite only transitions from the $m_j = 3/2$ levels, resulting in a polarization P of 1.

The band gap is much smaller than the electron affinity (difference between the vacuum level and the minimum of the conduction band), so that band gap light cannot cause electron emission. This is achieved by covering the surface with a Cs-O or a Cs-NF_3 layer, which reduces the electron affinity to such extent, that the surface becomes a negative affinity surface (Fig. 3.59c). This surface condition is very sensitive to residual gas and loss of Cs, so that the vacuum requirements are similar to those for the FE sources discussed above. In the 10^{-10} mbar range about daily regeneration is necessary but in the XHV range ($\leq 10^{-11}$ mbar) a lifetime of months has been achieved (T. Koshikawa, private communication).

The P values mentioned above are theoretical. In practice P is smaller for a variety of reasons. With unstrained GaAs cathodes it is usually between 20 and 35 %, in the best strained photocathodes up to 90 %. Another important parameter is

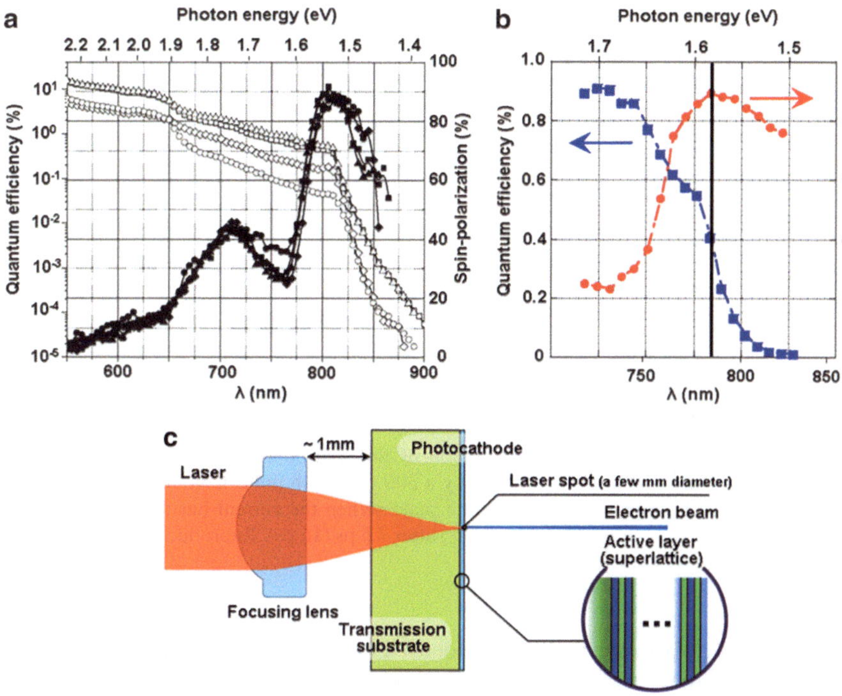

Fig. 3.60 Spin polarization *P* and quantum yield QE of (**a**) a front-illuminated strained AlInGaAs/AlGaAs superlattice on GaAs and (**b**) a back-illuminated strained GaAs/GaAsP superlattice on GaP. The best front-illuminated cathode achieved $P = 92$ % and QE = 0.85 %, which is better than the QE of the back-illuminated cathode. However, the photon beam can be focused much better on the strained layer in the second case as illustrated in (**c**), resulting in a much higher brightness of the source. (**a**) Reproduced from Ref. [285] with permission from the American Institute of Physics, © 2008. (**b**) Adapted from Ref. [286] with permission from the Japan Society of Applied Physics. (**c**) Reproduced from Ref. [101] with permission from T. Koshikawa

the quantum efficiency QE (number of electrons per photon). Typical values for unstrained cathodes at the photon energy of maximum polarization are 0.5 %, but values somewhat above 1 % have also been reported. At higher photon energies much higher QE values (up to 25 % in GaAs and up to 50 % in GaAsP [284]) can be obtained, however, with much lower polarization. For SPLEEM the QE is not so important because the light intensity can always be increased to obtain enough intensity, but high polarization is the main goal. Figure 3.60 shows some of the best data achieved to date. Figure 3.60a [285] shows *P* and QE from a front-illuminated strained AlInGaAs/AlGaAs superlattice source on a GaAs substrate, Fig. 3.60b [286] from a back-illuminated strained GaAs/GaAsP superlattice grown on a GaP substrate with a AlGaAs transparent inter-layer and Si_3N_4 antireflection coating, which is transparent for light with the energy of the band gaps of the superlattice grown on it. Figure 3.60c [101] shows schematically this photocathode, which presently is by far the most effective spin-polarized cathode despite its lower

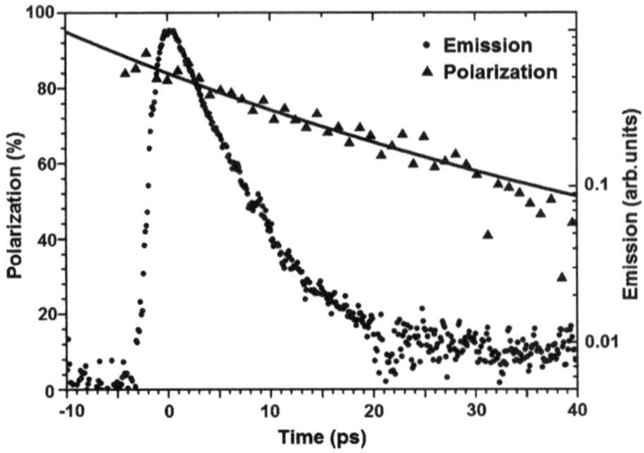

Fig. 3.61 Decay of the current and polarization of the spin-polarized electron pulse, excited from a front-illuminated cathode (Fig. 3.60a) by a 300 fs photon pulse from a Ti:sapphire laser. Note that the current is shown on a logarithmic scale. When the current has decreased to 10 % (2 %) of its maximum value P is still 54 % (24 %) after 7.5 ps (16 ps). Reproduced from Ref. [292] with permission from K. Aulenbacher

quantum efficiency because the laser is focused into a very small area. As a result it has an extremely high brightness of 2×10^7 A/cm^2 sr and also a very high spectral brightness, because of the narrow energy distribution at the low emission current needed in SPLEEM [287]. This cathode has been improved further, reaching slightly better P and QE values even without antireflection coating [287, 288]. Another potential spin-polarized source material is GaAsBi. Its band splitting is 60 mV already at 3 % Bi [289] and has been predicted theoretically [290] to increase to about 200 mV at 12 % Bi. The random distribution of Bi causes considerable band broadening [291], which has to be overcome for efficient spin selection. This material can be grown on GaAs as strained layer or with nitrogen incorporation strain-free. It has a smaller band gap as GaAs, so that it could be used for back-illuminated photocathodes.

For SPLEEM the high polarization is essential for good contrast between the images obtained with $\sigma+$ and $\sigma-$ polarized light. When the spin polarization P of the object is high, say $P \approx 0.9$, only one image, for example with $\sigma+$ light, needs to be taken. This allows real time image acquisition as in LEEM. Only at very weak magnetization or dominant structural contrast image subtraction will be necessary. Single image acquisition with pulsed laser illumination allows also pump-probe studies of faster processes. The time resolution is limited by charge and spin transport to the surface, which determines the decay time of the emission pulse caused by the laser pulse. Experiments with AlInGaAs/AlGaAs superlattice on GaAs photocathodes have shown, that the current decays initially with a decay time τ of 4 ps, followed by a slower decay with $\tau = 13$ ps, while the decay time of the spin polarization is much longer, 81 ps (Fig. 3.61 [292]). This is a developing field, driven mainly by the development of spin-polarized sources for high energy physics. The evolution in this field is, therefore, mostly documented in particle accelerator and free electron laser conference proceedings. In many of the instruments used in high energy physics spin polarization is not needed but high current in short pulses. For this purpose Cu, which has a very low QE (10^{-4} even with UV light) has been one of the main materials used because it is less sensitive to contamination than other materials with high QE such as Cs_2Te (QE = 0.07) or alkali antimonides (QE up to 0.46 [284]). The QE of Cu can be improved by a factor of 50 by coating with a CsBr layer [293] without losing its stability, but it is still inferior to that of the other materials, in particular GaAs and GaP surfaces with negative electron affinity [284]. For optimally Mg-doped GaN a QE value of even 0.72 has been reported with 5.4 eV illumination [294]. A comparison of these the electron sources can be found in Ref. [295]. For application in the field discussed in this book only GaAs, GaP, and GaN are of possible interest for pulsed sources. When illuminated with UV light the decay time of emission is significantly shorter [296] than shown in Fig. 3.61. When such a layer is grown on a substrate with larger band gap, so that illumination from the backside is possible as in the case of the spin-polarized cathode mentioned above, then time resolution could be improved below the ps range.

Concluding this section on electrons guns, a wide variety of sources is available and several possibilities exist for further improvement. For instruments, which have to tolerate rough treatment, the hexaborides probably are the best choice; for high resolution LEEM instruments without aberration correction—one of the FE sources; for LEEM with aberration correction—Schottky emitters or even hexaborides should be sufficient. For SPLEEM the back-illuminated III–V superlattice cathodes are the future and for pump-probe experiments back-illuminated III–V cathodes illuminated with UV light appear best suited.

3.2.9 Other Components (Image Detectors Vacuum System Including Airlock and Specimen Preparation Chamber, Electronics)

In this section we discuss briefly other components, which are essential for the operation of the instruments discussed in Sect. 3.1: beam deflectors, stigmators, apertures, specimen manipulators and sample holders, image detectors and vacuum system.

Beam deflectors are used to align the beam onto the optical axes of the lenses, beam separator, and energy filter. In magnetic microscopes they are outside the beam tube just like the lenses and consist usually of four coils, sometimes also of eight (to correct at the same time deviations from rotational symmetry of a nearby lens). In electrostatic microscopes the deflectors are by necessity in vacuum. They are usually four cylinder segments or rods, sometimes such as in the instrument shown in Fig. 3.10 also eight rods. Of foremost importance is the correction of the deviation of the objective lens from rotational symmetry, which is the main cause of its astigmatism. The coils or electrodes of the corrector must be as close as possible to the lens field for effective correction. This is the main reason for the pronounced asymmetry of the objective lens, which provides space for the corrector. The very first LEEM used only four coils but present instruments have eight coils or electrodes, which are connected to the power supplies in such a manner, as to allow independent correction of astigmatism and deflection in any direction.

Apertures are also indispensable parts of the instrument. One—the contrast aperture or angle-limiting aperture—is needed in an intermediate diffraction plane to select a small angular region for real space imaging, a second one—the selected area or field-limiting aperture—is needed in an intermediate image plane to select a small specimen region for reciprocal space imaging. In LEEM and MEM the desired area can also be selected by an aperture in the illumination system. The apertures are either small Pt discs with holes down to 5 μm, which are mounted on linear manual or motorized drives for interchanging holes sizes and for alignment, or iris apertures, whose opening can be adjusted continuously via a rotary feedthrough. The locations of these apertures in the various microscopes are indicated in Figs. 3.4, 3.7, 3.8, 3.10, 3.13, and 3.18. With the smallest apertures regions as small as 0.5 μm or less in diameter in real space and regions as small as 0.5 $Å^{-1}$ in diameter (including the factor 2π) in reciprocal space, can be selected.

Because of the spherical aberration of the objective lens great care has to be taken when correlating the reciprocal space image with small areas selected with the field-limiting aperture in an intermediate image plane. Similar but less serious problems are caused by the chromatic aberration of the objective lens and the dispersion of the beam separator. When area-selection is done with the illumination aperture the problems caused by the spherical aberration are avoided [297]. In addition to these apertures, instruments with a dispersive energy filter need an energy-selection slit, whose width and position is also controlled by a linear motion feedthrough.

Two more components determine to a large extent the image quality even with the best electron optics: the specimen manipulator and the image detection system. The specimen manipulator must fulfill many conditions. It must allow moving the specimen parallel to it surface (x, y) by several mm, so that all regions of the specimen can be imaged. It must allow tilting of the specimen, so that its surface normal is parallel to the optical axis of the objective lens and that the region being imaged stays within the field of view ("eucentric" tilt). The eucentric tilt is achieved by sliding the central part of the manipulator on a spherical calotte, whose center is in the center of the specimen surface. It should be noted in connection with these movements that x, y translations, which are not small compared to the diameter of the specimen holder, introduce a lateral field component, which destroys the rotational symmetry of the objective lens, so that tilt and/or astigmatism have to be corrected again. The specimen also has to be moved along the optical axis (z) for two reasons: (1) a fine movement for changing the field strength and focus position at fixed objective current/voltage settings and (2) a rough movement for specimen exchange, which in some instruments, such as those shown in Figs. 3.4, 3.8, 3.10,

Fig. 3.62 Computer-controlled sample manipulator. Two inchworms marked *x* and *y* shift the sample laterally, the other two inchworms move the sample holder on a spherical surface with the help of C clamps, thus providing tilt in two directions. Reproduced with permission from Ref. [28]. Copyright 1991 Elsevier

and 3.13 can be several cm. All these motions must be possible without endangering the stability of the specimen position relative to the objective lens due to vibrations or drift. The manipulator in Fig. 3.1, for example, did not fulfill these conditions and also had no tilt, which limited the resolution of the otherwise well-designed instrument. The manipulator of the first LEEM instrument, seen in Fig. 3.5 on top of the specimen chamber, allowed tilt up to 5°, all translations and even to swing out the specimen towards a side flange for deposition of thin films. Its stability allowed the first good LEEM images [27].

The more advanced electron optics in today's instrument requires better stability and ease of operation. The stability is achieved either by integrating the specimen manipulator with the objective lens [7–9] as seen in Fig. 3.9 or by firmly connecting the manipulator indirectly with the objective lens via a massive chamber as seen for example in Figs. 3.4, 3.7, and 3.14. In the first case the necessary motions are controlled by piezoelectric drives, in the second case either by external manual (Figs. 3.4, 3.7, and 3.14) or computer-controlled mechanical drives [298] or internally by piezoelectric drives as in Tromp's first two instruments, in which he used inchworms [28, 32] (Fig. 3.62 [28]). The piezoelectric drives in the more recent instruments [32, 33, 66, 299] are linear piezomotors (slip-stick drives) [300], capable of more force and longer travel. The relative merits of the two drive methods are obvious. Out-of-vacuum drives have to move the manipulator against atmospheric pressure, which causes considerable friction. Together with the play of the threads this makes precise positioning difficult. In-vacuum drives avoid this problem, so that much less force is needed and the electronic control of the piezoelectric drives allows precise specimen positioning in the 10–100 nm range. Another important aspect is specimen exchange. It should involve minimum friction but still allow stability once inserted in the manipulator. It can be done either from the backside as in Fig. 3.62 or from the side as it is done in the instruments shown in Figs. 3.4, 3.7, and 3.14. Once inserted, friction in UHV provides sufficient stability.

Additional features, which the specimen manipulator has to have for many experiments are specimen heating, specimen cooling, and temperature measurement. For magnetic measurements application of magnetic fields parallel and/or normal to the surface is needed. Other experiments, for example electromigration studies require in-plane currents in the sample, deformation experiments require the application of forces to the specimen. All these possibilities have to be incorporated with minimum loss of stability and alignment. Heating to intermediate temperatures, say up to 800 K, is usually achieved by radiation from filament behind the specimen, heating to high temperatures up to 2000 K or more, by bombardment with electrons with up to 1 keV energy or more from the same filament. In both cases the specimen drifts, which requires refocusing and usually also lateral repositioning. Cooling to low temperatures is more difficult, in particular when the specimen is at high voltage. In this case 50 K has been reached by incorporating a cooling liquid reservoir into the specimen cartridge holder and cooling with liquid He via a high voltage insulated transfer line (Altman, private communication). With liquid nitrogen 100 K can be reached in this manner. Cooling of samples at ground potential is simpler. With a liquid He-cooled manipulator 25 K has been reached within 30 min and 50 nm resolution at this temperature [301]. The specifications of a commercial instrument also claim 25 K [96]. With increasing interest in low temperature studies these limits will certainly be pushed down further. Without drift correction considerable waiting times for high resolution studies have to be accepted, however.

High temperatures can be most precisely measured with optical pyrometer provided the emissivity factor of the material and/or a calibration point in the temperature of interest is known. An example for a calibration temperature is the Si(111) $(7 \times 7) \leftrightarrow (1 \times 1)$ phase transition. However, in specimens, that are transparent in the sensitivity range of the pyrometer, the temperature of the hot metal support is measured, so that the surface temperature has to be calibrated differently. Thermocouples, spot welded to the cup, which holds the specimen in place, or in the case of metals directly to the sample edge, allow also good temperature measurement, provided the temperature gradients on the sample are small. The choice of the thermocouple material depends upon the maximum temperature used in experiments: WRe, PtRh, NiCr alloys etc. Also the specimen material has to be kept in mind because of the possibility of alloying with the thermocouple. For low temperatures some of these thermocouples are still useful but at the lowest temperatures accessible up to now Pt resistance thermometry is the choice.

For the study of magnetic processes magnetic fields $\boldsymbol{H}$ have to be applied to the specimen, either parallel or normal to the surface. With a carefully aligned magnetization coil behind the specimen, imaging with the field on is possible but $\boldsymbol{H}$-dependent refocusing is required because of the additional field. Weak $\boldsymbol{H}$ fields parallel to the surface still allow imaging but deflection and astigmatism have to be corrected. Window frame mounting of the specimen has been used successfully in spin-polarized low energy electron diffraction (SPLEED) [302] and found also to be useful in XMCDPEEM (Fig. 3.63a) [303]. With coils around the yoke magnetically soft specimens can also be imaged as a function of applied field with

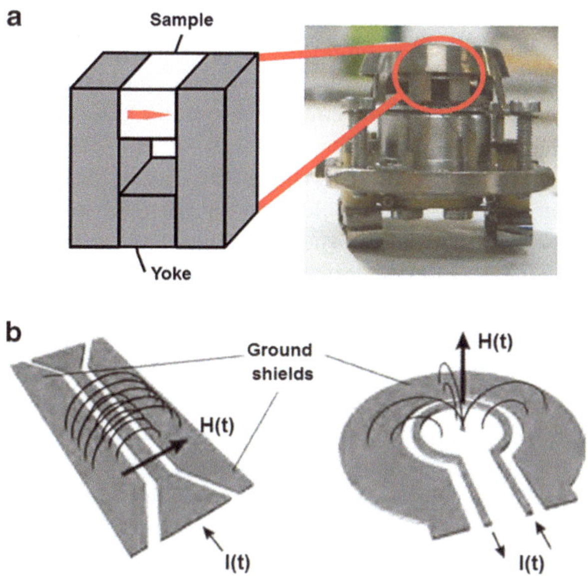

Fig. 3.63 Magnetization sample holders. (**a**) Magnetic window frame in a standard sample holder for in-plane magnetization. (**b**) In-plane (*left*) and out-of-plane (*right*) field pulse generation by 50 Ω linear and circular waveguides consisting of thin lithographically prepared Au or Cu stripes on SiO_2/Si. The sample is placed in close contact on these field generators. (**a**) Reprinted from Ref. [303]. Courtesy of International Business Machines Corporation, © 2011 International Business Machines Corporation. (**b**) Reproduced with permission from Ref. [80]. Copyright 2006 Elsevier

concurrent correction of deflection and astigmatism. In general the fields are too high for imaging with the field on and experiments are performed before and after application of the magnetic field or time-resolved in the pump-probe mode, with the field on in the pump phase and off in the imaging phase. Pump-probe experiments aim at the dynamics of magnetization processes and require, therefore, very short pump and probe pulses. Short pump pulses are obtained with miniaturized waveguide structures with 50 Ω impedance. Figure 3.63b [80] shows two examples, a stripline for generating a $\boldsymbol{H}$ field parallel to the surface and a microcoil for producing a field normal to the surface of the specimen. These patterns are produced by the usual lithography procedures used in microelectronics [304] on Si or GaAs substrates and consist either of Cu or Au. Typical widths and thicknesses are 10 μm and 500 nm. The magnetic specimen of interest is produced subsequently by the same methods along the center line of the stripline or in the center of the microcoil. The complete structure is mounted on standard specimen cartridges modified with high frequency connectors. For more information see the review by Schönhense et al. [80]. The structures discussed up to now are used to image the influence of external magnetic fields on the magnetic domain structure. For the study of spin current-induced magnetization changes such as the domain wall motion additional terminals are needed. The continuing interest in magnetic processes in magnetic nanostructures will very likely lead to the development of more sophisticated nanostructure supports.

Finally we have to consider one of the most important components, the image detector, whose relative importance compared to other components, is increasing with increasing demands on the microscopes. It should lose a minimum of the information contained in the signal arriving at the detector, in particular it should not deteriorate the resolution, it should be very sensitive but also have a high dynamic range, its signal/noise ratio should be high and for time-resolved studies the signal should have a short decay time or the read-out time should be short. All these conditions are analyzed in a very thorough manner by Herrmann [305]. Here we discuss only the instrumental aspects of image detection, the physical aspects are subject of Chap. 4. In pre-UHV emission electron microscopy the detector was either fine-grained film or photoplate within the microscope or a fine-grained fluorescent screen with a particle size between 1 and 10 μm plus film or photoplate outside the vacuum. Then film and photo plate were replaced by sensitive charge-coupled device (CCD) cameras, which lead to a considerable increase of the dynamic range and image acquisition rate. Slow-scan CCDs contributed to major progress in TEM [306, 307]. These detectors provided sufficient resolution even for the best electron optics.

The introduction of microchannelplate (MCP) image intensifiers in the early years of UHV microscopy did not change this situation much because their resolution was compatible with that of the electron optics at that time. However, other problems arose when the time-consuming photographic film, which was used to record the intensified image on the fluorescent screen, was replaced initially by low-light level TV cameras and later by CCD cameras: the low signal-to-noise ratio (SNR). This introduced two noise-increasing components into the detection system, which became to a certain extent also resolution-limiting with increasing resolution of the electron optics. Thus the sensitivity increase from the MCP was bought at the expense of the detective quantum efficiency $DQE = (SNR_{out}/SNR_{in})^2$ which decreased due to the increased output noise. This is of particular importance when the final image is a difference image of two images, for example in magnetic dichroism imaging or spectroscopic imaging without energy filter. The intensity difference between the two images is usually small, even near saturation of the detection system, so that the intensity in the difference is small, while the noise of the two original intense images is carried over to the difference image. For this reason it is important that the camera has a low dark current (e/pixel/s), a low read noise (e/pixel) and a high dynamic range. Table 3.3 [308] compares the properties of some cameras. While the electron multiplying CCD (EMCCD) has the lowest read noise, this advantage is partially lost by poorer other features such as smaller field of view. The scientific complementary metal oxide semiconductor (sCMOS) camera has also low read noise and in addition a very large dynamic range and readout speed, while the quantum efficiency (e/photon) is only slightly smaller than that of a CCD. This makes the sCMOS camera best suited for difference imaging.

Of course, before the light signal gets to the camera there is the fluorescence screen on the fiber optic window and the MCP. There is no information on fiber

Table 3.3 Comparison of image acquisition cameras [308]

Parameter	sCMOS (Neo)	Interline CCD	EMCCD
Sensor format (Mpixels)	5.5	1.4–4	0.25–1
Pixel size (μm)	6.5	6.45–7.4	8–16
Read noise	1 e^- @ 30 frames/s 1.4 e^- @ 100 frames/s	4–10 e^-	<1e^- (with EM gain)
Readout speed (max.) (Mpixel/s)	5500	25	~30
Quantum efficiency (max.)	0.57	0.6	90 % "back-illuminated" 65 % "virtual phase"
Dynamic range	30000:1 (@ 30 frames/s)	~3000:1 (@ 11 frames/s)	8500:1 (@ 30 frames/s with low EM gain)
Multiplicative noise	None	None	1.41× with EM gain (effectively halves the QE)

Courtesy of Laser Focus World

optic windows in the literature, which is relevant for the discussion here, except that the standard 6 μm fiber size does not reduce the resolution significantly and transfers the light efficiently to the camera, even if not connected directly to the fiber plate. The fluorescent screen material is usually P20 ((Zn,Cd)S:Ag) with particle sizes between 2 and 15 μm. It has an emission wavelength with maximum emission at 560 nm, which fits well to the sensitivity maximum of the CCD or CMOS camera, and a high efficiency but a long decay time (60 ms). For fast image acquisition the P43 ($Gd_2O_2S:Tb^{3+}$) phosphor with maximum emission wavelength 545 nm and 3 ms decay time is preferable. Light scattering within the layer limits the resolution beyond the particle size, so that the layer should be as thin as compatible with the thickness for optimum transfer of electron energy to light emission. No measurements for low energies are available but extrapolation of data for high energies [309] suggest a thickness in the μm range. With these dimensions in mind, the resolution degradation by the fluorescent screen is insignificant. Thus, in instruments without MCPs such as the ALS PEEM II the resolution of the primary detector is of the order of 1 μm, modified slightly by the diameter of the fibers in the fiber optics window.

The MCP and its combination with the fluorescent screen, however, cause a major deterioration of the resolution and of the DQE. Standard MCPs have a hexagonal pattern of channels with a diameter D of 10 μm, a periodicity ("pitch") P of 12 μm and a length to diameter ratio L/D of 40:1 or 60:1. MCPs with smaller D and L down to 2 and 3 are available too, with correspondingly higher resolution. The channels are tilted, typically by 8°, and two of them are stacked in a chevron style arrangement. With 2–2.4 kV applied across the plate pair a gain of 10^6–10^7 can be achieved. The secondary electron amplification in the channels produces from each incident electron a charge cloud at the ends of the channels, which spreads laterally somewhat between the plates At the exit from the second MCP the electrons in the charge cloud are accelerated on parabolic paths in the

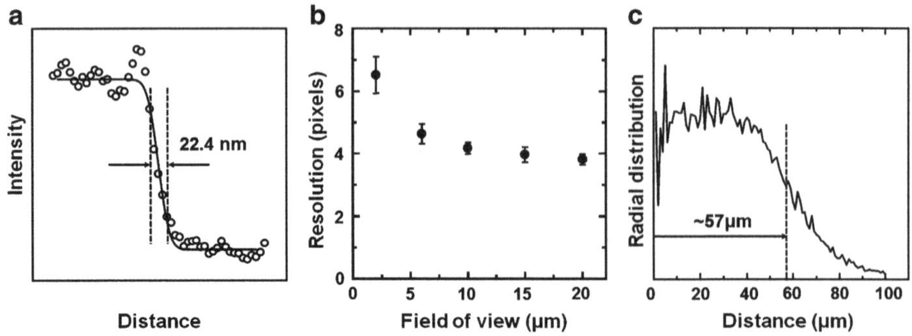

Fig. 3.64 Resolution limitation by a MCP-CCD image detection system. (**a**) The intensity profile of a sharp amplitude contrast object used for the resolution test. (**b**) Resolution in pixels as a function of the field of view. (**c**) Profile of the lateral electron distribution at the fluorescent screen from Monte Carlo calculations. Voltage across the channel plates 1.4 kV, distance between them 50 μm, voltage between channel plates and fluorescent screen 5.5 kV and distance between them 1 mm. The object is a Si(111) surface with a low Cu coverage, which forms a sharp boundary between Si(1 × 1) and Cu/Si"5 × 5" structure upon high-temperature annealing. (**a** and **c**) Courtesy of T. Yasue; (**b**) From Ref. [313]

quasi-homogeneous field between the exit plane of the channel plate and the fluorescent screen. Their initial angular and energy distribution determines then the resolution δ of the MCP-fluorescent screen combination, not the D, L values of the MCP. This has been realized already many years ago [310–312] but taken into account in surface microscopy with slow electrons only recently by Shimizu et al. [313]. They measured the resolution as a function of magnification. Figure 3.64 [313] shows some of the results. The 84 %/16 % width of the profile (a) of an amplitude contrast object is plotted in (b) as a function of the field of view in units of the number of pixels over which the width is spread. In an ideal detector (infinitely small pixel size) it would decrease linearly with increasing field of view but it soon saturates at 3.8 pixels, limited by the image detection system. The CCD camera was found to contribute 1.7 pixels so that $\delta_{MCP} = (\delta^2{}_{total} - \delta^2{}_{CCD})^{1/2} =$ 3.4 pixels, indicating a considerable spreading of the charge in the preceding MCP-fluorescent screen system. They simulated this system by Monte Carlo calculations using published energy distributions [314] and found that the spreading in the channel plates increases with decreasing voltage between them but is small compared to the spreading during the propagation of the electron cloud between the channel plate exit and the fluorescent screen shown in (c) (Yasue, private communication). These results are in agreement with the earlier work cited above and more recent reports, which show even stronger spreading, because of weaker fields between MCP and fluorescent screen and space charge effects [314–316]. An analysis in terms of the terminology of image formation theory [317] will be discussed in Chap. 4. In addition to the loss of resolution, the MCP also decreases the DQE because of the fluctuations of the signal amplification between different channels and its Poisson distribution, which creates significant noise. The resolution

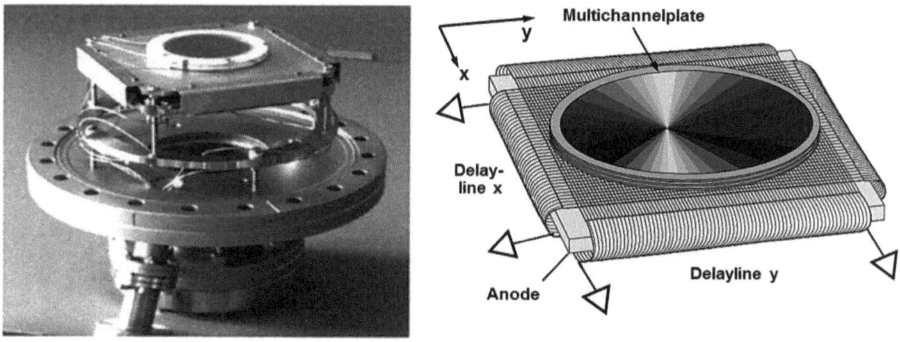

Fig. 3.65 Delay line single-electron detector for time-resolved imaging. A single electron hitting the channel plate produces a charge cloud at its exit, which produces an electromagnetic pulse in the *x*, *y* position of its impact. The difference between the arrival times of the *x* and *y* pulses at the ends of the delay lines determines the location of the electron on the channel plate. Coupling with the pulsed electron source allows time-resolved imaging. A spatial resolution of 47 μm has been reported in qualitative agreement with the dimensions of the charge cloud discussed before. Reproduced with permission from Ref. [78]. Copyright 2001 Elsevier

problem can in principle be solved by imaging with very large magnifications, say 1 μm field of view or less, but the longer image acquisition times awaken other resolution-limiting effects such as vibrations, electronic and specimen instabilities. The noise problem remains. If instead of the fluorescent screen and external camera a delay line detector (DLD) (Fig. 3.65) is used, as in the instruments described in Refs. [77–79], the MCP has less negative influence on resolution and noise—if any at all—because the resolution is determined by the distance between the wires in the DLD, which is comparable to the diameter of the charge cloud, and the MCP is operated in the saturation mode, so that every electron incident on the MCP produces an event in the DLD.

While the DLD eliminates the detour via the fluorescent screen to the final signal, it still involves the MCP. The ideal detector eliminates this step by acquiring the image directly on a chip inside the microscope. First attempts in this direction to detect slow electrons (<30 keV) with CCDs were made already in 1987 [318] for SEM, but serious efforts to develop such detectors started worldwide not until around 2000 [319–321]. More recently two other approaches have been taken: monolithic devices, in which electron sensor and signal processing are integrated from the beginning, and hybrid devices, in which electron sensor and signal processing are on separate chips, which are then bump-bonded with the flip-chip attachment technology. Both approaches have already led to commercially available products, the Gatan K2 camera [322] and the PANalytical PIXcel camera [323] for high energy electrons and for LEEM/PEEM, all of them very costly. Figure 3.66 (R.Tromp, private communication) illustrates this improvement in UVPEEM with the first commercially available direct electron detector [324]. Noncommercial versions of these detectors have already demonstrated excellent properties for

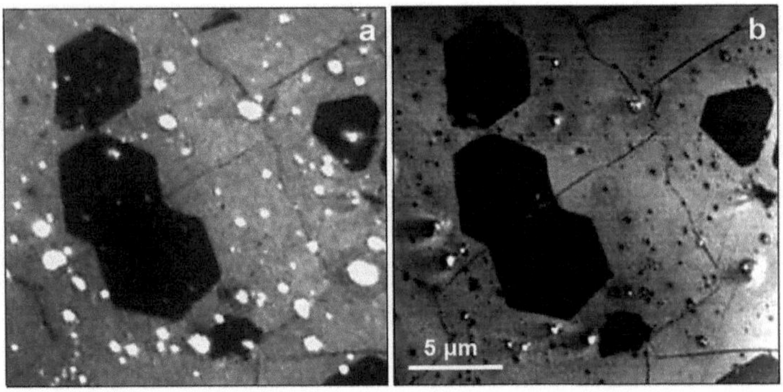

Fig. 3.66 Comparison of MCP/fluorescent screen/CCD (**a**) and direct electron detection (**b**) imaging. Graphene on Cu. Monolayer *bright*, bilayer *dark*. Courtesy of R. Tromp

imaging with electrons: the TEAM1k, a monolithic active pixel sensor (MAPS) detector, in a transmission microscope [325, 326] and the Medipix2, a hybrid detector, in LEEM [327, 328]. Like the CCDs used earlier these detectors are not UHV compatible and outgas strongly, so that differential pumping and removal or cooling during bakeout is necessary. A detailed account of direct image detection in electron microscopy has been given by Faruqi in 2011 [329] but the field is still in rapid development. For the time being MCP detectors will very likely remain the workhorse in LEEM/PEEM or their strong intensity enhancement has to be sacrificed for the benefit of a high DQE, which can be gained, for example with YAG:Ce single crystal screens [330, 306] with antireflection coating.

Vacuum is the next aspect which needs to be discussed. All modern LEEM and PEEM instruments are bakeable up to at least 150 °C, in critical regions, for example around cold FE gun or CsO-activated photocathodes, at least up to 200 °C in order to achieve sufficiently low pressures, in particular of water vapor, CO, CO_2, and hydrocarbons. Standard pumps are ion sputter pumps, supplemented in critical regions with nonevaporable sublimation getter pumps (SAES). Turbomolecular pumps, preferably magnetically levitated are used frequently when the highest resolution is not needed or when high gas pressures are used in the specimen region. At the usual field strength in the specimen region, pressures up to the middle 10^{-5} mbar range can be used but for the study of clean surfaces a pressure in the low 10^{-10} mbar or lower is required. As mentioned earlier, stable operation of spin-polarized guns needs XHV pressures, which can be achieved with the best pumps only after specific treatment of the surrounding surfaces (electropolishing to reduce surface area, coating with layers, to which residual gases bond only weakly and pre-baking at high temperatures to drive out dissolved gases in the material). In magnetic microscopes the specimen region is connected with the gun and the detector region with narrow tubes—the lenses are outside the vacuum—which allows large pressure differences between these regions. In order the keep the pressure in the specimen region low during specimen change the specimen is usually transferred to a pre-chamber with a base pressure comparable to that of the specimen chamber and from there to an airlock chamber. The pre-chamber serves at the same time as a preparation chamber for outgassing of the specimen holder and cleaning the specimen by sputtering or treatment in suitable gases, for example O_2 or H_2. Of course, in order to achieve the pressures and surface cleanness needed for well-defined experiments, all the usual procedures in UHV technology have to be followed.

A few final comments on the electronics of the instrument have to be made without going into detail. Electrostatic microscopes, in which voltage dividers supply the voltage to the object and various lenses from a common high voltage supply, can tolerate high voltage stabilities $\Delta U/U$ in the 10^{-4} range. In magnetic microscopes, in which all lenses are supplied by independent power supplies, the current stability $\Delta I/I$ has to be in the low 10^{-5} range, as does the high voltage stability, in order to reach the resolution possible with the objective lens. Clearly, aberration-corrected instruments, which aim at 1 nm resolution, require stabilities in the low 10^{-6} range, which are implemented in these instruments as described by Tromp [66]. All power supplies, including in particular also those for deflectors and stigmators, are computer-controlled, except in older instruments, in which cathode current and Wehnelt voltage are usually controlled manually.

In addition to the computer control of the electronics there are many other operations, which can be—and are in some instruments—computerized: specimen positioning with numerical readout, specimen heating with thermocouple or pyrometer control, aperture positioning, etc. Together with a variety of data (real and reciprocal images, spectra, etc.) acquisition and processing software packages commercial LEEM and PEEM systems have become convenient laterally-resolving surface and thin films analysis systems, whose results will be discussed in Chap. 5.

References

1. Tonner, B.P., Harp, G.R.: Photoelectron microscopy with synchrotron radiation. Rev. Sci. Instrum. **59**, 853–858 (1988)
2. Tonner, B.P., Harp, G.R.: Photoyield spectromicroscopy of silicon surfaces using monochromatic synchrotron radiation. J. Vac. Sci. Technol. A **7**, 1–4 (1989)
3. Tonner, B.P., Harp, G.R., Koranda, S.F., Zhang, J.: An electrostatic microscope for synchrotron radiation x-ray absorption microspectroscopy. Rev. Sci. Instrum. **63**, 564–568 (1992)
4. Bethge, H., Klaua, M.: Photo-emission electron microscopy of work function changes. Ultramicroscopy **11**, 207–214 (1983)
5. Engel, W., Kordesch, M.E., Rotermund, H.H., Kubala, S., von Oertzen, A.: A UHV-compatible photoelectron emission microscope for applications in surface science. Ultramicroscopy **36**, 148–153 (1991)
6. www.staibinstruments.com/.
7. Grzelakowski, K., Bauer, E.: A flange-on type low energy electron microscope. Rev. Sci. Instrum. **67**, 742–747 (1996)
8. Swiech, W., Fecher, G.H., Ziethen, C., Schmidt, O., Schönhense, G., Grzelakowski, K., Schneider, C.M., Frömter, R., Oepen, H.P., Kirschner, J.: Recent progress in photoemission

microscopy with emphasis on chemical and magnetic sensitivity. J. Electron. Spectros. Relat. Phenomena **84**, 171–188 (1997)
9. www.focus-gmbh.com.
10. www.elmitec.de.
11. Griffith, O.H., Rempfer, G.F., Lesch, G.H.: A High Vacuum Photoelectron Microscope for the Study of Biological Specimens. Scanning Electron Microscopy, vol. II, pp. 123–130. SEM Inc., AMF O'Hare, Chicago, IL (1981)
12. Rempfer, G.F., Skoczylas, W.P., Griffith, O.H.: Design and performance of a high-resolution photoelectron microscope. Ultramicroscopy **36**, 196–221 (1991)
13. Griffith, O.H., Lesch, G.H., Rempfer, G.F., Birrell, G.B., Burke, C.A., Schlosser, D.W., Mallon, M.H., Lee, G.B., Stafford, R.G., Jost, P.C., Marriott, T.B.: Photoelectron microscopy: a new approach to mapping organic and biological surfaces. Proc. Natl. Acad. Sci. U. S. A. **69**, 561–565 (1972)
14. Engel, W.: Entwicklung eines Emisssionsmikroskops höher Auflösung mit photoelektrischer, kinetischer und thermischer Elektronenauslösung. Ph.D. Thesis (1968)
15. Wegmann, L.: The photo-emission electron microscope: its technique and applications. J. Microsc. **96**, 1–23 (1972)
16. De Stasio, G., Capozi, M., Lorusso, G.F., Baudat, P.A., Droubay, T.C., Perfetti, P., Margaritondo, G., Tonner, B.P.: MEPHISTO: performance tests of a novel synchrotron imaging photoelectron spectromicroscope. Rev. Sci. Instrum. **69**, 2062–2066 (1998)
17. Watts, R.N., Liang, S., Levine, Z.H., Lucatorto, T.B., Polack, F., Scheinfein, M.R.: A transmission x-ray microscope based on secondary-electron imaging. Rev. Sci. Instrum. **68**, 3464–3476 (1997)
18. Huang, L.Y.: Röntgen-Bildwandler-Mikroskopie. Z. Phys. **149**, 225–253 (1957)
19. Anders, S., Padmore, H.A., Duarte, R.M., Renner, T., Stammler, T., Scholl, A., Scheinfein, M.R., Stöhr, J., Séve, L., Sinkovic, B.: Photoemission electron microscope for the study of magnetic materials. Rev. Sci. Instrum. **70**, 3973–3981 (1999)
20. Veneklasen, L.H.: Design of a spectroscopic low-energy electron microscope. Ultramicroscopy **36**, 76–90 (1991)
21. Bauer, E.: Photoelectron spectromicroscopy: present and future. J. Electron. Spectros. Relat. Phenomena **114–116**, 975–987 (2001)
22. Bauer, E.: Cathode lens electron microscopy: past and future. J. Phys. Condens. Matter **21**, 314001 (2009)
23. Bauer, E.: Low energy electron reflection microscopy. In: Breese, S.S.J. (ed.) Fifth Intern. Congr. Electron Microscopy 1962, p. D-11. Academic, New York (1962)
24. Turner, G., Bauer, E.: Design features of an ultrahigh-vacuum electron microscope. In: 22nd Ann. Meeting EMSA, Detroit 1964. J. Appl. Phys., p. 3080 (1964)
25. Bauer, E.: LEEM and UHV-PEEM: a retrospective. Ultramicroscopy **119**, 18–23 (2012)
26. Telieps, W.: Ein Ultrahochvakuum-Elektronenmikroskop zur Abbildung von Oberflächen mit langsamen reflektierten und emittierten Elektronen. Ph.D. Thesis, Technische Universität Clausthal (1983)
27. Telieps, W., Bauer, E.: An analytical reflection and emission UHV surface electron microscope. Ultramicroscopy **17**, 57–66 (1985)
28. Tromp, R.M., Reuter, M.C.: Design of a new photo-emission/low-energy electron microscope for surface studies. Ultramicroscopy **36**, 99–106 (1991)
29. Liebl, H., Senftinger, B.: Low-energy electron microscope of novel design. Ultramicroscopy **36**, 91–98 (1991)
30. Kolařík, V., Vašina, R., Mynář, M., Bejdák, T.: X-ray photoemission and low energy electron microscope. In: Frank, L., Čiampor, F. (eds.) 12th Eur. Congr. Electron Microscopy (EUREM 12), Brno 2000, pp. I 181–I 182. Czech. Soc. Electron Microscopy, Brno (2000)
31. Bauer, E., Koziol, C., Lilienkamp, G., Schmidt, T.: Spectromicroscopy in a low energy electron microscope. J. Electron. Spectros. Relat. Phenomena **84**, 201–209 (1997)

32. Tromp, R.M., Mankos, M., Reuter, M.C., Ellis, A.W., Copel, M.: A new low energy electron microscope. Surf. Rev. Lett. **5**, 1189–1197 (1998)
33. www.specs.de.
34. Tsuno, K.: Simulation of a Wien Filter as beam separator in a low energy electron microscope. Ultramicroscopy **55**, 127–140 (1994)
35. Sakai, Y., Kato, M., Masuda, S., Harada, Y., Ichinokawa, T.: Dynamical change of step and domain-structures on the Si(111) surface with heat-treatment observed by a low energy electron microscope. Surf. Sci. **336**, 295–302 (1995)
36. Sakai, Y., Kato, M., Masuda, S., Harada, Y., Ichinokawa, T.: Development of a low energy electron microscope with an energy analyzer. Surf. Rev. Lett. **5**, 1199–1211 (1998)
37. Griffith, O.H., Hedberg, K.K., Desloge, D., Rempfer, G.F.: Low-energy electron microscopy (LEEM) and mirror electron microscopy (MEM) of biological specimens: preliminary results with a novel beam separator system. J. Microsc. **168**, 249–258 (1992)
38. Skoczylas, W.P., Rempfer, G.F., Griffith, O.H.: A proposed modular imaging system for photoelectron and electron probe microscopy with aberration correction, and for mirror microscopy and low-energy electron microscopy. Ultramicroscopy **36**, 252–261 (1991)
39. Skoczylas, W.P., Rempfer, G.F., Griffith, O.H.: Electron optical benches for in-line and branched systems. A new bench designed for mirror-based aberration correction and low energy electron microscopy. Rev. Sci. Instrum. **65**, 3183–3193 (1994)
40. Adamec, P., Bauer, E., Lencová, B.: Compact low-energy electron microscope for surface imaging. Rev. Sci. Instrum. **69**, 3583–3787 (1998)
41. Adamec, P.: Compact Low Energy Electron Microscope for Surface Imaging. Ph.D. thesis, Masaryk University, Brno, Czech Republic (1997)
42. Mankos, M., Adler, D., Veneklasen, L., Munro, E.: Electron optics for high throughput low energy electron microscopy. Surf. Sci. **601**, 4733–4741 (2007)
43. Mankos, M., Spasov, V., Han, L., Kojima, S., Jiang, X., Harb, S., Grella, L., Czarnik, C.: Low energy electron microscopy for semiconductor applications. In: Adler, D.L., Bauer, E., Kellogg, G.L., Scholl, A. (eds.) Proc. MRS Spring Meeting: Symposium Advances and Applications of Surface Electron Microscopy, San Francisco 2008. MRS Proceedings, pp. W03–01 (2008)
44. Mankos, M., Adler, D., Veneklasen, L., Munro, E.: Electron optics for low energy electron microscopy. Phys. Procedia **1**, 485–504 (2008)
45. Mankos, M., Spasov, V., Munro, E.: Principles of dual-beam low-energy electron microscopy. In: Hawkes, P.W. (ed.) Advances in Imaging and Electron Physics, vol. 161, pp. 1–53. Academic, Burlington (2010)
46. Mankos, M.: Electron optics for dual-beam low energy electron microscopy. Nucl. Instrum. Methods Phys. Res. A **645**, 35–40 (2011)
47. Mankos, M., Shadman, K., N'Diaye, A.T., Schmid, A.K., Persson, H.H.J., Davis, R.W.: Progress toward an aberration-corrected low energy electron microscope for DNA sequencing and surface analysis. J. Vac. Sci. Technol. B **30**, 06F402, 12 pages (2012)
48. Mankos, M., Shadman, K.: A monochromatic, aberration-corrected, dual-beam low energy electron microscope. Ultramicroscopy **130**, 13–28 (2013)
49. Shimakura, T., Takahashi, Y., Sugaya, M., Ohnishi, T., Hasegawa, M., Ohta, H.: Mirror electron microscope for inspecting nanometer-sized defects in magnetic media. Microelectron. Eng. **85**, 1811–1814 (2008)
50. Kleinschmidt, H., Bostanjoglo, O.: Pulsed mirror electron microscope: a fast near-surface imaging probe. Rev. Sci. Instrum. **72**, 3898–3901 (2001)
51. Scherzer, O.: Ueber einige Fehler von Elektronenlinsen. Z. Phys. **101**, 593–693 (1936)
52. Scherzer, O.: Sphärische und chromatische Korrektur von Elektronen-Linsen. Optik **2**, 114–132 (1947)
53. Hawkes, P.W.: Aberration correction past and present. Philos. Trans. R. Soc. Lond. A **367**, 3637–3664 (2009)

54. Rose, H.H.: Historical aspects of aberration correction. J. Electron Microsc. **58**(3), 77–85 (2009)
55. Könenkamp, R., Jones, T., Elstner, J., Word, R.C., Rempfer, G., Dixon, T., Almaraz, L., Skoczylas, W.: Image properties in an aberration-corrected photoemission electron microscope. Phys. Procedia **1**, 505–511 (2008)
56. Könenkamp, R., Word, R.C., Rempfer, G.F., Dixon, T., Almaraz, L., Jones, T.: 5.4 nm Spatial resolution in biological photoemission electron microscopy. Ultramicroscopy **110**, 899–902 (2010)
57. Rose, H., Preikszas, D.: Outline of a versatile corrected LEEM. Optik **92**, 31–44 (1992)
58. Fink, R., Weiss, M.R., Umbach, E., Preikszas, D., Rose, H., Spehr, R., Hartel, P., Engel, W., Degenhardt, R., Wichtendahl, R., Kuhlenbeck, H., Erlebach, W., Ihmann, K., Schlogl, R., Freund, H.J., Bradshaw, A.M., Lilienkamp, G., Schmidt, T., Bauer, E., Benner, G.: SMART: a planned ultrahigh-resolution spectromicroscope for BESSY II. J. Electron. Spectros. Relat. Phenomena **84**, 231–250 (1997)
59. Wichtendahl, R., Fink, R., Kuhlenbeck, H., Preikszas, D., Rose, H., Spehr, R., Hartel, P., Engel, W., Schlogl, R., Freund, H.J., Bradshaw, A.M., Lilienkamp, G., Schmidt, T., Bauer, E., Benner, G., Umbach, E.: SMART: an aberration-corrected XPEEM/LEEM with energy filter. Surf. Rev. Lett. **5**, 1249–1256 (1998)
60. Hartel, P., Preikszas, D., Spehr, R., Müller, H., Rose, H.: Mirror corrector for low-voltage electron microscopes. In: Hawkes, P.W., Kazan, B. (eds.) Advances in Imaging and Electron Physics, vol. 120, pp. 41–133. Academic, San Diego, CA (2002)
61. Schmidt, T., Marchetto, H., Levesque, P.L., Groh, U., Maier, F., Preikszas, D., Hartel, P., Spehr, R., Lilienkamp, G., Engel, W., Fink, R., Bauer, E., Rose, H., Umbach, E., Freund, H.J.: Double aberration correction in a low-energy electron microscope. Ultramicroscopy **110**, 1358–1361 (2010)
62. Feng, J., Macdowell, A.A., Duarte, R., Doran, A., Forest, E., Kelez, N., Marcus, M., Munson, D., Padmore, H., Petermann, K., Raoux, S., Robin, D., Scholl, A., Schleuter, R., Schmidt, P., Stohr, J., Wan, W., Wei, D.H., Wu, Y.: An aberration-corrected photoemission electron microscope at the advanced light source. In: Proc. Eighth International Conference on Synchrotron Radiation Instrumentation 2003. AIP Conf. Proc., vol. 705, pp.1070–1073 (2004)
63. Feng, J., Forest, E., Macdowell, A.A., Marcus, M., Padmore, H., Raoux, S., Robin, D., Scholl, A., Schlueter, R., Schmidt, P., Stohr, J., Wan, W., Wei, D.H., Wu, Y.: An x-ray photoemission electron microscope using an electron mirror aberration corrector for the study of complex materials. J. Phys. Condens. Matter **17**, S1339–S1350 (2005)
64. Feng, J., Scholl, A.: Photoemission electron microscopy (PEEM). In: Hawkes, P.W., Spence, J.C.H. (eds.) Science of Microscopy, vol. 1, pp. 657–695. Springer, New York (2007)
65. MacDowell, A.A., Feng, J., DeMello, A., Doran, A., Duarte, R., Forest, E., Kelez, N., Marcus, M.A., Miller, T., Padmore, H.A., Raoux, S., Robin, D., Scholl, A., Schlueter, R., Schmid, P., Stöhr, J., Wan, W., Wei, D.H., Wu,Y.: Progress on PEEM3—an aberration corrected XRay photoemission electron microscope at the ALS. In: Choi, J.-Y., Rah, S. (eds.) Proc. Ninth International Conference on Synchrotron Radiation Instrumentation Korea 2006. AIP Conf. Proc., vol. 879, pp. 1341–1344. New York (2007)
66. Tromp, R.M., Hannon, J.B., Ellis, A.W., Wan, W., Berghaus, A., Schaff, O.: A new aberration-corrected, energy-filtered LEEM/PEEM instrument. I. Principles and design. Ultramicroscopy **110**, 852–861 (2010)
67. Tsuno, K., Yasue, T., Koshikawa, T.: Design of a mirror aberration corrector and a beam separator for LEEM. Appl. Surf. Sci. **256**, 1035–1041 (2009)
68. Schönhense, G., Spiecker, H.: Correction of chromatic and spherical aberration in electron microscopy utilizing the time structure of pulsed excitation sources. J. Vac. Sci. Technol. B **20**, 2526–2534 (2002)

69. Schönhense, G., Elmers, H.J.: PEEM with high time resolution—imaging of transient processes and novel concepts of chromatic and spherical aberration correction. Surf. Interface Anal. **38**, 1578–1587 (2006)
70. Ikuta, T.: Image restoration in coherent imaging system involving spherical aberration. J. Electron Microsc. **38**, 415–422 (1989)
71. Koshikawa, T., Shimizu, H., Amakawa, R., Ikuta, T., Yasue, T., Bauer, E.: A new aberration correction method for photoemission electron microscopy by means of moving focus. J. Phys. Condens. Matter **17**, S1371–S1380 (2005)
72. Möllenstedt, G., Gruner, H.: Monochromatisation of electron beams and improvement of emission images by means of an electron mirror. Optik **27**, 602–604 (1968)
73. Gruner, H., Gauckler, K.H., Möllenstedt, G.: Monochromatisierung von Elektronenstrahlen mittels Spiegelung und ihre Anwendung in der Emissions-Mikroskopie. Optik **33**, 255–269 (1971)
74. Merkel, M., Escher, M., Settemeyer, J., Funnemann, D., Oelsner, A., Ziethen, C., Schmidt, O., Klais, M., Schönhense, G.: Microspectroscopy and spectromicroscopy with photoemission electron microscopy using a new kind of imaging energy filter. Surf. Sci. **480**, 196–202 (2001)
75. Weber, N.B., Escher, M., Merkel, M., Oelsner, A., Schönhense, G.: Energy- and time-resolved microscopy using PEEM: recent developments and state-of-the-art. J. Phys. Conf. Ser. **100**, 072031, 4 pages (2008)
76. Spiecker, H., Schmidt, O., Ziethen, C., Menke, D., Kleineberg, U., Ahuja, R.C., Merkel, M., Heinzmann, U., Schönhense, G.: Time-of-flight photoelectron emission microscopy TOF-PEEM: first results. Nucl. Instrum. Methods Phys. Res. A **406**, 499–506 (1998)
77. Oelsner, A., Schmidt, O., Schicketanz, M., Klais, M., Schönhense, G., Mergel, V., Jagutzki, O., Schmidt-Böcking, H.: Microspectroscopy and imaging using a delay line detector in time-of-flight photoemission microscopy. Rev. Sci. Instrum. **72**, 3968–3974 (2001)
78. Schönhense, G., Oelsner, A., Schmidt, O., Fecher, G.H., Mergel, V., Jagutzki, O., Schmidt-Böcking, H.: Time-of-flight photoemission electron microscopy—a new way to chemical surface analysis. Surf. Sci. **480**, 180–187 (2001)
79. Oelsner, A., Krasyuk, A., Fecher, G.H., Schneider, C.M., Schönhense, G.: Image enhancement in photoemission electron microscopy by means of imaging time-of-flight analysis. J. Electron. Spectros. Relat. Phenomena **137–140**, 757–761 (2004)
80. Schönhense, G., Elmers, H.J., Nepijko, S.A., Schneider, C.M.: Time-resolved photoemission electron microscopy. In: Hawkes, P. (ed.) Advances in Imaging and Electron Physics, vol. 142, pp. 159–323. Academic, London (2006)
81. Oelsner, A., Rohmer, M., Schneider, C., Bayer, D., Schönhense, G., Aeschlimann, M.: Time- and energy resolved photoemission electron microscopy-imaging of photoelectron time-of-flight analysis by means of pulsed excitations. J. Electron. Spectros. Relat. Phenomena **178**, 317–330 (2010)
82. Lin, J., Weber, N., Wirth, A., Chew, S.H., Escher, M., Merkel, M., Kling, M.F., Stockman, M. I., Krausz, F., Kleineberg, U.: Time of flight-photoemission electron microscope for ultrahigh spatiotemporal probing of nanoplasmonic optical fields. J. Phys. Condens. Matter **21**, 314005, 7 pages (2009)
83. Castaing, R., Henry, L.: Filtrage magnétic des vitesses en microscopie électronique. C. R. Acad. Sci. Paris **255**, 76–86 (1962)
84. Reimer, L., Fromm, I., Rennekamp, R.: Operation modes of electron spectroscopic imaging and electron energy-loss spectroscopy in a transmission electron microscope. Ultramicroscopy **24**, 339–354 (1988)
85. Lanio, S., Rose, H., Krahl, D.: Test and improvement design of a corrected imaging magnetic energy filter. Optik **73**, 56–68 (1986)
86. Rose, H., Krahl, D.: Electron optics of imaging energy filters. In: Reimer, L. (ed.) Energy-Filtering Transmission Electron Microscopy, vol. 71, pp. 43–148. Springer, Berlin (1995)

87. Cazaux, J.: Microscope photoélectronique pour l'analyse chimique des surfaces. Rev. Phys. Appl. (Paris) **8**, 371–381 (1973)
88. Bauer, E., Franz, T., Koziol, C., Lilienkamp, G., Schmidt, T.: Recent advances in LEEM/PEEM for structural and chemical analysis. In: Rosei, R. (ed.) Proc. NATO Workshop on Chemical, Structural and Electronic Analysis of Heterogeneous Surfaces on Nanometer Scale, Trieste 1995, pp. 75–91. Kluwer, Dordrecht (1997)
89. Lilienkamp, G., Koziol, C., Schmidt, T., Bauer, E.: Cathode lens spectromicroscopy with a low-energy electron microscope. In: Thieme, J., Schmahl, G., Rudolph, D., Umbach, E. (eds.) X-Ray Microscopy and Spectromicroscopy, pp. III 25–III 34. Springer, Berlin (1998)
90. Schmidt, T., Heun, S., Slesak, J., Diaz, J., Prince, K.C., Lilienkamp, G., Bauer, E.: SPELEEM: combining LEEM and spectroscopic imaging. Surf. Rev. Lett. **5**, 1287–1296 (1998)
91. Tonner, B.P., Dunham, D., Droubay, T., Pauli, M.: A photoemission microscope with a hemispherical capacitor energy filter. J. Electron. Spectros. Relat. Phenomena **84**, 211–229 (1997)
92. Liebl, H.: Stigmatic sector-field energy analyzer without second-order angular image aberrations. Nucl. Instrum. Methods Phys. Res. A **292**, 537–540 (1990)
93. Grzelakowski, K.: Electron optical aspects of the dual-emission electron microscope. J. Phys. Condens. Matter **17**, S1351–S1362 (2005)
94. Escher, M., Weber, N., Merkel, M., Ziethen, C., Bernhard, P., Schönhense, G., Schmidt, S., Forster, F., Reinert, F., Krömker, B., Funnemann, D.: NanoESCA: a novel energy filter for imaging x-ray photoemission spectroscopy. J. Phys. Condens. Matter **17**, S1329–S1338 (2005)
95. Krömker, B., Escher, M., Funnemann, D., Hartung, D., Engelhard, H., Kirschner, J.: Development of a momentum microscope for time resolved band structure imaging. Rev. Sci. Instrum. **79**, 053701, 7pp (2008)
96. www.omicron.de.
97. Rose, H.: The retarding Wien filter as a high performance imaging filter. Optik **77**, 26–34 (1987)
98. Tsuno, K.: Electron-optical analysis of a retarding Wien filter for electron spectroscopic imaging. Rev. Sci. Instrum. **64**, 659–666 (1993)
99. Marx, G.K.L., Gerheim, V., Schönhense, G.: Multipole WIEN-filter for a high-resolution X-PEEM. J. Electron. Spectros. Relat. Phenomena **84**, 251–261 (1997)
100. Tromp, R.M., Hannon, J.B., Fujikawa, Y., Berghaus, A., Schaff, O.: A simple energy filter for LEEM/PEEM instruments. J. Phys. Condens. Matter **21**, 314007, 13 pages (2009)
101. Yamamoto, N., Jin, X.G., Mano, A., Ujihara, T., Takeda, Y., Okumi, S., Nakanishi, T., Yasue, T., Koshikawa, T., Ohshima, T., Saka, T., Horinaka, H.: Status of the high brightness polarized electron source using transmission photocathode. J. Phys. Conf. Ser. **298**, 012017, 6 pages (2011)
102. Bauer, E.: Spin-polarized low energy electron microscopy. In: Amelinckx, S., Van Dyck, D., Van Landuyt, J.F., Van Tendeloo, G. (eds.) Handbook of Microscopy, pp. 751–759. VCH Verlagsges, Weinheim (1997)
103. Duden, T., Bauer, E.: A compact electron spin polarization manipulator. Rev. Sci. Instrum. **66**, 2861–2864 (1995)
104. Grzelakowski, K., Duden, T., Bauer, E., Poppa, H., Chiang, S.: A new surface microscope for magnetic imaging. IEEE Trans. Magn. **30**, 4500–4502 (1994)
105. Bauer, E.: SPLEEM. In: Hopster, H., Oepen, H.P. (eds.) Magnetic Microscopy of Nanostructures, pp. 111–136. Springer, Berlin (2005)
106. Bauer, E.: Spin-polarized low-energy electron microscopy. In: Van Tandeloo, G., Van Dyck, D., Pennycook, S. (eds.) Handbook of Nanoscopy, pp. 697–707. Wiley-VCH, Weinheim (2012)
107. Scheinfein, M.R.: Second order transfer matrixes for inhomogeneous field Wien filters including spin-precession. Optik **82**, 99–113 (1989)

108. Kohashi, T., Matsuyama, H., Koike, K.: A spin rotator for detecting all three magnetization vector components by spin-polarized scanning electron microscopy. Rev. Sci. Instrum. **66**, 5537–5543 (1995)
109. Kohashi, T., Konoto, M., Koike, K.: A spin rotator for spin-polarized scanning electron microscopy. Rev. Sci. Instrum. **75**, 2003–2007 (2004)
110. Kronast, F., Schlichting, J., Radu, F., Mishra, S.K., Noll, T., Dürr, H.A.: Spin-resolved photoemission microscopy and magnetic imaging in applied magnetic fields. Surf. Interface Anal. **42**, 1532–1536 (2010)
111. Tusche, C., Ellguth, M., Ünal, A.A., Chiang, C.-T., Winkelmann, A., Krasyuk, A., Hahn, M., Schönhense, G., Kirschner, J.: Spin resolved photoelectron microscopy using a two-dimensional spin-polarizing electron mirror. Appl. Phys. Lett. **99**, 032505, 3 pages (2011)
112. Lencová, B.: Electrostatic lenses. In: Orloff, J. (ed.) Handbook of Charged Particle Optics, pp. 161–208. CRC Press, Boca Raton, FL (2008)
113. Rempfer, G.F., Nadakavukaren, K.K., Griffith, O.H.: Topographical effects in emission microscopy. Ultramicroscopy **5**, 437–448 (1980)
114. Recknagel, A.: Theorie des elektrischen Elektronenmikroskops für Selbststrahler. Z. Phys. **117**, 689–708 (1941)
115. Recknagel, A.: Zur Theorie des Elektronenmikroskops für Selbststrahler. Jahrbuch der AEG-Forschung **9**, 1–7 (1942)
116. Boersch, H.: Die Verbesserung des Auflösungsvermögens im Emissions-Elektronenmikroskop. Z. Tech. Phys. **23**, 129–130 (1942)
117. Bauer, E.: Optical properties of the uniform electric field. J. Appl. Phys. **35**, 3079 (1964)
118. Bauer, E.: The resolution of the low energy electron reflection microscope. Ultramicroscopy **17**, 51–56 (1985)
119. Cruise, D.R., Bauer, E.: Optical properties of a typical electrostatic immersion lens. J. Appl. Phys. **35**, 3080 (1964)
120. Davisson, C.J., Calbick, C.J.: Electron lenses. Phys. Rev. **42**, 580 (1932)
121. Lenc, M., Müllerova, I.: Electron optical properties of a cathode lens. Ultramicroscopy **41**, 411–417 (1992)
122. Lenc, M., Müllerova, I.: Optical properties and axial aberration coefficients of the cathode lens in combination with a focusing lens. Ultramicroscopy **45**, 159–162 (1992)
123. Wang, Y.L., Lai, M.Y., Shao, Z.: On the optical properties of an electrostatic retarding field lens. J. Vac. Sci. Technol. A **11**, 406–411 (1993)
124. Tromp, R.M., Wan, W., Schramm, S.M.: Aberrations of the cathode objective lens up to fifth order. Ultramicroscopy **119**, 33–39 (2011)
125. Baker, P.O.: Automatic electron trajectory plotting using the electrolytic tank analogue. Br. J. Appl. Phys. **5**, 191–195 (1954)
126. Pizer, H.I., Yates, J.G., Sander, K.F.: An automatic electron trajectory tracer. J. Electron. Control **2**, 65–86 (1965)
127. Liebmann, G.: The field distribution in asymmetrical magnetic electron lenses. Proc. Phys. Soc. B **68**, 679–681 (1955)
128. Rempfer, G.F.: Unipotential electrostatic lenses: paraxial properties and aberrations of focal length and focal point. J. Appl. Phys. **57**, 2385–2401 (1985)
129. Soa, E.A.: Systematische Untersuchungen an elektrostatischen Immersionsobjekten. In: Jenaer Jahrbuch 1959, vol. I, pp. 115–153. Kommissionsverlag VEB Gustav Fischer, Jena (1959)
130. Cruise, D.R.: A numerical method for the determination of an electric field about a complicated boundary. J. Appl. Phys. **34**, 3477–3479 (1963)
131. Hawkes, P.W., Kasper, E.: Principles of Electron Optics, vol. 1, 2. Academic, London (1989)
132. Orloff, J. (ed.): Handbook of Charged Particle Optics. CRC Press, Boca Raton, FL (1997)
133. www.simion.com.
134. www.mebs.co.uk.

135. www.lencova.com.
136. Liebl, H.: Applied Charged Particle Optics. Springer, Berlin (2008)
137. Rose, H.H.: Geometrical Charged-Particle Optics. Springer, Berlin (2009)
138. Heddle, D.W.O.: Electrostatic Lens Systems. Institute of Physics Publ, Bristol (2000)
139. Chmelik, J., Veneklasen, L., Marx, G.: Comparing cathode lens configurations for low energy electron microscopy. Optik **83**, 155–160 (1989)
140. Bauer, E.: Low energy electron microscopy. Rep. Prog. Phys. **57**, 895–938 (1994)
141. Bauer, E.: LEEM and SPLEEM. In: Hawkes, P., Spence, J.C.H. (eds.) Science of Microscopy, pp. 605–656. Springer, New York (2007)
142. Orloff, J., Swanson, L.W.: An asymmetric electrostatic lens for field emission microprobe applications. J. Appl. Phys. **50**, 2494–2501 (1979)
143. Tromp, R.M.: Measuring and correcting aberrations of a cathode objective lens. Ultramicroscopy **111**, 273–281 (2011)
144. Bauer, E.: The possibilities for analytical methods in photoemission and low energy electron microscopy. Ultramicroscopy **36**, 52–62 (1991)
145. Feng, J., Padmore, H., Wei, D.H., Anders, S., Wu, Y., Scholl, A., Robin, D.: Modeling the acceleration field and objective lens for an aberration corrected photoemission electron microscope. Rev. Sci. Instrum. **73**, 1514–1517 (2002)
146. Bernheim, M.: Image acquisition with immersion objective lenses using electrons emitted with several tenths of an electron volt energies: towards high spatial resolution ESCA analysis. Ultramicroscopy **106**, 398–412 (2006)
147. Bernheim, M.: Evaluation of abberations of immersion objective lenses in relation to electron emission microscopy. Eur. Phys. J. Appl. Phys. **36**, 193–204 (2006)
148. Pang, A.B., Müller, T., Altman, M.S., Bauer, E.: Fourier optics of image formation in LEEM. J. Phys. Condens. Matter **21**, 314006, 10 pages (2009)
149. Gribi, M., Thürkauf, M., Villiger, W., Wegmann, L.: Ein 70 kV-Elektronenmikroskop mit kalter Kathode und elektrostatischer Linse. Optik **16**, 65–86 (1959)
150. Boersch, H., Hamisch, H., Ehrlich, W.: Oberflächenentladungen über Isolatoren im Vakuum. Z. Angew. Phys. **15**, 518–525 (1963)
151. Fitzgerald, J.P.S., Word, R.C., Könenkamp, R.: Aberrations in asymmetrical electron lenses. Ultramicroscopy **119**, 40–44 (2012)
152. Liebmann, G., Grad, E.M.: Imaging properties of a series of magnetic electron lenses. Proc. Phys. Soc. B **64**, 956–971 (1951)
153. Liebmann, G.: Magnetic electron microscope projector lenses. Proc. Phys. Soc. B **65**, 94–108 (1952)
154. Liebmann, G.: A unified representation of magnetic electron lens properties. Proc. Phys. Soc. B **68**, 737–745 (1955)
155. Hawkes, P.W. (ed.): Magnetic Electron Lenses. Springer, Berlin (1982)
156. Livingood, J.J.: The Optics of Dipole Magnets. Academic, New York (1969)
157. Archard, G.D., Mulvey, T.: Magnetic deflexion of electron beams without astigmatism. J. Sci. Instrum. **35**, 279–283 (1958)
158. Mankos, M., Kolařik, V., Veneklasen, L.H.: Electron-optical properties of multiple magnetic prism systems. Nucl. Instrum. Methods Phys. Res. A **298**, 189–198 (1990)
159. Kolařík, V., Mankos, M., Veneklasen, L.H.: Close-packed prism arrays for electron microscopy. Optik **87**, 1–12 (1991)
160. Tsuno, K., Munro, E., Rouse, J.: A beam separator of a close packed prism array for low energy electron microscopy. Nucl. Instrum. Methods Phys. Res. A **363**, 276–283 (1995)
161. Degenhardt, R.: Korrektur von Aberrationen in der Teilchenoptik mit Hilfe von Symmetrien. Ph.D. Thesis, Technische Universität Darmstadt (1992)
162. Kan, H.-C., Dürkop, T., Phaneuf, R.J.: Comparison of stigmatically focusing magnetic prisms of square versus round symmetries. J. Vac. Sci. Technol. B **20**, 2519–2525 (2002)
163. Kan, H.-C., Auerbach, D., Phaneuf, R.J.: Approach for investigating the astigmatism of a magnetic prism in low-energy electron microscopy. Rev. Sci. Instrum. **74**, 1008–1015 (2003)

164. Preikszas, D., Rose, H.: Corrected low-energy electron microscope for multi-method operation. In: Jouffrey, B., Colliex, C., et al. (eds.) Proc. 13th Internat. Congr. Electron Microscopy, Paris 1994, pp. 197–198. Les éditions de physique, Les Ulis (1994)
165. Müller, H., Preikszas, D., Rose, H.: A beam separator with small aberrations. J. Electron Microsc. **48**, 191–204 (1999)
166. Preikszas, D., Hartel, P., Spehr, R., Rose, H.: SMART electron optics. In: Frank, L., Čiampor, F. (eds.) Proc. 12th European Congr. Electron Microscopy (EUREM 12) Brno 2000, pp. 181–184. Czech. Soc. Electron Microscopy, Brno (2000)
167. Wu, Y.K., Robin, D.S., Forest, E., Schlueter, R., Anders, S., Feng, J., Padmore, H., Wei, D.H.: Design and analysis of beam separator magnets for third generation aberration compensated PEEMs. Nucl. Instrum. Methods Phys. Res. A **519**, 230–241 (2004)
168. Schmid, P., Feng, J., Padmore, H., Robin, D., Rose, H., Schlueter, R., Wan, W.: Correction and alignment strategies for the beam separator of the photoemission electron microscope 3 (PEEM3). Rev. Sci. Instrum. **76**, 023302, 14 pages (2005)
169. Wan, W., Feng, J., Padmore, H.A.: A new separator design for aberration corrected photoemission electron microscopes. Nucl. Instrum. Methods Phys. Res. A **564**, 537–543 (2006)
170. Rose, H., Plies, E.: Correction of aberrations in electron optical systems with curved axes. In: Hawkes, P. (ed.) Image Processing and Computer-Aided Design in Electron Optics, pp. 344–369. Academic, London (1973)
171. Lichte, H., Möllenstedt, G.: Measurement of the roughness of supersmooth surfaces using an electron mirror interference microscope. J. Phys. E Sci. Instrum. **12**, 941–944 (1979)
172. Rempfer, G.F.: A theoretical study of the hyperbolic electron mirror as a correcting element for spherical and chromatic aberration in electron optics. J. Appl. Phys. **67**, 6027–6040 (1990)
173. Rempfer, G., Mauck, M.S.: Correction of chromatic aberration with an electron mirror. Optik **92**, 3–8 (1992)
174. Rempfer, G.F., Desloge, D.M., Skoczylas, W.P., Griffith, O.H.: Simultaneous correction of spherical and chromatic aberrations with an electron mirror: an electron optical achromat. Microsc. Microanal. **3**, 14–27 (1997)
175. Zworykin, V.K., Morton, G.A., Ramberg, E.G., Hillier, J., Vance, A.W.: Electron Optics and the Electron Microscope. John Wiley, New York (1945)
176. Ramberg, E.G.: Aberration correction with electron mirrors. J. Appl. Phys. **20**, 183–186 (1949)
177. Rempfer, G.F., Mauck, M.S.: Aberration-correcting properties of the hyperbolic electron mirror. In: Bailey, G.W. (ed.) Proc. 43rd Annual Meeting Electron Microscopy Society America, San Francisco 1985, pp. 132–133. San Francisco Press, San Francisco, CA (1985)
178. Shao, Z., Wu, X.D.: A study on hyperbolic mirrors as corrector. Optik **84**, 51–54 (1990)
179. Fitzgerald, J.P.S., Word, R.C., Könenkamp, R.: Adaptive aberration correction using a triode hyperbolic electron mirror. Ultramicroscopy **111**, 1495–1503 (2011)
180. Shao, Z., Wu, X.D.: Adjustable four-electrode electron mirror as an aberration corrector. Appl. Phys. Lett. **55**, 2696–2697 (1989)
181. Preikszas, D.: Korrektur des Farb- und Öffnungsfehlers eines Niederspannungs-Elektronenmikroskops mit Hilfe eines Elektronenspiegels. Ph.D. thesis, Technische Hochschule Darmstadt (1995)
182. Preikszas, D., Rose, H.: Correction properties of electron mirrors. J. Electron Microsc. **1**, 1–9 (1997)
183. Wan, W., Feng, J., Padmore, H.A., Robin, D.S.: Simulation of a mirror corrector for PEEM3. Nucl. Instrum. Methods Phys. Res. A **519**, 222–229 (2004)
184. Wang, L., Munro, E., Rouse, J., Liu, H.: Simulation of electron mirrors by the differential algebraic method. Phys. Procedia **1**, 297–304 (2008)
185. Schramm, S.M., van der Molen, S.J., Tromp, R.M.: Intrinsic instability of aberration-corrected electron optical systems. Phys. Rev. Lett. **109**, 163901, 5 pages (2012)

186. Fitzgerald, J.P.S., Word, R.C., Könenkamp, R.: Simultaneous and independent adaptive correction of spherical and chromatic aberration using an electron mirror and lens combination. Ultramicroscopy **115**, 35–40 (2012)
187. Hawkes, P. (ed.): Aberration-Corrected Electron Microscopy. Advances in Imaging and Electron Physics. Elsevier, Chichester (2008)
188. Rose, H.: Outline of an ultracorrector compensating for all primary chromatic and geometrical aberrations of charged-particle lenses. Nucl. Instrum. Methods Phys. Res. A **519**, 12–27 (2004)
189. Haider, M., Müller, H., Uhlemann, S., Zach, J., Loebau, U., Hoeschen, R.: Prerequisites for a Cc/Cs-corrected ultrahigh-resolution TEM. Ultramicroscopy **108**, 167–178 (2008)
190. Haider, M., Hartel, P., Müller, H., Uhlemann, S., Zach, J.: Current und future aberration correctors for the improvement of resolution in electron microscopy. Proc. R. Soc. A **367**, 3665–3682 (2009)
191. Tsuno, K., Ioanoviciu, D., Martinez, G.: Third-order aberration theory of Wien filters for monochromators and aberration correctors. J. Microsc. **217**, 205–215 (2005)
192. Liu, H., Wang, L., Rouse, J., Munro, E.: Design and optimization of multi-pole lens and Wien filter systems. Nucl. Instrum. Methods Phys. Res. A **645**, 300–306 (2011)
193. Erni, R.: Aberration-Corrected Imaging in Transmission Electron Microscopy. Imperial College Press, London (2010)
194. Liu, X.D., Tang, T.T.: Aberration analysis of Wien filters and design of an electron energy-selective imaging system. Nucl. Instrum. Methods Phys. Res. A **363**, 254–260 (1995)
195. Terauchi, M., Tanaka, M., Tsuno, K., Ishida, M.: Development of a high energy resolution electron energy-loss spectroscopy microscope. J. Microsc. **194**, 203–209 (1999)
196. Niimi, H., Kato, M., Tsutsumi, T., Kawasaki, T., Matsudaira, H., Suzuki, S., Chun, W.-J., Kitajima, Y., Kudo, M., Asakura, K.: Development of imaging energy analyzer using multipole Wien filter. Appl. Surf. Sci. **241**, 131–134 (2005)
197. Niimi, H., Chun, W.-J., Suzuki, S., Asakura, K.: Aberration-corrected multipole Wien filter for energy-filtered x-ray photoemission electron microscopy. Rev. Sci. Instrum. **78**, 0637101–6371018 (2007)
198. Niimi, H., Kato, M., Kawasaki, T., Miyamoto, T., Suzuki, S., Chun, W.-J., Kudo, M., Kawahara, N., Doi, M., Tsukamoto, K., Asakura, K.: Development of in-lab energy-filtered X-ray photoemission electron microscope using air-core-coil-type multipole Wien filter. Surf. Sci. **601**, 4742–4747 (2007)
199. Asakura, K., Niimi, H., Kato, M.: Energy filtered X-ray photoemission electron microscopy. In: Hawkes, P.W. (ed.) Advances in Imaging and Electron Physics, vol. 162, pp. 1–43. Academic, London (2010)
200. Yasufuku, H., Ohminami, Y., Tsutsumi, T., Asakura, K., Kato, M., Sakai, Y., Kitajima, Y., Iwasawa, Y.: Observation of energy-filtered image for X-ray photoemission electron microscopy (EXPEEM) using a retarding Wien-filter energy analyzer. Chem. Lett. **8**, 842–843 (2002)
201. Yasufuku, H., Yoshikawa, H., Kimura, M., Vlaicu, A.M., Kato, M., Kudo, M., Fujikata, J., Fukushima, S.: On the wide-energy-range tuning of x-ray photoemission electron microscope optics for the observation of the photoelectrons excited by several keV x-rays. Rev. Sci. Instrum. **77**, 033702, 6 pages (2006)
202. Yamaguchi, Y., Takakusagi, S., Sakai, Y., Kato, M., Asakura, K., Iwasawa, Y.: X-ray photoemission electron microscopy XPEEM as a new promising tool for the real-time chemical imaging of active surfaces. J. Mol. Catal. A **141**, 129–137 (1999)
203. Takakusagi, S., Kato, M., Sakai, Y., Fukui, K., Asakura, K., Iwasawa, Y.: Development of an X-ray photoemission electron microscopy system with multi-probes, and its application to surface imaging at static and dynamic states. J. Microsc. **200**, 240–250 (2000)
204. Imhof, R.E., Adams, A., King, G.C.: Energy and time resolution of the 180° hemispherical electrostatic analyser. J. Phys. E Sci. Instrum. **9**, 138–142 (1976)

205. Zouros, T.J.M., Benis, E.P.: The hemispherical deflector analyser revisited. I. Motion in the ideal 1/r potential, generalized entry conditions, Kepler orbits and spectrometer basic equation. J. Electron. Spectros. Relat. Phenomena **125**, 221–248 (2002)
206. Zouros, T.J.M., Benis, E.P.: The hemispherical deflector analyser revisited. I. Motion in the ideal 1/r potential, generalized entry conditions, Kepler orbits and spectrometer basic equation. Erratum. J. Electron. Spectros. Relat. Phenomena **142**, 175–176 (2005)
207. Zouros, T.J.M., Benis, E.P., Chatzakis, I.: Optimization of the energy resolution of an ideal ESCA-type hemispherical analyzer. Nucl. Instrum. Methods Phys. Res. B **235**, 535–539 (2005)
208. Zouros, T.J.M.: Theoretical investigation of the energy resolution of an ideal hemispherical deflector analyzer and its dependence on the distance from the focal plane. J. Electron. Spectros. Relat. Phenomena **152**, 67–77 (2006)
209. Zouros, T.J.M.: Theoretical investigation of the energy resolution of an ideal hemispherical deflector analyzer and its dependence on the distance from the focal plane. Erratum. J. Electron. Spectros. Relat. Phenomena **153**, 102–107 (2006)
210. Benis, E.P., Zouros, T.J.M.: The hemispherical deflector analyser revisited II. Electron-optical properties. J. Electron. Spectros. Relat. Phenomena **163**, 28–39 (2008)
211. Lilienkamp, G.: New aspects in LEEM and spectroscopic emission microscopy instrumentation. In: Frank, L., Čiampor, F. (eds.) Proc. 12th European Congr. Electron Microscopy (EUREM 12), Brno 2000, pp. I 177–I 180. Czech. Soc. Electron Microscopy, Brno (2000)
212. Tonner, B.P.: Energy-filtered imaging with electrostatic optics for photoelectron microscopy. Nucl. Instrum. Methods Phys. Res. A **291**, 60–66 (1990)
213. Grzelakowski, K.: A novel imaging energy filter for cathode lens electron microscopy. Ultramicroscopy **116**, 95–105 (2012)
214. Grzelakowski, K.: A flange on electron spectromicroscope with spherical deflector analyzer-simultaneous imaging of reciprocal and real spaces. Ultramicroscopy **130**, 29–35 (2013)
215. www.opticon-nanotechnology.com.
216. Legler, W.: Ein modifiziertes Wiensches Filter als Elektronenmonochromator. Z. Phys. **171**, 424–435 (1963)
217. Boersch, H., Geiger, J., Stickel, W.: Das Auflösungsvermögen des elektrostatisch-magnetischen Energieanalysators für schnelle Elektronen. Z. Phys. **180**, 415–424 (1964)
218. Plies, E., Marianowski, K., Ohnweiler, T.: The Wien filter: history, fundamentals and modern applications. Nucl. Instrum. Methods Phys. Res. A **645**, 7–11 (2011)
219. Kisker, E., Baum, G., Mahan, A.H., Raith, W., Reihl, B.: Electron field emission from ferromagnetic europium sulfide on tungsten. Phys. Rev. B **18**, 2256–2275 (1978)
220. Yasue, T., Suzuki, M., Tsuno, K., Goto, S., Arai, Y., Koshikawa, T.: Novel multipole Wien filter as three-dimensional spin manipulator. Rev. Sci. Instrum. **85**, 043701 (2014)
221. www.prolight.info/pdf_specs/OSRAM-MICROLITHOGRAPHY-HBO.pdf.
222. Zhu, H., Blackborow, P.: LDLS sheds light on analytical-sciences applications. Laser Focus World **47**(12), 53–59 (2011)
223. www.energetiq.com.
224. Feuerbacher, B., Fitton, B.: Experimental investigation of photoemission from satellite surface materials. J. Appl. Phys. **43**, 1563–1572 (1972)
225. Wegmann, L., Buhl, R., Dannöhl, H.-D., Graber, R., Grauer-Carstensen, E., Gribi, M.: Über einige neuere Ergebnisse aus der Photoemissions-Elektronen-Mikroskopie. In: Auwärter, M. (ed.) Ergebnisse der Hochvakuumtechnik und der Physik dünner Schichten, vol. II, pp. 111–130. Wiss. Verlagsges, Stuttgart (1971)
226. Nakagawa, T., Watanabe, K., Matsumoto, Y., Yokoyama, T.: Magnetic circular dichroism photoemission electron microscopy using laser and threshold photo-emission. J. Phys. Condens. Matter **21**, 314010, 6 pages (2009)
227. Jones, M.D., Massey, G.A., Habliston, D.L., Griffith, O.H.: Laser excitatiton in photoelectron microscopy. In: Pfefferkorn, G., Schur, K. (eds.) Proc. First Intern. Conf. Emission Electron

Microscopy/Germany. Beiträge zur elektronenmikroskopischen Direktabbildung von Oberflächen, Tubingen/Germany 1979, pp. 3–38. Verlag R.A. Remy, Münster (1979)
228. Massey, G.A., Jones, M.D., Johnson, J.C.: Nonlinear photoemission for viewing guided or evanescent waves. IEEE J. Quantum Electron. **17**, 1035–1041 (1981)
229. Massey, G.A., Jones, M.D., Johnson, J.C.: Measurement of laser photoelectron damage degradation at high current densities. IEEE J. Quantum Electron. **19**, 873–877 (1983)
230. Bostanjoglo, O., Weingärtner, M.: Pulsed photoelectron microscope for imaging laser-induced nanosecond processes. Rev. Sci. Instrum. **68**, 2456–2460 (1997)
231. Aeschlimann, M., Schmuttenmaer, C.A., Elsayed-Ali, H.E., Miller, R.J.D., Cao, J., Gao, Y., Mantell, D.A.: Observation of surface enhanced multiphoton photoemission from metal surfaces in the short pulse limit. J. Chem. Phys. **102**, 8606–8613 (1995)
232. Schmidt, O., Bauer, M., Wiemann, C., Porath, R., Scharte, M., Andreyev, O., Schönhense, G., Aeschlimann, M.: Time-resolved two photon photoemission electron microscopy. Appl. Phys. B **74**, 223–227 (2002)
233. Munakata, T., Masuda, T., Ueno, N., Abdureyim, A., Sonoda, Y.: Time-resolved photoemission microspectroscopy based on fs-VUV laser light. Surf. Sci. **507–510**, 434–440 (2002)
234. Mikkelsen, A., Schwenke, J., Fordell, T., Luo, G., Klünder, K., Hilner, E., Anttu, N., Zakharov, A.A., Lundgren, E., Mauritsson, J., Andersen, J.N., Xu, H.Q., L'Huillier, A.: Photoemission electron microscopy using extreme ultraviolet attosecond pulse trains. Rev. Sci. Instrum. **80**, 123703, 7 pages (2009)
235. Meyer zu Heringdorf, F.-J., Chelaru, L.I., Möllenbeck, S., Thien, D., Horn-von Hoegen, M.: Femtosecond photoemission microscopy. Surf. Sci. **601**, 4700–4705 (2007)
236. Höfer, A., Duncker, K., Kiel, M., Förster, S., Widdra, W.: Laser-excited PEEM using a fully tunable fs-laser system. IBM J. Res. Dev. **55**(4), 4.1–4.8 (2011)
237. Faure, J., Mauchain, J., Papalazarou, E., Yan, W., Pinon, J., Marsi, M., Perfetti, L.: Full characterization and optimization of a femtosecond ultraviolet laser source for time and angle-resolved photoemission on solid surfaces. Rev. Sci. Instrum. **83**, 043109, 7 pages (2012)
238. Buckanie, N.M., Göhre, J., Zhou, P., von der Linde, D., Horn-von Hoegen, M., Meyer zu Heringdorf, F.-J.: Space charge effects in photoemission electron microscopy using amplified femtosecond laser pulses. J. Phys. Condens. Matter **21**, 314003, 7 pages (2009)
239. Litvinenko, V.N., Burnham, B., Madey, J.M.J., Park, S.H., Wu, Y.: Duke storage ring UWVUV FEL: status and prospects. Nucl. Instrum. Methods Phys. Res. A **375**, 46–52 (1996)
240. Ade, H., Yang, W., English, S.L., Hartman, J., Davis, R.F., Nemanich, R.J.: A free electron laser photoemission electron microscope system (FEL PEEM). Surf. Rev. Lett. **5**, 1257–1268 (1998)
241. De Ninno, G., Trovo, M., Danailov, M., Marsi, M., Karantzoulis, E., Diviacco, B., Walker, R.P., Bartolini, R., Dattoli, G., Giannessi, L., Mezi, L., Couprie, M.E., Gatto, A., Kaiser, N., Gunster, S., Ristau, D.: The UV European FEL at ELETTRA: towards compatibility of storage ring operation for FEL and synchrotron radiation. Nucl. Instrum. Methods Phys. Res. A **507**, 274–280 (2003)
242. De Ninno, G., Allaria, E., Coreno, M., Curbis, F., Danailov, M.B., Karantzoulis, E., Locatelli, A., Menteş, T.O., Niño, M.A., Spezzani, C., Trovò, M.: Generation of ultrashort coherent vacuum ultraviolet pulses using electron storage rings: a new bright light source for experiments. Phys. Rev. Lett. **101**, 053902, 4 pages (2008)
243. Gregoratti, L., Menteş, T.O., Locatelli, A., Kiskinova, M.: Beam-induced effects in soft X-ray photoelectron emission microscopy experiments. J. Electron. Spectros. Relat. Phenomena **170**, 13–18 (2009)
244. Kong, X., Rowe, J.E., Nemanich, R.J.: Single molecule measurements with photoelectron emission microscopy. J. Vac. Sci. Technol. B **26**, 1461–1465 (2008)
245. Attwood, D.: Soft X-rays and Extreme Ultraviolet Radiation. Cambridge University Press, Cambridge (1999)
246. Willmott, P.: An Introduction to Synchrotron Radiation. John Wiley, Chichester (2011)

247. Clarke, J.A.: The Science and Technology of Undulators and Wigglers. Oxford University Press, Oxford (2004)
248. www.elettra.trieste.it/it/lightsources/elettra/elettra-beamlines/nanospectroscopy/beamline-description.html.
249. Cocco, D., Marsi, M., Kiskinova, M., Prince, K.C., Schmidt, T., Heun, S., Bauer, E.: A microfocussing VLS grating based beamline for advanced microscopy. In: Proc. SPIE Conference on EUV, X-Ray, Neutron Optics and Sources, Denver, Colorado 1999, pp. 271–279 (1999)
250. Locatelli, A., Bianco, A., Cocco, D., Cherifi, S., Heun, S., Marsi, M., Pasqualetto, M., Bauer, E.: High lateral resolution spectroscopic imaging of surfaces: the undulator beamline "Nanospectroscopy" at Elettra. J. Phys. IV (Paris) **104**, 99–102 (2003)
251. Locatelli, A., Menteş, T.O., Niño, M.Á., Bauer, E.: Image blur and energy broadening effects in XPEEM. Ultramicroscopy **111**, 1447–1454 (2011)
252. Harada, Y., Masuda, S., Ozaki, H.: Electron spectroscopy using metastable atoms as probes for solid surfaces. Chem. Rev. **97**, 1897–1952 (1997)
253. Morgner, H.: The characterization of liquid and solid surfaces with metastable helium atoms. In: Bederson, B., Walther, H. (eds.) Advances in Atomic, Molecular, and Optical Physics, vol. 42, pp. 387–488. Academic, San Diego, CA (2000)
254. Yasufuku, H., Okumura, M., Ibe, T., Okudaira, K., Harada, Y., Ueno, N.: Surface images of SiO_2/Si(100) pattern using electron emission microscopy with metastable atoms. Photons and low-energy electrons. Jpn. J. Appl. Phys. **40**, 2447–2450 (2001)
255. Wei, H., Beuermann, L., Helmbold, J., Borchardt, G., Kempter, V., Lilienkamp, G., Maus-Friedrichs, W.: Study of SrO segregation on $SrTiO_3$(100) surfaces. J. Eur. Ceram. Soc. **21**, 1677–1680 (2001)
256. Kamada, T., Sogo, M., Aoki, M., Masuda, S.: Mesoscopic pattern formation during initial oxidation of Ni(111) observed by electron emission microscopy. Surf. Sci. **602**, 724–732 (2008)
257. Yamamoto, S., Masuda, S., Yasufuku, H., Ueno, N., Harada, Y., Ichinokawa, T., Kato, M., Sakai, Y.: Study of solid surfaces by metastable electron emission microscopy: energy-filtered images and local electron spectra at the outermost surface layer of silicon oxide on Si(100). J. Appl. Phys. **82**, 2954–2960 (1997)
258. Wei, H., Maus-Friedrichs, W., Lilienkamp, G., Kempter, V., Helmbold, J., Gömann, K., Borchardt, G.: Surface structure of oxygen annealed donor doped SrTiO3(100) single crystals studied with spectroscopic electron microscopy. J. Electroceram. **8**, 221–228 (2002)
259. Lilienkamp, G., Wei, H., Maus-Friedrichs, W., Kempter, V., Marbach, H., Günther, S., Suchorski, Y.: Metastable impact electron emission microscopy of the catalytic H_2 oxidation on Rh(110). Surf. Sci. **532–535**, 132–136 (2003)
260. Lilienkamp, G., Suchorski, Y.: Metastable impact electron emission microscopy: principles and applications. Surf. Interface Anal. **38**, 378–382 (2006)
261. Reimer, L.: Transmission Electron Microscopy, 4th edn. Springer, Berlin (1997)
262. www.tedpella.com.
263. www.a-p-tech.com.
264. Mackie, W.A., Magera, G.G.: Defined emission area and custom thermal electron sources. J. Vac. Sci. Technol. B **29**, 06F601, 4 pages (2011)
265. Kagarice, K.J., Magera, G.G., Pollard, S.D., Mackie, W.A.: Cold field emission from HfC (310). J. Vac. Sci. Technol. B **26**, 868–871 (2008)
266. Forbes, R.G.: Physics of generalized Fowler-Nordheim-type equations. J. Vac. Sci. Technol. B **26**, 788–793 (2008)
267. Cook, B., Verduin, T., Hagen, C.W., Kruit, P.: Brightness limitations of cold field emitters caused by Coulomb interactions. J. Vac. Sci. Technol. B **28**, C6C74–C6C79 (2010)
268. Verduin, T., Cook, B., Kruit, P.: Influence of gun design on Coulomb interactions in a field emission gun. J. Vac. Sci. Technol. B **29**, 06F605, 4 pages (2011)

269. Swanson, L.W., Schwind, G.A.: Review of ZrO/W Schottky cathode. In: Orloff, J. (ed.) Handbook of Charged Particle Optics, pp. 1–28. CRC Press, Boca Raton, FL (2008)
270. Schwind, G.A., Magera, G., Swanson, L.W.: Comparison of parameters for Schottky and cold field emission sources. J. Vac. Sci. Technol. B **24**, 2897–2901 (2006)
271. Milne, W.I., Teo, K.B.K., Mann, M., Bu, I.Y.Y., Amaratunga, G.A.J., De Jonge, N., Allioux, M., Oostveen, J.T., Legagneux, P., Minoux, E., Gangloff, L., Hudanski, L., Schnell, J.-P., Dieumegard, L.D., Peauger, F., Wells, T., El-Gomati, M.: Carbon nanotubes as electron sources. Phys. Status Solidi A **203**(6), 1058–1063 (2006)
272. Nakahara, H., Kusano, Y., Kono, T., Saito, Y.: Evaluations of carbon nanotube field emitters for electron microscopy. Appl. Surf. Sci. **256**, 1214–1217 (2009)
273. de Jonge, N., Allioux, M., Oostveen, J.T., Teo, K.B., Milne, W.I.: Low noise and stable emission from carbon nanotube electron sources. Appl. Phys. Lett. **87**, 133118, 3 pages (2005)
274. de Jonge, N.: Carbon nanotube electron sources for electron microscopes. In: Hawkes, P.W. (ed.) Advances in Imaging and Electron Physics, vol. 156, pp. 203–233. Elsevier, Amsterdam (2009)
275. Oshima, C., Rokuta, E., Itagaki, T., Ishikawa, T., Cho, B., Kuo, H.S., Hwang, I.S., Tsong, T.T.: Demountable single-atom electron source. e-J. Surf. Sci. Nanotechnol. **3**, 412–416 (2005)
276. Ishikawa, T., Urata, T., Cho, B., Rokuta, E., Oshima, C., Terui, Y., Saito, H., Yonezawa, A., Tsong, T.T.: Highly efficient electron gun with a single-atom electron source. Appl. Phys. Lett. **90**, 143120, 3 pages (2007)
277. Chang, C.C., Kuo, H.S., Hwang, I.S., Tsong, T.T.: A fully coherent electron beam from a noble-metal covered W(111) single-atom emitter. Nanotechnology **20**, 115401, 6 pages (2009)
278. Hwang, I.-S., Kuo, H.-S., Chang, C.-C., Tsong, T.T.: Noble-metal covered W(111) single-atom electron sources. J. Electrochem. Soc. **157**, P7–P12 (2010)
279. Nakagawa, T., Rokuta, E., Murata, H., Shimoyama, H., Oshima, C.: Field ion microscopy of nanometer-size pyramid grown on a blunt end of tungsten tip. e-J. Surf. Sci. Nanotechnol. **10**, 12–16 (2012)
280. Nakagawa, T., Rokuta, E., Murata, H., Shimoyama, H., Oshima, C.: Fabrication of a single-atom electron source by noble-metal surface diffusion. J. Vac. Sci. Technol. B **31**, 02B105, 5 pages (2013)
281. Niu, Y.R., Altman, M.S.: Superparamagnetic limit to spin-polarized vacuum tunneling from Fe-coated W(001) tips. Appl. Phys. Lett. **95**, 203113–203115 (2009)
282. Niu, Y.R., Altman, M.S.: Spin polarized field emission from Fe and Co-coated W tips. Surf. Sci. **604**, 1055–1059 (2010)
283. Pierce, D.T.: Spin-polarized electron sources. In: Dunning, F.B., Hulet, R.G. (eds.) Atomic, Molecular and Optical Physics: Charged Particles. Methods in Experimental Physics, Part A, vol. 29, pp. 1–38. Academic, San Diego, CA (1995)
284. Suyama, M.: Latest status of PMTs and related sensors. In: International workshop on new photon-detectors PD07, Kobe University, Kobe, Japan 2007. Proceedings of Science, pp. 1–11
285. Mamaev, Y.A., Gerchikov, L.G., Yashin, Y.P., Vasiliev, D.A., Kuzmichev, V.V., Ustinov, V.M., Zhukov, A.E., Mikhrin, V.S., Vasiliev, A.P.: Optimized photocathode for spin-polarized electron sources. Appl. Phys. Lett. **93**, 081114, 3 pages (2008)
286. Jin, X., Ichihashi, F., Mano, A., Yamamoto, N., Takeda, Y.: Fourfold increase in quantum efficiency in highly spin-polarized transmission-type photocathode. Jpn. J. Appl. Phys. **51**, 108004, 2 pages (2012)
287. Jin, X., Yamamoto, N., Nakagawa, Y., Mano, A., Kato, T., Tanioku, M., Ujihara, T., Takeda, Y., Okumi, S., Yamamoto, M., Nakanishi, T., Saka, T., Horinaka, H., Kato, T., Yasue, T., Koshikawa, T.: Super-high brightness and high-spin-polarization photocathode. Appl. Phys. Express **1**, 045002, 3 pages (2008)

288. Jin, X., Matsuba, S., Honda, Y., Miyajima, T., Yamamoto, M., Utiyama, T., Takeda, Y.: Picosecond electron bunches from GaAs/GaAsP strained superlattice photocathode. Ultramicroscopy **130**, 44–48 (2013)
289. Francoeur, S., Seong, M.-J., Mascarenhas, A., Tixier, S., Adamcyk, M., Tiedje, T.: Band gap of GaAs1-xBix, $0 < x < 3.6\%$. Appl. Phys. Lett. **82**, 3874–3876 (2003)
290. Batool, Z., Hild, K., Hosea, T.J.C., Lu, X., Tiedje, T., Sweeney, S.J.: The electronic band structure of GaBiAs/GaAs layers: influence of strain and band anti-crossing. J. Appl. Phys. **111**, 113108, 7 pages (2012)
291. Kudrawiec, R., Kopaczek, J., Sitarek, P., Misiewicz, J., Henini, M., Novikov, S.V.: Unusual broadening of E0 and $E0 + \Delta SO$ transitions in GaAsBi studied by electromodulation spectroscopy. J. Appl. Phys. **111**, 066103, 3 pages (2012)
292. Gerchikov, L.G., Aulenbacher, K., Mamaev, Y.A., Riehn, E.J., Yashin, Y.P.: Spin polarized electron transport and partial localization of photoelectrons in highly doped photocathodes. J. Phys. Conf. Ser. **298**, 012013, 10 pages (2011)
293. Maldonado, J.R., Liu, Z., Dowell, H., Kirby, R.E., Sun, Y., Pianetta, P., Pease, F.: Robust CsBr/Cu photocathodes for the linac coherent light source. Phys. Rev. ST Accel. Beams **11**, 060702, 5 pages (2008)
294. Uchiyama, S., Takagi, Y., Niigaki, M., Kan, H., Kondoh, H.: GaN-based photocathodes with extremely high quantum efficiency. Appl. Surf. Lett. **86**, 103511, 3 pages (2005)
295. Cultrera, L.: Cathodes for photoemission guns. In: Proc. 2011 Particle Accelerator Conference, pp. 2099–2103, New York, NY (2011)
296. Bazarov, I.V., Dunham, B.M., Liu, X., Virgo, M., Dabiran, A.M., Hannon, F., Sayed, H.: Thermal emittance and response time measurements of a GaN photocathode. J. Appl. Phys. **105**, 083715, 4 pages (2009)
297. Tromp, R.M.: Selected-area diffraction and spectroscopy in LEEM and PEEM. Ultramicroscopy **120**, 73–7 (2012)
298. Yamamoto, I., Matsuura, N., Miyakubo, K., Yamada, T., Munakata, T.: High-precision sample stage for photoemission microscopy of organic films. J. Electron. Spectros. Relat. Phenomena **174**, 131–135 (2009)
299. Ellis, A.W., Tromp, R.M.: A versatile ultra high vacuum sample stage with six degrees of freedom. Rev. Sci. Instrum. **84**, 075112, 7 pages (2013)
300. www.smaract.de.
301. Doran, A., Church, M., Miller, T., Morrison, G., Young, A.T., Scholl, A.: Cryogenic PEEM at the advanced light source. J. Electron. Spectros. Relat. Phenomena **185**, 340–346 (2012)
302. Alvarado, S., Campagna, M., Hopster, H.: Surface magnetism of Ni (100) near the critical region by spin-polarized electron scattering. Phys. Rev. Lett. **48**, 51–54 (1982)
303. Yamaguchi, R., Terashima, K., Fukumoto, K., Takada, Y., Kotsugi, M., Miyata, Y., Mima, K., Komori, S., Itoda, S., Nakatsu, Y., Yano, M., Miyamoto, N., Nakamura, T., Kinoshita, T., Watanabe, Y., Manabe, A., Suga, S., Imada, S.: An XMCD-PEEM study on magnetized Dy-doped Nd-Fe-B permanent magnets. IBM J. Res. Dev. **55**(4), 12.1–12.6 (2011)
304. Heyderman, L.J., Kläui, M., Nöhammer, B., Vaz, C.A.F., Bland, J.A.C., David, C.: Fabrication of nanoscale magnetic ring structures and devices. Microelectron. Eng. **73–74**, 780–784 (2004)
305. Herrmann, K.-H.: Image recording in microscopy. In: Amelinckx, S., van Dyck, D., van Landuyt, J., van Tendeloo, G. (eds.) Handbook of Microscopy, vol. 2, pp. 885–921. VCH, Weinheim (1997)
306. Zuo, J.M.: Electron detection characteristics of slow-scan CCD camera. Ultramicroscopy **66**, 21–33 (1996)
307. Zuo, J.M.: Electron detection characteristics of a slow-scan CCD camera, imaging plates and film, and electron image restoration. Microsc. Res. Tech. **49**, 245–268 (2000)
308. Coates, C., Campillo, C.: CCDs lose ground to new CMOS sensors. Laser Focus World **47**, 40–45 (2011)

309. Fan, G.Y., Ellisman, M.H.: Optimization of thin-foil based phosphor screens for CCD imaging in TEM in the voltage range of 80–400 kV. Ultramicroscopy **66**, 11–19 (1996)
310. Wiza, J.L., Henkel, P.R., Roy, R.L.: Improved microchannel plate performance with a resistive anode encoder. Rev. Sci. Instrum. **48**, 1217–1218 (1977)
311. Edgar, M.L., Kessel, R., Lapington, J.S., Walton, D.M.: Spatial charge cloud distribution of microchannel plates. Rev. Sci. Instrum. **60**, 3673–3680 (1989)
312. Tremsin, A.S., Siegmund, O.H.W.: Spatial distribution of electron cloud footprints from microchannel plates: measurements and modeling. Rev. Sci. Instrum. **70**, 3282–3288 (1999)
313. Shimizu, H., Yasue, T., Koshikawa, T.: Image blur by micro channel plate in LEEM/PEEM. In: Proc. 6th Intern. Symp. Atomic Level Charact. New Mater. Devices (ALC'07), Kanazawa 2007. JSPS 141 Activity Report, pp. 203–206 (2008)
314. Koshida, N., Hosobuchi, M.: Energy distribution of output electrons from a microchannel plate. Rev. Sci. Instrum. **56**, 1329–1331 (1985)
315. Saito, M., Saito, Y., Asamura, K.: Spatial charge cloud size of microchannel plates. Rev. Sci. Instrum. **78**, 023302, 7 pages (2007)
316. Aase, J.G., Burchill, J.K., Knudsen, D.J., Hackett, J.P., Moffat, B.: Spatial resolution and relative brightness of a microchannel plate detector system with P20 and P43 phosphor screens. Opt. Eng. **50**, 064001, 6 pages (2011)
317. Moldovan, G., Matheson, J., Derbyshire, G., Kirkland, A.: Characterisation of a detector based on microchannel plates for electrons in the energy range 10–20 keV. Nucl. Instrum. Methods Phys. Res. A **596**, 402–408 (2008)
318. Daud, T., Janesick, J.R., Evans, K., Elliott, T.: Charge-coupled-devise response to electron beam energies of less than 1 keV up to 20 keV. Opt. Eng. **26**, 686–691 (1987)
319. Stearns, D.G., Wiedwald, J.D.: Response of charge-coupled devices to direct electron bombardment. Rev. Sci. Instrum. **60**, 1095–1103 (1989)
320. Horáček, M.: Charge-coupled device area detector for low energy electrons. Rev. Sci. Instrum. **74**, 3379–3384 (2003)
321. Horáček, M.: Modulation transfer function and detective quantum efficiency of electron bombarded charge coupled device detector for low energy electrons. Rev. Sci. Instrum. **76**, 093704, 6 pages (2005)
322. www.gatan.com/products/digital_imaging/products/K2-Direct-Detection.php.
323. www.panalyticalnow.com.
324. http://www.directelectron.com.
325. Battaglia, M., Contarato, D., Denes, P., Doering, D., Duden, T., Krieger, B., Giubilato, P., Gnani, D., Radmilovic, V.: Characterisation of a CMOS active pixel sensor for use in the TEAM microscope. Nucl. Instrum. Methods Phys. Res. A **622**, 669–677 (2010)
326. Krieger, B., Contarato, D., Denes, P., Doering, D., Gnani, D., Joseph, J., Schindler, S.: Fast, radiation hard, direct detection CMOS imagers for high resolution transmission electron microscopy. In: 2011 I.E. Nuclear Science Symposium, pp. 1946–1949 (2011)
327. van Gastel, R., Sikharulidze, I., Schramm, S., Abrahams, J.P., Poelsema, B., Tromp, R.M., van der Molen, S.J.: Medipix 2 detector applied to low energy electron microscopy. Ultramicroscopy **110**, 33–35 (2009)
328. Sikharulidze, I., van Gastel, R., Schramm, S., Abrahams, J.P., Poelsema, B., Tromp, R.M., van der Molen, S.J.: Low energy electron microscopy imaging using Medipix2 detector. Nucl. Instrum. Methods Phys. Res. A **633**, S239–S242 (2011)
329. Faruqi, A.R., McMullan, G.: Electronic detectors for electron microscopy. Q. Rev. Biophys. **44**, 357–390 (2011)
330. Schauer, R., Autrata, R.: Performance of YAG: Ce single crystal screens for TEM. In: Calderón Benavides, H.A., Yacamán, M.J. (eds.) Proc.14th Intern. Congr. Electron Microscopy, 1998, pp. 633–634. Institute of Physics Publishing, Bristol (1998)

Chapter 4
Theory of Image Formation

Abstract Full understanding of image formation requires the understanding of not only the resolution, discussed in Chap. 3 in terms of geometrical optics, but also the contrast. This is achieved in the wave-optical treatment of image formation, in which both aspects are intimately connected in the contrast transfer function formalism, a formalism which is applied in this chapter not only to the information transfer from the object to the detection system but also to its conversion in the detection system into the final image.

4.1 Introduction

In Chap. 2 we have seen how the electrons, which are used for imaging, are produced, and Chap. 3 has described the instruments used for imaging, their possibilities and limitations. With this background we can now address the problem of image formation. This may be divided into two steps: (1) the transfer of the information contained on the specimen plane to the image plane and (2) the image acquisition process, which in modern instruments is an electronic process while the former is an optical process. The optical process can be described either by geometrical optics as we have implicitly done already in Chap. 3 in connection with the calculation of the resolution, or by wave optics. Geometrical optics (ray optics or particle optics) is essential for following the electron on its way from the object plane to the image plane through the electric and magnetic fields of the electron-optical components of the microscope and gives us the aberrations caused by these components. However, it is insufficient for a complete understanding of the resolution and in particular of the contrast. This is already evident in the expression (3.5) for the resolution, in which one of the parts describes diffraction at the angle-limiting aperture (contrast aperture), clearly a wave-optical phenomenon. Conversely, wave optics needs the input from geometrical optics for the

E. Bauer, *Surface Microscopy with Low Energy Electrons*,
DOI 10.1007/978-1-4939-0935-3_4, © Springer Science+Business Media New York 2014

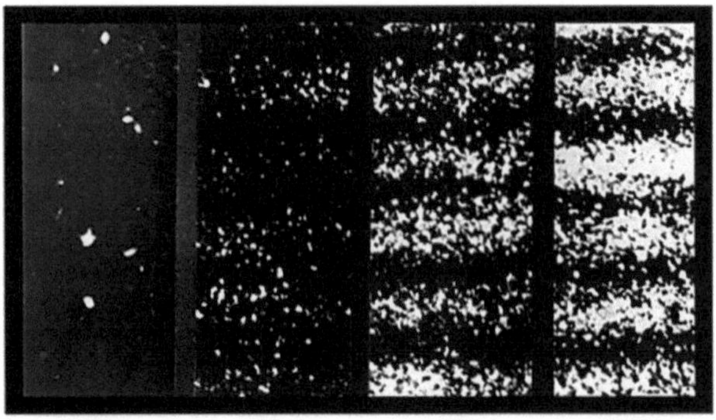

Fig. 4.1 Evolution of contrast and resolution with increasing electron dose. Biprism interference pattern. Reproduced with permission from ref. [1]. © IOP Publishing 2008. All rights reserved

influence of the optical elements, their focal properties, and aberrations. In this Chapter we take this input to be known and calculate resolution and contrast within the framework of wave optics, assuming the exit wave $\psi_{\mathrm{obj}}(\boldsymbol{r})$ at the specimen surface is known. This wave function is the superposition of many wavelets from individual electrons, each of which interferes with itself. This is evident at very low intensities at which the "image" consists of individual spots. A recognizable image develops only after summing over many such "images," that is over long image acquisition times, or at high intensities as illustrated in Fig. 4.1 [1].

With this in mind we can now follow the object wave on its way through the electron optics to the final image plane using the contrast transfer function theory. This method was originally developed in light microscopy in the middle of the twentieth century [2] and shortly thereafter for transmission and scanning electron microscopy as described by Hanszen [3]. Extensive references to this early work can be found in Hawkes's and Kasper's book on wave optics [4]. This book and several others [5–11] contain also a more detailed description of the transfer function theory than the outline given here. The first application of the transfer function theory to emission electron microscopy was made by Storbeck [12] and to LEEM by Müller [13]. An expanded version of Müller's thesis, which is difficult to access, was published by Pang et al. [14]. Schramm et al. [15] have extended the theory to higher-order aberrations, applied it to image formation in aberration-corrected LEEM and to PEEM and have done extensive numerical calculations illustrating the power of the optical transfer theory.

Before presenting contrast transfer in LEEM in terms of the contrast transfer function we briefly discuss Storbeck's work on PEEM. He studied the contrast transfer of electrons emitted with Maxwell energy distribution and $\cos^2\theta$ angular distribution and compared the wave-optical and geometric-optical intensity distribution in the image of small homogeneously emitting regions. At the low energies used in thermionic and threshold photoemission microscopy, apertures do not improve resolution, which is limited by the acceleration field of the objective lens. Therefore his calculations are for the case without aperture. Figure 4.2 shows some of his results for homogeneously emitting discs with radius r_0, normalized to $2\varepsilon_0/F$, on a non-emitting background. $\varepsilon_0 = E_0/e$, where E_0 is the energy of the maximum of

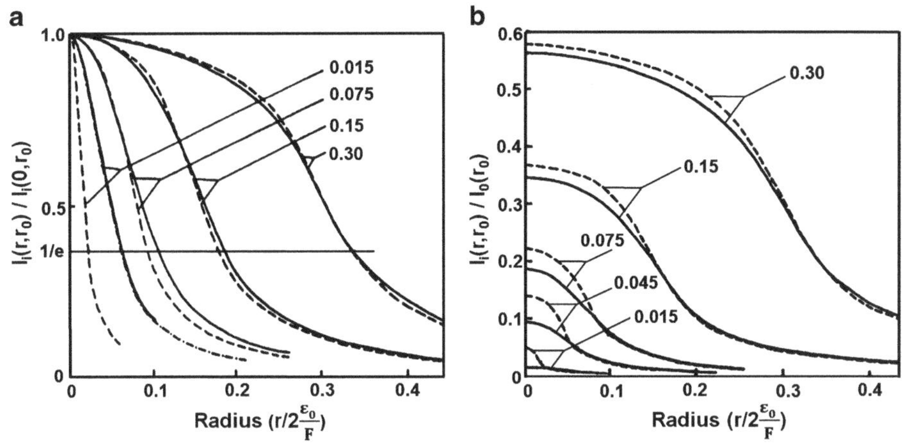

Fig. 4.2 Angular current density distribution in the images of small homogeneously emitting discs with the radii indicated by the curve parameters. *Solid lines*: wave-optical calculation, *dashed lines*: geometric-optical calculation. (**a**) Normalized to the intensity in the center of the disc in the image plane, (**b**) unnormalized. The radii are normalized to $2\varepsilon_0/F = 2lE_0/E$, where l is the length of the acceleration field and E_0 and E are the initial and final energies. Adapted from ref. [12] with permission from Wiley-VCH Verlag GmbH & Co. KGaA, Weinheim

the Maxwell distribution and F the field strength in V/m. All curves are for the optimum focus at $\varepsilon_0/F\lambda_0 = 30/4\pi$, where λ_0 is the wavelength corresponding to ε_0, and the radius r in the image plane is also normalized like r_0. The essential result is that the wave nature of the electron causes a broadening of the intensity distribution in the image plane compared to the geometrical optics distribution and that the broadening increases with decreasing emitting area as clearly seen in the curves normalized to the intensity in the center of the disc in (a). The curves in (b), which are not normalized, indicate a second aspect of the wave nature: the central part of the image is much less intense than according to geometric-optics. This has an important consequence: if the region around the emitting area also shows some emission, the area may not be recognizable. This shows the importance of wave-optics. It not only determines the resolution but also the contrast. The contrast in turn depends upon the signal to noise ratio S/N, which is strongly influenced by the image detection system. Therefore we have to discuss this in addition to the image transfer to the detector in order to understand the image acquired by it.

The need for a wave-optical treatment of the contrast in LEEM became obvious already in the first images published by Telieps and Bauer [16]. The contrast of monatomic steps on Mo(110) surface, which appeared only at certain energies, that is at certain wavelengths, could be explained only by Fresnel diffraction, similar to the light-optical experiment of Kastler [17]. He had observed beautiful interference patterns at the boundary between two transparent plates with a phase shift of $\lambda/2$ ($\Delta\varphi = \pi$) between them. In LEEM only the central minimum was observed and occasionally the first side minimum when the distance between two steps led to the superposition of the first minima. The aspirations to develop a wave-optical theory of contrast, indicated in Telieps's last publication [18] were not realized due to his untimely death. Another phenomenon, the quantum size effect contrast in thin Cu films on Mo(110), reported by Mundschau et al. [19] further called for a wave-optical theory, in addition to the diffraction contrast. However, all efforts during these early years of LEEM in Clausthal were concentrated on instrument development. Only towards the end of this period, when a mathematically gifted student wanted a more theoretical task for his M.S. thesis [13] the problem of the wave-optical theory of image formation in LEEM was taken up again. For the special case of the step contrast a wave-optical model in the spirit of Kastler's approach was developed independently by Chung and Altman [20], also motivated by experiment.

The influence of the imaging system on the information transfer from object to image can be described either in real space or in reciprocal space (Fourier space). In an ideal system the object details would be perfectly reproduced in the image plane at a larger scale given by the magnification and inverted or rotated, depending upon the type of lens. In a real system the object details are blurred in the image, a phenomenon which is described in real space by the point spread function PSF $h(\boldsymbol{r})$, sometimes also called impulse response function, in reciprocal space by the contrast transfer function (CTF) $H(\boldsymbol{q})$, which is related to $h(\boldsymbol{r})$ via the Fourier transformation

$$H(\boldsymbol{q}) = \int_{-\infty}^{+\infty} h(\boldsymbol{r}) \exp(-2\pi i \boldsymbol{q} \cdot \boldsymbol{r}) \mathrm{d}\boldsymbol{r}. \tag{4.1}$$

$\boldsymbol{q}$ has the dimension of inverse length and is called spatial frequency in analogy to the inverse relation between temporal frequency and time. When the waves emanating from the object are (partially) *coherent*—as in reflection microscopies (LEEM and MEM) and to some extent in energy-filtered PEEM—they can interfere on their way to the M-times magnified image. Therefore, the *wave functions* are blurred by the point spread function via the convolution with $h(\boldsymbol{r})$

$$\psi(\boldsymbol{r}') = \frac{1}{M} \int_{-\infty}^{+\infty} \psi(\boldsymbol{r}) h(\boldsymbol{r}' - \boldsymbol{r}) \mathrm{d}\boldsymbol{r} \equiv \psi(\boldsymbol{r}) \otimes h(\boldsymbol{r}). \tag{4.2}$$

The intensity distribution in the image is then

$$I(\boldsymbol{r}') = \psi(\boldsymbol{r}')\psi^*(\boldsymbol{r}') = \frac{1}{M^2} \left| \int_{-\infty}^{+\infty} \psi(\boldsymbol{r}) h(\boldsymbol{r}' - \boldsymbol{r}) \mathrm{d}\boldsymbol{r} \right|^2 = \frac{1}{M^2} |\psi(\boldsymbol{r}) \otimes h(\boldsymbol{r})|^2. \tag{4.3}$$

The simplest case of blurring is that caused by diffraction at the aperture in the back focal plane of the lens but the lens itself contributes significantly to the blurring too. With an ideal, aberration-free lens and ideal illumination (monochromatic plane wave) and a "point source" ($\psi(\boldsymbol{r}) = \delta(\boldsymbol{r})$) the intensity distribution in the image is $I(\boldsymbol{r}') = |h(\boldsymbol{r})|^2/M^2$, the well-known Airy pattern from light optics.

When the waves emanating from the object are *incoherent*, which is the case in emission microscopies, they cannot interfere and it is the *intensity* distribution which is modified by the optical system. The intensity distribution in the image is then (see e.g. [2], p. 484)

$$I(\boldsymbol{r}') = \frac{1}{M^2} I(\boldsymbol{r}) \otimes |h(\boldsymbol{r})|^2. \tag{4.4}$$

$|h(\boldsymbol{r})|^2$ is called the *incoherent* point spread function or impulse response function. These relations in real space have their counterparts in reciprocal space by taking their Fourier transforms. The convolution theorem states that the Fourier transform of the convolution of two functions is the product of their individual Fourier transforms. For example, Eq. (4.3) is in reciprocal space

$$\mathcal{F}\{I(\boldsymbol{r}')\} \equiv I(\boldsymbol{q}') = \mathcal{F}\left\{I(\boldsymbol{r}) \otimes |h(\boldsymbol{r})|^2\right\} = \mathcal{F}\{I(\boldsymbol{r})\} \bullet \mathcal{F}\left\{|h(\boldsymbol{r})|^2\right\}. \tag{4.5}$$

The Fourier transform of $|h(\boldsymbol{r})|^2$ in turn may be expressed by the self-convolution (autocorrelation function) of $H(\boldsymbol{q})$ (see e.g. ref. [1], p. 485)

$$\mathcal{F}\left\{|h(\boldsymbol{r})|^2\right\} = H(q) \times H^*(\boldsymbol{q}), \tag{4.6}$$

a relationship, which will be needed later in the discussion of emission microscopies. In the following we treat the information transfer in reciprocal space.

4.2 Wave Propagation: The Contrast Transfer Function

4.2.1 *Low Energy Electron Microscopy: The Wave Amplitude*

The transfer function approach taken by Müller [13] and in the subsequent work is an extension of the method used in transmission electron microscopy (TEM) to the case in which the specimen is not a weak phase object, that is in which the object wave

$$\psi_{\text{obj}}(\boldsymbol{r}) = \exp(\alpha(\boldsymbol{r}) - i\varphi(\boldsymbol{r})) \tag{4.7}$$

may not be expanded into a Taylor series keeping only the linear terms

$$\psi_{\text{obj}}(\boldsymbol{r}) \approx 1 + \alpha(\boldsymbol{r}) - i\varphi(\boldsymbol{r}) \tag{4.7a}$$

as it is generally done in TEM. Here $\boldsymbol{r}$ is the position vector in the object plane, which is normal to the optical axis. $\alpha(\boldsymbol{r})$ is the local reflection amplitude and $\varphi(\boldsymbol{r})$ the local phase shift upon reflection caused by the object topography and potential variations. LEEM and TEM differ also in another aspect: in TEM, at least in high resolution TEM (HRTEM) several diffracted beam contribute to the image intensity and the interference between these beams has to be taken into account via the so-called transmission cross-coefficient, which can make the transfer function nonlinear. In LEEM the image is formed by one diffracted beam—usually the specular reflected beam (bright field imaging) or a nonspecular diffracted beam (dark field imaging) and the much weaker waves in its surrounding. This simplifies the information transfer process through the lens and leads to a linear transfer function. Except for these differences the formalism developed for TEM can be transferred immediately to LEEM if the real object is replaced by the virtual object discussed in Chap. 3 and $\boldsymbol{r}$ is referred to this plane. Between the virtual object and the subsequent optical system the electrons have constant energy E just like in TEM. The illumination part of the system can then be treated like in TEM and the transmission in TEM is replaced by reflection. In particular the reduction of the large exit angles θ_0 at the real object, at which the exit energy is E_0, to the small exit angles θ at the virtual object ($\sin\theta/\sin\theta_0 = (E_0/E)^{1/2}$) allows now to use the paraxial approximation (small angle approximation). In addition we assume that all object points irrespective of their distance from the optical axis may be considered in the same manner as the point on the optical axis (isoplanatic approximation). The second approximation is well justified in LEEM, where the imaged object region is usually about 10 μm or less in diameter but may be marginal in MEM and PEEM where frequently much larger fields of view, for example with 100 μm diameter or more, are used.

The object wave is modified by a number of factors on its way to the final image plane. Important parts of the wave are cut off by the contrast aperture, the spherical and chromatic aberration of the objective lens deform the wave as does a defocus. In addition the illumination is not a plane wave but convergent, not monochromatic and there are high voltage and lens current or voltage instabilities. Furthermore image drift and vibrations relative to the objective lens may be present, which is always the case to some extent in a real system. All of these factors have to be taken into account in the contrast transfer function. We do not include optical elements beyond the objective lens such as the beam separator or energy filter because the factors just mentioned are the crucial ones. The aberrations of the beam separator have been found to be uncritical at the present state of the resolution of the image detection system [15].

In the absence of all these wave-deforming factors the imaging process is straightforward: the objective lens produces the diffraction pattern of the object in the back focal plane and the image in the image plane, using the complete wave field from the object. We assume a monochromatic plane wave at normal incidence which produces from each object point $\boldsymbol{r}$ reflected waves. When we expand the wave $\psi_{obj}(\boldsymbol{r})$ emitted from each object point $\boldsymbol{r}$ into plane waves $\exp(i\,2\pi\boldsymbol{k}\bullet\boldsymbol{r})$—with $|\boldsymbol{k}| = 1/\lambda$—then the wave fronts with wave vector $\boldsymbol{k}$ have a path difference Δs_k compared to the wave fronts along the optical axis ($\boldsymbol{k}_0$). It can be derived quite general from the optical path difference of the waves from two points O and P as

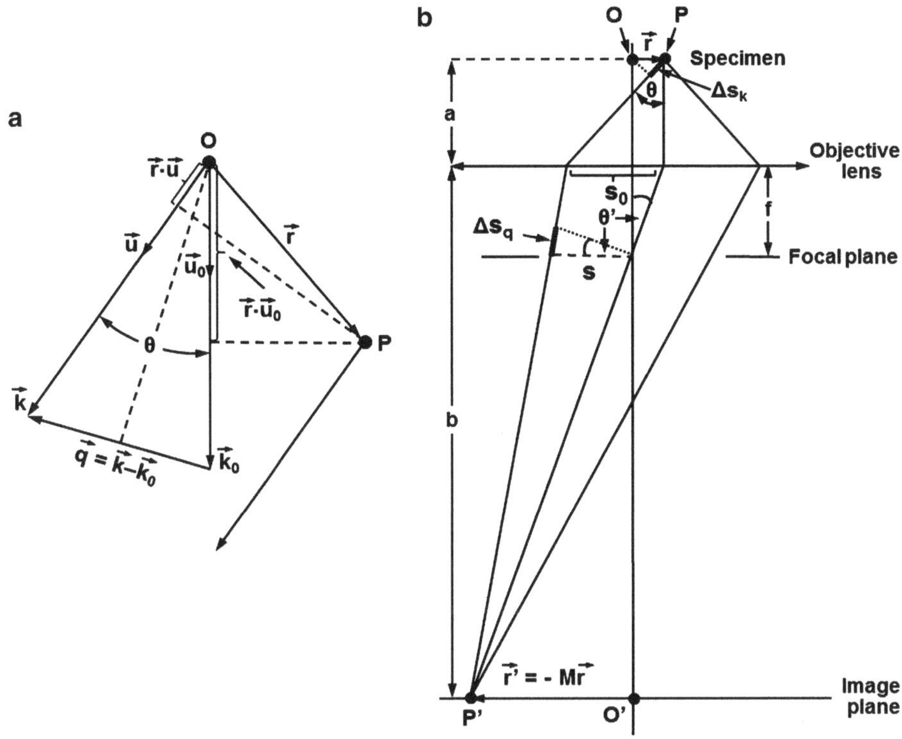

Fig. 4.3 (**a**) Path difference between waves scattered from two points (O and P) in directions $\boldsymbol{k}$ and $\boldsymbol{k}_0$. (**b**) Path difference between waves with directions $\boldsymbol{k}$ and $\boldsymbol{k}_0$ between back focal plane and image plane. The lateral dimensions are strongly enhanced compared to the longitudinal dimensions. In reality the angles are in the mrad range. For explanation see text. Adapted from ref. [7] with permission from Springer Science + Business Media

indicated in Fig. 4.3a: $\Delta s_k = \boldsymbol{r} \bullet \boldsymbol{u}_0 - \boldsymbol{r} \bullet \boldsymbol{u}$, where $\boldsymbol{u}_0 = \lambda \boldsymbol{k}_0$ and $\boldsymbol{u} = \lambda \boldsymbol{k}$ are unit vectors along $\boldsymbol{k}$ and $\boldsymbol{k}_0$, respectively. The corresponding phase difference is

$$\Delta\varphi_k = \frac{2\pi}{\lambda}\Delta s_k = -\frac{2\pi}{\lambda}\lambda(\boldsymbol{k} - \boldsymbol{k}_0) = -2\pi\boldsymbol{q} \bullet \boldsymbol{r}. \tag{4.8}$$

As seen in Fig. 4.3a $q = |\boldsymbol{q}| = |\boldsymbol{k} - \boldsymbol{k}_0| = 2k\ \sin(\theta/2) \approx \theta/\lambda$ for small angles. Because the waves from all object points with the same $\boldsymbol{k}$ direction intersect at the same point in the back focal plane with the phase difference $\Delta\varphi_k$ this plane displays the spatial frequency spectrum

$$F(\boldsymbol{q}) = \int \psi_{\mathrm{obj}}(\boldsymbol{r})\exp(i\Delta\varphi_k)\mathrm{d}A = \int \psi_{\mathrm{obj}}(\boldsymbol{r})\exp(-2\pi i\boldsymbol{q} \bullet \boldsymbol{r})\mathrm{d}\boldsymbol{r} \tag{4.9}$$

This is the Fourier transform $\widetilde{\psi}_{\mathrm{obj}}(\boldsymbol{r})$ of the object wave function, with the integration extending over all points from which electrons exit. $F(\boldsymbol{q})F^*(\boldsymbol{q})$ is the intensity distribution in the back focal plane, i.e. in crystalline materials the intensity distribution in the LEED pattern.

Beyond the back focal plane all waves with different $\boldsymbol{q}$—or equivalently different $\boldsymbol{k}$—from the same object point $\boldsymbol{r}$ have again phase differences $\Delta\varphi_{\boldsymbol{q}}$ when they are reunited in the Gaussian image plane in the image point at a distance $\boldsymbol{r}'$ from the optical axis with $|\boldsymbol{r}'| = M\,|\boldsymbol{r}|$, M being the magnification but $\boldsymbol{r}'$ pointing in a direction different from $\boldsymbol{r}$ in the case of a magnetic lens and opposite to $\boldsymbol{r}$ in an electrostatic lens. As indicated in Fig. 4.3b, when a/b is small then s in the focal plane is approximately equal to s_0 in the objective lens so that at high magnification $(a \approx f)$ $s \approx s_0 \approx a\theta \approx f\theta$. With $\theta \approx \lambda q$ (see above) and $\theta' \approx r/f$ we obtain for the path difference $\Delta s_{\boldsymbol{q}} \approx \theta' s \approx +\lambda qr$ and thus for the phase difference $\Delta\varphi_{\boldsymbol{q}} = 2\pi\Delta s_{\boldsymbol{q}}/\lambda \approx +2\pi\boldsymbol{q} \bullet \boldsymbol{r}$. In order to obtain the wave amplitude in the image point $\boldsymbol{r}'$ we have to integrate over all $\boldsymbol{q}$:

$$\psi_{\text{im}}(\boldsymbol{r}') = \frac{1}{M}\int F(\boldsymbol{q})\exp(+2\pi i\boldsymbol{q} \bullet \boldsymbol{r})\mathrm{d}\boldsymbol{q} = \frac{1}{M}\psi_{\text{obj}}(\boldsymbol{r}). \tag{4.10}$$

Thus $\psi_{\text{im}}(\boldsymbol{r}')$ is the inverse Fourier transform of the Fourier transform $F(\boldsymbol{q}) = \widetilde{\psi}_{\text{obj}}(\boldsymbol{q})$ of the object wave $\psi_{\text{obj}}(\boldsymbol{r})$, which reproduces the object wave with the amplitude reduced by the linear magnification M. The image intensity $I(\boldsymbol{r}') = \psi_{\text{im}}(\boldsymbol{r}')\ \psi^*_{\text{im}}(r')$ is then $\psi_{\text{obj}}(\boldsymbol{r})\psi^*_{\text{obj}}(\boldsymbol{r})/M^2$. The factor $1/M^2$ ensures conservation of the intensity during the magnification process.

This simple 1:1 ratio is no longer true when an aperture with radius $r_{\text{A}} = 1/q_{\max}$ is inserted into the back focal plane because this eliminates waves whose $\boldsymbol{q}$ vectors intersect the focal plane beyond the aperture diameter. We have therefore to insert in Eq. (4.10) an aperture function

$$A(\boldsymbol{q}) = \begin{cases} 1 \text{ for } |\boldsymbol{q}| < q_{\text{A}} \\ 0 \text{ for } |\boldsymbol{q}| \geq q_{\text{A}} \end{cases} \tag{4.11}$$

which changes the image wave function to

$$\psi_{\text{im}}(\boldsymbol{r}') = \frac{1}{M}\int \widetilde{\psi}_{\text{obj}}(\boldsymbol{q})A(\boldsymbol{q})\exp(+2\pi i\boldsymbol{q} \bullet \boldsymbol{r})\mathrm{d}\boldsymbol{q}. \tag{4.12}$$

Thus even in absence of any wave distortions the image is no longer a true magnified reproduction of the object because large spatial frequencies are cut off by the aperture. This means that short spatial distances are eliminated, resulting in loss of resolution. For example for a point source the image is the Airy disc. This effect is included in the geometrical optics theory of image formation by the diffraction at the aperture in Eqs. (3.5 and 3.6), which limits the resolution.

Next we consider the wave front deformation by spherical and chromatic aberration as well as by defocus. In Chap. 3 we have seen that the aberrations displace the image point in the Gaussian image plane by a distance Δr given by

Eq. (3.5) relative to its position in the absence of aberrations. A defocus Δz_0 causes the additional displacement $\Delta z_0 \theta$. Neglecting astigmatism, other aberrations which destroy axial symmetry and off-axis aberrations we have

$$\Delta r = -\Delta z_0 \theta + C_c \rho \theta + C_3 \theta^3 + C_{cc} \rho^2 \theta + C_{3c} \rho \theta^3 + C_5 \theta^5. \qquad (4.13)$$

From Δr the path difference between the distorted and undistorted wave is obtained by integration over θ ([7], p. 71, [21]):

$$\Delta s = -\frac{1}{2}\Delta z_0 \theta^2 + \frac{1}{2} C_c \rho \theta^2 + \frac{1}{4} C_3 \theta^4 + \frac{1}{2} C_{cc} \rho^2 \theta^2 + \frac{1}{4} C_{3c} \rho \theta^4 + \frac{1}{6} C_5 \theta^6. \qquad (4.13a)$$

Setting $\theta \approx \lambda q$ (see above) and dividing with λ we obtain the path difference in units of λ

$$\begin{aligned} \chi &= -\frac{1}{2}\Delta z_0 \lambda q^2 + \frac{1}{4} C_3 \lambda^3 q^4 + \frac{1}{6} C_5 \lambda^5 q^6 + \frac{1}{2} C_c \rho \lambda q^2 + \frac{1}{2} C_{cc} \rho^2 \lambda q^2 + \frac{1}{4} C_{3c} \rho \lambda^3 q^4 \\ &= \sum_{i=1}^{3} \chi_i + \sum_{i=4}^{6} \chi_i = \chi_c + \chi_{ic}. \end{aligned} \qquad (4.14)$$

Multiplication with 2π gives the phase shift $\Delta\varphi$. In Eq. (4.14) we have arranged the path differences in two groups. The first three generate the so-called geometric or coherent aberrations, which do not change the sharpness of the wave fronts. The second three are the chromatic aberrations which blur the wave fronts and destroy or reduce coherency. In addition we will also have to take into account that the object is not illuminated by a plane wave but by waves with wave vectors distributed over an angular region, which results in lateral image shifts resulting in a blurring of the wave fronts and loss of coherency. The first sum in Eq. (4.14), is called wave aberration function and

$$\exp\left(2\pi i\left(-\frac{1}{2}\Delta z_0 \lambda q^2 + \frac{1}{4} C_3 \lambda^3 q^4 + \frac{1}{6} C_5 \lambda^5 q^6\right)\right) = \exp(2\pi i \chi_c(q, \Delta z_0)\,) \qquad (4.15)$$

is the coherent contrast transfer function. If we abbreviate Eq. (4.15) with $W_c(\boldsymbol{q}, \Delta z_0)$ and the corresponding term for the incoherent contributions with $W_{ic}(\boldsymbol{q}, \rho)$ then the aberrations change (Eq. 4.12) to

$$\psi_{im}(\boldsymbol{r}') = \frac{1}{M}\int \tilde{\psi}_{obj}(\boldsymbol{q}) A(\boldsymbol{q}) W_c(\boldsymbol{q}, \Delta z_0) W_{ic}(\boldsymbol{q}, \rho) \exp(+2\pi i \boldsymbol{q} \bullet \boldsymbol{r}) d\boldsymbol{q}. \qquad (4.16)$$

Equation (4.16) contains only the effects of the spherical and of the chromatic aberration, of the intentional defocus, of the aperture, and of the energy spread-dependent incoherent contribution. Implicitly plane wave illumination is assumed. Although the goal of the illumination optics is to produce a point source in the back focal plane of the objective lens, so that the object is illuminated by a plane wave, in reality the beam has a finite diameter in the focal plane so that the object is illuminated by a nonparallel beam. This may be described by plane waves $\psi_t(\boldsymbol{k}')$ arriving from different directions $\boldsymbol{t}=\boldsymbol{k}'-\boldsymbol{k}$ within an illumination cone with a direction distribution $s(\boldsymbol{t})$ of the wave vectors, over which we have to integrate. $s(\boldsymbol{t})$ can to a good approximation be described by a Gaussian distribution function

$$s(\boldsymbol{t}) = \frac{1}{\sqrt{2\pi}q_{\text{ill}}}\exp\left(-\frac{1}{2}\left(\frac{\boldsymbol{t}}{q_{\text{ill}}}\right)^2\right), \tag{4.17}$$

where $q_{\text{ill}}=\alpha_{\text{ill}}/\lambda$ is the spatial frequency corresponding to the standard deviation α_{ill} of the angular distribution in the illumination cone. Because of the tilted illumination the tilt $\boldsymbol{t}$ enters $W(\boldsymbol{q})$ and the aperture function $A(\boldsymbol{q})$. When the angular range of the illumination cone is much smaller than the angular range accepted by the aperture in the back focal plane ($\boldsymbol{t}\ll\boldsymbol{q}$) so that $|\boldsymbol{q}+\boldsymbol{t}|<q_{\text{A}}$ then we can neglect $\boldsymbol{t}$ in $A(\boldsymbol{q}+\boldsymbol{t})$ and can write $A(\boldsymbol{q})A^*(\boldsymbol{q})=A(q)$ because of the circular aperture assumed in Eq. (4.11). Thus the influence of the angular distribution of the illumination has to be taken into account only in $W(\boldsymbol{q}+\boldsymbol{t})$.

In order to determine the influence of nonparallel illumination on the transfer function we assume that the phase changed caused by it is small compared to that from parallel illumination. Then χ_{c} $(\boldsymbol{q}+\boldsymbol{t})$ may be expanded into a Taylor series of $\boldsymbol{t}$ and only the linear terms in $\boldsymbol{t}$ retained [22]:

$$\chi_{\text{c}}(\boldsymbol{q}+\boldsymbol{t}) = \chi_{\text{c}}(\boldsymbol{q}) + \boldsymbol{t}\frac{\partial\chi_{\text{c}}}{\partial q}\bigg|_q + \dots. \tag{4.18}$$

From Eq. (4.14) we obtain

$$\partial\chi_{\text{c}}/\partial q = -\Delta z_0\lambda q + C_3\lambda^3q^3 + C_5\lambda^5q^5 \equiv u(q). \tag{4.19}$$

The contribution $\boldsymbol{t}u(q)$ to χ_{c} in Eq. (4.14) has to be weighted with the distribution function $s(\boldsymbol{t})$. This gives the additional exponential function

$$\int_{-\infty}^{+\infty} s(\boldsymbol{t})\exp(2\pi i t u)\big)\mathrm{d}\boldsymbol{t} = s(u), \tag{4.20}$$

which is the Fourier transform of the distribution function $s(\boldsymbol{t})$:

$$\tilde{s}(\boldsymbol{t}) \equiv s(u) = \exp\left(-2\pi^2q_{\text{ill}}^2u^2\right). \tag{4.21}$$

Inserting the expression (4.19) for u one finally obtains the function

$$s(q) = \exp\left(-2q_{\text{ill}}^2\left(C_5\lambda^5q^5 + C_3\lambda^3q^3 - \Delta z_0\lambda q\right)^2\right) \equiv E_{\text{S}}(q), \tag{4.22}$$

which, being a real function, decreases with increasing q. This so-called (transverse or spatial) source envelope function $E_S(q)$ is multiplied with the oscillating coherent transfer function $W_c(\boldsymbol{q}, \Delta z_0)$ (Eq. 4.15)

$$W_c(\boldsymbol{q}, \Delta z_0) = \exp(2\pi i \chi_c) = \exp\left(\pi i\left(-\Delta z_0 \lambda q^2 + \frac{1}{2} C_3 \lambda^3 q^4 + \frac{1}{3} C_5 \lambda^5 q^6\right)\right), \quad (4.23)$$

and therefore reduces its amplitude.

Next we consider the influence of instrument-related focus changes which have to be added to the intentional focus change Δz_0 in $W(\boldsymbol{q})$. They are caused by high voltage fluctuations ΔU and lens current fluctuations ΔI in magnetic lenses (or ΔU_L in electrostatic lenses). Lens current fluctuations cause focus fluctuations Δf_I and high voltage fluctuations cause focus fluctuations Δf_U. The influence of these focus fluctuations on the transfer function was first taken into account by Hanssen and Trepte [23] in transmission electron microscopy, together with the focus smearing Δf_E caused by the energy distribution in the illuminating beam, with a common chromatic aberration constant C_c as $\Delta z_c = C_c\ (\Delta f_I + \Delta f_U + \Delta f_E)$. In the more recent literature the joint effect of the Δf's on the focus smearing has generally been taken into account by $\Delta z_c = C_c\ (\Delta f_I^2 + \Delta f_U^2 + \Delta f_E^2)^{1/2}$ with $\Delta f_I = -2\Delta I/I$ and $\Delta f_U = \Delta U/U$, based on the approximation $f \sim U/I^2$ for the focal length of a weak magnetic lens. However $\Delta f_I = -2f\Delta I/I$ and $\Delta f_U = f\Delta U/U$, so that the expressions for Δz_c above implicitly assume that the focal length is of the same magnitude as the chromatic aberration constant $C_{cE} \equiv C_c$ describing the influence of the energy distribution ΔE via $\Delta f_E = C_c\ \Delta E/E$. For the magnetic immersion lenses used in the instruments described in this book the expression $f \sim U/I^2$ is valid only approximately because it is a compound lens consisting of a divergent lens (accelerating field) and a convergent lens (magnetic field). Numerical calculations for the objective lens, on which the contrast and resolution analysis of refs. [13, 14] is based, have indeed shown that $C_{cU} \approx 0.5\ C_{cI}$ is nearly an order of magnitude smaller than C_{cE}. Therefore in instruments with current and voltage stabilities of 10^{-5} to 10^{-6} and energy widths in the 10^{-1} to 1 eV range the influence of current and high voltage fluctuations on the focus smearing can be neglected. Schramm et al. [15] have shown that this is actually true also for the aberration-corrected case, even when all C_c's are set equal and even for the narrow energy distribution of a field emission gun. In practice other effects to be discussed later play a more important role. The influence of the focus smearing Δz_C on the transfer function has in the past been taken into account by expanding $\chi_{ic}\ (\Delta z_0 + \Delta z_c)$ in a Taylor series of Δz_c in the same manner as $\chi_c\ (\boldsymbol{q} + \boldsymbol{t})$ was expanded above as function of $\boldsymbol{t}$ [13, 14, 24]. Having neglected the contributions of high voltage and current fluctuations to Δz_c, only the contribution of the energy distribution remains. This, however, can be calculated directly without invoking $\Delta z_c \ll \Delta z_0$ as required in the expansion of χ_{ic} by evaluating the incoherent transfer function $W_{ic}(\boldsymbol{q}, \rho)$ taking the energy distribution ρ into account. This was shown by Schramm et al. [15], whose derivation we follow here.

We write for ρ explicitly ε/E and assume for the energy distribution ε a Gaussian distribution centered at the mean start energy of the electrons

$$f(\varepsilon) = \frac{1}{\sqrt{2\pi}\sigma}\exp\left(-\frac{1}{2}\left(\frac{\varepsilon}{\sigma}\right)^2\right) \tag{4.24}$$

with the standard deviation σ, which is connected to the FWHM ΔE via $\sigma = \Delta E/2\,(2\ln 2)^{1/2}$. In the evaluation of $w_{ic}(q,\ \rho) = \exp(2\pi i\chi_{ic}\ (q,\ \rho))$ $\rho = \varepsilon/E$ has to be weighted with this distribution. To illustrate the calculation we consider first the lowest-order term of χ_{ic} in Eq. (4.14)

$$\chi_{ic} = \frac{1}{2}C_c\rho\lambda q^2 + \frac{1}{2}C_{cc}\rho^2\lambda q^2 + \frac{1}{4}C_{3c}\rho\lambda^3 q^4. \tag{4.14a}$$

This leads to the envelope function

$$E_C^1(q) = \frac{1}{\sqrt{2\pi}\sigma}\int_{-\infty}^{\infty}\exp\left(-\frac{1}{2}\left(\frac{\varepsilon}{\sigma}\right)^2 + i\pi C_c\frac{\varepsilon}{E}\lambda q^2\right)\mathrm{d}\varepsilon. \tag{4.25}$$

If we call the coefficients of ε^2 and ε a^2 and b, respectively, then this integral and its solution becomes [25]

$$\int_{-\infty}^{\infty}\exp(-a^2x^2 \pm bx)\mathrm{d}x = \frac{\sqrt{\pi}}{a}\exp\left(\frac{b^2}{4a^2}\right). \tag{4.26}$$

Thus

$$E_C^1\left(q, \frac{\Delta E}{E}\right) = \exp\left(-\frac{(\pi C_c\lambda)^2}{16\ln 2}q^4\left(\frac{\Delta E}{E}\right)^2\right), \tag{4.27}$$

where we have used $\sigma = \Delta E/2(2\ln 2)^{1/2}$ in the exponent. Interestingly, the Δz_C expansion mentioned above gives the same result [13, 14]. In the same manner the total chromatic (or longitudinal) envelope function $E_{\text{Ctotal}}(q)$ can be computed using for χ_{ic} (Eq. 4.14a). In this case

$$a^2 = (2\sigma^2)^{-1} - i\pi C_{cc}\lambda q^2/E^2 \quad \text{and} \quad b = i\pi(C_c\lambda q^2 + C_{3c}\lambda^3 q^4/2)/E.$$

Inserting these expressions in Eq. (4.22) and expressing σ in the exponent again by ΔE finally yields

$$E_{\text{Ctotal}}(q) = E_{cc}(q)\exp\left[-E_{cc}^2(q)\frac{\pi^2}{16\ln 2}\left(\frac{\Delta E}{E}\right)^2(C_c\lambda q^2 + C_{3c}\lambda^3 q^4/2)^2\right] \tag{4.28}$$

with

$$E_{cc}(q) = \left(1 - i\frac{\pi}{4\ln 2}\left(\frac{\Delta E}{E}\right)^2 C_{cc}\lambda q^2\right)^{-1/2}. \tag{4.29}$$

$E_{cc}(q)$ is the chromatic envelope function if only the term with C_{cc} is taken into account in the integration. The influence of its imaginary part on the envelope is not immediately obvious but becomes clear by rewriting Eq. (4.29) in two forms, one useful for $E_{cc}(q)$ as prefactor, the other useful for $E_{cc}(q)$ in the exponent. Abbreviating the coefficient of i with z and using relation 1.622.3 of ref. [25], Eq. (4.29) can be rewritten as

$$(1 - iz)^{-1/2} = \left(1 + z^2\right)^{-1/4} \exp\left(i\frac{1}{2}\arctan z\right), \tag{4.30}$$

which for small values of z becomes $(1 - (z/2)) \exp(iz/2)$. Thus $E_{cc}(q)$ is a damped oscillatory function, whose damping and period depends upon $(\Delta E/E)^2 C_{cc}\lambda q^2$. Equation (4.29) can also be rewritten making use of the square root of a complex number theorem [26] as

$$\frac{1}{\sqrt{2}}\left[\frac{1}{(1+z^2)^{1/2}} + \frac{1}{(1+z^2)}\right]^{1/2} + i\frac{1}{\sqrt{2}}\left[\frac{1}{(1+z^2)^{1/2}} - \frac{1}{(1+z^2)}\right]^{1/2}, \tag{4.31}$$

so that $E_{cc}^2(q) = f(z) + ig(z)$ in the exponent also contributes to damping and oscillation, though with a faster decay with increasing $\Delta E/E$ and q as an expansion in z shows. Quantitative results require numerical calculations which have been made by Schramm et al. [15] and some of their results will be shown below.

Beforehand we will discuss briefly the so-called parasitic aberrations. They can be coherent or incoherent. Coherent parasitic aberrations are astigmatism or are caused by specimen tilt and illumination beam misalignment; incoherent parasitic aberrations are due to specimen drift and vibrations or stochastic electron–electron interactions at high current densities. Astigmatism destroys the axial symmetry of the system and contributes to $\chi(\boldsymbol{q}, \Delta z_0)$ an azimuth angle-dependent term $A_1 \lambda q^2 \sin(2\Phi)$ for the twofold astigmatism and $A_2 \lambda^2 q^3 \sin(3\Phi)$ for the threefold astigmatism [4]. Twofold astigmatism can be corrected with the stigmator and threefold astigmatism is of higher order and can be neglected compared to the other factors, which deteriorate image quality. Specimen drift and vibrations can be taken into account by additional envelope functions [27]. In well-designed and well-supported cathode lens electron microscopes the influence of vibrations is negligible compared to the other effects and drifts usually occur only upon temperature changes or charging, processes which can be avoided or minimized with proper experimental procedures. Stochastic electron–electron interactions (Boersch and Loeffler effects) in the objective lens region (object and back focal plane) can cause considerable image blurring at high current densities and currents [28, 29] as they occur in pulsed PEEM (synchrotron radiation or pulsed high-power laser-excited) but are negligible at the low currents and current densities used in LEEM, MEM, and conventional PEEM. Whether or not the results of the Monte Carlo calculations of these effects can be formulated in analytical form, which can be taken into account in the transfer function, has not been explored yet.

We discuss now the numerical results for the various contributions to the wave amplitude in the image plane, first for the non-aberration-corrected LEEM (nac-LEEM), subsequently for the aberration-corrected LEEM (AC-LEEM). Collating all contributions gives the complete contrast transfer function $H(q,\Delta z_0, \Delta E/E)$

$$\begin{aligned} H\left(q, \Delta z_0, \frac{\Delta E}{E}\right) &= A(q) W_c(q, \Delta z_0) E_S(q) E_C^1\left(q, \frac{\Delta E}{E}\right) \\ &= A(q)\left\{ \cos\left(\pi\left(-\Delta z_0 \lambda q^2 + \frac{1}{2} C_3 \lambda^3 q^4\right)\right)\right. \\ &\quad \left. + i \sin\left(\pi\left(-\Delta z_0 \lambda q^2 + \frac{1}{2} C_3 \lambda^3 q^4\right)\right)\right\} \\ &\quad \times \exp\left(2 q_{\text{ill}}^2 \left(C_3 \lambda^3 q^3 - \Delta z_0 \lambda q\right)^2\right) \times \exp\left(-\frac{(\pi C_c \lambda)^2}{16 \ln 2} q^4 \left(\frac{\Delta E}{E}\right)^2\right). \end{aligned} \tag{4.32}$$

Figure 4.4a (Pang, private communication) shows from top to bottom the various contributions for an objective lens as calculated in [13, 14] but with the parameters of one of the commercial instruments (Elmitec). The shaded region in the top two panels represents the q-region blocked by the aperture function $A(q)$. The oscillating curves in them show the real (Re) and imaginary (Im) part of the wave aberration function $W_c(q, \Delta z_0)$ (upper and lower panel, respectively) for in-focus ($z_0 = 0$, black lines) and strongly out-of-focus ($\Delta z_0 = 4(C_3\lambda)^{1/2}$, red lines). The following panel shows the source envelope $E_S(q\ )$ for the two focus values and the bottom panel the chromatic envelope $E_C^1(q, \Delta E/E)$ for $\Delta E/E = 3.75 \times 10^{-5}$. The real part of $W_c(q, \Delta z_0)$ shows that the contrast is transferred with the correct sign only up to a certain q value $q_{c0,}$ beyond which it changes sign and oscillates. A large defocus (red lines) reduces this value but as we will see later a small defocus can increase it somewhat by balancing the defocus term against the spherical aberration. $E_S(q)$ reduces the amplitude of the wave aberration function in the q range transmitted through the aperture very little but the chromatic envelope causes a strong damping. For comparison Fig. 4.4b shows the real and imaginary parts of the wave aberration function and Fig. 4.4c their damping by the source and chromatic envelope of the objective lens calculated by Schramm et al. [15] and used in the IBM/SPECS LEEM. The larger value of q_{c0} in the first lens is due to its smaller C_3 value. However, because of its larger C_c and ΔE values the chromatic envelope E_c reduces the q_{c0} below the value of the second lens. C_c decreases faster

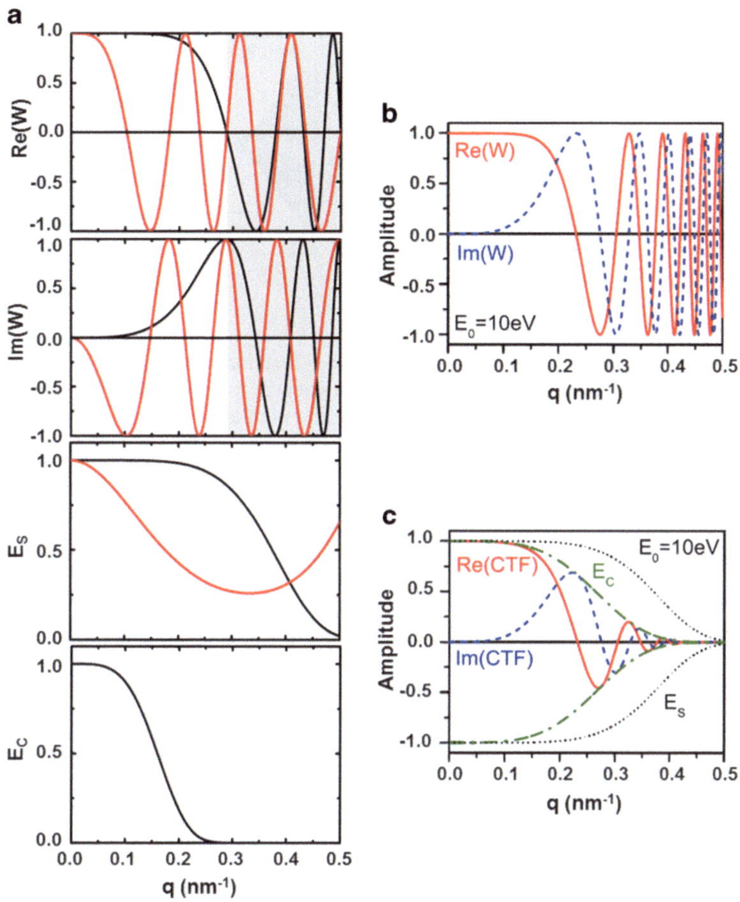

Fig. 4.4 (**a**) Real and imaginary part of the wave aberration function (Re(W), Im(W)), source envelope E_S and chromatic envelope E_C of the objective lens used in the Elmitec instrument. Parameters: $C_3=0.222$, $C_c=-0.106$, $\alpha_{ill}=\lambda q_{ill}=0.4$ mrad, $\Delta E=0.75$ eV, $E=20$ keV. *Black lines*: in focus ($\Delta z_0=0$), *red lines*: strongly defocused ($\Delta z_0=4(C_3\lambda)^{1/2}$). (**b**) Real and imaginary part of the wave aberration function of the objective lens used in the IBM/SPECS instrument and (**c**) their damping by source and chromatic envelopes. Parameters: $C_3=0.345$, $C_c=-0.075$, $\alpha_{ill}=\lambda q_{ill}=0.1$ mrad, $\Delta E=0.25$ eV, $E=15$ keV. Starting energy in all cases 10 eV. All aberration constants in m referred to the virtual image plane. (**a**) Courtesy A.B. Pang; (**b** and **c**) Adapted with permission from ref. [15]. Copyright 2012 Elsevier

with increasing energy than C_3 (see Fig. 3.30) so that each lens has its advantages and disadvantages, depending on start energy range.

These differences disappear in aberration-corrected systems and are replaced by the differences in the higher-order aberration constants. In the aberration corrected systems W_c, E_S, and E_C in Eq. (4.32) are now replaced with the corresponding terms (Eqs. 4.22, 4.23, and 4.28) with $C_s=C_3=0$. A comparison of Figs. 4.5a, b shows how aberration correction effects Re(W_c) at several start energies in the case of the

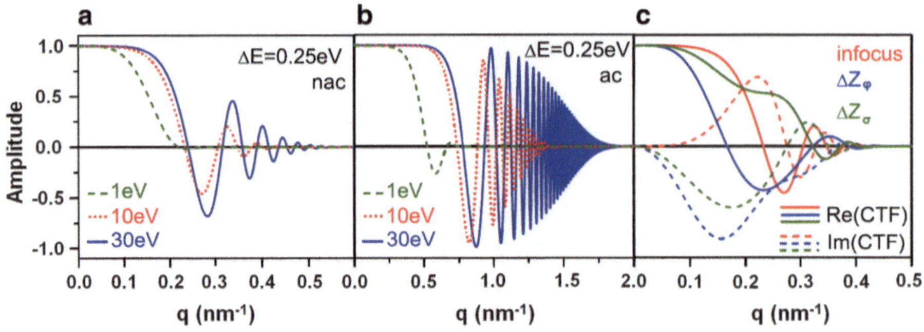

Fig. 4.5 (**a**) and (**b**) Real part of the contrast transfer function of the uncorrected (nac, **a**) and aberration-corrected (ac, **b**) IBM/SPECS LEEM instrument in focus condition for start energies of 1, 10, and 30 eV. (**c**) Real and imaginary part of the contrast transfer function of the same uncorrected LEEM instrument in focus, with focus optimized for imaginary part and real part of the contrast transfer function, Δz_φ and Δz_σ, respectively. For explanation of Δz_φ and Δz_σ see text. Reproduced with permission from ref. [15]. Copyright 2012 Elsevier

IBM/SPECS lens [15]. In the nac case q_{c0} is about 0.24 nm^{-1} nearly independent of start energy. Aberration correction shifts q_{c0} to much higher, more energy-dependent values, producing a much wider q range in which the wave amplitude is transferred with the same sign.

With the knowledge of the factors, which determine the transfer of the amplitude of the object wave function into the amplitude of the image wave function we can now address the question on how to optimize this transfer for given lens properties without and with aberration correction. The goal is to increase q_{c0} as much as possible and to minimize the oscillations of the wave aberration function. For TEM, which is concerned nearly exclusively with weak phase objects, it is the imaginary part of the CTF which has to be optimized. This was done long ago by Scherzer [21] who derived the so-called Scherzer (de)focus $\Delta z_\varphi = -c(C_3\lambda)^{1/2}$, where c varies between about 1 and 1.2, depending upon the detailed optimization criteria. Usually the Scherzer focus is written as $-1.2(C_3\lambda)^{1/2}$. c values such as $\sqrt{4/3}$ or $\sqrt{3/2}$, for which the argument of the imaginary part (the sin in Eq. (4.32)) at q_{c0} is $-120°$ or $-135°$, respectively, can be derived straightforward with somewhat different optimization criteria [10, 15]. The influence of the Scherzer focus Δz_φ for $c = \sqrt{3/2}$ on the imaginary part of the CTF is shown in Fig. 4.5c together with its influence on the real part of the CTF for the non-aberration-corrected case. q_{c0} of the imaginary part is clearly increased from 0.27 to 0.36 nm^{-1} by the defocus and so has the overall amplitude. On the other hand, q_{c0} of the real part has been reduced from 0.23 to 0.17 nm^{-1} and so has been the overall amplitude. This is a clear indication that the defocus is too large (see Fig. 4.4a) for the real part, which is important in LEEM because of the strong phase changes and the presence of not only phase contrast but also amplitude contrast. Therefore, a lower defocus has to be chosen. Using the same optimization criteria as for the imaginary part and $-45°$ for the argument of the cos at q_{c0}, the value $c = \sqrt{1/2}$ is obtained [15]. The resulting real and imaginary parts of the CTF are also plotted in Fig. 4.5c. With this defocus, $\Delta z_\sigma \approx 0.71(C_3\lambda)^{1/2}$, the q_{c0} value of the real part of the CTF is increased from the in-focus value of 0.23 to 0.32 nm^{-1} and the overall amplitude enhanced significantly, while the q_{c0} value of the imaginary part is not changed and the overall amplitude increased slightly compared to the in-focus case. The essential point of these considerations is that the Scherzer focus known from TEM is not suitable for LEEM in general but that a significantly lower value, as low as by a factor of 0.6, is optimal. For the aberration-corrected lens optimal defocus values can be derived, using the appropriate expressions for the CTF with $C_s = C_3 = 0$. Similar to the nac case focus optimization for amplitude contrast increases q_{c0} somewhat, while optimization for phase contrast reduces q_{c0} [15].

q_{c0} is obviously the largest q value at which the wave amplitude is transferred with the correct sign, though with strongly decreased amplitude. The intensity is even more reduced near q_{c0}. Nevertheless, q_{c0} is used to define the so-called point resolution. In TEM, where the wave aberration function is much less damped because of the much smaller aberration constants and smaller $\Delta E/E$, another resolution limit can be defined, the so-called information limit. The weak damping allows acquiring many oscillations and by proper computer processing contributions from very large q values can be included in the image formation. Information on these methods can be found in refs. [8, 10, 11]. The strong damping in LEEM reduces these possibilities significantly so that for all practical purposes the information limit coincides approximately with the point resolution.

4.2.2 *The Image Intensity*

In practice the resolution in LEEM is determined by analyzing the intensity distribution using model object structures which can be easily realized in experiment. Figure 4.6 [14] shows the simplest models, a phase object, and an amplitude object with a sudden change of phase φ and amplitude α between two regions with constant phase and amplitude, respectively. Physical realizations are steps on surfaces and step-free transitions between regions with strongly differing reflectivity. The intensity in the image, with the complete contrast transfer function $H(q,\Delta z_0, \Delta E/E)$ (Eq. 4.32) included is

$$
\begin{aligned}
I(\mathbf{r}') &= \psi_{\text{im}}(\mathbf{r}')\,\psi^*_{\text{im}}(\mathbf{r}') \\
&= \frac{1}{M^2}\left|\int \widetilde{\psi}_{\text{obj}}(\mathbf{q})H(q)\exp(2\pi i\mathbf{q}\cdot\mathbf{r})\mathrm{d}\mathbf{q}\right|^2 = \frac{1}{M^2}\left|\mathcal{F}^{-1}\{\widetilde{\psi}_{\text{obj}}(\mathbf{q})\cdot H(q)\}\right|^2,
\end{aligned}
\tag{4.33}
$$

where $\mathcal{F}^{-1}$ is the inverse Fourier transform. $\mathcal{F}^{-1}\{H(q)\}$ is the point spread function $h(\mathbf{r})$ (Eq. 4.1). Making use of the convolution theorem, equation 4.33 can be also written in terms of the point spread function as already mentioned earlier (Eq. 4.3)

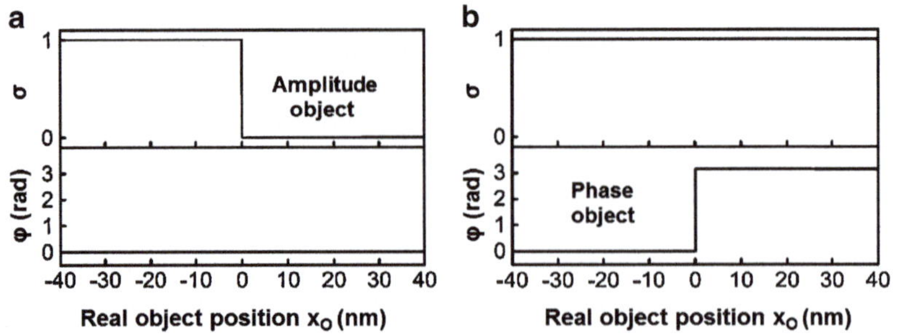

Fig. 4.6 Schematic of the one-dimensional amplitude object (**a**) and phase object (**b**) used in the calculations, illustrated for a 1/0 pure amplitude object (amplitude jump $1 \rightarrow 0$) and a 0/3 pure phase object (phase jump $0 \rightarrow \pi$). Reproduced with permission from ref. [14]. © IOP Publishing 2009. All rights reserved

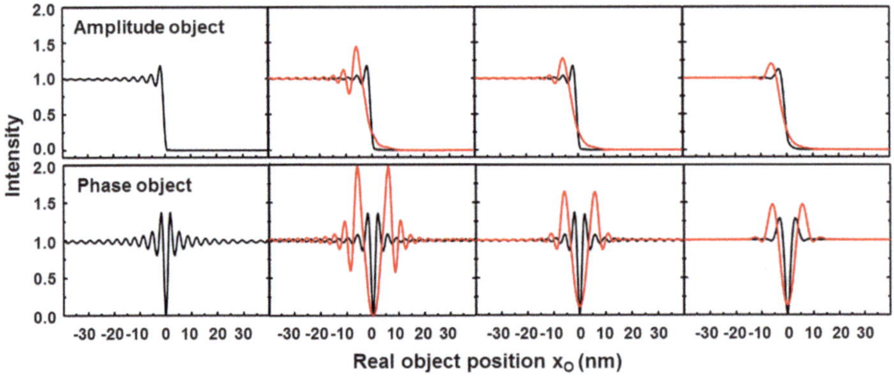

Fig. 4.7 Evolution of the intensity profile of a 1/0 amplitude object (*top*) and a 1/0 phase object ($\varphi = \pi$, *bottom*) with the number of terms taken into account in the contrast transfer function. From *left* to *right*: aperture function only, with coherent wave aberration function W_c included, with source spread function E_S included and with chromatic envelope function E_C included. A large energy width of about 1.4 eV was used in E_C. *Black curve*: in focus, *red curve*: with very strong defocus ($\Delta z_0 = 4(C_3\lambda)^{1/2}$). Reproduced with permission from ref. [14]. © IOP Publishing 2009. All rights reserved

$$I(\boldsymbol{r}') = \frac{1}{M^2} \left|\psi_{\mathrm{obj}}(\boldsymbol{r}) \otimes h(\boldsymbol{r})\right|^2.$$

Figure 4.7 [14] illustrates how the image of the objects shown in Fig. 4.6 evolves with the number of terms taken into account for in-focus and for a very large defocus $\Delta z_0 = 4(C_3\lambda)^{1/2}$ and the lens parameters of a commercial microscope (Elmitec). The abscissa is the real object position. Going from left to right we see first the diffraction at the aperture, then the influence of the wave aberration, which is minor in focus but dramatic at large defocus, where it increases period and amplitude of the intensity oscillations. The source envelope also does not cause

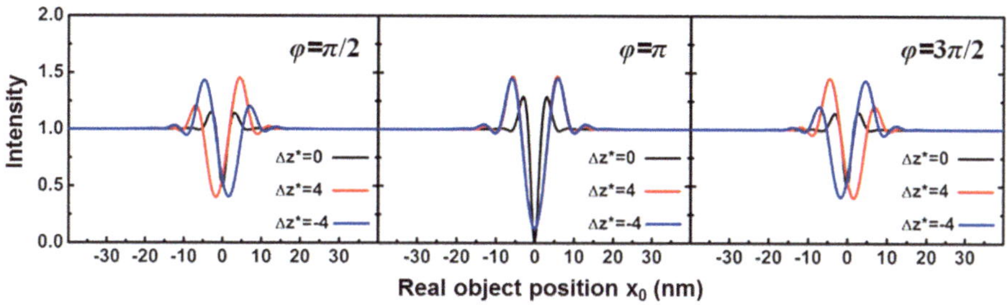

Fig. 4.8 Dependence of the intensity of a phase object upon phase difference ($\varphi = \pi/2$, π and $3\pi/2$) and focus ($\Delta z = 0$ (*black*), $+4(C_3\lambda)^{1/2}$ (*red*), and $-4(C_3\lambda)^{1/2}$ (*blue*). The large energy width of about 1.4 eV in E_C suppresses the asymmetry seen in focus at small ΔE values. With smaller energy widths, e.g. 0.25 eV, the intensity distribution of the phase object becomes asymmetric also in focus. Defocusing allows distinguishing between upper and lower side of the step via the intensity and position shift. The lower side of the step is on the right side. Adapted with permission from ref. [14]. © IOP Publishing 2009. All rights reserved

significant changes in focus, but dampens the oscillation amplitudes out of focus. Finally, the chromatic envelope has a strong effect not only at large defocus but also in-focus due to the strong decrease of q_{c0}, which it causes as seen in Fig. 4.4. This illustrates the importance of a small C_c and a small ΔE in uncorrected systems.

Of particular interest is the step contrast which is frequently observed in LEEM. Figure 4.8 [14] shows for the same lens parameters as used in Fig. 4.7 how the phase difference $\varphi = (2\pi/\lambda_0)2a$ between the two sides of the step with height a influences the intensity distribution around the step in focus and strongly out-of-focus [13, 14]. In-focus the step profile is symmetric at large energy widths, out-of-focus asymmetric. The asymmetry depends upon the phase difference, is maximal when $\varphi = (2n+1)\pi/2$, zero for $\varphi = (2n+1)\pi$ and for $\varphi = 2n\pi$ the contrast disappears. This allows within limits determination of the step height. Distinction between up and down side of the step is possible via the sign dependence of the defocus, which allows identification of ad-islands and vacancy pits. The physical appearance of the contrast produced by a step in non-aberration-corrected LEEM as seen in [16, 20], which is simulated approximately by these calculations, is illustrated in Fig. 4.9 for a step with a phase difference of $\pi/2$ for in-focus and defocus of $\pm 2(C_3\lambda)^{1/2}$ [13]. The asymmetry is clearly visible as well as the lateral shift of the step position at high defocus seen in Fig. 4.8.

The intensity profiles in Figs. 4.7 and 4.8 allow one to define resolution limits, for example for the amplitude object the distance between the positions at which the intensity has 90 % and 10 % of the maximum value (90 %/10 % criterion) or for the amplitude object the distance between the central minimum and the first maximum. The second criterion is similar to the frequently used Rayleigh criterion, which is defined by the distance between the maximum of one Airy diffraction disc from one point source and the first zero of the Airy disc from the second source. The amplitude resolution criterion gives approximately the same value as the 90 %/10 % phase resolution criterion, 5.5 nm for $\Delta E = 0.5$ eV, $E_0 = 10$ eV in the Elmitec lens, but the

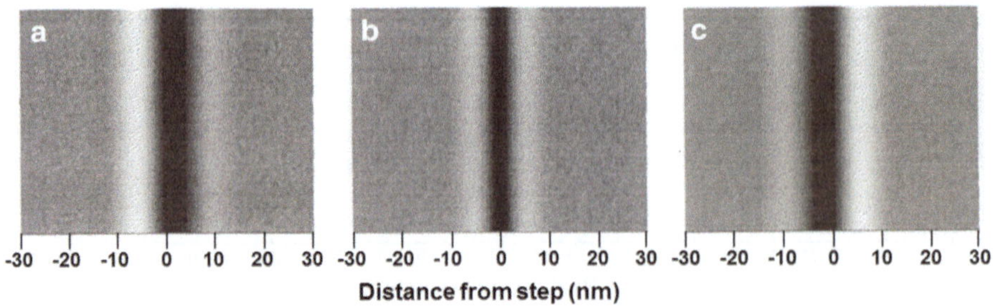

Fig. 4.9 Contrast of step with a height producing a phase difference $\varphi = \pi/2$ between the two sides of the step: in focus (**b**) and at defocus $\Delta z_0 = -2(C_3\lambda)^{1/2}$ (**a**) and $+2(C_3\lambda)^{1/2}$ (**c**). A large energy width of about 1.4 eV was used in the calculations. The lower side of the step is on the right side. Modified from ref. [13]

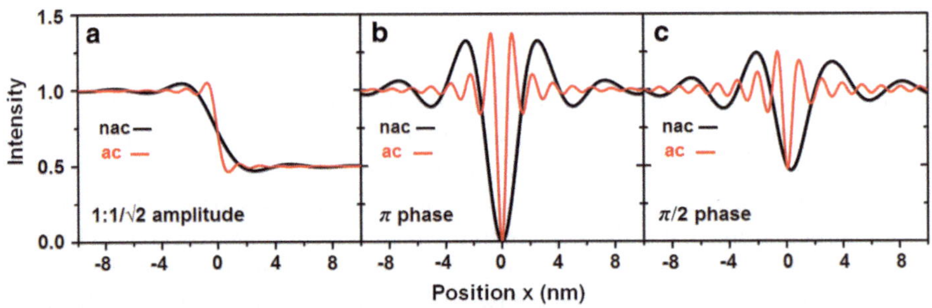

Fig. 4.10 Comparison of amplitude and phase step contrast profiles between uncorrected (nac) and corrected (ac) lens of the IBM/SPECS objective lens. (**a**) Reflectivity step 100 %/50 %, (**b**) phase step π, and (**c**) phase step $\pi/2$. Adapted with permission from ref. [15]. Copyright 2012 Elsevier

widely used 84 %/16 % criterion a significantly lower value, 3.0 nm [13, 14]. Thus, caution should be used when comparing resolution values.

The resolution improvement which can be achieved with smaller C_c and $\Delta E/E$ values and in particular with aberration correction is illustrated in Fig. 4.10 [15] for the IBM/SPECS lens ($W(q)$ shown in Figs. 4.4b, c, and 4.5). Its C_c and $\Delta E/E$ values are respectively 30 % and more than 70 % smaller than those used for Fig. 4.7 and accordingly the resolution as judged by the criteria used above is correspondingly better, 2.5 nm at the same starting energy as above (10 eV). Aberration correction reduces this value to 0.8 nm. The first side minima and side maxima are now clearly seen, in contrast to Fig. 4.7. Several more occur with aberration correction. Schramm et al. [15] define resolution somewhat different. They define amplitude contrast by the 84 %/16 % criterion but with the intensity difference taken at large distances from the contrast edge instead of the maxima close to it. Phase contrast is likewise defined by the FWHM of the intensity dip, with the maximum defined by the intensity far away from the phase step. They have calculated the so-defined resolution as function of q and plotted it as function of the corresponding aperture

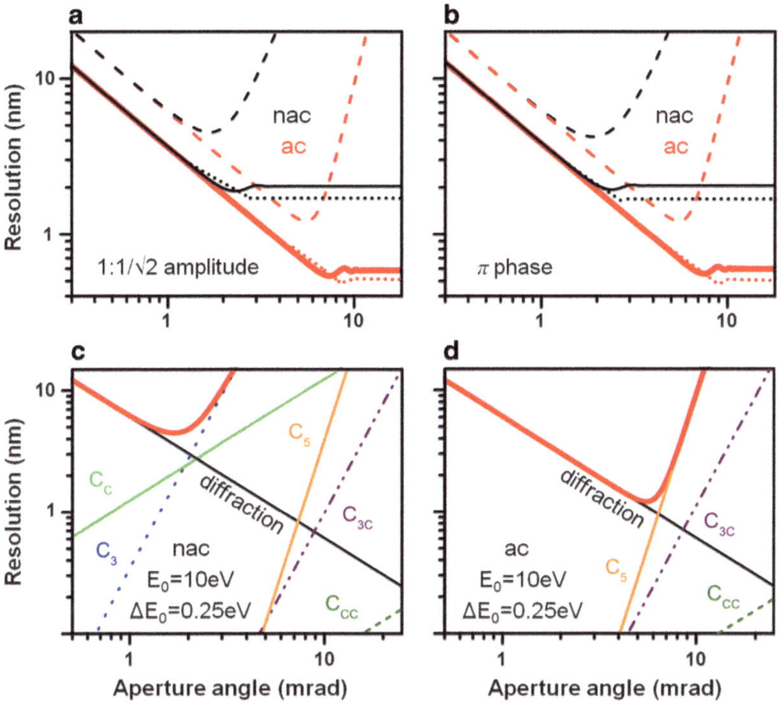

Fig. 4.11 Dependence of the resolution upon aperture angle. (**a** and **b**) as obtained from the amplitude and phase step profiles for a 100 %/50 % reflectivity step (**a**) and a π phase step (**b**) for an uncorrected (nac) and an aberration-corrected (ac) lens. *Solid lines*: wave-optical calculations with the contrast transfer function theory, *dashed lines*: geometric optics calculations. *Red* (*black*) with (without) aberration correction. The *dotted lines* are for the optimum defocus in the aberration-corrected system ($\Delta z^5_{A\Phi}$ in [15]). (**c** and **d**) Resolution calculated by geometric optics without aberration (**c**) and with aberration (**d**) correction. The contributions of the various aberrations in Eq. (4.34) are indicated. Adapted with permission from ref. [15]. Copyright 2012 Elsevier

angle $\theta_A = q_A \lambda$ for a 1:1/√2 amplitude object and for several phase objects. (The 1:1/√2 amplitude object (100:50 reflectivity) is more realistic than the 1:0 amplitude assumed in Figs. 4.6 and 4.7). This is shown in Fig. 4.11 [15] for (a) an amplitude and (b) a π phase object for the non-aberration-corrected (nac) and the aberration-corrected (ac) lens with solid lines and compared in (c) and (d) with the corresponding geometric optics results. In all cases resolution improves like $1/\theta_A$ as determined by diffraction at the aperture until aberrations determine the resolution. In the wave optics case it is the envelope function, specifically the chromatic envelope which suppresses larger q values as seen in the amplitude plots in Figs. 4.4c and 4.5a and in the weak oscillation of the ac curves. Enhanced plots show the oscillations out to about 12 mrad so that in principle contributions from larger q values could be included in the image formation but the contrast associated with them is too small to justify the effort. For all practical purposes the minima in Fig. 4.11a, b represent the resolution, larger apertures bring no improvement but increase only the background. The dotted curves show the improvement that can be achieved with the Scherzer type defocus discussed earlier.

In the geometrical optics calculation of the resolution (Eq. 3.5)

$$\delta = \left[\left(\frac{0.61\lambda}{\theta} \right)^2 + \left(C_3 \theta^3 \right)^2 + \left(C_5 \theta^5 \right)^2 + \left(C_c \frac{\Delta E}{E} \theta \right)^2 + \left(C_{cc} \left(\frac{\Delta E}{E} \right)^2 \theta \right)^2 + \left(C_{3c} \frac{\Delta E}{E} \theta^3 \right)^2 \right]^{1/2} \tag{4.34}$$

the $1/\theta_A$ decrease is terminated by the increase of the aberration terms C_3 in the nac case and C_5 in the ac case as seen in Figs. 4.11c, d. For larger $\Delta E/E$ values C_{3c} and C_{cc} terms also contribute to the resolution deterioration in the nac and ac case, respectively, in particular at lower starting energies.

Image quality does not depend only on resolution but also on contrast C, which can be defined in several ways. We follow here the choice of Schramm et al. [15]: for a α/β amplitude object $C_\sigma =$ (maximum intensity − far away intensity)/$|\alpha^2 - \beta^2|$, for a pure α/β phase object $C_\varphi =$ (maximum intensity − far away intensity)/far away intensity. They have calculated for a starting energy of 10 eV and $\Delta E = 0.25$ eV the in-focus contrast defined in this manner for a number of structural features with the following results: C_σ depends strongly upon the amplitude ratio α/β below a feature size of about 8 nm (2 nm) in the nac (ac) case. A bright feature on a dark background gives a somewhat better resolution than the inverse feature. In π:$-\pi$ phase objects the contrast is enhanced, up to 200 % in the ac case, for feature sizes of at about three times the resolution limit. In the case of $(2n+1)\pi/2$:$-(2n+1)\pi/2$ phase objects the contrast fluctuates with feature size below about 15 and 5 nm in the nac and ac case, respectively. These examples illustrate that image interpretation of features near the resolution limit becomes increasingly complicated with decreasing feature size. The resolution limits shown on Fig. 4.11a, b are for one-dimensional isolated linear sudden changes of amplitude or phase. In small features the intensity oscillations around the sudden changes overlap and smear out the intensity distribution. Extraction of the true object structure then requires model calculations.

Another question is to which extent does the Scherzer-like defocus influence not only resolution but also contrast. Calculations, again for the same energies, show [15] that in a $1{:}1/\sqrt{2}$ amplitude object the resolution is improved only by a few 0.1 nm but the contrast by about 5 %, that in a π phase object the contrast is independent of defocus, with even less resolution improvement than in the amplitude object. For a weak phase object ($\varphi = \pi/20$), in which the contrast in-focus is only about 2 %, the Scherzer Δz_φ focus increases the contrast to about 10 % (8 %) in the ac (nac) case but simultaneously causes a slight loss of resolution, in particular in the nac case, where it increases from about 1.5 to 2 nm. While these calculations are for a specific lens and specific operation conditions, their results can probably generalized in two respects: (1) the contrast of features with

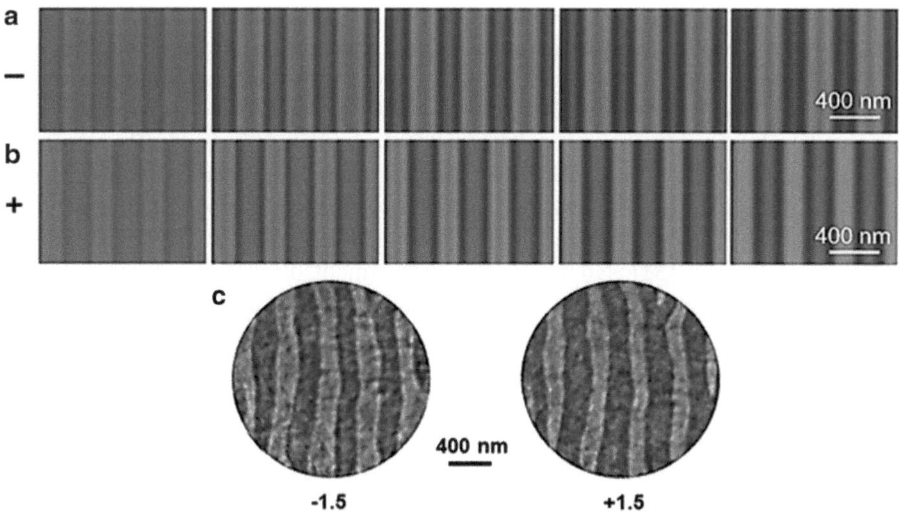

Fig. 4.12 Surface profile analysis with the contrast transfer function theory. (**a**) and (**b**) negative and positive defocus series. From *left* to *right*: 0.3 mA, 0.9 mA, 1.5 mA, 2.1 mA and 2.7 mA. (**c**) LEEM images at −1.5 and +1.5 mA defocus taken at 14 eV at room temperatures. The object is a 80 nm thick corrugated MnAs layer on GaAs(100). Adapted with permission from ref. [30]. Copyright 2013 Elsevier

dimensions near the resolution limit does not produce a 1:1 image of the object and (2) the main advantage of the Scherzer focus is not for resolution enhancement but for increasing contrast.

The object model calculations discussed up to now assume a planar surface. Real objects are in general not planar but approximate the models only to a certain extent. The best approximation for a pure phase object is the monatomic step on an otherwise perfect surface. However, if the height to width ratio is small and the electron energy not close to zero, objects with larger height differences may also be considered as flat. MnAs layers on GaAs are an example. Their surface consists of ridges and valleys with flat tops and bottoms of the same material. It shows no LEEM contrast in-focus but strong focus-dependent contrast out of focus. Pang et al. [30] have analyzed this contrast with the contrast transfer theory ("Fourier optics") assuming a square wave surface profile. Figure 4.12 [30] shows the calculated focus dependence and selected LEEM images taken at 14 eV for comparison. From the focus dependence of the contrast the average width of ridges and valleys could be determined with high accuracy to 172 and 229 nm but the height difference only to within a relatively large error to between about 1 and 2 nm in agreement with STM measurements. Accurate values require measurements at more than one energy.

A good amplitude object is a monolayer of a different material at a step of the same height. In this case the reflectivity difference involves in general not only amplitude but also phase. Thus the pure amplitude object is only an idealized approximation. In reality amplitude objects are three-dimensional features on a different substrate, which invalidates the assumption of a planar surface. As a consequence the electric field is no longer perpendicular to the macroscopic surface, resulting in different electron trajectories, enhanced yet by work function differences between substrate and the three-dimensional features. These effects become increasingly important with decreasing energy and determine the contrast in MEM. In LEEM the influence of surface roughness was briefly studied with a micropatterned periodic surface with a height/width ratio of 1:8 and modeled geometric-optically [31]. No wave-optical treatment of such rough surfaces has been made up to now within the framework of the phase contrast transfer theory.

4.2.3 Mirror Electron Microscopy

Contrast formation and resolution in MEM has been discussed from the early years of MEM in terms of geometrical optics as reviewed in several papers, e.g. [32, 33]. Geometric optics is also the basis of the most recent studies of MEM image interpretation [34–38], using different approaches, which have been compared in [37]. Contrast and resolution have been analyzed repeatedly also by wave optics [39–44]. The first attempt was made early on by Wiskott [39] who calculated the intensity distribution in the reflected wave in front of a twofold square periodic object in a homogeneous retarding field, modulated by the object structure (height or surface potential variations). With a maximum to minimum current density ratio of ¾ between the squares as criterion for lateral resolution he obtained a resolution of 8 nm at a field strength of 10 kV/mm. He did not calculate the propagation of the reflected wave to the entrance/exit of the field, which was done later by Gvosdover and Zel'dovich [42]. Working in Fourier space and including the propagation through the field via a transfer function $S(q)$ they analyzed the intensity distribution in the wave field reflected from a point source such as a small pit or hillock. They defined the resolution Δx by the half width of the linear response function $G(x) = 1/2\pi \int S(q)\exp(iqx)\mathrm{d}q$ at $x = 0$ for the point z_0 at which the wave changes direction above the surface. This definition gives $\Delta x = 1.25\sqrt{3/4\pi}(l\lambda)^{1/2} \approx 0.61(l\lambda)^{1/2}$ where l is the length of the field and λ the wavelength of the electron wave entering/leaving the field. For $l = 2$ mm and 20 keV electrons (i.e. a field strength of 10 kV/mm) this gives $\Delta x \approx 80$ nm, an order of magnitude larger than Wiskott's value but in good qualitative agreement with the best experimental results. This illustrates the importance of the transfer of the contrast from the object to the exit from the field which is taken into account by the transfer function.

Building on the earlier studies, Lichte's work [43, 44] is formulated in a language close that that used in the discussion of the LEEM transfer function and includes experimental results for comparison with theory. Therefore, we will discuss it here in more detail. Lichte considers an object with cos-shaped height (or potential) distribution $h(x) = h_0 \cos(kx)$ with $k = 2\pi q$. This modulates the homogeneous z-dependent potential $\Phi_0(z)$ of the flat object in the mirror field of length l with the potential

$$\Phi(x,z) = T(k,z)h(x) = \frac{\sinh(k(l-z))}{\sinh kl} h_0 \cos kx. \tag{4.35}$$

Using the Fourier presentation of the object function $h(x)$

$$\tilde{h}(k) = \int_{-\infty}^{+\infty} h(x)\exp(ikx)\mathrm{d}x, \tag{4.36}$$

$\Phi(x,z)$ can be written quite general as

$$\Phi(x,z) = \frac{1}{\sqrt{2\pi}} \int_{-\infty}^{+\infty} \tilde{h}(k)T(k,z)\exp(-ikx)\mathrm{d}k. \tag{4.37}$$

This potential causes in addition to the phase shift φ_0 caused by the homogeneous mirror field ($h(x) = \text{const}$), which depends only upon its length, at the exit from this field an additional x-dependent phase shift

$$\varphi(x) = \frac{2\pi}{\lambda}\frac{1}{\sqrt{2\pi}} \int_{-\infty}^{+\infty} \tilde{h}(k)P(k,z_0)\exp(-ikx)\mathrm{d}k, \tag{4.38}$$

where the phase transfer function $P(k,z_0)$ is obtained from $T(k,z)$ via integration from the wave return point z_0 in front of the object to the exit from the field at l at the potential U:

$$P(k,z_0) = \frac{\sqrt{l-z_0}}{U} \int_{z_0}^{l} \frac{T(k,z)}{\sqrt{z-z_0}}\mathrm{d}z == \sqrt{\pi}\exp(-kz_0)\frac{erf\left(\sqrt{kl}\right)}{\sqrt{kl}} \tag{4.39}$$

for $k > 2\pi/l$. For $k \to 0$ $P(k,z_0) \to 4/3$. $P(k,z_0)$ is the analog to the contrast transfer function in LEEM. There is no source envelope and no chromatic envelope because the incident wave has been assumed to be a plane, monochromatic wave.

In Fig. 4.13a [43] $P(k,z_0)$ is plotted as a function of k for several z_0 values. A comparison of the scales with those of the wave aberration function in LEEM (Figs. 4.4 and 4.5) shows that the transfer function decreases to very low values not somewhere at $q \leq 1\ \text{nm}^{-1}$ but at $10 \leq 2\pi q \leq 100\ \mu\text{m}^{-1}$ for $z_0 = 0.2\ \mu\text{m}$, the smallest practical value, where the deflections are strongest. z_0 cannot be chosen arbitrarily small because of the energy distribution of the electrons: faster electrons come closer to the surface before being reflected than slower ones. The difference in the

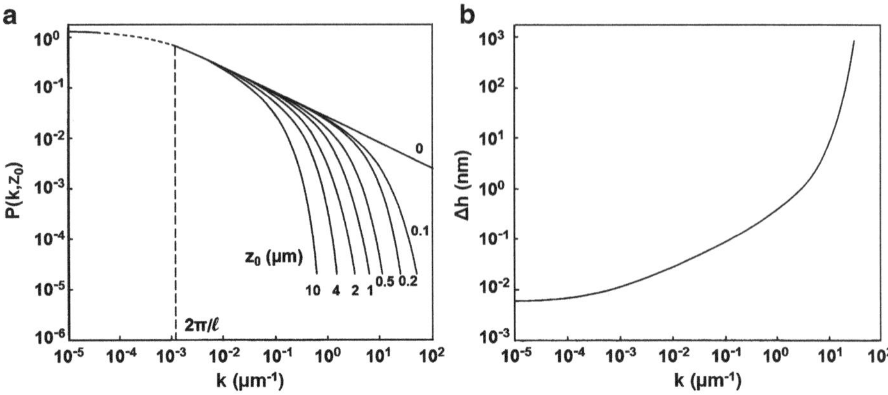

Fig. 4.13 Phase transfer function $P(k, z_0)$ for various return point values z_0 (**a**) and height resolution function $\Delta h(k)$ (**b**) in mirror electron microscopy as function of the spatial frequency $k = 2\pi q$. Adapted from ref. [43] with permission. Copyright 1980 Urban and Fischer (Elsevier) (Netherlands)

lateral deflection (phase modulation) depends upon distance. Averaging over the energy-dependent deflections decreases the contrast. Lichte has estimated that for an energy width $\Delta E = 0.3$ eV and a field strength of 5 kV/mm sufficient contrast can be obtained for $k = 60\ \mu\text{m}^{-1}$. This corresponds to a lateral resolution of $\Delta x \approx 100$ nm, which is of the same magnitude as that obtained in ref. [42]. The limits of the depth resolution Δh as a function of the spatial frequency $k = 2\pi q$, assuming a phase shift accuracy limit of $2\pi/10$ due to the energy width ΔE, are shown in Fig. 4.13b [43]. For $k = 10\ \mu\text{m}^{-1}$ $\Delta h = 0.8$ nm in good agreement with the value given by Gvosdover and Zel'dovich [42]. These authors have also briefly discussed the contrast and resolution in MEM of magnetic microfields in front of the surface, which are described by the magnetic vector potential $\boldsymbol{A}$. They found the same lateral resolution as in the electrostatic case and a minimum detectable magnetic flux $\Phi_0 \approx \hbar/ec$.

Lichte checked his theoretical results semi-quantitatively with an experimental set-up more sophisticated than today's standard LEEM systems except for the lack of UHV: a field emission electron gun, a Castaing-Henry magnetic prism as beam separator, an electrostatic four-electrode cathode objective lens and after the beam separator an intermediate and projective lens system. In addition an electrostatic bi-prism for interferometry was inserted before and after the beam separator. The deformation of the phase of the wave field beyond that of the accelerating field is negligible so that the final image should allow testing the calculations. Figure 4.14a–c [44] show the results. (a) is an image of a grating with 200 nm period, illustrating the lateral resolution. (b) is from a pattern of 2.5 nm high Au squares with 12 μm periodicity, whose high contrast gives an idea of the good depth resolution. (c) is a strongly defocused image of an edge of a groove, showing up to 30 Fresnel fringes. Imaging the diffraction pattern of a grating by focusing

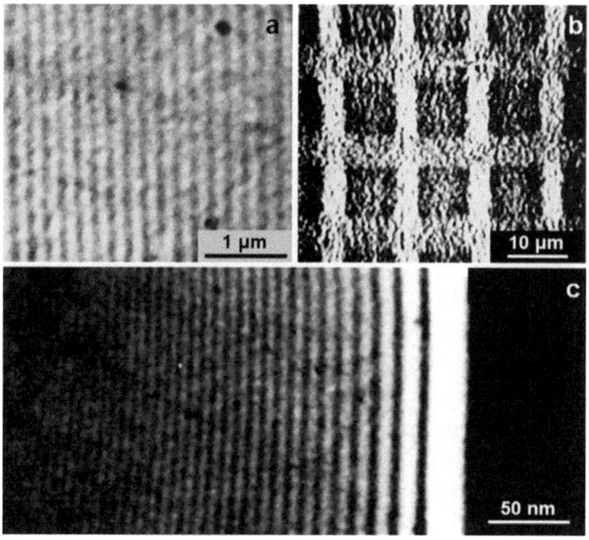

Fig. 4.14 Mirror electron microscope images characterizing (**a**) the lateral resolution using a 200 nm grating, (**b**) the depth resolution using 12 μm period Au squares with 2.5 nm height and (**c**) Fresnel diffraction pattern of the edge of a grove obtained by strong defocusing. Adapted from ref. [44] with permission. Copyright 1983 The Japan Society of Applied Physics

with the intermediate lens onto the back focal plane of the objective lens he also determined the z_0 dependence of the contrast via the number of diffraction maxima. Using the bi-prisms for beam splitting and superposition he could measure the step heights of two monolayer high steps on KBr, the local roughness of highly polished glass surfaces with an rms roughness of 0.27 nm and potential differences of 1 mV. Thus while the lateral resolution of MEM is not significantly better than 100 nm, the depth resolution is better than 1 nm, which makes MEM very sensitive to height variations or equivalent surface potential variations on a lateral scale of more than 10 nm.

4.2.4 Emission Electron Microscopy

Reflection electron microscopy is a partially coherent imaging method, with the degree of coherence determined by the size of the electron source and the energy distribution of the emitted electrons. As a consequence the information transfer from object to image is determined by the amplitude of the reflected wave. This wave is the superposition of the spherical waves scattered from *all* atoms, whose coherence is determined by the *spatial* and *temporal* coherence of the incident wave field as discussed in Sect. 4.2.1.

In emission microscopy the situation is different. When no energy filter is used as in standard UVPEEM or in the XAS imaging modes, such as in XANES, NEXAFS or X-ray dichroism imaging, the wide energy distribution makes the imaging process incoherent because of lack of *temporal* coherence. When the emitted electrons are energy-filtered to an energy width comparable with that discussed in LEEM (Sect. 4.2.1) then the imaging process is *temporal* coherent and *spatial* coherence becomes decisive. Here two extreme cases have to be considered: electrons strongly localized on atoms, such as the core electrons, and electrons strongly spread across the whole crystal, such as the electrons in the valence band. In the case of photoemission the incident EM wave interacts with all electrons but the core electron that produces the image point can come from only *one* atom, in contrast to LEEM, where the waves from *all* atoms contribute to the image point and to the diffraction spot in the LEED pattern. Thus, it is only the scattering of the spherical wave of the emitted photoelectron, which produces the photoelectron diffraction (PED) pattern and contributes to image formation. The spatial coherence is determined by the IMFP of the electron and is therefore in general only a few atomic distances. Therefore the PED pattern is much more diffuse than the LEED pattern and the *spatial* coherence is so small that the imaging process may be considered to be incoherent. At the other extreme, i.e. strongly delocalized electrons, *emission* can occur from any *equivalent* region in the crystal similar to *scattering* from any *equivalent* atom in the case of LEEM/LEED, resulting in large spatial coherence as evident in band structure images in $\boldsymbol{k}$-space ("$\boldsymbol{k}$-space images"). Despite the large spatial coherence no contrast appears in the real image because of the delocalization of the emission point, unless variations in the band structure or electronic density of state, which produce *nonequivalent* regions, occur within the resolution limit of the instrument. An example of such *nonequivalent* regions will be encountered in Sect. 6.1 (e.g. Fig. 6.4), where local variations of the graphene–substrate distance cause corresponding changes in the density of states at the Γ and K points.

From these considerations it follows that in general PEEM and other emission microscopies may be considered to be incoherent processes, even when coherence is seen in $\boldsymbol{k}$ space. As a consequence it is not the amplitude but the intensity whose transfer to the image has to be considered. The intensity distribution in the image is now given by Eq. (4.4) instead of the expressions (4.3) for the intensity in LEEM (normalized to $M = 1$)

$$I(\boldsymbol{r}') = \left|\psi_{\text{obj}}(\boldsymbol{r})\right|^2 \otimes |h(\boldsymbol{r})|^2. \tag{4.40}$$

In order to make use as much as possible of the results obtained in connection with the discussion of coherent imaging it is useful to go to reciprocal space via Eq. (4.5) and use Eq. (4.6) to calculate $\mathcal{F}\{|h(\boldsymbol{r})|^2\}$ from the contrast transfer function $H(q)$ and finally return to real space via the inverse Fourier transform, yielding the intensity distribution (Eq. 4.40).

The contrast transfer function $H(q)$ in emission microscopy differs in two respects from that used up to now in that there is no illuminating electron beam. Therefore there is (1) no source envelope and (2) the energy distribution ΔE in the chromatic envelope is determined now by the emission process. Schramm et al. [15] have taken the energy distribution into account by integrating $H(\boldsymbol{q},\varepsilon) \times H^*(\boldsymbol{q},\varepsilon)$, weighted with the energy distribution $N(\varepsilon)$, over ε. For UVPEEM they assumed a

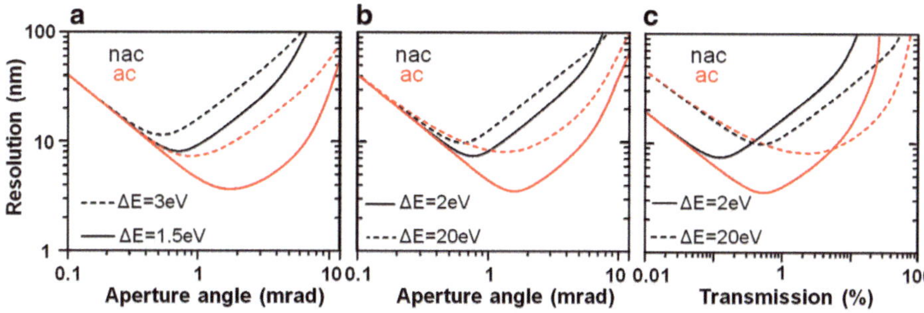

Fig. 4.15 Resolution and transmission in photoemission electron microscopy without (*black lines*) and with (*red lines*) aberration correction. (**a**) UV PEEM with Gaussian energy distribution, (**b** and **c**) X-ray-excited secondary electron emission. A $\cos\theta_0$ angular distribution is assumed in both cases. In (**a**) a mean energy of 1 eV, in (**b** and **c**) the energy of the maximum of the energy distribution has been used in the calculation of the chromatic damping. The energy range in (**b** and **c**) is limited either to a window of 2 eV (*solid lines*) or extends from 0 to 20 eV (*dashed lines*). Courtesy of S. Schramm

$\cos\theta$ angular and Gaussian energy distributions 1.5 and 3 eV wide with maxima at 1 eV. They calculated the intensity profiles of a $1{:}1/\sqrt{2}$ amplitude object, using the aberration coefficients for 1 eV. The resolution derived from the 84 %/16 % intensity profile criterion for the two energy widths and for the aberration-corrected (ac) and non-aberration-corrected (nac) case is shown in Fig. 4.15a [15]. Aberration correction improves the resolution at $\Delta E = 1.5$ (3) eV by a factor of 2.2 (1.6) and the transmission at the optimum aperture by a factor of 4.8 (2.9). These improvements are far below the expectations for energy-filtered imaging published in the early years of aberration correction. For example, *geometric-optical* calculations by Schmidt et al. [45] for the SMART microscope predicted for a mean emission energy of 2 eV and an energy width of 1 eV a resolution improvement by a factor of 5 and a 25-fold increase of the transmission by aberration correction. Even more improvements were predicted for an emission energy of 10 eV and an energy width of 2 eV: with the optimum aperture the resolution should improve from 14 to 1.6 nm and simultaneously the transmission should increase by a factor of 67. At a constant resolution of 14 nm the transmission should increase even 500-fold. These examples show that resolution and transmission depend strongly on the energy width of the electrons used for imaging, i.e., whether or not an energy filter is used, and to some extent on the calculation method.

Schramm et al. [15] calculated for their microscope the resolution and transmission also for imaging with secondary electrons with $\cos\theta$ angular distribution, specifically for X-ray excited secondary electrons from metals, for which Henke et al. [46] had measured an energy distribution $N(\varepsilon) \sim \varepsilon/(\varepsilon + W)^4$, where W is the work function. The aberration constants for 1 eV electrons were here taken too. They took energies up to 20 eV and the energy-dependent cut-off of the angular distribution by the aperture into account. The results are shown in Fig. 4.15b, c [15] for imaging with electrons up to 2 eV energy and with all electrons up to 20 eV energy.

For $\Delta E = 2$ eV (20 eV) aberration correction improves the resolution by a factor of 2.1 (1.2) and the transmission at the optimum aperture by a factor of about 4.5 in both cases. If resolution improvement is forfeited by increasing the aperture, keeping the resolution at the value of the uncorrected system (9.8 nm (nac) or 7.4 nm (ac)), the transmission increases 14-fold and 30-fold. These numbers can be compared with those of Feng et al. [47] for the PEEM2 and planned PEEM3 microscopes at ALS, which can be extracted from Fig. 3.39. They studied an electrostatic tetrode objective lens, aberration-corrected with a SMART type corrector (see Sect. 3.2.4), and used the same angular and energy distribution as Schramm et al. [15] but ray tracing. They obtained resolution and transmission improvements at optimum aperture by factors of 3 and 4.5, comparable to those of Schramm et al. [15] but a much larger transmission improvement of more than a factor of 70 when they optimized transmission, keeping resolution at the nac value of 13 nm. When a resolution of 50 nm is accepted the calculations of Feng et al. [47] predict 100 % transmission.

What all these calculations indicate is that aberration correction can increase in PEEM the resolution by a factor of 2–3 and transmission with the optimum aperture by a factor of about 5. A significant transmission improvement (though less than 100-fold), which is important in XPEEM, can be obtained at the expense of resolution. The best resolution obtainable with aberration correction is predicted to be about 4 nm, the best transmission at the optimum aperture 0.5–2 %.

4.3 Through-Focus Series Image Improvement

In Sect. 4.2.1 we have seen in Fig. 4.5 that the amplitude of the wave aberration function $W(q,\Delta z_0)$ [(Eq. 4.15), neglecting the fifth-order term] can be increased over a wider range of the spatial frequency q than in-focus by defocusing (Scherzer focus), thus increasing the q-band width and the resolution. Additional "pass bands," within which $W(q,\Delta z_0)$ does not go through 0 but has a relatively high amplitude, occur in higher Δz_0 ranges defined by the condition [8, 10, 11]

$$C_3\lambda^3 q^4 - 2\Delta z_0 \lambda q^2 = (8n+3)/2,$$

with $n = 0$ giving the Scherzer focus. Taking images in the various n regions and adding them up with proper weighting in the computer allows approaching the information limit, which is determined by the envelope functions. Another approach is the defocus-modulation imaging processing method, which makes use of all spatial frequencies by adding the oscillations of $W(q,\Delta z_0)$ with the same sign and proper filtering [48, 49]. The second reference gives the theoretical background of this method in terms of the transfer function. Both approaches and others have been successfully used in transmission electron microscopy before the advent of instruments with electron-optical aberration correction. In cathode lens microscopy this approach is impractical because the spatial and temporal envelope functions caused by convergent illumination and large chromatic aberration limit the useful spatial frequency range so much that these methods are not applicable. Once instruments with better illumination systems become available, so that the corresponding envelope functions cause less damping of the wave aberration function, they may become suitable for LEEM and MEM.

A more promising approach to improving the resolution in cathode lens microscopy has its origin in the desire in light microscopy to increase the depth of focus. In light microscopy the aberrations are well corrected and the apertures are large, so there is no need for aberration correction but for increasing the depth of focus which is small because of the large aperture θ_A. While the lateral resolution is given by $\Delta x \approx 0.61\lambda/\sin\theta_A$, the depth resolution is $\Delta z \approx \lambda/\sin\theta_A^2$ ([2], p. 441). This results in a kind of uncertainty relation $\Delta z\ \Delta x^{-2} \approx 1/\lambda$ [50]. With $\sin\theta_A \approx 1$ the depth resolution is limited to $\Delta z \approx \lambda$, but object features within this range are sharp, those outside of this range contribute a background with increasing diffuseness with increasing thickness. The diffuse background is dominated in spatial frequency space by the low end of the q spectrum and can be filtered out with suitable filters. A three-dimensional transfer function theory for this situation was developed already in 1967 by Frieden [51] and in 1972 by Häusler [50], who developed also the methodic: adding (incoherently) images taken at different focus and filtering them (coherently) with a filter whose frequence transparency is proportional to the inverse transfer function. Applying this approach to some demonstration objects produced convincing results.

Today three-dimensional imaging with several versions of through-focus methods is standard in light optics ([52] Chap. 14, pp. 329–399) and many filtering methods, commercial instruments, and software packages are available. More recently, these extended depth-of-field or "optical sectioning" methods have been applied also in high resolution electron microscopy. In scanning TEMs (STEMs) with fifth-order spherical aberration (C_5) correction, which are equipped with field emission guns, such large aperture angles (e.g. $\theta_A \approx 33$ mrad) can be used now that the depth of focus is only a few nm [53]. Imaging of rough samples, such as nanoparticles on rough substrates without loss of lateral resolution is done now also with through-focus series imaging. This produces a "z-stack" of images which is processed by one of the many methods available from light microscopy. An example is the study of Pt-Co nanoparticles on carbon support particles in an aberration-corrected STEM, in which 31 images with 30 nm defocus steps were taken [54]. The lateral image shifts upon focus change, which are difficult to avoid, were corrected by cross correlation. Several processing methods were found to produce comparably sharp images with high lateral resolution over the full depth range. For completeness, old discussions of the depth of focus in cathode lens microscopy can be found in [55–57].

In cathode lens microscopy with its large aberrations the situation is somewhat different from aberration-free light-optical or aberration-corrected high-energy electron microscopy whose goal is to reconstruct from a series of images a three-dimensional image or a two-dimensional projection thereof. In cathode lens

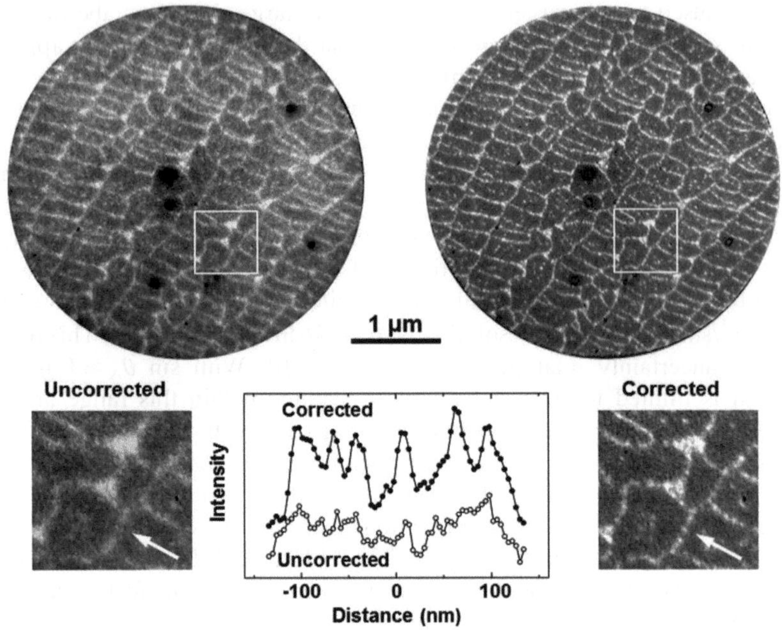

Fig. 4.16 Through-focus series image improvement. *Top*: *left*: image taken in focus; *right*: spatial frequency-filtered through-focus series image. *Bottom*: enlargements of the areas indicated in the *top images* and profiles along the *lines* marked by *arrows*. The object is a Si(111) surface. The step edges and domain walls are decorated with Pb particles, which are also found on the terraces. Illumination with focused 100 W high-pressure Hg arc lamp. *Top*: Reproduced with permission from ref. [60]. © IOP Publishing (2010). All rights reserved. *Bottom*: Courtesy of T. Yasue

microscopy an aberration-corrected image should be constructed from electrons with different energies and from electrons emitted or reflected into different angular cones $\theta\ldots\theta + \mathrm{d}\theta$, all of which are focused in different z-planes. Full correction of chromatic and spherical aberration is not possible but a significant reduction can be achieved. Inasmuch as chromatic aberration limits the resolution at the low energies used in most imaging methods the emphasis must be on its correction. For TEM a chromatic aberration correction method was proposed by Van Dyck and Coene in 1987 [58] and successfully implemented [59]. Koshikawa et al. [60] have used Ikuta's method of defocus modulation [48, 49] in an attempt to improve the resolution in LEEM to better than 10 nm. This failed because of instrumental limitations and the strong damping of the contrast transfer function by the source and chromatic envelope functions. However, a significant improvement of UV-PEEM images was obtained by processing the image stack with a suitable spatial frequency filters as illustrated in Fig. 4.16 [60]. Forty one images were taken without contrast aperture in 0.02 mA steps around the in-focus current of 1698 mA and added incoherently without weighting. A spatial frequency filter, in which frequencies above 49 $\mu\mathrm{m}^{-1}$ were weighted by a factor of 10 compared to frequencies below 2.4 μm^{-1}, with a linear interpolation of the weighting factor between these two values and a high frequency limit of about 194 μm^{-1}, was found to produce the best result. Not only was the resolution improved to the detector-limited value (12.5 nm) but also the tails of the object features were suppressed. Similar improvements by through-focus series imaging were found in energy-filtered XPEEM with an energy widow of 0.25 eV width. The 84 %/16 % resolution of an amplitude contrast object taken in focus with a small aperture was improved from 32 to 23 nm by through-focus series imaging with a five times larger aperture, simultaneously with an increase of the signal to noise ratio (Yasue, private communication). Unfortunately these preliminary efforts in through-focus "aberration correction" were terminated by redirection of the research program and lack of synchrotron radiation beamtime. However they clearly show the potential of this method at least for PEEM, in particular when full use is made of the experience available from optical microscopy. In LEEM some improvement can be expected too, provided that a number of instrumental preconditions are fulfilled: field emission electron gun, stable object manipulator, aberration-free energy filter in SPELEEM instruments, high-stability power supplies, improved detectors, to name the most important ones.

4.4 Information Transfer in the Image Acquisition System

Most image acquisition systems in surface electron microscopes consist of a double channel plate electron multiplier and fluorescent screen in vacuum and of a lens outside the vacuum, which images the fluorescent screen onto a CCD or CMOS detector. In-vacuum direct electron detectors have also become available but their price and limited vacuum compatibility will delay their widespread use until these problems are solved. Therefore, mainly the properties of the first-mentioned systems will be discussed. In Sect. 3.2.9 this was done already in real space. Here the image acquisition system will be considered in reciprocal (spatial frequency) space. First we have to consider what arrives at the first detector surface. It is the wave field of the electron originating from the object, modified by the lenses and the aperture. In the wave field the electron is delocalized over many pixels. De Jong and Van Dyck [61] described this by a detector envelope function, which acts in a manner similar to the other envelope functions. It has the effect that the point resolution δ_{eo} of the preceding electron optics is deteriorated by the primary detection process to $\delta_D > \delta_{eo}$.

δ_D is now the electron-optical image on the detector. The pixel size in all pixelated detectors, whether MCP, CCD or CMOS, corresponds in reciprocal space to a spatial frequency. For optimum transfer of the electron-optical spatial frequency spectrum the electron-optical magnification has to be adapted to the spatial frequency of the detector. This is the case when the minimum detector frequency (maximum pixel size) is two times larger than the maximum frequency (minimum feature size) in the electron-optical image on the detector. This is the famous Nyquist theorem. In optical microscopy a factor of 2.3 is used for better low-pass filtering, which may be appropriate here too. The spatial frequency of the detector is measured either in lines/mm or line pairs/mm (lp/mm). Sometimes the frequency is also expressed in units of the maximum detector frequency (minimum pixel size), called Nyquist (frequency) so that 1 Nyquist = 1/(2 pixels).

The performance of a given detector can be characterized by three quantities: the contrast transfer function (CTF), also called modulation transfer function (MTF), the noise power spectrum (NPS) and the detective quantum efficiency (DQE) as a function of the spatial frequency. The first describes the degree to which the intensity of a given frequency component of the input signal is reproduced in the output, that is the dependence of the contrast upon feature size. It is measured with black/white signals such as a sharp edge, which contains a wide frequency spectrum, sharp line pairs with varying spacing (CTF) or with sinusoidal gratings (MTF). The contrast transfer and modulation transfer are defined by image modulation/object modulation, where modulation = $(I_{max} - I_{min})/(I_{max} + I_{min})$. The noise power spectrum describes the spatial frequency dependence of the mean-square noise per unit frequency interval. Finally, the detective quantum efficiency quantifies the degree of information transfer between input and output via the S/N ratios, normalized to DQE($q = 0$), as a function of spatial frequency q: DQE(q) = $(\mathrm{S/N}(q)_{in})^2/(\mathrm{S/N}(q)_{out})^2$. These three functions are related via

$$\mathrm{DQE}(q) = N\frac{\mathrm{MTF}(q)^2}{\mathrm{NPS}(q)},$$

where N is the number of electrons per unit area [62]. The three functions are shown in Fig. 4.17a–c [63] for the MCP-fluorescent screen-CCD detector discussed above for 10, 15, and 20 keV electrons. The CTF is decreasing rapidly with q to 10 % at 8.86 pixels corresponding to a resolution of 113 μm, which was attributed to charge spreading in the fluorescent screen. However as seen in Fig. 3.63c, this resolution loss is already caused by the charge spreading between MCP and fluorescent screen. All three quantities are remarkably improved by direct electron detection as shown in Fig. 4.17d–f [64] for 10 and 20 keV electrons. The improvement by direct electron detection is significant, in particular if the larger q range is taken into account. The state of the art around 2011 has been reviewed by Faruqi and McMullan [65] but the field has been in rapid development since.

While the information transfer as a function of the spatial frequency q is well described theoretically and for some image detection systems also experimentally, only limited work has been done to describe the performance of the complete microscope in this manner. The contrast transfer theory for partially coherent imaging is well-developed now but experimental data are available up to the time of writing only for incoherent imaging with XPEEM [66, 67]. Figure 4.18a (Doran et al., unpublished) shows the measured contrast as function of spatial frequency of the PEEM-2 instrument at ALS for different field strengths at the specimen (field length 2 mm) and apertures. The image detection system is a

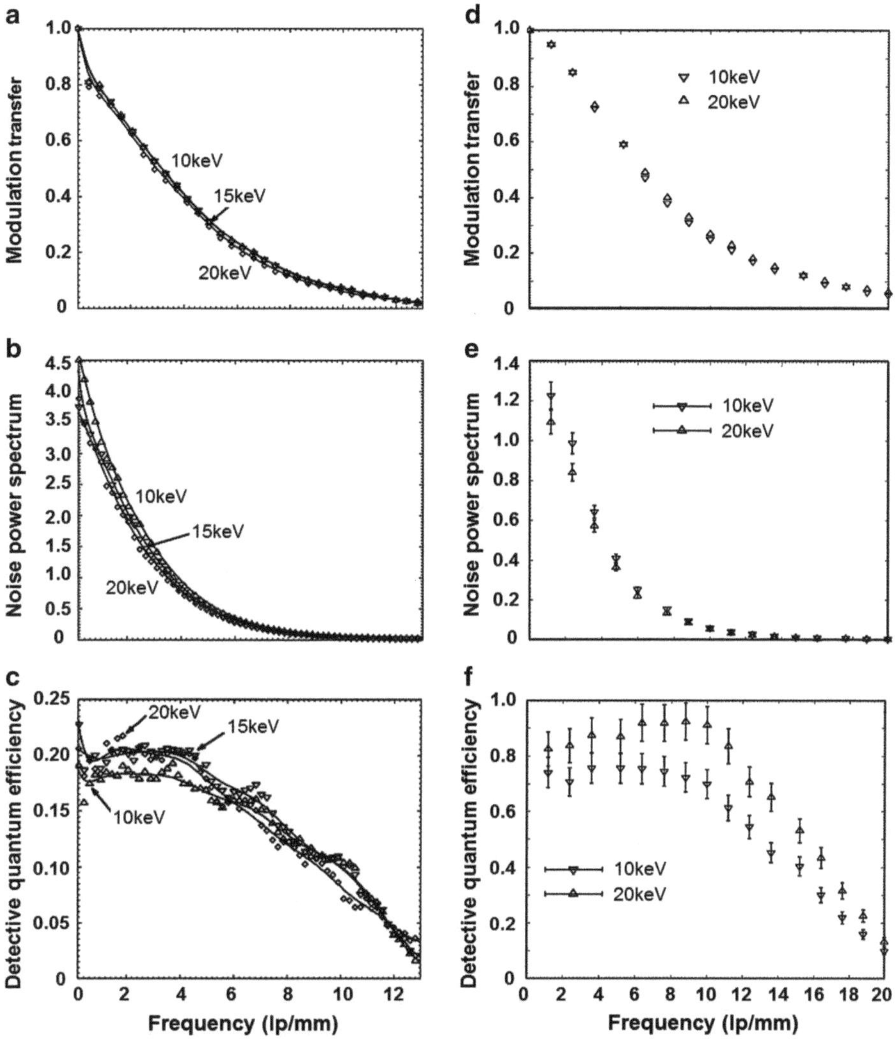

Fig. 4.17 Functions characterizing the spatial frequency dependence of the information transfer in the image detector. From *top* to *bottom*: modulation transfer function (MTF), noise power spectrum/electron/unit area (NPS), detective quantum efficiency (DQE) as function of spatial frequency. *Left* (**a–c**): for a system consisting of MCP, fluorescent screen, and CCD. *Right* (**d–f**): for a direct detection device (DDD). Note the different scales. (**a–c**) Reproduced with permission from ref. [63]. Copyright 2008 Elsevier. (**d–f**) Reproduced with permission from ref. [64]. Copyright 2009 Elsevier

fluorescent screen directly fiber-optically coupled to a cooled CCD detector, thus avoiding the image spreading connected with the MCP detection. The strong effect of the field strength at the specimen and the aperture size are clearly visible. In Fig. 4.18b [66] the measurements for a 12 μm aperture and 20 kV final energy

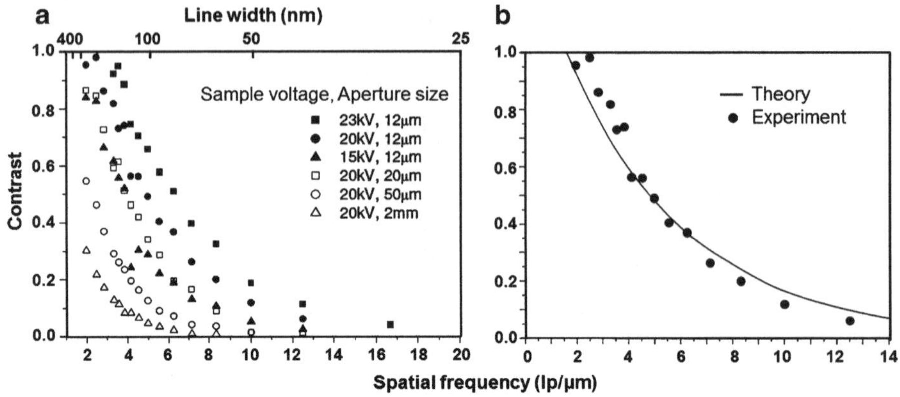

Fig. 4.18 Contrast transfer function of the complete PEEM-2 microscope (electron optics and image detection system). (**a**) As a function of high voltage at constant aperture and as a function of aperture at constant high voltage, (**b**) comparison between theory and experiment. Sample: 16 nm thick Ni pattern on Si, imaged in XPEEM at Ni L_3 resonance. (**a**) Courtesy of J. Feng; (**b**) Reproduced from ref. [66] with permission from Springer Science + Business Media

are compared with the calculated MTF, which was obtained by Fourier transformation of the PSF, derived from ray tracing, showing the good agreement between experiment and theory. Hopefully, future instrument specifications will not simply be in terms of edge resolution but rather give also CTF and DQE.

4.5 Summary and Outlook

Surface microscopy with slow electrons has reached a state of sophistication, in particular due to aberration correction, that geometric optics no longer is sufficient to describe the imaging process. The wave-optical treatment of the imaging process has made the strong interconnection between resolution and contrast amply evident as illustrated, for example, by the contrast transfer theory calculations of Schramm et al. [15] for small amplitude and phase features. Also the limitations of the conventional image detectors, which have been used now for half a century, the MCP/fluorescent screen/CCD detector, have become clearly evident. Reaching the wavelength-limited resolution is now no longer an electron-optical problem but a challenge to solve other instrument problems such as mechanical and electronic stability and image acquisition system properties.

Further resolution improvement beyond the about 1 nm, which is theoretically achievable with present aberration correction at low energies, seems possible by imaging at higher energies. For example, going in LEEM from 3 to 30 eV could improve the wave length-limited resolution by a factor of three. Inasmuch as the reflectivity decreases in general considerably with energy, this will require longer image acquisition times and thus better overall instrument stability and possibly through focus series to determine the focus. At the same time the inelastic background has to be suppressed by energy filtering or with image filtering with filters which suppress the low energy part of the spatial frequency spectrum. The latter will also have to be done to eliminate the thermal diffuse background.

Image processing has been explored much less than in TEM and promises image improvement, whether it is high-frequency noise or low-frequency background suppression. In particular, aberration "correction" via through focus series and image deconvolution can provide an alternative to electron-optical aberration correction in PEEM in environments, which do not fulfill the stability requirements needed for aberration-corrected systems. This is particularly relevant for instruments at synchrotron radiation beamlines and other multi-user facilities.

Further resolution improvements in LEEM can be expected when the incident beam is monochromatized, which is becoming routine now in high resolution TEM and STEM. This will shift the zero of the chromatic envelope to larger spatial frequency values and allow using regions of the contrast transfer function beyond the point resolution with the image processing methods used in high resolution TEM for resolution improvement. However, before the resolution can be improved significantly further C_5 will have to be corrected and not only extreme mechanical stability but in particular also electronic stability will be necessary because of the intrinsic instability of the aberration correction as pointed out by Schramm et al. [68].

References

1. Lichte, H., Lehmann, M.: Electron holography-basics and applications. Rep. Prog. Phys. **71**, 016102, 46 pages (2008)
2. Born, M., Wolf, E.: Principles of Optics, 6th edn. Cambridge University Press, Cambridge (1980)
3. Hanszen, K.J.: The optical transfer theory of the electron microscope: fundamental principles and applications. Adv. Opt. Electron Microsc. **4**, 1–84 (1971)
4. Hawkes, P.W., Kasper, E.: Wave Optics. Principles of Electron Optics, vol. 3. Academic, London (1994)
5. Williams, C.S., Becklund, O.A.: Introduction to the Optical Transfer Function. John Wiley, New York (1989)
6. Horiuchi, S.: Fundamentals of High-Resolution Transmission Electron Microscopy. North-Holland, Amsterdam (1994)
7. Reimer, L.: Transmission Electron Microscopy. Springer, Berlin (1997)
8. Spence, J.C.H.: High-Resolution Electron Microscopy, Fourth edn. Oxford University Press, Oxford (2013)
9. Goodman, J.R.: Introduction to Fourier Optics, 3rd edn. Roberts and Company, Greenwood Village, CO (2005)
10. Williams, D.B., Carter, C.B.: Transmission Electron Microscopy, vol. 3. Springer, New York (2009)
11. Erni, R.: Aberration-Corrected Imaging in Transmission Electron Microscopy. Imperial College Press, London (2010)
12. Storbeck, F.: Zur Kontrastübertragung im Emissions-Elektronenmikroskop. Ann. Phys. **23**, 208–227 (1969)
13. Müller, T.: Bildentstehung im LEEM. M.S. Thesis, Technical University Clausthal (1995)

14. Pang, A.B., Müller, T., Altman, M.S., Bauer, E.: Fourier optics of image formation in LEEM. J. Phys. Condens. Matter **21**, 314006, 10 pages (2009)
15. Schramm, S.M., Pang, A.B., Altman, M.S., Tromp, R.M.: A Contrast Transfer Function approach for image calculations in standard and aberration-corrected LEEM and PEEM. Ultramicroscopy **115**, 88–108 (2012)
16. Telieps, W., Bauer, E.: An analytical reflection and emission UHV surface electron-microscope. Ultramicroscopy **17**, 57–65 (1985)
17. Kastler, A.: Un système de franges de diffraction a grand contraste. Rev. Opt. **29**, 307–314 (1950)
18. Telieps, W.: Surface imaging with LEEM. Appl. Phys. A **44**, 55–61 (1987)
19. Mundschau, M., Bauer, E., Swiech, W.: Initial epitaxial growth of Cu on Mo(011) by low energy electron microscopy and photoemission electron microscopy. J. Appl. Phys. **65**, 581–584 (1989)
20. Chung, W.F., Altman, M.S.: Step contrast in low energy electron microscopy. Ultramicroscopy **74**, 237–246 (1998)
21. Scherzer, O.: The theoretical resolution limit of the electron microscope. J. Appl. Phys. **20**, 20–29 (1949)
22. Frank, J.: The envelope of electron microscope transfer functions for partially coherent illumination. Optik **38**, 519–536 (1973)
23. Hanßen, K.-J., Treptke, L.: Der Einfluss von Strom- und Spannungsschwankungen, sowie der Energiebreite der Strahlelektronen auf Kontrastübertragung und Auflösung des Elektronenmikroskops. Optik **32**, 519–538 (1971)
24. Wade, R.H., Frank, J.: Electron microscope transfer function for partially coherent axial illumination and chromatic defocus spread. Optik **49**, 81–92 (1977)
25. Gradshteyn, I.S., Ryzhik, I.M.: Table of Integrals, Series and Products, p. 337. Academic, Oxford (2007)
26. Mostowski, A., Stark, M.: Introduction to Higher Algebra, p. 93. Pergamon Press, Oxford (1964)
27. Frank, J.: Nachweis von Objektbewegungen im lichtoptischen Diffraktogramm von elektronenmikroskopischen Aufnahmen. Optik **30**, 171–180 (1969)
28. Mankos, M., Adler, D.: Electron–electron interactions in cathode objective lenses. Ultramicroscopy **93**, 347–354 (2002)
29. Yasue, T., Nakaguchi, A., Hashimoto, M., Mentes, T.O., Locatelli, A., Bauer, E., Koshikawa, T.: Effects of Coulomb interaction between photoelectrons on XPEEM. In: Proceedings of the 6th International Symposium on Atomic Level Characterizations for New Materials and Devices (ALC'07), Kanasawa, Japan 2007, pp. 207–211. JSPS 141 Activity Report (2008)
30. Pang, A.B., Pavlovska, A., Däweritz, L., Locatelli, A., Bauer, E., Altman, M.S.: LEEM image phase contrast of MnAs stripes. Ultramicroscopy **130**, 7–12 (2013)
31. Kan, H.-C., Phaneuf, R.J.: Focusing of low energy electrons by submicrometer patterned structures in low energy electron microscopy. J. Vac. Sci. Technol. B **19**, 1158–1163 (2001)
32. Luk'yanov, A.E., Spivak, G.V., Gvozdover, R.S.: Mirror electron microscopy. Sov. Phys. Usp. **16**, 529–552 (1974)
33. Nepijko, S.A., Sedov, N.N.: Aspects of mirror electron microscopy. In: Hawkes, P.W. (ed.) Adv. Imaging Electron Phys, vol. 102, pp. 273–323. Academic, San Diego (1997)
34. Kennedy, S.M., Zheng, C.X., Tang, W.X., Paganin, D.M., Jesson, D.E.: Laplacian image contrast in mirror electron microscopy. Proc. Roy. Soc. A Math. Phys. Eng. Sci. **466**, 2857–2874 (2010)
35. Kennedy, S.M., Zheng, C.X., Tang, W.X., Paganin, D.M., Jesson, D.E.: Laplacian image contrast in mirror electron microscopy. Addendum. Proc. Roy. Soc. A Math. Phys. Eng. Sci. **467**, 3332–3341 (2011)
36. Kennedy, S.M., Zheng, C.X., Tang, W.X., Paganin, D.M., Jesson, D.E.: Caustic imaging of gallium droplets using mirror electron microscopy. Ultramicroscopy **111**, 356–363 (2011)

37. Kennedy, S.M., Jesson, D.E., Paganin, D.M.: Laplacian and caustic imaging theories of MEM work-function contrast. IBM J. Res. Dev. **55**, 3.1–8 (2011)
38. Kennedy, S.M., Hjort, M., Mandl, B., Marsell, E., Zakharov, A.A., Mikkelsen, A., Paganin, D. M., Jesson, D.E.: Characterizing the geometry of InAs nanowires using mirror electron microscopy. Nanotechnology **23**, 125703, 9 pages (2012)
39. Wiskott, D.: Zur Theorie des Auflicht-Elektronenmikroskops. II. Wellenmechanische Elektronenoptik in der Umgebung des Objects. Optik **13**, 481–493 (1956)
40. Bok, A.B.: A Mirror electron microscope. Ph.D. thesis, Technical University of Delft (1968)
41. Hermans, A.J., Petterson, J.A.: A quantum mechanical treatment of the mirror electron microscope. J. Eng. Math. **4**, 141–154 (1970)
42. Gvosdover, R.S., Zel'dovich, B.Y.: The quantum theory of the image contrast formation of electrical and magnetic microfields in mirror electron microscopy. J. Microsc. **17**, 107–130 (1973)
43. Lichte, H.: Ein Elektronen-Auflicht-Interferenzmikroskop zur Präzisionsmessung von Unebenheiten und Potentialunterschieden auf Oberflächen. Optik **57**, 35–67 (1980)
44. Lichte, H.: Coherent electron optical experiments using an electron mirror. In: Proceedings of the International Symposium Foundations of Quantum Mechanics, Tokyo 1983, pp. 29–38 (1983)
45. Schmidt, T., Groh, U., Fink, R., Umbach, E., Schaff, O., Engel, W., Richter, B., Kuhlenbeck, H., Schlögl, R., Freund, H.-J., Bradshaw, A.M., Preikszas, D., Hartel, P., Spehr, R., Rose, H., Lilienkamp, G., Bauer, E., Benner, G.: XPEEM with energy-filtering: advantages and first results from the SMART project. Surf. Rev. Lett. **9**, 223–232 (2002)
46. Henke, B.L., Smith, J.A., Attwood, D.T.: 0.1-10-keV x- ray-induced electron emissions from solids: Models and secondary electron measurements. J. Appl. Phys. **48**, 1852–1866 (1977)
47. Feng, J., Forest, E., Macdowell, A.A., Marcus, M., Padmore, H., Raoux, S., Robin, D., Scholl, A., Schlueter, R., Schmidt, P., Stöhr, J., Wan, W., Wei, D.H., Wu, Y.: An x-ray photoemission electron microscope using an electron mirror aberration corrector for the study of complex materials. J. Phys. Condens. Matter **17**, S1339–S1350 (2005)
48. Ikuta, T.: Image restoration in coherent imaging system involving spherical aberration. J. Electron Microsc. **38**, 415–422 (1989)
49. Taniguchi, Y., Ikuta, T., Shimizu, R.: Mathematical background of defocus-modulation image processing. Optik **96**, 129–135 (1994)
50. Häusler, G.: A method to increase the depth of focus by two step image processing. Opt. Commun. **6**, 38–42 (1972)
51. Frieden, B.R.: Optical transfer of the three-dimensional object. J. Opt. Soc. Am. **57**, 56–66 (1967)
52. Wu, Q., Merchant, F.A., Castleman, K.R.: Microscope image processing. Elsevier, Amsterdam (2008)
53. Borisevich, A.Y., Lupini, A.R., Pennycook, S.J.: Depth sectioning with the aberration-corrected scanning transmission electron microscope. Proc. Natl. Acad. Sci. U. S. A. **103**, 3044–3048 (2006)
54. Hovden, R., Xin, H.L., Müller, D.A.: Extended depth of field for high-resolution scanning transmission electron microscopy. Microsc. Microanal. **17**, 75–80 (2011)
55. Bartz, G.: Zur Tiefenschärfe der elektronenoptischen Abbildung mittels elektrostatischer Immersionsobjektive. Optik **17**, 135–140 (1960)
56. Soa, E.A.: Bemerkungen zur Tiefenschärfe im Emissions-Elektronenmikroskop. Exp. Tech. Phys. **22**, 463–466 (1974)
57. Rempfer, G.F., Nadakavukaren, K.K., Griffith, O.H.: Depth of field in emission microscopy. Ultramicroscopy **5**, 449–457 (1980)
58. Van Dyck, D., Coene, W.: A new procedure for wave function restoration in high resolution electron microscopy. Optik **72**, 125–128 (1987)
59. Thust, A., Coene, W.M.J., Op de Beeck, M., Van Dyck, D.: Focal-series reconstruction in HRTEM: simulation studies on non-periodic objects. Ultramicroscopy **64**, 211–230 (1996)

60. Koshikawa, T., Shimizu, H., Amakawa, R., Ikuta, T., Yasue, T., Bauer, E.: A new aberration correction method for photoemission electron microscopy by means of moving focus. J. Phys. Condens. Matter **17**, S1371–S1380 (2005)
61. de Jong, A.F., van Dyck, D.: Ultimate resolution and information in electron microscopy: II. The information limit of transmission electron microscopes. Ultramicroscopy **49**, 66–80 (1993)
62. Meyer, R.R., Kirkland, A.I.: Characterisation of the signal and noise transfer of CCD cameras for electron detection. Microsc. Res. Tech. **49**, 269–280 (2000)
63. Moldovan, G., Matheson, J., Derbyshire, G., Kirkland, A.: Characterisation of a detector based on microchannel plates for electrons in the energy range 10–20 keV. Nucl. Instrum. Meth. Phys. Res. A **596**, 402–408 (2008)
64. Matheson, J., Moldovan, G., Clark, A., Prydderch, M., Turchetta, R., Derbyshire, G., Kirkland, A., Allinson, N.: Characterisation of a monolithic active pixel sensor for electron detection in the energy range 10–20 keV. Nucl. Instrum. Meth. Phys. Res. A **608**, 199–205 (2009)
65. Faruqi, A.R., McMullan, G.: Electronic detectors for electron microscopy. Quart. Rev. Biophys. **44**, 357–390 (2011)
66. Feng, J., Scholl, A.: Photoemission electron microscopy (PEEM). In: Hawkes, P.W., Spence, J.C.H. (eds.) Science of Microscopy, vol. 1, pp. 657–695. Springer, New York (2007)
67. Doran, D., Church, M., Miller, T., Morrison, G., Young, A.T., Scholl, A.: Cryogenic PEEM at the advanced light source. J. Electron Spectrosc. Relat. Phenom. **185**, 340–346 (2012)
68. Schramm, S.M., van der Molen, S.J., Tromp, R.M.: Intrinsic instability of aberration-corrected electron optical systems. Phys. Rev. Lett. **109**, 163901, 5 pages (2012)

Chapter 5
Applications in Surface Science

Abstract In this chapter we illustrate the applications of surface electron microscopy with slow electrons by examples selected from the extensive list of references, which should be accessible to most readers. The list has no claim to completeness, in particular because some important results have been published only in conference proceedings, which are hard to find, and some in the open literature may have been overlooked. Many application examples have already been mentioned in Chap. 3 in connection with the description of the methods and will not be repeated here. The examples are ordered primarily according to the material studied, an imperfect ordering system because certain phenomena are studied using several materials. This will become evident already in the first section, which is concerned with the microstructure of surfaces, studied on metals, semiconductors, and insulators. All methods, which contribute to the understanding of a given phenomenon will be discussed in the corresponding section to illustrate their complementarity whenever existing. Every method has its strengths and limitations and accordingly will dominate the various chapters. For example, LEEM is ideally suited for the study of the microstructure of surfaces and of processes that influence it, while of only limited use for the study of the surface composition. On the other hand, XPEEM gives little information on the microstructure but is powerful in chemical characterization. Some of the other methods such as MIEEM provide very limited information but will nevertheless be mentioned whenever they have been used in a given context.

5.1 Surface Microstructure

The microstructure of surfaces was one of the first applications of LEEM, using the two main contrast mechanisms, phase contrast and diffraction contrast. These show monoatomic steps, step bunches, facets, etch pits, the intersection of glide planes

E. Bauer, *Surface Microscopy with Low Energy Electrons*,
DOI 10.1007/978-1-4939-0935-3_5, © Springer Science+Business Media New York 2014

and screw dislocations with the surface and other surface features. The processes which determine the microstructure are imaged dynamically while heating, during sublimation, growth, electromigration, or other surface treatments such as annealing of the radiation damage caused by ion, electron, or photon bombardment. From these images the phenomenological quantities which characterize these processes and the microstructure can be extracted. The atomic aspects such as the nature of the diffusing species or the interactions between the adsorbed atoms, of course, are not accessible because of the limited resolution of LEEM. For example, scanning tunneling microscopy (STM) studies have clearly shown that the high temperature (1×1) phase of the Si(111) surface contains 10 % more atoms than the ideal (1×1) plane [1] and that these atoms form a two-dimensional gas of magic clusters [2], which play an important role in the step dynamics of the (1×1) phase and in the transition to the low-temperature (7×7) phase. In principle LEED could give some information on the adsorbed species but even high resolution diffractometer studies showed only a diffuse (2×1) structure and a high background around the (00) spot, superimposed on sharp (1×1) reflections, which indicate a well-ordered (1×1) structure [3]. The diffuse features are compatible with the magic clusters but do not allow extracting size and structure from the data.

With these limitations in mind we discuss in the following the information which can be and has been obtained from LEEM studies of the surface microstructure starting with metal surfaces, followed by Si surfaces, and ending with compound surfaces. Initially dynamic processes were recorded with video cameras on video tape or intermittently with still cameras. This made analysis tedious and led mostly to qualitative results. CCD cameras, data acquisition with computers, and the development of data analysis software eliminated this hurdle, allowing extraction of quantitative results from the images.

5.1.1 Metals

The first metal surface studied with LEEM was the Mo(110) surface [4]. Figure 5.1 shows round monoatomic sublimation steps and straight monoatomic glide steps introduced by glide during rapid cooling. This was the first demonstration of phase contrast, which was analyzed using Kastler's calculations of Fresnel diffraction from two half-planes shifted by half a wavelength relative to each other [5]. Following Telieps' original work Mundschau et al. studied surface step phenomena qualitatively not only with LEEM but also with PEEM (Fig. 5.2), making use of the preferential condensation of Cu at steps to decorate them and of the work function difference between Cu and Mo [6–10]. Other early work was concerned with the influence of steps on the orientation of reconstruction domains of Au(100), using dark field LEEM [11, 12] and with the influence of stress on reconstruction of Pb(110) [13–15]. Step structure studies continued in the late 1990s in Flynn's group [16–22] with thick epitaxial Mo(110) and Nb(110) layers grown and annealed on $Al_2O_3\,(11\bar{2}0)$ surfaces. These studies were concerned in particular with faceting of slightly misoriented

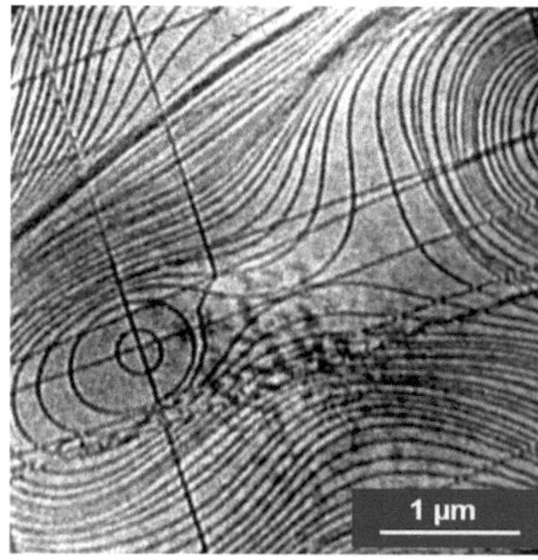

Fig. 5.1 LEEM image of a clean Mo(110) surface taken with 14 eV electrons. The spots are defects in the channel plate image intensifier. Reproduced with permission from Ref. [4]. Copyright 2010 Elsevier

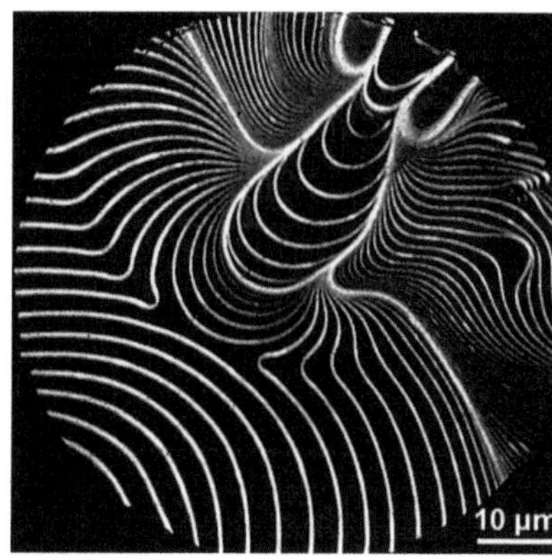

Fig. 5.2 PEEM image of a Mo(110) surface. Monoatomic steps are decorated with Cu deposited at 700 K. Mercury high-pressure arc lamp illumination with $h\nu \approx 5$ eV. From Ref. [6]

surfaces, resulting from the slight miscut of the substrate, and with the defects introduced by the misfit with the substrate. Microstructure changes caused by sputtering at elevated temperatures were later studied on Pd(111) [23] and Pt(111) [24] surfaces. Some of the references ([19, 22]) contain a wealth of information on the influence of carbon and oxygen on the surface microstructure, for example oxygen-induced domain formation.

The studies mentioned up to now gave a good qualitative understanding of the microstructure of single crystals surfaces and the factors upon it depends, but a quantitative understanding requires a detailed study of the step structures as a function of time and temperature and analysis of the experimental results in terms of theoretical models. These have a long history, starting from the famous Burton-Cabrera-Frank (BFC) model [25], and have been reviewed repeatedly, e.g. in Refs. [26–29], where the derivations of the formulas used below in the data analysis can be found. Both equilibrium and non-equilibrium situations have been studied, predominantly with two methods: step edge fluctuations of more or less straight steps and shrinking/growth of more or less circular two-dimensional islands. The first method, called step fluctuation spectroscopy, was pioneered by Bartelt et al. [30] for the

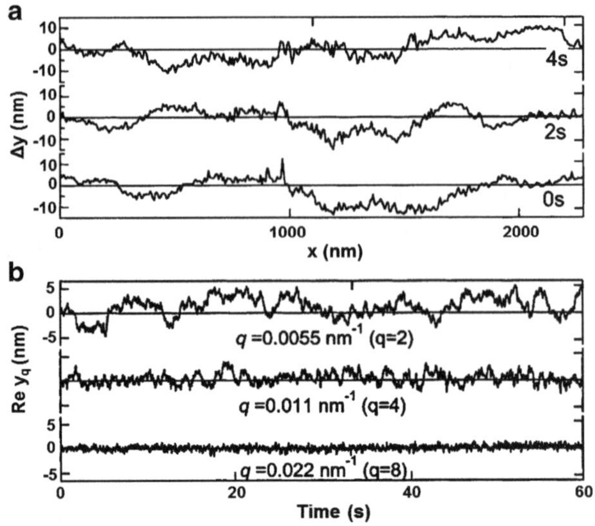

Fig. 5.3 Fluctuations of a straight step on a Mo(110) surface. (**a**) Fluctuations $y(x, t_i)$ at three moments t_i separated by 2 s. (**b**) Fluctuations of the Fourier amplitudes $y_q(t)$ with different spatial frequency q. 1 s = 30 images. Reproduced with permission from Ref. [32]. Copyright 2003 Elsevier

analysis of step fluctuations on the Si(111) surface at 900 °C observed with high energy reflection electron microscopy (REM). Later it was applied by Ondrejcek et al. in Flynn's group in LEEM studies of the microstructure of close-packed metal surfaces (Mo(110), Nb(110), Pt(111), Pd(111), Au(111), and Ni(111)) [31–43].

In step fluctuation spectroscopy the fluctuations $y(x, t)$ around a step (Fig. 5.3a) are extracted from many LEEM images (typically several hundreds) and Fourier-analyzed:

$$y(x,t) = \sum_q y_q(t)\exp(iqx) \quad (0 \le x \le L) \tag{5.1}$$

with $y_q = y^*_{-q}$, $q = 2\pi q/L$ (q integer) and L the step length, which is limited by the image diameter and the magnification. Figure 5.3b shows $y_q(t)$ for three q values, indicating a rapid decrease with decreasing wavelength (increasing q). The temperature determines the amplitudes of the fluctuations q, whose energies U_q add up to the excess free energy over that of the non-fluctuating step:

$$U = \sum_q U_q = \frac{L\tilde{\beta}}{2}\sum_q q^2 |y_q|^2, \tag{5.2}$$

where

$$\tilde{\beta} = \beta + \mathrm{d}^2\beta/\mathrm{d}\Theta^2 \tag{5.3}$$

is the step stiffness, which is a measure for the resistance to step fluctuations, and $\beta(\theta, T)$ is the step free energy or line tension per unit length. In thermal equilibrium on average $U_q = kT/2$ so that from Eq. (5.2)

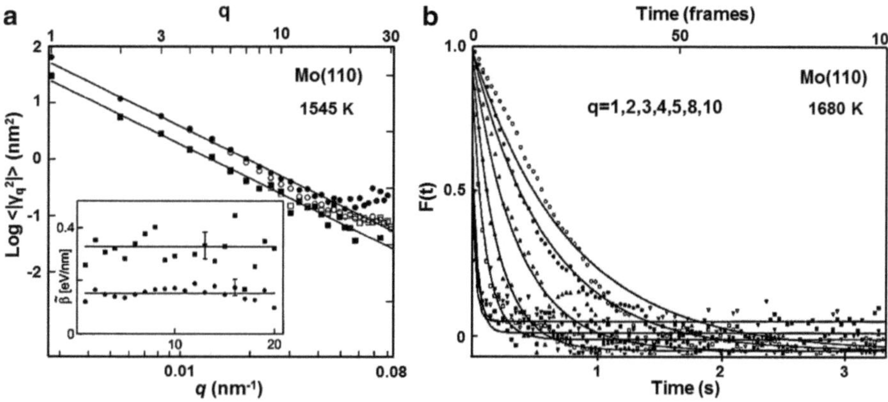

Fig. 5.4 (**a**) Fourier amplitudes of the fluctuations of straight steps with different orientations on a Mo(110) surface at 1545 K as a function of q. At large q values the amplitudes are so small that corrections are needed. The differing symbols indicate uncorrected and corrected values. The *inset* shows the step stiffness deduced from these data via Eq. (5.4). (**b**) Time dependence of the time correlation of the Fourier amplitudes $F(t' - t)$ for several q values at 1680 K. The relaxation time at this temperature is too short for $q \geq 8$ for the determination of τ_q. Reproduced with permission from Ref. [32]. Copyright 2003 Elsevier

$$|y_q|^2 = k_B T / L\tilde{\beta} q^2. \tag{5.4}$$

Thus by determining $|y_q|^2$ as a function of q^2, $\tilde{\beta}$ can be determined as a function of step orientation and temperature as illustrated in Fig. 5.4a for two step orientations on Mo(110). Up to about $q = 10$ the data fit very well the q^{-2} dependence but at larger q the y_q values decrease into the noise level. Nevertheless the fitting range is large enough to give step stiffness values with a small error bar. Measurements of several step orientations show a strong angular dependence of the step stiffness which can be described by $\tilde{\beta} = 0.26 \pm 0.03 \cos 2\theta \,\mathrm{eV/mm}$, with $\theta = 0°$ in the [001] direction, a consequence of the twofold symmetry of this surface. On the Nb(110) surface $\tilde{\beta}$ shows a stronger angular dependence at about 1600 K: $\tilde{\beta} = 0.32 \pm 0.19 \cos 2\theta \,\mathrm{eV/mm}$. On the threefold symmetric close-packed fcc(111) surfaces $\tilde{\beta}$ was found in the case of Au to vary from 0.3 eV/nm to about 0.1 eV/nm almost up to the melting point T_m, but to be isotropic in the case of Ni within the limits of error. Except for Ni, $\tilde{\beta}$ decreased only little with temperature in the temperature range studied, $0.4 \leq T/T_m \leq 0.96$, varying with metal (Mo, Nb, Pd, Ni, Pt, Au). Thus, although some common trends were found, there are significant differences between the close-packed surfaces. On the Nb(110) surface another unrelated phenomenon was observed, the movement of steps by a moving edge dislocation intersecting the surface. The dislocation pulls the step along its path, deforms it until the step tension becomes too large, leaves it behind, and the process continues with the next step. From the step shape at the step-dislocation separation point the "plucking" force acting on the dislocation of 75 pN could be derived [39].

Beyond the determination of the step stiffness the fluctuations provide information on the kinetics of the fluctuations, which are caused by the locally varying attachment and detachment of atoms. The kinetics can be extracted from the q-dependent relaxation of the various modes

$$\langle y_q(t')\rangle = y_q(t)\exp\{-(t'-t)/\tau_q\} \tag{5.5}$$

The relaxation time τ_q is extracted from the time correlation functions of the different modes $y_q(t)$

$$F_q(t'-t) = \frac{y_q(t')y_q^*(t) + \text{c.c.}}{2|y_q|^2}\exp\{-(t'-t)/\tau_q\}. \tag{5.6}$$

Figure 5.4b shows an example of such a fit for several q values of a step on Mo (110), from which τ_q is obtained. As expected, the relaxation time decreases rapidly to the image acquisition time with increasing q, limiting the q range. Nevertheless the fit gives a clear q^{-2} dependence of τ_q.

Inasmuch as the step edge fluctuations are caused by attachment/detachment of atoms they give information on the corresponding processes, such as surface diffusion or diffusion from the bulk. Assuming that the two processes are additive [34]

$$\tau_q^{-1} = \left(D_\text{b} + \frac{2qa}{\pi}D_\text{s}\right)\frac{\pi A^2 q^2}{\Omega k_\text{B}T}\tilde{\beta}, \tag{5.7}$$

where D_b and D_s are the temperature-dependent diffusion coefficients for bulk and surface diffusion, a is the spacing of the surface planes, A is the surface area per atom and Ω is the atomic volume. A plot of $\left(\tau_q q^2\tilde{\beta}\right)^{-1}\Omega k_\text{B}T/\pi A^2$ as a function of q gives an effective diffusion coefficient $D_\text{eff} = D_\text{b} + (2qa/\pi)D_\text{s}$, whose slope gives D_s and whose value at $q = 0$ gives D_b. The plot of D_eff in Fig. 5.5 shows pure surface diffusion below about 1450 K (D_eff (0) = 0) but an increasing contribution of bulk diffusion D_eff (0) at temperatures above about half of the melting temperature, increasing with increasing temperature. Using this method the diffusion coefficients for the close-packed surfaces mentioned above have been determined. The results for D_s are compiled in Fig. 5.6 and in Table 5.1 together with the step stiffness. These and general theoretical consideration of the mass transport processes involved [44–46] have led Flynn to the proposal of the "standard metal" model of diffusion with parameters relative to T_m, which is indicated in the inset of Fig. 5.6, common to many fcc(111) surfaces with minor deviations.

The second method using surface steps to extract quantitative information on surface energetics and kinetics is based on the growth or shrinking of two-dimensional islands under non-equilibrium conditions. Such conditions exist on flat surfaces at hills or shallow valleys consisting of a series of concentric

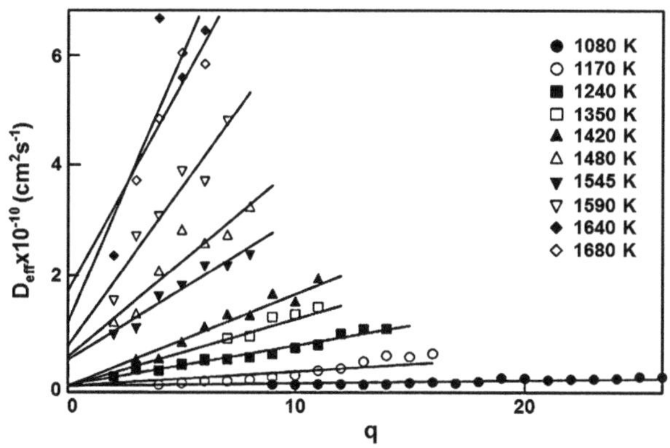

Fig. 5.5 Effective diffusion coefficients $D_{\rm eff} = D_{\rm B} + (2qa/\pi)D_{\rm S}$ on the Mo(110) surface as a function of q for various temperatures. At high temperatures the fast relaxation causes large uncertainties even for small q. Reproduced with permission from Ref. [41]. Copyright 2006 Elsevier

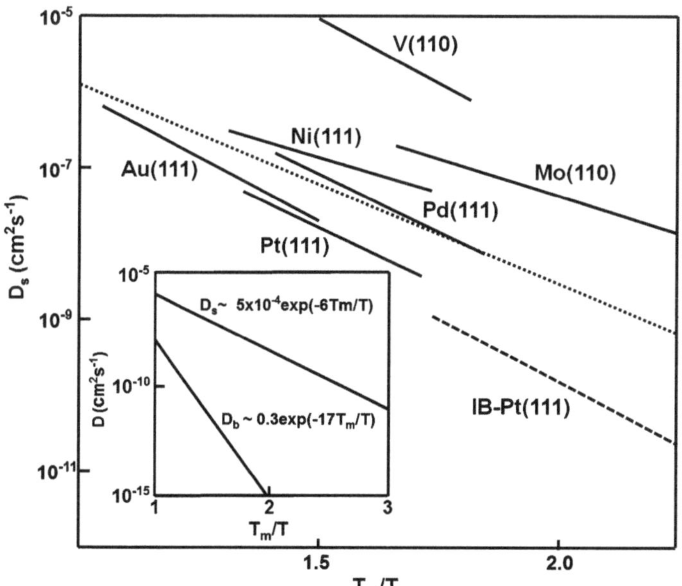

Fig. 5.6 Surface diffusion coefficients for the surfaces studied with step fluctuation spectroscopy as a function of T_m/T. The *dotted line* is the "standard metal" surface diffusion coefficient, which is shown again in the *inset* together with the bulk "standard metal" diffusion coefficient. The *dashed line* marked IB-Pt(111) is from data obtained from an ion bombardment-modified Pt(111) surface. New figure based on data in Refs. [36–43]

Table 5.1 Step stiffness and diffusion parameters

Surface	$\widetilde{\beta}$ (eV/nm)	α_S (cm^2/s)	E_S (eV)	α_B (cm^2/s)	E_B (eV)
Ni(111)	0.41	1×10^{-4}	0.65	1.4	2.45
Pd(111)	0.27	3×10^{-3}	1.2	0.2	2.75
Pt(111)	0.21	5×10^{-4}	1.2	0.8	2.75
Au(111)	0.17	1.7×10^{-3}	0.87	6×10^{-2}	1.95
Mo(110)	0.27	4×10^{-4}	1.13	0.4	3.1
Nb(110)	0.32				
V(110)	0.66	0.8	1.43		

The step stiffness $\widetilde{\beta}$ values are average values at $T/T_m = 0.6$. Only Au, Mo, and Nb show a significant θ dependence and Ni a significant temperature dependence. Some pre-exponentials α and activation energies E have large error bars and are not listed therefore. Compiled from data in Refs. [36–42]

monoatomic terraces. Driving force is the minimization of the step free energy by reducing the step length. In the study of metal surfaces with LEEM the method was first applied by McCarty [47] to a NiAl(110) surface. He noticed that all islands shrunk at the same rate, which is difficult to understand if the shrinking would be accomplished by surface diffusion from the edges of the smaller islands to the edges of the larger islands. In order to understand the material transport mechanism he oscillated the temperature and found that the normalized island area fluctuations dA/dtA were independent of island size, which means that the material transport occurred locally at the island edge. In the temperature range of his experiment ($0.53 \leq T/T_m \leq 0.63$) there is already a significant temperature-dependent vacancy concentration in the bulk, which changes during the temperature fluctuations by vacancy creation or annihilation at the surface causing the surface steps to move. From the area fluctuations as a function of temperature the vacancy formation energy could be determined. Superimposed on the oscillating area changes is a monotonic area decrease due to the isothermal smoothing, which depends upon the temperature around which the oscillations occur. From the temperature dependence of this area decrease an activation energy of 2.54 eV was obtained, which gives an activation energy of 1.9 eV for the rate of vacancy exchange between the bulk and surface steps, taking into account the vacancy formation energy of 0.64 eV.

Further studies of two-dimensional island growth/shrinking have shown that the changes of the island size with time after sudden temperature changes can occur either by volume or surface diffusion. On the Pt(111) surface it was found that volume diffusion was responsible for the island size change in the temperature range studied ($0.52 \leq T/T_m \leq 0.62$) and a bulk vacancy formation energy of 1.54 eV was derived from the temperature dependence of the changes [48]. On the other hand surface diffusion was deduced from the isothermal and post-temperature jump island shrinking on the Rh(100) surface in a similar temperature range ($0.54 \leq T/T_m \leq 0.65$) with an unusually high energy for creation and diffusion of the active species [49]. Apparently the difficulties of interpreting such data have stopped further work along this line.

In the studies discussed up to now the step changes involved no changes in the total number of atoms in the crystal. Useful information on surface microstructure and energetics can also be obtained from step studies in which this is not the case, i.e., during deposition and sublimation. An example is the determination of the Ehrlich-Schwoebel barrier, i.e., the energy E_{ES} needed to transfer an atom from the upper side of a step to the lower side on the terrace (where it is incorporated in the step). This energy can play an important role in growth at low temperatures and can be determined by step flow studies. These have been carried out by exposing concentric monatomic terraces on flat hills on a Cu(100) surface at temperatures between 360 and 400 K to a constant Cu atom flux and measuring the time dependence of the terrace radii. At these temperatures the thermal adatom concentration is negligible compared to the concentration produced by the incident flux. Assuming steady state conditions and precisely known Cu flux a value of 125 meV was obtained [50]. Surface step studies of the phenomenon opposite to condensation give useful information on the dynamics of sublimation, which proceeds via island decay, nucleation and growth of vacancy islands and in addition via formation and rotation of double spirals when dislocation loops intersect the surface. The analysis of these processes for a Cr(100) surface showed that no mass transfer with the bulk was involved and gave activation energies in agreement with the heat of sublimation [51].

A powerful method, which allows both addition and removal of atoms to the crystal in the same experimental setup was developed in Flynn's group [24] but unfortunately used only for a short period due to termination of funding. It uses bombardment of the surface by "self-ions," that is ions of the same material as the crystal, at elevated temperatures at which point defects anneal out rapidly forming two-dimensional structures, either vacancy islands or adatom islands, depending upon if the sputter rate (incoming flux/outgoing flux) is smaller or larger than 1 (Fig. 5.7). The transition between these two modes was found to occur for Pt at 245 eV ion energy. By proper choice of ion current density and temperature large flat regions up to 10 μm in diameter can be prepared, on which single or multiple adatom or vacancy islands can be nucleated and grown in a controlled manner by changing temperature, ion current density, or energy. This is illustrated in Fig. 5.7 [52] for the influence of energy. These processes have been studied in detail theoretically and experimentally [52–57]. Alternately, on stepped surfaces a high density of ion bombardment-produced two-dimensional islands can coalesce with the existing steps by annealing at higher temperatures and produce strongly oscillating steps, whose decay can be analyzed by step fluctuation spectroscopy over a wider temperature range than in the experiments described above. In this manner the curve marked IB-Pt(111) in Fig. 5.6 was obtained. Concluding the subject step motions a study of the high temperature behavior of steps on *i*-Al-Pd-Mn quasicrystal surfaces should be mentioned. Two sets of steps were found, which cross each other forming terraces with identical surface composition in contrast to the alternating composition in the bulk. The steps move with different velocities in opposite directions when the temperature is changed due to addition or removal of atoms from the bulk and have different step height. For details see Ref. [58].

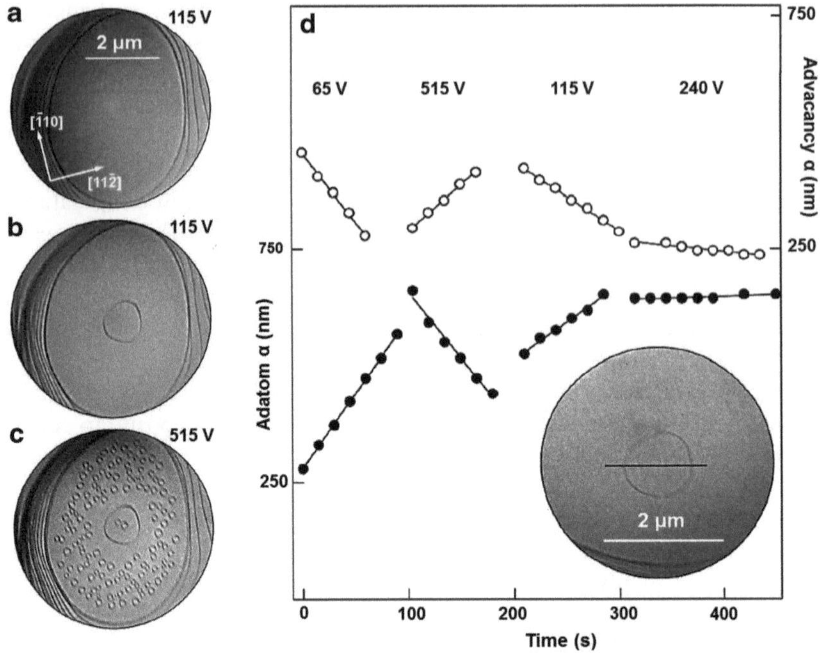

Fig. 5.7 Surface modification of a Pt(111) surface by bombardment with Pt^- ions at 1005 K. The single terrace with 5 μm diameter in (**a**) was formed initially from a preexisting hill by bombardment with 115 eV ions. Bombardment of this terrace with 115 eV ions created the adatom island shown in (**b**). Subsequent bombardment with 515 eV islands produced many vacancy islands and caused shrinking of the adatom island as shown in (**c**). (**d**) Illustrates how the size of an adatom or vacancy island can be varied at 1070 K by changing the ion energy. Adapted from Ref. [52] by permission of Taylor and Francis Group, LLC

Summarizing the quantitative studies of the surface microstructure, it is evident that only a few metals have been studied over a limited T/T_m range. The temperature range is limited towards the top by the vapor pressure, towards the bottom by the magnitude of the step changes that can be detected by LEEM. The results show to some extent common features such as in the surface diffusion behavior of fcc(111) surfaces ("standard metal") or the transition to bulk diffusion at high temperatures as dominating process, which determines the surface microstructure. Significant deviations from this trends have, however also been observed such as in the case of V(110) and Rh(100). Other unexpected results are the strong angle dependence of the step stiffness on Au(111) up to close to the melting point in contrast to the weak dependence on other fcc(111) surfaces even at lower temperatures as well as the weak temperature dependence of the step stiffness on most surfaces except Ni(111). Much work is still necessary for a comprehensive picture to emerge. In any case, LEEM has been demonstrated to be a powerful method for the study of surface energetics and kinetics of metals.

5.1.2 Semiconductors

Semiconductors have been one of the most studied materials with LEEM, mostly the Si(111) surface because of its interesting phase transition and the Si(100) surface because of its technological importance, followed by other Si surface orientations, Ge, GaAs and other compound semiconductors of practical importance such as SiC, GaN and other binary compounds. In the following only results of interest for the microstructure will be discussed while aspects of more practical interest will be described in Sect. 6.3.

5.1.2.1 Si(111)

The Si(111) was the first semiconductor surface studied with LEEM, mainly in order to settle a problem of the $(7 \times 7) \leftrightarrow (1 \times 1)$ transition on this surface, which was much discussed in the early 1980s. LEED studies by Bennett and Web [59] had originally favored a second order phase transition but Witt in a very careful precision LEED diffractometer study [3], which was not accepted for publication, excluded a second order transition because of the absence of critical scattering. On the other hand reflection electron microscopy (REM) studies led originally [60] to the conclusion that the transition is first order but a later REM study [61] indicated that the transition is second order. REM studies are hampered by the strong foreshortening of the image and dynamical diffraction problems, which make contrast interpretation difficult. In LEEM neither problem exist because of the normal incidence of the beam and the easy distinction between (1×1) and (7×7) structure due the difference in the $I(V)$ curves, even for the (00) beam, so that dark field imaging is not needed for contrast formation. Therefore, in a detailed study of the kinetics of the phase transition with LEEM, Telieps, and Bauer showed unambiguously that the transition is first order and determined the shape, orientation, nucleation, growth, and decay of the triangular (7×7) domains [62, 63]. Depending upon heating and cooling history many microstructural features were obtained, some of which are reported in later topical reviews [64–67]. Examples are shown in Fig. 5.8 [66, 67]. The influence of surface defects was also studied, though mainly on Si(100) [68]. Si(111) served later for a demonstration of the possibilities of a commercial instrument (JEOL) [69].

The early work in Bauer's group [62–65, 68, 70, 71] was more or less exploratory, looking qualitatively at the phase transition, determining the shape and orientation of the (7×7) domains, their primary nucleation at steps as a function of undercooling and step orientation and the secondary nucleation and growth, finally ending with the surface completely covered with (7×7) domains separated by domain boundaries. The growth of these domains was later studied by Phaneuf et al. [72], who found that there was no anisotropy in the boundary energy and that grain growth was not purely driven by boundary curvature as predicted by theory. They also studied the phase transition on vicinal Si(111) surfaces with the result

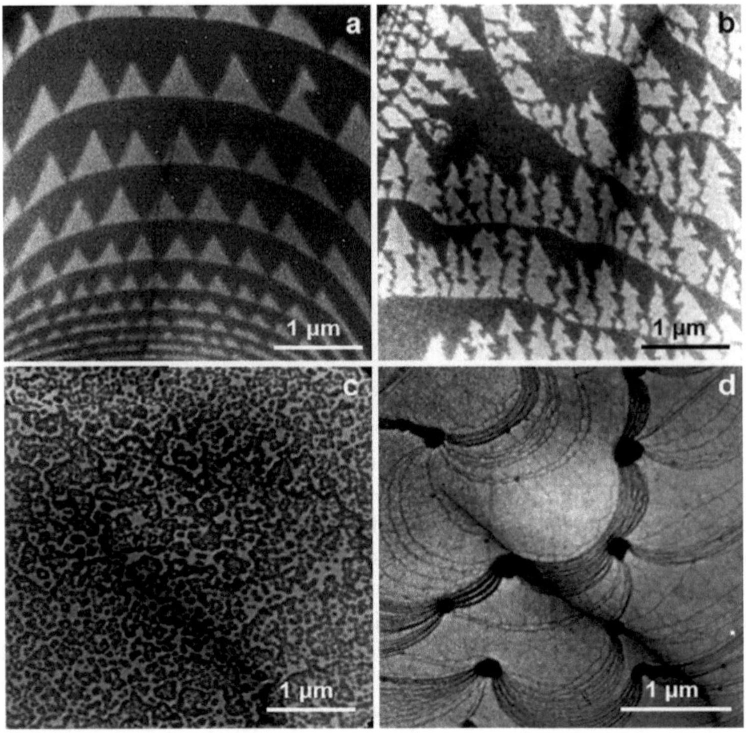

Fig. 5.8 LEEM images of a Si(111) surface after different treatments. (**a**) Slowly cooled a few degrees below the transition temperature. (**b**) Rapidly quenched to low temperatures. (**c**) Annealed for several hours at about 1200 K. (**d**) Cooled to room temperature after many annealing and cooling cycles, which have led to the formation of SiC crystals which pin the steps. Bright region have (7 × 7) structure, dark regions have (1 × 1) structure (except the *black dots* in (**d**)). Electron energy 10–11 eV. The *dark diffuse line* seen in these and some later images is a defect in the channel plate image intensifier. (**a**) Reproduced from Ref. [67] with permission from the American Institute of Physics, © 1991; (**b–d**) from Ref. [66] with permission from Springer Science + Business Media

that the surface broke up very rapidly into step bunches and (7 × 7) structure-covered (1 × 1) terraces, stopping spontaneously at a well-defined width. This theoretically unexpected result of finite facet width was attributed to elastic interactions caused by the terrace boundaries [73, 74]. A closer examination of the influence of the quenching rate revealed at high rates a spinodal pattern, which they attributed to the crossing of the surface free energies of the (7 × 7) and (1 × 1) structures as a function of miscut, resulting in a spinodal region [75].

With exception of the work reported in Ref. [69] all studies discussed up to now were made with the first LEEM instrument, which was used for the study of many other surfaces. Once newer LEEM instruments became available in the early to middle 1990s, in particular in semiconductor industry laboratories such as IBM Yorktown Heights and later in the NTT Basic Research Laboratories, in-depth studies of Si

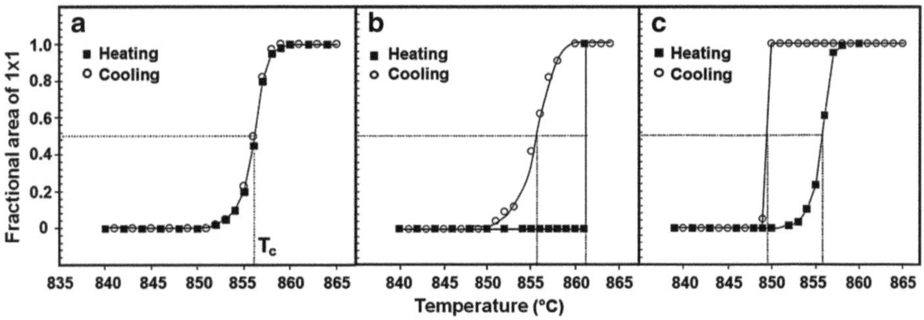

Fig. 5.9 Fractional area of the (1 × 1) regions on a Si(111) surface during very slow heating and cooling to achieve equilibrium. (**a**) In a region with 0.3–0.6 μm step spacing; (**b**) in a vacancy island with 4 μm average diameter; (**c**) on a mesa with 15 μm diameter. Adapted with permission from Ref. [85]. Copyright 2001 Elsevier

surfaces started, both on Si(111) and Si(100). They may be roughly divided into two groups, one with the main attention focused on the (7 × 7)-(1 × 1) transition, the other focusing on surface diffusion in general and on its involvement in the phase transition.

We discuss first the former group of studies, which were done predominantly by Hannon et al. [76–84] but others have also contributed to this subject [85, 86]. They were directed on the one hand to test the validity of standard phase transition theory using the (7 × 7) ↔ (1 × 1) transition as test object, on the other to understand this transition within the framework of existing theory. Both aspects and the results obtained up to 2002 are thoroughly discussed in two excellent review papers [87, 88] so that they will be sketched here only briefly. The most striking aspect of the transition is the relatively wide coexistence range around the critical temperature T_c, whose values varies somewhat from author to author due to the difficulty of accurate absolute temperature measurements in LEEM. The phase coexistence range can be seen on large step-free regions, on which (homogeneous) nucleation is not pre-emptied by (heterogeneous) nucleation at the upper side of the steps, where the nucleation barrier is lower. On the step-free regions considerable overheating, undercooling, and hysteresis occurs as shown in Fig. 5.9b, c, with a wide coexistence range, in particular upon cooling, in contrast to stepped regions as shown in Fig. 5.9a. This coexistence region allows now the study of the energetics and kinetics of domain nucleation, growth, and decay and to determine the physical parameters, which determine these processes.

Phase coexistence in two dimensions is usually determined by the competition between short range attractive and long range repulsive elastic interactions but electrostatic interactions can also produce coexistence when the work functions of the two phases differ sufficiently. Measurements of the relative size of (1 × 1) and (7 × 7) domains as function of temperature and terrace widths showed that in this system elastic interactions are more than an order of magnitude larger than electrostatic interactions, because of the low work function difference between the two phases as determined by the transition between mirror and LEEM mode of imaging [77]. The elastic interactions are caused by the surface stress difference between the

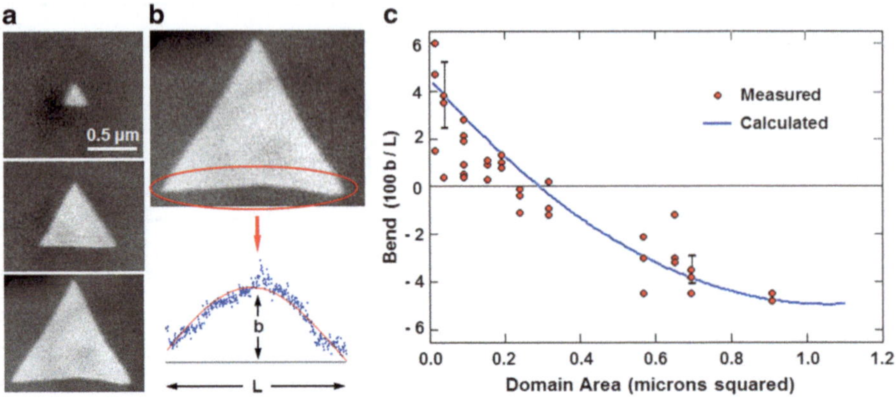

Fig. 5.10 Size dependence of the equilibrium shape of an isolated (7 × 7) domain on a large Si(111) terrace. (**a**) Observed domain shapes; (**b**) definition of the curvature ratio *b*/*L*; (**c**) measured *b*/*L* ratio and *b*/*L* ratio calculated with the parameters derived from the fit. Adapted from Ref. [82]. Permission from Nature Publishing Group (UK)

two phases. Fluctuation spectroscopy (similar to the step fluctuation spectroscopy described in Sect. 5.1.1) of phase boundaries, was used to determine the stiffness and the surface stress difference between the two phases [78].

Measurements of the size of stable isolated triangular (7 × 7) domains on a large terrace as a function of temperature above the critical temperature T_c showed at 5.3 K above T_c the smallest stable domain size with edge lengths $L_0 = 200$ nm, corresponding to a free energy difference $\Delta\gamma = 6.15$ μeV/(1 × 1) unit cell. This gives a very small entropy difference $\Delta S = \Delta\gamma/\Delta T = 0.013k_B$ per (1 × 1) unit cell which illustrates how delicate this transition is. A detailed analysis of the elastic interactions showed that the stable domain size is determined by elastic interactions at the phase boundaries [81]. A later study of stable isolated (7 × 7) domains on a very large terrace [82] showed that their equilibrium shape changed from convex to concave with increasing domain size as illustrated in Fig. 5.10a. Concave shapes had already been observed in the very early work [63] but had been attributed to growth shapes. The equilibrium, characterized by the ratio *b*/*L* in Fig. 5.10b, is plotted in Fig. 5.10c as a function of domain area *A*. The experimental data can be well fitted by an azimuthal angle θ dependence of the boundary energy $\beta(\theta) = \beta_0 + \beta_1(1 - \cos(3\theta))/2$ with $\beta_0 = 15.5$ meV/nm, $\beta_1 = 3.2$ meV/nm, assuming the surface stress difference of 30 meV/Å^2 determined in the earlier work. Medhekar et al. [83] have generalized these calculations and found that the concave shape is actually metastable. Figure 5.11a shows the experimental equilibrium distribution of domains, Fig. 5.11b the calculated domain shape of an isolated domain and Fig. 5.11c the domain distribution as a function of system parameters. As seen in (5.11c) the stable configurations of large domains are corner-connected subdomains. Their formation requires overcoming a notch nucleation barrier, which however is so small that these domains do form as seen in the equilibrium configurations on a large terrace (Fig. 5.11a). Apparently the tendency to form

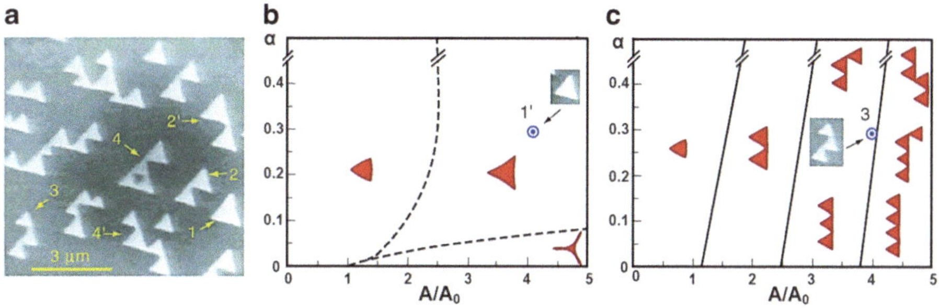

Fig. 5.11 (7 × 7) domain equilibrium structures of domain groups on a large Si(111) terrace. (**a**) LEEM image showing different equilibrium domain configurations. (**b** and **c**) Morphological phase diagrams of the equilibrium configurations. The *shapes* in (**b**) develop continuously with size but become metastable with increasing size, those in (c) have to overcome nucleation barriers indicated by lines but are lower in energy. $A_0 = 0.18\ \mu m^2$, α is the anisotropy parameter related to the parameters of the preceding analysis by $\alpha = \beta_1/2\beta_0$. For the (7 × 7) domains $\alpha = 0.28$. Adapted with permission from Ref. [83]. Copyright 2007 by the American Physical Society

interconnected domains above a given domain size is pre-emptying another phenomenon often seen in pattern formation in the presence of long range elastic interactions: arrangement of the (7 × 7) regions in a periodic pattern of stripes or droplets (if the domain boundary energy were isotropic), depending upon coverage. A theoretical study of this phenomenon showed that also the triangular (7 × 7) domains should order in this manner. The fact that this was not observed in experiment was attributed earlier to a too weak elastic driving force or too slow kinetics [80] but obviously it is the process described in [83].

While a considerable amount of work has been done on the (quasi)equilibrium aspects of the (7 × 7) structure, which gave considerable insight in the energetics, the kinetics of the (7 × 7) ↔ (1 × 1) transition was studied much less. Telieps and Bauer had done only a few measurements of the dependence of nucleation and growth rate upon undercooling without detailed analysis [63]. Hannon et al. measured the decay rate of superheated (7 × 7) domains and found a linear decay of the domain size with time, independent of the size of the domains [76]. This implies a steady supply of atoms needed to convert the (7 × 7) structure into the (1 × 1) structure, which has a higher atomic density. They explained this with continuous vacancy-adatom creation in the (1 × 1) region surrounding the (7 × 7) domains, with the adatoms diffusing to the (7 × 7) domains and the vacancy to the steps, thus decreasing the size of the (7 × 7) domains and causing the steps to retreat. The activation energy for vacancy-adatom generation was estimated to be in the range 3.2–3.8 eV, much higher than the sum of the activation energy for adatom formation by detachment from a step E_{ad} and of the energy barrier for surface diffusion E_{diff}, which varies between 1.1 and 1.9 eV on the pure (1 × 1) surface [96]. It is well established that the (1 × 1) surface is covered in equilibrium by about 0.25 monolayers of fast moving atoms (or clusters). This adatom coverage is continuously replenished by supply from the steps causing their retreat. Apparently this process is suppressed at steps whose upper level has (7 × 7) structure as will be seen later. A detailed theoretical study of the decay of the (7 × 7) structure with the phase field method has shown that it proceeds in two steps: first the corner holes and dimer chains are broken, then the stacking fault is eliminated, which is the rate-limiting process [89]. The growth of (7 × 7) domains has also been studied with the phase field method and found to be in good agreement with experiment [84]. In particular the serrated domains formed at low growth rate and the pine tree-like shapes seen already in the first LEEM studies [62–65] and in a compilation of experiments in the coexistence range [86] are reproduced by this theory.

Keeping the atomic density difference between (7 × 7) and (1 × 1) structure and the presence of the self-adsorption layer in the (1 × 1) surface in mind it is immediate obvious that their relative surface coverage can be manipulated in the coexistence range by changing the equilibrium between the two phases, either by adding or removing Si atoms. This has been accomplished by exposing the surface to silane—increasing the number of Si atoms—or by exposing to oxygen—removing Si atoms from the adsorption layer. These chemical potential changes have also been described by effective free energy differences between the two phases [79]. Concluding the discussion of the phase transition studies, LEEM has revealed a fascinating scenario of two-dimensional crystal phenomena, which has also contributed to the understanding of phase transitions in general.

The second group of studies, surface diffusion processes, has been done mainly by Altman's group [90–96] and by Hibino et al. [97–99]. The goal of their studies was to understand the relative importance of surface diffusion, step attachment–detachment of atoms, and step barriers for the surface microstructure by studying the behavior of monatomic steps as a function of time and temperature. Most of the work was done by studying the decay of circular islands and holes (vacancy islands), which was already discussed in Sect 5.1.1, but step boundary fluctuation spectroscopy of circular and linear steps was used too. Hibino et al. used the decoration of the upper sides of steps by the (7 × 7) as step marker and were therefore limited to a temperature region near T_C while Altman's group made use of the step phase contrast, which allows a wider temperature range and excludes a possible influence of the (7 × 7) structure on step behavior.

The analysis of the experimental data is based as in the case of metals on the methods reviewed in Refs. [26–29]. Briefly, the size of an island decays with time at a fixed temperature as $\mathrm{d}A/\mathrm{d}t = A_0(t_0 - t)^{\alpha}$, where A_0 is the initial area, t_0 the total decay time at which the island disappears and α the decay exponent, whose value is determined by the processes involved in the island decay. The limiting processes are adatom surface diffusion and attachment–detachment. The diffusion rate is characterized by the diffusion constant D, the step attachment–detachment rate by the step attachment–detachment kinetic coefficient K (assuming symmetry with respect to the attachment direction) and the relative importance of the two processes by the kinetic length $d = D/K$. For two concentric circular islands theory predicts for the decay of the island/hole area A

$$\frac{dA}{dT} = -2\pi\sqrt{3}\Omega D n_{eq} \frac{\exp\left(\frac{\Omega\tilde{\beta}}{rk_B T}\right) - \exp\left(\frac{\Omega\tilde{\beta}}{Rk_B T}\right)}{d/r + d/R + \ln(r/R)} \quad . \tag{5.8}$$

Here r and R are the radii of the inner and outer island/hole, $\tilde{\beta}$ is the step stiffness, $\Omega = \sqrt{3}a^2/2$ the area per atom on the (111) lattice with lattice constant a. The prefactor $n_{eq} = 2\Omega^{-1}\exp(-E_{ad}/k_B T)$ is the adatom concentration in equilibrium with a straight step, the prefactor $D = \sqrt{3}\Omega\nu\exp(-E_{diff}/k_B T)$ is the diffusion constant and E_{ad} and E_{diff} are the energy needed for formation of an adatom by step detachment and the activation energy for surface diffusion, respectively. When $\beta a^2/kT \ll r \ll R$ Eq. (5.8) can be simplified for the limiting cases of $d = D/K$. For $K \ll D$ the island decay is attachment-detachment-limited (ADL) and Eq. (5.8) can be reduced to $dA/dt = -2\pi\tilde{\beta}n_{eq}\Omega^2 K/k_B T$, so that $A(t) \sim t$ ($\alpha = 1$). When $K \gg D$ then diffusing atoms can be attached/detached easily at the step and Eq. (5.8) can be reduced to $dA/dt = -2\pi c_{eq}\tilde{\beta}\Omega^2 D/k_B T r \ln R$. With $A = r^2\pi$ integration leads to $A(t) = \pi\left(3c_{eq}\tilde{\beta}\Omega^2 D/k_B T \ln R\right)^{2/3}(t_0 - t)^{2/3}$, i.e., $\alpha = 2/3$. With the concentric ring configuration the experiments are straightforward—the art is in the sample preparation, careful data acquisition, and analysis—and give α and the pre-exponential $n_{eq}\tilde{\beta}D$.

The results of the island decay measurements from many islands with R/r ratios ranging from 3 to 200, initial island and hole distributions and areas up to a few micrometer in the temperature region 1090–1150 K gave a decay exponent $\alpha = 0.76 \pm 0.06$ [97]. In this temperature region the rim of the steps was decorated by a band of (7 × 7) structure. An example of such an experiment is shown in Fig. 5.12a [97]. Measurements at temperatures in the absence of (7 × 7) step decoration gave values of 0.726 at 1143 K and 0.771 at 1183 K [94] and of 0.75 at 1163 K [95]. While the last three values suggest a temperature dependence, it appears that all values taken together congregate around a mean value of ¾ instead of 2/3. This problem is discussed in the cited publications based on Eq. (5.8) for the decay rate in terms of the variation of the kinetic length d, i.e., in terms of the competition between surface diffusion and step attachment–detachment.

When $\tilde{\beta}$ is known, then the sum of the energy for adatom formation by detachment from the step, E_{ad}, and of the surface diffusion energy, E_{diff}, can be obtained from the temperature dependence of the pre-exponential in Eq. (5.8). These energies are important parameters for the understanding of the mass transport on the surface. The island and hole shapes on Si(111) at high temperatures are more or less circular so that there is no anisotropy and $\tilde{\beta} \approx \beta$. Hibino et al. [97] assumed in the temperature range of their measurements (1090–1150 K) a constant value of 137 meV/Å, while Altman's group [96] measured $\tilde{\beta}$ by step fluctuation spectroscopy and found values decreasing from 68 meV/Å at 1145 K to 56 meV/Å at 1233 K. Where the temperatures of the two experiments overlap, the $\tilde{\beta}$ value of the

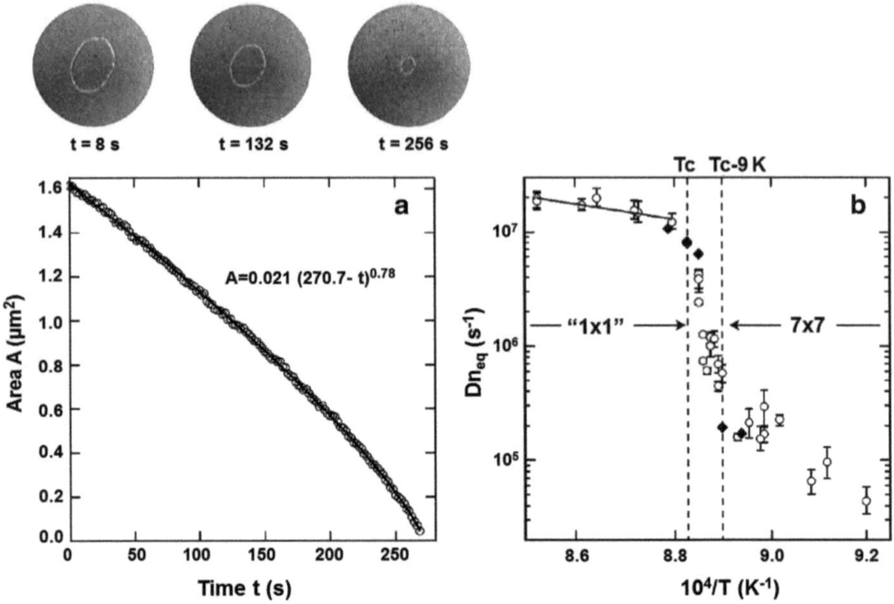

Fig. 5.12 Island decay on Si(111) surface. (**a**) *Top*: Snapshots of an isolated island on a 20 µm wide terrace at T_C (1133 K); *bottom*: area as a function of decay time with data fitted to $A_0(t_0 - t)^\alpha$. (**b**) Dn_{eq} from island decay times measured at many temperatures, assuming a step stiffness of 137 meV/Å. Adapted with permission from Ref. [97]. Copyright 2001 by the American Physical Society

(7 × 7) structure-decorated step [97] is by a factor of two larger than that of the pure (1 × 1) step [96]. If the larger $\tilde{\beta}$ value for the decorated step is correct then the (7 × 7) structure would cause considerable step stiffening. However it has to be kept in mind that $\tilde{\beta}$ and β values determined with different methods differ considerably.

Figure 5.12b shows the temperature dependence of Dn_{eq} derived from measurements of the island area decay (Fig. 5.12a) at many temperatures, using Eq. (5.8). $Dn_{eq} = 2\sqrt{3}\nu\ \exp(-(E_{ad} + E_{diff})/k_B T)$ decreases in a narrow temperature range by a factor of 50 from above T_C to below T_C, indicating a significant change of $E_{ad} + E_{diff}$ as the jump frequency ν is unlikely to vary much between the two surface structures. From the temperature dependence of Dn_{eq} above T_C the values $\nu = 1.1 \times 10^{13}\ s^{-1}$ and $E_{ad} + E_{diff} = 1.3$ eV were obtained [97]. From island decay measurements at higher temperatures, at which the steps are not decorated, values of $\nu = 2.59 \times 10^{13}\ s^{-1}$ and $E_{ad} + E_{diff} = 1.53$ eV [95, 96] were extracted. E_{ad} can be estimated from $n_{eq} \approx 0.2$ monolayers to be 0.2 or 0.23 eV, which gives 1.3 eV for E_{diff} for the (1 × 1) structure. Below T_C the scatter of the data is too large to allow such an analysis. The much smaller Dn_{eq} value on the (7 × 7) surface has, however, an important consequence for the decay of vacancy islands vs. adlayer islands: the rim of the vacancy islands is decorated by the (7 × 7) structure, which due to its lower Dn_{eq} reduces the rate at which vacancy islands are filled. This reduction acts as an effective downhill Ehrlich-Schwoebel barrier, for whose height an upper limit of 0.6 eV was estimated. No such barrier exists for the decay of adlayer islands. This phenomenon and the influence of n_{eq} differences in the decay of superheated and equilibrium vacancy and adlayer islands have been discussed in detail by Hibino et al. [98, 99].

At the higher temperatures studied by Altman's group, at which this (7×7) barrier at the step does not exist, the question of the step permeability P and of the distinction between diffusion-limited and attachment–detachment-limited step movement as characterized by the kinetic length $d = D/K$ can be studied more clearly [94, 95]. A numerical analysis of Eq. (5.8) for various d values as a function or r/R [95] shows that α depends upon the magnitude of d. For $d \approx 240a$, where a is the lattice constant, α is 0.77 from the largest r/R values considered (0.4) down to $r/R \approx 0.1$. When $d < 240a$, then α decreases with decreasing r/R but is still far away from 2/3 at $r/R = 0.5$, even at $d = 10a$, a value for a pure DL mechanism. When $d > 240a$, then α increases with decreasing r/R but still is very different from 1, the value for a pure attachment–detachment-limited process. The analysis of island decay at 1163 K gave $d = 75a$, so that diffusion is dominating the island decay. In the studies, described up to now, R was fixed. A study of the time dependence of the area of two concentric islands with radii r and $R > r$ on a larger disc allows determination of the step permeability P. Assuming zero detachment from the outer island R will increase until $r = 0$ because all atoms detached from the inner island will be attached to the outer when $P = 0$. When $P = 1$ R will remain constant because all atoms detached from the inner island will diffuse beyond the step of the outer island. A detailed analysis clearly indicated negligible permeability [95].

Island decay experiments at the highest temperatures studied (up to 1380 K) showed above 1200 K increasing deviations from the behavior expected on the basis of Eq. (5.8), which were attributed to desorption. Therefore this equation was generalized by incorporating desorption via the lifetime $\tau = \nu_0^{-1} \exp(E_{des}/k_B T)$ of the adsorbed atoms. The diffusion length before desorption is $x_s = \sqrt{D\tau} = (\sqrt{3}/2) a \exp[(E_{des} - E_{diff})/2k_B T]$, assuming the same value of ν_0 for diffusion and desorption. The inclusion of desorption gives excellent agreement between theory and experiment for island decay times from 1145 to 1380 K with the fit parameters $\nu_0 = 2.59 \times 10^{13}$ s^{-1}, $E_{ad} + E_{diff} = 1.53$ eV and $E_{des} - E_{diff} = 2.56$ eV. The last two values give a sublimation energy $E_{subl} = E_{ad} + E_{des} = 4.09$ eV in good agreement with literature, supporting the consistency of the values [96].

Summarizing the island decay studies, LEEM has provided considerable insight into the processes that control the microstructure of the Si(111) surface both in the phase transition region and above it. These studies were done at constant atom density except in those, in which sublimation occurred [95]. Additional information can be obtained from growth studies. Altman's group studied the competition between step flow growth and nucleation on the terraces on the (7×7) surface between 700 and 850 K [90–93]. As a first step the so-called critical terrace width for step flow growth, λ_c, defined by the last appearance of nuclei on the terrace (Fig. 5.13a) [92] was studied

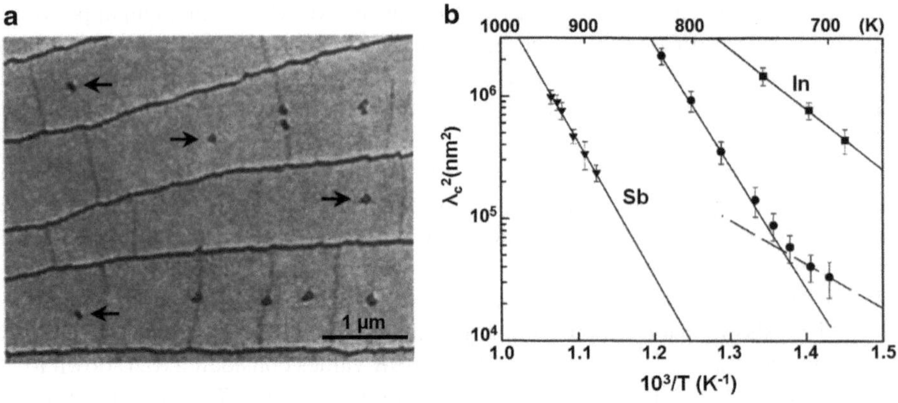

Fig. 5.13 Critical terrace width λ_c for step flow growth on the Si(111)-(7 × 7) surface. (**a**) LEEM image of the transition between two-dimensional nucleation on the terrace and step flow growth at 800 K and a flux of 0.015 Si monolayers/min. The islands determining λ_c are indicated by *arrows*. Energy 42.5 eV. (**b**) Critical terrace width as a function of 1/*T* for a flux of 0.1 monolayer/min. (**a**) Reprinted with permission from Ref. [92]. Copyright 2002 by the American Physical Society. (**b**) Reprinted with permission from Ref. [93]. Copyright 2012 World Scientific

for various step orientations, with the result shown by the unmarked central curve in Fig. 5.13b [93]. With a number of assumptions theory relates λ_c to the activation energy for surface diffusion E_{diff} and the binding energy E_{i*} of the critical nucleus consisting of i^* atoms by

$$\lambda_c^2 = A\ \exp[-(E_{diff} + E_{i^*}/i^*)/k_B T], \tag{5.9}$$

with the pre-exponential containing the flux, the jump frequency, the density of the critical nuclei and their size. Fitting the data in Fig. 5.13b to this expression gave for sum of the two energies $E = 2.05$ and 0.72 eV, for the pre-exponential 7.0×10^{19} and 5.0×10^9 nm^2 in the high and low temperature ranges, respectively. At high temperatures, at which nucleation occurs only at the (7 × 7) domain walls, nucleation was considered to be the dominating factor determining λ_c, while at low temperatures, at which nuclei form also within the domains, diffusion was concluded to be dominating, with a transition between these two processes at about 750 K [90, 91]. Adsorption layers can either enhance or impede step flow growth (surfactants or anti-surfactants) by changing surface diffusion and/or step attachment, which should have a strong influence on the critical terrace width. This is evident in the curves marked Sb and In in Fig. 5.13b: the pre-factor in Eq. (5.9) are strongly modified but the activation energy in the presence of Sb is similar to that on the pure (7 × 7) structure at high temperature, while that in the presence of In is similar to that in the low temperature range on the (7 × 7) surface. This suggests that In (Sb) suppresses (enhances) island nucleation on the terraces and enhances (suppresses) Si adatom attachment [93].

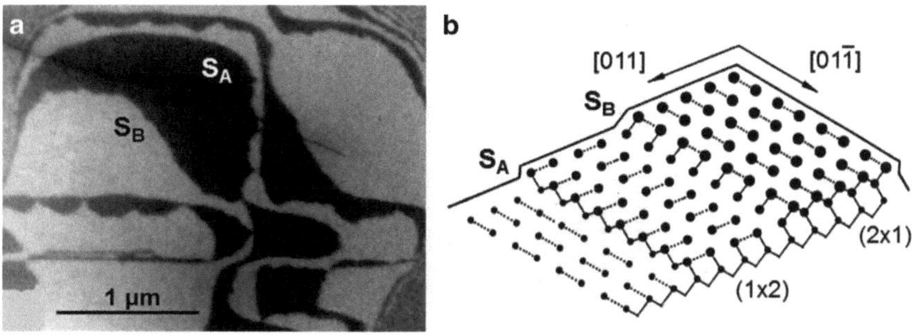

Fig. 5.14 Si(100)-(2 × 1) surface. (**a**) LEEM image showing the contrast between (2 × 1) and (1 × 2) reconstructed regions, smooth S_A and rough S_B steps, obtained by tilting the beam into one of the ⟨011⟩ directions. Electron energy 6 eV. (**b**) Schematic of Si(100) surface showing the orientation of the steps with respect to the dimer directions. From Ref. [101] with permission from Springer Science + Business Media, Copyright 1988

The location of the nuclei used in the determination of λ_c with respect to the steps bounding the terrace allows determination of the asymmetry of the step attachment–detachment kinetic coefficient *K*. If attachment is symmetric from bottom and top side of the step then nucleation will occur halfway in between the two steps, otherwise it is shifted towards the step with smaller *K*, provided that the steps are completely impermeable, i.e., incorporate all arriving atoms. Such shift towards the step-up side was actually observed at $\lambda_c = 950$ nm, 800 K, and a flux of 0.015 monolayers/min, indicating K_- (step-down) > K_+ (step-up) [92]. More data would be needed to determine how general these results are, whose explanation requires extensions of the Burton-Cabrera-Frank (BCF) theory. One example, in which the BCF theory completely describes step processes, is the growth and sublimation of Si(111) via spiral motion around screw dislocations. However for the same process on the Si(100) surface the theory has to be generalized [100].

5.1.2.2 Si(100)

The study of the Si(100) surface, the most important Si surface in semiconductor technology, began shortly after the nature of the (1 × 1) ↔ (7 × 7) phase transition on the (111) surface had been demonstrated. The first published images [65] showed very clear (2 × 1) structure contrast with slightly tilted illumination, which makes the (2 × 1) and (1 × 2) regions inequivalent. Steps in which the bonds are oriented along the step (type A) appeared smooth, steps with bonds perpendicular to the step (type B) appeared rough (Fig. 5.14) [101]. Depending upon cooling conditions from the high temperature annealing or flashing the roughness difference between the two steps can be much larger, indicating a strong temperature dependence of the stability of the two steps. Above 1300 K the steps were moving rapidly due to sublimation and at about 1350 K the steps disappeared. After Telieps's initial experiments Mundschau et al. studied the dependence of the microstructure of this surface

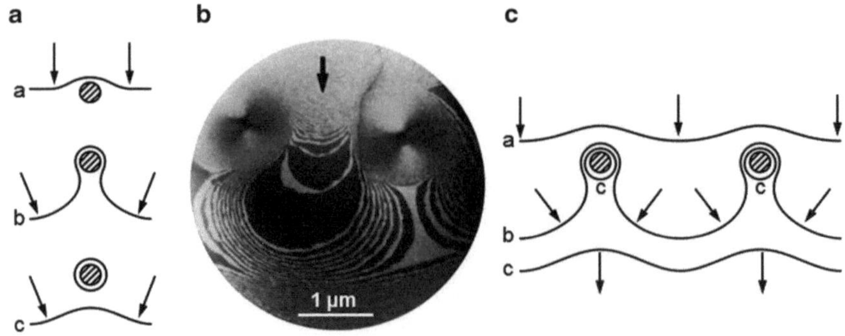

Fig. 5.15 Hillock formation on Si(100) upon extended sublimation caused by step pile-up at pinning centers. Electron energy 7 eV. The *sketches* indicate the formation process of the resulting step/terrace structure. Adapted with permission from Ref. [102]. Copyright 1989 Elsevier

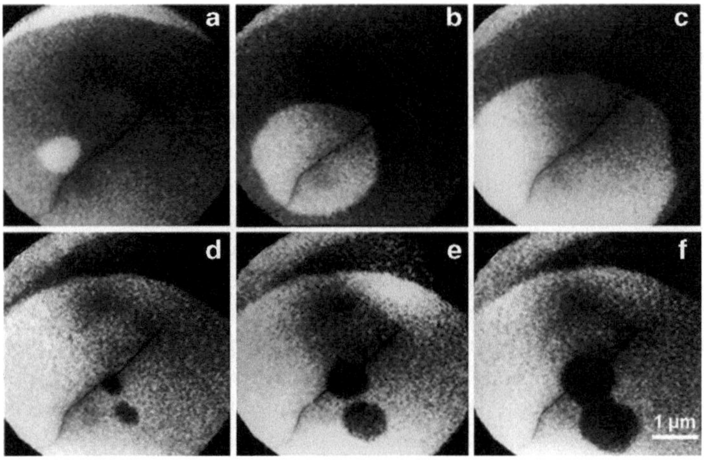

Fig. 5.16 LEEM video frames of the sublimation process via hole formation. Electron energy 5 eV. Adapted with permission from Ref. [103]. Copyright 1992 Elsevier

upon preparation conditions in considerable detail, in particular the influence of slip lines, partial and screw dislocations on step structure and dynamics [68] and the influence of step pinning centers such as SiC precipitates on the step dynamics during sublimation between 1100 and 1350 K [102]. Figure 5.15 [102] shows a later stage of the sublimation when already many steps have piled up in front of two pinning centers. In between these centers the steps have passed due to sublimation, forming sublimation hillocks. Sublimation via repeated Lochkeim formation at still higher temperature is illustrated in Fig. 5.16 [103]. Here the step free energy is isotropic, resulting in circular hole patterns.

The anisotropy of the step energy at low temperatures suggested by the different roughness of A and B steps after rapid cooling (Fig. 5.14) can be estimated from the

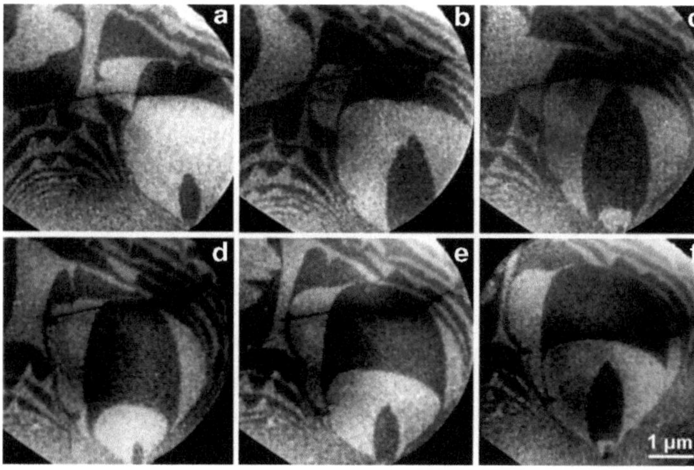

Fig. 5.17 Near equilibrium growth of Si on Si(100) with terrace nucleation at a defect at about 900 K and a deposition rate of 0.2 monolayers/min. LEEM video frames. Time elapsed between images: 180, 300, 480, 660, and 780 s. Reproduced with permission from Ref. [104]. Copyright 1991 Elsevier

shape of two-dimensional islands formed during very slow growth. Figure 5.17 [104] shows frames from a movie taken during deposition of Si under conditions at which no two-dimensional nucleation on terraces occurred, but new terraces started only at defects. Growth is clearly strongly anisotropic. Assuming that growth is close enough to equilibrium, a step energy ratio $\beta_A/\beta_B = 2.6$ would explain the observed terrace shapes, a value much smaller than the theoretical value of 15 available at that time, but in qualitative agreement with STM results. Of course, the conclusion implies that surface diffusion, step attachment anisotropies, and energy barriers have no influence on the shape [104]. The work in Bauer's group on clean Si surfaces, about which additional information can be found in some summaries [71, 103, 105, 106] turned after this study to metal layers on surfaces.

The study of clean Si(100) surfaces was continued by Tromp et al. [107–115] and extended to the determination of quantitative values of the parameters, which determine the surface microstructure similar to the work discussed in Sects. 5.1.1 and 5.1.2.1. In one of their first studies they investigated the step morphology of surfaces with a miscut of less than 0.1° as a function of miscut and found that the resulting step structure could be classified into four phases with decreasing miscut: double height steps, single height straight steps, single height wavy steps, and a hilly structure [107, 108], in agreement with theory according to which step undulations reduce the size of stress domains.

As the preceding images show, the surface morphology of the Si(100) surface as seen in the terrace-step shapes, is strongly temperature-dependent, much more than that of the Si(111) and the metal surfaces discussed beforehand. This makes step fluctuation spectroscopy an ideal tool for the determination of the step dynamics and energetics on Si(100). In a first experiment Bartelt et al. [109] measured step fluctuations (Eq. (5.1)) from 910 to 1480 K and analyzed the data via the correlation function (Eq. (5.6)) with the pre-exponential $A(q) = 2k_BT/L\widetilde{\beta}q^2$. The stiffness $\widetilde{\beta}_A$ of the S_A steps was found to be much higher than that of the S_B steps, $\widetilde{\beta}_B$, for example by a factor of 7, and the ratio $\widetilde{\beta}_A/\widetilde{\beta}_B$ strongly decreased with increasing temperature as expected. In this study also the ratio of the step free energies was determined from the shape of equilibrium shapes of islands produced by Si deposition and annealing. Between 1020 and 1330 K only elliptical shapes with temperature-dependent axis ratio were observed, but not the shape with sharp tips shown in Fig. 5.17. However, extrapolation of the linear increase of the β_A/β_B ratios derived from the equilibrium shape measurements and of those derived from the step stiffness to the lower temperature used in Fig. 5.17 [104] agrees well with the ratio reported in that study. Towards high temperatures there is still a significant anisotropy at 1330 K ($\beta_A/\beta_B \approx 0.7$). At all temperatures the step stiffness is considerably smaller than on the (111) surface, e.g. $\widetilde{\beta}_A$ by a factor of 8 compared to the values measured by Altman's group and $\widetilde{\beta}_B$ is also correspondingly smaller.

In an extension of this work Bartelt et al. [113] determined also another measure of the step stability, the step-edge mobility Γ, which is the rate at which adatoms attach/detach at step edges and which is related to the kink formation energy E_k via $\Gamma = \Gamma_0 \exp(-E_0/k_BT)$. In the case of attachment–detachment limited kinetics, which is indicated by the observed $1/q^2$ dependence of $\tau(q)$, Γ is related to the quantities obtained from step fluctuation spectroscopy $A(q)$ and $\tau(q)$ via $\Gamma = LA(q)/2\tau(q)$. In the temperature and q range studied (990–1370 K, 0.01–0.035 nm^{-1}) Γ was found to increase from 10^3 to about 10^5 nm^3/s and to be independent of q. From the slope of the Arrhenius plot of Γ a value of $E_0 = 1.45 \pm 0.15$ eV was derived, interestingly both for S_A and S_B steps. Thus the difference between the two steps is not caused by different attachment–detachment rates but by step stiffness and step free energy.

With steps well characterized, the interaction between steps and surrounding two-dimensional gas, i.e., the relative importance of step attachment/detachment kinetics and surface diffusion, was studied. This was done by analyzing the time dependence of the island areas in a high-density, non-equilibrium two-dimensional island distribution, the so-called Ostwald ripening, in a large step-free region at 940 K [110, 114]. The average area of *all* islands was found to increase linearly with time while that of the *smallest* islands decreased linearly with time. This confirms the earlier conclusion that the change of the island size is limited by attachment/detachment kinetics at the steps. The rate of change of the individual island areas $\partial A/\partial t$ is determined by the chemical potential difference between island and two-dimensional gas via

$$\frac{\partial A}{\partial t} = \frac{L\Gamma}{\Omega k_BT}(\mu_{is} - \mu_{ad}), \tag{5.10}$$

where L is the island circumference and Ω the area per atom. The chemical potential of the islands is determined by the Gibbs-Thomson equation, which relates it to the size and shape of the island. Therefore measurements of $\partial A/\partial t$ of individual islands

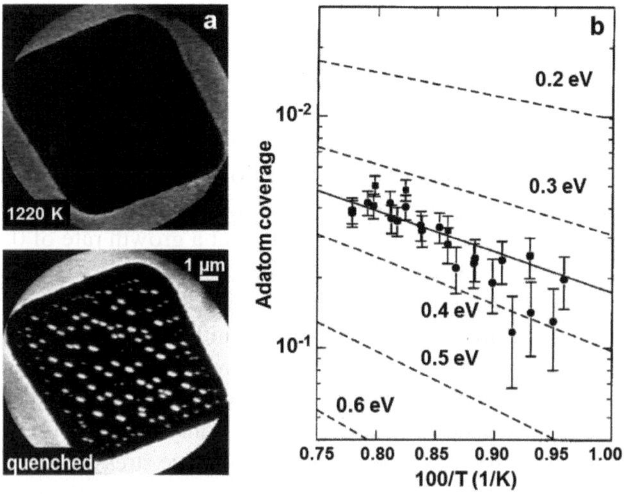

Fig. 5.18 Thermal adatom concentration on a step-free Si(100) terrace. (**a**) (1/2,0) dark-field LEEM images before and after quenching taken with 3.5 eV electron energy. The islands seen after quenching have formed from the two-dimensional adatom gas, which does not produce contrast in LEEM. (**b**) Adatom concentration as function of $1/T$ from measurements of the total island area coverage shown in (**a**). The *dashed lines* show the concentrations calculated for different formation energies. Adapted with permission from Ref. [115]. Copyright 1998 by the American Physical Society

give the chemical potential surrounding it. The chemical potential maps generated in this manner showed strong variations depending upon the island surrounding. Thus the chemical potential of the adatom gas is not uniform with corresponding consequences for the adatom diffusion coefficients.

Other interesting experiments addressed the critical nucleus size in homogeneous nucleation [111] and the thermal adatom concentration [115]. The first problem was studied using silane and disilane gas as supply of Si, a substrate temperature of 920 K, which is above the H desorption temperature, and growth rates from 0.1 to 2 monolayers/min. Under these conditions the adatom concentration from the incoming flux is only about 2 % of that of thermal adatom concentration, i.e., the supersaturation is very small and consequently the critical nucleus size N_c large. Careful analysis of the data, using classical homogeneous nucleation theory, gave $N_c \approx 650$ adatoms, generally accepted to be dimers. The thermal adatom concentration was studied between 1310 and 1560 K using large lithographically prepared [112] step-free regions with up to 10 µm edge length. Quenching from these temperatures to low temperature caused condensation of the two-dimensional adatom gas into many small two-dimensional crystals, whose total area coverage is a measure for the concentration c_T of the gas at the starting temperature T, excluding edge effects (Fig. 5.18) [115]. c_T was found to increase from 0.02 at 1310 K to 0.04 at 1560 K. This gave an activation energy for thermal adatom creation on terraces of 0.35 ± 0.05 eV, which fits best to the formation energy of a dimer instead of an atom.

Another method for producing large flat areas for measurements minimizing edge effects should be mentioned briefly too: Si deposition onto the sublimating surface, which allows higher temperatures without increase of the net sublimation rate. This increases the diffusion rate (D) without decreasing the adatom lifetime (τ), resulting in a larger diffusion length $\lambda = (D\tau)^{1/2}$. λ values from 10 to 30 μm at 1270 K have been observed under these conditions [116]. This method was used to study the change of the island shape with island size during Si deposition under quasi-equilibrium conditions at 960 and 1130 K at a growth rate of 0.1 monolayer/min from disilane gas in the absence of defects. To exclude edge effects, 10×15 μm^2 terraces were prepared lithographically. With increasing size the island shape changes from elliptical to "American football" shape, a shape similar to that in Fig. 5.17, and at the lower temperature the aspect ratio approaches the value derived from this figure. The size dependence was interpreted in terms of a theoretical model, which takes into account not only the step free energy in the calculation of the equilibrium shape but also the surface stress anisotropy. By fitting the aspect ratios measured at the two temperatures to the theory the step free energies and the surface stress anisotropy could be determined. This experiment clearly shows that whenever stress is present, it has to be taken into account in the interpretation of the equilibrium shape [117]. To complete the discussion of studies with patterned surfaces, the work of Blakely's group on photolithographically prepared biperiodic surfaces still needs to be mentioned. These surfaces consist initially of a square arrangement of flat top square pillars with a height of 0.1–0.25 μm and a periodicity of 4 μm, separated by flat trenches. When annealed at high temperatures surface diffusion and sublimation produce beautiful characteristic hill and valley structures whose shape can be explained in terms of the processes studied on flat surfaces and described above [118–120].

Silicon is frequently heavily boron-doped. When B diffuses to the surface it causes considerable tensile strain due to the short Si–B bond. The resulting changes in the surface stress anisotropy can cause the break-up of terraces into periodic step structures due to stress relaxation at the step edges. This is actually the case and has been the subject of several studies. In the first study in 1996 [121] of miscut surfaces the S_A steps were smooth and the S_B steps were rough at high temperatures ($\geq$1240 K) as on undoped surfaces. With decreasing temperature the S_B steps developed a periodic sawtooth shape, leading to triangular islands. The width of these islands at the smooth S_A steps decreased continuously with decreasing temperature, resulting in a striped phase, which could not be resolved any longer at 1160 K. The cause of this process was attributed to increasing stress anisotropy with increasing B segregation to the surface but a lowering of the S_A step energy by preferential B adsorption could not be excluded. Studies on larger terraces revealed a more complex behavior as illustrated in Fig. 5.19 [122]: striped phases on the same terrace orthogonal to each other, formation of elongated islands with temperature-dependent aspect ratio, which split to form new stripes and nucleation of stripes between existing stripes. These features were clarified later by more thorough studies of Hannon et al. [123, 124]. They derived the step stiffness from step fluctuation spectroscopy and came to the conclusion that the driving

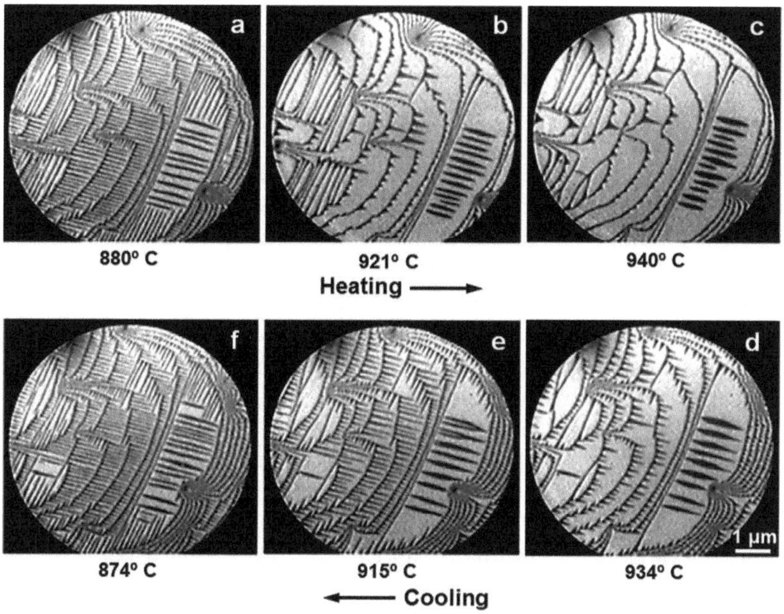

Fig. 5.19 Striped phase and step shape transition on B-doped Si(100). From (**a**) to (**f**) the crystal is cycled from high to low temperature and back again. Due to the different response of the thermocouple and the crystal to temperature changes the temperatures shown are only rough numbers. Adapted from Ref. [122] with permission from Springer Science + Business Media

force for stripe formation was not surface stress but a strong decrease of the free energy of S_A steps with decreasing temperature. The coexistence of orthogonal stripe domains was attributed to vacancy island formation. This suggests that a high vacancy concentration at high temperatures on the B-doped surface is necessary for the mass transport in the stripe formation process. In the studies just discussed pre-existing steps play an important role in the stripe formation.

The lithographically pre-patterned surfaces with large atomically flat surfaces, mentioned earlier, allow avoiding the step effects by adding the material needed for stripe formation from the vapor phase. Initial experiments using diborane as Si supply were only partially successful but deposition of 0.5 monolayers of Si atoms and annealing at 1010 K brought success [125]. This method was later perfected by using higher temperatures at which sublimation creates vacancies and Si deposition creates adatoms. By varying temperature and Si deposition rate a wide variety of microstructures could be formed, including perfect stripe patterns in regions with more than 5 μm diameter or edge length. Stripe pattern formation was limited to temperatures above 1150 K and attributed to stress interactions as in the original interpretation. Upon prolonged annealing at 1120 K the stripe pattern broke up into oval islands with sharp tips seen in Fig. 5.19. Their number and aspect ratio decreased with annealing time via Ostwald ripening, driven by minimization of the step free energy [126].

In connection with B doping it is interesting to note that segregation to the Si(111) surface has quite different consequences for the surface microstructure. No nanofacetting occurs but a ($\sqrt{3} \times \sqrt{3}$) structure is formed. On this surface Si grows between about 800 and 900 K in double layers in twin orientation with respect to the substrate, above about 950 K in parallel orientation. The twinning allows producing an artificial twinning superlattice but occurs only when the twinned crystallites are small. Small isolated crystals were observed even at 1100 K but as soon as they coalesce they convert into parallel orientation, which explains the loss of twinning in continuous films above about 950 K. This interesting phenomenon was attributed to the reduction of the step energy by twinning, which overcompensates the energy increase from the twin boundary energy [127].

Unless passivated with hydrogen Si is always covered with an at least 1 nm thick SiO_2 layer, which is usually removed by flashing in UHV to $\geq$1050 K. Slow removal leads to pit formation, whose mechanism and kinetics was thoroughly studied by Hibino et al. [128]. They found that pit growth proceeds via reaction of Si monomers with the edge of the pit with an activation energy of 4 eV. There have been several other studied of SiO_2 on Si(100) but they were aimed mainly at the understanding of contrast information in LEEM, PEEM, and MIEEM [129–131], of the interaction of light with extreme UV light [132] and at the determination of the thickness and composition of the layer [133]. Two other studies involving oxygen aimed at the understanding of the etching process by O_2 in the 10^{-8} mbar range. Below 850 K etching was found to occur randomly on the surface, producing vacancy clusters, above 850 K the vacancies diffused to the steps and caused effects similar to those seen in sublimation [134]. At higher temperatures (1000–1250 K) etching leads on large terraces to random vacancy island formation similar to what is shown in Fig. 5.16 during sublimation. The growth rate of the islands was found to depend upon their neighborhood, which shows that it is determined by vacancy diffusion from the environment. A study of the temperature dependence of the growth rate gave an effective activation energy of 2.0 eV for etching, an unexpectedly low value [135].

Finally some other studies involving the clean Si(100) surface and its interface with SiO_2 need to be discussed. They concern the influence of interfacial strain on the LEEM contrast and on the stability of SOI (silicon-on-insulator) films, which have been studied in detail by Lagally's group [136–140] and Müller's group [141–144]. The strain field originating at the film-substrate interface produces lattice distortions concentrated around dislocation lines which cause diffraction contrast in form of diffuse lines. These can be used to characterize the stress [136–138] and are particularly pronounced when the stress is enhanced by thermal shock [136]. When 10–20 nm thick Si(100) single crystalline films on a 150 nm thick amorphous SiO_2 layer on a Si(100) surface are annealed at 1100–1200 K they break up into self-organized nanocrystals ("dewetting"), leaving voids in the Si film behind [139, 140]. This process has been studied with LEEM, in particular with dark field LEEM. Figure 5.20 [142] illustrates the void formation process. The speckle in the image is due to the contrast between small crystals with (2×1) and (1×2) orientation. The break-up forms a pattern with

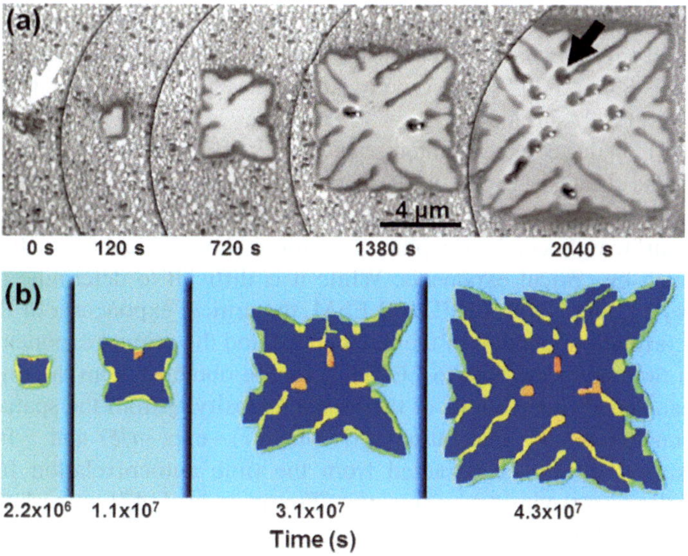

Fig. 5.20 Dewetting of a 22 nm thick SiO layer at 1100 K. (**a**) Dark-field LEEM images taken with 7.8 eV electrons during the dewetting process. The *white arrow* marks the nucleation of the void at a defect, the *black arrow* a nanocrystal distorted by charging. (**b**) Kinetic Monte Carlo simulation of the process shown in (**a**). Adapted with permission from Ref. [142]. © IOP Publishing & Deutsche Physikalische Gesellschaft (2011). All rights reserved

fourfold symmetry with agglomerated Si at the rim surrounding the exposed SiO_2. With increasing dewetting the Si nanocrystals form fingers aligned along the ⟨001⟩ directions. The details of the kinetics and energetics have been studied in considerable detail [141–144].

5.1.2.3 Other Si Surfaces

Very little work has been done with other surfaces because they are not practically important as the (100) surface or interesting as the (111) surface. The (110) surface has been studied by Swiech et al. [145, 146], motivated by the phase transition from its (16 × 2) reconstruction to the (1 × 1) structure at 1005 K and to understand step energetics and dynamics on a Si surface with lower symmetry. The phase transition was found to have a similar coexistence region as the Si(111) (7 × 7) ↔ (1 × 1) transition, with large aspect ratio (16 × 2) domains coexisting with (1 × 1) regions. The results were analyzed in the same manner as in Ref. [77]. The step stiffness was obtained from island decay measurements of islands on large terraces, which were produced by laser texturing. The islands are elliptical and the decay of their areas between 780 and 982 °C occurred with constant aspect ratio and followed the law $dA/dt = A_0(t_0 - t)^\alpha$ discussed earlier with $\alpha = 0.65 \pm 0.03$. This means that diffusion is isotropic in this temperature range and the decay is diffusion-limited. From the shape of the islands the following values for the step energy β and step stiffness $\widetilde{\beta}$ were derived (in units of meV/Å): for the long axis 13.1 and 5.4, respectively, for the short axis 8.4 and 20.5, respectively, i.e., values between those cited earlier for the (111) and (100) surfaces.

The Si(311) surface was of interest because LEED and X-ray diffraction measurements had shown a second order phase transition from a (3 × 1) reconstruction to the (1 × 1) structure at 693 °C, in contrast to the first order transition on the (111) and (110) surfaces. Second order phase transitions show critical fluctuations and are characterized by critical exponents. While it is difficult to determine the critical exponents β, γ, and ν with LEED in LEEM, the critical exponent α of the specific heat, the correlation length ξ of the fluctuations and the critical exponent z of their relaxation time are accessible to LEEM. α can be obtained from the time average of the mean square fluctuations of the image intensity, ξ from the spatial autocorrelation function of the fluctuations $g(r)$ via $g(r) \sim \exp(-r/\xi)$ and z from $\tau \sim \xi^z$. The relaxation time τ is extracted from the time autocorrelation function of the fluctuations $c(t)$ via $c(t) \sim \exp(-t/\tau)$. Tromp et al. [147] embarked successfully on this ambitious endeavor and obtained $z = 1.9 \pm 0.3$. The ξ values were limited in one direction $\left(\left[2\overline{33}\right]\right)$ to 80 nm by steps, while in the perpendicular direction ($\left[01\overline{1}\right]$) the instrument resolution gave an upper limit of 40 nm. α could be determined only with large uncertainty. With improved instrumentation, for example with aberration-corrected systems and better data acquisition, the accuracy of such experiments could be improved significantly.

5.1.3 Other Inorganic Semiconductors

Surprisingly little work has been done on the microstructure of other clean semiconductor surfaces. Only for Ge(100), GaAs(100), GaN(0001), and SiC(0001) are limited results available. On Ge(100) Poelsema's group studied the (2 × 1) ↔ (1 × 1) phase transition, for which laterally averaging studies had given conflicting interpretations such as dimer breakup, step proliferation, or domain wall proliferation without dimer breakup. They interpreted their LEED (1/2,0) beam intensity measurements in the LEEM system in terms of dimer breakup followed by step proliferation, with the dimer concentration decreasing from 100 % at 900 K to 0 at 1100 K [148]. Extension of the LEED measurements to the (00) intensity and to the (00) and (1/2, 0) halfwidths supported this initial interpretation and lead to a new model of the phase transition, which assumes vibrating non-interacting dimers randomly distributed over lattice sites [149]. Although the detailed atomistic picture is not clear, the phenomenological aspects are in general supported by more recent phase field model calculations [150] with two exceptions: (1) while the dimer concentration as expressed by the order parameter decreases initially continuously with increasing temperature, it drops abruptly to zero at about 50 %

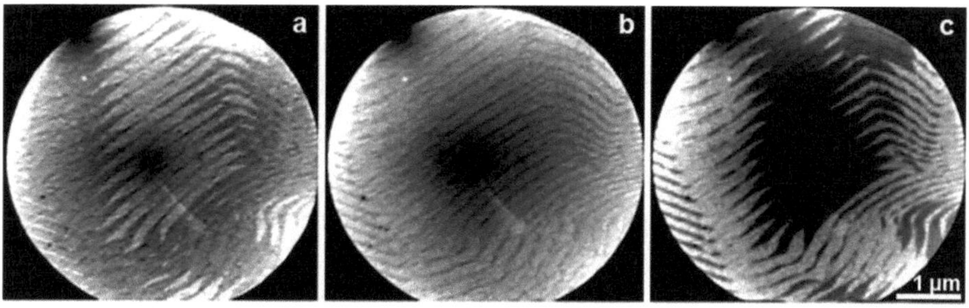

Fig. 5.21 LEEM images of a clean 6H-SiC(0001) surface showing domains with different surface terminations. (a) 13.0 eV; (b) 14.5 eV and (c) 15.5 eV. Reproduced with permission from Ref. [155]. Copyright 2002 Elsevier

concentration and (2) it shows considerable hysteresis upon cooling. This phenomenon possibly is not visible in experiment because of surface heterogeneity.

In the case of GaAs(100) only the decomposition [151] and congruent evaporation [152] has been studied, except for a brief MEM/LEEM inspection of a surface with a highly doped overlayer by the author. GaAs decomposes above 900 K in As, which sublimates and Ga, which forms droplets. The first paper [151] served to control droplet formation and find preconditions for congruent evaporation, the second [152] studied congruent evaporation by compensating the loss of As with an incoming As flux.

The GaN(0001) surface was studied only briefly in connection with GaN homoepitaxy. The surface was not from a bulk single crystal but from a 1.4 μm thick GaN film produced by OMVPE (Organo-Metallic Vapor Phase Epitaxy) on a AlN buffer layer on 6H-SiC(0001), the only GaN material available at that time. After cleaning with atomic nitrogen from a radiofrequency discharge at about 675 °C a step and terrace structure or spiral structure was obtained, depending upon substrate used [153]. The surface can be either Ga or N terminated. Distinction between different terminations is possible with dark field imaging with diffracted beams such as the (10) or (11) beams [154].

While GaN is of practical interest mainly for solar cells, SiC is not only important as substrate for nitride solar cells but also by itself as a material for high-power, high-frequency, high-temperature electronics. More recently it has become also important as template for graphene. SiC comes in a variety of polytypes, differing in the SiC bilayer stacking. They are characterized by their repetition frequency n and symmetry (H, C for hexagonal and cubic), with 4H-SiC and 6H-SiC the most used types. The first LEEM studies were made by Pavlovska et al. [155] in connection with GaN heteroepitaxy on 6H-SiC(0001) crystals, which had been etched at about 1600 °C in H_2. In this study the various LEED patterns previously reported were correlated with LEEM, which showed a large variety of structures, depending upon heat treatment. Some of the images are shown in Fig. 5.21 [155]. The contrast differences between different regions are attributed to different stacking sequences. Imaging of larger areas with FEL-PEEM revealed beautiful patterns attributed to screw dislocations [156].

More detailed LEEM studies of the SiC(0001) surface were made later in connection with its application for the growth of graphene [157–159]. In these studies the crystal was cleaned in disilane at a much lower temperature (700–800 °C), which produced larger terraces and apparently fewer stacking faults. Both 4H- and 6H-SiC(0001) surfaces were studied. On the 6H-SiC surface the same sequence of LEED and LEEM patterns as reported previously [155] was found. High resolution images of the step roughening observed previously and the subsequent growth of the $(6\sqrt{3} \times 6\sqrt{3})$-R30° layer, the "buffer layer" in graphene growth, gave a much deeper insight in the graphitization of the surface than the earlier work [155]. Another study explored the process of polytype formation during homoepitaxial growth on the 4H-SiC. Upon annealing cubic phase monolayer islands formed, which grew upon further annealing, displacing the hexagonal phase and causing step bunching and thus increasing surface roughness, a technologically detrimental effect. This phenomenon was interpreted in terms of the various surface terminations possible on the 6H-SiC(0001) surface, based on first principle calculations of their surface energy [158]. The "softer" cleaning with silane compared to H_2 produces apparently wider terraces which made local LEED from single 200 nm wide terraces possible. The wider terraces allowed μLEED of the 4H-SiC(0001) surface and its atomic structure with a dynamical LEED analysis of the $I(V)$ curves [159].

This surface was also subject of a MEM study aiming at the understanding of tilted stacking faults in a homoepitaxial layer, which are detrimental in electronic devices. While pure MEM could not image them due to surface charge fluctuations, simultaneous illumination with light with energy slightly above the band gap clearly revealed the stacking faults. This was attributed to charge generation and transport in the layer, eliminating the surface charge. For contrast between stacking fault and surrounding, generation and/or transport must obviously be different [160].

5.1.4 Other Inorganic Compounds

Amongst the many inorganic materials that have been studied by surface microscopy with slow electrons, only two of them have been subject of thorough surface microstructure studies, titanium nitride and titanium oxide. Many compounds are not suitable for these methods, either because of charging or radiation damage by electrons and photons, such as halides and many oxides, or are not available in sufficiently perfect, large, and clean crystals, or are not of sufficient practical interest. Therefore the discussion of inorganic compounds in this section is limited to TiN and TiO_2, with a few comments on CeO_2.

TiN is a corrosion resistant, hard refractory compound (melting point $T_m = 3200$ K) with NaCl structure. Changes of its surface microstructure occur at similar high temperatures as on the refractory metals discussed in Sect. 5.1.1 and the experimental procedures and goals are similar. In contrast to the alkali halides the

(111) surface is the most stable surface and has been studied on thick epitaxial layers, whose surface structure is believed to be representative for that of bulk crystals. Like in the case of metals and semiconductors TiN surfaces had been studied before at lower temperatures (1050–1250 K) with STM. LEEM served to extend the measurements on the (111) surface to higher temperatures (1500–1700 K) at which different processes can determine the surface microstructure but at which sublimation is still negligible. Step–step interactions, step permeability, step mobility, and the possible contributions of bulk mass transport processes were deduced from the decay of conical stacks of bilayer-height islands [161]. The details of the decay process showed high step permeability, strong temperature-dependent step–step repulsion and attachment–detachment limited decay. From the temperature dependence of the island area decay an activation energy of 2.8 ± 0.3 eV was obtained, which is much smaller than that for bulk diffusion so that surface diffusion is the rate-controlling process. In an extension of this study the anisotropy of the step mobility Γ, which is a measure for the rate of attachment–detachment of atoms at the step, was deduced from island decay rate data. Γ is intimately related to the step energy β because the attachment–detachment probability increases with β, so that Γ is orientation-dependent when β is direction-dependent. Theory shows that $\Gamma(\varphi) \sim \beta(\varphi)\mathrm{d}A/\mathrm{d}t$. With the proportionality factor and $\beta(\varphi)$ known from other experiments, Γ has been obtained from the angular dependence of the decay rate of the inequivalent $\langle 1\bar{1}0 \rangle$ steps of the strongly rounded triangular islands at 1560 K [162].

While the last two studies showed surface diffusion as the dominating mass transport mechanism, a study of dislocations terminating at surfaces [163] revealed that in the same temperature range bulk diffusion was causing microstructure changes as illustrated in Fig. 5.22 by the evolution of a screw dislocation (a) and a dislocation loop (b) intersection with the surface. In (a) the spiral step has made one full turn in the image sequence, the dislocation core has climbed one unit step height normal to the surface and then the process repeats itself with constant unwinding frequency ω. In (b) the spiral steps originating at the two end of the dislocation loop rotate in opposite direction and form a complete loop, which subsequently expands while the previous step repeats itself with constant period τ. A comparison with other step movements showed that the spiral in (a) is moving inward and that loop in (b) bounds a vacancy island. Because there was no sublimation this was explained by vacancy diffusion from the bulk. This was supported by measurements of ω at several temperatures T and plotting $\omega(T) = \omega_0 \exp(-E_d/kT)$, which gave $E_d = 4.5 \pm 0.2$ eV and $\omega_0 = 10^{12}\ \mathrm{s}^{-1}$. The high value of E_d is typical for bulk processes such as point defect formation and diffusion, so that the change of the surface microstructure in this case is due to vacancy diffusion along the dislocations [163]. In order to ensure that sublimation made a negligible contribution to spiral step formation the experiments were repeated as a function of N_2 pressure up to 5×10^{-7} mbar. While ω was found to decrease slightly with time from cycle to cycle, E_d had the value mentioned before, independent of N_2 pressure [164]. Of course this phenomenon is limited to materials which can sustain a high concentration of vacancies.

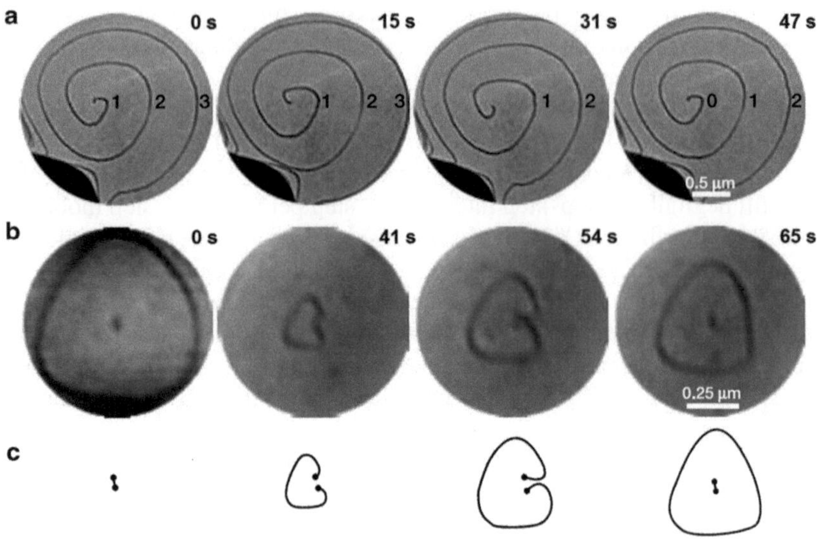

Fig. 5.22 Nucleation and growth of steps from a screw dislocation at 1688 K (**a**) and from a dislocation loop at 1653 K (**b**) on TiN(111) during annealing in N_2.without deposition and evaporation. The LEEM images show the time evolution of the bilayer high steps. The step edges are pinned at the core(s) of the dislocation(s), which in (**b**) have opposite sign. The loop generation process observed in (**b**) is shown schematically in (**c**). Adapted from Ref. [163]. Permission from Nature Publishing Group (UK)

TiO_2 also has a significant homogeneity range though not as wide as TiN, so that surface microstructure changes due to mass transport between volume and surface can be expected at elevated temperatures. Actually, changes of the microstructure of the (110) surface of the rutile phase of TiO_2 were studied before those of TiN. In a series of experiments McCarty and Bartelt developed a very comprehensive picture of the processes occurring on this surface at temperatures between about 690 and 1230 K as a function of the deviation from exact stoichiometry [165–169]. The specific problem addressed was the transition between the (1×1) and the (1×2) surface terminations, which served as probe for the mass transport and can be easily distinguished in LEEM at selected energies. The crystals were heated in UHV to several high temperatures to reduce the oxygen content from 2 to values between 1.9977 and 1.9958. The phase transition kinetics was studied under several experimental conditions: isothermal annealing after quenching, heating, cooling, oxidation after reduction for several surface geometries such as step density, pits, and multilayer islands. Of the wealth of results only a few can be mentioned. The phase transition is always first order. Under isothermal conditions it occurs mainly via mass redistribution across the surface, upon heating or cooling mass flow between surface and bulk is dominating except on wide terraces, on which also the first process is involved. During oxidation of oxygen-deficient crystals surface restructuring occurred via periodic transitions between (1×1) and (1×2)

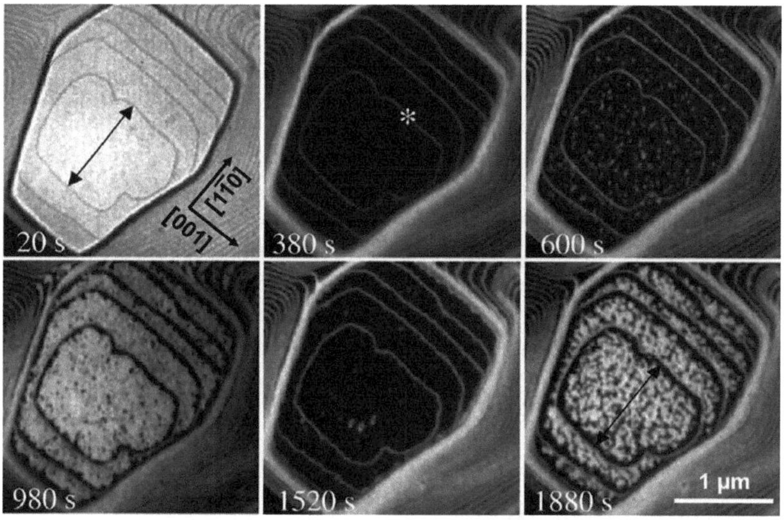

Fig. 5.23 Bright field LEEM images taken during the oxidation of slightly reduced a TiO_2 (110) surface in 1×10^{-6} mbar at 547 °C, starting from a (1 × 1) structure. At the electron energy used for imaging the (1 × 1) structure appears *bright*, the (1 × 2) *dark* and the surface oscillates between these two structures due to transport of Ti from the bulk and reaction with oxygen without step movement at these specific experimental conditions (see *double arrows*). Adapted with permission from Ref. [168]. Copyright 2003 Elsevier

termination, either in the step flow growth mode without nucleation (at high temperatures) or with two-dimensional nucleation and layer-by-layer growth (at low temperatures), with the details depending on the terrace size. Figure 5.23 [168] illustrates the oxidation of an oxygen-deficient crystal at 547 °C in 1×10^{-6} mbar O_2. The surface oscillates with time between (1 × 1) termination (bright) and (1 × 2) termination (dark). For the many interesting details the reader is referred to the original publications.

In the work just discussed the (1 × 2) structure was intimately connected with oxygen deficiency in the bulk and produced by cooling an oxygen-deficient crystal. In another approach Menteş et al. [170] produced the (1 × 2) on stoichiometric TiO_2(110) surfaces by electron-stimulated desorption (ESD) at about 720 K, at which the surface vacancies order in the (1 × 2) structure. In contrast to the added row model of this structure, which is generally accepted when produced in the manner discussed before, the (1 × 2) structure produced by ESD was interpreted in terms of the missing row model, based on a number of indirect indications. A later comparative analysis of the LEED *I*(*V*) curves of the (1 × 2) structure produced by the two methods, however, suggested that the ESD-generated (1 × 2) structure is also of the added row type. The (1 × 2) structure has also been produced by photon-stimulated desorption with synchrotron radiation [171] and studied with LEED and LEEM. Light from a frequency doubled Ti:sapphire laser was also used to produce the (1 × 2) structure in a UVPEEM study [172].

5.2 Adsorption

Adsorption layers can form either from the gas phase or by diffusion from the bulk. They are frequently the precursors of the growth of three-dimensional phases but differ so much from them that they will be treated here as a separate phase. Their interaction with the surface imposes usually a structure on them different from the bulk, in particular also a different electronic structure, and vice versa modifies also the properties and structure of the substrate, for example by reconstruction or by more extensive changes such as alloying and faceting. Many of the results presented in this section have been reported in connection with film growth but are included here for the reasons just mentioned. Organic adsorbates, however, will be discussed in connection with organic layers.

5.2.1 Adsorption on Metals

5.2.1.1 Nonmetallic Adsorbates

The adsorption of O_2 on metal surfaces is a good example of the application of the various methods and of the variety of phenomena, which can be studied with them. The simplest phenomenon is the work function change $\Delta\Phi$ caused by adsorption, which can be studied well with UVPEEM via the change of the emission current caused the work function change. PEEM has been used to image surface diffusion of oxygen on Ni [173], to study the influence of adsorption on various surface orientations of polycrystalline Cu and Ti surfaces [174] and to extract some information on the various binding states of oxygen on Pd(100) [175]. Faceting upon oxygen adsorption at high temperatures is a well-known process on refractory metal surfaces but only one attempt was made to study it with LEEM, with limited success because of the facet size on the crystal studied, Ir(210), were at the resolution limit of the instrument used [176]. LEEM played a limited but important role in some LEED studies. Examples are the structure analysis of the Ag(111)-(4×4)O structure, which required imaging for selection of a region with a minimum number of defects for microspot $I(V)$ analysis [177] and studies of the strain in the W(110)-p(2×1)O structure as a function of p(2×1)O island size [178] and temperature [179], which combined with first principle calculations allowed to determine the associated stress. In this work it was necessary to select step-free regions for quantitative spot position and shape analysis. Figure 5.24 [179] illustrates the usefulness of LEEM not only for this selection but shows also that adsorbate domains are not necessarily bounded by surface steps.

LEEM plays a central role whenever obtaining the desired information requires surface-sensitive imaging. This is the case, for example, for the understanding of the later stages of oxygen adsorption on W(100) at high temperatures. This system had been studied earlier by LEED and other techniques. However, many of the

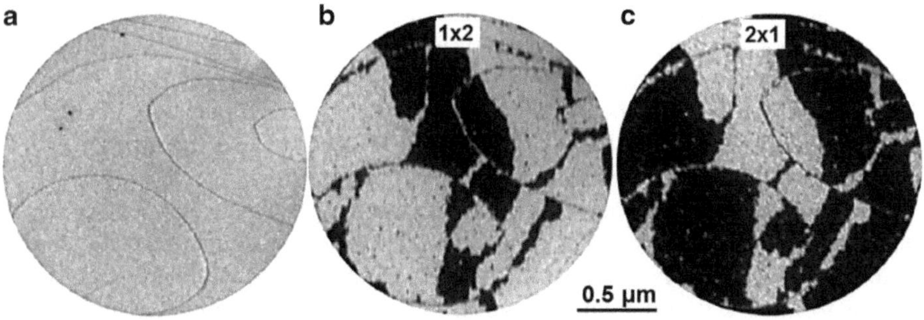

Fig. 5.24 LEEM images taken with 35 eV electrons of a W(110) surface covered with 1/2 monolayer oxygen with (1 × 2) structure. (**a**) Bright field image, (**b**) and (**c**) dark-field images taken with the (0,1/2) and the (1/2,0) beams of the two domains of the (1 × 2) structure, respectively. The domains extend frequently across the monatomic steps. Adapted with permission from Ref. [179]. Copyright 2010 by the American Physical Society

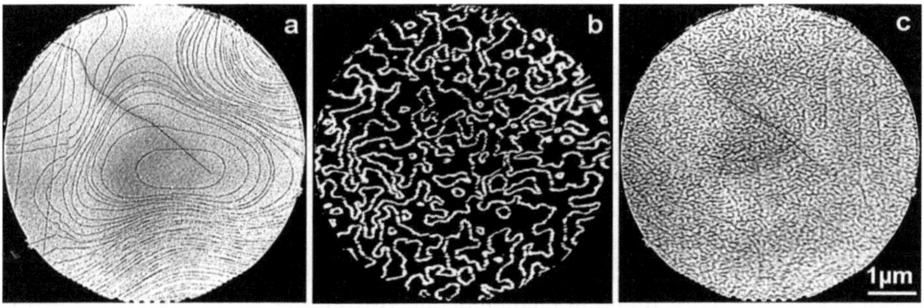

Fig. 5.25 Bright field LEEM images taken from a study of the oxidation of the W(100) surface at 1060 K in 2.5×10^{-8} mbar O_2. (**a**) Clean surface before oxidation (7.0 eV), (**b**) after exposure to 5.6×10^{-6} mbar s O_2 (16 eV), (**c**) after completion of the oxidation (16 eV). The *dark regions* in (**b**) are from two domains with a (5 × 1) structure, the white chains decorate the domain walls and have p(2 × 2) structure. In (**c**) monatomic steps are visible again. Adapted with permission from Ref. [181]. Copyright 1996 Elsevier

structural changes occurring with increasing coverage, in particular between the (5 × 1) pattern at 1.0 ML and the saturation of the two-dimensional layer with p(2 × 1) and p(2 × 2) structure at about 1.25 ML could not be explained using laterally averaging methods. The first LEEM studies [180, 181] provided an explanation (Fig. 5.25). The clean surface with monatomic steps is shown in Fig. 5.25a. At the beginning of the transition small islands, identified by LEED as p(2 × 1) islands, grew rapidly into random chains of small islands, which were unrelated to the step structure of the substrate (Fig. 5.25b). They became more intense and grew laterally over the surface while simultaneously developing a p(2 × 2) structure. When this structure completely covered the surface at 1.25 ML the substrate steps became visible again (compare Fig. 5.25a, c). Dark field LEEM with (1/2,0) spots indicated that the chains grew at the domain boundaries between

(5 × 1) and (1 × 5) domains, initiating the transition from a chemisorbed layer to a two-dimensional oxide. A detailed analysis of the LEEM images showed that the two (5 × 1) domains formed a two-dimensional percolating system [180, 181]. The formation of the chemisorbed oxygen layer and of the two-dimensional oxide was studied also briefly with LEEM on the Ru(0001) surface as a preliminary to CO oxidation on this surface [182].

On the close-packed refractory bcc (110) surfaces oxygen induces at high coverage and high temperatures complex superstructures with two domains corresponding to the twofold symmetry of the surface, which can be imaged with dark field LEEM similar to the p(2 × 1) domains shown in Fig. 5.25. On well-oriented surfaces the two domains form with equal probability but on Nb(110) films grown on Al_2O_3 $(11\bar{2}0)$ this symmetry is broken in a temperature-dependent manner due to the strain in the films caused by the anisotropic thermal expansion of the substrate [16]. Near the disordering temperature a stripe pattern was observed which was attributed to the interaction of the strain introduced by the substrate with the anisotropic surface stress of the film [183].

Most of the work on CO adsorption was connected with CO oxidation which will be discussed in Sect. 5.2.1.2 but CO alone or with H_2 has been studied too. Rotermund et al. [184–186] developed in the early 1990s a method for studying CO surface diffusion by desorbing CO locally from a saturated CO layer on Pt(110) and Pd(111) using a laser pulse and recording the subsequent filling of the region with reduced CO coverage with PEEM via work function contrast. This method, called laser-induced thermal desorption (LITD) was later applied in significantly improved version to the study of CO diffusion on Pt(111) using the LEEM intensity at the electron energy of maximum sensitivity to CO for local coverage measurement [187]. The work function change deduced from the MEM–LEEM transition energy served only to check the accuracy of the LEEM calibration. A detailed analysis of the time dependence of the coverage profile after LITD, which desorbed 0.06 ML in the center from an initial coverage of 0.31 ML, allowed to determine the coverage dependence of the CO diffusion coefficient between 0.25 and 0.31 ML. Beyond the demonstration of the use of LEEM for this type of measurement a comparison of the LEEM images before and after LITD showed that the laser pulse needed for desorption of a measurable amount of CO caused considerable surface damage, an effect which would be difficult to be seen with PEEM [187]. Another study of CO adsorption with LEEM was concerned with the interaction with hydrogen on Pd(111). Here LEEM served to illustrate the displacement of adsorbed hydrogen from the surface to below the first Pd layer, making use of low energy reflectivity differences. The structural details were obtained from $I(V)$ curves and first principle calculations [188].

Sulfur adsorption by segregation from the bulk is an unwelcome process encountered with many metals, in particular also with Cu. In a LEEM study of intentional S adsorption on a well cleaned Cu(111) surface an interesting phenomenon was discovered: Cu surface self-diffusion was enhanced several orders of magnitude by less than 0.01 ML of S. This was deduced from monolayer island decay and coarsening, similar to the diffusion studies on clean surfaces discussed in Sect. 5.1 [189].

As a final example of the use of imaging in adsorption studies the PEEM study of Xe on graphite should be mentioned. Xe monolayer adsorption required cooling to 60 K and for photoemission a windowless low-pressure hydrogen discharge source was needed. The photo ionization cross section of Xe for the 10.2 eV photons from this source is very high, producing high contrast between graphite and Xe. From variations of the contrast, shape, and size of wave-like patterns as a function of Xe exposure, temperature and during desorption, conclusions regarding the phase of the Xe layer were drawn [190]. LEEM combined with diffraction, of course, would be much better suited for such a study.

5.2.1.2 Coadsorption and Reaction: Catalysis

Imaging of the coadsorption and reaction of gases on metal surfaces has long been and still is an exciting playground for surface microscopists and a gold mine for surface chemists. It reveals a large variety of time-dependent patterns depending upon fine details of the partial pressure of the adsorbed gases and the temperature, which are summarized under the term spatiotemporal patterns. They include periodic plane waves, soliton waves, spherical waves, spiral waves, turbulent and chaotic states, and other configurations, which under the same conditions vary from crystal plane to crystal plane. The field was started in Ertl's group in the late 1980s, first with scanning PEEM [191] and has reached a high level of sophistication since. Initially PEEM was the main method because the first LEEM used in these studies had some technical problems which made the reflection methods LEEM and MEM difficult to use. Later LEEM and XPEEM made important contributions too in contrast to MIEEM, which was also tried [192]. The basics of surface reactions and the early phase of imaging have been reviewed by Imbihl and Ertl [193] and later developments are covered in several reviews [194–197]. The literature in this field is too extensive so that we will restrict ourselves to a few general remarks, a few examples, and the most recent literature.

CO oxidation is by far the most intensely studied reaction followed by the water formation reaction ($H_2 + O_2$) and the reaction between NO and H_2. The low index planes of Pt ((110), (100), (111)) are usually used for the CO oxidation, the Rh (110), and (111) surfaces for the other two reactions. Most of the work has been done and is being done with PEEM—a beautiful example of this work is shown in Fig. 5.26 [198]—but LEEM is increasingly used too and in cases, in which chemical information is essential, also XPEEM. It has to be kept in mind that the substrate is not static but rather is modified more or less by the reaction occurring on it. These changes are in general below the resolution limit of PEEM but in some cases occur over large enough regions to be detectable. An example is the formation of subsurface oxygen in the CO oxidation on Pt(110) in excess oxygen, which occurs over large enough regions to become visible via the more positive dipole

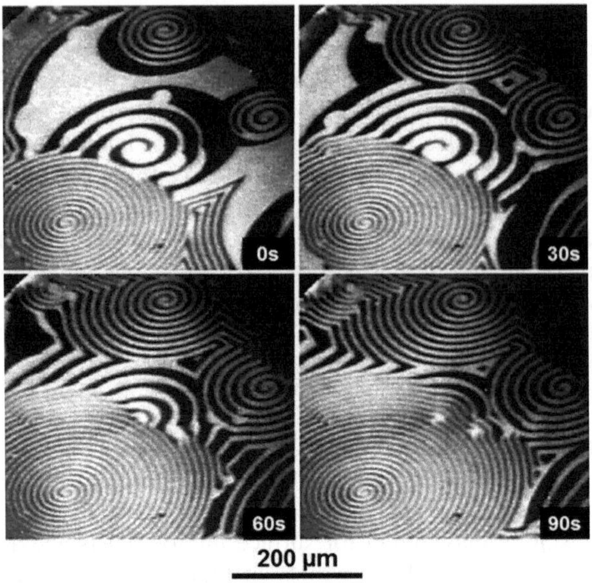

Fig. 5.26 UVPEEM images of the evolution of $O_2 + CO$ reaction spirals on a Pt(110) surface at 448 K, 4×10^4 mbar O_2 and 4.3×10^{-5} mbar CO. The *spirals* start at defects and have different rotation periods and wave lengths. The image sequence shows how the faster rotating spirals with shorter wave length annihilate the slower one in the center. Adapted from Ref. [198] with permission from the American Institute of Physics, © 1993

moment [199]. In general, however, LEEM is the method of choice as illustrated by the same reaction on the same surface [200].

LEEM images not only the microstructure but via µLEED also the local crystal structure. This is particularly useful when—as is often the case—the reacting species produce characteristic LEED patterns. In this case dark-field imaging with LEED spots of the various species make LEEM a chemically sensitive imaging method. The fast image acquisition of LEEM allows following the dynamics of each reactant, which is not possible with XPEEM because of its longer image acquisition time. As an example Fig. 5.27a–d shows dark field LEEM snapshots of the propagation fronts in the $NO + H_2$ reaction on Rh(110) [201]. The images are taken with LEED spots corresponding to different compositions. Following the bright regions with time gives a vivid picture of how the reaction proceeds in space. The time sequence can be recorded quantitatively by selecting a small region of the crystal with the field limiting aperture and recording the intensities of the various LEED spots. This is illustrated in Fig. 5.27e. The time scale is obtained from the speed of the reaction fronts in the LEEM video [201].

In cases in which the species involved in the surface reaction do not have characteristic LEED patterns XPEEM is called for. This can be done in the XANES-PEEM or threshold ionization PEEM mode, which requires resetting the photon energy for each species but has high intensity from the high secondary electron yield and does not

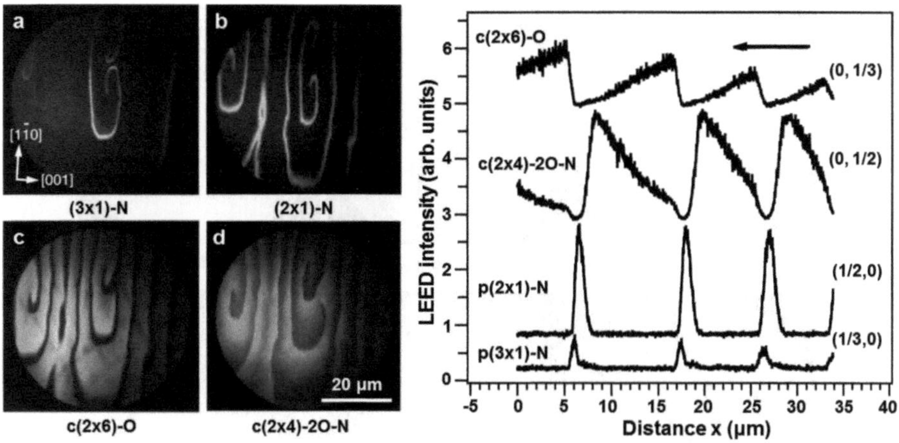

Fig. 5.27 LEEM and LEED study of chemical reaction waves in the NO + H_2 reaction on Rh (110). (**a–d**) Dark-field images taken with diffraction beam characteristic for the various phases occurring in the reaction system about 1 min apart at 550 K, 3.5×10^{-7} mbar NO and 2×10^{-6} mbar H_2. (**e**) LEED intensities of the characteristic diffraction beams acquired simultaneously during propagation of the reaction wave at 530 K, 3.2×10^{-7} mbar NO and 4.5×10^{-6} mbar H_2. Adapted with permission from Ref. [201]. Copyright 2000 Elsevier

need an energy filter. In instruments equipped with an energy filter it is more convenient to image with characteristic photoelectrons because only one photon energy is needed, which is chosen so that all species are ionized with sufficient probability. An example, in which XPEEM was needed for chemical analysis, is the multicomponent system of the water formation reaction on Rh(110) partially covered with Au and Pd, in which Au is passive and Pd active in the reaction [202, 203, 207]. The question arises how these two species are distributed. During reaction the lamellar structures shown in Fig. 5.28 [202, 203] form, which spread again upon reduction and become asymmetric in width upon oxidation. The composition of the various regions in the various states can be determined with XPEEM. In the state depicted in Fig. 5.28 it is evident that the oxygen is adsorbed in the Rh regions ((c) and (f)), that Au (a) and Pd (b) form alloy lamellas, which have a high reflectivity in LEEM, and that the higher work function in the oxygen regions (f) cause a higher reflectivity in MEM (d) [203].

Most work in the past was concerned with the three reactions on clean single crystal surfaces discussed up to now. More recently the attention has turned to the influence of catalytically active and/or passive adsorbates on these reactions as illustrated in the previous example and to catalysis on clean polycrystalline surfaces. The first development, advanced mainly by Locatelli, Kiskinova, Imbihl et al. [202–214], is motivated by the desire to tailor the activity and selectivity of the basic catalyst with co-adsorbates, the second route, followed mainly in Schlögl's group [215–219], by simulating pure real catalysts, which always expose several surface orientations. Depending upon the reaction system and the instrument used in the study, the experiments have been performed between 500 and

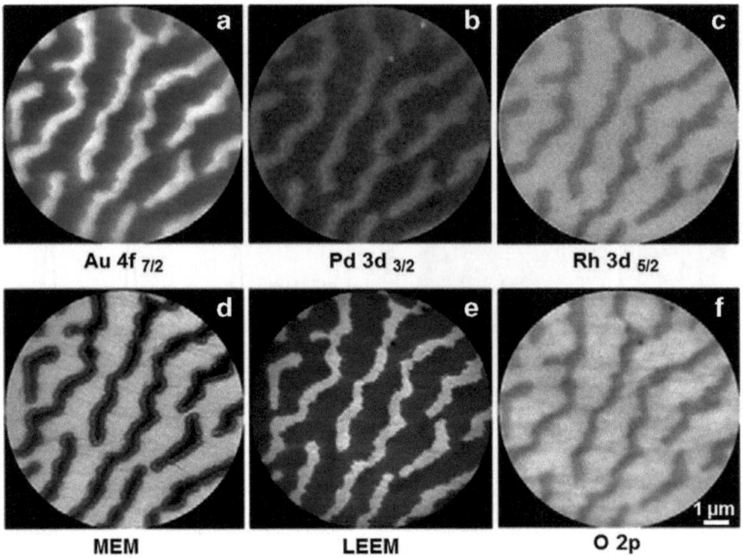

Fig. 5.28 XPEEM, LEEM, and MEM images of the stationary patterns resulting from the water formation reaction on the Rh(110) surface originally uniformly covered with 1/2 monolayer of AuPd. Reaction parameters: 880 K, $p_{H2}/p_{O2} = 0.45$, 5.3×10^{-7} mbar total pressure. The XPEEM images taken with the photoelectrons shown *below* the images show that Au and Pd form alloy stripes separated by oxygen-covered Rh regions, which are bright in MEM because the higher work function. The LEEM image taken with 8 eV electrons shows diffraction contrast. The $[1\bar{1}0]$ direction is normal to the stripe direction. Adapted with permission from Ref. [202]. Copyright 2006 American Chemical Society and with permission from Ref. [203]. © IOP Publishing 2008. All rights reserved

1000 K at pressures ranging from 10^{-7} to 10^{-5} mbar, in the simple PEEM system used in [215–219] up into the low 10^{-4} mbar range.

The initial studies of the first group looked at the influence of an inhibitor species, Au, on the $O_2 + H_2$ reaction on Rh(110) [204, 205] and were then extended to include also an activator species, Pd, simulating a bimetallic catalyst [202, 206, 207]. The lamellar structure shown in Fig. 5.28 is only one of many found at different stages of the reaction. Another effort aimed at the understanding of the influence of coadsorbed K, which is known to be a strong promoter in practically important reactions, on pattern formation, in this case in the $NO + H_2$ reaction on Rh(110) [208, 209, 220]. As in the case of Au and Pd, phase separation was observed, in this case not into pure K regions but into K + O regions due to the strong K–O interaction. Similar redistributions of catalytically active coadsorbates, called reactive phase separation, on single crystal surfaces have been observed in the $O_2 + H_2$ reaction on Rh(111) with vanadium oxide, also an industrially important catalyst in form of VO_x ($x \approx 3$) [210–212]. A more complicated situation was encountered in the $O_2 + H_2$ reaction on a bimetallic Rh(111)-Ni catalyst prepared by alloying the pure Rh surface with an unknown amount of Ni. Here the reaction produced many small

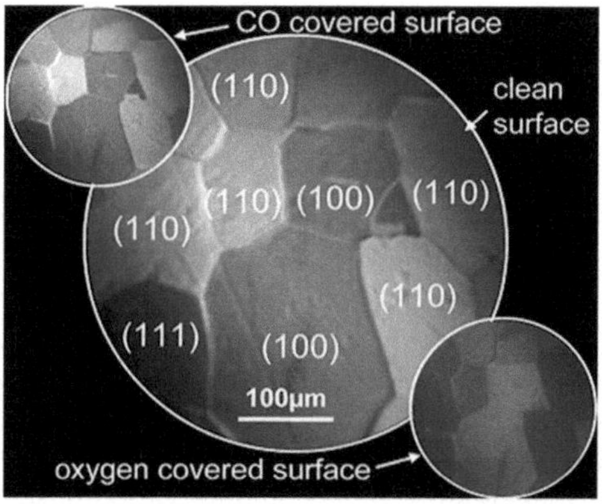

Fig. 5.29 UVPEEM images of a clean, CO-covered, and O-covered polycrystalline Pt foil. The contrast varies not only with surface orientation but also slightly between surfaces of the same orientation due to step density differences, which influence the work function. These differences influence also adsorption as seen in the strong intensity differences between different (110) surfaces upon CO adsorption. Reproduced from Ref. [216] with permission from Springer Science + Business Media

three-dimensional NiO crystals, which acted as obstacles to the propagation of the spiral wave excited by the addition of Ni. By reducing the pressure and increasing the H_2:O_2 ratio the NiO crystals could be reduced again leaving an O-free alloy surface. Thus the NiO crystallites, believed to be catalytically inactive in this reaction, acted as reservoir of Ni, which could be brought to the surface by coadsorption of a proper amount of oxygen [213, 214]. An even more complex situation was found when VO_x and Ni were present simultaneously in this reaction on this surface [214], which illustrates the challenges of simulating catalysis on real surfaces.

The second route in recent catalysis experiments, the study of reactions on polycrystalline surfaces, allows comparing different surfaces under identical conditions. Pt is particularly well suited for this purpose because polycrystalline foils with regions, which are sufficiently large for PEEM studies, can be prepared easily. The regions with different crystal orientation are usually tilted somewhat relative to each other so that PEEM is the preferred imaging method. In LEEM this tilt requires realignment or moving the contrast aperture when moving from grain to grain, which makes a quantitative comparison of the various surfaces difficult. The CO oxidation is an ideal reaction for this study because of the many results from single crystal studies, which can be used for comparison, in particular for the determination of the local coverage via the image intensity. Figure 5.29 illustrates how the contrast in the PEEM images of such a foil changes from orientation to orientation and with adsorbate [216]. Though the spatial resolution is poor, simultaneous data acquisition from several orientations is fast. This allows efficient determination of different reaction states such as high and low reactivity and bistability regions [215–218]. The method was also applied to a Si-contaminated polycrystalline Pd foil. The adsorbed oxygen enhanced Si segregation to the surface, resulting in silicone oxide formation, which drastically inhibited CO oxidation and its dependence on crystal orientation [219].

Recently other reactions have been added to the three classical ones discussed up to now, the $NH_3 + O_2$ reaction on Rh(110) [221] and the reaction of ethylene with O_2 on clean and Ag submonolayer-covered Ni(111) [222]. In the first case an interesting reaction front modulation was found. In the second experiment Ag increased the reactivity and shifted the stability of the reaction product to higher temperatures.

Of course, from the practical point of view all these studies suffer from the "pressure gap" which separates the low-pressure experiments from the high-pressure conditions of technical catalysis and which is frequently used to question the usefulness of these studies. Nevertheless these experiments make important contributions to the fundamental understanding of catalysis, in particular when complementary imaging and diffraction methods such as UVPEEM, LEEM, LEED, and XPEEM are combined.

5.2.1.3 Metallic Adsorbates

Metal adsorption on a clean metal surface under UHV conditions was imaged already in 1985 by Bethge et al. [223]. They studied the diffusion of Ag at submonolayer coverage on polycrystalline Ni with UVPEEM and correlated image intensity with Ag coverage using Auger electron spectroscopy, which also was used to characterize the cleanness of the surface. Although the first LEEM image of a metal adsorbate on a metal, Ag on Mo(110), was published in the same year it served mainly to demonstrate the usefulness of LEEM for such studies [4]. In the late 1980s the interest turned to submonolayer Cu and Au films on Mo(110) and W(110). Mundschau et al. imaged step decoration and step flow of Cu on Mo(110) with LEEM and PEEM [7], studied with LEEM the misfit flip transition [71, 103, 224] and the step roughening [8, 225] in the Cu double layer on Mo(110). On the W (110) surface the first two Cu monolayers have a different structure despite nearly identical lattice constants of W and Mo. Their formation, structure, and stability were studied by Altman et al. [226, 227]. A more detailed study of the evolution of these layers by Koshikawa et al. [228, 229] revealed additional features before the formation of three-dimensional crystals. Contrary to Cu, Au forms on Mo(110) a striped phase in the sub-monolayer range [71, 225, 230]. Later this growth mode was studied also on W(110) briefly by Duden [231] and thoroughly by de la Figuera et al. [232]. These studies stimulated the work of Menteş et al. and Stojić et al. on stress-induced stripe formation in Pd/W(110) [233, 234] and its stabilization by coadsorption of oxygen [235]. Figure 5.30 shows a few examples of submonolayer stripe formation of metals on bcc(110) surfaces. On Mo(110) Au stripes grow in the $[1\bar{1}2]$ directions (a), on W(110) in the $[1\bar{1}0]$ direction (b, c) [231] and Pd stripes on W(110) grow in the $[1\bar{1}0]$, $[1\bar{1}2]$, or [001] directions, depending upon oxygen coverage (d–f) [235]. The comparison of (b) and (c) shows that stripe formation is

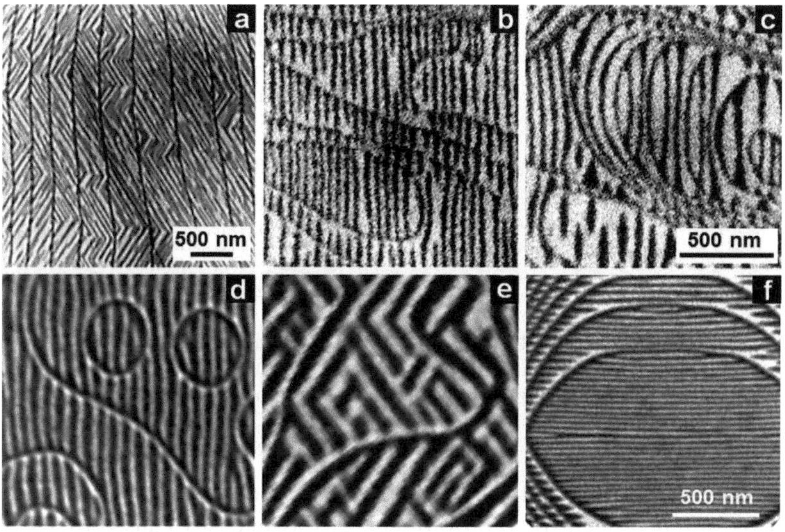

Fig. 5.30 LEEM images of stripes phases at submonolayer metal coverages (**a**) Au/Mo (110), stripes parallel to $\langle 11\bar{2}\rangle$; (**b** and **c**) Au/W (110), 0.3 monolayers at 1045 and 1015 K, respectively; (**d–f**) Pd/W(110) with 0, 0.2, and 0.3 monolayer oxygen coverage and 0.4–0.5 monolayers Pd, (**d**, **e**) at 1240 K, (**f**) quenched to room temperature. The dark metal stripes are in (**a**) and (**e**) parallel to $[1\bar{1}2]$, in (**b–d**) parallel to $[1\bar{1}0]$ and in (**f**) parallel to [001]. The *curved lines* are monatomic steps. (**a**) E. Bauer, M. Mundschau et al., unpublished; (**b**, **c**) Adapted from Ref. [231] with permission from T. Duden; (**d–f**) adapted from Ref. [235] with permission from Europhysics Letters

very temperature-dependent. The transition from stripes to elongated islands is similar to that on B-doped Si(100) shown in Fig. 5.19 [122], although in opposite direction. The mechanism is however the same: anisotropic surface stress and temperature-dependent boundary free energy. In contrast to Si(100), Au on W (110) shows no long range order in the striped phase as determined by LEED diffractometry [236]. Thus, long range order is not a necessary condition for stripe formation. This is in accord with the conclusion drawn from the detailed theoretical analysis [232] that the stripes are in the diffuse-interface limit of surface stress domains. A similar situation was found in the Pd on W(110) study [233]. The discovery that coadsorption of oxygen can be used to select orientation, width, and periodicity of metal stripe patterns [235] has opened the door to surface stress engineering of linear nanostructures and has actually already been used to produce ferromagnetic nanowires (see Chap. 7). Surface stress-induced pattern formation has been studied theoretically in several newer papers [237–239] in which also references to the pioneering work in this field can be found. A theoretical analysis of the influence of the elastic anisotropy of the substrate can be found in [240, 241]. The conditions for the formation of the herringbone structure seen in Fig. 5.30e are discussed in [237, 238].

Up to a temperature-dependent coverage metal atoms form initially a two-dimensional gas before nucleation of two-dimensional crystals. The two

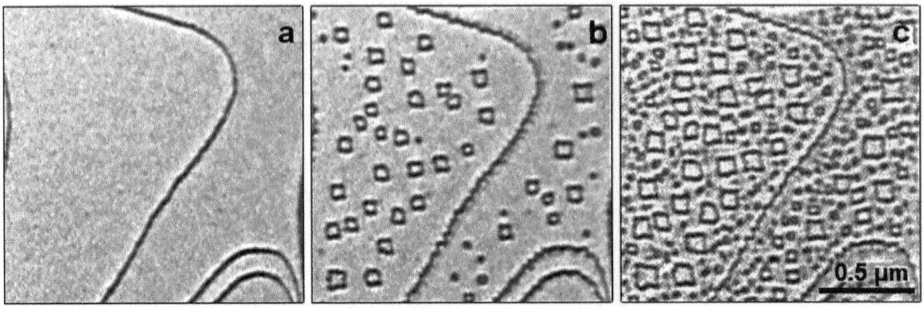

Fig. 5.31 LEEM images of the two-dimensional alloy formation of Ag on Mo(100) at 860 K. (**a**) 0, (**b**) 0.16, and (**c**) 0.38 monolayers Ag. For explanation see text. The *dark lines* are monoatomic steps. Electron energy 6.0 eV. Adapted with permission from Ref. [249]. Copyright 2007 Elsevier

phases can be distinguished easily via the difference of the work function, the LEED background intensity, or also by the LEEM intensity caused by the attenuation of the (00) beam intensity due to the diffuse scattering of the adatoms. de la Figuera et al. [242] have used the latter method to determine the gas—condensate boundary for Ag on W(110) and in addition the Ag desorption energy, both in good agreement with earlier work function and thermal desorption measurements. Yasue et al. [243, 244] used the diffuse scattering contrast to understand abnormalities in the step contrast of W(110) covered by Pb gas. Little use has been made otherwise of this potential of LEEM and via the work function contrast in PEEM.

While the close-packed bcc(110) surfaces are relative stable in the presence of adsorbates, more open low index bcc surfaces are not. An extreme response to metal adsorption is encountered on the (111) surfaces. From field emission microscopy studies it has been known for some time that the W(111) surface facets upon Au and Pd adsorption or annealing at high temperatures into {112} planes, which have a lower surface energy. Later STM studies showed that also other metals have the same effect. Contrary to STM, LEEM, and LEED have the advantage that the faceting can be studied in real time as a function of metal coverage or temperature but have been used for only one metal, Pt [245, 246]. The bright field imaging and limited resolution used, however, did not deliver much information beyond that obtained from STM.

The bcc(100) metal surfaces are more stable than the (111) surfaces but can also rearrange easily as illustrated by the reconstruction caused by hydrogen adsorption. Many metal atoms produce at half a monolayer coverage c(2×2) structure, which had been attributed originally to atoms on top of the surface with (1×1) periodicity but later to two-dimensional ordered alloys. Altman's group has extended the original LEED work to LEEM studies of Cu, Ag, Au, and Pd on Mo(100) and combined them with first principle calculations, which led to a deep insight in the kinetics and energetics of two-dimensional alloy formation [247–249]. While most of the information was extracted from LEED, no $I(V)$ structure analysis was needed to demonstrate alloy formation because of the evolution of the LEEM images with coverage. This is illustrated in Fig. 5.31 for the case of Ag [249], which is representative for the other atoms. In order to form a two-dimensional alloy there must be an exchange between adsorbed and substrate atoms. This leads to alloy monolayer islands on top of the Mo surface and alloy regions in the Mo surface with the same composition, which are visible by step contrast. If pure c(2 × 2) islands would form on top of the (1 × 1) Mo surface, they would show diffraction contrast at least at some energies. The islands reach the maximum density at 0.5 monolayers. Al lower coverage step faceting and an island-denuded zone around the steps is seen (Fig. 5.31b) suggesting the two-dimensional alloying occurs also by place exchange at the steps. The many aspects of this alloying–dealloying process are discussed in great detail in the papers cited.

A second metal on which surface alloying has been studied thoroughly is Cu. For comparison with the bcc(100) surface just discussed we start with the (100) surface, which, being fcc, is more densely packed. The apparently simplest adsorbed metal is Pb because it does not form bulk alloys with Cu. However, on the surface a variety of two-dimensional alloy phases exist. The LEEM studies of Kellogg's group [250–252] have revealed the evolution of the various phases as function of Pb coverage and temperature and are an excellent demonstration of the power of LEEM in surface microstructure analysis. They have shown that the three superstructure phases known from previous LEED studies involve initially alloying followed by two stages of dealloying, each with specific LEED pattern and microstructure, which has led to a new interpretation of the alloying process. Too many images are needed to illustrate this process so that the reader is referred to the original publications, in particular to Ref. [250].

While Pb does not form a bulk alloy with Cu because of its much larger atomic diameter, which is also at least partially responsible for the complex surface alloy, Pd is completely miscible with Cu. Here a quite different question was addressed with LEEM and dynamical theory LEED $I(V)$ analysis: how does the alloy composition evolve during Pd deposition at temperatures, at which no diffusion into the bulk occurs? [253, 254]. Because of the different backscattering cross sections of Cu and Pd the reflectivity of the surface is very sensitive to the Pd distribution normal to the surface so that it can be extracted from a careful $I(V)$ analysis. While Pd is incorporated all over the surface, including at the steps, Cu atoms, which have been displaced on the terraces by Pd, not only form islands but also diffuse to the steps, where the largest changes occur in a narrow region close to the steps. Understanding these changes requires LEEM images with the highest lateral resolution as a function of energy for local $I(V)$ analysis of the (00) beam. With the usual trial and error approach in LEED—dynamical LEED $I(V)$ calculations for model distributions and comparison with experimental $I(V)$ curves—it was found that at a Pd coverage of about 0.5 monolayer most Pd was in the second layer, while the top layer consisted almost completely of Cu ("buried Pd layer"). The Pd distribution around the step indicated effective Pd–Pd repulsion, resulting in step overgrowth, which was identified as the major cause of compositional heterogeneity in thin film growth.

The studies just discussed were made at a temperature (200 °C) at which diffusion into the bulk can be neglected. Pd diffusion into the bulk was addressed in a subsequent LEEM study [255] by annealing the buried surface alloy layer with c(2 × 2) structure at temperatures up to 370 °C. From the image intensity decay at an energy characteristic for this structure as function of temperature an activation energy for diffusion from the alloy into the bulk of only 1.6 eV was obtained. A dynamic LEED *I*(*V*) analysis of the image intensity changes near the steps gave a Pd distribution similar to that in an ordered Cu_3Pd phase known from the bulk. Another LEEM study of the system Pd-Cu(100), combined with extensive first principle and kinetic Monte Carlo calculations looked at the influence of the buried Pd c(2 × 2) layer on the surface self-diffusion on the Cu(100) surface [256]. In this study the decay of Cu monolayer islands deposited onto the Pd-Cu surface alloy was studied as a function of Pd concentration up to the saturation of the c(2 × 2) structure and analyzed in a manner similar to the methods briefly discussed in Sect. 5.1.2.1. The results are illustrated by the following data. At 240 °C the product of the diffusion coefficient *D* and the concentration *c* of the diffusing species at 0.06 and 0.4 Pd monolayers decreases to 50 and 10 % of that on the Pd-free surface, respectively, while the activation energy for surface diffusion increases from 0.83 eV for the clean surface by about 20 % for a Pd coverage of 0.4 monolayers. The results for the lowest Pd coverage, 0.06 monolayers, could be modeled well by surface vacancy diffusion, impeded by the interaction with the Pd atoms buried in the second layer.

On the Cu(111) surface even more interesting phenomena occur. Starting with Sn, the formation of the Sn-Cu alloy, bronze, has been found in LEEM studies [257] to proceed in a rather unusual manner. Already at low coverages Sn forms clusters consisting of several hundred thousand atoms and identified by STM as two-dimensional crystalline Sn islands, which move more or less randomly across the surface, leaving an alloy trace behind. Their speed at 270 K is about 0.8 nm/s independent of particle size and increases by more than two orders of magnitude within 60 K. From the temperature dependence of the speed an activation energy of 0.9 eV was obtained. The driving force of the particle motion was attributed to repulsion between the Sn in the clusters and the Sn atoms in the alloy track left behind the clusters. This experiment is a nice demonstration of the application of the fast image acquisition (30 frames/s) which LEEM allows even at high magnification.

The most exciting phenomenon discovered on Cu surfaces upon alloying is the self-organization of submonolayer Pb films on the two-dimensional Pb-Cu alloy on the Cu(111) surface [258, 259]. The alloy contains 0.22 Pb monolayers (in units of the Cu(111) packing density) and has no long range order. On top of it pure Pb grows up to a coverage of 0.56 monolayers, which corresponds to a slightly compressed Pb(111) monolayer, before three-dimensional Pb crystals form. At partial coverages the theoretically predicted scenario for two-phase systems with competing short range and long range interactions in an elastically isotropic system was found as shown in Fig. 5.32 [258]: the transition from a Pb droplet phase (a) to a striped phase (b) to an anti-droplet (hole) phase (c) with increasing coverage. Here the Pb islands/stripes represent one phase, the uncovered alloy regions the other. The width of the stripes and their periodicity decreased strongly with increasing temperature as expected based on their decreasing boundary free energy and the

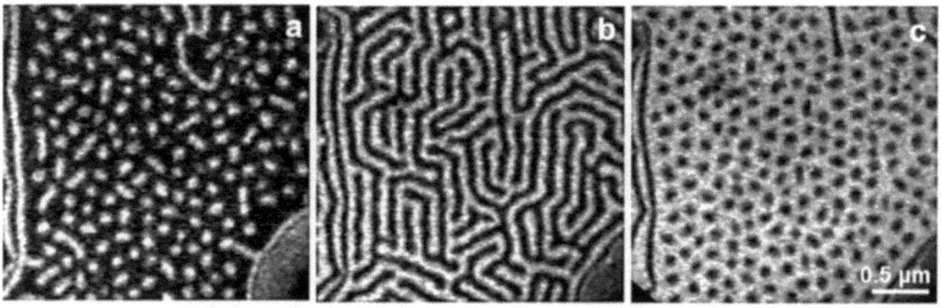

Fig. 5.32 LEEM images of domain pattern evolution in two-dimensional Pb-Cu alloys on Cu (111) at 673 K. The Pb coverages are in (**a**) 0.33, in (**b**) 0.39, and in (**c**) 0.48 monolayers. Electron energy 18–20 eV. For explanation see text. Adapted from Ref. [258] with permission from Nature Publishing Group (UK)

islands elongated. The phases form surprisingly fast, keeping the size of the islands (50–200 nm diameter) in mind, which requires a significant driving force. By analyzing the magnitude of the fluctuations of the island distances the surface stress difference between the two phases was identified as source of this force and found to be 1.2 N/m independent of temperature [260]. The temperature dependence of the boundary free energy was determined by the fluctuation spectroscopy described in Sect. 5.1 of the boundaries of the striped phase, resulting in values decreasing from 22 meV/nm at 600 K to 10 meV/nm at 650 K [261]. Other LEEM studies, which have been made on this system, concerned the shape transition of Pb droplets and anti-droplets caused by the surface stress difference between them and the alloy phase [262], the effect of the Pb in the alloy layer on Cu surface diffusion as measured via the island decay rate [263] and the mechanism responsible for the fast movement of the large Pb droplets during self-assembly [264]. This last study came to the surprising conclusion that the mechanism not only involved fast transport of Pb atoms but in addition also diffusion of Cu atoms *through* the Pb islands, mediated by a high vacancy concentration in the islands.

Another usual phenomenon seen in submonolayer metal adsorption on a metal surface is the snake-like growth of two-dimensional Pd islands on Ru(0001). This growth mode has also been explained with the help of first principle calculations by slight alloying, with slow growth in alloy regions surrounding the sides of the snakes and fast growth along unalloyed surface along the head of the snake over which Pd diffuses easily. The details are complicated but the messages are clear: there are other mechanisms for pattern formation than surface stress phenomena and partial surface alloying can suppress step flow growth [265]. A final example in which LEEM was used for studying metal adsorption on a metal concerned the herringbone and triangular dislocation patterns of Ag, Au and AgAu monolayers, and double layers on Ru(0001). Ag monolayers developed two different herringbone structures with different coverage, Au and AgAu monolayers only one. In the double layer the dislocations of Ag and Au formed an interconnected network, that of AgAu was more complex [266].

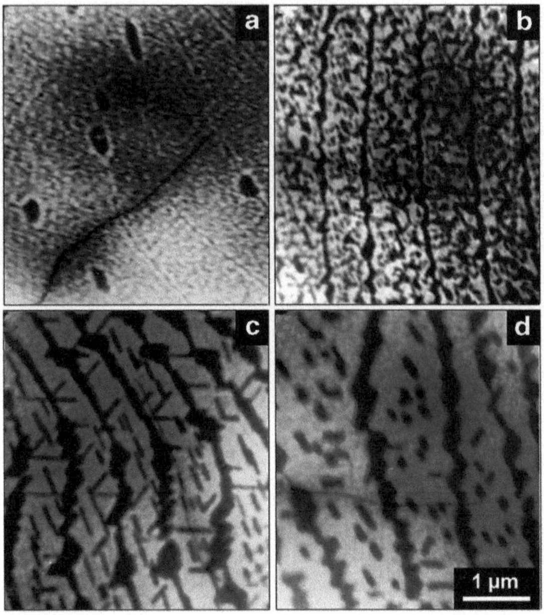

Fig. 5.33 Frames from a LEEM video of the growth of Au on Si(111) at 620 K (**a**), 750 K (**b**), 880 K (**c**) and 1000 K (**d**). The Au coverage is approximately 0.15 monolayers in all images. LEED shows a (5 × 2) pattern. Nucleation and growth occurs preferentially at the steps but on the terraces (5 × 2) domains grow too elongated along the $\langle 1\bar{1}0 \rangle$ direction of the Si(111) surface. Adapted with permission from Ref. [267]. Copyright 1991 Elsevier

5.2.2 Adsorption on Semiconductors

5.2.2.1 Adsorption of Metals

Au on Si(111) was first studied by Telieps and Bauer. They used the three (5 × 2) superstructure beams closest to the (00) beam in dark-field imaging to distinguish the three domains. With PEEM they used the work function difference between different surface phases (($\sqrt{3} \times \sqrt{3}$)-R30°, (5 × 2)) to image their propagation during diffusion [4]. The first systematic video-LEEM study of the evolution of the three Au surface phases at various temperatures by Swiech et al. [267] showed a wide range of the morphologies, a strong influence of steps on nucleation, domain growth of the various phases and the phase transitions between them. Figure 5.33 [267] illustrates the growth of the (5 × 2) phase at various temperatures for the same coverage. At the lowest temperature many small (5 × 2) regions form except for isolated larger regions at steps. Diffusion of Au was studied later with the same coverage edge shift method used by Telieps in more detail in a SPELEEM, with PEEM complemented by XPEEM [268]. Starting from the edge of a region with monolayer coverage the diffusion coefficients derived from the edge displacement with time at 985 K were found to be significantly higher on the (5 × 2) surface than on the (7 × 7) surface. In another surface diffusion study [269] Au microspheres were deposited on the hydrogen-terminated surface. Upon hydrogen desorption the spheres converted into Au silicide particles surrounded by the (7 × 7) structure. The spreading of the (√3 × √3)-R30° structure from these particles over the surface was measured from 800 to 930 K where the higher coverage (6 × 6) structure is disordered. The spreading rate of the edge of the (√3 × √3)-R30° structure was found to be linear over the complete temperature range studied, independent of distance travelled, with spreading rates increasing from less than 0.1 to 50 nm/s in this temperature range. This indicates that it is not the diffusion rate of Au atoms on top of the (√3 × √3)-R30° structure but the rate of the structural rearrangement occurring during the propagation of the Au front at the border between the (√3 × √3)-R30° and the (7 × 7) regions, which determines the spreading rate [269].

The adsorption of Ag on Si(111) was studied to a lesser extent. The first paper [270] clarified some apparent discrepancies of earlier LEED studies concerning the evolution of the (√3 × √3)-R30° and (3 × 1) structures. Later a spectroscopic PEEM study with a SPELEEM showed that the Ag 3d core level of the Ag in the (√3 × √3)-R30° structure was shifted 0.5 eV [271]. PEEM measurements of the spreading of Ag from a Ag island at temperatures at which both (√3 × √3)-R30° and (3 × 1) structures exist at their respective coverages, could be explained with diffusion coefficients from other studies and suggested that diffusion in the (√3 × √3)-R30° region is fast, slows down in the (3 × 1) region and speeds up again in the (7 × 7) region [272].

Cu adsorption on Si(111) was studied only in the context of three-dimensional silicide formation with LEEM [230, 273]. The two-dimensional ("5 × 5") phase ((5.42 × 5.42) according to LEED) was found to nucleate at steps but to spread across both bordering terraces, to disorder above 850 K by dissolution in the bulk and to reappear upon cooling in the same manner. Later work [274, 275] looked at the formation of the adsorption layer in more detail, also at lower temperature at which nucleation occurred at the (7 × 7) domain boundaries and on the domains themselves. Hydrogen adsorption was found to suppress the formation of the ("5 × 5") phase.

Adsorption of the transition metals Ni and Co does not produce superstructures visible in LEED but at high temperatures an "impurity-stabilized" (1 × 1) structure, which consists of a two-dimensional gas of ring clusters according to STM studies. This "(1 × 1)-RC" structure is in a temperature and cooling rate-dependent competition with the (7 × 7) structure, with the (7 × 7) regions growing and disappearing at the steps. Formation of the "(1 × 1)-RC" structure requires thermal activation so that at lower temperatures rather three-dimensional silicide crystals grow in the sea of the (7 × 7) structure [276]. A (√7 × √7)-R19.1° structure with a Co coverage of 1/7 monolayers was seen in LEEM via diffraction contrast and the phase diagram of the three surface phases (7 × 7), "(1 × 1)-RC" and (√7 × √7)-R19.1° could be determined [277–280].

Aluminum forms initially a (√3 × √3)-R30° on Si(111) like Ag but at higher coverage a (9.4 × 9.4) structure. LEEM studies of a surface covered with many Si islands showed that the (9.4 × 9.4) structure started to grow from the island edges, consuming the islands and then to complete the layer with the Si between the islands. From the evolution of the surface morphology an Al coverage of 0.68 monolayers at the completion of this layer could be determined. The (9.4×9.4) periodicity agreed well with this coverage assuming misfit accommodation via a dislocation network with this periodicity [281]. A more detailed LEEM–LEED study [282] of this adsorption system revealed that it is more complex: there are two $(\sqrt{3} \times \sqrt{3})$-R30° structures, a relatively unstable $(\sqrt{7} \times \sqrt{7})$ structure and the (9.4×9.4) structure with coverages of 0.25, 0.33, 0.43, and 0.68 monolayers in addition to a high temperature disordered phase. From studies at many temperatures and coverages the phase diagram was established and the interesting evolution and stability of the various phases deduced.

Pb adsorption on Si(111) is of particular interest because of the subsequent growth of three-dimensional Pb crystals with quantized thicknesses, which requires significant mass transport in the adsorption layer. This process was studied with LEEM using the LITD method for two phases, the low-temperature amorphous wetting layer and the α-$(\sqrt{3} \times \sqrt{3})$-R30° phase with 4/3 monolayer coverage [283]. The measurements at 186 and 300 K showed identical unusual desorption profiles differing only in the time scales. As in the case of Au on Si(111) the profile moved at a constant rate. Detailed analysis suggested a convection-like mass transport, which was attributed to diffusion of thermally activated atoms on top of the wetting layer.

Adsorption of In on Si(111) had been studied extensively with STM and laterally averaging methods before the LEEM studies, which showed many but not all of the previously reported phases. Their evolution with increasing coverage, temperature, and other parameters is too complicated to be sketched here. Therefore the reader is referred to the original work [284, 285]. A spectroscopic PEEM study [286] showed a chemical shift of the In 3d peaks of 0.5 eV to higher binding in the $(\sqrt{3} \times \sqrt{3})$-R30° and the $(\sqrt{31} \times \sqrt{31})$ structures, no shift in the In double layer and two peaks in the (4×1) structure, one unshifted, the other shifted as in the previous structures, indicating two binding sites as expected from the LEEM–LEED studies [284, 285]. Spectroscopic XPEEM combined with LEEM and LEED has also been used to look at the interaction of Sb and Ag with the $(\sqrt{3} \times \sqrt{3})$-R30° and the $(\sqrt{31} \times \sqrt{31})$ structures. The data were interpreted in terms of displacement and intermixing processes [287–289]. Finally, Ga adsorption on Si(111) was subject of a LEEM study, which showed the successive replacement of the (7×7) structure by the $(\sqrt{3} \times \sqrt{3})$-R30°-Ga structure with nice domain formation [290].

Metal adsorption on Si(100) surfaces has been studied much less extensively. The adsorption of Au has attracted most of the attention because the faceting it induces on slightly misoriented (100) surfaces at high temperatures. Several methods were combined to understand this process: bright-field and dark-field LEEM, LEED, STM for structural and spectroscopic PEEM for chemical characterization. Medium energy ion scattering served for absolute Au coverage determination [291–295]. 4° miscut surfaces consisting of 4 nm wide terraces separated by double steps, Au deposition rates of about 1.2×10^{-3} monolayers/s and temperatures between 1000 and 1150 K were used in all experiments. Driven by surface

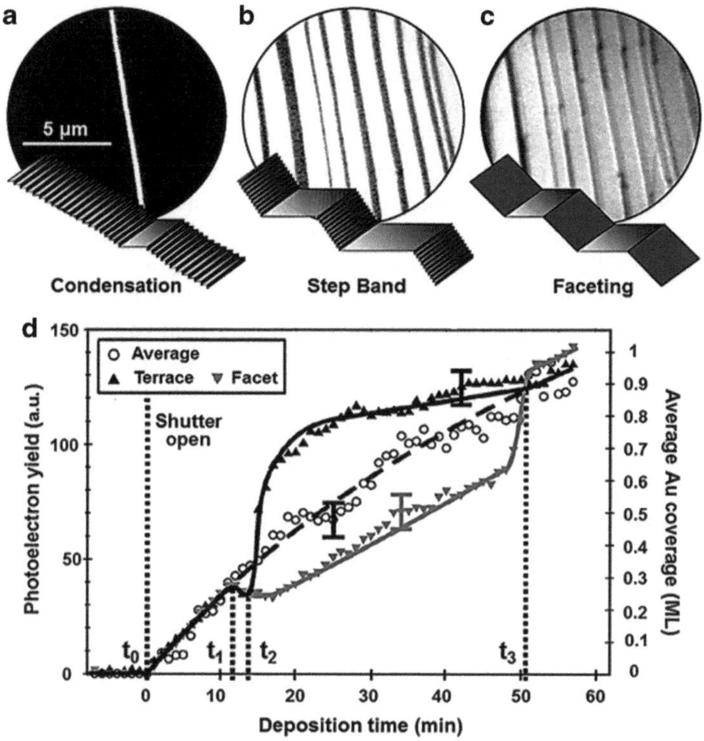

Fig. 5.34 Faceting of a vicinal Si surface 4° off (100) with increasing Au deposition at 850 °C. (**a–c**) LEEM images and sketches of the surface topography at three stages of the faceting process. The (100) surface is perpendicular to the electron beam so that (100) regions appear *bright* and tilted regions are *dark*. (**a**) (100) terrace nucleation during initial condensation stage, (**b**) step bunching connected with lateral terrace growth, (**c**) (119) facet formation from step bunches. (**d**) Evolution of the Au 4f μXPS signal from terraces and step bunches/facets and average 4f signal with deposition time illustrating the Au coverage difference between the different surface regions. (**a–c**) Reproduced with permission from Ref. [294]. Copyright 2001 Elsevier. (**d**) Reproduced with permission from Ref. [295]. Copyright 2001 by the American Physical Society

energy minimization, the Au lattice gas-covered surface breaks first up into (100) terraces covered with a (5 × 3.2) adsorption layer and step bunches which transform into (911) facets with increasing coverage. A few steps of the interesting dynamics, in which Au-decorated terraces spread with speeds up to 100 μm/s over the surface and reach aspect ratios of 10000:1, are shown in Fig. 5.34 [294, 295]. These experiments, which were performed in the first SPELEEM in Elettra, Trieste, are a good demonstration of the power of combining LEEM with spectroscopic PEEM.

Ag adsorption on Si(100) at high temperatures leads to a (2 × 3) superstructure whose composition and structure has been controversial. The solution of this controversy is another feat achieved by combining dark-field LEEM with LEED. With the same method mentioned already in connection with Al on Si(111), i.e., Si island deposition before Ag deposition, the amount of Si needed for completion of the (2 × 3) structure was determined. The amount of Ag was obtained from MEIS measurement, resulting in a 1:1 composition with three atoms each per unit mesh. From the analysis of the LEED patterns a detailed structural model was obtained [281, 296]. Other studies of Ag adsorption focused on surface diffusion on the (100) surface and some of its vicinals with PEEM [297, 298]. The spreading of the front of the (2 × 3) region originating from an Ag island was measured, similar to diffusion measurements mentioned earlier and activation energies were determined. On the 4° miscut surface an anisotropy of these energies parallel and perpendicular to the step direction of 0.7 eV was found. Ag diffusion on (100), 4° miscut (100), (911), and (111) surfaces is compared in Ref. [299].

Adsorption of In on Si(100) was studied with dark-field LEEM from room temperature up to 700 °C [300]. Below 600 °C and coverages up to 0.6 monolayers no dramatic changes in surface morphology were seen, only the transition from the (2 × 1) structure to the (4 × 3) structure with increasing coverage with varying degree of order. At higher temperatures, which required higher In fluxes because of desorption from the surface, major changes in the morphology occurred, similar to those seen with As adsorption, which will be discussed below: massive displacements of Si atoms, leading to island and hole formation and finally to progressive etching. Obviously In is much more aggressive than Au and Ag and so may be other non-noble metal adsorbates such as Sn.

A final interesting result concerning metal adsorption is the observation that Cs adsorption allows to image the (1 × 2) and (2 × 1) domains on Si(100) using linear polarized light. Between 0.3 and somewhat above 0.5 monolayer coverage the work function is low enough for photoemission excitation with 2.33 eV laser light, but not from states along the surface normal as the density functional theory calculations in this study show. However electrons from higher lying off-normal states are transmitted and allow imaging. The calculations identify the electrons as originating from π^* anti-bonding Si dimer states, to which Cs has donated electrons without destroying the (2 × 1) periodicity, and explain the polarization dependence of the contrast [301].

Other surfaces on which metal atom adsorption has been studied in connection with its possible surfactant effect are the (311) and (211) surfaces. On the (311) surface, which is one of the equilibrium planes of Si, Ag adsorption causes at saturation of the monolayer faceting with a periodicity of about 40 nm into (111) facets covered with the Ag(√3 × √3)R30° superstructure and (511) facets covered with a (n × 2)-like superstructure [302]. Saturation of the (311) surface with Ga also causes faceting with about the same periodicity into (211) facets with (1 × 6) reconstruction and (511) facets with (1 × 4) reconstruction [303, 304]. At lower coverages intermediate phases form, preferentially at the steps. In all cases temperatures around 500 °C were used. On the non-equilibrium (211) surface, which in the clean state consists of (111) facets with (7 × 7) structure and (5,5,12) facets with (1 × 2) structure, the adsorption of Ga and In was studied [305, 306]. In both cases adsorption initially reduces the faceting and then induces new facets with new superstructures. Both the facets on the clean and on the Ga and In covered surfaces are, however, too narrow to be resolvable with the LEEM instruments used so that all information had to be obtained from LEED.

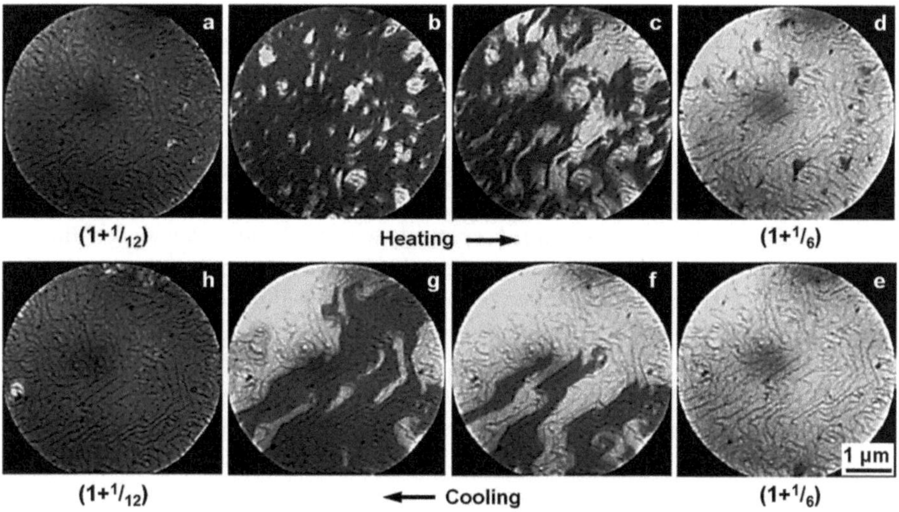

Fig. 5.35 Frames from a LEEM movie of the "(1 + 1/12)" to "(1 + 1/6)" phase transition of the Ga double layer on GaN(0001). (**a–d**) Heating and (**e–h**) cooling through the transition. *Dark*: "(1 + 1/12)" structure, *bright* "(1 + 1/6)" structure. The step structure is not influenced by the phase transition. Electron energy 11 eV. Adapted with permission from Ref. [307]. Copyright 2001 Elsevier

An interesting adsorption system on a compound semiconductor, Ga on GaN (0001), was studied by Pavlovska and Bauer [307]. Ga forms a double layer, which is stable up to about 680 °C and shows three phase transitions, a first order transition at 225 °C and two continuous transitions at higher temperatures. The LEED patterns connected with the first order transition shown in the LEEM images in Fig. 5.35 [307] indicate a material exchange between first and second layer, in which the first layer expels atoms into the second layer and becomes pseudomorphic upon heating; upon cooling it converts to the same misfitting packing density as the second layer. This transition is similar to the misfit phase transition in the Cu double layer mentioned earlier [224]. The importance of the Ga double layer for the growth of smooth GaN layers will be discussed below.

5.2.2.2 Nonmetallic Adsorbates

In view of the interesting surface structure of Si(111) and the practical importance of the Si(100) surface remarkably little work on the interaction of nonmetallic elements with these surface has been done, oxygen excepted. Arsenic has been studied on both surfaces. On the (100) surface As was found to displace Si atoms *on* the terraces at high temperatures (900 K) resulting in Si monolayer island formation. The driving force was identified to be due to a change of the surface stress anisotropy from the clean to the As-covered surface [107]. On the Si(111) surface MEM studies of As adsorption at low temperatures (<750 °C) also show displacive adsorption on the terraces but with increasing temperature and thus decreasing As coverage, when steps become the preferential adsorption sites, Si is increasingly displaced from the steps. This was attributed to surface stress relaxation on terrace regions near the steps [308].

Two more complicated adsorbates have been studied too: CCl_4 [309] and benzoic acid (BA) (C_6H_5COOH) [310]. In the first study UVPEEM was used to compare the reactive adsorption on the Si(111)-(7 × 7), the Ag(111) and the Si(√3 × √3)-R30°-Ag surface. The coverage was determined separately with XPS under identical conditions. The reactivity decreased in the same sequence, with dissociate adsorption on the first two surfaces but molecular adsorption on the last, which was interpreted in terms of the different electronic surface structure. The second study addressed the question of the influence of OH adsorption on the bonding of benzoic acid to the Si(100) surface with XPEEM and XPS. A patterned surface with OH patches and carbon-contaminated SiO_x regions was formed. In order to distinguish between adsorption on OH and C-contaminated regions 4-nitrobenzoic acid (4-NBA) was used and the N K absorption edge used for imaging, which clearly showed adsorption only in the OH regions.

Concluding the section adsorption on semiconductors some studies on Ge should be mentioned briefly. Apparently, no nonmetallic adsorbates have been studied and only a few metals on Ge(111). Adsorption of Ag has been the subject of a low-resolution LEEM–LEED study, which showed mainly the evolution of the various adsorption phases with coverage and temperature [311]. In a Pb adsorption layer a novel phase transformation was found at 189 °C at a coverage of 1.30 at which the high coverage (√3 × √3)R30° phase and the (1 × 1) phase coexist due to thermal fluctuations of rather large domains between the two phases without nucleation ("spontaneous domain switching") [312].

5.3 Film Growth and Structure

In this section we will discuss not only the growth and properties of continuous films but also of "nanostructures" such as nanowires, quantum dots, and droplets on the clean and wetting layer-covered surfaces described in the previous sections.

Included are not only studies of purely scientific interest but also work motived by technological problems.

5.3.1 Films on Semiconductors

5.3.1.1 Metal Films

In the growth of Au on Si(111) three-dimensional particles form before the two-dimensional (6 × 6) structure. Below the eutectic temperature (632 K) they have hexagonal cross section, above this temperature they are round (liquid silicide

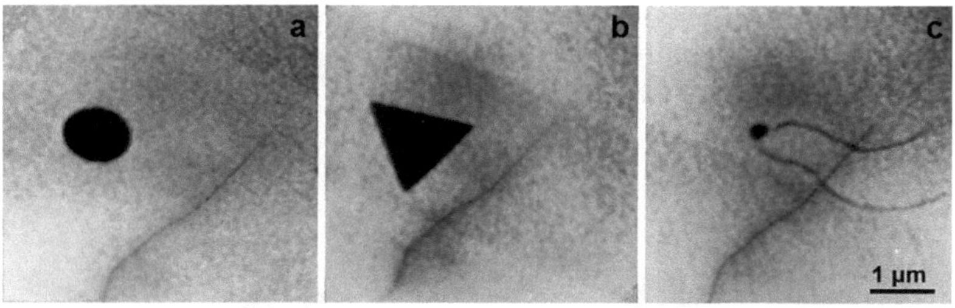

Fig. 5.36 Frames from a LEEM video of the temperature dependence of the shape of Au-Si eutectic particles on Si(111). (**a**) At the eutectic temperature (650 K), (**b**) at 900 K, (**c**) at 1150 K after most Au had evaporated during the particle movement across the surface. The transition between (**a**) and (**b**) is completely reversible. Adapted with permission from Ref. [267]. Copyright 1991 Elsevier

droplets), which are mobile and suppress (6 × 6) formation in the tracks they leave behind during their migration across the surface. These droplets show an interesting reversible wetting transition to larger triangular-shaped particles, which have {112} interfaces with the substrate (Fig. 5.36a, b) [267]. With increasing temperature the speed of the droplets increases strongly so that at 1150 K, where Au sublimates, their motion can be followed with the help of the tracks they leave behind (Fig. 5.36c) [267]. The droplets move perpendicular to the surface steps uphill leaving a step-free track behind. A later study on a better equilibrated surface with lower step density showed that the droplets formed right after completion of the (6 × 6) structure at the steps [313]. Droplet formation is also briefly discussed in [314].

The growth of Ag on Si(111) and Si(100) has been studied in much more detail, starting with studies of crystal shape oscillations [315] and shape transition induced by stress relaxation, leading to nanowire formation [316]. Tromp et al. imaged the interface between Ag crystals and the Si(111) substrate making use of the strain field and the long inelastic mean free path of slow electrons in Ag [317]. With the growth of Ag on Si(111) covered with a ($\sqrt{3} \times \sqrt{3}$)R30°-Sb layer his group also gave the first demonstration of the surfactant effect in thin film growth. This effect reduces the tendency to three-dimensional growth, which is very pronounced in the growth of Ag films [318]. Altman's group determined the three-dimensional shape of Ag crystals via the facet spot movements on Si(111) [319] and studied the influence of In co-deposition on their growth shape, which they attributed to the modification of the Ehrlich-Schwoebel barrier [320]. A PEEM study of the shape and orientation of Ag crystals from 200 to 640 °C showed that up to 240 °C the crystals had (100) and (111) orientations, from 240 to about 600 °C all crystals had (111) orientation with polygonic shape, above 600 °C with triangular shape [321]. In a different context, Ag films with different thickness and agglomeration were studied in an effort to understand the emission process as a function of photon energy and power [322].

On the Si(100) surface Ag crystals were observed to grow at about 250 and 300 °C in a bamboo-like manner along the ⟨011⟩ directions, with non-uniform

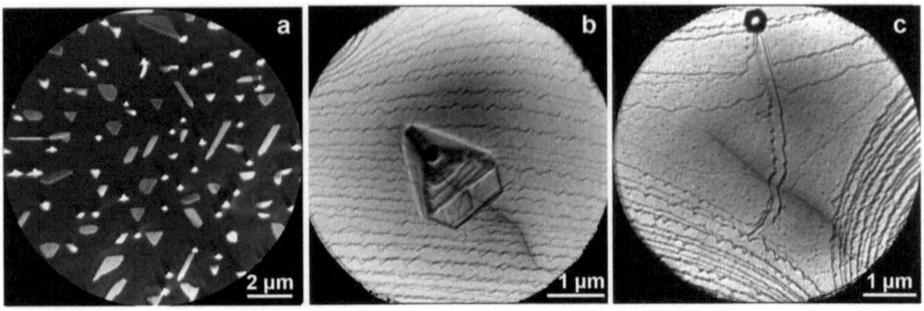

Fig. 5.37 Growth of Cu silicide on Si(111). (**a**) PEEM image showing *rods* and *triangles* which are flat (*no shadows*) and smaller three-dimensional particles (with *shadows*). (**b**) LEEM image of a triangular particle and (**c**) three-dimensional particle after reactive migration at high temperature. Adapted from Ref. [273] with permission from the American Institute of Physics, © 1989

width and height. This excludes a stress-induced mechanism as cause for the elongated growth. Instead a high density of nuclei along certain $\langle 011 \rangle$ directions followed by connection between the crystals during growth was proposed [323]. In PEEM studies at higher temperatures, around 600 °C, wires with uniform cross section have been reported, together with compact crystals. The ratio of wires to compact crystals increased with increasing deposition time and strongly with increasing temperature and miscut from 0° to 4° [324]. Specific studies have been made for a 4° miscut surface [325] and an Au-pre-facetted 4° miscut surface [326], in the last one also with LEEM. The wire growth has been attributed mainly to the anisotropic diffusion on the stepped surfaces.

The growth of Cu on Si(111) was already studied in the early years of LEEM but little has been published about it [225, 230, 273]. Irrespective whether the surface has initially the (7 × 7) structure or the hydrogen-terminated δ-(7 × 7) structure, the crystals grown on this surface were Cu silicides in various shapes. This is illustrated in the PEEM image of Fig. 5.37a [273] from a film grown at about 850 K. Figure 5.37b shows one of the triangular islands seen in PEEM, illustrating how much more information is contained in a LEEM image. Heating slightly above this temperature causes melting of the small polyhedral crystals, cooling below 850 K crystallization again. At high temperature the smaller particles migrate over the surface in the step-up direction leaving pronounced reaction tracks behind (Fig. 5.37c), similar to the Au silicide particles shown earlier. In the later studies of Cu growth on the hydrogen-terminated surface qualitatively the same distribution of particle shapes and sizes was found. The surface planes of the most frequent shape were found from the facet spot movements to be high index planes as expected for growth shapes [274, 275, 327].

Even less has been published on the growth of Co on Si(111) than for Cu. Some results can be found in [71, 225, 328]. Like Cu, Co forms a silicide, $CoSi_2$ that is very stable and has a higher surface energy than Si. Unlike Cu, Co does not form a

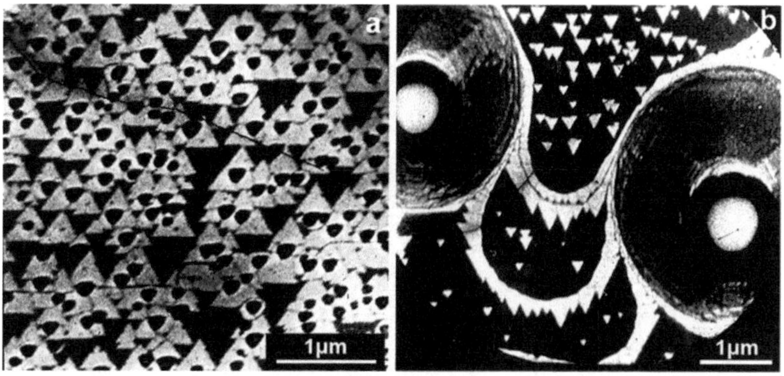

Fig. 5.38 Co on /Si(111) (**a**) $CoSi_2$ crystals (*small dark triangles*) on a Si(111) surface surrounded by regions with (7 × 7) (*bright triangles*) and (1 × 1) (*dark areas in between*) structure. Energy 10 eV. (**b**) Hillock formation on Si(111) due to Si sublimation suppression by $CoSi_2$ islands. Energy 10 eV. (**a**) Reproduced from Ref. [328] with permission from Springer Science + Business Media. (**b**) Reproduced from Ref. [67] with permission from the American Institute of Physics, © 1991

wetting layer as soon as enough Si has been deposited for nucleation of three-dimensional crystals. These crystals are surrounded by regions with (7 × 7) and (1 × 1) structure (Fig. 5.38a) [328]. Depending upon deposition and annealing temperature a wide variety of growth shapes were observed. When heated to temperatures at which Si sublimates, large $CoSi_2$-topped hillocks form on the Si surface with (1 × 1) and (7 × 7) because of the lower vapor pressure of $CoSi_2$ (Fig. 5.38b) [67]. A later study found that the two-dimensional structures at low Co coverages [277, 278, 280] indeed disappear once silicide islands have nucleated, even at very low coverage, apparently having been absorbed into the islands. The influence of SiO_2 layer on Si(100) on the formation of $CoSi_2$ [329] and $FeSi_2$ [330] has also been studied with LEEM and XPS in a SPELEEM, respectively.

The growth of Pb on Si(111) was subject of a study [331] aimed at the understanding and control of the three basic growth modes of thin films (Volmer-Weber, Stranski-Krastanov and Frank-van der Merwe described in Refs. [332–334]). On the Pb wetting layer on the Si(7 × 7) surface three-dimensional crystals form on the terraces and preferentially at the steps. Despite significant lateral growth and coalescence the film is still not continuous at 6 monolayers because of the simultaneous strong growth normal to the surface (Fig. 5.39a) [331]. Normal growth was reduced strongly by growing not directly on the (7 × 7) structure but on an Au wetting layer, which acts as an interfactant, encouraging lateral growth. This produced a continuous Pb film already at one monolayer, which continued to grow quasi-monolayer-by-monolayer via step flow and two-dimensional island formation up to the largest thickness of 5 ML studied (Fig. 5.39b [331]). Thus the interfactant converted the growth mode from Stranski-Krastanov to Frank-van der Merwe. A similar but less effective action could be achieved by growing on a Ag($\sqrt{3} \times \sqrt{3}$)R30° wetting layer. The effect of

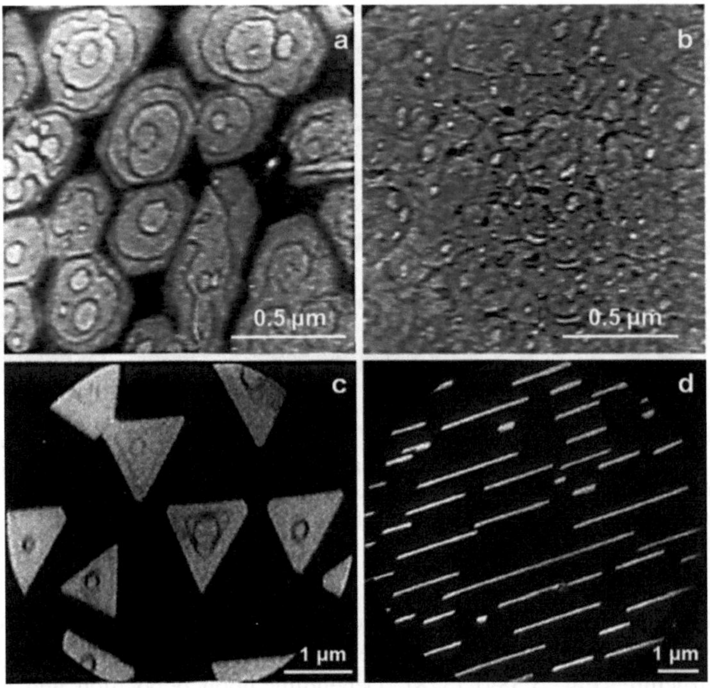

Fig. 5.39 LEEM images of the growth of Pb on several Si surfaces at room temperature. (**a** and **b**) 5 monolayers on the S(111)-(7 × 7) and on the Si(111)-(6 × 6)Au surface, respectively, grown at room temperature. Electron energy 8 eV. (**c**) Pb islands after deposition of 12 Pb monolayers on the Si(100)-c(4 × 4)Pb surface, grown at room temperature. Electron energy 5 eV. (**d**) 6.6 monolayers on the Si(755) surface with 0.2 pre-adsorbed Au monolayers. 276 K. Electron energy 7.1 eV. (**a**, **b**) Adapted with permission from Ref. [331]. Copyright 2000 by the American Physical Society. (**c**) Reproduced with permission from Ref. [336]. Copyright 1994 by the American Physical Society. (**d**) Reproduced with permission from Ref. [338]. Copyright 2001 Elsevier

the interfactant was attributed to misfit stress relief, increase of the nucleation rate and decrease of surface diffusion [331, 335].

On the Si(100) surface Pb grows also in the Stranski-Krastanov mode, with the first crystals forming at room temperature already before completion of the wetting layer. At a deposition rate of 1 monolayer/min the density of the particles is about $0.2/\mu m^2$ so that they have a large supply region from their surroundings. This and the much faster diffusion on the wetting layer than on top of the Pb crystals leads to the formation of rims on the crystal as shown in Fig. 5.39c [336]. The resulting "ashtrays" fill up in the later stages of the growth when the crystals become so large that the direct supply of atoms from the vapor phase dominates that from the wetting layer [336]. On 4° miscut (100), (110), and high index surface such as (533) or (755) Pb also grows in the Stranski-Krastanov mode but with a completely different growth shape, mesoscopic wires [337, 338]. Length to width ratios of up to

130 with temperature-dependent widths were obtained but in general there is a wide distribution of lengths as illustrated in Fig. 5.39d [338] for a film on a (755) surface after deposition of nearly 7 monolayers at 276 K. The wires are not flat but have (111) and (100) side faces and tips. Anisotropic stress and one-dimensional diffusion were invoked as driving force for the formation of the wires.

The growth of In on Si(111) is the most complex one of all metals, which have been studied on this surface. The multitude of thickness- and temperature-dependent two-dimensional phases has already been discussed. Three-dimensional crystals grow on the double layer predominantly with (100) orientation, which can be explained by the very small misfit with the centered square unit mesh of the ($\sqrt{7} \times \sqrt{3}$) structure of the double layer. The (100) surface of the crystals is reconstructed. Deconstruction and melting occur at the same temperature as in the bulk [284, 285].

Several metals including Co and Ni are known to form silicide nanowires on Si surfaces in a certain temperature range, at least on those which a $\langle 110 \rangle$ direction in the surface [339]. A LEEM study of the growth and stability of $TiSi_2$ nanowires [340] showed that they form on Si(111) at about 850 °C and become unstable above 900 °C. In FEL-PEEM studies [341] their growth on the same surface was studied as a function of deposition time at 1150 °C. The wires were found to develop from isometric particles at a critical size, after which the width remained constant while the length increased linearly with deposition time as expected for stress-induced wire formation [316]. Short depositions produced only isometric crystals [342]. On Si(100) $ErSi_2$ nanowires developed when a 10–20 Å thick Er film deposited at 750 °C was annealed at 800 °C, with linearly increasing length and constant width as a function of annealing time, suggesting the same stress effect as in $TiSi_2$. Annealing at 1050 °C caused only Ostwald ripening of the smaller particles [343]. Thus the formation of silicide nanowires is limited to a narrow temperature range.

When the temperature during deposition of silicide-forming metals is above or close to the bulk eutectic temperature then the silicide particles migrate across the surface. This has been seen in LEEM first with Au on Si(111) [267] and with Cu on Si(111) [273] but not studied in detail. However, it was noticed that the particles moved uphill perpendicular to the steps and the driving force was assumed to be dissolution of Si on the uphill side of the particle and expulsion on its downhill side. Detailed studies of the formation and migration of liquid silicide droplets were later performed with FEL-PEEM [344] and LEEM [345, 346], using Pt on Si (100). In these experiments a temperature gradient was assumed as driving force, with Si dissolution on the high temperature side and expulsion on the low temperature side of the particle. In the FEL-PEEM studies the migration rate increased independent of particle size (2–7 μm diameter) from 0.8 μm/s at 1085 ° C to 3.6 μm/s at 1210 °C, yielding an activation energy of 0.59 eV for migration. The later LEEM studies found that only larger particles moved normal to the step direction. Smaller particles are guided by the dragging force of the steps along the step direction. In all cases migration occurs from lower towards higher temperature regions on the substrate.

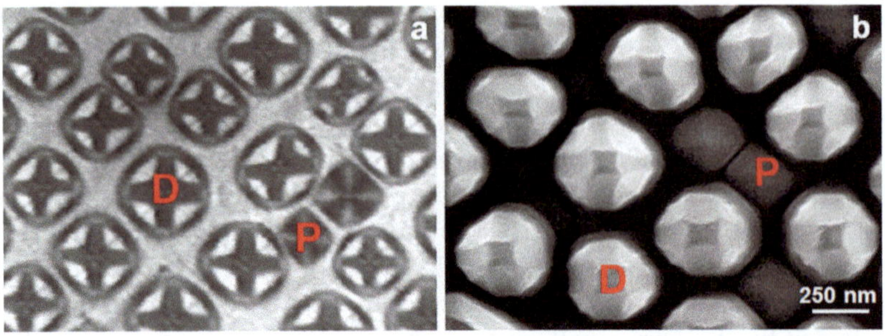

Fig. 5.40 (**a**) LEEM image, taken with 4 eV electrons, and (**b**) SEM image of Ge_xSi_{1-x} islands grown on Si(100) by exposure to a silane–germane mixture at 650–700 °C, showing the transition from *pyramidal* shapes (P) to *domes* shapes (D). From Ref. [350] with permission by the American Association for the Advancement of Science

5.3.1.2 Ge on Si

The importance of Ge-Si nanostructures in electronics and optoelectronics [347] has made the growth and structure of Ge on Si and of Si on GeSi [348] an interesting subject for LEEM and XPEEM. The technologically important (100) surface has been studied by Tromp's group [349–353] and Lagally's group [354–359], in part jointly [360]. The first study [349] compared Ge growth on the clean and As-terminated surface at 630 °C and found that on the clean surface three-dimensional islands developed after three monolayers while As induced at every level a high nucleation rate leading to quasi-monolayer-by-monolayer growth (surfactant-mediated growth). The high nucleation rate was attributed to exchange between As and Ge dimers during growth. Intermixing of Ge and Si occurs already during the formation of the first monolayer via Ge partially displacing Si on the terraces and both Ge and displaced Si diffusing to the steps [361]. A later study [353] provided detailed insight into the further intermixing. At higher temperatures (>750 °C) the GeSi film roughens already at 2 monolayers or less without involving the substrate with a thickness-dependent roughening temperature in between that of Si(100) and Ge(100) [360].

The study of the growth of technologically important $Ge_x\ Si_{1-x}$ layers brought some interesting results. Growth by exposing the Si(100) surface at 650–700 °C to Si_2H_6 and Ge_2H_6 at growth rates of several monolayers/min leads after a few monolayers to roughness, which develops into mounds with long range order with a Ge concentration-dependent period. While the period is independent of thickness, the slopes of the mounds increase reaching a final angle of 11° and azimuthal orientation, resulting in pyramids bounded by {105} faces [351, 352, 357, 358]. With further increasing thickness the pyramids coarsen, decrease in density, and transform into circular domes bounded predominantly by {311} and {15,3,23} surfaces [350, 352]. Figure 5.40 [350] shows this final stage in the LEEM image (a) and from a similar film in the SEM image (b) for comparison. All these processes, the roughening, mound formation, and transition to the domes are strain-induced and occur without nucleation. When these layers are imbedded in Si by Si_2H_6 exposure at 650 °C the islands underwent considerably shape changes ending up with large (100) top faces [355, 356].

The initial growth of Ge on Si(111) was studied already in 1993 on the Sb (√3 × √3)R30° surface during simultaneous Ge and Sb exposure and for comparison on the clean (7 × 7) surface, in both cases at 775 K [318]. On the (7 × 7) surface three-dimensional crystals formed after the completion of four two-dimensional layers, the first two with (7 × 7), the second two with (5 × 5) structure. On the Sb-covered surface growth proceeded by periodic two-dimensional island formation up to the highest coverage studied, demonstrating the surfactant action of Sb. No information on alloying of Ge with Si could be given at this early state of the art. Such information was obtained nearly 20 years later by combining LEEM with XPS in a SPELEEM instrument. Here only the first monolayer was studied at lower temperatures than in the original work. A detailed analysis of the Ge 3d/Si 2p intensity ratio showed that at 200 °C very little if any Ge-Si exchange occurred, while at 270 °C the data indicated considerable intermixing which decreased with coverage. Combining this with the evolution of the island sizes led to the conclusion that steps play the crucial role in intermixing [362], confirming earlier interpretations based on LEEM observations only. Another XPS-XPEEM [363] study looked at a phenomenon at the other extreme of Ge coverage: the dependence of the shape of islands upon their size up to lateral dimensions of tens of microns and thickness up to 500 nm. With increasing size a transition from compact to dendritic-looking islands occurred, which was attributed to strain relief at the island perimeters. Ge 2p and Si 1s XPEEM images clearly indicated a composition gradient from island center to edge but no quantitative information could be extracted [363]. Some information on the composition was obtained in a series of SPELEEM studies of crystals of various shapes grown at lower temperatures (460–560 °C) [364–366]. A LEEM study that looked at the distance distribution between three-dimensional Ge crystals found a substantial tendency to self-ordering without a significant influence of the size of the crystals on their distance [366].

Finally some studies of the influence of adsorbed metal layers on the growth of Ge on Si(111) should be mentioned. Growth on a surface incompletely covered with the Ga(√3 × √3)R30° structure produced preferred nucleation of Ge on the Ga-covered regions [367]. The situation is different when the surface is covered by a Ag(√3 × √3)R30° structure. On this surface large 3 monolayer thick islands grow after some intermediate stages, spread across the surface and coalesce, forming large flat regions, in contrast to the compact pyramidal crystals on the clean surface. Thus, Ag acts like a surfactant, though in a manner completely different from those mentioned earlier. In the case of As and Sb the surfactant reduces surface diffusion, which produced a large island density, while Ag enhances the surface diffusion, which drastically reduces the island density [368]. For a review of the processes involved in Ge growth on Si(111) see Ref. [369].

The growth of Ge on clean and metal adsorbate-covered (11h) surfaces ($h = 2, 3$) has also been studied with LEEM and LEED [370–372]. The (113) surface is the third most stable clean Si surface and is (3 × 2) or (3 × 1) reconstructed, depending

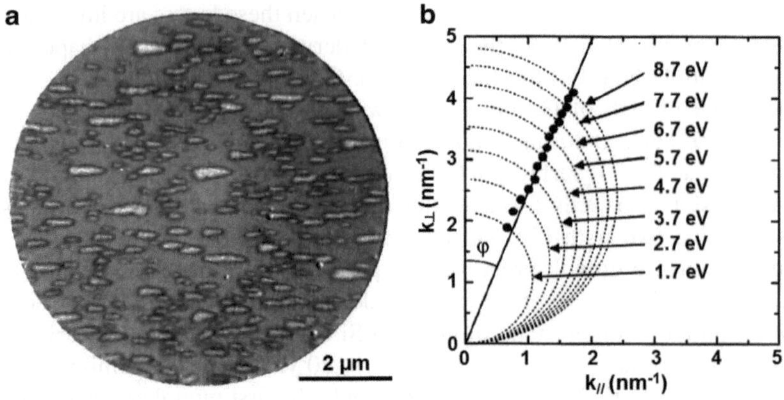

Fig. 5.41 (**a**) LEEM image after deposition of 36 Ge monolayers on Si(113) at about 560 °C, showing 3D islands. Electron energy 1.7 eV. (**b**) Reciprocal lattice section in an azimuth in which one of the facet spots move with energy. The *dotted circles* are the Ewald spheres for the indicated energies. φ is the tilt angle of the facet. Adapted from Ref. [370] with permission from the American Institute of Physics, © 2002

upon temperature while the (112) surface breaks up into (111) and (337) facets. On the clean (113) surface Ge was found to form initially an about 4.4 monolayer thick wetting layer followed by the formation of three-dimensional clusters. These develop with increasing temperature and thickness into larger crystals, which are strongly elongated along the $[33\overline{2}]$ direction and bounded by many facets (Fig. 5.41) [370]. On the clean (112) surface growth also proceeds via a wetting layer and three-dimensional islands. Their number density decreases and size increases with increasing thickness, from which—like in the growth on the (113) surface—activation energies for surface processes were derived [305].

Pre-adsorption of metals has a strong influence on the growth, at least when the adsorption layer is saturated. On the Ga-saturated (113) surface, which consists of (112) and (115) facets, Ge grows with a much higher nucleation density than on the clean, non-faceted surface with a particle distribution, which is closely correlated with the facet structure and periodicity [367, 372]. On the non-faceted Ga-saturated (112) surface with (6×1) reconstruction Ge grows in rod-shaped crystals with the long axis parallel to the $[1\overline{1}0]$ direction instead of the isometric crystals seen on the clean faceted surface [305]. Pre-adsorption of In has a similar influence on the growth of Ge on the Si(112) surface, which after In adsorption is also flat and is (3.5×1) reconstructed. After completion of the wetting layer a high density of crystals grow, initially extended along the $[1\overline{1}0]$ direction but transforming with increasing thickness and temperature into triangles with their apex in the $[11\overline{1}]$ direction. They are bounded by (112) top faces and (111) and {130} side faces [306]. Finally, on the Ag-pre-adsorbed, faceted (113) surface growth proceeds in a similar manner with Ge crystals elongated along the facet direction $[1\overline{1}0]$ [302].

Summarizing this section, Ge on Si, having been studied originally intensely because of its importance in semiconductor electronics, has developed in an interesting playground for the study of the influence of surface orientation, reconstruction, faceting, and adsorbates on epitaxy. Many detailed results of the structure and kinetics have been obtained which can be found in the original literature. For comparison with the growth on Si a brief LEEM study of the growth on GaAs(100) should be mentioned [373]. On this surface Ge grows at 420 °C layer-by-layer from small two-dimensional islands, which become anisotropic with increasing size and increasing temperature (450–480 °C). At higher temperatures growth proceeds via step flow until at 540 °C growth becomes unstable due to As desorption and surface roughening.

5.3.1.3 Other Films on Semiconductors

The growth of CaF_2 on Si(111) was studied by Tromp et al. from the sub-monolayer range [374] to thick films [375] in two temperature ranges. Below about 700 °C they found an interesting de-wetting transition: initially monolayer CaF_2 islands formed but once second layer islands nucleated on top of the first layer, the first layer contracted to form bilayer islands. This was interpreted as follows: below this temperature CaF_2 does not dissociate, diffuses over large distances on the (7 × 7) structure and CaF_2 bonding to the substrate is weaker than the bonding between the first and second layer. Above 700 °C CaF_2 dissociates and forms a double layer with Ca bonded to Si and F on top of Ca, with one F atom being desorbed, possibly as SiF_x molecule during the restructuring from the (7 × 7) to the (1 × 1) structure in the adsorption process. On top of this wetting layer initially isolated two-dimensional CaF_2 islands grow without crossing the steps but the third layer crosses the steps and leads to step flow growth up to a critical thickness of about 30 Å, above which the misfit strain is released by formation of a dislocation network. Using different energies and focus the authors were able to correlate the interfacial dislocations with the surface structure. In combination with transmission microscopy they could determine the evolution of the dislocation structure in considerable detail. It should be noted that electron energies below 10 eV were used in these studies in order to avoid electron stimulated desorption of F, which starts at about 25 eV due to ionization of the Ca 2p level followed by interatomic Auger transitions.

Another interesting film growth process is that of silicon nitride on Si(111). It had been studied extensively with lateral averaging methods with contradicting results before laterally resolving studies gave some insight in the growth process. In the first such study the nitride layer was grown by reaction with ammonia in the low 10^{-7} mbar range [376], in the second one [377] nitrogen from a radio frequency (RF) gas discharge was used for nitridation in the low 10^{-5} mbar range. In both cases LEEM and LEED were used, in the first case in addition also STM. Saturation of the surface with NH_3 followed by annealing to 1210 K led to triangular islands with (8 × 8) structure which decorated the steps and domain walls of the original

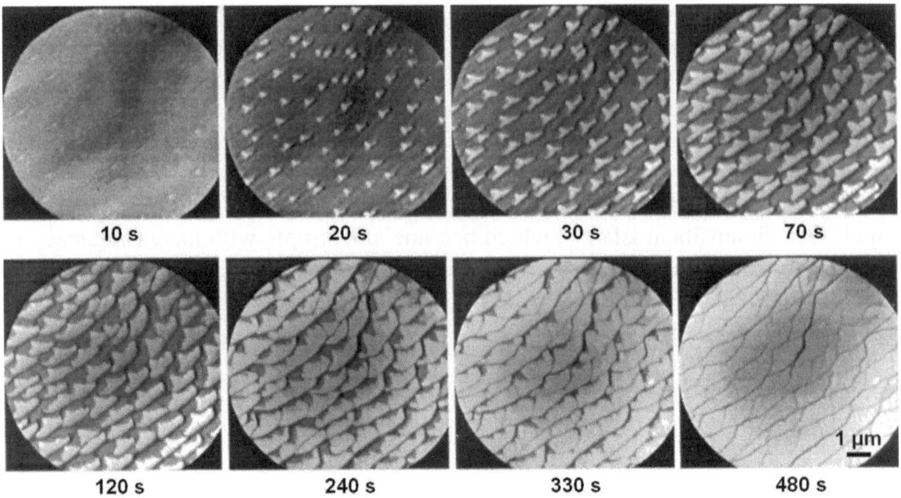

Fig. 5.42 Frames from a LEEM video taken during the growth of the Si-N (8 × 8) structure on a 0.5° miscut Si(111) surface exposed to 1×10^{-7} mbar ammonia at 1270 K, showing widening of terraces and step bunching as growth proceeds. Exposure times (**a–h**) from 10 to 480 s. Electron energy 3.6 eV. Adapted from Ref. [376] with permission. Copyright 1995 by the American Physical Society

(7 × 7) structure, with a total nitride coverage of 0.25 monolayers. During reaction at 1180 K, i.e., above the (7 × 7) ↔ (1 × 1) transition, a high density of concave triangular (8 × 8) islands forms, on which second layer islands form before the first layer covers the surface completely. At 1260 K the nucleation and growth of the (8 × 8) nitride islands is indistinguishable from that of the (7 × 7) Si islands during the (1 × 1) ↔ (7 × 7) transition. On slightly miscut surfaces the (8 × 8) islands are elongated along the steps and with increasing coverage pronounced step bunching occurs as seen in Fig. 5.42 [376]. This is much more pronounced on surfaces with larger miscut as seen in STM.

Once the (8 × 8) wetting layer is completed no three-dimensional Si nitride crystals form above 1200 K at the NH_3 pressures used. They do nucleate and grow occasionally, however, before completion of the wetting layer at surface defects. On a heavily contaminated surface, which evolves after many growth and desorption cycles, they grow with a high density, in particular in the electron-irradiated region. This is attributed to elution of adsorbed gases from the walls by the NH_3 and dissociation of these gases on the Si surface, enhanced by electron irradiation. At the higher pressures during reaction with nitrogen from the gas discharge this phenomenon is even more pronounced. The LEED patterns of the three-dimensional crystals are compatible with the (111)-oriented high-pressure phase of Si_3N_4 (γ-Si_3N_4). If contracted by 1 % it would fit perfectly in a 2:1 ratio in parallel orientation to the substrate but it is expanded 4.7 % and rotated several degrees relative to the substrate. In thicker films these crystals are also observed in exact parallel orientation with a ($\sqrt{3} \times \sqrt{3}$) superstructure. There are still many details of the high temperature growth of Si nitride, which are not understood yet despite of its technological importance.

Another practically important film on semiconductor surfaces is GaN on SiC (0001) or on thick GaN layers grown on various substrates. Similar to the Si nitride growth studies GaN growth is another example of in situ studies at high pressures of reactive gases and their limitations due to gas-induced contamination. The homoepitaxial growth of GaN has been studied with LEEM and LEED on GaN (0001) [153, 155, 378] and on SiC(0001) [155, 379–381] using more complex sources. Ga was evaporated from a BN crucible and nitrogen was supplied either from a RF plasma source as in the growth of Si nitride or from a supersonic He jet source seeded with up to 10 % NH_3. The motivation for the second choice was the belief that hyperthermal NH_3 would be more reactive and allow growth at lower temperature. Operation pressures were around 1×10^{-5} mbar with the RF discharge source and in the middle 10^{-8} mbar range with the jet source. During homoepitaxial growth on a GaN(0001) layer with ($\sqrt{3} \times \sqrt{3}$) superstructure at around 670 °C with a Ga/NH_3 ratio above 2, a two-dimensional layer formed from a high density of two-dimensional islands, completely restoring the original surface configuration. At lower Ga excess the surface was etched and no growth occurred [153, 378]. A more detailed study showed that the Ga double layer mentioned earlier, which is liquid around 670 °C [307] is necessary for continuous growth and has to be maintained during growth by proper adjusting the Ga/N ratio. This type of growth may be considered as quasi-liquid phase epitaxy. If the Ga/N ratio is too high Ga droplets form, if it is too low the surface becomes rough [155, 307].

Similar considerations apply also to the growth of GaN on SiC(0001) [155, 379–381]. The most extensive information is contained in the last study [155], which combined LEEM and LEED with cross-sectional high resolution TEM, and in the preceding one [381], which also corrects misinterpretations in [380]. The essential results are as follows. GaN grows on SiC(0001) both with activated nitrogen from the RF plasma discharge and the NH_3-seeded He jet source three-dimensionally, independent of Ga/N ratio. Ga excess produces flat epitaxial crystals with (0001) top faces and $\{10\bar{1}1\}$ and $\{10\bar{1}0\}$ side faces while with N excess the $\{10\bar{1}1\}$ faces are dominating and the (0001) face is absent, leading to a rough surface. The polar (0001) surface is not an equilibrium plane and appears only because it is stabilized by Ga adsorption. Substrate defects and N excess can cause growth of the cubic phase of GaN and stacking faults. Activated N is better suited for GaN film growth than NH_3, at least below 700 °C.

PEEM with the spontaneous radiation from a FEL has also been used in the study of GaN films, in this case, however, not for understanding of the growth but of inversion domain boundaries between Ga and N terminated surfaces [382]. The boundary was obtained by growing part of the film directly on sapphire, which gave N-face termination, the other part on an AlN nucleation layer, which resulted in Ga-face termination. Photoemission from the N-face started at 4.9 eV with emission from the conduction band, from the Ga-face at 6.3 eV from the valence band. The about 2 μm wide inversion domain boundary region showed strong emission above 4.9 eV from the space charge region between the two terminations, predominantly due to emission of electrons in an accumulation layer in the conduction band. Interestingly, the AFM and PEEM images suggest that the Ga-terminated surface is rougher than the N-terminated surface, in contrast to the LEEM, LEED, and TEM studies mentioned above.

5.3.1.4 Nanostructures and Droplets on Semiconductors

Nanostructures, when truly deserving of "nano" are at the limits of microscopes without aberration correction. Therefore mostly structures ranging from 50 nm to several 100 nm have been studied. A few structures, in general prepared ex situ, will be discussed here: dots, rings, and wires. An example for dots are the so-called InAs quantum dots on GaAs(100). They were grown either randomly epitaxial in the Stranski-Krastanov growth mode on a Se-terminated surface at 200 °C [383], on a Si doped surface at 540 °C [384, 385] or on a prepatterned GaAs surface [386]. After growth they were capped with As for transfer to the SPELEEM, decapped in it and studied with LEEM, XPEEM, μXPS, and LEED with the goal to determine the degree of intermixing between InAs and GaAs. This is a difficult experiment for two reasons: (1) the dots are surrounded by the wetting layer which contains both Ga and In and (2) the spherical aberration of the objective lens deflects electrons from the surrounding of the dots into the dots and vice versa. This makes quantitative chemical analysis on the nanoscale difficult. Nevertheless useful information can be extracted with this limitation in mind. The situation becomes more complicated in the analysis of quantum rings, which evolve when InAs quantum dots are overgrown with a thin GaAs layer and subsequently annealed properly [387, 388]. The transition from dot to ring was studied by XPEEM, LEEM, AFM, and low-temperature photoluminescence spectroscopy as a function of annealing time. The combination of AFM and photoluminescence spectroscopy showed a clear correlation with the shape of the dots and the rings which could be explained by the distribution of In and Ga as deduced from XPEEM [389]. The field of quantum dots and rings is reviewed in Ref. [390].

A quite different system, which demonstrates very nicely the power of XPEEM with core level photoelectrons is the formation of Ge nanoparticles on an array of Au discs on a hydrogen-terminated Si(100) surface after it had been exposed to air (Fig. 5.43) [391]. The air exposure caused Au-catalyzed local oxidation. Annealing this surface in UHV at 600 °C agglomerates the Au in the oxidized regions and desorbs the hydrogen between them, producing clean regions. When Ge is deposited on this surface at 450 °C it forms three-dimensional crystals only in the clean regions and Au diffuses from the oxidized regions onto the Ge, where it produces a much stronger Au signal than the large Au particles in the oxidized regions. Thus by oxidizing the surface Au not only produces the regions in which no Ge condenses but also has a major influence on the formation of the Ge crystals to which it migrates during their growth.

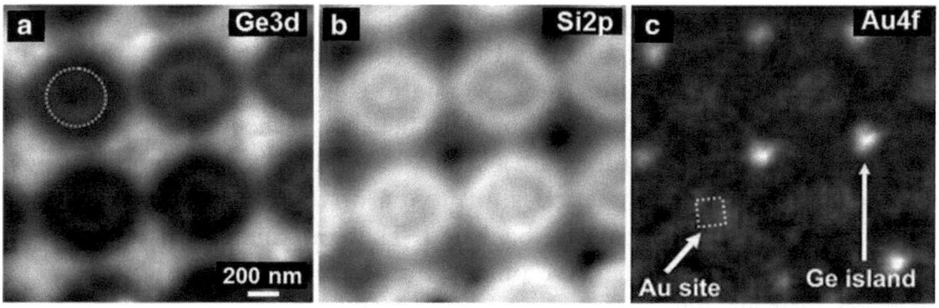

Fig. 5.43 Core level XPEEM images of Au-SiO_2-induced Ge nanopyramid arrays. The regular pattern of Au islands, one of which is indicated by the *circle* in (**a**), was deposited on hydrogen-terminated Si(100) and locally oxidized during transfer through air (**b**). Ge grew subsequently on this surface in UHV at 400–600 °C only in the clean areas between the SiO_2 regions while most of the Au diffused into the Ge regions, catalyzing the growth of Ge into pyramids (**c**). Photon energy 195 eV. Adapted with permission from Ref. [391]. Copyright 2007 American Chemical Society

The third group of semiconductor nanostructures, nanowires, cannot be studied directly with LEEM or XPEEM when grown perpendicular to the substrate. These nanostructures have to be removed from their substrate and remounted flat on another substrate for microscopic study. An example is the secondary electron PEEM, core photoelectron PEEM, and MEM study of InP nanowires, whose ends had been heavily n-doped. The combination of these methods gave detailed information on doping, surface chemistry, and morphology of wires [392]. Another example is the study of InAs nanowires, which had been grown through a SiO_x mask on InAs(111) and transferred onto a Si(111) substrate for imaging with MEM. A careful analysis of through focus series images allowed a precise determination of the diameter and shape of the wires [393]. LEEM and XPEEM have also contributed directly to the understanding of the formation of these nanowires by studying the formation of the holes in the SiO_x mask, from which they grow, as a function of temperature. By comparing these results with the number and distribution of long nanowires grown in a separate system, the growth was established to occur via a vapor–liquid–solid mechanism starting from liquid In droplets formed in the holes. The oxide layer immobilizes the droplets on the surface and restricts their size [394].

Studies of other nanostructures, oxide patches on GaAs [395, 396] and GaAs/AlAs/GaAs [397–399] heterostructures on GaAs(100), ranging from several 100 nm to a few micrometer, showed one of the limitations of XPEEM and μXPS, photon stimulated desorption. The nanostructures were produced by anodic oxidation in humid atmosphere by AFM with a voltage of 10 to several 10 V between tip and sample. A detailed μXPS analysis with Ga 3d, As 3d, and Al 2p photoelectrons combined with AFM height measurements showed a complex decomposition process of the nanostructures as a function of irradiation time and thickness, resulting in Al enrichment of the near surface region.

The formation, growth, shape, migration, coalescence, and evaporation of Ga droplets on GaAs and GaP surfaces has been studied in much more detail than the corresponding processes of liquid metal-Si eutectic droplets discussed in Sect. 5.3.1.1, using MEM, LEEM and UV-PEEM [151, 152, 308, 400–405]. The droplets form during Langmuir (free) evaporation above the congruent evaporation temperature 625 °C of GaAs (638 °C of GaP) because of the higher evaporation rate of As (and P). They grow (decrease) in size at higher (lower) temperatures and move across the surface in a stick–slip motion driven by the chemical potential difference between their front and backside. The combination of experiment and theory resulted in a comprehensive picture of the thermodynamics and kinetics of droplets on surfaces, which is summarized in [405].

Metal droplets are used in the droplet epitaxy of nanowires. Au droplets have been studied to understand the growth of GaAs nanowires [406] and Ge nanowires [407]. In the first brief study LEEM and LEED confirmed that a Au wetting layer was necessary to stabilize the Au droplets for nanowire growth. The second very carefully executed study combined several complementary methods (STM, SEM, RHEED, TEM, and energy-dispersive X-ray spectroscopy) with XAS-PEEM, core level PEEM, and XPD in order to obtain deeper insight in the nanowire formation process. Of the many results obtained in this manner only a few salient features will be mentioned here. The Au nanodroplets were produced by annealing a 0.8–1.2 nm thick film deposited at room temperature on a well-cleaned Ge(111) surface at or somewhat above the eutectic temperature (350–400 °C), which resulted in alloy particles with sizes ranging from 5 to 200 nm, surrounded by a Au wetting layer. The Au 4f XPS spectrum from the particles differs little from that of pure Au while that from the wetting layer is shifted nearly 0.5 eV due to the strong bonding to the substrate. AFM and cross-sectional TEM images show that the particles are sitting on pedestals. The corresponding Ge and Au EDX data show that with increasing annealing temperature and increasing time an increasing amount of Ge is incorporated into the particles via material transport across the surface. Thus, vertical Ge nanowire growth is achieved even without an external Ge supply.

Nanowire/nanocolumn formation was studied for InN, starting from In droplets produced by annealing a 2.3 monolayer thick In film on Si(111) at 480 °C [408]. AlN nanodots, formed by ex situ nitridation of 30 nm thick Al islands with 90 nm × 130 nm lateral dimensions were subject of another study. The islands were produced by electron beam lithography and reactive ion etching on SiO_2/Si(111) [409]. In both approaches the metals were nitrided in about 7×10^{-6} mbar NH_3, in the first case at about 480 °C, in the second case by heating in steps from 680 to 800 °C. In the first study the combination of MEM, Si 2p XPEEM, In, Si and N XPS and AFM lead to the conclusion that not only the In droplets were nitrided, leading to InN nanowire formation, but also the In-covered Si surface, which caused In aggregation into droplets and partial In desorption. The second study was done with MEM, LEEM, XPEEM in the absorption mode and Al and Si in the core level mode as well as with Al, Si, and N XPS and found that nitridation did not change the structural order of the pattern but decreased the size of the individual nanopattern and made it more corrugated. Also the surrounding SiO_2 layer was partially nitrided.

Summarizing this subsection, many of the studies reported in it are at the resolution limit of LEEM and XPEEM so that these methods served more or less to provide useful complementary information to that obtained with other methods such as STM, AFM, SEM, and TEM, in part with considerable, in part with limited success. Aberration-corrected instruments should make this kind of studies more successful.

5.3.2 Films on Metals

5.3.2.1 Metal Films

This subsection discusses the growth of metals on metals beyond the initial adsorption state, which was the subject of Sect. 5.2.1.3. In all film/substrate pairs studied the surface energy of the film material was higher than that of the substrate so that films grew in the Stranski-Krastanov mode (Refs. [332–334]) by forming first a two-dimensional adsorption or wetting layer before the growth of three-dimensional crystals. Under suitable growth conditions (temperature and deposition rate) a metastable wetting layer can be grown beyond the stable adsorption layer in the quasi-monolayer-by-monolayer (Frank-van der Merwe) growth mode before the layers in excess of the wetting layer coalesce into three-dimensional crystals. Many of the films had been studied previously with lateral averaging methods but the imaging methods discussed here have given a much deeper insight into their growth and microstructure. Most of the work has been done on close-packed surfaces of metals with high surface energy and high melting point, the bcc(110) surfaces of W and Mo, the hcp(0001) surface of Ru and the fcc(111) surface of Ni, but growth on bcc(100) and (111) and fcc(100) surfaces has also been studied. The film materials studied include Cu, Ag, Pd, Cr, Mg, Pb, and Sb as well as Fe, Ni, and Co. The latter will be discussed in Chap. 7 in connection with their ferromagnetic properties.

The growth of Cu on Mo(110) was the first system studied [230, 410] except for a cursory study of Ag on Mo(110) in the first LEEM paper [4]. In this study it had already been noticed that at high temperatures flat three-dimensional crystals, once nucleated on a terrace, grew rapidly along it due to supply of atoms diffusing rapidly on the wetting layer. Mundschau et al. [230, 410] found that crystals on different terraces developed different thickness due to step-limited growth as manifested by different reflectivities, caused by quantum size effects (QSE). The QSE effects in the growth of Cu on W(110) were studied in detail by Altman et al. [411] and analyzed in terms of a Koster-Kronig band structure model. The combined LEEM-UVPEEM studies of the same system by Koshikawa's group [228, 412] found that also the PEEM intensity was layer-dependent and not simply determined by the work function but strongly influenced by the electronic structure of the film and in the wetting layer also of the substrate.

Ag grows on W(110) above two monolayers quasi-monolayer-by-monolayer with a distorted (111) plane parallel to the substrate. Chung et al. [413] have

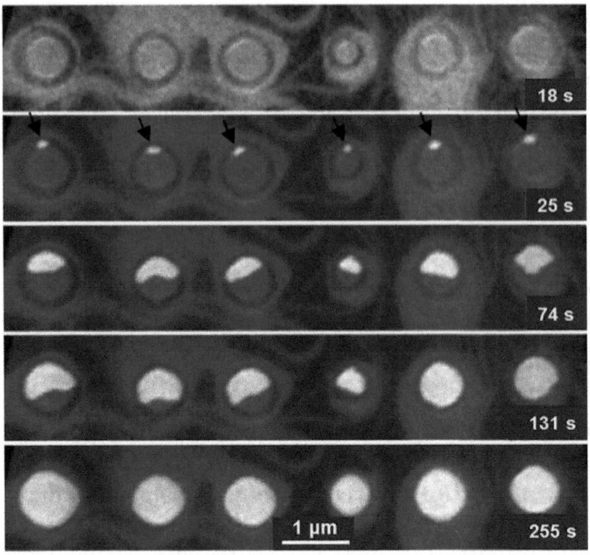

Fig. 5.44 LEEM images of the growth of Ag on W(110) with pits at 290 °C. Nucleation of three-dimensional crystals at the step bunch bordering the pits occurs shortly after completion of the wetting layer. Thereafter the crystals grow rapidly across the pits due to rapid diffusion of Ag atoms on the wetting layer surrounding the pits and fill the pits. Adapted with permission from Ref. [416]. Copyright 2006 American Chemical Society

analyzed the resulting QSE effects by a dynamical theory analysis and—combined with first principle calculations—determined interlayer and film-substrate spacings. From the same experimental data Altman [414] determined the electronic band structure above the vacuum level using the phase accumulation model [415]. A quite different study of the growth of Ag on W(110) made use of the high diffusion rate of atoms on the wetting layer, the nucleation on steps and the tendency of growth over downhill steps. McCarty [416] deposited Ag on a W (110) surface on which 200 nm diameter holes had been produced by focused ion beam milling. At 290 °C crystals nucleate only at the step bunches of the holes and grow until they fill the holes (Fig. 5.44). At lower temperatures, e.g. at 110 °C, Ag crystals form everywhere but annealing at high temperature, e.g. 540 °C, transports the Ag into the holes. This procedure provides a method for producing three-dimensional metal nanostructures of controlled shape and position, not only of Ag on W(110) but also in other Stranski-Krastanov growth systems.

Pd on W(110) has been subject of a brief study of the influence of the layer thickness on the valence band spectrum with µXPS [417] and of a µLEED $I(V)$ study of the structure of different layer stacking sequences resulting in twin formation [418]. The growth, structure, and electronic structure of Mg on W(110) up to about 450 °C has been studied in great detail in a SPELEEM instrument with LEEM, LEED, and XPS from the submonolayer range to 12 monolayers [419]. Between 390 and 400 °C quasi-monolayer-by-monolayer growth occurs on top of the stable wetting layer, whose structure evolves similar to that of Pb submonolayers [420]. XPS showed clear differences between the bonding in the two-dimensional gas phase, the wetting layer, and thick films. The most detailed LEEM/LEED studies of metals on bcc(110) surfaces was made by McCarty's group. Santos et al. [421] studied the first three monolayers and thick films of Cr on W(110) with LEEM and LEED $I(V)$ measurements and derived their interlayer spacings with dynamical LEED analysis. Comparison with first principle calculations showed that these spacings were compatible only with antiferromagnetic ordering. McCarty et al. [422] made a very thorough study of the dewetting of thick Cr layers and of the dewetting during the transition from 3 to 4 monolayers, which gave deep insight into the energetics and kinetics of growth and dewetting.

Dewetting was also the subject of a study of Ag and Cu films on Ru(0001), which showed the importance of steps for the formation of three-dimensional crystals via downhill migration without the need of nucleation of two-dimensional islands on top of the crystal [423]. Other studies of the growth of Ag on Ru(0001) addressed the question how twin boundaries between crystals with different layer stacking sequence are eliminated during growth [424] and the decrease of the circular dichroism in the valence band photoemission with increasing thickness from 1 to 3 monolayers [425]. LEEM and LEED studies showed that Pd can be grown on Ru (0001) pseudomorphic quasi monolayer-by-monolayer up to 6 monolayers at 750 K, with twin formation-induced monatomic substrate steps. Interlayer spacings were obtained from $I(V)$ data for these layers and for a thick film [418]. In a similar but more detailed study of the initial layer-by-layer growth of Co on Ru(0001) at 500 K only the first monolayer was found to be pseudomorphic, but quasi-monolayer-by-monolayer growth occurred also in the next few laterally relaxed monolayers. The island shapes reflected the stacking sequence of the layers whose spacings were determined via LEED $I(V)$ analysis [426]. Contrary to Co, quasi-monolayer-by-monolayer growth was observed in a combined LEEM, LEED, and STM study of Mg on Ru(0001) up to 10 monolayers between room temperature and 390 K. Similar to Pb and Mg on W(110) the two-dimensional gas below one monolayer is compressed with increasing coverage into an ordered lattice. The spacing was found to decrease continuously to the lattice constant of a slightly compressed Mg(0001) plane at the completion of the monolayer. With increasing thickness stack faults develop [427].

On Ni(111) the growth of Ag, Au, and Pb has been studied with LEEM and LEED. In the case of Ag [428] and Au [429] growth on a 2 monolayer thick wetting layer occurred in the Stranski-Krastanov mode with multilayer islands at the temperatures studied. Ag, grew above 700 K along step edges before very thick crystals formed. The wetting layer had already bulk lateral periodicity, with two temperature dependent azimuthal orientations. LEED $I(V)$ revealed a large film-substrate spacing. The study of Pb on Ni(111) focused on the influence of QSE effects on the thickness of the Pb islands, their movement across steps [430] and the rapid break-up of these metastable islands into hemispherical droplets above the growth temperature [431]. Interesting structural changes in the wetting layer were reported [430].

Summarizing the results of the studies of metals on densely packed surfaces of high surface energy metals, they all grow as expected in the Stranski-Krastanov mode, with wetting layers differing in thickness and structure depending upon the binding energy of the deposited metals and their misfit with the substrate. With proper growth conditions (temperature, deposition rate, terrace size) metastable growth in the quasi-Frank-van der Merwe mode has been achieved in most cases, sometimes up to 10 monolayers, which makes these films ideal model systems for the study of QSE effects. The LEEM studies have given detailed insight in the growth and structure, in particular in combination with STM, which were not accessible to the laterally averaging studies used in the past, such as island size, shape, and extended defects.

Similar phenomena were found in some less densely packed surfaces such as the (100) surface of bcc crystals. Although in some cases two-dimensional alloying occurs in the submonolayer range such as in Cu, Ag, and Au layers (see Sect. 5.2.1.3), pseudomorphic wetting layers form before dewetting occurs leading to three-dimensional crystals on top of a thinner wetting layer. For example, in the case of Cr on W(100) the wetting layer consists of three pseudomorphic layers at growth temperatures between 575 and 840 K but once islands of the forth layer form the somewhat compressed third layer dewetts resulting in three-dimensional crystals and a double layer [432]. Ag on Fe(100) turned out to be a particular interesting system with quasi-monolayer-by-monolayer growth at 300 K up to 18 monolayers thanks to the small lattice misfit, which allows nearly strain-free pseudomorphic growth. The first pure LEEM study of this system [433] showed well-pronounced QSE effects (Fig. 5.45a). Upon annealing to 500 K the 3 and the 4 monolayer thick films broke up into 2 monolayer thick regions (the wetting layer) and 5 monolayer thick regions (Fig. 5.45b), which are QSE-stabilized. During growth at temperatures above 400 K these preferred thicknesses also occur ("electronic growth" of the 5 monolayer thick regions). Kinetic effects in the growth accompanied the electronic growth. A later study [434] combined with detailed LEED spot profile analysis brought deeper insight in the mechanism of the thermal decomposition process into the electronically stabilized thicknesses. The originally flat films were found to have pinholes which were assumed to facilitate vacancy and adatom island nucleation. This study is a good illustration of the power of combining LEEM with quantitative LEED spot-profile analysis, but combining simple LEED with LEEM is frequently also helpful for the understanding of growth processes studied with LEEM. An example is the growth of Pb on Cu(100). In LEEM three-dimensional crystals with two azimuthal orientations are seen on top of the compressed Pb two-domain wetting layer. LEED allowed to correlate the orientations of the crystals with those of the domains [250]. All systems discussed up to now involve heteroepitaxial growth. However, once the influence of the substrate and possible QSE effects have faded away, growth occurs very similar to homoepitaxial growth, which has been studied theoretically by Tersoff et al. [435] and illustrated with the growth of Ag on Ag(111). They have developed a growth "phase diagram" for homoepitaxial growth, which predicts at high temperatures and high growth rates step flow growth, at lower growth rates with decreasing temperature layer-by-layer, multilayer, and reentrant

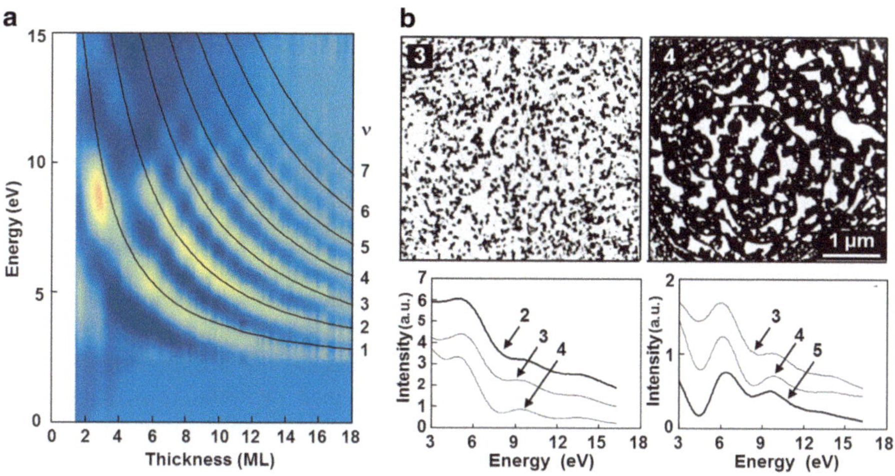

Fig. 5.45 Quantum size effects in Ag films on Fe(100). (**a**) LEEM intensity as a function of energy and thickness measured during deposition at 300 K. (**b**) *Top*: LEEM images of initially uniform 3 (*left*) and 4 monolayer (*right*) thick films after annealing to 500 K. *Bottom*: local intensity spectra in the *bright* regions (*left*) and *dark* regions (*right*) of the two films (*thin lines*) compared to spectra of the 2 and 5 monolayer thick uniform films (*thick lines*). Adapted with permission from Ref. [433]. Copyright 2004 by the American Physical Society

layer-by-layer growth. Finally an example, Sm on Ag [436], should be mentioned yet in which the film does not break up but massive alloying occurs upon annealing above 400 °C as determined by XPS, accompanied by a strong decrease of the intensity in UVPEEM. Ultraviolet photoelectron spectroscopy (UPS) showed that the valence of Sm changes with coverage and temperature in a complex manner. Thus while XPS, UPS, and PEEM give some useful information little can be said about growth and structure without LEEM and LEED, which are the essential methods in this field.

Compared to the studies of metal film on metal surfaces discussed up to now and on semiconductor surfaces (Sect. 5.3.1.1) little work has been done on insulator surfaces. The structural changes of thick (50–135 nm) Mo films [19, 22] and thick (50–500 nm) Nb films [16, 17, 21] with (110) orientation on $\left(11\bar{2}0\right)$ oriented Al_2O_3 surfaces at 1070–1200 K have been studied extensively with LEEM (Sect. 5.1.1) but the films were grown ex situ so that no information on growth was obtained. During growth studies it has to be kept in mind that the surface energy of insulator surfaces usually is lower than that of the film material. As a consequence, growth occurs in the Volmer-Weber mode described in Ref. [332], page 146 of Ref. [333] and page 171 of Ref. [334]. In a cathode lens electron microscope the grazing angle of incidence of the vapor beam is usually between 15° and 30°. Unless the surface mobility of the arriving atoms is low, three-dimensional crystals form, which shadow the surface behind the incident vapor so that the film is initially discontinuous and rough. Only once it has become continuous and thick enough so that it does not break up during annealing, the annealing can lead to surface smoothing, as it was done in the Nb and Mo films after transfer into the LEEM system. Another problem is that insulating substrates charge under the electron or photon beam, making imaging impossible. This problem can frequently alleviated by heating, producing some ionic or electronic conductivity but the increased surface mobility leads to larger crystals, which makes it more difficult to grow continuous films.

These problems can be reduced or even avoided by depositing a suitable nucleation layer, which ensures a high nucleation rate of the film and strong bonding to the substrate, or by surface modification so that the film material wets the substrate. An example of the second method is electron- or photon-stimulated desorption of oxygen on oxide surfaces via interatomic Auger transitions. The oxygen vacancies act as high binding energy sites for the arriving metal atoms and increase the nucleation density or can even lead to initial two-dimensional growth. An illustration of this approach is the study of the initial growth of Au on $TiO_2(110)$ with MEM, μLEED, and μXPS [170, 171]. In electron-irradiated regions, in which the surface has oxygen vacancy rows, Au grows in these rows, on unirradiated regions in three-dimensional crystals. In order to avoid charging, the crystal had to be heated to about 150 °C.

5.3.2.2 Inorganic Compound Films

Oxide films are the most-studied nonmetallic films because of their importance in corrosion and catalysis. Films grown simply by oxidizing bulk material ("homo-oxide films"), by oxidizing deposited films or by "reactive deposition" of metals in oxygen at low pressures ("hetero-oxide films") have been studied. The first group includes oxide films on Ni, Cu, Ag, Ru, and NiAl. The oxidation of a Ni(111) surface has been studied as a function of oxygen dose and temperature with PEEM, MIEEM (metastable impact electron emission microscopy) and MIES (metastable impact electron spectroscopy) [437, 438]. A more thorough study with LEEM and μLEED determined the structure by comparing the (00) I(V) curves with band structure calculations [439]. NiO patches were observed already during room temperature oxidation after saturation of the chemisorption layer, with O_2 exposure- and temperature-dependent patch size and density. The more detailed structural studies gave a more complex picture: at room temperature a continuous NiO layer 2–3 monolayers thick formed, while at 750 K a small number of (111)-oriented NiO crystals appear which spread across the surface leaving essentially clean Ni between them. Upon oxidation of the Cu(100) surface at 870 K (111)-oriented Cu_2O crystals were observed already at an oxygen coverage of less than 2.6 monolayers on an ordered chemisorption layer [440].

Ag does not oxidize in molecular oxygen and forms only a chemisorbed oxygen layer when exposed to the more reactive NO_2. However, during irradiation with electrons—and to a certain extent also with photons—a thick Ag_2O film can be grown, which is initially crystalline but becomes amorphous with increasing thickness. This process has been studied thoroughly with LEEM, LEED, MEM,

XPEEM, and valence band and core level μXPS and has been used for micropatterning of Ag(111) and polycrystalline Ag [441, 442]. The oxidation of Ru, whose oxide plays an important role in catalysis, has also been studied in O_2 with UVPEEM and in NO_2 with LEEM, LEEM-*I*(*V*), μLEED, and valence band XPS. The study in O_2 [443, 444], which covered a wide temperature and pressure range (10^{-5}–10^{-1} mbar, 300–1100 K) found three phases and determined the phase diagram. One of them, characterized by rod-shaped particles in three azimuthal orientations, existed only in a narrow pressure-dependent temperature range and was attributed to RuO_2 crystals. The later studies in NO_2 [182, 445] identified these particles via LEED clearly as (110)-oriented RuO_2 crystals. LEEM-*I*(*V*) and valence band μXPS allowed the conclusion that the three coexisting phases at high temperature were chemisorbed O on top of Ru, O–Ru–O trilayer regions, and RuO_2, a nice demonstration of the power of SPELEEM.

The probably most interesting oxidation layer is that on NiAl(110), which has been studied extensively by MCarty's group with LEEM. Usually the surface is exposed at about 550 K to a high O_2 dose ($>10^3$ mbar s). This produces an amorphous layer, which crystallizes into an about 5 Å thick κ-Al_2O_3 film in two azimuthal orientations upon annealing to 1000–1200 K. Dark-field LEEM imaging with LEED spots of the two domains showed that the domain size increases strongly with annealing temperature, while imaging with a NiAl LEED beam reveals oxide-free islands in the film whose density (size) decreases (increases) simultaneously. Step bunches acted as sinks for oxide-free regions, which spread onto the terraces at the highest temperatures studied ($\approx$1300 K). At low oxygen doses annealing at 1200 K lead to step decoration by the oxide [47]. Exposure to low O_2 doses (4–20×15^{-6} mbar s) at 800−1200 K produced quite different features: thin rods aligned parallel to the [001] direction of the substrate, accompanied with κ-Al_2O_3 islands above 950 K. From the temperature dependence of the growth rate of the rods an activation energy for growth of 1.2 eV was derived. Steps were found to slow the growth, annealing above the growth temperature to shrink the rows accompanied by the growth of the islands [446]. The islands themselves developed internal translation domain boundaries, attributed to stress relief, which could be clearly imaged by defocusing [447]. In the absence of O_2 the κ-Al_2O_3 islands were found to rapidly decrease with increasing temperature. A detailed analysis showed that the O_2 pressure necessary to stabilize the oxide film is much higher than that of bulk Al_2O_3, which shows that the film is kinetically stabilized far from equilibrium [448]. These studies are a good illustration of the advantage of LEEM over STM when information can be obtained only from the study of large areas and from kinetics.

Oxidation of deposited films has been studied only seldom. One example is the oxidation of one and two monolayer thick pseudomorphic Cr films on W(100) between 375 and 740 K at very low pressures [432]. One monolayer was found to be stable below 630 K but to dewet above 790 K, which resulted in three-dimensional Cr_2O_3 clusters surrounded by a penetrated (2 × 2) oxide structure on W. Oxidation of the two monolayer film caused phase separation already below 630 K but the crystals were not oxidized. This illustrates the complexity of the oxidation process

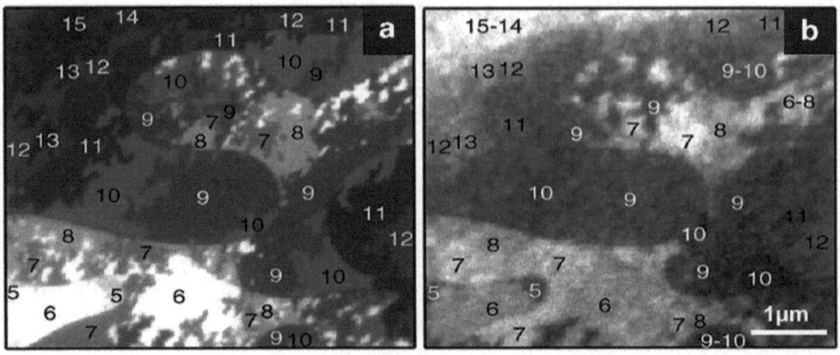

Fig. 5.46 Quantum size effect in the oxidation of thin Mg films on W(110). (**a**) LEEM image, taken with 1.3 eV electrons, (**b**) XPEEM image, taken with the chemically shifted Mg 2p photoelectrons of MgO. The thickness of the various regions of the film, indicated by the *numbers*, has been determined during the growth via the QSE contrast in the LEEM image. The thickness of the oxide layer formed by oxidation of the completed film with about 10^{-6} mbar s O_2 at 50 °C is proportional to the *brightness* in (**b**). Adapted with permission from Ref. [449]. Copyright 2004 by the American Physical Society

when the substrate oxidizes more easily than the film. The second example is the oxidation of Mg and Al films on W(110), which show well-pronounced QSE effects up to more than ten monolayers thanks to their quasi-monolayer-by-monolayer growth near room temperature. Aballe et al. [449, 450] combined LEEM with XPEEM, 2p core electron, and valence band XPS and found that the oxide layer thickness correlated with the QSE effects seen in LEEM. Figure 5.46 [449] shows that the thickness of the oxide layer on 7 and 14 monolayer thick regions of the Mg film is largest, indicating a higher oxidation rate than at other thickness. The maximum at 7 monolayers correlated well with the maximum of the density of states at the Fermi level. The experimental results for Al films were compared with first principle calculations of the electronic structure of the films and showed that the oxidation rate is correlated with the electronic decay length into the vacuum, which enhances the reaction probability before the O_2 molecule is reflected from the surface [450]. This study shows the importance of combining LEEM with XPEEM and spectroscopies with first principle calculations for the fundamental understanding of surface processes.

Oxide films produced by reactive growth have been studied for several materials on several surfaces: Fe oxides on Ru(0001) and Pt(111); Ce oxides on Ru(0001), W(110), and Re(0001); and Ti oxides on Pt(111). Four Fe oxides have been grown: FeO, Fe_3O_4, γ-Fe_2O_3, and α-Fe_2O_3, which are interesting because of their magnetic properties and as supports in catalysis. In the first study [451] films were grown at about 900 K in 10^{-6} mbar O_2 and annealed at about 1200 K, which resulted in large flat triangular crystals surrounded by a FeO wetting layer as identified with μLEED. In a subsequent study [452] films grown in this manner were studied also with μXPS and XMCD, which identified the crystals as Fe_3O_4 and their thickness as about 1 nm. Oxidizing these films with the more reactive NO_2 gas converted the Fe_3O_4 islands into Fe_2O_3 based on μXPS without changing the wetting layer [453]. μLEED gave some information in these studies too but only XPS and XMCD allowed a reliable identification of the crystals. Growth on Pt(111) was studied with the same combination of methods and found to be very similar (Bauer et al, unpublished).

Ceria (CeO_2) is an important oxide in catalysis because of its easy oxygen exchange in the bulk. CeO_2 growth on Ru(0001) has been studied with LEEM and in particular with LEED at temperatures between 575 °C and 1000 °C at oxygen pressures from 2×10^{-7} to 6×10^{-7} mbar in the highly oxidized state and at 1×10^{-8} mbar at 360 °C in a low oxidized state [454]. In all cases the film grew in the Volmer-Weber mode with triangular (111)-oriented crystals in the expected manner (decreasing particle density and increasing crystal size with increasing temperature) in a few azimuthal orientations surrounded by a two-domain Ru-p(2×2)-O layer. Annealing of films grown at low temperature (460 °C) in vacuum did not produce the p((2×2)-pattern and ended at 1000 °C with CeO_2 in parallel orientation to the substrate. No reaction between layer and substrate was observed. The situation is different for growth on W(110). In this case the film was grown only at one temperature and pressure (870 K, 5×10^{-7} mbar) and was featureless in LEEM so that analysis had to be based only on LEED and XPS [455]. After saturation of the high temperature oxygen superstructure of W(110) a characteristic LEED pattern developed which could be explained equally well with epitaxial CeO_2 and Ce_6WO_{12}, both in (100) orientation. Distinction between the two explanations was achieved with detailed XPS studies, in particular of Ce 3d spectra and fitting them with the known spectra for Ce^{3+}, characteristic for Ce_6WO_{12}, and Ce^{4+}, characteristic for CeO_2. From the thickness dependence of the relative contributions of the two components it was concluded that the layer consisted up to about 0.9 nm of Ce_6WO_{12}, on top of which CeO_2 grew epitaxial. Thus, in this system CeO_2 layer grows via an interfacial reaction layer.

Ce oxide layers can also be grown by first depositing Ce at room temperature in UHV and subsequent annealing in O_2. This method was used for growth on Re (0001) by depositing about 0.8 monolayers Ce and annealing in 1×10^{-7} mbar O_2 to 850 K. LEEM and LEED showed then flat (111)-oriented crystals with slightly expanded in-plane lattice constants in four azimuthal orientations. Re 4f and O 1s XPS indicated that the Re surface was oxidized and that the Ce oxide crystals were slightly reduced. The degree of reduction was determined by valence band μXPS of the islands, resonantly excited at the Ce^{3+} and Ce^{4+} binding energies, resulting in nearly uniform composition $CeO_{1.64}$ of all islands [456].

Finally a LEEM/μLEED study of Ti oxide films grown on Pt(111) in 5×10^{-5} Pa O_2 at 450 °C and at room temperature followed by annealing should be mentioned yet [457]. Movies taken during growth and annealing showed the evolution of a number of complex phases with thickness and temperature. Ti diffusion into the substrate and electron stimulated oxygen desorption were considered to be involved in these transitions. Such processes have to be considered in all oxide studies. This is also true for films grown ex situ. Unpublished studies (Bauer et al) of thick epitaxial (111)-oriented CeO_2 films on YSZ have shown in addition other problems: diffusion of substrate impurities (Al, Si) into the film upon annealing and strong defect generation in XPEEM by photon-stimulated dissociation.

Hydride films are of interest, in particular MgH_2 because of its large storage capability. The growth and thermal decomposition of MgH_2 films on Ru(0001) was studied [458] by exposing up to 10 monolayer thick Mg films grown between 300 and 430 K to atomic hydrogen and annealing the resulting MgH_2 films up to 500 K. In LEEM, taken near the mirror mode because of the destructive effect of electrons with energies above a few eV, a low density of nuclei was observed right after the start of the H_2 exposure from which large crystals with different azimuthal orientations developed with increasing exposure. Their structure could not be determined because of the destructive beam effect. Thermal decomposition was thickness-dependent and started around 450 K as deduced from MEM and thermal desorption of H_2. In the thickest film it was completed at 480 K [458]. This system is an example for an extreme case of radiation damage, limiting the number of electron and photon beam methods useful for characterization.

Another film growth method is thermal decomposition of molecules, which contain the constituents of the film. BN films have been grown in this manner by decomposition of borazine $(BHNH)_3$ on the Ru(0001) and on the Fe(110) surface at about 800 °C in the low 10^{-7} mbar range. The study on the Ru(0001) surface [459] looked at the reaction of the BN film, called h-BN nanomesh because of the pattern of the hexagonal film seen in STM, with O_2. The combination of MEM, LEED, and Ru 3d, B 1s and N 1s XPS showed that the removal of BN started at defects and occurred in 5×10^{-8} mbar O_2 at about 750 °C within several minutes with a well-defined two-step reaction front, leaving a Ru surface with a (2×2) LEED pattern behind. In the second study [460], the h-BN monolayer was grown on an epitaxial Fe film on W (110). Because of the twofold symmetry of this surface the hexagonal BN lattice grew in two orientations as seen in LEED and dark-field LEEM. Accompanying STM studies showed a one-dimensional "washboard" corrugation in agreement with the periodicity of the LEED superstructure spots. The varying film-substrate contact in this corrugation was also supported by comparing N 1s NEXAFS and N 1s XPS spectra of BN on substrates with different film-substrate contact [460].

Summarizing the section on inorganic compound films, this is a field in which the combination of structural and spectroscopic information is in general needed to obtain a full understanding of the film. The films are generally heterogeneous so that good resolution in LEEM and μLEED are required for structural analysis and μXPS for chemical identification. In poorly ordered systems as encountered here in some cases MEM and PEEM can give some information but XPS is the main tool for analysis. As a result, SPELEEM systems at synchrotron radiation facilities dominate this field.

5.3.3 *Organic Films*

The development of organic thin film semiconductor devices around the turn of the century such as organic field-effect transistors (OFETs) and organic light-emitting diodes has made the study of organic thin films an important application field of cathode lens electron microscopy and associated techniques. Film morphology, crystallinity, and the orientation of the molecules in the crystals determine to a large extent the properties of the devices, together with the interface of the organic film with the electrodes (metal or semiconductor) of the device. For example charge transport is very sensitive to crystallinity and best perpendicular to the molecular axis while light emission is polarized parallel to it. The contact resistance depends strongly on the molecule/electrode material interaction. Many of the crystals can grow in different polymorphic forms. All these features are controlled by film deposition and interface conditions, which make surface characterization and in situ deposition essential tools for device optimization.

AFM and STM are frequently used in this endeavor but LEEM, UVPEEM, XPEEM combined with LEED are major tools too, in particular when combined with linear polarized NEXAFS, which allows determination of the orientation of the molecules. Of course these experiments are limited by the requirement that the molecules do not break up during sublimation and condensation. Radiation damage is another limiting factor, not only at the high flux densities in XPEEM but also in UVPEEM. However, there are many aromatic molecules, which satisfy these requirements. The first studies of organic films compared the possibilities of different imaging modes (UVPEEM, MIEEM, LEEM) using the square chloroaluminum phthalocyanine molecule ($C_{32}H_{18}AlClN_8$) on MoS_2. No contrast was seen in LEEM [461] but weak contrast between first and second layer in PEEM and to a lesser extent in MIEEM [461, 462]. The second layer appeared brighter, which led to the conclusion that emission is not determined by the work function but by the electronic structure. Structure and growth could not be studied because of limited resolution.

No more PEEM work was done on phthalocyanine films until much later when more sophisticated equipment became available. In one study [463] a better PEEM instrument with tunable laser excitation was combined with high resolution UPS in a separate system to correlate image contrast with work function and electronic structure of up to two monolayers thick Pb phthalocyanine films on graphite. A second study [464], interesting from the methodical point of view, combined laterally averaging angle-dependent linear polarized NEXAFS in an XPS system with laterally resolved NEXAFS at fixed angle in a PEEM. The goal was to determine the molecular orientation of silicon phthalocyanine chloride via NEXAFS resonance absorption spectra. This molecule has a Si atom in the center to which a Cl atom is attached on each side of the molecule normal to the molecular plane. At the Si K absorption edge at 1840 eV there are two 1s–σ^* transitions separated by about 2 eV, one to the in-plane Si–N bond, the other one to the Si–Cl bond normal to the plane. After determining the intensity ratio of these two transitions as function of angle in the angle-dependent NEXAFS measurements the authors could determine the tilt angle of the molecules on the microscopic level in PEEM by analyzing the local intensity in a series of spectroscopic images around the absorption edge. This method was applied to a study of five monolayers thick patterned films on a Au surface [464].

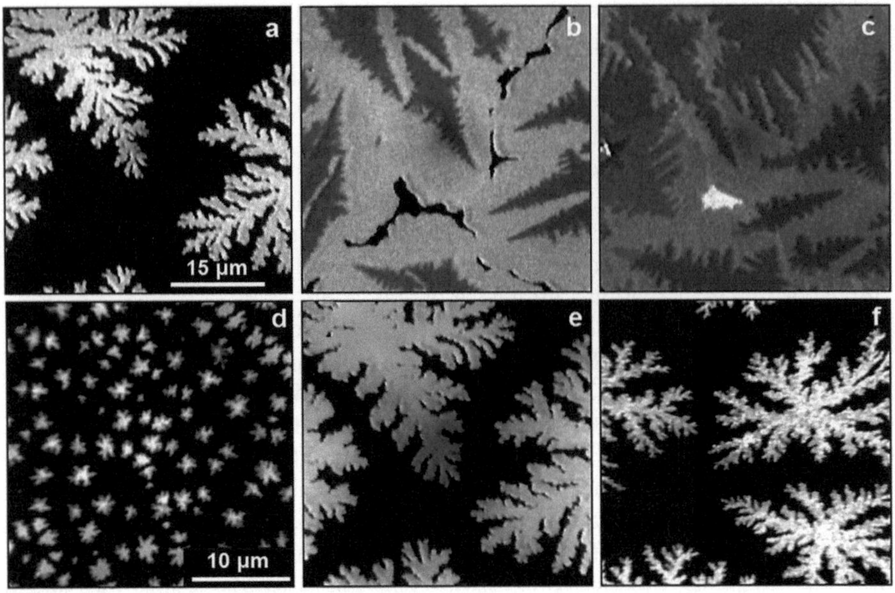

Fig. 5.47 UVPEEM images of the growth of pentacene on Si at room temperature. (**a–c**) growth on the clean Si(100)(2 × 1) surface: (**a**) one layer, (**b**) two layers, (c) three layers. (**d–f**) Growth on different surface terminations: (**d**) on plasma-cleaned oxide, (**e**) on the clean (2 × 1) surface, (**f**) on a well-ordered self-assembled cycloxane (C_6H_{10}) layer, which saturates the dangling bonds and suppresses the flat-lying pentacene wetting layer present in (**e**). Adapted from Ref. [466] with permission from Springer Science + Business Media

The first studies of the growth of organic films were made in Tromp's group, stimulated by the development of pentacene transistors at IBM. The linear molecule pentacene ($C_{22}H_{14}$) can be easily sublimated from purified crystals at 200–300 °C. Meyer zu Heringdorf et al. [465, 466] deposited it on Si(100) surfaces and studied film growth with UVPEEM. In order to minimize radiation damage they reduced exposure time per image to about 1 s in 60 s. Figure 5.47 [466] shows some of their results. The top row (a–c) shows the dendritic quasi-monolayer-by-monolayer growth on the clean (2 × 1) surface on top of a wetting layer of flat lying molecules. µLEED showed that the individual islands are single crystals with random azimuthal orientation. From the rate dependence of the nucleation rate a critical nucleus size of six molecules was deduced. The bottom row (d–f) shows the influence of the surface condition on the growth. A comparison of the images shows that the number density of crystals is an order of magnitude smaller on the clean and cyclohexane-passivated surface than on the oxidized surface, resulting in a much smaller number of grain boundaries, which are detrimental to charge transport. A later combined PEEM, LEEM, and LEED study confirmed the beneficial effects of passivating layers of two other organic adsorbates and found that the pentacene crystals grew epitaxial on top of them to very large grain sizes [467].

While the early work concentrated on the technologically important (100) surface, later work in Sakurai's group studied pentacene growth on Si(111) surfaces and other surfaces with hexagonal surface nets, starting with the hydrogen-terminated Si(111) surface [468, 469]. These studies combined tilted-beam LEEM with μLEED, which showed complex anisotropic dendritic polycrystalline growth with the (001) plane parallel (molecules perpendicular to the surface) to the substrate and the *b* axis of the two-dimensional pentacene unit cell somewhat rotated relative to the $\langle 1\bar{1}0\rangle$ and $\langle 11\bar{2}\rangle$ directions of the substrate. The *b* axis was the main growth direction of both the main branch and of the side branches of the dendrites. Thus the direction of *b* determined together with the density gradient in the two-dimensional molecular gas phase the shape of the crystals. The crystals grew directly on the H-terminated surface from a small number of nuclei radially outward and not on a wetting layer as on the clean (111) or (100)-(2 × 1) surface.

The strong influence of the substrate on the growth mode is even more dramatically demonstrated by the comparison with the growth on the Bi(0001) surface [470, 471] on which pentacene grows from local defects in the step flow mode instead dendritic, with twinned domains. This is attributed to the fact that every second molecular row parallel to the *a* direction in the *ab* plane of pentacene coincides with the Bi atomic rows, resulting in only one-dimensional misfit. The preferred growth direction was perpendicular to these rows as in the dendritic growth on the H-terminated Si(111) surface although the bonding is strongest along the row direction (*a* axis). Similar growth behavior was found on the Bi-terminated surface ("$\alpha\sqrt{3}$-Bi-Si(111)" [471, 472]. This shows that the preferential growth direction is determined by the molecular structure and bonding, which determines the crystal anisotropy. The detailed analysis of the data showed also a significant reorientation barrier for the transition from the diffusing molecules which are parallel to the substrate to the crystal in which they are upright.

Pentacene growth was also studied on several other surfaces with hexagonal surface net: on a fullerene (C_{60}) film on the Bi(0001) surface [473] on the (0001) graphite surface [474], on the BN monolayer on Ru(0001) [475] and on the ($\sqrt{3} \times \sqrt{3}$)R30°-Ag/Si(111) surface [476]. On the fullerene surface, crystals with flat lying molecules formed initially dendritic islands, on top of which crystals with nearly perpendicular molecules in several azimuthal orientations grew as on the surfaces mentioned before. On graphite growth on the wetting layer occurred also mainly with nearly perpendicular molecules but continued partially also with flat lying molecules. The crystal size increased strongly with deposition temperature and all crystals had preferred azimuthal alignments. Growth on BN starts already in the submonolayer range with nearly perpendicular molecules when the density of the flat-lying molecules reaches a critical density. Large two-dimensional crystals with several azimuthal orientations form, similar to the growth on the Bi(0001) surface. In the second monolayer the nucleation density is much higher and the islands have fractal shapes. This study was combined with angle-dependent NEXASF spectroscopy, which clearly showed the orientation of the molecules. A e ($\sqrt{3} \times \sqrt{3}$)R30°-Ag/Si(111) surface: the molecules were lying flat on the substrate not only in the wetting layer but also in the several 10 nm thick crystals growing on it. Summarizing the growth on surfaces with hexagonal surface nets, the results show the strong influence of the molecule-substrate interaction, which determines the orientation of the molecules, the diffusion process, and the resulting island shape, but also influences the growth on top of the first layer. In all cases the film grows with preferred azimuthal orientations and orientation-dependent growth rates.

The results obtained from the growth studies on well-defined hexagonal surfaces are of fundamental interest and allow some general suggestions for optimizing organic thin film devices but practical devices usually are grown on SiO_2 and polycrystalline metal surfaces. The problems encountered on such surfaces have also been studied with LEEM and PEEM. The LEEM study [477] addressed the question how self-assembled monolayers, which are frequently used to improve the performance of OFETs by reducing the trapping sites at the film-substrate interface, influence the growth. It found that hexamethyldisilazane ($(CH_3)_3SiNHSi(CH_3)_3$) and octadecyltrichlorosilane ($CH_3(CH_2)_{17}SiCl_3$) adsorption on SiO_2 caused spontaneous aggregation deleterious to OFET performance due to the decrease and loss of the conduction path. This effect was attributed to the lower surface energy of the layers compared to that of pentacene, causing Volmer-Weber growth.

A PEEM study [478] was concerned with the problems, which can occur at the pentacene/electrode contact. Au contact patterns were grown on a SiO_2 layer on a Si (111) wafer and pentacene was grown on this patterned surface. PEEM showed that a denuded zone ("groove") at the SiO_2–Au boundary developed during deposition, which persisted up to thick films, also when the SiO_2 surface was precovered with a $((CH_3)_3SiNHSi(CH_3)_3$ self-assembled (SAM) monolayer as in the preceding LEEM study. However when the Au surface was covered with a octanethiol ($CH_3(CH_2)_7SH$) SAM no groove formed during growth. Conductivity measurements showed that this caused an increase of the charge-carrier mobility by orders of magnitude due to the elimination of the high resistance region at the contact. The groove formation was attributed to the high nucleation and growth rate of pentacene crystals with flat-lying molecules on the clean Au surface compared to that on SiO_2. This reduced the molecule concentration in the surroundings of Au, while on the octanethiol-covered Au surface pentacene grew in the same manner as on SiO_2 due to the lowering of the surface energy by the SAM, thus eliminating the concentration gradient. This experiment is a nice example of the usefulness of UVPEEM for the understanding of technological problems. Such problems stimulated also a LEEM study of the growth of 6,13-pentacenequinone ($C_{22}H_{12}O_2$) on Si(111), which forms upon oxidation of pentacene and causes deterioration of OFETs exposed to air [479]. The study revealed a very complex growth process. Initially an amorphous wetting layer forms followed by the nucleation and growth of compact, amorphous islands, from which long curved feather-like crystals grow with the *ab* plane parallel to the substrate. They have left-handed and right-handed curvature, related to the tilt direction of the nearly upright molecules, and the in-plane orientation changes continuously with the curvature. During very slow growth large rectangular crystals formed, which indicates that the feather-like growth is determined by kinetics.

While pentacene was at the center of the interest in organic films, some others have been studied, at least briefly. A UVPEEM study of the growth of PTCDA ($C_{24}H_8O_6$) on the Ag(111) surface looked at the influence of the surface morphology on the film structure [480]. In the first monolayer the nucleation rate in stepped regions was much larger than in flat regions, in the second monolayer the difference was much smaller and in the third monolayer three-dimensional crystals formed, indicating Stranski-Krastanov growth. Anthracene ($C_{14}H_{10}$) was found in standard UVPEEM studies to grow on Si(111) in the same manner as pentacene with a wetting layer of flat-lying molecules on which dendritic crystal layers grow with nearly perpendicular molecules [481]. 2PPE imaging with 400 nm fs laser light ($h\nu = 3.1$ eV) gave much deeper insight, in particular by making use of the polarization of the light [482]. 3.1 eV is the energy of the lowest lying exciton level, which can be excited only if the $\boldsymbol{E}$ vector has a component along the $\boldsymbol{b}$ axis. Thus the intensity of the two-electron photoemission depends upon the orientation of the $\boldsymbol{b}$ axis relative to the $\boldsymbol{E}$ vector and allows determination of the azimuthal orientation of the crystals. This method of orientation determination, which is based on the selective excitation effect in a *crystal*, is different from the orientation determination with linear-polarized NEXAFS, which gives the orientation of a *molecule* via the dependence of the excitation probability of π and σ orbitals on the relative orientation of $\boldsymbol{E}$ and the orbital orientation under resonance condition.

Diindenoperylene ($C_{32}H_{16}$) films on polycrystalline, (111), and (100) Au surfaces were subject of a series of UVPEEM, XPEEM, and LEEM studies combined with linear polarized NEXAFS in a separate system by Casu et al. [483–486]. Their results can be summarized as follows: the films grow in the Stranski-Krastanov mode with flat-lying molecules at sub-monolayer coverage and thereafter with fractal crystals in which the molecules are tilted (relative to the surface) by angles increasing from 23° to 46° with thickness from 0.6 to 3.6 nm. LEEM contrast in the three-dimensional crystals was attributed to QSE effects.

The initial growth, structure, and thermal stability of 4,4-biphenyldicarboxylic-acid (BDA, $C_{14}H_{12}O_4$) on Cu(100) was studied in detail with LEEM and μLEED in Poelsema's group [487–491]. These studies have given deep insight into the complex processes, which can occur in the submonolayer range. Nucleation of two-dimensional crystals with neighboring flat-lying molecules in orthogonal orientation occurred rapidly after a considerable incubation time (t_i), followed by lateral growth without much further nucleation until renewed nucleation occurred at a critical crystal size (2×10^4 nm^2) during growth at room temperature. The renewed nucleation at $1.5t_i$ was attributed to stress build-up in the existing crystals, which made further growth energetically unfavorable. During growth at 448 K further growth was not suppressed indicating the absence of stress but at this temperature the crystals decayed with a decay exponent of 0.6. LEED showed that the crystals were epitaxial, with the long axis of the molecules parallel to the $\langle 110 \rangle$ directions of Cu [487]. Further LEEM intensity studies allowed determining the phase boundary between the two-dimensional gas phase and the crystal + gas coexistence region as a function of temperature from which a cohesive energy of 0.35 eV was derived based on a lattice gas model. The analysis gave also a large entropy contribution, which was attributed to the large size and shape of the molecule and considered to be responsible for the high density in the gas phase relative to that in the crystals. From island decay studies diffusion constants were derived, steps did not act as limiting diffusion barriers [488]. A detailed study of the formation and decay of crystal nuclei between 297 and 332 K gave critical island sizes between 400 and 600 nm^2 and a Gibbs free energy for nucleation of 4.0 eV [489]. Another interesting observation, found by combining LEEM with μLEED, was the formation of a compressed phase inside large crystals during growth above 370 K, in four rotational domains and 14 % higher density than the uncompressed region. It was stabilized by the incoming molecular flux but disappeared without it [490]. At high temperature (410 K) growth was strongly influenced by steps: once a crystal had formed on a terrace it continued to grow only along the terrace. This is similar to the growth of metals on close-packed metal surfaces, as illustrated by Cu on W (110). At high temperature steps are no diffusion barriers so that atoms/molecules can easily surmount them get incorporated in the crystal while growth on neighboring terraces requires a nucleation event. There is, however, a difference between metals and large elongated molecules: they have to arrive at the growth front in the correct orientation or have to rotate, which results in a roughening of the growth front as observed [491].

Another interesting phenomenon, which has been found in BDA films on Au(111) in the submonolayer range, is the formation of three different crystalline phases, depending upon growth or annealing temperature [492]. The crystals grown at room temperature are needle-shaped and transform into compact ramified crystals upon annealing above about 330 K with a somewhat different crystal structure. Growth above about 330 K produces compact crystals with well-defined straight edges and a completely different structure. In all structures the molecules are lying flat. The transition between the first two structures has been attributed to loss of the H at the carboxyl group at the ends of the molecules ("deprotonation"), which changes the lateral interactions between them, causing the structural changes. This suggests that on the Cu(100) surface, with which the molecule interact much stronger, significant deprotonation should occur too. Another substrate with which BDA interacts more weakly than with Cu is graphene. On graphene supported on Ir(111) the crystals nucleated at room temperature at the wrinkles and protrusions of the graphene layer and grew across it in azimuthal orientations determined by the wrinkle direction. During growth at 370 K no crystals formed on graphene but decorated the substrate at the edges of the graphene flakes. During annealing of room temperature-grown films the crystals disappeared above 350 K but nucleated again upon cooling. This indicates that no sublimation occurred and molecules diffused back from the surrounding Ir onto the graphene. No second layer growth was observed [493].

Para-sexiphenyl (6P, $C_{22}H_{12}O_2$) is another model molecule, whose growth was studied with several imaging methods in Poelsema's group with LEEM and μLEED on graphene and the supporting Ir(111) surface in the temperature range from 240 K

[494, 495] to 405 K [496]. 6P forms initially an epitaxial layer of flat-lying molecules. This converts into an alternating row of flat-lying and upright (edge-on) epitaxial layer, which nucleates at the wrinkles of the graphene and completes the first layer. This molecular arrangement is maintained with further growth. At 240 K this occurs in a quasi-monolayer-by-monolayer mode as evidenced by characteristic changes of the image intensity, suggesting quantum size contrast, up to the largest thickness studied (4.35 monolayers) but at 320 K the crystals grow as needles across the surface, starting from wrinkles again. Above 380 K the molecules diffuse onto the Ir surface and form near the graphene edge crystals with upright standing molecules. At lower temperatures these crystals grow all over the Ir surface with ramified shape.

Fleming et al. [497] studied the growth of the same molecule (6P) on the Cu(110) (2×1)-O surface from 25 to 150 °C with UVPEEM. It shows some similarities to that on graphene but also distinct differences. The mode of film growth, which probably occurs also in many other organic films, is illustrated in Fig. 5.48 [497]. The initially flat-lying molecules tilt upward forming a monolayer of tilted molecules, which is stable up to 220 °C. On this monolayer a second layer grows, also with initially flat-lying molecules which turn upright with increasing coverage to form a second monolayer of molecules tilted into the opposite direction. This layer is metastable and breaks up into needles along the Cu $\left[1\bar{1}0\right]$ direction, which grow in length and thickness with increasing deposition time, initially by the supply of molecules from the second monolayer until dewetting is completed. From a detailed study and analysis of the evolution of the image intensity and the needle dimensions as a function of deposition rate, temperature and annealing temperature, combined with AFM measurements and results from other studies the authors have developed a very detailed model of the growth of these films.

Another film/substrate system studied with PEEM in connection with STM is α-sexithiophene (α-6T, $C_{24}H_{14}S_6$) on Ag(110) [498–500]. The dependence of the PEEM intensity upon deposition time showed the formation of two monolayers. With STM they were identified to consist of flat-lying molecules, which were partially aligned in the first layer but perfectly aligned in the double layer. On top of this wetting layer a third layer formed in which needle-shaped crystals grew across the surface, partially depopulating the third layer. As judged by the coverage of needles their height was estimated to be 100 nm after deposition of 33 monolayers. Thin platelets were observed too occasionally. The use of D_2 and He I light ($h\nu = 6.4$ and 21.2 eV, respectively) did not produce significant information in addition to imaging with the Hg lamp but the polarization had a strong influence on the emission, in particular from the needles. Local UPS spectra obtained by imaging with a retarding grid energy filter and varying the retarding voltage showed little difference between wetting layer and needles suggesting the same molecular orientation in both of them. The spectrum of the platelets, however, was very different and assigned to nearly upright standing molecules.

Summarizing the in situ work on molecular layers, cathode lens electron microscopy methods have made important contributions although most of the

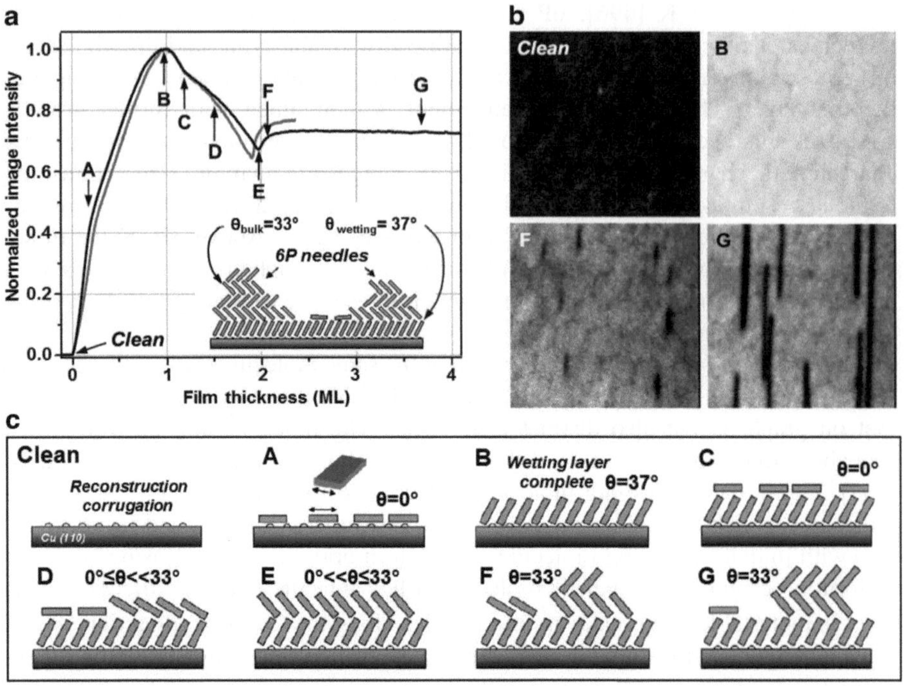

Fig. 5.48 (**a**) Two typical plots of the photoelectron intensity as a function of the thickness of para-sexiphenyl (6P) films deposited at 140 °C at a rate of about 1/3 monolayer/min. Unless otherwise stated, all discussions in the text refer to the *darker curve*. The change in work function from clean substrate (4.8 eV) to 1 monolayer at point B is about −0.4 eV. The evaporator shutter was continuously open. The deposition rate had no significant influence on the trends of the curves. *Inset*: schematic illustrating the structure of the various layers that assemble. (**b**) Frames from the PEEM video taken at the points marked in (**a**) at the *darker curve*. (**c**) Schematic diagrams illustrating the orientation and packing of the molecules that explain the photoelectron intensity at the *points* marked in (**a**). Only the molecular end-on view is shown; the short axis of the 6P molecule is illustrated by a *double-headed arrow*. The packing angle θ is the angle between the molecular short axis and the surface. The O-induced corrugated surface reconstruction is indicated by the O rows which are normal to the plane of the drawing and have a distance of 5.1 Å. Adapted with permission from Ref. [497]. © IOP Publishing 2009. All rights reserved

understanding of the molecular structure and orientation has been obtained by other methods such as STM, AFM, Fourier transform infrared spectroscopy, UPS, to name the most important ones. Cathode lens electron microscopy allows the study of the kinetics of film growth, from which energetic parameters can be extracted, at least in the early growth stages. It is indispensable in the later growth stages, when three-dimensional crystals form. Used alone, it gives an incomplete picture. Therefore, in most of the studies mentioned the interpretation of the results relies more or less heavily on information from the other methods mentioned above. Although the general picture of growth and structure of molecular layers, that has emerged, is similar to what is known from the growth of atomic layers, it is strongly modified by the shape and bonding of molecules. Many questions are still open, whose answer will require systematic studies of more molecule-substrate combinations over a wider temperature range, with improved methods such as excitation with polarized tunable light in PEEM or addition of $I(V)$ measurements to μLEED.

The organic film studies discussed up to now were made in situ with PEEM, LEEM, and μLEED but XPEEM and XPS were used only occasionally. There are two other groups of organic films, prepared ex situ, in which XPEEM and XPS are the main laterally resolving and only chemically resolving surface characterization methods: Langmuir-Blodgett films and polymer films. The interest in these fields is in phase-separating two- or more components systems. Two-component Langmuir-Blodgett film samples are prepared by spreading a mixture of mutually immiscible surfactant molecules on water and transferring the Langmuir-Blodgett film onto a Si wafer or other substrate. This is done by submerging the wafer below the film and pulling it from the water, followed by drying before transfer into the PEEM system. Chemical contrast in imaging is achieved via NEXAFS spectromicroscopy at the C K edge making use of the energy difference between the absorption resonances of different bonds, for example of the C 1s $\rightarrow \sigma^*_{C\text{–}F}$ resonance at $h\nu = 282.8$ eV in a perfluorinated region and of the C 1s $\rightarrow \sigma^*_{C\text{–}H}$ resonance at $h\nu = 287.9$ eV in a hydrocarbon-rich region. Excitation at these two photon energies allows a clear distinction between the corresponding regions. This method has been used very successfully, together with AFM, in studies of phase-separated Langmuir-Blodgett films of fluor-free and fluorinated organic acids [501, 502] and even of more complex systems, in which the spectra of the two components were not clearly separated but differed only in width [503].

Polymers were studied for a variety of reasons, ranging from technical to medical applications. A polymer was actually the first non-biological organic material studied with cathode lens microscopy [504, 505]. The subject of this study was a polyimide film which is used to orient the molecules in the liquid crystal layer of flat panel displays. In order to achieve this, the molecules in the polyimide film have to be oriented by rubbing parallel to the direction of rubbing. As the intensity of the NEXAFS resonances depends upon the angle between E and the orientation of the final state orbital (σ^* or π^*) the intensity of the resonance peaks allows determination of the orientation of the molecules as discussed before. The image series ("stacks") taken in a 50 eV wide photon energy range around the C K absorption edge with the rubbing trace parallel and perpendicular to the in-plane component of E showed clear intensity differences at the π^* resonance. From these intensity differences the minimum pressure needed for orientation of the molecules could be determined. In a later study [506] molecular orientation was produced by irradiation with linear polarized 4.8 eV laser irradiation, which preferentially decomposes polymer chains that are parallel to E. As a result the surface layer, in which decomposition occurs is enriched with molecules aligned perpendicular to E. In this study not only the resonance at the C K edge was used as before but also resonances at the N K edge and the O K edge, with the latter showing the largest dichroism.

Semiconducting polymers are also important in electronics. High mobility is usually attributed to high intrachain and interchain order in the film. This is achieved by annealing the film, which causes considerable changes in the film morphology such as grain boundary formation and break-up into separate ordered regions, resulting in a deterioration of the conductivity. A UVPEEM study of these morphology changes in 18 nm thick poly(3-hexylthiophene) films on Si upon annealing at 170 and 220 °C [507] indeed showed dramatic changes in the image contrast connected with the grain growth. They were interpreted to differences of the work function and ionization potential caused by variations of the molecule orientation. A study of a doped polypyrrole film on an insulating fluorinated ethylene propylene substrate, in part covered with a Cu grid pattern, combined XPEEM with NEXAFS and some AFM studies [508]. The film was found to be chemically homogeneous in XPEEM images at the absorption edges of C, N, and S. NEXAFS did not show dichroism, indicating that the molecules were not ordered. The study was hampered by considerable radiation damage by the high photon flux density of the undulator beamline used in these experiments. This example illustrate one of the limitations of cathode lens microscopy.

The polymer films studied most intensely with XPEEM consist of two immiscible polymers, which phase separate upon annealing and are used mainly for competitive adsorption studies on the phase-separated mixtures. XPEEM is frequently combined with scanning transmission X-ray microscopy (STXM), which gives the structure of the bulk material for comparison with the surface structure seen in XPEEM. The first combination studied was polystyrene (PS) and poly(methylmethacrylate) (PMMA) [509], a system also used later for protein adsorption [510]. Figure 5.49 [510] illustrates this kind of studies for an annealed PS:PMMA film on Si. The NEXAFS spectrum of the two polymers is very different as seen in the absorption spectra (b) so that the two components can be distinguished easily by imaging at the corresponding resonance transitions, C 1s $\rightarrow \pi^*_{C-C}$ of PS at 285.1 eV and of PMMA at 288.4 eV, as shown in the NEXAFS images. Quantitative information has been extracted by detailed image analysis, taking into account several possible artifacts. This is of particular importance in studies of protein adsorption on these polymer blends because of the small signal from the proteins and the small difference of their resonance energies from those of the substrate. For example, the resonance of human serum albumin protein is shifted only 0.25 eV relative to that of PMMA. Nevertheless preferred adsorption of the protein at the interfacial PS/PMMA region could be established with a resolution of better than 100 nm [510–512]. Other polymer combinations and protein adsorption were studied too, for example polystyrene (PS)-polylactide (PLA) blends [513] and PS blended with a copolymer of polymethylmethacrylate and polyacrylic acid (PMMA-*b*-PAA) [514]. In both studies the XPEEM and XSTM were combined with AFM in order to obtain also information on the film morphology. The first one was concerned with the dependence of the morphology and chemical composition upon the polymer ratio and annealing conditions (time and temperature) and found that the AFM images did not correspond to the XPEEM images. The second one studied the

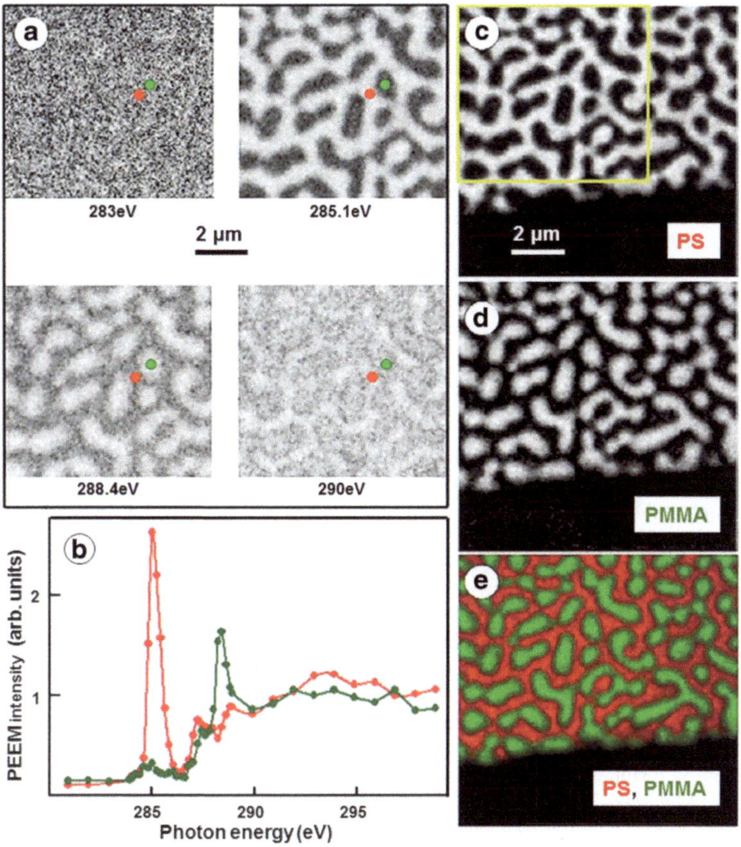

Fig. 5.49 (**a**) X-PEEM images at 283, 285.1, 288.4, and 290 eV, 4 of the 40 images in the C 1s region of an annealed 28:72 (w/w) PS:PMMA blend thin film spun cast on native oxide Si (the average surface composition is 57:43). (**b**) Spectra from the indicated spots. (**c** and **d**) Component maps of PS and PMMA derived by singular value decomposition of the C 1s image sequence. (**e**) Color coded composite map (*red*: PS; *green*: PMMA). Adapted with permission from Ref. [510]. Copyright 2004 Elsevier

adsorption of positively charged human serum albumin and a negatively charged peptide. Quantitative analysis showed that the thickness of the adsorbate film was strongly correlated with its charge, which suggests that adsorption is driven by electrostatic interactions with the negatively charged surface during adsorption. A quite different method of two-component polymer pattern formation was used to study the adsorption of the protein ubiquitin. It used two superimposed polymer films. The water-soluble top film was patterned by electron beam lithography, which formed cross-linked patterns, which remained after washing. Protein was found to adsorb only on the cross-linked polymer [515]. For a review of the work on polymer films with XPEEM and XSTM see Hitchcock et al. [516].

References

1. Yang, Y.-N., Williams, E.D.: High atom density in the "1 × 1" phase and origin of the metastable reconstructions on Si(111). Phys. Rev. Lett. **72**, 1862–1865 (1994)
2. Ong, W.J., Tok, E.S.: Real time dynamics of Si magic clusters mediating phase transformation: Si(111)-(1 × 1) to (7 × 7) reconstruction revisited. Surf. Sci. **606**, 1037–1044 (2012)
3. Witt, W.: Untersuchungen 2-dimensionaler Phasenübergänge mit einem hochauflösenden LEED-Diffraktometer. Ph.D. thesis, Technische Universität Clausthal (1984)
4. Telieps, W., Bauer, E.: An analytical reflection and emission UHV surface electron microscope. Ultramicroscopy **17**, 57–66 (1985)
5. Kastler, A.: Un systeme de franges de diffraction à grand contraste. Rev. Opt. **29**, 307–314 (1950)
6. Bauer, E., Mundschau, M., Swiech, W., Telieps, W.: Low energy electron microscopy. In: Goodhew, P.J., Dickinson, H.G. (eds.) Proc. 9th European Congress on Electron Microscopy (EUREM 88), 1988. IOP Publ., Inst. Phys. Conf. Ser. No 93, vol. 1, Chapter 4, pp. 213–218. York, England (1988)
7. Mundschau, M., Bauer, E., Swiech, W.: Photoemission microscopy and atomic steps on Mo (011). Surf. Sci. **203**, 412–422 (1988)
8. Mundschau, M., Bauer, E., Swiech, W.: Modification of atomic steps by adsorbates observed by low energy electron microscopy and photoemission microscopy. Catal. Lett. **1**, 405–412 (1988)
9. Mundschau, M., Bauer, E., Swiech, W.: Defects on the surface of Mo{011} observed by low-energy electron microscopy. Phil. Mag. A **59**, 217–226 (1989)
10. Mundschau, M., Bauer, E., Swiech, W.: The role of ledges in vapor/solid phase transformations observed by low-energy electron microscopy and photoemission electron microscopy. Metall. Trans. **22A**, 1311–1315 (1991)
11. Telieps, W., Mundschau, M., Bauer, E.: Dark field imaging with LEEM. Optik **77**, 93–97 (1987)
12. Telieps, W., Mundschau, M., Bauer, E.: Surface domain structure of reconstructed Au(100) observed by dark field low energy electron microscopy. Surf. Sci. **225**, 87–96 (1990)
13. Altman, M.S., Chiang, S., Statiris, P., Gustafsson, T., Bauer, E.: Stress-induced microfacetted reconstructions of the Pb(110) surface. In: Xi, X.D., Tong, S.Y., van Hove, M.A. (eds.) The Structure of Surfaces IV, pp. 183–191. World Scientific Publ, Singapore (1994)
14. Altman, M.S., Bauer, E.: The reconstruction of the Pb(110) surface studied by low energy electron microscopy. Surf. Sci. **344**, 51–64 (1995)
15. Altman, M.S.: Evidence of a Pb(110) (2x2) reconstruction. Surf. Sci. **344**, 65–69 (1995)
16. Swiech, W., Appleton, R.S., Wiemeyer, B.D., Flynn, C.P.: LEEM investigation of refractory metals grown on insulating substrates: Nb(110) on sapphire (1120). Surf. Rev. Lett. **5**, 1221–1231 (1998)
17. Flynn, C.P., Świech, W.: Periodic states in the constrained equilibrium of vicinal Nb (011) miscut along [100]. Phys. Rev. Lett. **83**, 3482–3485 (1999)
18. Mundschau, M., Swiech, W., Durfee, C.S., Flynn, C.P.: Slip propagation in epitaxial Mo (011) studied by low-energy electron microscopy. Surf. Sci. **440**, L831–L834 (1999)
19. Swiech, W., Mundschau, M., Flynn, C.P.: Characterization of single crystal films of molybdenum (011) grown by molecular beam epitaxy on sapphire (1120) and studied by low-energy electron microscopy. Surf. Sci. **437**, 61–74 (1999)
20. Swiech, W., Mundschau, M., Flynn, C.P.: Interfacial defects in thin refractory metal films imaged by low-energy electron microscopy. Appl. Phys. Lett. **74**, 2626–2628 (1999)
21. Flynn, C.P., Swieçh, W., Appleton, R.S., Ondrejcek, M.: Nanofaceting of vicinal Nb(011). Phys. Rev. B **62**, 2096–2107 (2000)
22. Swiech, W., Ondrejcek, M., Appleton, R.S., Durfee, C.S., Mundschau, M., Flynn, C.P.: LEEM investigations of bcc metals grown heteroepitaxially on sapphire. In: Bentley, J.,

Allen, C., Dahmen, U., Petrov, I. (eds.) Proc. MRS Symposium advances in materials problem solving with the electron microscopy, 2001, pp. 27–38. MRS Symposium Proceedings. Boston, USA (2001)
23. Ondrejcek, M., Swiech, W., Rajappan, M., Flynn, C.P.: Ripples formed in the sputter erosion of Pd(111). J. Phys. Condens. Matter **15**, L735–L742 (2003)
24. Swiech, W., Rajappan, M., Ondrejcek, M., Sammann, E., Burdin, S., Petrov, I., Flynn, C.P.: Real-time imaging of surface evolution driven by variable-energy ion irradiation. Ultramicroscopy **108**, 646–655 (2008)
25. Burton, W.K., Cabrera, N., Frank, F.C.: The growth of crystals and the equilibrium structure of their surfaces. Phil. Trans. Royal Soc. London Ser A Math. Phys. Sci. **243**, 299–358 (1951)
26. Jeong, H.-C., Williams, E.D.: Steps on surfaces: experiment and theory. Surf. Sci. Rep. **34**, 171–294 (1999)
27. Giesen, M.: Step and island dynamics at solid/vacuum and solid/liquid interfaces. Prog. Surf. Sci. **68**, 1–153 (2001)
28. Flynn, C.P.: Step-edge fluctuations on crystal surfaces. Phys. Rev. B **66**, 155405, 13 pages (2002)
29. Misbah, C., Pierre-Louis, O., Saito, Y.: Crystal surfaces in and out of equilibrium: a modern view. Rev. Mod. Phys. **82**, 981–1040 (2010)
30. Bartelt, N.C., Goldberg, J.L., Einstein, T.L., Williams, E.D., Heyraud, J.C., Metois, J.J.: Brownian-motion of steps on Si(111). Phys. Rev. B **48**, 15453–15456 (1993)
31. Ondrejcek, M., Swiech, W., Yang, G., Flynn, C.P.: Low energy electron microscopy studies of steps on single crystal thin films of refractory metals. J. Vac. Sci. Technol. B **20**, 2473–2477 (2002)
32. Ondrejcek, M., Swiech, W., Durfee, C.S., Flynn, C.P.: Step fluctuations and step interactions on Mo(011). Surf. Sci. **541**, 31–45 (2003)
33. Flynn, C.P., Ondrejcek, M., Swiech, W.: Capillary waves and thermodynamics of multisteps on Pt(111). Chem. Phys. Lett. **378**, 161–166 (2003)
34. Ondrejcek, M., Swiech, W., Yang, G., Flynn, C.P.: Crossover from bulk to surface diffusion in the fluctuations of step edges on Pt(111). Phil. Mag. Lett. **84**, 69–77 (2004)
35. Ondrejcek, M., Swiech, W., Flynn, C.P.: Studies of step stiffnesses and relaxation on Pt(111), Pd(111) and Mo(011). Surf. Sci. **566**, 160–164 (2004)
36. Ondrejcek, M., Rajappan, M., Swiech, W., Flynn, C.P.: Step fluctuation spectroscopy of Au (111) by LEEM. Surf. Sci. **574**, 111–122 (2005)
37. Ondrejcek, M., Swiech, W., Rajappan, M., Flynn, C.P.: Fluctuation spectroscopy of step edges on Pt(111) and Pd(111). Phys. Rev. B **72**, 085422, 12 pages (2005)
38. Ondrejcek, M., Rajappan, M., Swiech, W., Flynn, C.P.: Low energy electron microscopy investigations of kinetics and energetics on clean close-packed metal surfaces. J. Phys. Condens. Matter **17**, S1397–S1406 (2005)
39. Ondrejcek, M., Swiech, W., Flynn, C.P.: From step fluctuations to nanostructure energetics on Nb(011). Phil. Mag. **86**, 1387–1400 (2006)
40. Ondrejcek, M., Rajappan, M., Swiech, W., Flynn, C.P.: Step fluctuation studies of surface diffusion and step stiffness for the Ni(111) surface. Phys. Rev. B **73**, 035418, 8 pages (2006)
41. Ondrejcek, M., Swiech, W., Flynn, C.P.: Surface mass diffusion and step stiffness on an anisotropic surface: Mo(011). Surf. Sci. **600**, 4673–4678 (2006)
42. Ondrejcek, M., Rajappan, M., Swiech, W., Flynn, C.P.: Surface mass diffusion and step stiffness on V(011). J. Appl. Phys. **100**, 083523, 6 pages (2006)
43. Rajappan, M., Swiech, W., Ondrejcek, M., Flynn, C.P.: Surface mass diffusion over an extended temperature range on Pt(111). J. Phys. Condens. Matter **19**, 226006, 8 pages (2007)
44. Flynn, C.P.: Point defect reactions at surfaces and in bulk metals. Phys. Rev. B **71**, 085422, 16 pages (2005)
45. Flynn, C.P.: Coupling among bulk, surface and edge diffusion. Phys. Rev. B **73**, 155417, 12 pages (2006)

46. Flynn, C.P.: Exact linear response of reacting thermal defects driven by creation processes. Phys. Rev. B **75**, 134106, 11 pages (2007)
47. McCarty, K.F., Nobel, J.A., Bartelt, N.C.: Vaccine in solids and the stability of surface morphology. Nature **412**, 622–625 (2001)
48. Poelsema, B., Hannon, J.B., Bartelt, N.C., Kellogg, G.L.: Bulk-surface vacancy exchange on Pt(111). Appl. Phys. Lett. **84**, 2551–2553 (2004)
49. Kellogg, G.L., Bartelt, N.C.: Surface-diffusion-limited island decay on Rh(001). Surf. Sci. **577**, 151–157 (2005)
50. Gerlach, R., Maroutian, T., Douillard, L., Martinotti, D., Ernst, H.J.: A novel method to determine the Ehrlich-Schwoebel barrier. Surf. Sci. **480**, 97–102 (2001)
51. Tang, S.J., Kodambaka, S., Swiech, W., Petrov, I., Flynn, C.P., Chiang, T.C.: Sublimation of Atomic Layers from a Chromium Surface. Phys. Rev. Lett. **96**, 126106, 4 pages (2006)
52. Rajappan, M., Swiech, W., Ondrejcek, M., Flynn, C.P.: Surface sink action during irradiation for Pt^- on Pt(111) by LEEM. Phil. Mag. **87**, 4501–4518 (2007)
53. Flynn, C.P., Swiech, W., Ondrejcek, M., Rajappan, M.: Surface islands nucleated by a beam of energetic self-ions. Phys. Rev. B **77**, 045406, 7 pages (2008)
54. Flynn, C.P., Ondrejcek, M., Swiech, W.: Synthesis of pans and mesas using a beam of self-ions. J. Phys. Condens. Matter **20**, 395001, 8 pages (2008)
55. Flynn, C.P., Swiech, W., Ondrejcek, M.: Universal growth of islands driven by ion beams: theory and measurements on Pt(111). Phys. Rev. B **78**, 075420, 9 pages (2008)
56. Ondrejcek, M., Swiech, W., Petrov, I., Rajappan, M., Flynn, C.P.: LEEM investigations of surfaces using a beam of energetic self-ions. Microsc. Res. Tech. **72**, 197–207 (2009)
57. Swiech, W., Ondrejcek, M., Flynn, C.P.: LEEM investigations of ion beam effects on clean metal surfaces: quantitative studies of the driven steady state. J. Phys. Condens. Matter **21**, 314021, 10 pages (2009)
58. Sato, Y., Ünal, B., Lograsso, T.A., Thiel, P.A., Schmid, A.K., Duden, T., Bartelt, N.C., McCarty, K.F.: Periodic step arrays on the aperiodic i-Al-Pd-Mn quasicrystal surface at high temperature. Phys. Rev. B **81**, 161406, 4 pages (2010)
59. Bennett, P.A., Webb, M.W.: The Si(111) 7x7 to "1x1" transition. Surf. Sci. **104**, 74–104 (1981)
60. Osakabe, N., Tanishiro, Y., Yagi, K., Honjo, G.: Direct observation of the phase transition between the (7x7) and (1x1) structures of clean (111) silicon surfaces. Surf. Sci. **109**, 353–366 (1981)
61. Tanishiro, Y., Takayanagi, K., Yagi, K.: On the phase transition between the (7x7) and (1x1) structures of the silicon (111) surface studied by reflection electron microscopy. Ultramicroscopy **11**, 95–102 (1983)
62. Telieps, W., Bauer, E.: Kinetics of the (7x7) (1x1) transition on Si(111). Ber. Bunsenges. Phys. Chem. **90**, 197–200 (1986)
63. Telieps, W., Bauer, E.: The (7x7)↔(1x1) phase transition on Si(111). Surf. Sci. **162**, 163–168 (1985)
64. Bauer, E., Telieps, W.: Low energy electron microscopy. In: Scanning Microscopy, pp. 99–108. Scanning Microscopy Int., Chicago (1987)
65. Telieps, W.: Surface imaging with LEEM. Appl. Phys. A **44**, 55–61 (1987)
66. Bauer, E., Mundschau, M., Swiech, W., Telieps, W.: Low energy electron microscopy (LEEM) and photoemission microscopy (PEEM) of semiconductor surfaces. In: Cherns, D. (ed.) NATO Advanced research workshop on the evaluation of advanced semiconductor materials by electron microscopy, 1988. NATO ASI series B, vol. 203, pp. 283–294. Plenum Publ. Corp. Bristol, England (1989)
67. Bauer, E., Mundschau, M., Swiech, W., Telieps, W.: Low energy electron microscopy of semiconductor surfaces. J. Vac. Sci. Technol. A **9**, 1007–1013 (1991)
68. Mundschau, M., Bauer, E., Telieps, W., Swiech, W.: Atomic step and defect structure on surfaces of Si(100) and Si(111) observed by low energy electron microscopy. Phil. Mag. A **61**, 257–280 (1990)

69. Sakai, Y., Kato, M., Masuda, S., Harada, Y., Ichinokawa, T.: Dynamical change of step and domain-structures on the Si(111) surface with heat-treatment observed by a low energy electron microskope. Surf. Sci. **336**, 295–302 (1995)
70. Telieps, W., Bauer, E.: Low energy electron microscopy (LEEM). In: van der Veen, J.F., Hove, M.A. (eds.) Structure of surfaces. Springer series in surface science, vol. 11, pp. 53–57. Springer, Berlin (1988)
71. Bauer, E., Mundschau, M., Swiech, W.: Low energy electron microscopy of nanometer scale phenomena. J. Vac. Sci. Technol. B **9**, 403–406 (1991)
72. Phaneuf, R.J., Bartelt, N.C., Williams, E.D., Swiech, W., Bauer, E.: LEEM Investigations of the domain growth of the (7x7) reconstruction on Si(111). Surf. Sci. **268**, 227–237 (1992)
73. Phaneuf, R.J., Swiech, W., Bartelt, N.C., Williams, E.D., Bauer, E.: LEEM investigations of orientational phase separation on vicinal Si(111) surfaces. Phys. Rev. Lett. **67**, 2986–2989 (1991)
74. Williams, E.D., Phaneuf, R.J., Bartelt, N.C., Swiech, W., Bauer, E.: The role of surface stress in the facetting of stepped Si(111) surfaces. MRS Symp. Proc. **1992**, 219–227 (1992)
75. Phaneuf, R.J., Bartelt, N.C., Williams, E.D., Swiech, W., Bauer, E.: Crossover from metastable to unstable facet growth on Si(111). Phys. Rev. Lett. **71**, 2284–2287 (1993)
76. Hannon, J.B., Hibino, H., Bartelt, N.C., Swartzentruber, B.S., Ogino, T., Kellogg, G.L.: Dynamics of the silicon (111) surface phase transition. Nature **405**, 552–554 (2000)
77. Hannon, J.B., Meyer zu Heringdorf, F.-J., Tersoff, J., Tromp, R.M.: Phase coexistence during surface phase transitions. Phys. Rev. Lett. **86**, 4871–4874 (2001)
78. Hannon, J.B., Tromp, R.M.: Phase boundary fluctuations on Si(111). J. Vac. Sci. Technol. A **19**, 2596–2600 (2001)
79. Hannon, J.B., Tersoff, J., Reuter, M.C., Tromp, R.M.: Influence of supersaturation on surface structure. Phys. Rev. Lett. **89**, 266103, 4 pages (2002)
80. Hannon, J.B., Tersoff, J., Tromp, R.M.: The stability of triangular 'droplet' phases on Si(111). J. Cryst. Growth **237–239**, 181–187 (2002)
81. Hannon, J.B., Tersoff, J., Tromp, R.M.: Surface stress and thermodynamic nanoscale size selection. Science **295**, 299–301 (2002)
82. Thayer, G.E., Hannon, J.B., Tromp, R.M.: Shape and stability of self-assembled surface domains. Nat Mater. **3**, 95–98 (2004)
83. Medhekar, N.V., Shenoy, V.B., Hannon, J.B., Tromp, R.M.: Metastability in 2D self-assembling systems. Phys. Rev. Lett. **99**, 156102, 4 pages (2007)
84. Medhekar, N.V., Shenoy, V.B., Hannon, J.B., Tromp, R.M.: Self-assembling surface stress domains far from equilibrium. Appl. Phys. Lett. **91**, 253101, 4 pages (2007)
85. Hu, C.W., Hibino, H., Ogino, T., Tsong, I.S.T.: Hysteresis in the (1 x 1)-(7 x 7) first-order phase transition on the Si(111) surface. Surf. Sci. **487**, 191–200 (2001)
86. Hibino, H., Homma, Y., Hu, C.W., Uwaha, M., Ogino, T., Tsong, I.S.T.: Structural and morphological changes on surfaces with multiple phases studied by low-energy electron microscopy. Appl. Surf. Sci. **237**, 51–57 (2004)
87. Tromp, R.M., Hannon, J.B.: Thermodynamics of nucleation and growth. Surf. Rev. Lett. **9**, 1565–1593 (2002)
88. Hannon, J.B., Tromp, R.M.: Low-energy electron microscopy of surface phase transitions. Annu. Rev. Mater. Res. **33**, 263–288 (2003)
89. Xu, Y.-C., Liu, B.-G.: Two-speed phase dynamics in the Si(111) (7x7) - (1x1) phase transition. Phys. Rev. Lett. **100**, 056103, 4 pages (2008)
90. Altman, M.S.: The role of diffusion on step flow growth. In: Tringides, M.C. (ed.) Proc. NATO ASI conference on surface diffusion: atomistic and collective processes rhodes, Greece 1996. Plenum Press, NATO ASI series, vol. B360, pp. 227–233, New York (1997)
91. Altman, M.S., Chung, W.F., Franz, T.: LEEM determination of critical terrace widths for Si/Si(111) step flow growth. Surf. Rev. Lett. **5**, 27–30 (1998)
92. Chung, W.F., Altman, M.S.: Kinetic length, step permeability, and kinetic coefficient asymmetry on the Si(111) (7x7) surface. Phys. Rev. B **66**, 075338, 5 pages (2002)

93. Chung, W.F., Bromann, K., Altman, M.S.: The transition to step flow growth on the clean and surfactant covered Si(111) surface studied by in-situ LEEM. Int. J. Mod. Phys. B **16**, 4353–4362 (2002)
94. Man, K.L., Tang, W.X., Jin, X.F., Altman, M.S.: Kinetic regime of step motion on the Si(111) (1x1) surface. Surf. Interface Anal. **38**, 1632–1635 (2006)
95. Man, K.L., Pang, A.B., Altman, M.S.: Kinetic length and step permeability on the Si(111) (1x1) surface. Surf. Sci. **601**, 4669–4674 (2007)
96. Pang, A.B., Man, K.L., Altman, M.S., Stasevich, T.J., Szalma, F., Einstein, T.L.: Step line tension and step morphological evolution on the Si(111) (1x1) surface. Phys. Rev. B **77**, 115424, 12 pages (2008)
97. Hibino, H., Hu, C.W., Ogino, T., Tsong, I.S.T.: Decay kinetics of two-dimensional islands and holes on Si(111) studied by low-energy electron microscopy. Phys. Rev. B **63**, 245402, 8 pages (2001)
98. Hibino, H., Hu, C.-W., Ogino, T., Tsong, I.S.T.: Diffusion barrier caused by 1x1 and 7x7 on Si(111) during phase transition. Phys. Rev. B **64**, 245401–245405 (2001)
99. Hibino, H., Watanabe, Y., Hu, C.-W., Tsong, I.S.T.: Thermal decay of superheated 7x7 islands and supercooled "1x1" vacancy islands on Si(111). Phys. Rev. B **72**, 245424, 6 pages (2005)
100. Hannon, J.B., Shenoy, V.B., Schwarz, K.W.: Anomalous spiral motion of steps near dislocations on silicon surfaces. Science **313**, 1266–1269 (2006)
101. Bauer, E., Telieps, W.: Emission and low energy reflection electron microscopy. In: Howie, A., Valdre, U. (eds.) Surface and Interface Characterization by Electron Optical Methods, vol. 191 NATO ASI Series B: Physics, pp. 195–233. Plenum Press, New York (1988)
102. Mundschau, M., Bauer, E., Telieps, W., Swiech, W.: Atomic steps on Si(100) and step dynamics during sublimation studied by low energy electron microscopy. Surf. Sci. **223**, 413–423 (1989)
103. Bauer, E.: Low energy electron microscopy of surface processes. Appl. Surf. Sci. **60/61**, 350–358 (1992)
104. Swiech, W., Bauer, E.: The growth of Si on Si(100): a video-LEEM study. Surf. Sci. **255**, 219–228 (1991)
105. Bauer, E., Mundschau, M., Swiech, W., Telieps, W.: Low energy electron microscopy of surface processes. Vacuum **41**, 5–10 (1990)
106. Bauer, E., Mundschau, M., Swiech, W., Telieps, W.: Low energy electron microscopy and surface processes on clean Si(111) and Si(100). In: Bringans, R.D., Feenstra, R.M., Gibson, J.M. (eds.) MRS symposium atomic scale structure of interfaces, vol. 159, pp. 225–233, Boston, Massachusetts 1989. Materials Research Society (1990)
107. Tromp, R.M., Dernier van der Gon, A.M., Reuter, M.C.: Surface stress as a driving force for interfacial mixing. Phys. Rev. Lett. **68**, 2313–2316 (1992)
108. Tromp, R.M., Reuter, M.C.: Step morphologies on small-miscut Si(001) surfaces. Phys. Rev. B **47**, 7598–7601 (1993)
109. Bartelt, N.C., Tromp, R.M., Williams, E.D.: Step capillary waves and equilibrium island shapes on Si(001). Phys. Rev. Lett. **73**, 1656–1659 (1994)
110. Theis, W., Bartelt, N.C., Tromp, R.M.: Chemical potential maps and spatial correlations in 2D-island ripening on Si(001). Phys. Rev. Lett. **75**, 3328–3331 (1995)
111. Theis, W., Tromp, R.M.: Nucleation in Si(001) homoepitaxial growth. Phys. Rev. Lett. **76**, 2770–2773 (1996)
112. Tanaka, S., Umbach, C.C., Blakely, J.M., Tromp, R.M., Mankos, M.: Fabrication of arrays of large stepfree regions on Si(001). Appl. Phys. Lett. **69**, 1235–1237 (1996)
113. Bartelt, N.C., Tromp, R.M.: Low-energy electron microscopy study of step mobilities on Si(001). Phys. Rev. B **54**, 11731–11740 (1996)
114. Bartelt, N.C., Theis, W., Tromp, R.M.: Ostwald ripening of two-dimensional islands on Si(001). Phys. Rev. B **54**, 11741–11751 (1996)

115. Tromp, R.M., Mankos, M.: Thermal adatoms on Si(001). Phys. Rev. Lett. **81**, 1050–1053 (1998)
116. Nielsen, J.-F., Pelz, J.P., Hibino, H., Hu, C.-W., Tsong, I.S.T.: Enhanced terrace stability for preparation of step-free Si(001)-(2x1) surfaces. Phys. Rev. Lett. **87**, 136103, 4 pages (2001)
117. Zielasek, V., Liu, F., Zhao, Y., Maxson, J.B., Lagally, M.G.: Surface stress-induced island shape transition in Si(001) homoepitaxy. Phys. Rev. B **64**, 201320 (2001)
118. Tanaka, S., Umbach, C.C., Blakely, J.M., Tromp, R.M., Mankos, M.: Atomic step distributions on annealed periodic Si(001) gratings. J. Vac. Sci. Technol. A **15**, 1345–1350 (1997)
119. Tanaka, S., Bartelt, N.C., Umbach, C.C., Tromp, R.M., Blakely, J.M.: Step permeability and the relaxation of biperiodic gratings on Si(001). Phys. Rev. Lett. **78**, 3342–3345 (1997)
120. Blakely, J.M., Tanaka, S., Tromp, R.M.: Atomic step dynamics on periodic semiconductor surface structures. J. Electron. Microsc. **48**, 747–752 (1999)
121. Jones, D.E., Pelz, J.P., Hong, Y., Bauer, E., Tsong, I.S.T.: Striped phase and temperature dependent step shape transition on highly B-doped Si(001)-(2x1) surfaces. Phys. Rev. Lett. **77**, 330–333 (1996)
122. Bauer, E.: LEEM and SPLEEM. In: Hawkes, P., Spence, J. (eds.) Science of Microscopy, vol. I, chapter 8, pp. 606–656. Springer, New York (2007)
123. Hannon, J.B., Bartelt, N.C., Swartzentruber, B.S., Hamilton, J.C., Kellogg, G.L.: Step faceting at the (001) surface of boron doped silicon. Phys. Rev. Lett. **79**, 4226–4229 (1997)
124. Hannon, J.B., Swartzentruber, B.S., Kellogg, G.L., Bartelt, N.C.: LEEM measurements of step energies at the (001) surface of heavily boron-doped silicon. Surf. Rev. Lett. **5**, 1159–1165 (1998)
125. Nielsen, J.-F., Pelz, J.P., Hibino, H., Hu, C.-W., Tsong, I.S.T., Kouvetakis, J.: Controlled striped phase formation on ultraflat Si(001) surfaces during diborane exposure. Appl. Phys. Lett. **79**, 3857–3859 (2001)
126. Ermanoski, I., Bartelt, N.C., Kellogg, G.L.: Self-assembly of defect-free nanostripe arrays on B-doped Si(001). Phys. Rev. B **83**, 205432, 5 pages (2011)
127. Hibino, H., Watanabe, Y.: Growth of twinned epitaxial layers on Si(111)$\sqrt{3}$x$\sqrt{3}$-B studied by low-energy electron microscopy. Jpn. J. Appl. Phys. **44**, 358–364 (2005)
128. Hibino, H., Uematsu, M., Watanabe, Y.: Void growth during thermal decomposition of silicon oxide layers studied by low-energy electron microscopy. J. Appl. Phys. **100**, 113519, 8 pages (2006)
129. Yamamoto, S., Masuda, S., Yasufuku, H., Ueno, N., Harada, Y., Ichinokawa, T., Kato, M., Sakai, Y.: Study of solid surfaces by metastable electron emission microscopy: energy-filtered images and local electron spectra at the outermost surface layer of silicon oxide on Si(100). J. Appl. Phys. **82**, 2954–2960 (1997)
130. Yasufuku, H., Okumura, M., Ibe, T., Okudaira, K., Harada, Y., Ueno, N.: Surface images of SiO_2/Si(100) pattern using electron emission microscopy with metastable atoms, photons and low-energy electrons. Jpn. J. Appl. Phys. **40**, 2447–2450 (2001)
131. Fukidome, H., Tanaka, K., Yoshimura, M., Ueda, K., Guo, F.-Z., Kinoshita, T., Kobayashi, K.: Microscopically controlled oxidation of H/Si(100) by lateral surface electric field studied by emission electron microscopies. Surf. Sci. **601**, 4675–4679 (2007)
132. Heun, S., Kremmer, S., Ercolani, D., Wurmbauer, H., Teichert, C.: Behavior of SiO_2 nanostructures under intense extreme ultraviolet illumination. J. Appl. Phys. **97**, 104333, 4 pages (2005)
133. de la Peña, F., Barrett, N., Zagonel, L.F., Walls, M., Renault, O.: Full field chemical imaging of buried native sub-oxide layers on doped silicon patterns. Surf. Sci. **604**, 1628–1636 (2010)
134. Wurm, K., Kliese, R., Hong, Y., Rottger, B., Wei, Y., Neddermeyer, H., Tsong, I.S.T.: Evolution of surface morphology of Si(100)-(2x1) during oxygen adsorption at elevated temperatures. Phys. Rev. B **50**, 1567–1574 (1994)
135. Hannon, J.B., Bartelt, M.C., Bartelt, N.C., Kellogg, G.L.: Etching of the Si(001) surface with molecular oxygen. Phys. Rev. Lett. **81**, 4676–4679 (1998)

136. Sutter, P., Lagally, M.G.: Quantitative determination of dislocation-induced strain at the surface of (001) silicon-on-insulator. Phys. Rev. Lett. **82**, 1490–1493 (1999)
137. Euaruksakul, C., Li, Z., Savage, D.E., Lagally, M.G.: X-ray absorption spectroscopy of strained-Si nanomembranes. ECS Trans **6**, 257–262 (2007)
138. Seo, S., Euaruksakul, C., Savage, D.E., Lagally, M.G., Evans, P.G.: Nanostructure formation in the initial roughening of a thin silicon sheet. Phys. Rev. B **81**, 041302, 4 pages (2010)
139. Flack, F.S., Yang, B., Huang, M., Marcus, M., Simmons, J., Castellini, O.M., Eriksson, M.A., Liu, F., Lagally, M.G.: Pattern formation on silicon-on-insulator. In: Chason, E., Gilmer, G. H., Huang, H., Wang, E. (eds.) Proc. MRS symposium KK—kinetics-driven nanopatterning on surfaces, vol. 849, pp. KK1.3.1–KK1.3.9, Boston, Massachusetts, USA 2004. Materials Research Society (2005)
140. Yang, B., Zhang, P., Savage, D.E., Lagally, M.G., Lu, G.-H., Huang, M., Liu, F.: Self-organization of semiconductor nanocrystals by selective surface faceting. Phys. Rev. B **72**, 235413, 4 pages (2005)
141. Bussmann, E., Cheynis, F., Leroy, F., Müller, P.: Thermal instability of silicon-on-insulator thin films measured by low-energy electron microscopy. In: Henrion, G., Belmahi, M., Andrieu, S. (eds.) Proc. innovations in thin film processing and characterisation (ITFPC 2009), Nancy, France 2009. IOP Conf. Series: Materials Science and Engineering, vol. 12, pp. 1–4. Institute of Physics, Bristol, England (2010)
142. Bussmann, E., Cheynis, F., Leroy, F., Müller, P., Pierre-Louis, O.: Dynamics of solid thin-film dewetting in the silicon-on-insulator system. New J. Phys. **13**, 043017, 9 pages (2011)
143. Cheynis, F., Bussmann, E., Leroy, F., Passanante, T., Müller, P.: Dewetting dynamics of silicon-on-insulator thin films. Phys. Rev. B **84**, 245439, 9 pages (2011)
144. Leroy, F., Cheynis, F., Passanante, T., Müller, P.: Dynamics, anisotropy, and stability of silicon-on-insulator dewetting fronts. Phys. Rev. B **85**, 195414, 6 pages (2012)
145. Swiech, W., Schwarz-Selinger, T., Cahill, D.G.: Phase coexistence and morphology at the Si(110) surface phase transition. Surf. Sci. **519**, L599–L603 (2002)
146. Watanabe, F., Kodambaka, S., Swiech, W., Greene, J.E., Cahill, D.G.: LEEM study of island decay on Si(110). Surf. Sci. **572**, 425–432 (2004)
147. Tromp, R.M., Theis, W., Bartelt, N.C.: Real-time microscopy of two-dimensional critical fluctuations: disordering of the Si(113)-(3x1) reconstruction. Phys. Rev. Lett. **77**, 2522–2525 (1996)
148. van Vroonhoven, E., Zandvliet, H.J.W., Poelsema, B.: (2x1)-(1x1) phase transition on Ge(001): dimer breakup and surface roughening. Phys. Rev. Lett. **91**, 116102, 4 pages (2003)
149. van Vroonhoven, E., Zandvliet, H.J.W., Poelsema, B.: A quantitative evaluation of the dimer concentration during the (2x1)-(1x1) phase transition on Ge(001). Surf. Sci. **574**, L23–L28 (2005)
150. Xu, Y.-C., Liu, B.-G.: Phase-field-crystal modelling of the (2x1)-(1x1) phase transitions of Si (001) and Ge(001) surfaces. J. Phys. D Appl. Phys. **42**, 035402, 5 pages (2009)
151. Tersoff, J., Jesson, D.E., Tang, W.X.: Decomposition controlled by surface morphology during langmuir evaporation of GaAs. Phys. Rev. Lett. **105**, 035702, 4 pages (2010)
152. Zhou, Z.Y., Zheng, C.X., Tang, W.X., Jesson, D.E., Tersoff, J.: Congruent evaporation temperature of GaAs(001) controlled by As flux. Appl. Phys. Lett. **97**, 121912, 3 pages (2010)
153. Pavlovska, A., Torres, V.M., Edwards, J.L., Bauer, E., Smith, D.J., Doak, R.B., Tsong, I.S.T., Thomson, D.B., Davis, R.F.: Homoepitaxial GaN layers studied by low-energy electron microscopy, atomic force microscopy and transmission electron microscopy. Phys. Status Solidi A **176**, 469–473 (1999)
154. Maxson, J.B., Perkins, N., Savage, D.E., Woll, A.R., Zhang, L., Kuech, T.F., Lagally, M.G.: Novel dark-field imaging of GaN {0001} surfaces with low-energy electron microscopy. Surf. Sci. **464**, 217–222 (2000)
155. Pavlovska, A., Bauer, E., Smith, D.J.: In situ studies of the role of excess Ga on the growth morphology of thin GaN layers. Surf. Sci. **496**, 160–178 (2002)

156. Hartman, J.D., Naniwae, K., Petrich, C., Nemanich, R.J., Davis, R.F.: Photo-electron emission and atomic force microscopies of the hydrogen etched 6H-SiC(0001) surface and the initial growth of GaN and AlN. Appl. Surf. Sci. **242**, 428–436 (2005)
157. Hannon, J.B., Tromp, R.M.: Pit formation during graphene synthesis on SiC(0001): in situ electron microscopy. Phys. Rev. B 77, 241404, 4 pages (2008)
158. Hannon, J.B., Tromp, R.M., Medhekar, N.V., Shenoy, V.B.: Spontaneous formation and growth of a new polytype on SiC(0001). Phys. Rev. Lett. **103**, 256101, 4 pages (2009)
159. Sun, J., Hannon, J.B., Tromp, R.M., Pohl, K.: Highly uniform step and terrace structures on SiC(0001) surfaces. IBM J. Res. Dev. **55**, 7, 6 pages (2011)
160. Hasegawa, M., Ohno, T.: Non-destructive observation of in-grown stacking faults in 4H-SiC epitaxial layer using mirror electron microscope. J. Appl. Phys. **110**, 073507, 6 pages (2011)
161. Kodambaka, S., Israeli, N., Bareno, J., Swiech, W., Ohmori, K., Petrov, I., Greene, J.E.: Low-energy electron microscopy studies of interlayer mass transport kinetics on TiN(111). Surf. Sci. **560**, 53–62 (2004)
162. Bareno, J., Kodambaka, S., Khare, S.V., Swiech, W., Petrov, I., Greene, J.E.: Orientation-dependent mobilities from analyses of two-dimensional TiN(111) island decay kinetics. Thin Solid Films **510**, 339–345 (2006)
163. Kodambaka, S., Khare, S.V., Swiech, W., Ohmori, K., Petrov, I., Greene, J.E.: Dislocation-driven surface dynamics on solids. Nature **429**, 49–52 (2004)
164. Kodambaka, S., Bareño, J., Khare, S.V., Świech, W., Petrov, I., Greene, J.E.: Nucleation and growth kinetics of spiral steps on TiN(111): an in situ low-energy electron microscopy study. J. Appl. Phys. **98**, 034901, 5 pages (2005)
165. McCarty, K.F., Bartelt, N.C.: Role of bulk thermal defects in the reconstruction dynamics of the TiO_2 (110) surface. Phys. Rev. Lett. **90**, 046104, 4 pages (2003)
166. McCarty, K.F., Bartelt, N.C.: The 1x1/1x2 phase transition of the TiO_2(110) surface–variation of transition temperature with crystal composition. Surf. Sci. **527**, L203–L212 (2003)
167. McCarty, K.F., Bartelt, N.C.: Spatially resolved dynamics of the TiO_2(110) surface reconstruction. Surf. Sci. **540**, 157–171 (2003)
168. McCarty, K.F.: Growth regimes of the oxygen-deficient TiO_2(110) surface exposed to oxygen. Surf. Sci. **543**, 185–206 (2003)
169. McCarty, K.F., Bartelt, N.C.: Crystal growth rate limited by step length - the case of oxygen-deficient TiO_2 exposed to oxygen. J. Cryst. Growth **270**, 691–698 (2004)
170. Menteş, T.O., Locatelli, A., Aballe, L., Pavlovska, A., Bauer, E., Pabisiak, T., Kiejna, A.: Surface modification of oxides by electron-stimulated desorption for growth-mode control of metal films: experiment and density-functional calculations. Phys. Rev. B **76**, 155413, 14 pages (2007)
171. Locatelli, A., Pabisiak, T., Pavlovska, A., Menteş, T.O., Aballe, L., Kiejna, A., Bauer, E.: One-dimensional Au on TiO_2. J. Phys.: Condens. Matter **19**, 082202, 8 pages (2007)
172. Xiong, G., Joly, A.G., Beck, K.M., Hess, W.P.: Laser-induced oxygen vacancy formation and diffusion on TiO_2(110) surfaces probed by photoemission electron microscopy. Phys. Status Solidi C **3**, 3598–3602 (2006)
173. Bethge, H., Klaua, M.: Photo-emission electron microscopy of work function changes. Ultramicroscopy **11**, 207–214 (1983)
174. Przychowski, M.D., Marx, G.K.L., Fecher, G.H., Schönhense, G.: A spatially resolved investigation of oxygen adsorption on polycrystalline copper and titanium by means of photoemission electron microscopy. Surf. Sci. **549**, 37–51 (2004)
175. Huang, W., Zhai, R., Bao, X.: Investigation of oxygen adsorption on Pd (100) with defects. Appl. Surf. Sci. **158**, 287–291 (2000)
176. Ermanoski, I., Swiech, W., Madey, T.E.: Nucleation and coexistence of nanometer-scale facet domains on O/Ir(210). Surf. Sci. Lett. **592**, L299–L304 (2005)
177. Reichelt, R., Günther, S., Wintterlin, J., Moritz, W., Aballe, L., Mentes, T.O.: Low energy electron diffraction and low energy electron microscopy microspot I/V analysis of the (4x4)O structure on Ag (111): surface oxide or reconstruction? J. Chem. Phys. **127**, 134706, 8 pages (2007)

178. Menteş, T.O., Stojić, N., Binggeli, N., Niño, M.A., Locatelli, A., Aballe, L., Kiskinova, M., Bauer, E.: Strain relaxation in small adsorbate islands: O on W(110). Phys. Rev. B **77**, 155414, 9 pages (2008)
179. Stojić, N., Menteş, T.O., Binggeli, N., Niño, M.A., Locatelli, A., Bauer, E.: Temperature dependence of surface stress across an order–disorder transition: p(1x2)O/W(110). Phys. Rev. B **81**, 115437, 7 pages (2010)
180. Altman, M.S., Bauer, E.: The reaction of oxygen with the hot W(001) surface studied by low-energy electron microscopy. J. Vac. Sci. Technol. A **9**, 659–660 (1991)
181. Altman, M.S., Bauer, E.: LEEM/LEED investigation of reconstruction and initial oxidation of the W(001). Surf. Sci. **347**, 265–279 (1996)
182. Flege, J.I., Hrbek, J., Sutter, P.: Structural imaging of surface oxidation and oxidation catalysis on Ru(0001). Phys. Rev. B **78**, 165407, 5 pages (2008)
183. Ondrejcek, M., Appleton, R.S., Swiech, W., Petrova, V.L., Flynn, C.P.: Thermally Activated Stripe Reconstruction Induced by O on Nb (011). Phys. Rev. Lett. **87**, 116102, 4 pages (2001)
184. Rotermund, H.H., Nettesheim, S., von Oertzen, A., Ertl, G.: Observation of surface diffusion of adsorbates on metal surfaces by photoemission electron microscopy (PEEM). Surf. Sci. Lett. **275**, L645–L649 (1992)
185. Rotermund, H.H.: Investigation of dynamic processes in adsorbed layers by photoemission electron microscopy (PEEM). Surf. Sci. **283**, 87–100 (1993)
186. von Oertzen, A., Rotermund, H.H., Nettesheim, S.: Investigation of diffusion of CO adsorbed on Pd(111) by a combined PEEM/LITD technique. Chem. Phys. Lett. **199**, 131–137 (1992)
187. Yim, C.M., Man, K.L., Xiao, X., Altman, M.S.: Low-energy electron microscopy of CO/Pt (111) surface diffusion by nonequilibrium coverage profile evolution. Phys. Rev. B **78**, 155439, 9 pages (2008)
188. Cerdá, J.I., Santos, B., Herranz, T., Puerta, J.M., de la Figuera, J., McCarty, K.F.: CO-assisted subsurface hydrogen trapping in Pd(111) films. J. Phys. Chem. Lett. **3**, 87–91 (2012)
189. Ling, W.L., Bartelt, N.C., Pohl, K., de la Figuera, J., Hwang, R.Q., McCarty, K.F.: Enhanced self-diffusion on Cu(111) by trace amounts of S: chemical-reaction-limited kinetics. Phys. Rev. Lett. **93**, 166101, 4 pages (2004)
190. Przychowski, M.D., Wiechert, H., Marx, G.K.L., Schönhense, G.: Real-space observation of xenon adsorption and desorption kinetics on graphite (0001) by photoemission electron microscopy. Surf. Sci. **541**, 46–58 (2003)
191. Rotermund, H.H., Ertl, G., Sesselmann, W.: Scanning photoemission microscopy of surfaces. Surf. Sci. **217**, L383–L390 (1989)
192. Lilienkamp, G., Wei, H., Maus-Friedrichs, W., Kempter, V., Marbach, H., Günther, S., Suchorski, Y.: Metastable impact electron emission microscopy of the catalytic H_2 oxidation on Rh(110). Surf. Sci. **532–535**, 132–136 (2003)
193. Lmbihl, R., Ertl, G.: Oscillatory kinetics in heterogeneous catalysis. Chem. Rev. **95**, 697–733 (1995)
194. Rotermund, H.H.: Imaging of dynamic processes on surfaces by light. Surf. Sci. Rep. **29**, 265–364 (1997)
195. Imbihl, R.: Non-linear dynamics in catalytic reactions. In: Hasselbrink, E., Lundquist, B. (eds.) Handbook of Surface Science, vol. 3, pp. 341–428. Elsevier, Amsterdam (2008)
196. Imbihl, R.: Nonlinear dynamics on catalytic surfaces: the contribution of surface science. Surf. Sci. **603**, 1671–1679 (2009)
197. Rotermund, H.H.: Real time imaging of catalytic reactions on surfaces: past, present and future. Surf. Sci. **603**, 1662–1670 (2009)
198. Nettesheim, S., von Oertzen, A., Rotermund, H.H., Ertl, G.: Reaction diffusion patterns in the catalytic CO-oxidation on Pt(110): front propagation and spiral waves. J. Chem. Phys. **98**, 9977–9985 (1993)
199. Sadeghi, P., Dunphy, K., Punckt, C., Rotermund, H.H.: Inversion of pattern anisotropy during CO oxidation on Pt(110) correlated with appearance of subsurface oxygen. J. Phys. Chem. C **116**, 4686–4691 (2012)

200. Wei, H., Lilienkamp, G., Imbihl, R.: Surface topographical changes and chemical wave patterns in catalytic CO oxidation on Pt(110). Chem. Phys. Lett. **389**, 284–288 (2004)
201. Schmidt, T., Schaak, A., Guenther, S.G., Ressel, B., Bauer, E., Imbihl, R.: In situ imaging of structural changes in a chemical wave with low energy electron microscopy: the system Rh(110)/NO + H_2. Chem. Phys. Lett. **318**, 549–554 (2000)
202. Locatelli, A., Mentes, T.O., Aballe, L., Mikhailov, A., Kiskinova, M.: Formation of regular surface-supported mesostructures with periodicity controlled by chemical reaction rate. J. Phys. Chem. B **110**, 19108–19111 (2006)
203. Locatelli, A., Bauer, E.: Recent advances in chemical and magnetic imaging of surfaces and interfaces by XPEEM. J. Phys. Condens. Matter **20**, 82202–82224 (2008) (and references therein)
204. Locatelli, A., Heun, S., Kiskinova, M.: Direct observation of reaction-induced lateral redistribution of sub-monolayers of Au deposited on a Rh(110) surface. Surf. Sci. **566–568**, 1130–1136 (2004)
205. Locatelli, A., Sbraccia, C., Heun, S., Baroni, S., Kiskinova, M.: Energetically driven reorganization of a modified catalytic surface under reaction conditions. J. Am. Chem. Soc. **127**, 2351–2357 (2005)
206. Locatelli, A., Aballe, L., Menteş, T.O., Kiskinova, M., Bauer, E.: Photoemission electron microscopy with chemical sensitivity: SPLEEM methods and applications. Surf. Interface Anal. **38**, 1554–1557 (2006)
207. Locatelli, A., Aballe, L., Menteş, T.O., Guo, F.Z., Kiskinova, M.: A spectro-microscopic study of the reactive phase separation of Au + Pd and O on Rh(110). Surf. Sci. **601**, 4663–4668 (2007)
208. Hong, L., Uecker, H., Hinz, M., Qiao, L., Kevrekidis, I.G., Günther, S., Mentes, T.O., Locatelli, A., Imbihl, R.: Mass transport of alkali metal with pulses in a surface reaction. Phys. Rev. E **78**, 055203(R), 4 pages (2008)
209. Hong, L., Hesse, M., Imbihl, R.: Patterns in the NO + H_2 reaction on Rh(110) modified by coadsorbed potassium. J. Phys. Chem. C **113**, 4174–4183 (2009)
210. Lovis, F., Hesse, M., Imbihl, R.: Redistribution of supported vanadium oxide catalysts by pattern formation. Catal. Lett. **136**, 171–176 (2010)
211. Lovis, F., Imbihl, R.: Self-organization of ultrathin vanadium oxide layers on a Rh(111) surface during a catalytic reaction. Part I: a PEEM study. J. Phys. Chem. C **115**, 19141–19148 (2011)
212. Lovis, F., Hesse, M., Locatelli, A., Mentes, T.O., Nino, M.A., Lilienkamp, G., Borkenhagen, B., Imbihl, R.: Self-organization of ultrathin vanadium oxide layers on a Rh(111) surface during a catalytic reaction. Part II: a LEEM and spectromicroscopy study. J. Phys. Chem. C **115**, 19149–19157 (2011)
213. Lovis, F., Smolinsky, T., Locatelli, A., Niño, M.Á., Imbihl, R.: Chemical waves and rate oscillations in the $H_2 + O_2$ reaction on a bimetallic Rh(111)/Ni catalyst. J. Phys. Chem. C **116**, 4083–4090 (2012)
214. Imbihl, R.: Chemical selforganization of composite catalysts during catalytic reactions. J. Electron. Spectrosc. Rel. Phenom. **185**, 347–355 (2012)
215. Suchorski, Y., Spiel, C., Vogel, D., Drachsel, W., Schlögl, R., Rupprechter, G.: Local reaction kinetics by imaging: CO oxidation on polycrystalline platinum. Chem. Phys. Chem. **11**, 3231–3235 (2010)
216. Spiel, C., Vogel, D., Suchorski, Y., Drachsel, W., Schlögl, R., Rupprechter, G.: Catalytic CO oxidation on individual (110) domains of a polycrystalline Pt foil: local reaction kinetics by PEEM. Catal. Lett. **141**, 625–632 (2011)
217. Vogel, D., Spiel, C., Suchorski, Y., Urich, A., Schlögl, R., Rupprechter, G.: Mapping the local reaction kinetics by PEEM: CO oxidation on individual (100)-type grains of Pt foil. Surf. Sci. **605**, 1999–2005 (2011)

218. Vogel, D., Spiel, C., Suchorski, Y., Trinchero, A., Schlögl, R., Grönbeck, H., Rupprechter, G.: Local catalytic ignition during CO oxidation on low-index Pt and Pd surfaces: a combined PEEM, MS, and DFT study. Angew. Chem. Int. Ed. **51**, 10041–10044 (2012)
219. Vogel, D., Budinska, Z., Spiel, C., Schlögl, R., Suchorski, Y., Rupprechter, G.: Silicon oxide surface segregation in CO oxidation on Pd: an in situ PEEM, MS and XPS study. Catal. Lett. **143**, 235–240 (2013)
220. Günther, S., Liu, H., Menteş, T.O., Locatelli, A., Imbihl, R.: Spectromicroscopy of pulses transporting alkali metal in a surface reaction. Phys. Chem. Chem. Phys. **15**, 8752–8764 (2013)
221. Rafti, M., Uecker, H., Lovis, F., Krupennikova, V., Imbihl, R.: Traveling interface modulations in the $NH_3 + O_2$ reaction on a Rh(110) surface. Phys. Chem. Chem. Phys. **14**, 5260–5264 (2012)
222. Rettew, R.E., Meyer, A., Senanayake, S.D., Chen, T.-L., Petersburg, C., Flege, J.I., Falta, J., Alamgir, F.M.: Interactions of oxygen and ethylene with submonolayer Ag films supported on Ni(111). Phys. Chem. Chem. Phys. **13**, 11034–11044 (2011)
223. Bethge, H., Krajewski, T., Lichtenberger, O.: Investigation of ultra-thin Ag films on Ni with the photoelectron emission microscope. Ultramicroscopy **17**, 21–30 (1985)
224. Tikhov, M., Stolzenberg, M., Bauer, E.: New type of two-dimensional phase transition. Phys. Rev. B **36**, 8719–8724 (1987)
225. Bauer, E., Mundschau, M., Swiech, W., Telieps, W.: Surface Studies by low energy electron microscopy (LEEM) and conventional UV photoemission electron microscopy. Ultramicroscopy **31**, 49–57 (1989)
226. Altman, M.S., Luo, E.Z., Chung, W.F., Orr, B.G.: Growth kinetics and structure of ultrathin copper films on the W(110) surface studied by LEEM. In: Zangwill, A. (ed.) Proc. MRS symposium evolution of epitaxial structure and morphology, vol. 399, pp. 523–528, Boston, Massachusetts, USA 1995. MRS Symposium Proceedings Series, Pittsburgh (1996)
227. Luo, E.Z., Cai, Q., Chung, W.F., Orr, B.G., Altman, M.S.: Competing desorption pathways during epitaxial growth: LEEM investigation of Cu/W(110) heteroepitaxy. Phys. Rev. B **54**, 14673–14678 (1996)
228. Koshikawa, T., Shimizu, H., Nakaguchi, A., Yasue, T., Bauer, E.: Dynamic measurements of ultra thin film growth by using LEEM and PEEM. In: Proc. atomic level characterizations for new materials and devices (ALC'05) Big Island, Hawaii, USA 2005, pp. 653–657. JSPS 141 Activity Report (2006)
229. Shimizu, H., Nakaguchi, A., Yasue, T., Koshikawa, T., Bauer, E.: Dynamic observation of structure change on double layer Cu/W(110) by means of LEEM and selected area LEED. In: Proc. atomic level characterizations for new materials and devices (ALC'05) Big Island, Hawaii, USA 2005, pp. 694–697. JSPS 141 Activity Report (2006)
230. Mundschau, M., Bauer, E., Telieps, W., Swiech, W.: In situ studies of epitaxial growth in the low energy electron microscope. Surf. Sci. **213**, 381–392 (1989)
231. Duden, T.: Entwicklung und Anwendung der Polarisationsmanipulation in der Niederenergie-Elektronenmikroskopie. Ph.D. thesis, Technical University Clausthal, Germany (1996)
232. de la Figuera, J., Leonard, F., Bartelt, N.C., Stumpf, R., McCarty, K.F.: Nanoscale Periodicity in Stripe-Forming Systems at High Temperature: Au/W(110). Phys. Rev. Lett. **100**, 186102, 4 pages (2008)
233. Menteş, T.O., Locatelli, A., Aballe, L., Bauer, E.: Stress Induced Stripe Formation in Pd/W (110). Phys. Rev. Lett. **101**, 085701, 4 pages (2008)
234. Stojić, N., Menteş, T.O., Binggeli, N.: Self-organization in Pd/W(110): interplay between surface structure and stress. J. Phys. Condens. Matter **25**, 355010, 12 pages (2013)
235. Menteş, T.O., Stojić, N., Locatelli, A., Aballe, L., Binggeli, N., Nino, M.A., Kiskinova, M., Bauer, E.: Stress engineering at the nanometer scale: two-component adlayer stripes. Europhys. Lett. **94**, 38003, 6 pages (2011)

236. Georgiev, N.: LEED-diffractometry of phase transitions in chemisorbed layers and on single crystal surfaces. Ph.D. thesis, Technische Universität Clausthal, Germany (1996)
237. Gao, Y.F., Lu, W., Suo, Z.: A mesophase transition in a binary monolayer on a solid surface. Acta Mater. **50**, 2297–2308 (2002)
238. Lu, W., Suo, Z.: Symmetry breaking in self-assembled monolayers on solid surfaces: anisotropic surface stress. Phys. Rev. B **65**, 085401, 9 pages (2002)
239. Gao, Y.F., Suo, Z.: The orientation of the self-assembled monolayer stripes on a crystalline substrate. J. Mech. Phys. Solids **51**, 147–167 (2003)
240. Lu, W., Suo, Z.: Symmetry breaking in self-assembled monolayers on solid surfaces. II. Anisotropic substrate elasticity. Phys. Rev. B **65**(205418, 7 pages) (2002)
241. Léonard, F., Bartelt, N.C., Kellogg, G.L.: Effects of elastic anisotropy on the periodicity and orientation of striped stress domain patterns at solid surfaces. Phys. Rev. B **71**, 045416, 7 pages (2005)
242. de la Figuera, J., Bartelt, N.C., McCarty, K.F.: Electron reflectivity measurements of Ag adatom concentrations on W(110). Surf. Sci. **600**, 4062–4066 (2006)
243. Yasue, T., Amakawa, R., Shimizu, H., Nakaguchi, A., Koshikawa, T., Bauer, E.: Anomalous step contrast in LEEM during growth of Pb on W(110). In: Proc. atomic level characterizations for new materials and devices (ALC'05) Big Island, Hawaii, USA 2005, pp. 680–683. JSPS 141 Activity Report (2006)
244. Yasue, T., Amakawa, R., Shimizu, H., Nakaguchi, A., Koshikawa, T., Bauer, E.: Step contrast reversal in LEEM during Pb deposition on W(110). J. Phys. Condens. Matter **21**, 314024, 5 pages (2009)
245. Pelhos, K., Hannon, J.B., Kellogg, G.L., Madey, T.E.: LEEM investigation of the faceting of the Pt covered W(111) surface. Surf. Sci. **432**, 115–124 (1999)
246. Pelhos, K., Madey, T.E., Hannon, J.B., Kellogg, G.L.: Nucleation and growth of the platinum-covered W(111) surface. Surf. Rev. Lett. **6**, 767–774 (1999)
247. Wu, D., Lau, W.K., He, Z.Q., Feng, Y.J., Altman, M.S., Chan, C.T.: Ordered alloying of Pd with the Mo(100) surface. Phys. Rev. B **62**, 8366–8375 (2000)
248. Man, K.L., Feng, Y.J., Altman, M.S.: Formation kinetics of the Mo(100)-Ag c(2 x 2) surface alloy. Phys. Rev. B **74**, 085420, 11 pages (2006)
249. Man, K.L., Feng, Y.J., Chan, C.T., Altman, M.S.: Vibrational entropy-driven dealloying of Mo(100) and W(100) surface alloys. Surf. Sci. **601**, L95–L101 (2007)
250. Kellogg, G.L., Plass, R.: The relationship between the growth shape of three-dimensional Pb islands on Cu(100) and the domain orientation of the underlying c(5√ 2 x √ 2)R45 structure. Surf. Sci. **465**, L777–L782 (2000)
251. Plass, R., Kellogg, G.L.: Surface morphology changes during Pb deposition on Cu(100): evidence for surface alloyed Cu(100)-c(2 x 2) Pb. Surf. Sci. **470**, 106–120 (2000)
252. Kellogg, G.L., Plass, R.A.: Mesoscopic scale observations of surface alloying, surface phase transitions, domain coarsening, and 3D island growth: Pb on Cu(100). Surf. Rev. Lett. **7**, 649–655 (2000)
253. Hannon, J.B., Sun, J., Pohl, K., Kellogg, G.L.: Origins of nanoscale heterogeneity in ultrathin films. Phys. Rev. Lett. **96**, 246103, 4 pages (2006)
254. Sun, J., Hannon, J.B., Kellogg, G.L., Pohl, K.: Local structural and compositional determination via electron scattering: heterogeneous Cu(001)-Pd surface alloy. Phys. Rev. B **76**, 205414, 10 pages (2007)
255. Bussmann, E., Sun, J., Pohl, K., Kellogg, G.L.: Palladium diffusion into bulk copper via the Cu(100) surface. J. Phys. Condens. Matter **21**, 314016, 7 pages (2009)
256. Bussmann, E., Ermanoski, I., Feibelman, P.J., Bartelt, N.C., Kellogg, G.L.: Buried Pd slows self-diffusion on Cu(001). Phys. Rev. B **84**, 245440, 10 pages (2011)
257. Schmid, A.K., Bartelt, N.C., Hwang, R.Q.: Alloying at surfaces by the migration of reactive two-dimensional islands. Science **290**, 1561–1564 (2000)
258. Plass, R., Last, J.A., Bartelt, N.C., Kellogg, G.L.: Nanostructures: self-assembled domain patterns. Nature **412**, 875 (2001)

259. Plass, R., Bartelt, N.C., Kellogg, G.L.: Dynamic observations of nanoscale self-assembly on solid surfaces. J. Phys. Condens. Matter **14**, 4227–4240 (2002)
260. van Gastel, R., Plass, R., Bartelt, N.C., Kellogg, G.L.: Thermal motion and energetics of self-assembled domain structures: Pb on Cu(111). Phys. Rev. Lett. **91**, 055503, 4 pages (2003)
261. van Gastel, R., Bartelt, N.C., Feibelman, P.J., Leonard, F., Kellogg, G.L.: Relationship between domain-boundary free energy and the temperature dependence of stress-domain patterns of Pb on Cu(111). Phys. Rev. B **70**, 245413, 6 pages (2004)
262. van Gastel, R., Bartelt, N.C., Kellogg, G.L.: Reversible shape transition of Pb islands on Cu (111). Phys. Rev. Lett. **96**, 36106, 4 pages (2006)
263. Anderson, M.L., Bartelt, N.C., Feibelman, P.J., Swartzentruber, B.S., Kellogg, G.L.: The effect of embedded Pb on Cu diffusion on Pb/Cu(111) surface alloys. Surf. Sci. **600**, 1901–1908 (2006)
264. Anderson, M.L., Bartelt, N.C., Feibelman, P.J., Swartzentruber, B.S., Kellogg, G.L.: How Pb-overlayer islands move fast enough to self-assemble on Pb-Cu surface alloys. Phys. Rev. Lett. **98**, 096106, 4 pages (2007)
265. Rougemaille, N., El Gabaly, F., Stumpf, R., Schmid, A.K., Thürmer, K., Bartelt, N.C., de la Figuera, J.: Labyrinthine island growth during Pd/Ru(0001) heteroepitaxy. Phys. Rev. Lett. **99**, 106101, 4 pages (2007)
266. Ling, W.L., Hamilton, J.C., Thurmer, K., Thayer, G.E., de la Figuera, J., Hwang, R.Q., Carter, C.B., Bartelt, N.C., McCarty, K.F.: Herringbone and triangular patterns of dislocations in Ag, Au, and AgAu alloy films on Ru(0001). Surf. Sci. **600**, 1735–1757 (2006)
267. Swiech, W., Bauer, E., Mundschau, M.: A low energy electron microscopy study of the system Si(111)-Au. Surf. Sci. **253**, 283–296 (1991)
268. Slezak, J., Ondrejcek, M., Chvoj, Z., Chab, V., Conrad, H., Heun, S., Schmidt, T., Ressel, B., Prince, K.C.: Surface diffusion of Au on Si(111): a microscopic study. Phys. Rev. B **61**, 16121–16128 (2000)
269. Ferralis, N., El Gabaly, F., Schmid, A.K., Maboudian, R., Carraro, C.: Real-time observation of reactive spreading of gold on silicon. Phys. Rev. Lett. **103**, 256102, 4 pages (2009)
270. Dernier van der Gon, A.W., Tromp, R.M.: Phase formation sequence and domain structure: a low-energy-electron-microscopy study of the Si(111)-($\sqrt{3}$ x$\sqrt{3}$) Ag surface. Phys. Rev. Lett. **69**, 3519–3522 (1992)
271. Hashimoto, M., Guo, F.Z., Suzuki, M., Ueda, M., Matsuoka, Y., Kinoshita, T., Kobayashi, K., Shin, S., Oura, M., Takeuchi, T., Saito, Y., Matsushita, T., Yasue, T., Koshikawa, T.: Mapping of chemical bonding states of Ag/Si(111) with synchrotron radiation photo emission electron microscopy. Surf. Interface Anal. **40**, 1772–1776 (2008)
272. Wall, D., Lohmar, I., Roos, K.R., Krug, J., Horn-von Hoegen, M., Meyer zu Heringdorf, F.-J.: Imaging diffusion fields on a surface with multiple reconstructions: Ag/Si(111). New J. Phys. **12**, 103019, 8 pages (2010)
273. Mundschau, M., Bauer, E., Telieps, W., Swiech, W.: Initial epitaxial growth of copper silicide on Si(111) studied by low energy electron microscopy and photoemission electron microscopy. J. Appl. Phys. **65**, 4747–4752 (1989)
274. Yasue, T., Koshikawa, T., Jalochowski, M., Bauer, E.: Dynamic LEEM observation of Cu nanostructure formation processes on Si(111) with hydrogen. Surf. Rev. Lett. **7**, 595–599 (2000)
275. Yasue, T., Koshikawa, T., Jalochowski, M., Bauer, E.: Dynamic observations of the formation of thin Cu layers on clean and hydrogen-terminated Si(111) surfaces. Surf. Sci. **480**, 118–127 (2001)
276. Bennett, P.A., Lee, M.Y., Parikh, S.A., Wurm, K., Phaneuf, R.J.: Surface phase transformations in the Ni/Si(111) system observed in real time using low-energy electron microscopy. J. Vac. Sci. Technol. A **13**, 1728–1732 (1995)
277. Phaneuf, R.J., Hong, Y., Horch, S., Bennett, P.A.: Two dimensional phase separation for Co adsorbed on Si(111). Phys. Rev. Lett. **78**, 4605–4608 (1997)

278. Phaneuf, R.J., Bennett, P.A.: LEEM studies of phase separations and surface depletion of Co and Ni on Si(111). Surf. Rev. Lett. **5**, 1179–1188 (1998)
279. Phaneuf, R.J., Bennett, P.A., Marsi, M., Gunther, S., Gregoratti, L., Casalis, L., Kiskinova, M.: Equilibration of ring-cluster surface phases and silicide islands for Co adsorbed on Si(111). Surf. Sci. **431**, 232–241 (1999)
280. Phaneuf, R.J., Hong, Y., Horch, S., Bennett, P.A.: Observations of reversible and irreversible structural transitions of cobalt on Si(111) with LEEM. Micron **30**, 13–20 (1999)
281. Tromp, R.M., Michely, T.: Atomic-layer titration of surface reactions. Nature **373**, 499–501 (1995)
282. Michely, T., Reuter, M.C., Tromp, R.M.: Al on Si(111): phase diagram and atomic mechanisms. Phys. Rev. B **53**, 4105–4108 (1996)
283. Man, K.L., Tringides, M.C., Loy, M.M.T., Altman, M.S.: Anomalous Mass Transport in the Pb Wetting Layer on the Si(111) Surface. Phys. Rev. Lett. **101**, 226102, 4 pages (2008)
284. Pavlovska, A., Bauer, E., Giesen, M.: Low energy electron microscopy study of In on Si(111). J. Vac. Sci. Technol. B **20**, 2478–2491 (2002)
285. Pavlovska, A., Bauer, E.: Surface reconstructions in two and three dimensions: In on Si(111). Surf. Interface Anal. **37**, 110–114 (2005)
286. Shimizu, H., Nakaguchi, A., Yasue, T., Koshikawa, T., Guo, F-Z., Tamenori, Y., Matsushita, T., Kinoshita,T., Kobayashi, K., Bauer, E.: Surface chemical analysis on the selected area of In/Si(111) by means of LEEM, LEED and SR-XPEEM. In: Proc. atomic level characterizations for new materials and devices (ALC'05) Big Island, Hawaii, USA 2005, pp. 710–713. JSPS 141 Activity Report (2006)
287. Nakaguchi, A., Guo, F.Z., Hashimoto, M., Ueda, M., Yasue, T., Matsushita, T., Tamenori, Y., Kinoshita, T., Kobayashi, K., Koshikawa, T.: Sb on In/Si(111) processes with dynamically observable LEEM, selected area LEED and chemically analyzed SR-XPEEM. Surf. Interface Anal. **38**, 1773–1776 (2006)
288. Ueda, M., Guo, F-Z., Nakaguchi, A., Suzuki, M., Hashimoto, M., Matsuoka, T., Kinoshita, T., Kobayashi, K., Shin, S., Oura, M., Takeuchi, T., Saito, Y., Yasue, T., Koshikawa, T.: Observation of the growth process Sb on In/Si(111) by observation of the growth process Sb on In/Si(111) by the synchrotron radiation XPEEM and LEEM. In: Proc. atomic level characterizations for new materials and devices (ALC'07), Kanasawa, Japan 2007, pp. 216–219. JSPS 141 Activity Report (2008)
289. Matsuoka, Y., Guo, F-Z., Hashimoto, M., Suzuki, M., Ueda, M., Yasue, T., Kinoshita, T., Kobayashi, K., Shin, S., Oura, M., Takeuchi, T., Saito, Y., Koshikawa, T.: Ag on In/Si(111) processes by means of LEEM, selected area XPS and chemically analyzed SR -XPEEM. In: Proc. atomic level characterizations for new materials and devices (ALC'7), Kanasawa, Japan 2007, pp. 212–215. JSPS 141 Activity Report (2008)
290. Schmidt, T., Gangopadhyay, S., Flege, J.I., Clausen, T., Locatelli, A., Heun, S., Falta, J.: Self-organized 2D nanopatterns after low-coverage Ga adsorption on Si(111). New J. Phys. **7**, 193, 11 pages (2005)
291. Horn-von Hoegen, M., Meyer zu Heringdorf, F.-J., Kaehler, D., Schmidt, T., Bauer, E.: Adsorption induced giant faceting of vicinal Si(100). Thin Solid Films **336**, 16–21 (1998)
292. Meyer zu Heringdorf, F.-J., Kaehler, D., Schmidt, T., Bauer, E., Copel, M., Minoda, H., Horn-von Hoegen, M.: Giant faceting of vicinal Si(001) induced by Au adsorption. Surf. Rev. Lett. **5**, 1167–1178 (1998)
293. Horn-von Hoegen, M., Meyer zu Hennigdorf, F.-J., Hild, R., Zahl, P., Schmidt, T., Bauer, E.: Au-induced giant faceting of vicinal Si(100). Surf. Sci. **433–435**, 475–480 (1999)
294. Meyer zu Heringdorf, F.-J., Hild, R., Zahl, P., Schmidt, T., Ressel, B., Heun, S., Bauer, E., Horn-von Hoegen, M.: Local Au coverage as driving force for Au induced faceting of vicinal Si(001): a LEEM and XPEEM study. Surf. Sci. **480**, 103–108 (2001)
295. Meyer zu Heringdorf, F.-J., Schmidt, T., Heun, S., Hild, R., Zahl, P., Ressel, B., Bauer, E., Horn-von Hoegen, M.: Spatial variation of Au coverage as driving force for nanoscopic pattern formation. Phys. Rev. Lett. **86**, 5088–5091 (2001)

296. Michely, T., Reuter, M.C., Copel, M., Tromp, R.M.: Symmetry, structure, and step induced ordering of the Si(001)-(2 x 3)Ag surface. Phys. Rev. Lett. **73**, 2095–2098 (1994)
297. Roos, K.R., Roos, K.L., Lohmar, I., Wall, D., Krug, J., Horn-von Hoegen, M., Meyer zu Heringdorf, F.-J.: Real-time view of mesoscopic surface diffusion. Phys. Rev. Lett. **100**, 016103, 4 pages (2008)
298. Wall, D., Roos, K.R., Horn-von Hoegen, M., Meyer zu Heringdorf, F.-J.: LEEM/PEEM study of anisotropic diffusion fields in the Ag/Si(001) system. In: Adler, D.L., Bauer, E., Kellogg, G.L., Scholl, A. (eds.) Proc. MRS symposium advances and applications of surface electron microscopy, San Francisco 2008. MRS Proceedings, vol. 1088, pp. W05–04 (2008)
299. Sindermann, S., Wall, D., Roos, K.R., Horn-von Hoegen, M., Meyer zu Heringdorf, F.-J.: Anisotropy of Ag diffusion on vicinal Si surfaces. e-J. Surf. Sci. Nanotech. **8**, 372–376 (2010)
300. McKay, H.A., Feenstra, R.M.: Low energy electron microscopy of indium on Si(001) surfaces. Surf. Sci. **547**, 127–138 (2003)
301. Thien, D., Kury, P., Horn-von Hoegen, M., Meyer zu Heringdorf, F.-J., van Heys, J., Lindenblatt, M., Pehlke, E.: Domain sensitive contrast in photoelectron emission microscopy. Phys. Rev. Lett. **99**, 196102, 4 pages (2007)
302. Speckmann, M., Schmidt, T., Locatelli, A., Menteş, T.O., Niño, M.Á., Falta, J.: Silver: a novel growth catalyst for Ge nanoislands on Si(113). Phys. Status Solidi RRL **3**, 305–307 (2009)
303. Clausen, T., Schmidt, T., Flege, J.I., Falta, J., Locatelli, A., Mentes, T.O., Heun, S., Guo, F.Z.: Real-time low-energy electron microscopy study of Ga adsorption and facet array formation on Si(113). e-J. Surf. Sci. Nanotech. **3**, 379–383 (2005)
304. Schmidt, T., Clausen, T., Flege, J.I., Gangopadhyay, S., Locatelli, A., Mentes, T.O., Guo, F.Z., Heun, S., Falta, J.: Adsorbate induced self-ordering or germanium nanoislands on Si(113). New J. Phys. **9** (392), 1–13 (2007)
305. Speckmann, M., Schmidt, T., Flege, J.I., Sadowski, J.T., Sutter, P., Falta, J.: Temperature dependent low energy electron microscopy study of Ge island growth on bare and Ga terminated Si(112). J. Phys. Condens. Matter **21**, 314020, 7 pages (2009)
306. Speckmann, M., Schmidt, T., Flege, T.J., Falta, J.: In adsorption on Si(112) and its impact on Ge growth. IBM J. Res. Dev. **55**, paper 11, 8 pages (2011)
307. Pavlovska, A., Bauer, E.: Wetting and reactive thin film growth. Surf. Sci. **480**, 128–136 (2001)
308. Zheng, C.X., Tang, W.X., Jesson, D.E.: Relief of surface stress at steps during displacive adsorption of As on Si(111). Appl. Phys. Lett. **100**, 201602, 4 pages (2012)
309. Yao, Y., Fu, Q., Tan, D., Bao, X.: A comparative study of CCl_4 reactions on Ag and Si surfaces by in situ ultraviolet photoemission electron microscopy. J. Phys. Condens. Matter **21**, 314014, 7 pages (2009)
310. Ihm, K., Han, J.H., Kim, B., Chung, S., Hwang, C.-C., Kang, T.-H., Kim, K.-J., Jung, Y.J., An, K.-S.: Selective adsorption of benzoic acid species on patterned OH/Si(100) surface. J. Appl. Phys. **100**, 043510, 5 pages (2006)
311. Mullet, C.H., Chiang, S.: High temperature growth of Ag phases on Ge(111). J. Vac. Sci. Technol. A **31**, 020602, 5 pages (2013)
312. Sato, Y., Chiang, S., Bartelt, N.C.: Spontaneous domain switching during phase separation of Pb on Ge(111). Phys. Rev. Lett. **99**, 096103, 4 pages (2007)
313. Hibino, H., Watanabe, Y.: Arrangement of Au-Si alloy islands at atomic steps. Surf. Sci. **588**, L233–L238 (2005)
314. Hannon, J.B., Kodambaka, S., Ross, F.M., Tromp, R.M.: The influence of the surface migration of gold on the growth of silicon nanowires. Nature **440**, 69–71 (2006)
315. Tersoff, J., Dernier van der Gon, A.W., Tromp, R.M.: Shape oscillations in growth of small crystals. Phys. Rev. Lett. **70**, 1143–1146 (1993)
316. Tersoff, J., Tromp, R.M.: Shape transition in growth of strained islands: spontaneous formation of quantum wires. Phys. Rev. Lett. **70**, 2782–2785 (1993)

317. Tromp, R.M., Denier van der Gon, A.W., LeGoues, F.K., Reuter, M.C.: Observation of buried interfaces with low energy electron microscopy. Phys. Rev. Lett. **71**, 3299–3302 (1993)
318. Denier van der Gon, A.W., Tromp, R.M., Reuter, M.C.: Low energy electron microscopy studies of Ge and Ag growth on Si(111). Thin Solid Films **236**, 140–145 (1993)
319. Tang, W.X., Man, K.L., Huang, H., Woo, C.H., Altman, M.S.: Growth shapes of Ag crystallites on the Si(111) surface. J. Vac. Sci. Technol. B **20**, 2492–2495 (2002)
320. Man, K.L., Tang, W.X., Huang, H., Altman, M.S.: Kinetic Limitations in Two-and Three-Dimensional Growth. In: Chason, E., Gilmer, G.H., Huang, H., Wang, E. (eds.) MRS Symposium KK—kinetics-driven nanopatterning on Surfaces, vol. 849, pp. KK5.3.1–KK5.3.9, Boston, Massachusetts, USA 2004. Materials Research Society (2005)
321. Wall, D., Tikhonov, S., Sindermann, S., Spoddig, D., Hassel, C., Horn-von Hoegen, M., Meyer zu Heringdorf, F.-J.: Shape, orientation, and crystalline composition of silver islands on Si(111). IBM J. Res. Dev. **55**, Paper 9, 6 pages (2011)
322. Gloskovskii, A., Valdaitsev, D., Nepijko, S.A., Schönhense, G., Rethfeld, B.: Coexisting electron emission mechanisms in small metal particles observed in fs-laser excited PEEM. Surf. Sci. **601**, 4706–4713 (2007)
323. Li, B.Q., Swiech, W., Venables, J.A., Zuo, J.M.: A LEEM study of bamboo-like growth of Ag crystals on Si(001) surfaces. Surf. Sci. **569**, 142–148 (2004)
324. Wall, D., Sindermann, S., Roos, K.R., Horn-von Hoegen, M., Meyer zu Heringdorf, F.-J.: The influence of anisotropic diffusion on Ag nanowire formation. J. Phys. Condens. Matter **21**, 314023, 4 pages (2009)
325. Roos, K.R., Roos, K.L., Horn-von Hoegen, M., Meyer zu Heringdorf, F.-J.: High temperature self-assembly of Ag nanowires on vicinal Si(001). J. Phys. Condens. Matter **17**, S1407–S1414 (2005)
326. Meyer zu Heringdorf, F.-J., Roos, K.L., Wiethoff, C., Horn-von Hoegen, M., Roos, K.R.: Growth of Ag nanowires on Au-pre-facetted 4° vicinal Si(001). Surf. Sci. **602**, 1852–1857 (2008)
327. Yasue, T., Koshikawa, T., Jalochowski, M., Bauer, E.: LEEM observation of formation of Cu nano-islands on Si(111) surface by hydrogen termination. Surf. Sci. **493**, 381–388 (2001)
328. Bauer, E.: Low energy electron microscopy. In: Vanselow, R., Howe, R. (eds.) Chemistry and Physics of Solid Surfaces VIII, pp. 267–288. Springer, Berlin (1990)
329. Maeda, F., Hibino, H., Suzuki, S., Kobayashi, Y., Watanabe, Y., Guo, F.Z.: Surface reactions of Co on SiO_2 thin layer/Si substrate studied by LEEM and PEEM. e-J. Surf. Sci. Nanotech. **4**, 155–160 (2006)
330. Maeda, F., Hibino, H., Suzuki, S., Guo, F.Z.: Oxide-mediated formation of α-$FeSi_2$ on Si (001) studied by X-ray adsorption near edge structure analysis using SPELEEM. Surf. Interface Anal. **40**, 1747–1750 (2008)
331. Schmidt, Th., Bauer, E.: Interfactant-mediated quasi-Frank-van der Merwe growth of Pb on Si(111). Phys. Rev. B **62**, 15815–15825 (2000)
332. Bauer, E.: Phänomenologische Theorie der Kristallabscheidung an Oberflächen I. Z. Kristallogr. **110**, 372–394 (1958)
333. Venables, J.A.: Introduction to Surface and Thin Films Processes. Cambridge University Press, Cambridge, UK (2000). p. 146
334. Ibach, H.: Physics of Surfaces and Interfaces. Springer Verlag, Berlin (2006). p. 171
335. Schmidt, Th., Bauer, E.: Influence of interfactants on thin metal film growth. Surf. Sci. **480**, 137–144 (2001)
336. Li, L., Koziol, C., Wurm, K., Hong, Y., Bauer, E., Tsong, I.S.T.: Surface morphology of Pb overlayers grown on Si(100)-(2x1). Phys. Rev. B **50**, 10834–10842 (1994)
337. Jalochowski, M., Bauer, E.: Self-assembled parallel mesoscopic Pb-wires on Au-modified Si(533) substrates. Prog. Surf. Sci. **67**, 79–97 (2001)
338. Jalochowski, M., Bauer, E.: Growth of metallic nanowires on anisotropic Si substrates: Pb on vicinal Si(001), Si(755), Si(533) and Si(110). Surf. Sci. **480**, 109–117 (2001)

339. Bennett, P.A., He, Z., Smith, D.J., Ross, F.M.: Endotaxial silicide nanowires: a review. Thin Solid Films **519**, 8434–8440 (2011)
340. Bennett, P.A., Ashcroft, B., He, Z., Tromp, R.M.: Growth dynamics of titanium silicide nanowires observed with low-energy electron microscopy. J. Vac. Sci. Technol. B **20**, 2500–2504 (2002)
341. Yang, W.-C., Ade, H., Nemanich, R.J.: Shape stability of $TiSi_2$ islands on Si (111). J. Appl. Phys. **95**, 1572–1576 (2004)
342. Yang, W.-C., Zeman, M., Ade, H., Nemanich, R.J.: Attractive migration and coalescence: a significant process in the coarsening of $TiSi_2$ islands on the Si(111) surface. Phys. Rev. Lett. **90**, 136102, 4 pages (2003)
343. Fitting, L., Zeman, M.C., Yang, W.-C., Nemanich, R.J.: Influence of strain, surface diffusion and Ostwald ripening on the evolution of nanostructures for erbium on Si(001). J. Appl. Phys. **93**, 4180–4184 (2003)
344. Yang, W.-C., Ade, H., Nemanich, R.J.: Stability and dynamics of Pt-Si liquid microdroplets on Si(001). Phys. Rev. B **69**, 045421, 8 pages (2004)
345. Bennett, P.A., Chobanian, J., Flege, J.I., Sutter, E., Sutter, P.: Surface thermomigration of nanoscale Pt-Si droplets on stepped Si(100). Phys. Rev. B **76**, 125410, 6 pages (2007)
346. Sutter, P., Bennett, P.A., Flege, J.I., Sutter, E.: Steering liquid Pt-Si nanodroplets on Si(100) by interactions with surface steps. Phys. Rev. Lett. **99**, 125504, 4 pages (2007)
347. Shiraki, Y., Usami, N. (eds.): Silicon-Germanium (SiGe) Nanostructures: Production, Properties and Applications in Electronics. Woodhead Publishing, Cambridge, UK (2011)
348. Aqua, J.-N., Berbezier, I., Favre, L., Frisch, T., Ronda, A.: Growth and self-organization of SiGe nanostructures. Phys. Rep. **522**, 59–189 (2013)
349. Tromp, R.M., Reuter, M.C.: Local dimer exchange in surfactant-mediated epitaxial growth. Phys. Rev. Lett. **68**, 954–967 (1992)
350. Ross, F.M., Tromp, R.M., Reuter, M.C.: Transition states between pyramids and domes during Ge/Si island growth. Science **286**, 1931–1934 (1999)
351. Tromp, R.M., Ross, F.M., Reuter, M.C.: Instability-driven SiGe island growth. Phys. Rev. Lett. **84**, 4641–4644 (2000)
352. Tromp, R.M., Ross, F.M.: Advances in in situ ultra-high vacuum electron microscopy: growth of SiGe on Si. Annu. Rev. Mater. Sci. **30**, 431–449 (2000)
353. Hannon, J.B., Copel, M., Stumpf, R., Reuter, M.C., Tromp, R.M.: Critical role of surface steps in the alloying of Ge on Si(001). Phys. Rev. Lett. **92**, 216104, 4 pages (2004)
354. Sutter, P., Mateeva, E., Lagally, M.G.: Si growth on partially relaxed Ge islands. J. Vac. Sci. Technol. B **16**, 1560–1563 (1998)
355. Sutter, P., Lagally, M.G.: Embedding of nanoscale 3D SiGe islands in a Si matrix. Phys. Rev. Lett. **81**, 3471 (1998)
356. Sutter, P., Mateeva, E., Sullivan, J.S., Lagally, M.G.: Low-energy electron microscopy of nanoscale three-dimensional SiGe islands on Si(100). Thin Solid Films **336**, 262–270 (1998)
357. Sutter, P., Lagally, M.G.: Nucleationless three-dimensional island formation in low-misfit heteroepitaxy. Phys. Rev. Lett. **84**, 4637–4640 (2000)
358. Sutter, P., Lagally, M.G.: Nucleationless island formation in SiGe/Si(100) heteroepitaxy. Mater. Sci. Eng. B **89**, 45–48 (2002)
359. Huang, M., Ritz, C.S., Novakovic, B., Yu, D., Zhang, Y., Flack, F., Savage, D.E., Evans, P.G., Knezevic, I., Liu, F., Lagally, M.G.: Mechano-electronic superlattices in silicon nanoribbons. ACS NANO **3**, 721–727 (2009)
360. Maxson, J.B., Savage, D.E., Liu, F., Tromp, R.M., Reuter, M.C., Lagally, M.G.: Thermal roughening of a thin film: a new type of roughening transition. Phys. Rev. Lett. **85**, 2152–2155 (2000)
361. Tromp, R.M.: Surface stress and interface formation. Phys. Rev. B **47**, 7125–7127 (1993)
362. Moutanabbir, O., Ratto, F., Heun, S., Scheerschmidt, K., Locatelli, A., Rosei, F.: Dynamic probe of atom exchange during monolayer growth. Phys. Rev. B **85**, 201416(R), 5 pages (2012)

363. MacLeod, J.M., Lipton-Duffin, J.A., Lanke, U., Urquhart, S.G., Rosei, F.: Shape transition in very large germanium islands on Si(111). Appl. Phys. Lett. **94**, 103109, 3 pages (2009)
364. Ratto, F., Rosei, F., Locatelli, A., Cherifi, S., Fontana, S., Heun, S., Szkutnik, P.-D., Sgarlata, A., De Crescenzi, M., Motta, N.: Composition of Ge(Si) islands in the growth of Ge on Si(111). Appl. Phys. Lett. **84**, 4526–4528 (2004)
365. Ratto, F., Rosei, F., Locatelli, A., Cherifi, S., Fontana, S., Heun, S., Szkutnik, P.-D., Sgarlata, A., De Crescenzi, M., Motta, N.: Composition of Ge(Si) islands in the growth of Ge on Si(111) by x-ray spectromicroscopy. J. Appl. Phys. **97**, 043516, 8 pages (2005)
366. Ratto, F., Locatelli, A., Fontana, S., Kharrazi, S., Ashtaputre, S., Kulkarni, S.K., Heun, S., Rosei, F.: Diffusion dynamics during the nucleation and growth of GeSi nanostructures on Si(111). Phys. Rev. Lett. **96**, 096103, 4 pages (2006)
367. Schmidt, Th., Flege, J.I., Gangopadhyay, S., Clausen, T., Locatelli, A., Heun, S., Falta, J.: Alignment of Ge nanoislands on Si(111) by Ga-induced substrate self-patterning. Phys. Rev. Lett. **98**, 066104, 4 pages (2007)
368. Schmidt, Th., Speckmann, M., Falta, J., Mentes, T.O., Nino, M.A., Locatelli, A.: Ge growth on partially and entirely Ag covered Si(111). e-J. Surf. Sci. Nanotech. **8**, 221–226 (2010)
369. Wang, Y.-L., Guo, H.-M., Qin, Z.-H., Ma, H.-F., Gao, H.-J.: Toward a detailed understanding of Si(111)-7 $\times$ 7 surface and adsorbed Ge nanostructures: fabrications, structures, and calculations. J. Nano Mater. **2008**, 874213, 19 pages (2008)
370. Yasue, T., Koshikawa, T., Bauer, E.: LEEM and LEED study of the growth of Ge on the Si(113) surface. J. Vac. Sci. Technol. B **20**, 2496–2499 (2002)
371. Clausen, T., Schmidt, T., Flege, J.I., Locatelli, A., Mentes, T.O., Heun, S., Guo, F.Z., Falta, J.: Temperature dependent low energy electron microscopy.study of Ge growth on Si(113). Appl. Surf. Sci. **252**, 5321–5325 (2006)
372. Schmidt, T., Flege, J.I., Speckmann, M., Clausen, T., Gangopadhyay, S., Locatelli, A., Mentes, T.O., Heun, S., Guo, F.Z., Sutter, P., Faltaa, J.: From nanoislands to nanowires: Growth of germanium on gallium-terminated silicon surfaces. Phys. Stat. Sol. A **206**, 1718–1722 (2009)
373. Falta, J., Reuter, M.C., Tromp, R.M.: Growth modes of Ge on GaAs(001). Appl. Phys. Lett. **65**, 1680–1682 (1994)
374. Tromp, R.M., Reuter, M.C.: Kinetic Instability in the Growth of CaF_2 on Si(111). Phys. Rev. Lett. **73**, 110–113 (1994)
375. Tromp, R.M., LeGoues, F.K., Reuter, M.C.: Strain relief during growth: CaF_2 on Si(111). Phys. Rev. Lett. **74**, 2706–2709 (1995)
376. Bauer, E., Wei, Y., Muller, T., Pavlovska, A., Tsong, I.S.T.: Reactive crystal growth in two dimensions: silicon nitride on Si(111). Phys. Rev. B **51**, 17891–17901 (1995)
377. Bauer, E., Pavlovska, A., Tsong, I.S.T.: In situ growth studies by low energy electron microscopy (LEEM) and low electron diffraction (LEED). In: Bailey, G.W., Dimlich, R.V.W., Alexander, K.B., McCarty, J.J., Pretlow, T.P. (eds.) Proc. MSA meeting microscopy and microanalysis '97, Cleveland, Ohio 1997. Microscopy and Microanalysis, vol. 3, suppl. 2, pp. 611–612 (1997)
378. Pavlovska, A., Torres, V.M., Bauer, E., Doak, R.B., Tsong, I.S.T., Thomson, D.B., Davis, R.F.: Low-energy electron microscopy observations of GaN homoepitaxy using a supersonic source. Appl. Phys. Lett. **75**, 989–991 (1999)
379. Pavlovska, A., Bauer, E., Torres, V.M., Edwards, J.L., Doak, R.B., Tsong, I.S.T., Ramachadran, V., Feenstra, F.M.: In situ real-time studies of GaN growth on 6H-SiC (0001) by low energy electron microscopy (LEEM). J. Cryst. Growth **189/190**, 310–316 (1998)
380. Hu, C., Smith, D.J., Doak, R.B., Tsong, I.S.T.: Morphological control of gan buffer layers grown by molecular beam epitaxy on 6H-SiC (0001). Surf. Rev. Lett. **7**, 565–570 (2000)
381. Pavlovska, A., Bauer, E.: Low energy electron microscopy studies of the growth of GaN on 6H-SiC(0001). Surf. Rev. Lett. **8**, 337–346 (2001)

382. Yang, W.C., Rodriguez, B.J., Park, M., Nemanich, R.J., Ambacher, O., Cimalla, V.: Photoelectron emission microscopy observation of inversion domain boundaries of GaN-based lateral polarity heterostructures. J. Appl. Phys. **94**, 5720–5725 (2003)
383. Heun, S., Watanabe, Y., Ressel, B., Bottomley, D., Schmidt, T., Prince, K.C.: Core-level photoelectron spectroscopy from individual heteroepitaxial nanocrystals on GaAs(001). Phys. Rev. B **63**, 125335, 8 pages (2001)
384. Biasiol, G., Heun, S., Golinelli, G.B., Locatelli, A., Mentes, T.O., Guo, F.Z., Hofer, C., Teichert, C., Sorba, L.: Surface compositional gradients of InAs/GaAs quantum dots. Appl. Phys. Lett. **87**, 223106, 3 pages (2005)
385. Biasiol, G., Heun, S., Golinelli, G.B., Locatelli, A., Mentes, T.O., Guo, F.Z., Sorba, L.: Surface concentration mapping of InAs/GaAs quantum dots. In: Jantsch, W., Schaffler, F. (eds.) AIP Conf. Proc., pp. 49–50 (2007)
386. Biasiol, G., Baranwal, V., Heun, S., Prasciolu, M., Tormen, M., Locatelli, A., Mentes, T.O., Nino, M.A., Sorba, L.: Composition uniformity of site-controlled InAs/GaAs quantum dots. J. Cryst. Growth **323**, 176–179 (2011)
387. Biasiol, G., Magri, R., Heun, S., Locatelli, A., Mentes, T.O., Sorba, L.: Surface compositional mapping of self-assembled InAs/GaAs quantum rings. J. Cryst. Growth **311**, 1764–1766 (2009)
388. Magri, R., Heun, S., Biasiol, G., Locatelli, A., Mentes, T.O., Sorba, L.: Surface compositional profiles of self-assembled InAs/GaAs quantum rings. In: Caldas, M.J., Studart, N. (eds.) AIP Conf. Proc., pp. 3–4 (2010)
389. Baranwal, V., Biasiol, G., Heun, S., Locatelli, A., Mentes, T.O., Niño, M., Sorba, L.: Kinetics of the evolution of InAs/GaAs quantum dots to quantum rings: a combined x-ray, atomic force microscopy, and photoluminescence study. Phys. Rev. B **80**, 155328, 7 pages (2009)
390. Biasiol, G., Heun, S.: Compositional mapping of semiconductor quantum dots and rings. Phys. Rep. **500**, 117–173 (2011)
391. Robinson, J.T., Ratto, F., Moutanabbir, O., Heun, S., Locatelli, A., Mentes, T.O., Aballe, L., Dubon, O.D.: Gold-catalyzed oxide nanopatterns for the directed assembly of Ge island arrays on Si. Nano Lett. **7**, 2655–2659 (2007)
392. Hjort, M., Wallentin, J., Timm, R., Zakharov, A.A., Andersen, J.N., Samuelson, L., Borgstrom, M.T., Mikkelsen, A.: Doping profile of InP nanowires directly imaged by photoemission electron microscopy. Appl. Phys. Lett. **99**, 233113, 3 pages (2011)
393. Kennedy, S.M., Hjort, M., Mandl, B., Marsell, E., Zakharov, A.A., Mikkelsen, A., Paganin, D.M., Jesson, D.E.: Characterizing the geometry of InAs nanowires using mirror electron microscopy. Nanotechnology **23**, 125703, 9 pages (2012)
394. Mandl, B., Stangl, J., Hilner, E., Zakharov, A.A., Hillerich, K., Dey, A.W., Samuelson, L., Bauer, G., Deppert, K., Mikkelsen, A.: Growth mechanism of self-catalyzed group III-V nanowires. Nano Lett. **10**, 4443–4449 (2010)
395. Ercolani, D., Lazzarino, M., Mori, G., Ressel, B., Sorba, L., Locatelli, A., Cherifi, S., Ballestrazzi, A., Heun, S.: GaAs oxide desorption under extreme ultraviolet photon flux. Adv. Funct. Mater. **15**, 587–592 (2005)
396. Mori, G., Lazzarino, M., Ercolani, D., Sorba, L., Heun, S., Locatelli, A.: Desorption dynamics of oxide nanostructures fabricated by local anodic oxidation nanolithography. J. Appl. Phys. **97**, 114324, 8 pages (2005)
397. Mori, G., Lazzarino, M., Ercolani, D., Biasiol, G., Sorba, L., Heun, S., Locatelli, A.: Evidence of material mixing during local anodic oxidation nanolithography. J. Appl. Phys. **98**, 114303, 6 pages (2005)
398. Mori, G., Lazzarino, M., Ercolani, D., Biasiol, G., Locatelli, A., Sorba, L., Heun, S.: X-ray induced variation of the chemistry of GaAs/AlAs oxide nanostructures. Nucl. Instrum. Method Phys. Res. B **246**, 39–44 (2006)
399. Lazzarino, M., Mori, G., Sorba, L., Ercolani, D., Biasiol, G., Heun, S., Locatelli, A.: Chemistry and formation process of Ga(Al)As oxide during local anodic oxidation nanolithography. Surf. Sci. **600**, 3739–3743 (2006)

400. Jesson, D.E., Tang, W.X.: Surface electron microscopy of Ga droplet dynamics on GaAs (001). Microsc Sci Technol Appl Educ **4**, 1608–1619 (2010)
401. Tersoff, J., Jesson, D.E., Tang, W.X.: Running droplets of gallium from evaporation of gallium arsenide. Science **324**, 236–238 (2009)
402. Hilner, E., Zakharov, A.A., Schulte, K., Kratzer, P., Andersen, J.N., Lundgren, E., Mikkelsen, A.: Ordering of the nanoscale step morphology as a mechanism for droplet self-propulsion. Nano Lett. **9**, 2710–2714 (2009)
403. Tang, W.X., Jesson, D.E., Pavlov, K.M., Morgan, M.J., Usher, B.F.: Ga droplet morphology on GaAs(001) studied by Lloyd's mirror photoemission electron microscopy. J. Phys. Condens. Matter **21**, 314022, 5 pages (2009)
404. Zhou, Z.Y., Tang, W.X., Jesson, D.E., Tersoff, J.: Time evolution of the Ga droplet size distribution during Langmuir evaporation of GaAs(001). Appl. Phys. Lett. **97**, 191914, 3 pages (2010)
405. Tang, W.X., Zheng, C.X., Zhou, Z.Y., Jesson, D.E., Tersoff, J.: Ga droplet surface dynamics during Langmuir evaporation of GaAs. IBM J. Res. Dev. **55**(4), 10:1–10:7 (2011)
406. Hilner, E., Mikkelsen, A., Eriksson, J., Andersen, J.N., Lundgren, E.: Au wetting and nanoparticle stability on GaAs(111)*B*. Appl. Phys. Lett. **89**, 251912, 3 pages (2006)
407. Hajjar, S., Garreau, G., Josien, L., Bubendorff, J.L., Berling, D., Mehdaoui, A., Pirri, C., Maroutian, T., Renard, C., Bouchier, D., Petit, M., Spiesser, A., Dau, M.T., Michez, L., Le Thanh, V., Menteş, T.O., Niño, M.Á., Locatelli, A.: Morphology and composition of Au catalysts on Ge(111) obtained by thermal dewetting. Phys. Rev. B **84**, 125325, 10 pages (2011)
408. Qi, B., Ólafsson, S., Göthelid, M., Gislason, H.P., Agnarsson, B.: Photoemission and low energy electron microscopy study on the formation and nitridation of indium droplets on Si (111)7 $\times$ 7 surfaces. Thin Solid Films **531**, 61–69 (2013)
409. Qi, B., Ólafsson, S., Zakharov, A.A., Göthelid, M., Agnarsson, B., Gislason, H.P.: Study of spatial homogeneity and nitridation of an Al nanopattern template with spectroscopic photoemission and low energy electron microscopy. Appl. Surf. Sci. **264**, 349–357 (2013)
410. Mundschau, M., Bauer, E., Swiech, W.: Initial epitaxial growth of Cu on Mo(011) by low energy electron microscopy and photoemission electron microscopy. J. Appl. Phys. **65**, 581–584 (1989)
411. Altman, M.S., Chung, W.F., Liu, C.H.: LEEM phase contrast. Surf. Rev. Lett. **5**, 1129–1141 (1998)
412. Shimizu, H., Nakaguchi, A., Yasue, T., Koshikawa, T., Bauer, E.: Abnormal contrast of PEEM images on Cu/W(110). In: Proc. atomic level characterizations for new materials and devices (ALC'05), Big Island, Hawaii, USA 2005, pp. 702–709. JSPS 141 Activity Report (2006)
413. Chung, W.F., Feng, Y.J., Poon, H.C., Chan, C.T., Tong, S.Y., Altman, M.S.: Layer spacings in coherently strained epitaxial metal films. Phys. Rev. Lett. **90**, 216105, 4 pages (2003)
414. Altman, M.S.: Low energy electron microscopy of quantum well resonances in Ag films on W(110). J. Phys. Condens. Matter **17**, 1305–1310 (2005)
415. Smith, N.V., Brookes, N.B., Chang, Y., Johnson, P.D.: Quantum-well and tight-binding analyses of spin-polarized photoemission from Ag/Fe(001) overlayers. Phys. Rev. B **49**, 332–338 (1994)
416. McCarty, K.F.: Deterministic positioning of three-dimensional structures on a substrate by film growth. Nano Lett. **6**, 858–861 (2006)
417. Aballe, L., Barinov, A., Locatelli, A., Heun, S., Cherifi, S., Kiskinova, M.: Spectromicroscopy of ultrathin Pd films on W(110): interplay of morphology and electronic structure. Appl. Surf. Sci. **238**, 138–142 (2004)
418. Santos, B., Puerta, J.M., Cerda, J.I., Herranz, T., McCarty, K.F., de la Figuera, J.: Structure of ultrathin Pd films determined by low-energy electron microscopy and diffraction. New J. Phys. **12**, 023023, 21 pages (2010)

419. Aballe, L., Barinov, A., Locatelli, A., Mentes, T.O., Kiskinova, M.: Initial stages of heteroepitaxial Mg growth on W(110): early condensation, anisotropic strain, and self-organized patterns. Phys. Rev. B **75**, 115411, 9 pages (2007)
420. Amakawa, R., Shimizu, H., Yasue, T., Koshikawa, T., Bauer, E.: Dynamic observation of nano-structure growth by LEEM and PEEM. In: Proc. 4th international symposium on atomic level characterization of new materials and devices (ALC'03), Kauai, Hawaii 2003, pp. 573–576. JSPS 141 Activity Report (2004)
421. Santos, B., Puerta, J.M., Cerda, J.I., Stumpf, R., von Bergmann, K., Wiesendanger, R., Bode, M., McCarty, K.F., de la Figuera, J.: Structure and magnetism of ultra-thin chromium layers on W(110). New J. Phys. **10**, 013005, 16 pages (2008)
422. McCarty, K.F., Hamilton, J.C., Sato, Y., Saá, A., Stumpf, R., de la Figuera, J., Thürmer, K., Jones, F., Schmid, A.K., Talin, A.A., Bartelt, N.C.: How metal films de-wet substrates-identifying the kinetic pathways and energetic driving forces. New J. Phys. **11**, 043001, 38 pages (2009)
423. Ling, W.L., Giessel, T., Thürmer, K., Hwang, R.Q., Bartelt, N.C., McCarty, K.F.: Crucial role of substrate steps in de-wetting of crystalline thin films. Surf. Sci. **570**, L297–L303 (2004)
424. Ling, W.L., Bartelt, N.C., McCarty, K.F.: Twin boundaries can be moved by step edges during film growth. Phys. Rev. Lett. **95**, 166105, 4 pages (2005)
425. Mascaraque, A., Mentes, T.O., Mentes, T.O., McCarty, K.F., Marco, J.F., Schmid, A.K., Locatelli, A., de la Figuera, J.: Valence band circular dichroism in non-magnetic Ag/Ru (0001) at normal emission. J. Phys. Condens. Matter **23**, 305006, 5 pages (2011)
426. El Gabaly, F., Puerta, J.M., Klein, C., Saa, A., Schmid, A.K., McCarty, K.F., Cerda, J.I., de la Figuera, J.: Structure and morphology of ultrathin Co/Ru(0001) films. New J. Phys. **9**, 80, 23 pages (2007)
427. Herranz, T., Santos, B., McCarty, K.F., de la Figuera, J.: Real-space study of the growth of magnesium on ruthenium. Surf. Sci. **605**, 903–911 (2011)
428. Meyer, A., Flege, J.I., Rettew, R.E., Senanayake, S.D., Schmidt, T., Alamgir, F.M., Falta, J.: Ultrathin silver films on Ni(111). Phys. Rev. B **82**, 085424, 9 pages (2010)
429. Hashimoto, M., Umezawa, K., Yasue, T., Koshikawa, T.: Growth of Au on Ni(111) Surface studied with low energy electron microscopy. In: Proc. 8th international symposium on atomic level characterization of new materials and devices (ALC'11) Seoul, Korea 2011, pp. 300–302. JSPS 141 Activity Report (2012)
430. Bollmann, T.R.J., van Gastel, R., Zandvliet, H.J.W., Poelsema, B.: Quantum size effects on surfaces without a projected bandgap: Pb/Ni(111). New J. Phys. **13**, 103025, 16 pages (2011)
431. Bollmann, T.R.J., van Gastel, R., Zandvliet, H.J.W., Poelsema, B.: Anomalous decay of electronically stabilized lead mesas on Ni(111). Phys. Rev. Lett. **107**, 136103, 4 pages (2011)
432. Man, K.L., Guo, Q., Altman, M.S.: Growth and oxidation of Cr films on the W(100) surface. Surf. Sci. **600**, 1060–1070 (2006)
433. Man, K.L., Qiu, Z.Q., Altman, M.S.: Kinetic limitations in electronic growth of Ag films on Fe(100). Phys. Rev. Lett. **93**, 236104, 4 pages (2004)
434. Man, K.L., Qiu, Q., Altman, M.S.: Quantum size effect-driven thermal decomposition of Ag films on Fe(100) in the presence of pinhole growth morphological defects. Phys. Rev. B **81**, 045426, 10 pages (2010)
435. Tersoff, J., Denier van der Gon, A.W., Tromp, R.M.: Critical island size for layer-by-layer growth. Phys. Rev. Lett. **72**, 266–269 (1994)
436. Juel, M., Samuelsen, B.T., Kildemo, M., Raaen, S.: Valence variations of Sm on polycrystalline Ag. Surf. Sci. **600**, 1155–1159 (2006)
437. Sogo, M., Kamada, T., Aoki, M., Masuda, S.: Initial oxidation of Ni(111) observed by electron emission microscopy: PEEM and MEEM. Surf. Sci. **601**, 3988–3991 (2007)
438. Kamada, T., Sogo, M., Aoki, M., Masuda, S.: Mesoscopic pattern formation during initial oxidation of Ni(111) observed by electron emission microscopy. Surf. Sci. **602**, 724–732 (2008)

439. Flege, J.I., Meyer, A., Falta, J.: Self-limited oxide formation in Ni(111) oxidation. Phys. Rev. B **84**, 115441, 10 pages (2011)
440. Devlin, C.L.H., Sato, Y., Chiang, S.: Morphology of the Cu_2O surface oxide phase formed on Cu(100) at high temperature. J. Appl. Phys. **105**, 123534, 5 pages (2009)
441. Günther, S., Reichelt, R., Wintterlin, J., Barinov, A., Menteş, T.O., Niño, M.Á., Locatelli, A.: Chemical patterning of Ag(111): spatially confined oxide formation induced by electron beam irradiation. Appl. Phys. Lett. **93**, 233117, 3 pages (2008)
442. Günther, S., Bocklein, S., Reichelt, R., Wintterlin, J., Barinov, A., Mentes, T.O., Nino, M.A., Locatelli, A.: Surface patterning of silver using an electron- or photon-assisted oxidation reaction. Chem. Phys. Chem. **11**, 1525–1532 (2010)
443. Böttcher, A., Krenzer, B., Conrad, H., Niehus, H.: Mesoscopic-scale pattern formation induced by oxidation of Ru(0001). Surf. Sci. Lett. **466**, L811–L820 (2000)
444. Böttcher, A., Krenzer, B., Conrad, H., Niehus, H.: Mesoscopic-scale growth of oxygen-rich films on Ru(0001) investigated by photoemission electron microscopy. Surf. Sci. **504**, 42–58 (2002)
445. Flege, J.I., Sutter, P.: Nanoscale analysis of Ru(0001) oxidation using low-energy and photoemission electron microscopy. J. Phys. Condens. Matter **21**, 314018, 6 pages (2009)
446. Pierce, J.P., McCarty, K.F.: Self-assembly and dynamics of oxide nanorods on NiAl(110). Phys. Rev. B **71**, 125428, 11 pages (2005)
447. McCarty, K.F., Pierce, J.P., Carter, C.B.: Translation-related domain boundaries form to relieve strain in a thin alumina film on NiAl (110). Appl. Phys. Lett. **88**, 141902, 4 pages (2006)
448. Pierce, J.P., Bartelt, N.C., Stumpf, R., McCarty, K.F.: Stability of ultrathin alumina layers on NiAl(110). Phys. Rev. B **77**, 195438, 7 pages (2008)
449. Aballe, L., Barinov, A., Locatelli, A., Heun, S., Kiskinova, M.: Tuning surface reactivity via electron quantum confinement. Phys. Rev. Lett. **93**, 196103, 4 pages (2004)
450. Aballe, L., Barinov, A., Stojić, N., Binggeli, N., Menteş, T.O., Locatelli, A., Kiskinova, M.: The electron density decay length effect on surface reactivity. J. Phys. Condens. Matter **22**, 015001, 6 pages (2010)
451. Santos, B., Loginova, E., Mascaraque, A., Schmid, A.K., McCarty, K.F., de la Figuera, J.: Structure and magnetism in ultrathin iron oxides characterized by low energy electron microscopy. J. Phys. Condens. Matter **21**, 314011, 10 pages (2009)
452. Monti, M., Santos, B., Mascaraque, A., de la Fuente, O.R., Niño, M.Á., Menteş, T.O., Locatelli, A., McCarty, K.F., Marco, J.F., de la Figuera, J.: Magnetism in nanometer-thick magnetite. Phys. Rev. B **85**, 020404, 5 pages (2012)
453. Monti, M., Santos, B., Mascaraque, A., de la Rodriguez Fuente, O., Niño, M.Á., Menteş, T. O., Locatelli, A., McCarty, K.F., Marco, J.F., de la Figuera, J.: Oxidation pathways in bicomponent ultrathin iron oxide films. J. Phys. Chem. C **116**, 11539–11547 (2012)
454. Kaemena, B., Senanayake, S.D., Meyer, A., Sadowski, J.T., Falta, J., Flege, J.I.: Growth and morphology of ceria on ruthenium (0001). J. Phys. Chem. C **117**, 221–232 (2013)
455. Skala, T., Tsud, N., Niño Orti, M.Á., Menteş, T.O., Locatelli, A., Prince, K.C., Matolın, V.: In situ growth of epitaxial cerium tungstate (100) thin films. Phys. Chem. Chem. Phys. **13**, 7083–7089 (2011)
456. Grinter, D.C., Yim, C.-M., Pang, C.L., Santos, B., Menteş, T.O., Locatelli, A., Thornton, G.: Oxidation state imaging of ceria island growth on Re(0001). J. Phys. Chem. C **117**, 16509–16514 (2013)
457. Agnoli, S., Menteş, T.O., Niño, M.Á., Locatelli, A., Granozzi, G.: A LEEM/micro-LEED investigation of phase transformations in TiOx/Pt(111) ultrathin films. Phys. Chem. Chem. Phys. **11**, 3727–3732 (2009)
458. Herranz, T., McCarty, K.F., Santos, B., Monti, M., de la Figuera, J.: Real space observations of magnesium hydride formation and decomposition. Chem. Mater. **22**, 1291–1293 (2010)
459. Goriachko, A., Zakharov, A.A., Over, H.: Oxygen-etching of h-BN/Ru(0001) nanomesh on the nano- and mesoscopic scale. J. Phys. Chem. C **112**, 10423–10427 (2008)

460. Vinogradov, N.A., Zakharov, A.A., Ng, M.L., Mikkelsen, A., Lundgren, E., Mårtensson, N., Preobrajenski, A.B.: One-dimensional corrugation of the h-BN monolayer on Fe(110). Langmuir **28**, 1775–1781 (2012)
461. Yasufuku, H., Okumura, M., Kera, S., Okudaira, K.K., Harada, Y., Ueno, N.: PEEM and MEEM of chloroaluminum phthalocyanine ultrathin film on MoS_2. J. Electron. Spectrosc. Rel. Phenom. **114–116**, 1025–1030 (2001)
462. Yasufuku, H., Ibe, T., Okumura, M., Kera, S., Okudaira, K.K., Harada, Y., Ueno, N.: Diffusion of chloroaluminum phthalocyanine on MoS_2 surface detected by photoemission electron microscopy and metastable electron emission microscopy. J. Appl. Phys. **90**, 213–216 (2001)
463. Yamamoto, I., Matsuura, N., Mikamori, M., Yamamoto, R., Yamada, T., Miyakubo, K., Ueno, N., Munakata, T.: Imaging of electronic structure of lead phthalocyanine films studied by combined use of PEEM and Micro-UPS. Surf. Sci. **602**, 2232–2237 (2008)
464. Baba, Y., Sekiguchi, T., Shimoyama, I., Honda, M., Hirao, N., Narita, A., Deng, J.Z.: Real-time observation on surface diffusion and molecular orientations for phthalocyanine thin films at nanometer spacial resolution. Surf. Sci. **603**, 2612–2618 (2009)
465. Meyer zu Heringdorf, F.-J., Reuter, M.C., Tromp, R.M.: Growth dynamics of pentacene thin films. Nature **412**, 517–520 (2001)
466. Meyer zu Heringdorf, F.-J., Reuter, M.C., Tromp, R.M.: The nucleation of pentacene thin films. Appl. Phys. A **78**, 787–791 (2004)
467. Weidkamp, K.P., Tromp, R.M., Hamers, R.J.: Epitaxial growth of large pentacene crystals on Si(001) surfaces functionalized with molecular monolayers. J. Phys. Chem. C **111**, 16489–16497 (2007)
468. Sadowski, J.T., Sazaki, G., Nishikata, S., Al-Mahboob, A., Fujikawa, Y., Nakajima, K., Tromp, R.M., Sakurai, T.: Single-nucleus polycrystallization in thin film epitaxial growth. Phys. Rev. Lett. **98**, 046104, 4 pages (2007)
469. Nishikata, S., Sazaki, G., Sadowski, J.T., Al-Mahboob, A., Nishihara, T., Fujikawa, Y., Suto, S., Sakurai, T., Nakajima, K.: Polycrystalline domain structure of pentacene thin films epitaxially grown on a hydrogen-terminated Si(111) surface. Phys. Rev. B **76**, 165424, 10 pages (2007)
470. Al-Mahboob, A., Sadowski, J.T., Nishihara, T., Fujikawa, Y., Xue, Q.K., Nakajima, K., Sakurai, T.: Epitaxial structures of self-organized, standing-up pentacene thin films studied by LEEM and STM. Surf. Sci. **601**, 1304–1310 (2007)
471. Al-Mahboob, A., Fujikawa, Y., Sadowski, J.T., Hashizume, T., Sakurai, T.: Formation of giant crystalline grain via delayed growth process driven by organic molecular anisotropy. Phys. Rev. B **82**, 235421, 6 pages (2010)
472. Al-Mahboob, A., Sadowski, J.T., Fujikawa, Y., Nakajima, K., Sakurai, T.: Kinetics-driven anisotropic growth of pentacene thin films. Phys. Rev. B **77**, 035426, 6 pages (2008)
473. Al-Mahboob, A., Sadowski, J.T., Fujikawa, Y., Sakurai, T.: The growth mechanism of pentacene-fullerene heteroepitaxial films. Surf. Sci. **603**, L53–L56 (2009)
474. Liu, H.W., Al-Mahboob, A., Fujikawa, Y., Fukui, N., Hitosugi, T., Hashizume, T., Xue, Q.K., Sakurai, T.: Pentacene growth on graphite investigated by low-energy electron microscope. J. Cryst. Growth **312**, 967–970 (2010)
475. Ng, M.L., Preobrajenski, A.B., Zakharov, A.A., Vinogradov, A.S., Krasnikov, S.A., Cafolla, A.A., Mårtensson, N.: Effect of substrate nanopatterning on the growth and structure of pentacene films. Phys. Rev. B **81**, 115449, 7 pages (2010)
476. Wall, S., Thien, D., Meyer zu Heringdorf, F.-J.: The interplay of topography and energy dissipation in pentacene thin films. J. Electron. Spectrosc. Rel. Phenom. **185**, 436–440 (2012)
477. Yoshikawa, G., Sadowski, J.T., Al-Mahboob, A., Fujikawa, Y., Sakurai, T.: Spontaneous aggregation of pentacene molecules and its influence on field effect mobility. Appl. Phys. Lett. **90**, 251906, 3 pages (2007)

478. Tsuruma, Y., Al-Mahboob, A., Ikeda, S., Sadowski, J.T., Yoshikawa, G., Fujikawa, Y., Sakurai, T., Saiki, K.: Real-time observation and control of pentacene film growth on an artificially structured substrate. Adv. Mater. **21**, 4996–5000 (2009)
479. Al-Mahboob, A., Sadowski, J.T., Nishihara, T., Fujikawa, Y., Xue, Q.K., Nakajima, K., Sakurai, T.: Mechanism of two-dimensional chiral growth of 6,13-pentacenequinone thin films on Si(111). Surf. Sci. **601**, 1311–1318 (2007)
480. Marchetto, H., Groh, U., Schmidt, T., Fink, R., Freund, H.J., Umbach, E.: Influence of substrate morphology on organic layer growth: PTCDA on Ag(111). Chem. Phys. **325**, 178–184 (2006)
481. Buckanie, N.M., Meyer zu Heringdorf, F.-J.: Photoemission electron microscopy study of anthracene growth on Si(111). Surf. Sci. **601**, 1701–1704 (2007)
482. Buckanie, N.M., Meyer zu Heringdorf, F.-J.: Exciton sensitive microscopy of anthracene monolayers on Si(111). Org. Electron. **10**, 446–452 (2009)
483. Casu, M.B., Biswas, I., Nagel, M., Nagel, P., Schuppler, S., Chassé, T.: Photoemission electron microscopy of diindenoperylene thin films. Phys. Rev. B **78**, 075310, 7 pages (2008)
484. Casu, M.B., Schuster, B.-E., Biswas, I., Nagel, M., Nagel, P., Schuppler, S., Chassé, T.: From interfaces to surfaces: soft x-ray spectromicroscopy investigations of diindenoperylene thin films on gold. J. Phys. Condens. Matter **21**, 314017, 5 pages (2009)
485. Casu, M.B., Schuster, B.-E., Biswas, I., Raisch, C., Marchetto, H., Schmidt, T., Chassé, T.: Locally resolved core-hole screening, molecular orientation, and morphology in thin films of diindenoperylene deposited on Au(111) single crystals. Adv. Mater. **22**, 3740–3744 (2010)
486. Casu, M.B., Savu, S.-A., Hoffmann, P., Schuster, B.-E., Menteş, T.O., Niño, M.A., Locatelli, A., Chassé, T.: Direct observation of step-edge barrier effects and general aspects of growth processes: morphology and structure in diindenoperylene thin films deposited on Au(100) single crystals. CrystEngComm. **13**, 4139–4144 (2011)
487. Khokhar, F.S., van Gastel, R., Schwarz, D., Zandvliet, H.J.W., Poelsema, B.: A low energy electron microscopy study of the initial growth, structure, and thermal stability of 4,4′-biphenyldicarboxylic acid domains on Cu(001). J. Chem. Phys. **135**, 124706, 7 pages (2011)
488. Schwarz, D., van Gastel, R., Zandvliet, H.J.W., Poelsema, B.: Phase transformations of 4,4’-biphenyldicarboxylic acid on Cu(001). Phys. Rev. B **85**, 235419, 10 pages (2012)
489. Schwarz, D., van Gastel, R., Zandvliet, H.J.W., Poelsema, B.: Size fluctuations of near critical nuclei and gibbs free energy for nucleation of BDA on Cu(001). Phys. Rev. Lett. **109**, 016101, 4 pages (2012)
490. Schwarz, D., van Gastel, R., Zandvliet, H.J.W., Poelsema, B.: Formation and decay of a compressed phase of 4,4′-biphenyldicarboxylic acid on Cu(001). Phys. Chem. Chem. Phys. **15**, 5007–5016 (2013)
491. Schwarz, D., van Gastel, R., Zandvliet, H.J.W., Poelsema, B.: Growth anomalies in supramolecular networks: 4,4′-biphenyldicarboxylic acid on Cu(001). Phys. Rev. Lett. **110**, 076101, 5 pages (2013)
492. Schwarz, D., van Gastel, R., Zandvliet, H.J.W., Poelsema, B.: In situ observation of a deprotonation-driven phase transformation: 4,4′-biphenyldicarboxylic acid on Au(111). J. Phys. Chem. C **117**, 1020–1029 (2013)
493. Khokhar, F.S., van Gastel, R., Poelsema, B.: Role of topographical defects in organic film growth of 4,4′-biphenyldicarboxylic acid on graphene: a low-energy electron microscopy study. Phys. Rev. B **82**, 205409, 8 pages (2010)
494. Hlawacek, G., Khokhar, F.S., van Gastel, R., Poelsema, B., Teichert, C.: Smooth growth of organic semiconductor films on graphene for high-efficiency electronics. Nano Lett. **11**, 333–337 (2011)
495. Hlawacek, G., Khokar, F.S., van Gastel, R., Teichert, C., Poelsema, B.: Diffusion and submonolayer growth of para-sexiphenyl on Ir(111) and Ir(111)-supported graphene. IBM J. Res. Dev. **55**(4), 15:1–15:7 (2011)

496. Khokhar, F.S., Hlawacek, G., van Gastel, R., Zandvliet, H.J.W., Teichert, C., Poelsema, B.: The influence of substrate temperature on growth of para-sexiphenyl thin films on Ir{111} supported graphene studied by LEEM. Surf. Sci. **606**, 475–480 (2012)
497. Fleming, A.J., Netzer, F.P., Ramsey, M.G.: Nucleation and 3D growth of para-sexiphenyl nanostructures from an oriented 2D liquid layer investigated by photoemission electron microscopy. J. Phys. Condens. Matter **21**, 445003, 11 pages (2009)
498. Wagner, T., Fritz, D.R., Zeppenfeld, P.: Standing and flat lying alpha-6T molecules probed by imaging photoelectron spectroscopy. Org. Electron. **12**, 442–446 (2011)
499. Wagner, T., Fritz, D.R., Zeppenfeld, P.: α-6T on Ag(110): the formation of the wetting layer. Synth. Met. **161**, 2006–2010 (2011)
500. Wagner, T., Fritz, D.R., Zeppenfeld, P.: Probing organic nanostructures by photoelectron-emission microscopy. Appl. Surf. Sci. **267**, 26–29 (2013)
501. Christensen, S., Lanke, U.D., Haines, B., Qaqish, S.E., Paige, M.F., Urquhart, S.G.: Structural and compositional mapping of a phase-separated Langmuir-Blodgett monolayer by X-ray photoelectron emission microscopy. J. Electron. Spectrosc. Rel. Phenom. **162**, 107–114 (2008)
502. Qaqish, S.E., Urquhart, S.G., Lanke, U., Brunet, S.M.K., Paige, M.F.: Phase separation of palmitic acid and perfluorooctadecanoic acid in mixed Langmuir-Blodgett monolayer films. Langmuir **25**, 7401–7409 (2009)
503. Leung, B.O., Hitchcock, A.P., Won, A., Ianoul, A., Scholl, A.: Imaging interactions of cationic antimicrobial peptides with model lipid monolayers using X-ray spectromicroscopy. Eur. Biophys. J. **40**, 805–810 (2011)
504. Cossy-Favre, A., Diaz, J., Anders, S ., Padmore, H.A., Liu, Y., Samant, M., Stöhr, J., Brown, H., Russell, T.P.: Photoelectron emission microscopy and its application to the study of polymer surfaces. In: Proc. third international school and symposium on synchrotron radiation in natural science, Jaszowiec, Poland 1996. Acta Physica Polonica A **91** (5) 10 pages (1997)
505. Cossy-Favre, A., Díaz, J., Liu, Y., Brown, H.R., Samant, M.G., Stöhr, J., Hanna, A.J., Anders, S., Russell, T.P.: X-PEEM study on surface orientation of stylized and rubbed polyimides. Macromolecules **31**, 4957–4962 (1998)
506. Zharnikov, M., Ouchi, Y., Hasegawa, M., Scholl, A.: X-ray absorption spectromicroscopy study of UV-photoinduced surface modification and anisotropy in polyimide films. J. Phys. Chem. B **108**, 859–863 (2004)
507. Kanai, K., Miyazaki, T., Wakita, T., Akaike, K., Yokoya, T., Ouchi, Y., Seki, K.: Lateral inhomogeneity in the electronic structure of a conjugated poly(3-hexylthiophene) thin film. Adv. Funct. Mater. **20**, 2046–2052 (2010)
508. Kappen, P., Hale, P.S., Brack, N., Prissanaroon, W., Pigram, P.J.: X-PEEM/NEXAFS and AFM of polypyrrole and copper micro-patterns on insulating fluoropolymer substrates. Appl. Surf. Sci. **253**, 1473–1479 (2006)
509. Morin, C., Ikeura-Sekiguchi, H., Tyliszczak, T., Cornelius, R., Brash, J.L., Hitchcock, A.P., Scholl, A., Nolting, F., Appel, G., Winesett, D.A., Kaznacheyev, K., Ade, H.: X-ray spectromicroscopy of immiscible polymer blends: polystyrene-poly(methyl methacrylate). J. Electron. Spectrosc. Rel. Phenom. **121**, 203–224 (2001)
510. Morin, C., Hitchcock, A.P., Cornelius, R.M., Brash, J.L., Urquhart, S.G., Scholl, A., Doran, A.: Selective adsorption of protein on polymer surfaces studied by soft X-ray photoemission electron microscopy. J. Electron. Spectrosc. Rel. Phenom. **137–140**, 785–794 (2004)
511. Li, L., Hitchcock, A.P., Robar, N., Cornelius, R., Brash, J.L., Scholl, A., Doran, A.: X-ray microscopy studies of protein adsorption on a phase-segregated polystyrene/polymethyl methacrylate surface. 1. Concentration and exposure-time dependence for albumin adsorption. J. Phys. Chem. B **110**, 16763–16773 (2006)
512. Li, L., Hitchcock, A.P., Cornelius, R., Brash, J.L., Scholl, A., Doran, A.: X-ray microscopy studies of protein adsorption on a phase segregated polystyrene/polymethylmethacrylate surface. 2. Effect of pH on site preference. J. Phys. Chem. B **112**, 2150–2158 (2008)

513. Leung, B.O., Hitchcock, A.P., Brash, J.L., Scholl, A., Doran, A.: Phase segregation in polystyrene-polylactide blends. Macromolecules **42**, 1679–1684 (2009)
514. Leung, B.O., Hitchcock, A.P., Cornelius, R.M., Brash, J.L., Scholl, A., Doran, A.: Using X-PEEM to study biomaterials: protein and peptide adsorption to a polystyrene-poly(methyl methacrylate)-b-polyacrylic acid blend. J. Electron. Spectrosc. Rel. Phenom. **185**, 406–416 (2012)
515. Ceccone, G., Leung, B.O., Perez-Roldan, M.J., Valsesia, A., Colpo, P., Rossi, F., Hitchcock, A.P., Scholl, A.: X-ray spectromicroscopy study of ubiquitin adsorption to plasma polymerized microstructures. Surf. Interface Anal. **42**, 830–834 (2010)
516. Hitchcock, A.P., Stöver, H.D.H., Croll, L.M., Childs, R.F.: Chemical mapping of polymer microstructure using soft X-ray spectromicroscopy. Aust. J. Chem. **58**, 423–432 (2005)

Chapter 6
Applications in Other Fields

Abstract This chapter first discusses studies of graphene, whose intriguing physical properties and promising technological applications has attracted widespread attention. An area of similar interest, both from the point of view of basic physics and technology, plasmon physics is addressed next. While in the graphene studies all available imaging methods have been used with great success, the study of plasmons has been a domain of PEEM. Surface microscopy with slow electrons has also been used in a wide range of more technology-oriented applications ranging from minerals to electronic devices, and these are discussed next, followed by a short section describing the limited amount or work done in biology. A study of an unknown material, in which many methods had to be combined, completes this chapter.

6.1 Graphene

6.1.1 Introduction

What silicon was towards the end of the twentieth century graphene became at the beginning of the twenty-first century. The strong interest in graphene was initiated by the discovery of the outstanding electronic properties of ultrathin graphite, consisting of only a few graphene layers, by Novoselov et al. [1] and de Heer's group [2], which promised a new age in electronics. In the years after these discoveries graphene has been found to have a multitude of other fascinating properties with possible widespread applications. These are described in a book [3] and many topical review papers, of which only a few are referenced here [4–8]. Although much of the information on graphene, which can be found in these reviews, has been obtained with laterally averaging methods, STM, and other probe microscopies, cathode lens electron microscopy has also made significant contributions to this field. In the few

E. Bauer, *Surface Microscopy with Low Energy Electrons*,
DOI 10.1007/978-1-4939-0935-3_6, © Springer Science+Business Media New York 2014

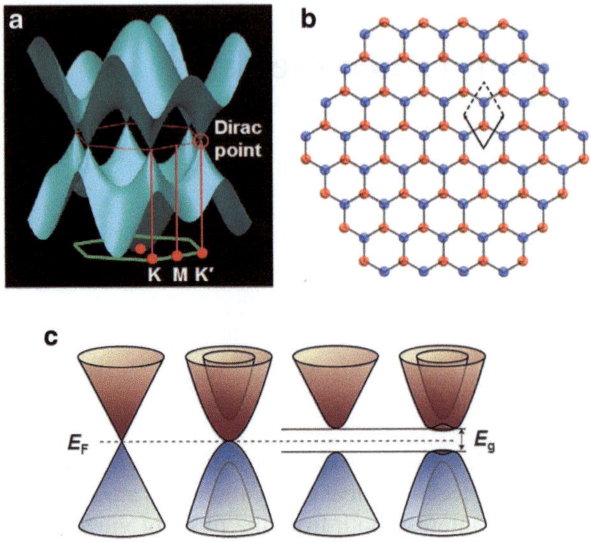

Fig. 6.1 (**a**) Reciprocal space. Electronic band structure of flat free-standing graphene. The occupied and unoccupied cones intersect in the Dirac point at the Fermi level. The high symmetry points of the reciprocal lattice are shown at the bottom. (**b**) Real space lattice of graphene showing the two sublattices and the unit cell. (**c**) Dirac cones of graphene. From *left* to *right*: free-standing single layer and bilayer, electron-doped single layer and bilayer. (**a**, **b**) Adapted with permission from Ref. [6]. Copyright 2012 Elsevier. (**c**) Reproduced from Ref. [5] with permission form Wiley-VCH Verlag GmbH & Co. KGaA, Weinheim

years since the first LEEM [9], XPEEM, and μARPES [10] publications in 2008 so many publications have appeared that mostly only those published after 2011 will be discussed here. Previous work can be found in the very thorough review by Man and Altman [11]. Another more recent review [12] contains some additional references to work published in 2012. First a brief introduction to the aspects of graphene which are relevant for cathode lens electron microscopy has to be given.

The unusual properties of graphene are a consequence of its band structure, which is shown in Fig. 6.1a [6] for the case of a flat free-standing graphene monolayer in vacuum. It has apparent hexagonal symmetry with six cone pairs near the Fermi energy E_F at the K points of the reciprocal space. The symmetry is only apparent because the unit cell in real space contains two atoms (Fig. 6.1b) [6]. This introduces a pseudo-spin state, which has important consequences for the transport properties of graphene. It has been studied with ARPES with different polarization directions [13, 14] but not in the work with imaging instruments discussed here. The linear dispersion $E(k) = \hbar v_F k$ near E_F is shown for one of the K points in the left side of Fig. 6.1c [5]. According to theory the Fermi velocity v_F is about 10^8 cm/s, in good agreement with experiment [10]. The dispersion is similar to that of light and, therefore, called Dirac dispersion based on his theory of light, and the intersection of the two cones is called Dirac point. The occupied states below E_F are formed from the π bonds between the atoms, the unoccupied

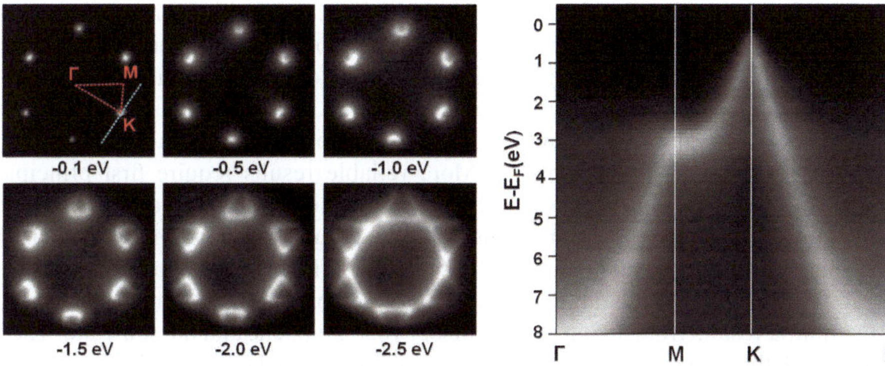

Fig. 6.2 Angle-resolved photoemission spectra (ARPES) of single layer graphene suspended on 5 μm diameter cylindrical cavities etched to a depth of 500 nm into a SiO_2 support. *Left side*: ARPES at several photoemission energies below the Fermi level excited with 50 eV photons. The spectra were taken in a SPELEEM with an energy and wave vector resolution of 0.3 eV and 0.05 $Å^{-1}$, respectively (horizontal cuts through the band structure shown in Fig. 6.1b). *Right side*: vertical cuts through the band structure along the ***k*** directions indicated in the first spectrum on the left side. The spectra are diffuse because the layer is not flat but undulated. Most ARPES data in the literature show cuts along the *blue line*. Courtesy of A. Locatelli

states are π* states. A second graphene monolayer introduces a second band and destroys the linear dispersion near E_F (Fig. 6.1c). When graphene is in contact with some other material as is the case in most experiments, electrons are transferred to or from graphene, shifting E_F relative to the band structure, depending upon the contact material. This is indicated in Fig. 6.1c for the monolayer and the double layer for the case of electron transfer to graphene (n-doping). These shifts can be studied very well in cathode lens microscopy with μARPES [10] as a function of type and amount of contact material, degree of contact, and temperature, though not with the high energy resolution as in laterally averaging high energy resolution spectrometers. ARPES is done by imaging at fixed photon energy the angular distribution of photoelectrons in the back focal plane of the objective. At fixed energy an ARPES pattern is obtained as illustrated for several emission energies near the Fermi energy in Fig. 6.2 (left) [A. Locatelli, private communication] for a free-standing graphene layer. The band structure $E(\boldsymbol{k})$ is obtained from patterns taken at many energies and is usually plotted for specific ***k*** directions as shown in Fig. 6.2 (right) [15]. While this is in general done with synchrotron radiation in SPELEEM instruments, ARPES has also been demonstrated with He II radiation in an energy filtered PEEM instrument [16].

Cathode lens microscopy is also uniquely suited for the study of the growth of graphene over a wide temperature range and the changes occurring upon adsorption, intercalation, and other in situ processes. μLEED allows structure analysis of small regions in films consisting of several graphene layers ("few-layer graphene"), which often occur because of thickness variations. In particular, information on the interface between the graphene layer and the substrate can be extracted from the

$I(V)$ curves by comparison with model calculations. To a first approximation this can be achieved by considering the few-layer graphene as a quantum well and analyzing the intensity oscillations with the phase accumulation model by identifying the number of minima n in the reflectivity with the number n of layers. While this is useful for thickness determination, extraction of layer spacings requires, however, a number of assumptions. More reliable results require first principle calculations similar to those developed by Krasovskii [17] for the transmission and reflection of slow electrons from surfaces, which require fewer assumptions such as the shape of the surface potential (see Sect. 2.3.4). This has been done by Feenstra et al. [18, 19], who found out that n layers produce only $n - 1$ minima and that the nth minima results from an interface state between graphene and the substrate. They compared their experimental $I(V)$ curves of graphene layers on Cu with their theoretical $I(V)$ curves and found best agreement with a substrate–graphene spacing of 3.25 Å [20]. This value is within the range of the values from several first-principle calculations, which range from 3.09 to 3.26 Å [21]. The most reliable experimental data, however, do not come from LEED $I(V)$ curves in LEEM instruments but from X-ray standing wave measurements, for example for graphene on Ir(111) [22].

Cathode lens microscopy has been used to study exfoliated free-standing graphene, graphene grown on SiC, and graphene grown on metals. Except for a brief demonstration of the usefulness of dark field XPEEM using photoelectrons at the Dirac point [23], work on free-standing graphene has been reviewed by Man and Altman [11]. Therefore only supported graphene will be discussed in the following. Two aspects will be considered: (1) growth and microstructure and (2) intercalation and other aspects, studied by all methods available in SPELEEM instruments. For the study of aspect (1) LEEM and μLEED are sufficient but μARPES is also useful, for aspect (2) in addition XPS is needed.

6.1.2 *Graphene on SiC*

6.1.2.1 Growth and Microstructure

SiC has many polytypes differing in the stacking of the SiC bilayers, the most used substrates being hexagonal 6H-SiC and 4H-SiC single crystals because they are commercially available. Some work has also been done on cubic 3C-SiC, which is grown only on the laboratory scale as epitaxial layers on Si(111), Si(100), and 4H-SiC and 6H-SiC. Growth on the hexagonal crystals has been studied both on the (0001) and the $(000\overline{1})$ surfaces, which are terminated by Si and C atoms, respectively. Graphene growth is achieved by thermal decomposition of the SiC surface. Because of the different surface terminations the growth on all these surfaces shows characteristic differences. These depend of course also on the surface defect structure such as steps and growth conditions such as temperature, ambient pressure, and gas. Nearly all groups working with cathode lens microscopy have been and are studying graphene growth. Their work has already produced a rather detailed understanding of the growth processes [24, 25]. The main goal is to understand the conditions which lead to reproducible desired electrical properties such as high electron mobility, doping, and band gap.

While the early studies were done in situ, in most of the later work the graphene layers were prepared ex situ for at least two reasons. First of all, at the high temperatures needed for graphene growth at a reasonable rate (>1000 °C) continuous observation is difficult because of the outgassing of the sample holder and the close-by objective lens so that heating has to be done in steps, losing one of the strengths of LEEM, dynamic studies. Secondly, it was recognized already early on that thermal decomposition at high Ar pressures produces much better films. This is due to the reduction of the Si desorption rate, which shifts the strongly non-equilibrium decomposition process in vacuum towards quasi-equilibrium as clearly demonstrated in a LEEM study of the silane pressure dependence of the decomposition process [26]. On all Si-terminated surfaces studied the thermal decomposition process proceeds via Si–C bond breaking at the steps, Si desorption and C diffusion on the terraces. Graphene formation is preceded by growth of a carbon rich graphene-like "buffer" layer, identified by its superstructure LEED pattern, which nucleates at the steps and grows preferentially along the steps until is covers the surface completely [26–28], providing a template for the growth of graphene. However, growth of nanographene clusters on the buffer layer has also been reported in a study, which combined LEEM with STM [29]. With the dominant role of steps, miscuts are favorable for growth by minimizing the pit formation and thickness fluctuations encountered on surfaces with large step spacings but at the same time create small graphene domain sizes [30, 31], which are undesirable for electronic purposes. Thickness fluctuations can be suppressed by micropatterning the SiC surface in narrow ($\leq$1 μm) stripes parallel to the $[1\bar{1}00]$ direction [32]. However most studies have been made with well-oriented surfaces (miscut $\leq 0.06°$) and pit formation minimized by shifting growth conditions towards equilibrium using high-pressure noble gas or silane environment during thermal decomposition. These studies have given a widely accepted picture of the growth and microstructure of graphene on the Si-terminated 6H-SiC(0001) and the 4H-SiC(0001) surfaces (see [11]).

Growth and microstructure on the C-terminated surfaces however remained controversial although it had been studied also early on. Agreement exists that under identical growth conditions the graphene layer grows faster and is less well ordered than on the Si-terminated surfaces, independent of temperature and environment, and that during growth in vacuum no interface layer is formed. Convincingly established is also that growth proceeds by forming new layers below the already existing graphene layers [28]. Open questions are the stacking of the layers and the structure of the interface of graphene layers grown not in free thermal decomposition, i.e., in high-pressure noble gas or an incoming Si flux.

The state of understanding around 2011 has been summarized by Hicks et al. [33] but experiments in Zakharov's group [34, 35] have in the meantime put the generally accepted azimuthal misaligned stacking of the layers in question. The misalignment was deduced in the past from LEED studies, which averaged over *laterally* misaligned domains. LEED in a LEEM instrument allows selecting submicron regions by either the illumination aperture or the field-limiting aperture. Making use of this possibility Johansson et al. [34, 35] could select individual domains up to 6 monolayers thick and show that no additional diffraction spots appeared. ARPES patterns from individual domains accordingly also consisted of only one sextet of Dirac cones. Their dispersion near the Dirac point suggested AA stacking of the bilayers in contrast to the AB stacking in layers on the Si-terminated surface and in apparent disagreement with theory. Another interesting observation should be mentioned yet. Domains of equal thickness as judged by the QSE oscillations differed considerable in brightness in LEEM images taken at energies close to the LEEM–MEM transition. At constant thickness the work function and surface potential should be the same on all domains. As just mentioned, the stacking sequence is the same in every domain. Therefore the cause of the brightness difference has to be sought in other differences connected with the azimuthal alignment of the domains, most likely at the interface. Different brightness levels at constant thickness due to work function differences have also been observed in XPEEM threshold images. In this case these changes were also attributed to different interfaces, in this case to different interface periodicity although not seen in LEED in this study [36]. An interfacial layer has been observed, however, in graphene films grown in 1 atm of Ne at 1630 °C and 10^{-6}–10^{-4} Torr of disilane at 1200–1400 °C [37–39]. LEED shows that this layer has a superstructure similar to the buffer layer on the Si-terminated surface but is rotated by a different angle relative to the substrate. It was suggested that this buffer layer plays a role in the rotational misalignment of the domains. However the misalignment was also observed without this layer in films grown at 1 atm of Ar, i.e., under similar thermal decomposition rate-reducing conditions as in the case of Ne and silane [35]. Thus understanding of the azimuthal misalignment and interfacial structure in the growth on C-terminated SiC surfaces still needs further work. Some information on the effect of the azimuthal misalignment between layers might be obtained by producing it artificially by superimposing an ex situ prepared graphene monolayer on a graphene layer on SiC. This has been done successfully on a Si-terminated surface with a graphene monolayer grown on a Cu foil [40, 41].

Graphene formation on the 3C-SiC surfaces has been studied very little with cathode lens microscopy despite the desire to integrate graphene into silicon technology, which can be achieved only via 3C-SiC. This is due to the fact that at the low temperatures compatible with Si the cubic 3C structure is the stable polytype of SiC and forms therefore upon carbonization of Si. The first cathode lens microscopy studies of graphene growth on 3C-SiC were made on layers grown on Si(111) [42] and on Si(100) [43], using temperatures between 1150 and 1350 °C. On the 3C-SiC(111) surface graphene grew in the same manner as on the 6H-SiC

(0001) surface as evidenced by LEED and ARPES but the surface was rough, consisting of submicron triangular terraces, reflecting the roughness of the epitaxial 3C-SiC(111) layer [42]. The 3C-SiC(100) surface consisted of micron-sized antiphase domains caused by the growth on the Si(100) surface, with height differences of about 10 nm so that the graphene layer grown on it was far from flat [43]. On these domains the graphene layers grew in two different azimuthal orientations as expected for the growth of a hexagonal layer on a square substrate. XPS indicated that the interface between graphene and the substrate is abrupt, similar to the graphene/SiC$(000\overline{1})$ interface, so that no insulating buffer layer is formed in contrast to the (111) orientation. In contrast to graphene on SiC$(000\overline{1})$ LEED did not show the azimuthal variation of the layer stacking seen on that surface. With growth on Si(100), miscut in one of the [011] directions, the antiphase domains might be eliminated or at least reduced significantly so that large graphene domains might be grown. In any case considerable research is still necessary before the hope of integration of graphene-based devices with Si-CMOS devices [44] can be fulfilled.

The problems encountered with 3C-SiC(111) grown on Si(111) caused by the misfit, thermal expansion and growth temperature limitations are avoided by 3C-SiC(111) growth on SiC(0001) surfaces, which produces nearly strain-free layers. This has been done first on 4H-SiC(0001) [45] and subsequently on 6H-SiC(0001) [46]. In the first study only the buffer layer, sometimes called zero-layer graphene, was grown and converted into quasi free-standing monolayer graphene (QFMLG) by intercalation with hydrogen, a process which will be discussed in the next section. Several μm wide QFMLG areas, separated by step bunches, could be prepared in this manner. The second study compared graphene formation on the three hexagonal SiC surfaces under identical growth conditions. A LEEM study found the following: when the 3C surface was covered completely with a single graphene layer and only locally with small bilayer regions, already 10 % of the 6H surface was covered with by a bilayer and about 40 % of the 4H surface by a bilayer with small trilayer regions. These layer coverage differences could be simply due to growth rate differences but were attributed to differences in the spatial homogeneity of the C atom supply, which is determined by the step height distribution. AFM studies showed that the step height distribution was narrow on the 3C surface, with 1 SiC bilayer step dominating while the 4H surface had the widest step height distribution with 2 and 4 SiC bilayers dominating. As the C supply is determined by the Si sublimation rate, which changes with step height, the C concentration is most homogenous on the 3C surface while on the 4H surface large local fluctuations can occur leading to more inhomogeneous growth. However, the higher growth rate on the 3C surface is hard to understand in view of the lower Si sublimation rate from 1 SiC bilayer high steps. Whatever the mechanism, $50 \times 50\ \mu m^2$ single layer graphene was grown on 3C-SiC(111) and if this preferred lateral growth occurs in subsequent layers this surface is far better suited for homogeneous few-layer growth than the 4H surface and somewhat better than the 6H surface.

6.1.2.2 Intercalation

Intercalation is a process in which a foreign material is inserted between two layers, in this case between the graphene layer and the substrate. On SiC surfaces it is used to change the bonding of the graphene layer to the substrate, which determines not only the stress but equally important the charge transfer. Both play an important role for the position of the Fermi level and the transport processes, which motivated many intercalation studies. Starting with hydrogen many atoms have already been intercalated: Li, Na, O, F, Si, Ge, and Au. Most of these systems have been studied in cathode lens microscopy systems because these allow not only getting information on the electronic structure via μARPRES but also chemical and structural information from the same regions via μXPS, LEEM, and μLEED. The ARPES studies show that the Dirac point E_D of the single graphene layer on top of the buffer layer is located at −0.45 eV below the Fermi level E_F. E_D of the single graphene layer formed from the buffer layer by co-intercalation with H and Li [47] is −1.25 eV below E_F, while intercalation with F shifts E_D up as far as +0.79 eV above E_F [48]. Although Li causes a downward shift [49] and H, O, Si, Ge all cause upward shifts of E_D, there is no simple relationship with the electronegativity of the inserted atoms. The position of E_D depends also sensitively on the preparation conditions, mainly due to the degree of intercalation. The shifts are usually attributed to the saturation of the dangling bonds of the substrate but atoms on top of the layer can also cause doping. Another possible cause is the intercalation-caused modification of the polarization charge connected with the spontaneous polarization of the 6H and 4H SiC(0001) surfaces [50]. A review of the understanding of intercalation around 2012 can be found in [51].

A reliable identification of intercalation requires additional methods such as LEEM, LEED $I(V)$, XPS, or Raman spectroscopy. Most of these intercalation studies were made with SPELEEM instruments at synchrotron radiation facilities, foremost at the MAXLab, which allow combining the most relevant methods except Raman spectroscopy. LEED gives the clearest and simplest criterion: intercalation eliminates the characteristic LEED pattern of the buffer layer. LEEM $I(V)$ curves show the increase of the number of graphene layer when the buffer layer is converted into a graphene layer during the intercalation. The chemical shifts in XPS show the changes in C and Si bonding and the redistribution of the doping atoms, which occur upon intercalation. Hydrogen intercalation has been studied most extensively, starting with the work of Riedl et al. [52], in which ARPES was done with HeII radiation and XPS with a separate spectrometer. Figure 6.3 shows some of their results. The graphene layer, which had been intercalated with hydrogen at atmospheric pressure between 600 and 1000 °C, consisted of regions with different thickness as seen in the $I(V)$ curves. The local $I(V)$ curves (Fig. 6.3a) taken from LEEM image sequences illustrate the increase of the number of minima from before (a1) to after (a2) intercalation. The C 1s (b) and Si 2p (c) XPS spectra, excited with 450 eV and 140 eV photons, respectively, show how the graphene bilayer, which was formed from the single layer by the conversion of the buffer layer, reverts back again onto a single graphene

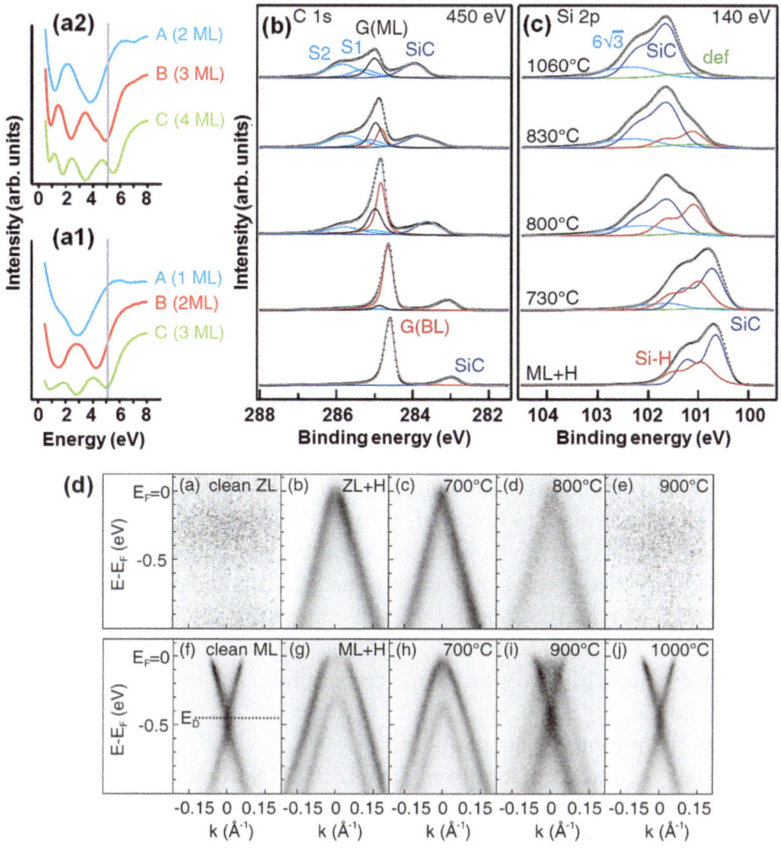

Fig. 6.3 Hydrogen intercalation of graphene on Si(0001). The graphene layer has regions with different thickness as seen in the number of minima of the LEEM *I*(*V*) curves in (**a1**, without hydrogen) and (**a2**, with hydrogen). Hydrogen intercalation (**a1** → **a2**) increases the number of layers in all regions by converting the buffer layer into a graphene layer. The C 1s (**b**) and Si 2p (**c**) XPS spectra show the changes occurring in a bilayer (BL) produced by intercalation of a monolayer (ML) with increasing temperature ending in a monolayer after complete desorption of hydrogen at 1060 °C. The experimental data (*black curves*) are fitted by several colored curves corresponding to characteristic bonding states. The changes in the band structure as measured in ARPES along the *blue line* in Fig. 6.2 upon hydrogen treatment and subsequent annealing steps are shown in (**d**) for an initial buffer layer (*top row*) and an initial monolayer on top of the buffer layer (*bottom row*). Adapted with permission from Ref. [52]. Copyright 2009 by the American Physical Society

layer with increasing temperature due to the desorption of hydrogen bonded to Si at the interface (Si–H). In (d) one can see the corresponding changes in the electronic structure, in the top row for the case that no graphene layer was present before intercalation (clean "zero layer"), i.e., only the buffer layer, in the bottom row for the case of a single graphene layer (clean monolayer) on the buffer layer before intercalation. The formation of the graphene bilayer Dirac cone in the first case and of the second graphene bilayer Dirac cone in the second case is clearly visible as is their disappearance with the loss of hydrogen with increasing temperature. These results have been extended and refined with better defined initial surface conditions in subsequent studies [53–56, 45].

Li intercalation was studied with the same methods for "zero-layer" (buffer layer only) and monolayer graphene surfaces. Partial intercalation was found to occur already at room temperature, full intercalation at about 350 °C and stability up to 450 °C. The Dirac point was shifted by to about 1.0 eV below E_F. Similar to hydrogen, LEED and the LEEM $I(V)$ data showed the conversion of the buffer layer into a graphene layer. In XPS the Si 2p peak was shifted about 2 eV to lower binding energy, which was attributed to Li bonding to Si on the SiC surface, thus creating a dipole layer [57, 58, 49]. When a hydrogen-intercalated graphene monolayer (Fig. 6.3d, bottom middle) is exposed to Li the band structure is changed dramatically: E_D is shifted from near E_F downward by about 1.25 eV. The XPS spectra indicate several Si and Li binding states and the LEEM $I(V)$ curve has an unusual shape. The results were attributed in part to charge transfer from Li on top of the graphene layer, in part to below it, where it displaces some of the interfacial hydrogen and creates Li–Si bonds [47]. Na intercalation of a graphene film consisting of single layer and bilayer regions was studied with LEEM, μLEED, μXPS, and XPEEM with the following results: intercalation occurs in both regions, beginning at the boundaries between them, but is very inhomogeneous. In contrast to Li, which intercalates directly at room temperature, Na intercalation is strongly promoted by heating to 100 °C and also by electron and photon beam exposure. The heavier alkali do not intercalate at all [59].

In the first oxygen intercalation study in a LEEM instrument [60] intercalation of a graphene monolayer on the (0001) surface was proved by the disappearance of the buffer layer LEED pattern, by the appearance of a pronounced plasmon energy loss, which is characteristic for the π electron system of graphene and graphite, and by depth profiling with medium energy ion scattering, which revealed a strong oxygen peak below the graphene layer, and corresponding changes in the C 1s and Si 2p XPS spectra. In a later SPELEEM study of oxygen intercalation below a single graphene layer the LEEM $I(V)$ curves showed only weak signs of buffer layer conversion. In LEED the buffer layer pattern was still visible but in addition also a $(\sqrt{3} \times \sqrt{3})$-R30° structure, which was attributed to an ordered oxygen layer below the graphene layer covering 1/5 of the graphene honeycomb centers. Although ARPES showed a shift of E_D towards E_F, no second cone indicating formation of a bilayer was seen, only a broadening of the Dirac cone branches. While the limited energy resolution (0.25 eV) may have smeared out the two cones, a more likely interpretation is that only partial intercalation had occurred, due to the softer oxidation compared to that in the first experiment (10^{-4} Torr at 500 °C for 6 h compared to 1 atm at 250 °C for 5 s) [61]. The $(\sqrt{3} \times \sqrt{3})$-R30° structure was also found to replace the buffer LEED pattern after uncontrolled oxidation of graphene layers on the (0001) surface, which had been prepared by heating in disilane at high temperatures (see Sect. 6.1.2.1). The LEEM $I(V)$ curves also indicated addition of another graphene layer connected with the disappearance of the buffer pattern [37–39]. Thus is appears that in both cases only partial intercalation has occurred but the results are too limited to conclude this firmly.

Si intercalation has been subject of much more thorough studies. The first one looked at the intercalation of a single graphene layer, combining LEEM, LEED, and XPEEM in a SPELEEM instrument with C 1s, Si 2p, and valence band PES in a high resolution electron spectrometer [62]. In contrast to Li, which started to intercalate already at room temperature, 800 °C are needed for Si intercalation as evidenced by the disappearance of the LEED pattern of the buffer layer, the changes in the C 1s, Si 2p, and valence band spectra and in the ARPES data. For the temperature range, in which the intercalation is stable, the different methods give different values: K point ARPES at least up to 1000 °C and core level spectra at least up to 900 °C. In LEED the buffer layer spots appear already at 1000 °C, together with $(\sqrt{3} \times \sqrt{3})$-R30° spots. Above 1300 °C all methods show that the surface has returned to its original state. The $(\sqrt{3} \times \sqrt{3})$-R30° pattern was tentatively attributed to SiC formation on top of the graphene layer. However, it could also be explained in the same manner as in the case of oxygen intercalation (see above) because it is difficult to extract the Si location from the XPS data. Intercalation of a graphene bilayer was the subject of a second study, which used mainly C 1s and Si 2p XPS, combined with LEED [63]. A careful analysis of the XPS data revealed that up to 6.3 Si monolayers could be intercalated at 1020 °C. 0.2–0.5 Si monolayer stayed on top of the graphene layer over the complete Si coverage range studied, showed some ordering and could be desorbed only above 1400 °C.

A study of Ge intercalation of a surface terminated by the buffer layer, using the same combination of methods as the first Si intercalation study, revealed a number of interesting phenomena [64]. At 720 °C 2 monolayers of Ge were intercalated, leading to p-type doping ($E_D > E_F$), at 920 °C only 1 monolayer was intercalated, leading to n-type doping. At intermediate temperatures 1 and 2 monolayer thick regions coexisted so that lateral p–n junctions were formed. LEEM movies shed some light on the mechanism of this complex intercalation process and Ge 3d core level PEEM images as well as LEEM $I(V)$ curves contributed also to the understanding of this intercalation system. This study is a nice demonstration of the power of the combination the methods used.

6.1.3 Graphene on Metals

6.1.3.1 Introduction

Similar to graphene on SiC, which was first interpreted as a graphite layer [65], many researchers have produced unintentionally graphene layers on metal surfaces during surface cleaning without recognizing its significance. What was a nuisance in the twentieth century has become a gold mine for many researchers in the twenty-first century. Graphene on metals has been subject of numerous studies with are described in detail in the review by Batzill [4] and more condensed by

Bartelt and McCarty [66] in addition to the references given at the beginning of this chapter. Here only a brief introduction will be given and more recent developments sketched. Graphene on metals is usually grown either by direct deposition of elemental carbon, by segregation of carbon dissolved in the metal or by decomposition of adsorbed hydrocarbons. The first method is universal and exactly controllable; the second is limited to metals with sufficient carbon solubility, the third one requires catalytic activity. This makes the last two material-specific. Depending on the metal and the formation conditions single or multiple layers can be grown. The graphene–metal bonding ranges from weak, for example on the (111) faces of Al, Cu, Ag, and Au, to strong such as on the (111) faces of Ni, Co, Re, and Ru, resulting in a wide variety of film structures and morphology. The mismatch between the atomic distance in the graphene layer (2.46 Å) and the atomic distances in the closest packed substrate planes, which range from 2.49 Å for Ni to 2.89 Å for Ag determines the azimuthal alignment of the layer, its distance from and bonding strength to the surface. Misfitting layers usually consist of regions with good fits and bad fits with correspondingly varying layer–substrate distances and bonding strength. This produces large unit cells of the composite system, clearly visible in STM and LEED. For these regions first principle calculations, which assume fixed positions of the C atoms with respect to the substrate [67, 21] are of little help for the experimentalist in the interpretation of his data. Rather theory has to start from the experimental superstructure cell and relax the atomic positions by minimizing the total energy. As an example, graphene on Ru(0001) shows in experiment an approximate 12×12 supercell in substrate units. Density functional theory calculations for a 12×12 supercell give after relaxation a distance between the graphene layer and the topmost Ru layer varying from 2.15 to 3.70 Å [68] with according bond strength variations, while first principle calculations based on fixed C and Ru positions would give only one distance and binding energy. In the following recent results obtained mainly with cathode lens microscopy but also borrowing some results from other methods, will be discussed, first growth and microstructure, then intercalation.

6.1.3.2 Growth and Microstructure

The best studied metals are Ru(0001), mainly by the groups of McCarty, Sutter, and Bao, who looked at different aspects of growth and structure, and Ir(111), which was thoroughly studied by McCarty's group. Other groups have contributed too to the understanding of these graphene-on-metal systems, which are well described in the review papers cited earlier. Additional insight comes from studies aimed at modifying the growth conditions. One method used Ar sputtering followed by annealing at temperatures above 750 °C, which leads to the formation of subsurface Ar bubbles from the implanted Ar. The bubbles cause a local surface strain visible in MEM as little bulges. Graphene growth on such a surface, either by C segregation from the substrate of by exposure to C_2H_4, is significantly faster than on the unstrained surface and grows radially outward from the center of the bubble, irrespective of the step directions on the surface. This is in contrast to growth via segregation on the normal, unsputtered surface, on which growth along the steps is about twice as fast as normal to them [69]. Another method for controlling graphene growth is to limit the amount of C in the metal which is available for segregation, either in growth by pure segregation or by dissociation of hydrocarbons from the gas phase. This has been achieved by replacing the single crystal substrate with an epitaxial film, which is C-free when deposited under clean conditions. It can also be doped with carbon up to a thickness- and temperature-dependent limit, which is significant lower than the bulk solubility at the corresponding temperature [70]. This limited solubility has been attributed to the same phenomenon described in the preceding study, strain from trapped Argon, which was used for sputter cleaning. Actually the LEEM images of graphene islands obtained by segregation from a 1 μm thick Ru(0001) film on sapphire are very similar to those in the previous study. More detailed information on C in this film can be found in [70].

A quite different growth and structure control was adopted in a study of the influence of alloying Ru, to which graphene bonds very strongly, with Pt, with which graphene interacts only weakly [68]. By changing the Ru:Pt ratio in the surface it should be possible to tune the interaction and the structure and growth behavior continuously. RuPt surface alloys with 12, 19, and 39 % Pt were formed by deposition at 300 K followed by annealing at 1300 K. Graphene was grown on the clean surface and on the alloys by exposure to C_2H_4 at 1050 K up to more than 2 monolayers while recording the growth continuously by LEEM and XPS in a SPELEEM instrument. The C 1s peak of graphene on Ru(0001) consists of a strong and a weak binding energy peak corresponding the regions of small and large distances of the undulating layer from the metal surface. With increasing Pt content the ratio of strong to weak peak height decreased systematically as qualitatively expected because of the weaker bonding to Pt. The accompanying density functional theory calculations mentioned already earlier, which were made for 0, 10, 20, and 50 % Pt coverage showed that the shortest graphene–substrate distance increases accordingly from 1.55 Å above Ru atoms to 3.45 Å above the Pt atoms. The two-dimensional local C–Ru distance and charge transfer plots resulting from these calculations are a good illustration of the overriding influence of the electronic structure on the formation of graphene on metals.

Most of the graphene growth studies have been done on surfaces with hexagonal surface lattice. The question how growth occurs on a substrate with different symmetry was addressed in a study by Locatelli et al. of graphene growth on Ir (100), combining bright and dark field LEEM and XPEEM with LEED, ARPES, STM, and density functional theory calculations [71]. They grew graphene islands with several μm size by exposing the surface to C_2H_4 at about 800–850 °C. LEEM showed structureless islands at temperatures down to about 500 °C. Below 500 °C oriented striations formed in them, which at room temperature covered about 25 % of the island area as seen in Fig. 6.4a. The LEED pattern at room temperature (Fig. 6.4b) is a pronounced coincidence pattern, which is dominated by the dark

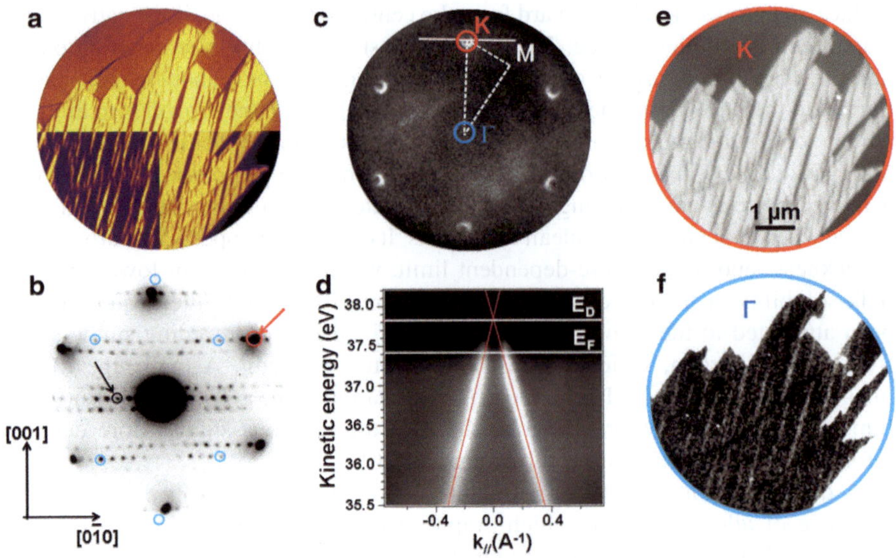

Fig. 6.4 Graphene monolayer island on Ir(100). (**a**) LEEM images taken with the (00) beam (*upper half*), with the superstructure beam marked *black* in the LEED pattern (**b**) (*lower right quadrant*), both with 12 eV electrons, and with the first order graphene beam marked *red* in (**b**) with 50 eV electrons. The LEED pattern (**b**) was taken with 38 eV electrons; the *blue circles* in it indicate the position of the Ir diffraction spots. (**c**) ARPES pattern at E_F and (**d**) band structure along the normal to ΓK through K (*white line* in **c**), photon energy 40 eV, (**e**) and (**f**) dark field and bright field XPEEM images taken with photoelectrons emitted in the K and Γ directions at E_F, respectively, marked *red* and *blue* in (**c**). Adapted with permission from Ref. [71]. Copyright 2013 American Chemical Society

stripes as seen in the dark field LEEM image in the left lower quarter of Fig. 6.4a, taken with one of these spots. Dark field images of the bright regions obtained with the strong first order graphene spots reproduce exactly the same structure and contrast observed in the bright field (right lower quarter of Fig. 6.4a). Only above 500 °C, after the dark regions have disappeared, can the LEED pattern of the bright regions be identified clearly as being incommensurate with the substrate.

Laterally resolved XPS spectra extracted from a series of XPEEM images show in both regions a strong C 1s core level peak characteristic of weakly bound graphene but in the dark regions in addition a weak satellite at higher binding energy, indicating stronger bonding. In ARPES (Fig. 6.4c, d) pronounced Dirac cones can be seen, with E_D shifted 0.42 eV above E_F. Dark field XPEEM with the electrons around the K point (Fig. 6.4e) clearly show that the Dirac cones exist only in the regions which appear bright in LEEM, while imaging with photoelectrons emitted normal to the surface, i.e., in the Γ direction (bright field imaging) (Fig. 6.4f) shows intensity only in the stripes, which was attributed to hybridization with Ir states near E_F. Combining all these results with the STM results and the first principle calculations resulted in the following structure model. At room temperature graphene forms on Ir(100) a layer consisting of wide flat regions with a large distance from the substrate (maximal about 3.8 Å), which is quasi-freestanding and supports a slightly energy-shifted Dirac cone band structure, and narrow regions with a minimum graphene–Ir distance of 2.1 Å. In these narrow regions, which contain only 11 % of the atoms, the strong interaction with the substrate suppresses the Dirac cone structure. The graphene–Ir distance variations produce a one-dimensional rippling with a lateral periodicity of 21 Å and amplitude of about 1.7 Å. It forms upon cooling from the flat state at the growth temperature as a consequence of the different expansion coefficients of film and substrate and is reversible. This experiment is a very good demonstration of how a complicated system can be unraveled by making use of many of the operation modes available in a SPELEEM instrument.

In a study of the structure and stability of graphene on Re(0001) [72] LEEM and µLEED played only a supporting role, demonstrating that graphene was not rotated relative to the substrate over the complete film and that it formed a (9×9) coincidence structure in Re lattice units, which served as the unit cell for density functional theory calculations. These produced a strongly corrugated graphene with C–Re distances predominantly between 2.1 and 2.4 Å but also distances up to 3.8 Å. The same calculations also predicted accordingly a C 1s XPS doublet with the strong high binding energy peak dominating in good agreement with experiment. Upon annealing this peak, which is connected with the short C–Re distances decreased much faster with temperature than the low energy binding peak as expected because the stronger interaction leading to a faster dissolution in Re, which has a high C solubility and forms also a carbide.

LEEM and µLEED has also been used to characterize graphene crystals up to mm size grown ex situ on Pt at 1040 °C with atmospheric pressure CVD at low CH_4/H_2 ratio. Although these crystals developed many pits in UHV, their electron mobility was very high (7100 cm^2/V s at a hole concentration of 2×10^{11} cm^{-2}) as measured in a field-effect transistor after removal from Pt and mounting it on SiO_2/Si [73]. An in situ LEEM and µLEED study of the growth of graphene via C segregation from Pd(111) between 970 and 660 °C addressed nucleation, growth, layer stacking, and work function of graphene. Heterogeneous nucleation, followed by multilayer growth below the existing layer was found and characterized by LEEM $I(V)$ curves up to 8 monolayer thick regions. Many azimuthal orientations of the graphene islands were found with irregular layer stacking. The work function of bilayers varied with both the azimuthal orientation of the first and second layer within 0.07 eV and was higher than that of the single layer which could be explained by first principles calculations [74].

Ni(111) is a particular surface for graphene growth because its interatomic distance is very close to that of the C distances in graphene and the bulk of the crystal is a convenient source of C because its high solubility. By decreasing the temperature from high temperatures and thus decreasing the C solubility single layer, bilayer, and trilayer graphene can be grown systematically. This was done in the first LEEM and µLEED study of this system [75–77] by recording graphene single layer growth at 1125 K and bilayer and trilayer growth at 1050 K. All layers were found to grow in a carpet-like fashion across the steps, starting from a nucleation point outside the 100 μm wide field of view, but with vastly different growth rate (about 10 μm/s, 1 μm/s and 0.1 μm/s for the single layer, bilayer and trilayer, respectively), and growth front shapes. While the growth front of the single layer was relatively straight over large distances and independent of the surface morphology, the bilayer growth front was strongly influenced by the step distribution and was, therefore dendritic-like. This is a clear indication that growth occurred below the first layer. Interestingly, the trilayer, which started to grow already before completion of the bilayer had again straight growth fronts reflecting the anisotropy of the boundary energy of graphene. With the growth of the second and third layer the substrate step-induced wrinkling of the layer decreased [75]. μLEED showed a (1 × 1) pattern for all thicknesses with different azimuthal orientations in the bilayer and trilayer regions [T. Yasue, private communication]. Thus this relatively straightforward study showed that large graphene sheets without apparent grain boundaries can be grown in a simple manner.

Shortly after this study it was shown that single layer graphene can also be grown by thermal decomposition [78] using C_2H_4 at a pressure of 10^{-6} Torr, however only within a narrow temperature window of about 550–600 °C, with higher temperatures leading to dissolution of C in the bulk and lower temperatures to carbide formation. LEEM showed that the nucleation density was much higher than during segregation growth at high temperatures and that the growth front was dendritic. No LEED results were reported so that nothing can be said about the orientation. In contrast to growth on Ni(111) *single crystals* a quite different growth behavior has been reported for growth at temperatures of about 550–750 °C by exposure to 5×10^{-4} C_2H_4 on initially carbon-free epitaxial Ni(111) *films* on yttria-stabilized zirconia [79]. Not only graphene islands with parallel orientation but also islands rotated by about $17 \pm 7°$ with respect to the substrate nucleated and grew significantly faster. Upon cooling a second layer grew only below the rotated first layer. This was attributed to the weaker bonding of graphene in these layers, in which many C atoms are in unfavorable binding sites compared to the layer in parallel orientation, which allows easier transport and nucleation.

For practical applications graphene growth on single crystal surfaces, either of bulk crystals or thin films is unrealistic. The observation made on the Ni(111) surface that at high temperatures nucleation is very localized, suggests that grain boundary-free graphene could also be grown on polycrystalline foils. This requires that the nucleation probability on one grain orientation is much higher than on all others and that growth from this nucleus spreads across the other grains without change of the azimuthal orientation at the beginning of the growth. This possibility was studied with LEEM and LEED for growth via surface segregation in situ [80] and ex situ after heating at 900 °C in H_2 and Ar atmosphere containing CH_4 [81]. In the first study a 100 μm thick Ni foil with strong (111) texture was doped with C by heating in 10^{-3} Pa C_6H_6 for 30 min and graphene growth was recorded at 1150 K similar to the study on the Ni(111) single crystal discussed before. At this temperature only a single layer grows. The LEEM movies clearly showed that graphene, which had nucleated outside the field of view spreads rapidly across all grains and µLEED showed that the orientation of the layer did not change from grain to grain. Rapid quenching to room temperature suppressed the formation of a second layer. Therefore the hope that large single crystalline single layer graphene regions can be grown on suitably prepared polycrystalline Ni was fulfilled. The second study, however, suggests that this is not true for multilayer graphene films.

Similar to Ni, the solubility of C in Fe is also high and so is the tendency to carbide formation. Nevertheless single layer graphene has been grown on Fe(110) by thermal decomposition of hydrocarbons, using the same method described before: a hydrocarbon pressure which produces C supersaturation, an at least initially C-free thin film and a growth temperature low enough to avoid dissolution of C in the film but high enough to avoid the competing carbide formation. This was done by exposing an about 20 nm thick epitaxial Fe film on W(110) to $2–4 \times 10^{-6}$ mbar C_3H_6 at 520–600 °C [82]. This resulted in a monodomain graphene layer, with a complex moiré LEED pattern indicating a long period modulation in the $[1\overline{1}0]$ direction and a shorter period modulation in the [001] direction, which was confirmed by STM. First principles theory calculations for the corresponding supercell (37.5×17.2 Å^2) showed maximum and minimum graphene–Fe distances of 3.03 Å and 2.12 Å, respectively, with less than 10 % of the atoms at distances above 2.7 Å. The short distances are connected with strong bonding. This is also evident in the C 1s XPS spectrum, which shows only the strong binding energy peak. Thus the situation is opposite to that found on Ru(100) where the weak binding energy was dominating, involving only 11 % of the atoms and where most of the C atoms had a large distance from the metal surface.

Graphene growth on polycrystalline Cu has been developed already in 2010 to such an extent that touch-screen panels with 30 in. diagonal dimensions could be produced after removal from the Cu foil [83]. However, for active electronic devices such as field effect transistors the carrier mobilities were still low and irreproducible. As Cu is the most promising substrate because of the weak interaction with graphene and the availability in highly oriented foil efforts have continued to understand and to improve the optimum formation conditions of graphene on Cu. Most of this work has been done with other imaging methods such as STM, SEM, and TEM because practical growth methods use high pressures unsuitable for in situ studies. Post-growth LEEM, and LEED have also been used, in particular to understand the differences between the growth on the (100) surface, which is the dominating surface in the cold-rolled and annealed Cu foil, and the (111) surface [84, 85]. The first study [84] compared the structure of graphene layers grown under identical conditions on (111)- and (100)-oriented epitaxial Cu films on MgO. Graphene was grown under optimized conditions (CH_4:Ar:H_2 ratio and flow rate) at atmospheric pressure at 500 °C to completion of a single layer and rapidly quenched to room temperature. On the (111) surface graphene was found to grow in a single orientation over an area as large as 1 mm with negligible defects, on the (100) surface in predominantly 2 µm-sized domain orientations with many defects connected with the domain boundaries. The second study [85] compared the structure of graphene also grown from CH_4 in Ar + H_2 at 500 °C on a

(111)-oriented epitaxial film on sapphire with the structure of graphene on a Cu foil with (100) texture. On the (111) surface the same structure was found as previously. On the foil the individual grains showed the dominant domain structure with two rotated domains described above, but due to the many different azimuthal orientations in the foil a corresponding number of different grain boundaries. Field effect transistor mobility measurements of the graphene films mounted on SiO_2/Si after removal from Cu showed that the mobility of the films grown on the foil was an order of magnitude lower than that grown on the (111)-oriented surface. Thus, these LEEM and LEED studies indicate that the growth of high-mobility graphene on Cu foils faces a fundamental problem: the many domain boundaries caused by the at least two domains on the individual grains and the many grain orientations reduce the mobility. A detailed bright and dark field LEEM study combined with μLEED [86] has shown a further complication: the graphene growth process in the CH_4/H_2 mixture causes faceting of the imperfectly (100)-oriented Cu grains into (100) and (410) planes and (n11) facets. The resulting steps influence the azimuthal orientation of the graphene. As a consequence additional orientations are introduced and the graphene layer, though continuous, becomes strongly micro-corrugated. The great advantage of close atomic distance match and weak interaction between graphene and Cu apparently can only be used on well-oriented (111)-oriented Cu, which might require growth on suitably oriented wafers such as Si(111), sapphire (0001), or SiC(0001).

Another surface, with which graphene interacts only weekly is Au(111). While Cu has a very small mismatch with graphene, Au has a large one and one could expect a quite different growth behavior if lattice mismatch would play an important role. Apparently this seems not to be the case as a comparison of in situ growth on Au(111) [87] with that on Cu(111) [88] under similar growth conditions shows. In both cases graphene was grown by deposition from a carbon rod heated by electron bombardment with the substrate at temperatures between 770 and 940 °C (Au) and 690 and 900 °C (Cu). Nucleation started preferentially at steps with higher nucleation density on Au than on Cu but the growth shape on both surfaces was dendritic with distinct lobes. After completion of 1 monolayer no second graphene layer was formed. All crystals were (111)-oriented with several azimuthal orientations at the lower temperatures but only parallel orientation at the higher temperatures. ARPES showed that the Dirac point deviated only slightly from E_F, indicating very weak interaction between the graphene layer and Au, which explains the apparent insensitivity of growth and structure to misfit.

6.1.3.3 Intercalation

When a graphene-covered metal surface is exposed to a material (atom or molecule), which interacts more strongly with the metal than with graphene, it intercalates, provided it has access to the graphene–metal interface in order to minimize the total energy of the system material–graphene–metal surface. This is true for any material with sufficient resident time and diffusion rate on the graphene or on the only partially covered metal surface so that the approach to thermodynamic equilibrium is not kinetically limited. Access to the interface is provided by defects in the graphene layer at complete coverage and at the edges of graphene islands at partial coverage. Although at first sight counterintuitive because of the short C-metal distance, intercalation is particularly favorable in the case of strong graphene–metal interaction as most of the out-of-plane bonding charge of graphene is shifted towards the metal, weakening bonding to adsorbates on top of graphene. At the same time the electron configuration of the metal surface responsible for the strong interaction with graphene also promotes bonding of the intercalant. In the case of weak graphene–metal interaction the "empty" space between graphene and metal cries for being filled. Thus, independent of the strength of the graphene–metal, intercalation is a general phenomenon but most work has been devoted to intercalation of strongly bound systems.

The first intercalation studies were made on a weakly interacting system, graphene on Ni(111), already around 2000, using Cu, Ag and Au as intercalant, with laterally averaging methods. More recently graphene–Ni(111) intercalation with Al and Sn has also been studied but not with cathode lens microscopy, which, however has been used in a number of intercalation studies on Ru(0001). Oxygen and NO_2 intercalation has been reviewed already [11, 12]. Here only the work on metal, Si and Si intercalation, followed by oxidation, will be discussed. In metal intercalation metals with extreme atomic diameters were chosen: Ni ($d = 2.49$ Å) and Pb ($d = 3.50$ Å), which have strong and weak interaction with Ru, respectively. These differences suggest a completely different intercalation behavior. Combined LEEM, PEEM, and μLEED studies indeed confirm this. Pb cannot be intercalated into a complete graphene layer but only into graphene islands after the surrounding free Ru surface is covered by a complete Pb adsorption layer. Intercalation occurs via the edges of the graphene islands [89, 90]. To the contrary, Ni is intercalated directly into graphene islands as evident in well-resolved LEEM images in many small intercalation islands within large graphene islands. The difference between the two intercalation systems is clearly visible in the PEEM images produced with high-pressure Hg arc lamp light with increasing temperature of annealing, which is needed for intercalation (Fig. 6.5) [90]. In the case of Ni intercalation the contrast of the graphene island changes homogeneously with increasing temperature; (the small intercalation islands are not resolved in these low magnification images). Pb intercalation clearly starts at the boundary of the graphene island in an inhomogeneous manner moving towards the center with increasing temperature. Apparently more than 1 Pb monolayer was deposited leading the agglomeration of Pb in excess of 1 monolayer except near to the graphene island, where Pb is consumed by the intercalation [90]. The high density and random distribution of Ni intercalation islands suggests that Ni atoms penetrate through the graphene monolayer, apparently supporting the expectation mentioned before.

Another small atom for which this should be true is Si. A in situ laser PEEM study of its intercalation process, combined with μLEED, STM, and first principles theory studies, has led to a specific intercalation mechanism. Intercalation occurred at 900 K.

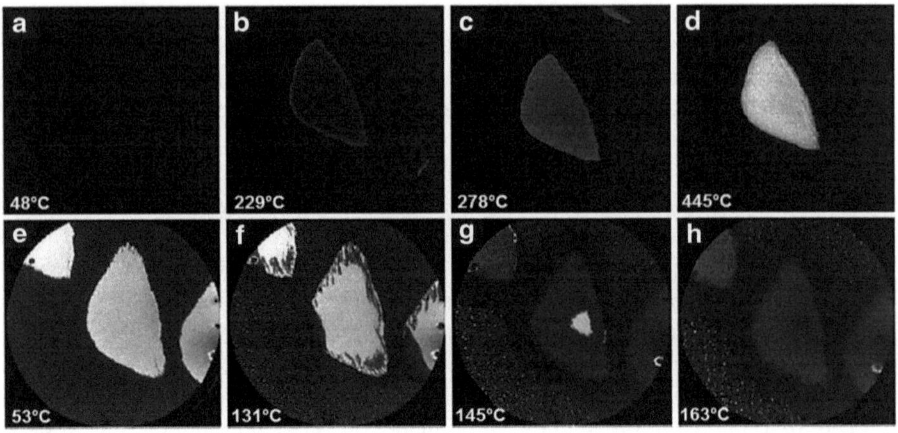

Fig. 6.5 Intercalation of graphene islands on Ru(0001) by Ni (*top row*) and Pb (*bottom row*) with increasing annealing temperature. UVPEEM images with 50 μm field of view. Ni deposition increases the work function of graphene by about 1.2 eV to a value similar to that of Ni on Ru so that there is no contrast at the beginning (**a**) while Pb has little influence on the work function of graphene on Ru(0001), which is about 3.3 eV and causes high emission from the islands (**e**). Intercalation with increasing temperature (**a–d** and **e–h**) decreases the work function in the case of Ni but increases it in the case of Pb, causing corresponding changes in the emission intensity. For details of these changes seen see text. Adapted with permission from Ref. [90]. Copyright 2013 Elsevier

PEEM showed the same behavior as in the case of Ni, but the calculations produced a more complicated process than the one considered before [91]. This process involves several steps: Si adsorption on top of the graphene layer followed by penetration through it, and rotation of the Si–C dimer so that finally the C atom is on top again leaving Si strongly bound to Ru. In this model, direct intercalation through the graphene layer is not so much a question of the size of the atom but rather determined by the strength of its interaction with carbon and should be applicable also to other atoms with strong bonds to C, including the Ni intercalation discussed before. The next question is: can the intercalated Si be oxidized to form an insulating SiO_2 layer so that the graphene layer is quasi-free-supported with high carrier mobility? The earlier oxygen intercalation experiments, reviewed in [11, 12] had shown that oxygen can be intercalated below about 650 K via the edges of graphene islands. More recent laterally averaging studies [92] report that oxygen intercalation of Si-intercalated graphene on Ru(0001) is possible also on continuous graphene layers leading to an SiO_2 interfacial layer as thick as 7 monolayers formed by about 4 monolayers of Si intercalated to a large part in form of Ru silicides. The graphene layer had been grown at 1100 K by exposure to C_2H_4 at low pressure until completion of monolayer coverage. Si intercalation had been performed by Si deposition at 720 K and the complete process (graphene growth, Si and O intercalation) was recorded continuously with C 1s, Ru $3d_{5/2}$, Si 2p, and O 1s XPS up to a total oxygen dose of about 2×10^{-5} L. What is missing is microscopic evidence that the film is fully continuous.

Before completing this section a comment concerning the work function of graphene systems should be made yet. Graphene and its intercalation can cause large work function changes. This fact was already used in early PEEM studies [93] of graphene formation by segregation from Ru(0001) and its removal by oxidation and later again in the PEEM Pb and Ni intercalation studies discussed above [90]. Quantitative work function values of graphene-covered surfaces have been determined by Starodub et al. [94] by measuring the temperature dependence of the thermionic emission current and energy of the emitted electrons. The work functions were obtained from fits to the Richardson-Dushman equation and the Maxwell-Boltzmann equation, respectively. Graphene/Ru(0001), a strongly bound system, was found to have a work function of 3.3 eV. In the graphene/Ir(111) system, which is more weakly bound, the work function depended on the azimuthal orientation of the graphene layer, yielding two values 4.2 and 4.6 eV. All values are lower than the work function of Ru(0001), providing good contrast in PEEM. These results are in agreement with the picture that stronger bonding, i.e., stronger charge transfer causes stronger work function reduction.

Summarizing this section, it is evident that cathode lens microscopy has already made significant contributions to the understanding of graphene growth, structure and properties but it is also clear that there are still many open questions, as illustrated by the last example, waiting to be answered with the methods available in LEEM/PEEM and in particular in SPELEEM instruments. In particular the dynamic observation of growth, intercalation or of changes with temperature on the mesoscopic scale are best or only accessible with these methods as is the combination of spectroscopic microscopy with spectroscopy. Thus graphene will remain a popular and important subject of cathode lens microscopy for some time to come.

6.2 Plasmons

6.2.1 Introduction

In Chap. 2 we have seen that electron emission can be strongly enhanced when the incident light can excite surface plasmons or surface plasmon polaritons due to the strong field enhancement by the plasmon radiation. Here we discuss now the contributions by PEEM to the understanding of this phenomenon and of its applications. In PEEM the surface is usually exposed to a DC field of about 10^5 V/cm which is small compared to the local AC plasmon fields of the order 10^7 V/cm. However, surface asperities, on which plasmons can be excited, enhance also the DC field considerably, up to by about a factor of 10^3 on very sharp asperities. Diamond field emitter films are an example. Thus, while in general the high intensity of the small spots ("hot spots") seen in PEEM from particles resulting from excitation by light, can be attributed to the influence of plasmon fields, sometimes DC field enhancement may be involved too. Imaging with light off the plasmon resonance can easily distinguish between the two contributions to the intensity. Another effect, not known from imaging with conventional light sources, which emit incoherent light, is Fresnel diffraction of the coherent laser light at the asperities, a problem also encountered in synchrotron radiation PEEM. The light intensity modulation in the diffraction fringes causes a corresponding variation of the photoemission intensity in the background surrounding the asperities. The oscillation period of these fringes is usually large compared to that of the fringing field of the plasmon oscillation surrounding the asperity so that they can be easily distinguished.

With these comments in mind we can now review what PEEM has contributed up to now to plasmon physics. Two commercial instruments have been used in all experiments, one (Focus) with an angle of incidence of 65°, the other (Elmitec) with an angle of incidence of 74°, angles, which give in p-polarization a large field component normal to the surface. PEEM gives the spatial distribution of the plasmon field strength, with energy filtering also the photoelectron energy distribution, and when used in the pump-probe mode also the time dependence of the processes involved in the emission process. The initial observations of hot spots were more or less of incidental nature and showed mainly the usefulness of PEEM, in particular with time-of-flight energy resolution, for the study of hot spots [95–98]. The first studies specifically aiming at photo emission via the plasmon channel appeared in 2005. Cinchetti and Schönhense [99] compared the energy spectra of photoelectrons emitted from a Ag nanoparticle film with $\approx$ 50 nm average particle size with those from a continuous film, using $<$200 fs pulses of frequency-doubled p-polarized light from a Ti:sapphire laser with an energy of 3.1 eV and a fluence of 6.4 $\mu J/cm^2$, incident at 65°. In 2PPE they found that the emission from the nanoparticle film sample, integrated over the nearly 2 eV wide spectrum, was 70 times higher than that of the continuous film. This is much less than expected on the basis of earlier laterally averaging measurements of films but the photon energy was significantly off the maximum of the plasmon resonance and the continuous film between the nanoparticles also showed some roughness causing plasmon-enhanced emission. Another experiment by Cinchetti et al. [100] aimed specifically at the shape dependence of the plasmon resonance and its near field by imaging Ag moon crescents with 400 nm diameter, using the same conditions as in the experiment just described. Strong plasmon-field enhanced electron emission was found inside the Ag crescents but the resolution was not sufficient to show the calculated field enhancement distribution at $h\nu = 3.1$ eV.

6.2.2 Linear Structures

While the preceding experiments were concerned with localized surface plasmons Kubo et al. [101] in a landmark study aimed at propagating surface plasmons (PSP) with their associated electromagnetic field (surface plasmon polariton (SPP)) by exciting plasmon waves in a grating, also at $\theta = 65°$ angle

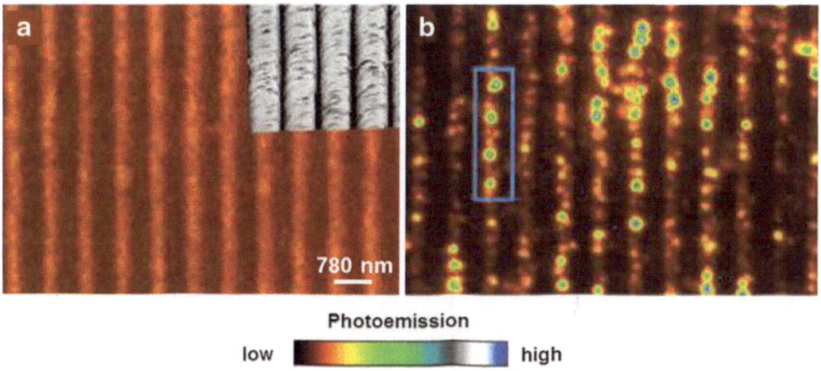

Fig. 6.6 PEEM images of the identical region of the 780 nm period Ag grating obtained (**a**) with light from a high-pressure Hg arc lamp (254 nm, 1PPE) and (**b**) with p-polarized 400 nm fs laser excitation (2PPE). The SEM image in the *inset* in (**a**) shows the true configuration of the grating and surface roughness—unresolved in PEEM—from which the localized surface plasmon modes, the "hot spots" seen in the 2PPEEM image (**b**) are excited. Adapted with permission from Ref. [101]. Copyright 2005 American Chemical Society

of incidence. At this large angle of incidence the k-matching condition via diffraction at a grating with period a

$$k_{E//} = (\omega/c)\sin\theta + 2\pi/a = k_{SPP} = (\omega/c)[\varepsilon/(1+\varepsilon)]^{1/2} \tag{6.1}$$

is not fulfilled. Here ω is the angular frequency of light, ε the real part of the dielectric function, k_{SPP} the wave number of the SPP, which depends on the period of the grating, and $k_{E//}$ the wavenumber of the diffracted incident wave parallel to the grating. Therefore the SPP could not be excited. The grating was prepared by deposition of a 400 nm thick Ag film on a quartz substrate on which a linear grating with meander profile and 780 nm period was etched. This method produced the periodic array of Ag stripes shown in the Hg arc lamp image (Fig. 6.6a), which is dominated by the ridge of the stripes. The inserted SEM image, which is free from these surface potential modulation effects and has better resolution, shows the true widths of the stripes and considerable roughness (<10 nm rms). The roughness is uncorrelated between the different lines so that only local SPs (LSPs) were excited as seen in the image taken with 3.10 eV photons in the 2PPE mode (Fig. 6.6b). The sophisticated experimental set-up of the authors, based on many years of experience with fs time-resolved 2PPE studies of the electron dynamics in metals [102], allowed them to study the dynamics of the LSP excitation. They could compress the pulse duration to 10 fs and use the pump-probe method with a minimum delay time interval of 50 as. By varying the delay time between pulses they measured the evolution of the intensity of individual hot spots via interferometric time resolution PEEM (ITR-PEEM) continuously. They took images of the type shown in Fig. 6.6b in steps of phase differences of $\pi/2$ between the two pulses and made a movie from

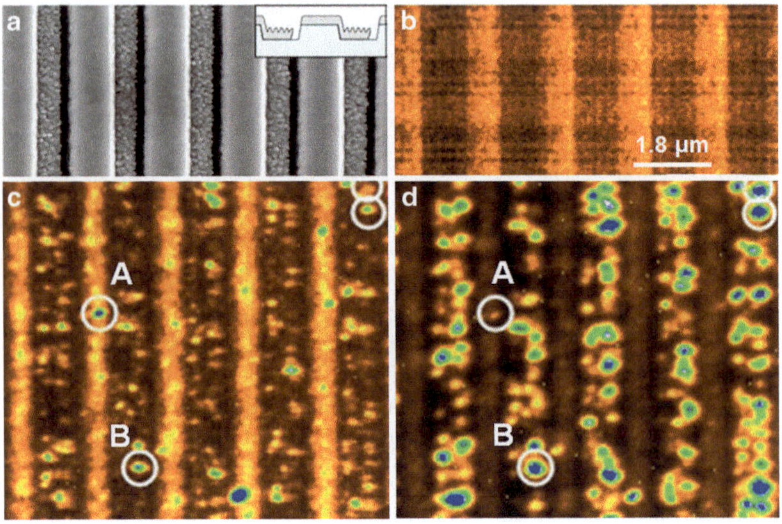

Fig. 6.7 Images of the 1.8 µm period Ag grating. (**a**) SEM image, (**b**) PEEM image excited with light from a high-pressure Hg arc lamp (1PPE), (**c**) and (**d**) excited with p-polarized fs laser light with 386 nm and 402 nm wavelength (2PPE), respectively. Note the shift by one half period between (**c**) and (**d**). There is no SPP wave in (**d**) and only LSPs are excited. The two rows of emission centers from the Ag particles in between the Ag ribbons is due to the fact that there are two antinodes in the total field of light wave and SPP wave, exciting LSP emission. Adapted with permission from Ref. [103]. © IOP Publishing 2007. All rights reserved

them. These measurements gave detailed insight into the dephasing behavior of the LSPs through coupling between the externally imposed SP mode and other SP modes or through decay into single-electron excitations. In continuation of these studies Kubo et al. [103] applied this method to a 1.8 µm period grating, which fulfills approximately the k matching condition Eq. (6.1) for 400 nm light, and investigated also the wavelength dependence of the emission pattern. The SEM image of the grating (Fig. 6.7a) shows that in contrast to the 780 nm period grating (Fig. 6.6a) the Ag layer on the higher parts of the grating is now rather smooth. Now also the lower parts of the grating are visible but the Ag layer is grainy similar to the layer on the 780 nm grating. The Hg arc lamp image (Fig. 6.7b) confirms the overall smoothness of the upper level. In the 2PPE imaging mode with the frequency-doubled light of the Ti:sapphire laser the image changes considerably with the wavelength of the incident light as shown in Fig. 6.7c, d for $\lambda = 386$ nm and 402 nm, respectively. Apparently for $\lambda = 386$ nm the in-phase condition Eq. (6.1) for exciting a SPP is at least approximately fulfilled, although this should be more the case for $\lambda = 402$ nm, keeping the variation of ε with λ in mind. In the in-phase condition the combined field of incident wave and SPP excite LSP emission on the relative smooth top of the grating (A in Fig. 6.7c). This case is illustrated in Fig. 6.8a. At $\lambda = 402$ nm predominantly LSP excitation in the Ag nanoparticles in the rougher lower part of the grating occurs, with much higher photoemission cross

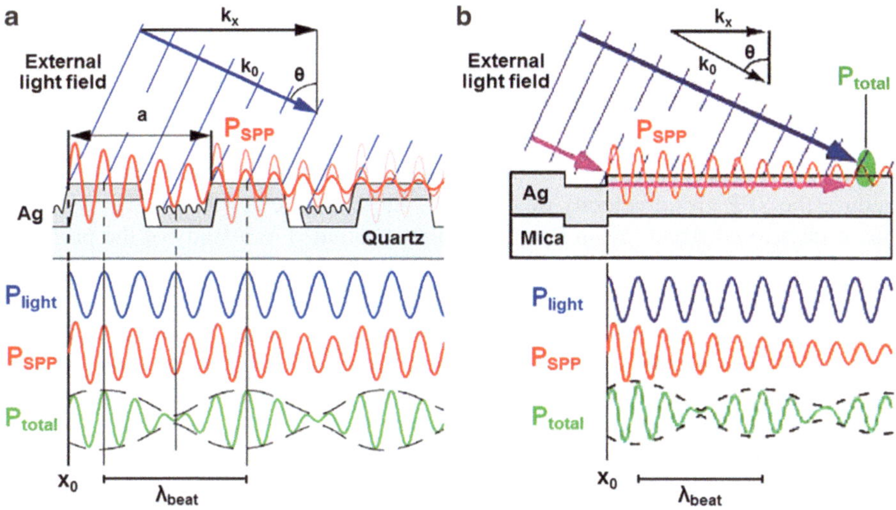

Fig. 6.8 Schematic of the excitation of propagating surface plasmons (surface plasmon polaritons). (**a**) 1.8 μm period Ag grating. The SPPs are excited at the edge of each grating mesa and interfere constructively to form the effective polarization field P_{SPP}, modulated by the grating periodicity. This field interferes with the polarization field P_{light} of the incident light, resulting in the total polarization field P_{total} with a beat length λ_{beat}. The case of resonance excitation ($\lambda = na$) is shown. (**b**) Flat Ag film with edge. The SPP wave is excited by the incident light at the edge and decays with distance. It interferes with the constant amplitude light wave, which produces a decaying beat pattern with wave length λ_{beat} due to the different wave lengths of the two waves. The higher field strength at the antinodes leads to plasmon-enhanced photoemission. (**a**) Reproduced with permission from Ref. [103]. © IOP Publishing 2007. All rights reserved. (**b**) Reproduced with permission from Ref. [104]. Copyright 2007 American Chemical Society

section. The study of the wavelength dependence of individual LSP emission centers such as B in Fig. 6.7d showed that they had different resonance wavelengths ranging from 399 to 409 nm, corresponding to different particle sizes. The amplitude of the SPP mode decreases while traveling along the surface due to electron emission, radiation, and resistive losses but is restituted after each period by excitation from the incident light. This example illustrates the complications with polycrystalline gratings, which at first sight appear to be ideal for surface plasmon polariton studies.

This experience leads to the next experiment, the study of SPP generation at the edge of a very smooth film and propagation along the film as schematically shown in Fig. 6.8b. Kubo et al. [104] implemented this by depositing a mirror-like Ag film on cleaved mica which had occasional cleavage steps. At these steps the momentum change of the incident wave allowed coupling to surface plasmon polaritons. Because the $\omega(k)$ relations of the incident light and the SPP do not match perfectly there is a phase velocity difference between the two waves propagating along the film, which causes corresponding field maxima and minima visible in the intensity of the electron emission. Inasmuch as there is no reconstitution of the SPP amplitude as in the previous case, the SPP is damped continuously. Accounting for this damping requires going beyond the simplified k matching condition of Eq. (6.1) with real k and replacing it with a complex k, whose imaginary part describes the damping and can be derived from the complex dielectric constant. Again using the ITR-PEEM method the authors could extract from the damping a SPP lifetime of 12 fs and from a careful analysis of the various interference processes they could deduce the SPP group velocity of 1.83×10^6 m/s = 0.61c at 400 nm. The work on one-dimensional structures in Petek's group culminated in a study of the propagation of SPPs excited at nanometer slits with widths ranging from 80 to 142 nm. They were prepared by focused-ion-beam milling of the Si(100) substrate on which a 80 nm thick Ag film was deposited after cleaning the surface by high temperature flashing. The interference between incident wave and SPP wave was found to increase with slit width, which detailed modeling showed to be due to increasing constructive coupling between the two fields [105].

Other aspects of SPP propagation in linear patterns were studied with Au films on Si, which were cesiated to reduce the work function enabling two photon-photoemission (2PPE) using the 800 nm light of the Ti:sapphire laser. One experiment looked at the efficiency of short gratings consisting of only a few periods for SPP excitation. For their specific experiment (wavelength, angle of incidence, grating material, and dimensions) the authors found that the coupling efficiency into the SPP mode did not improve significantly when the grating length was larger than the damping length of the SPP. In another experiment the interference between SPP and light pulses propagating in opposite direction was studied by illuminating a 2.25 μm wide and 140 nm high gold bar on a 60 nm thick Au layer. This bar acted as a source for SPP wave packet emission in the direction opposite of the incident light pulse. By varying the delay time between pump and probe pulse the propagation of the SPP could be studied in detail [106]. The modification of the SPP on a Au film by an organic nanofiber was subject of another interesting study. The 0.04 μm high, 1 μm wide fiber was normal to the edge of a 60 μm thick Au film at which the SPP pulses were excited by the laser pulses. The fiber produced a pronounced periodic pattern parallel to the fiber, which was attributed to dielectric modulation of the surface induced by the fiber [107].

6.2.3 Nanostructures

Other ITR-PEEM studies, also at 65° angle of incidence, looked at the plasmon dynamics in small Ag discs and disc pairs prepared by electron beam lithography [108–110]. The discs had a diameter of 200 nm, a height of 50 nm and the gap between the discs in the pair was 150 nm, too large for dipole–dipole interactions to be of influence. Studies without ITR [111] showed in images taken with SEM and Hg high pressure arc lamp light PEEM no variations between the discs but the 2PPE images obtained with the frequency-doubled Ti:sapphire showed considerable fluctuations, indicating the extreme sensitivity of plasmon excitation to small

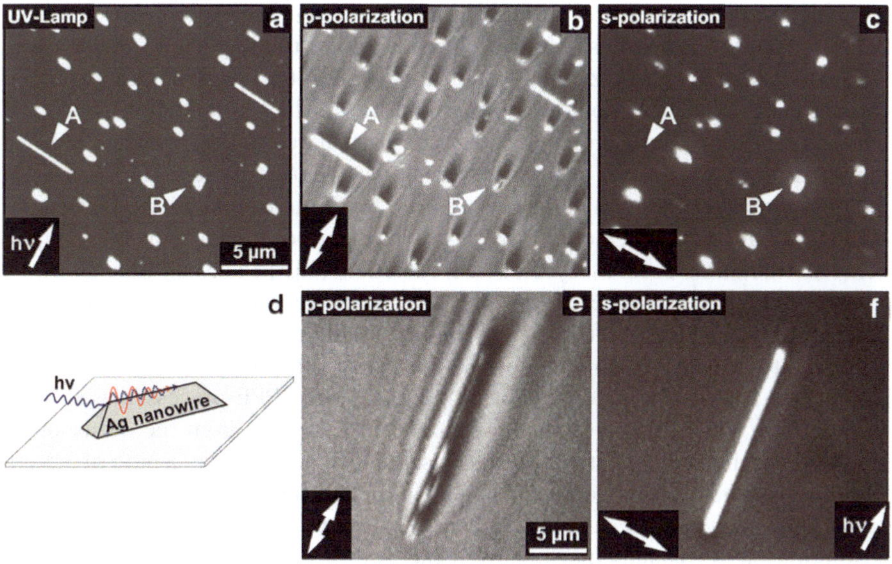

Fig. 6.9 PEEM images of Ag crystals on Si(100) illuminated (**a**) with light from a high-pressure Hg arc lamp ($h\nu = 4.9$ eV, 1PPE), (**b**) and (**c**) with p- and s-polarized fs laser light ($h\nu = 3.1$ eV, 2PPE). A *single arrow* indicates the direction of the incident light, a *double arrow* the in-plane component of the electric field. (**d**) illustrates the shape of the Ag rods and the wave fields (*blue* laser pulse, *red* SPP). (**e**) and (**f**) are images of a rod taken with p- and s-polarized fs laser light ($h\nu = 3.1$ eV, 2PPE). (**a–c**) Reproduced with permission from Ref. [112]. Copyright 2006 by the American Physical Society. (**d–f**) Adapted with permission from Ref. [113]. Copyright 2007 Elsevier

heterogeneities. This was also evident within individual 200 nm diameter discs by changing the phase delay between pump and probe pulse in steps of 0.13 fs. This allowed to follow the shift of the local field of the LSP excited by the incident wave across the disc via the enhanced electron emission due to the locally enhanced field [15].

While the experiments discussed up to now used nanostructures prepared by electron beam lithography, Chelaru et al. [112–114] prepared their samples in situ by Ag deposition at about 600 °C on clean Si(100), which results in the formation of three-dimensional islands and epitaxial quasi-onedimensional nanowires on a Ag wetting layer as shown in the PEEM image (Fig. 6.9a). The nanowires are believed to be roof-shaped as indicated in Fig. 6.9d with widths and heights of 150–300 nm and lengths up to tens of microns and grow in two orthogonal directions. This allows a very good demonstration of the polarization and shape dependence of SP excitation in nanostructures. The authors used frequency doubled pulses from a 800 nm Ti:sapphire laser with 40 fs pulse length, 1 nJ maximum pulse energy at an angle of incidence of 74° and p- and s-polarization. Figure 6.9b, c show that 3.1 eV photons both with p- and s-polarization can excite strong 2PPE from the three-dimensional islands. Compared to the 1PPE from the Hg arc lamp (Fig. 6.9a) the photoelectron yield fluctuates strongly as already seen in the experiments described before. From the nanowires, which are perpendicular to the plane of incidence no emission is seen in s-polarization. On the other hand, if the nanowires are parallel to the plane of incidence, homogeneous 2PPE occurs in s-polarization (Fig. 6.9f) while in p-polarization a pronounced beating pattern due to the mismatch between the exciting wave ($k_{E//}$) and the surface plasmon wave (k_{SPP}) is seen (Fig. 6.9e). The interference pattern around the nanowires and three-dimensional islands is caused by Fresnel diffraction features similar to what has been seen with other highly coherent radiation such as synchrotron radiation [115]. The asymmetry of the fringes in Fig. 6.9e is due to fact that the nanowire was not aligned precisely parallel to the plane of incidence. The much higher background intensity in p-polarization than in s-polarization is very likely due to excitation of a surface state in the Ag wetting layer similar to the 2PPE and 3PPE processes used in the surface state spectroscopy studies mentioned in Chap. 2. From the decay of the amplitude of the beating pattern the authors deduced a SPP decay length of 3.5 μm. The beat period was in good agreement with the values of $\lambda_{E//} = 416$ nm and $\lambda_{SPP} = 359$ nm calculated from ε of Ag. Later studies of the influence of C_{60} adsorption on Ag islands on Si showed that the SPP wavelength decreased with increasing coverage due to the change of the adsorption-induced change of the dielectric constant; also the SPP was strongly damped [116]. More on these studies can be found in [117].

A different approach to the study of plasmon-induced PEEM was taken by Gloskovskii et al. [118, 119]. They deposited thin Ag stripes with different mass thickness ranging from continuous films to small Ag clusters and measured the energy distribution and irradiation power-dependence of the emitted electrons for different photon energies and mass thicknesses. The photon energy was varied over the full tuning range of the Ti:sapphire laser. Both fundamental ("IR," $h\nu = 1.45$–1.65 eV, 80 fs pulses, 750 mW) and frequency doubled ("UV," $h\nu = 2.9$–3.3 eV, 200 fs pulses, 80 mW) light were used for excitation with maximum power densities of 40 and 9 MW/cm^2 at the sample. Pulse duration, average power, and delay time in the pump-probe experiments were varied. Two of the many results of their extensive studies are (1) that with UV excitation via a 2PPE process nanoclusters (sizes in the few nm range) emitted more strongly than islands in the 10–100 nm range; (2) that the observations of IR-excited emission could not be explained by a pure 3PPE process but required transient heating of the electron gas ("thermal-assisted nPPE"). For the many other results the reader is referred to the original papers.

Several other studies have brought also some insight into this phenomenon. Two of them [120, 121] combined PEEM with STM to characterize the size distribution of Ag clusters formed on defect sites on graphite produced by sputtering. Then they studied the 2PPE emission intensity of individual clusters and cluster distributions as well as the polarization dependence, with maximum intensity occurring around 3.3 eV and p-polarization. The comparison between STM and 2PPEEM images revealed the high sensitivity of 2PPE excitation to cluster size and shape. Only clusters, for which the plasmon resonance condition was fulfilled, were visible while in 1PEE emission excited by Hg arc lamp light, which is far off resonance, the cluster density was comparable with that determined with STM.

In another study, in this case of a Cu(100) surface with defects, 2PPE-PEEM was combined with LEEM for surface characterization [122]. They found that both p- and s-polarized 3.17 eV laser light showed the same number of defects as Hg arc lamp light, though with some of the defects 10^3 times brighter with p-polarization than with Hg arc lamp light and that some of the defects were brighter with s-polarization than with p-polarization. Changing the polarization caused some defects to disappear and others to appear. Illumination with 1.65 eV laser light of about six times higher power density showed fewer of the defects which have been seen with 3.17 eV light. Defects smaller than 0.5 μm (as seen by LEEM) were brighter in p- than in s-polarization. While emission with 3.17 eV excitation was shown to be a 2PPE process via its illumination power dependence, the features seen with 1.65 eV excitation require a 3PPE process. It should be noted that the photon energies used were far away from the Cu surface plasmon energy of about 2.2 eV and that the emitting surface features are much larger than those discussed before. DC field enhancement at the defects was excluded by changing the DC field strength. More detailed studies on single crystal surfaces with well-defined surface features should shed some light at the excitation process.

A better defined system consists of polystyrene latex spheres on mica with 465 μm average diameter, coated with 50 nm Ag. This produced Ag hemispheres surrounded by a flat Ag film with an average grain height of 2.3 nm and a rms roughness of 0.66 nm as determined by atomic force microscopy. The specific goal of this experiment was to determine the near field enhancement of nanoparticles relative to a flat surface by measuring the 2PPE yield Y excited by 3.15 eV light with p-polarization at 74° angle of incidence [123]. The enhancement factor, defined as $\mathrm{EF} = Y_{\mathrm{Particle}}/Y_{\mathrm{Film}}$ was found to have a mean value of 16, with some particles having values above 100. SEM images showed that particles in the first group had smooth surfaces while those with EF above 30 had asperities and up to six times higher yields. Thus even with particle diameters of about 500 nm plasmon excitation enhances the short range field strength by a factor of 4 compared to a planar surface.

The excitation modes of well-defined individual Au nanostructures were studied by Douillard's group, initially of nanodots [124] and nanorods [125], using 80 fs IR (138–1.72 eV) pulses with p-polarization parallel to the rods. The rods were prepared by electron beam lithography and were 35 nm high, 30 nm wide and varied in length from 90 to 925 nm in steps of 25 nm. The results gave clear evidence that the rods acted as short-range SPP resonators, with standing waves for rod lengths below the attenuation length of the short-range SPP. Another study [126] aimed at the excitation modes of well-defined Au stars. These are particles with a near-spherical core with less than 100 nm diameter from which 5–8 short arms stick outward. They end in tips with a radius of curvature of typically 5 nm, resulting in a strong field enhancement. They were prepared by wet chemistry instead electron beam lithography but otherwise the conditions were the same as in the preceding studies: deposition on an indium tin oxide (ITO) coated glass substrate with sufficient distances so that no interaction between them was possible; shape and orientation could be easily identified with SEM and LEEM. PEEM images were taken with Ti:sapphire light with wavelengths ranging from 720 to 940 nm, 125 MW/cm^2 power density, 75° angle of incidence, and different polarization directions. Correlation with the LEEM image showed unambiguously that the emission regions were located at the tips of the arms. The emission changed between the various arms with the direction of the polarization of the light relative to the tip orientation. The yield dependence upon laser power indicated a 3PPE process, which became possible in the high-field regions of the tips. These results were obtained with an instrument resolution of 20 nm. The localized surface plasmon resonances seen on the Au stars are much clearer evident and can be much easier analyzed on triangular nanostructures. This was done with electron beam lithography-prepared Au prisms 50 nm thick with edge length varying from 100 to 300 nm. The same experimental conditions as in the previous study were used. The results showed convincingly that the various localized plasmon resonances of these prisms could be selected by choosing the polarization direction [127]. Könenkamp's group extended the study of the dependence of the location of the resonances upon the polarization, using Y-shaped Au nanostructures with 400 nm lateral dimensions on ITO, which were prepared by ion beam milling. The Y-shaped structures were illuminated at an angle of incidence of 60° in the symmetry plane of the Y. When the polarization was switched +/−45° with respect to the plane of incidence the hot spots switched accordingly between the two arms of the Y, with the by far highest emission at the two ends. This led to the suggestion to use such structures for plasmonic routing [128].

The first application of an aberration-corrected PEEM instrument [129] to plasmon physics has already brought an interesting result. The authors studied a 200 nm thick smooth Au film with random nanostructures on a ITO film on glass with Ti:sapphire laser light varying the energy from 1.37 to 1.60 eV and the laser power from 100 to 1000 mW. The power dependence of the yield clearly indicated a 3PPE process. The emission occurred in many hot spots, which could be located by simultaneous illumination with an Hg arc lamp to be in narrow gaps between the Au islands. The emission area was smaller than the gap between the Au islands so that it clearly came from the ITO, apparently induced by energy transfer from the absorbing Au film. The intensity was about 6000 times higher than that from ITO film regions far away from the gaps in the Au film. The wavelength dependence of the hot spot excitation showed typical resonances expected for excitation of Au surface plasmons, varying from hot spot to hot spot, which implies constructive overlap of the plasmon fields in the gap between the Au islands. This phenomenon was studied subsequently in more detail as a function of the polarization direction of the incident beam with the result that the hotspot intensity in the gaps of coupled metal junctions was highest for s-polarization normal to the gap direction. This was attributed to the fact that due to momentum conservation s-polarized light does not generate propagating modes, leading to hot spot formation, while p-polarization excites propagating modes [130]. Similar effects could be achieved with circular polarized light, with the excitation probability depending on the helicity of the light [131]. Awada et al. [127] studied Au films ranging from isolated to interconnected clusters to nearly continuous films at wavelengths from 800 to 1040 nm in steps of 10 nm. The found many hot spots with up to $N = 6$ excitation resonances at different frequencies. The probability for N resonances to occur at a given site decayed exponentially with N. The highest intensities were observed near the percolation threshold but hot spots occurred in all three film configurations.

All these studies of granular Au films suggest strong local plasmon generation between narrow-spaced Au particles. This phenomenon has been brought to light very clearly in a study by Schertz et al. [132, 133]. They prepared their sample with the procedure previously used by Hutter et al. in optical studies of surface plasmon resonances [134]: dipping an Au-layer-covered substrate into a solution of 2-aminoethanethiol (cysteamine) and depositing Au nanoparticles on top of this surface. The cysteamine formed a self-assembled monolayer, which produced a gap of 0.8 nm (0.4 nm in [134]) between the Au particles and the planar Au film. In contrast to the earlier laterally averaging measurements [134] UVPEEM could measure the plasmon-induced electron emission from individual particles with different shapes and also the energy distribution. The unusually strong dependence of the electron yield Y upon the incident laser intensity P, $Y \sim P^{5.6}$ was incompatible with a 3PP emission process and was attributed to the enhancement of the field between particle and film by the plasmon near-field. A comparison of the dependence of the yield upon the strength of the laser field, $Y \sim E^{11.2}$, with the slope of the Fowler-Nordheim equation $I(E)$ gave a resulting effective field E_{eff} of about 5 V/nm, which explains the high yield as a plasmonic field emission process. The observed energy distribution of the electrons and their strong dependence on the laser field strength could be attributed to their acceleration in the strongly inhomogeneous field between particle and film. These results suggest that the hot spots observed in the semicontinuous films discussed before may have a similar origin. This is in fact suggested by the dependence of the angular distribution of the intensity of the hot spots upon the angle between polarization and gap direction between Au particles reported in [130]. Although there is a considerable scatter of the data, they are concentrated on curve, which can be fitted better with $\cos^{10}\theta$ than with $\cos^{6}\theta$. This suggests that the emission does not occur via a 3PP process but that it is also plasmon field assisted.

6.2.4 *Complex Wave Fields*

All the experiments described up to now were done with linear polarized light pulses, with one exception, in which also circular polarization was used [131]. PEEM images of the specimen were taken with different polarizations, mostly with p-polarization (because of its strong $\boldsymbol{E}$ field component normal to the surface conducive to surface plasmon excitation on flat surfaces) but also with s-polarization and in some cases with tilted polarization planes. Around 2000

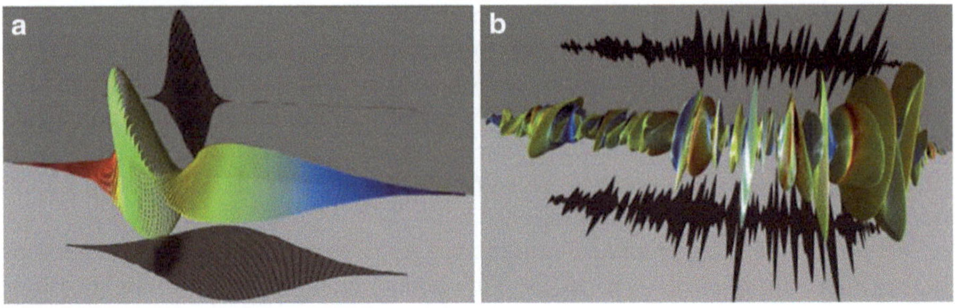

Fig. 6.10 Quasi-three-dimensional electric field presentation of a fs polarization-shaped light pulse, in which the momentary intensity, frequency and polarization change as functions of time from left to right. The frequency is indicated by the color. The electric field amplitude is shown by the size and shape of the ellipses whose horizontal and perpendicular projections are shown by the *black shadows*. The pulse shaping is achieved by fast computer control of the relative amplitudes of the two *E* field components. (**a**) Simple pulse (**b**) Complex pulse. Adapted from Ref. [136] with permission from Springer Science + Business Media

Brixner et al. [135, 136] developed a method which allows changing the polarization (ellipticity, orientation angle) *within* each fs pulse and in addition also amplitude and frequency. Examples of such fs pulses are shown in Fig. 6.10 in a quasi-three-dimensional presentation. The shadows are the projections of the momentary ***E*** vector onto two orthogonal planes, indicating the degree of ellipticity and the orientation of the ellipse, the color indicates the frequency. The pulse is at the beginning and at the end polarized in the horizontal direction. In between a vertical field component with a maximum at ¼ of the pulse length is added, producing a rotating elliptical polarization. Simultaneously a frequency compression ("red to blue light") is made. This method, called femtosecond polarization pulse shaping opened up a new era in plasmon PEEM because it allowed switching in the fs range between various plasmon excitation states such as in the stars mentioned above. Aeschlimann and coworkers [137–141] have applied this method successfully with an adaptive learning loop [142] to fs switching via surface plasmon excitation in well-defined nanostructures. They used various two-dimensional nanopatterns produced by electron beam lithography: a Y-shaped array of Ag discs with 180 nm diameter and 30 nm height [137], Ag discs with six rectangular arms in radial direction ("suns") [138], pairs of opposing triangles ("bowtie nanoantennas") [139, 140] and finally a corrugated Ag surface [141]. As an illustration we discuss here the work on bowtie nanoantennas shown in Fig. 6.11 [139, 140]. The 25 fs light pulses with 795 nm center wavelength from the 80 MHz Ti:sapphire laser are sent through the polarization pulse shaper described in Refs. [135, 136] and are focused to about 100 μm on the specimen at an angle of 65°. The bowtie consists of two 40 nm thick Au triangles on an ITO layer on glass with side lengths of 350 and 100 nm distance between their tips as seen in the SEM image (a). The image taken with Hg arc lamp light (b) shows homogeneous intensity. Images (c) and (d) are taken with p- and s-polarized light, respectively in the 3PPE mode and show the regions of maximum plasmon-induced photoemission in the two polarization

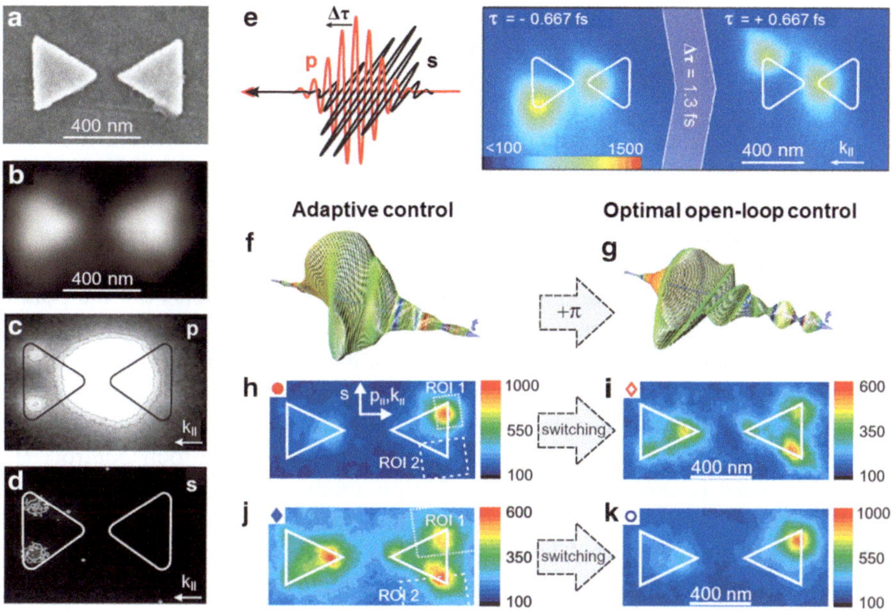

Fig. 6.11 Local surface plasmon switching by pulse shape control using two opposing Au prisms ("bowtie" structure). (**a**) SEM image of the structure, (**b**) PEEM image excited with light from a high-pressure Hg arc lamp (1PP), (**c**) 3PP-PEEM excited with 25 fs Ti:sapphire p-polarized laser pulses, showing strong emission from the gap between the prisms ("bowtie antenna"), centered on the tip of the *right prism*, and two *hot spots* at the opposing corners of the *left prism*. 3PP emission excitation with s-polarization (**d**) produces only weak emission from these two corners. (**e**) shows how the surface plasmon excitation conditions can be changed by the phase shift between two perpendicular polarized laser pulses. A relative phase shift of $+\pi/2$ and $-\pi/2$ (left and right circular polarization) shifts the emission between the corners of the *left prism*. (**f–k**) illustrate the power of the pulse shaping shown in Fig. 6.10. Using different pulse shapes ("adaptive" and "optimal open-loop" control) allow to move the $\boldsymbol{E}$ field with optimum frequency, polarization direction and field strength for surface plasmon excitation between the various corners of the prisms without causing excitation in the gap. (**a–e**) Adapted with permission from Ref. [139]. Copyright 2011 by the American Physical Society. (**f–k**) Reproduced with permission from Ref. [140]. © IOP Publishing & Deutsche Physikalische Gesellschaft 2012. All rights reserved

directions of the light coming from the right hand side. With p-polarization the maximum intensity is centered at the tip of the triangle but has much higher intensity in the gap compared to that in the corners of the opposite triangle, which are the only emission spots in s-polarization. If p- and s-polarized pulses of equal height are superimposed the excitation pattern is quite different and depends upon the phase shift between them as illustrated in Fig. 6.11e. In the left image the phase shift was $+\pi/2$, in the right image $-\pi/2$, corresponding to left and right circular polarized light. The experimental results agree very well with simulations of the polarization- and phase shift dependence of the local field of the incident wave [143].

It is evident that very rapid switching between different field concentrations is possible with fs polarization pulse forming. This is illustrated in Fig. 6.11f–k [140].

Here more complex pulse shapes than the one in Fig. 6.11e were used as shown in the quasi-three-dimensional plots of the time evolution of the ***E*** vector in the pulse, similar to pulse presentation in Fig. 6.10. The images were produced by coherent superposition of two pulses, a polarization-shaped pump pulse with 795 nm wavelength and a frequency-doubled unmodified probe pulse capable of 2PPE excitation with s-polarization. The first one can cause only emission via a 3PPE process while the second one produces homogeneous emission from the triangles. The emission sites are determined by the field concentration produced by the pump pulse but the emission intensity is the result of the combined action of pumps and probe pulse. The images Fig. 6.11h–k were obtained in the following manner. First the polarization pattern of the pump pulse was optimized for excitation of one corner, using adaptive feedback from the intensity of the spot into the polarization pulse shaper ("adaptive control" (h).) Then the phase of one of the polarization components was shifted by π to change the pulse shape in such a manner that the strongest field enhancement occurred in the opposite corner, ("optimal open-loop control" (i)) (similar to what is shown in Fig. 6.11e). With another polarization-shaped pulse another excitation pattern can be formed (j) and switched with optimal open-loop control in a different pattern. The location of the emission centers, i.e., of the high-field regions is determined either by the fit of the incident polarization to the local dipole or by the constructive interference of the local fields produced by the two polarization components [143]. In the example shown here it is the second mechanism. It should be noted that PEEM is used here as a high-resolution field sensor and not as an imaging method of extended lateral structures. It is expected that the optical near-field control by manipulation of the two incident polarization components in fs pulses will become important in ultrafast nanooptics and other applications. Not all nanooptical surface processes excited by polarization-shaped pulses are as ultrafast as those seen in these nanopatterns. On corrugated Ag surfaces Aeschlimann et al. have observed long-lived coherent modes, whose origin still need clarification [141].

Returning from possible applications to fundamentals, two proposed PEEM imaging methods involving surface plasmons should be mentioned yet: coherent two-dimensional nanoscopy, proposed by Aeschlimann et al. and applied to a corrugated Ag surface [144] and attosecond nanoplasmonic field microscopy, proposed by Stockman et al. [145]. The application range of the first method and the feasibility of the second one are still open but a few comments on the second one are appropriate in view of its potential power. Its basic idea is to produce the high near fields of plasmonic nanostructures with intense waveform-controlled, plasmon-creating light pulses and probe the fields with attosecond XUV pulses, synchronized with the plasmon-creating pulses, in an energy-resolving PEEM. The XUV photons produce in addition to secondary electrons, which have a wide energy spread, monochromatic photoelectrons whose energy is much larger than that of the electrons from the low energy photon pulse. Their energy distribution is determined by the time-dependent magnitude of the near field and estimated to be about 10 eV wide. Specifically the authors have considered the following parameters: 93 eV, 170 as XUV pulses with a repetition rate of 3 kHz and 10^6 photons per pulse, focused into a spot with diameter 2 μm. At a photon energy of 93 eV the only photoelectrons, which could be used for energy resolved imaging are the 4p and 5p electrons of Ag and Au, respectively, because the 4d and 5d are broad valence states. The Ag 4p binding energies are about 64 and 58 eV, those of Au 5p electrons are 74 and 57 eV. Therefore the photoelectrons resulting from the ionization of these states will be superimposed on the much stronger secondary electron background and selecting them for imaging will be difficult if possible at all. If attosecond photon pulses with much higher energies could be achieved then the chances are much better. For example, with 300 eV photons 4f photoelectrons could be produced with high ionization cross section and at their energy of about 215 eV the secondary electron background is small. However, at 215 eV the transmission of PEEM instruments is already small and the low repetition rate will require excessive image acquisition times unless the electron yield per pulse is very high. In this case space charge effects very likely will prevent imaging with the desired resolution. Thus, the interesting idea may face some insurmountable obstacles.

6.2.5 Limitations

The detrimental effects of space charges at low pulse repetition rate have been demonstrated dramatically in spectroscopic PEEM studies of Ag islands on Si(111) using 800 nm and 200 nm pulses with 100 kHz and 250 kHz repetition rate, respectively [146]. With the 800 nm,100 kHz pulses the average laser power was varied from 10 to 100 mW, with the 200 nm, 250 kHz pulses from 0.3 to 103 μW. Imaging with 800 nm light requires a 3PPE process, while with 200 nm direct excitation (1PPE) is sufficient. At low average power, for example at 33 mW and 89 nW average power with 800 nm and 200 nm light, respectively, no image distortions occur but with higher power, e.g., with 228 mW and 212 nW at these wavelengths, the images are severely distorted. Interestingly, no contrast was observed between Ag and Si with the 1.55 eV and 6.2 eV pulsed laser illumination in the 3PPE and 1PPE mode, respectively, while the Hg arc lamp (1PPE) with photon energy up to 5.2 eV shows strong contrast. Only the edges of the Ag crystals were visible, in part because of the higher light intensity due to the large angle of incidence (74°), in part due to hot spots from localized surface plasmons. Local electron spectroscopy of a region containing a hot spot as a function of average laser power showed very little change in the full width at half maximum (FWHM) of the energy distribution up to 45 mW and 1–2 μW for 800 nm and 200 nm illumination, respectively, provided that the angular distribution of the photoelectrons was limited by the contrast aperture. At higher power the FWHM increased from 0.5 to 1.4 eV at 90 mW with 800 nm excitation and from 1.5 eV to about 5.5 eV at 20 μW with 200 nm excitation. The broadening did not occur towards higher energy, which could have been attributed to acceleration in the increasing higher plasmonic field. Rather, the distribution broadened more towards lower energy. This and the much stronger energy broadening when the angular distribution was not limited clearly indicate space charge effects.

Space charge effects have never been seen with Hg arc lamp or with cw laser illumination because of the continuous illumination of the specimen. To achieve the same average power with pulsed illumination a much higher power per pulse is needed and consequently much higher photoelectron current densities per pulse. This is the reason why space charge effects set in much earlier at low pulse repetition rates than at high rates and makes the 80 MHz Ti:sapphire laser uniquely suitable for pulsed laser PEEM.

Ending this section on plasmon PEEM, whose progress has been reviewed by Douillard and Charra [147, 148], the question may be asked: is PEEM really necessary for the understanding of localized and propagating surface plasmons? Many of the results of PEEM can be and have been obtained with laterally averaging measurements, either by electron emission or with optical methods, in particular combined with imaging methods for the characterization of the microstructure of the specimen, such as TEM, SEM, AFM, or STM. With these methods a 1:1 correlation between image and the laterally averaging plasmon characterization method is difficult, except scanning tunneling luminescence, which is easily combined with STM [53]. In PEEM plasmon characterization and microstructure measurement are intimately connected, giving a direct correlation between them. The limitations of past PEEM work compared to that done in combination with parallel external imaging methods are also clear: PEEM has had less resolution and requires UHV. The resolution, in the past 50 or 20 nm, depending upon instrument, will be below 10 nm in aberration-corrected instruments in the future, which will be sufficient for most if not all studies. The need of UHV may be inconvenient but also provides opportunities for in situ preparation of well-defined specimens and their modification during the measurements. An example is the change of the length of Ag nanowires on Si by annealing, which was used in Ref. [108]. Thus PEEM will continue to play an important role in surface plasmon physics and its applications. For recent reviews of this field see Refs. [149–153].

6.3 Technological Applications

This section discusses a wide variety of subjects which have been studied with cathode lens microscopy ranging from mineralogy to electronics, but to a lesser extent than the other applications covered in this chapter, in order to illustrate its wide application range.

6.3.1 General Materials Applications

Minerals. XANES microscopy at the Si $L_{2,3}$ edge has been used to image quartz (SiO_2) inclusions in zircon ($ZrSiO_4$) crystals [154]. The thickness and chemical composition of synthetic lamellar clay mineral flakes was studied with Si 2p

XPEEM and μXPS [155]. Useful and worthless minerals are commonly separated by froth flotation, in which the hydrophobic mineral is brought to the surface by the air bubbles in the water while the hydrophilic, usually more oxidized mineral stays in suspension. The oxidation of a sample consisting of chalcopyrite ($CuFeS_2$), pentlandite ($(Fe,Ni)_9S_8$), and pyrrhotite ($Fe_{10}S_{11}$) regions was studied with Fe $L_{2,3}$ absorption XPEEM and spectroscopy. The latter mineral was found to be much more oxidized than the former two, in agreement with the floating practice of Cu and Ni extraction from mixed ores consisting of these minerals [156]. Another study addressed the mechanism of pyrite (FeS_2) oxidation and leaching with S 2p SPEM and Fe $L_{2,3}$ absorption PEEM imaging and spectroscopy but was mainly a demonstration experiment [157].

Metallurgy. One of the last studies with the Balzers K3 Metioscope concerned the grain shrinking connected with boundary migration during grain growth of a TiAlZr alloy at 1100 °C. From the UVPEEM videos the shrinking kinetics could be extracted quantitatively [158]. More recently UVPEEM was used in a study of the thermally induced martensitic phase transition in CuZnAl and NiTiCu shape memory alloys. A strong decrease of the image intensity caused by a corresponding work function increase was observed during the transition from the low temperature martensitic phase to the high temperature austenitic phase [159, 160]. A somewhat unusual application of XPEEM is the study of Ti-Al phase formation during high energy ball milling of $NaAlH_4$ with $TiCl_3$ powder. $NaAlH_4$ (sodium alanate) is used for hydrogen storage and Ti in intimate contact in a $TiAl_x$ complex catalyzes the hydrogen exchange rate. Ball milling produces this contact and the question posed was how this contact is formed and what the ionization level of Al is. To answer it absorption spectra were taken around the Ti L, Al K, and Na K absorption edges and peak to pre-edge ratio images were taken of powders supported on Au, resulting in a clear picture of the process [161].

Tribology. Friction, being a problem of two surfaces and of the lubricant on them, seems to be well suited for studies with the methods discussed in this book. Early UVPEEM work looked at the influence of lubricant adsorption and the influence of the lubricant on the wear caused by contact between metals. Complementary chemical information on the changes of the lubricant caused by heating and electron bombardment was obtained by high resolution vibrational energy loss spectroscopy [162, 163]. In a XPEEM study of magnetic recording media simultaneous chemical and structural information was later obtained of the tribochemical processes between the slider of the recording head, coated with diamond like carbon, and the disc, coated with amorphous hard carbon. The lubricant was a perfluoropolyether. Selected area XAS showed strong differences between the regions within and outside of the contact in the C K-edge and F K-edge NEXAFS spectra, indicating oxidation of the lubricant and removal of fluorine from the disc in the contact area [164–166]. A similar approach was used to study the tribochemical effects of the additive zinc dialkyl dithiophosphate (ZDDP) to automotive oil on steel surfaces during frictional contact. In this case P L-edge XPEEM and XANES indicated differences in the polyphosphate spectrum between regions inside and outside of the contact region; the corresponding data of the

O K-edge showed clearly that the region outside the contact area was oxidized in contrast to the oxide-free contact area [167]. These examples illustrate that XPEEM is well-suited for the study of tribology problems.

Carbon, diamond. High density diamond-like carbon films have been studied extensively because of their hardness and diamond films for their electron emission properties. In the early 1990s in situ carbon growth by low pressure chemical vapor deposition (CVD) was studied with UVPEEM on polycrystalline Mo [168], on Mo (100) [169] and on Mo(310) [170]. In another experiment diamond films, which had been grown in a separate chamber on Mo(100), the oxidation and dissolution at high temperature in the substrate was observed with UVPEEM and XPS [171]. XANES-spectromicroscopy and μXANES allowed a more detailed analysis of the distribution of various C bonding states (sp^2 and sp^3) in C and diamond films by exciting resonance emission from the corresponding final states at about 285 eV and 290 eV, respectively [172, 173]. In a very detailed study of the surface conductivity of undoped CVD diamond films on the (100) surface of synthetic diamonds, complemented by several spectroscopic measurements, UVPEEM with Hg and Xe light was used to examine the correlation between conductivity, band structure, and surface defects [174].

Most of the emission microscopy work on CVD diamond and carbon films was made to understand and optimize electron emission from these films by imaging the distribution of the emission regions, starting with the work in Kordesch's group on field emission from a diamond film [175]. This film was subsequently also imaged with XPEEM together with other diamond emitters [176]. Another study compared SEM, UVPEEM, and field emission images of clean and hydrogenated films and measured the emission from the images as a function of photon energy and applied field [177]. Nemanich's group investigated several aspects of electron emission form diamond films using PEEM, field emission, thermionic emission (TEEM), thermionic field emission, SEM, and combinations of these methods: the correlation between morphology and field emission [178], the influence of temperature on the thermionic emission from nitrogen-doped films [179], the thermionic emission properties of Si tips coated with nitrogen-doped diamond [180], field-enhanced thermionic emission from sulfur-doped films on a Mo disc and on Si(100) [181, 182] and thermionic emission from nitrogen-doped homoepitaxial diamond films [183]. Nanocrystalline graphite films, consisting of flake-like crystals and nanotubes, were also studied by this group [184]. In a different study CVD was used to grow nitrogen-doped spherical nanostructures consisting of diamond nanoclusters, carbon nanotubes, and graphene nanowires/nano-onions, identified by TEM, transmission electron diffraction, and SEM, which showed strong field emission. C K edge NEXASF indicated predominant diamond character with some graphitic admixture [185].

6.3.2 Electronics

Electronic technology. Although Si-based electronics has moved long ago from research to routine production there are still areas to which cathode lens microscopy can make useful contributions. The small dimensions of field effect transistors

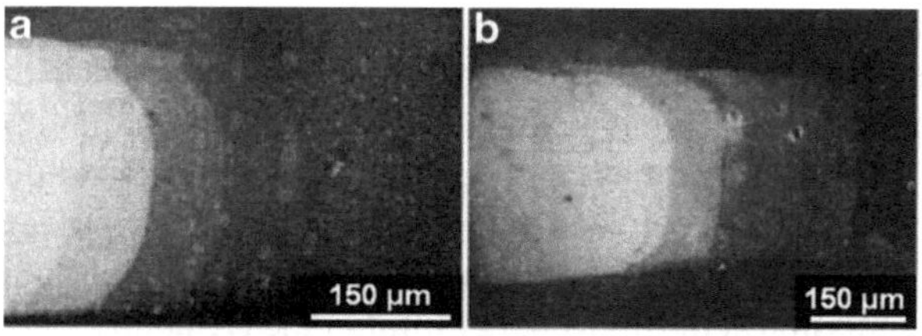

Fig. 6.12 Doping-dependent contrast in UVPEEM. (**a**) Boron implanted Si(100) in 8 dosis steps, (**b**) Zink implanted GaAs(100) in 8 dosis steps. Not all doping levels are visible in these images because of the poor quality of the original prints, in which the details of the original data got lost. For details see text. Modified from Ref. [192]

require very thin gate oxide layers with large dielectric constants, large band gap, and high thermal stability, which has stimulated studies of the thermal stability of several transition metal oxide films on Si(100) by UVPEEM and AFM [186] and of ZrO_2 films on Si(100) by AFM, LEEM, LEED, XPS, and AES [187]. The transition metal oxide films were grown in UHV by depositing the metals in oxygen ambient while the ZrO_2 films were produced by a sol–gel process, in both cases on a SiO_2 buffer layer. The two studies gave a rather comprehensive picture of the decomposition of these ultrathin layers (3–4 nm) at temperatures ranging from 700 to 1000 °C, depending upon film material and experimental conditions. The decomposition proceeded by SiO formation, diffusion, and sublimation, resulting in Si-transition metal oxide layer reaction, silicide formation, and agglomeration. In a brief XPEEM/XPS study the quality of the silicon oxide-Si interface of microcrystalline Si was compared with that of Si(100) with the result that Ca, K, Na, and hydrocarbon contamination reduced the quality of the former [188].

Ga ion implantation and ion beam milling in Si(100) can lead to diffusion of Ga during implantation or annealing in the subsequent processing steps. Diffusion at 200 °C after ion beam milling was studied with XPEEM and MEM, which showed strong local outdiffusion from the milled region [189]. Diffusion upon implantation at 400 °C was studied with UVPEEM. In this case not only a broadening of the doping profile was found but also additional intensity maxima between closely spaced doping lines [190, 191].

The contrast used in the study of the doping profile just mentioned has attracted considerable interest for many years. The first thorough investigation under UHV conditions was made by Driesel and Bethge [192] who used UVPEEM in the system shown in Fig. 3.1 to image an n-doped Si(100) surface, locally p-doped with 100 keV B ion bombardment (Fig. 6.12a) and a GaAs(100) surface p-doped with 50 keV Zn ion bombardment (Fig. 6.12b), in both cases with 10^{11}–10^{16} atoms/cm^2 in eight steps. They studied the influence of high temperature annealing and sputtering and discovered that adsorption of CH_4 strongly enhanced contrast and emission from the cleaned surface, which allowed clear distinction between 6(8) doping levels in the case of B(Zn) doping. The intensity was found to be proportional to the CH_4 coverage as determined by AES. Doping contrast studies with UVPEEM were later resumed again in Williams' group, now on surfaces covered with Si oxides of different thickness, addressing various questions. The first study [193], using a B(p) doping level of 10^{18} cm^{-3} in n-doped Si(100), concluded that the contrast between p- and n-doped regions is due to a difference in the Fermi level pinning, which is determined by the density n_S of surfaces states. The difference in the photoemission p-n contrast between oxide-covered and clean surface was attributed to this effect, which lowers the photoemission threshold of the oxide-covered surface so that emission at the high energy limit of the Hg lamp (5.1 eV) is enhanced. The second study looked at the B concentration dependence of the contrast and found a monotonic increase from 10^{18} to 10^{20} cm^{-3} with the Hg lamp [194]. In the next study [195] the dependence of the contrast upon photon energy was studied for the same doping level range using spontaneous FEL light between 4.5 and 5.2 eV in order to determine the origin of photoemission at low photon energies. In the final paper [196] the results are summarized and a general model of the doping-induced contrast is presented, based on surface state effects and band gap reduction.

This model was later repudiated by high-pass energy filtered UVPEEM measurements of Si(100) samples with p-doped patterns in an n-type substrate and via versa [197]. In both samples the p-n contrast was found to depend upon the photoelectron energy selected. This dependence was caused by a small shift of the energy distribution of the electrons emitted from the n-type regions towards higher energy relative to the p-type regions. However the energy-integrated intensity from the p-regions was higher than from the n-regions, an observation, which was later confirmed to be valid over a wide dopant concentration range [198]. These observations let to the conclusion that the emission difference between p-type and n-type regions is not related to the photoemission threshold but due the different absorption of hot electrons caused by electron–hole pair generation. This picture is supported by p-n contrast measurements in the elastic backscattering of electrons in scanning LEEM (SLEEM) measurements [199].

In all these studies the electrons had to pass through an oxide layer, whose influence can play an important role in the p-n contrast as already the very first UVPEEM studies [192] had shown. The importance of the oxide layer for the contrast was emphasized by SEM studies of oxidized and HF-cleaned samples [200]: on the cleaned samples emission from p-type regions was always stronger than from n-type regions, independent of incident electron energy while on the oxidized sample a contrast inversion occurred with increasing energy. The higher intensity in p-type regions on oxygen-free (H-terminated) Si(100) is also seen in other imaging modes, e.g., in XPEEM when imaging with very low energy secondary electrons (0.5 eV, but not with 5 eV) or in MEM. In LEEM with 3 eV electrons the p-type region was darker, which was attributed to damage caused by the implantation [201]. Spectroscopic XPEEM has brought more insight into the influence of the oxide layer on the contrast. High energy resolution Si 2p images show clear variations of the contrast with binding energy connected with the Si oxidation state (Si^0 to Si^{4+}) within the native oxide layer, both for p doping in n-doped substrates and via versa [202]. Ratio images taken at energies corresponding to different binding states produce very high contrast [203]. These two papers give a detailed analysis of the various factors, which influence the contrast.

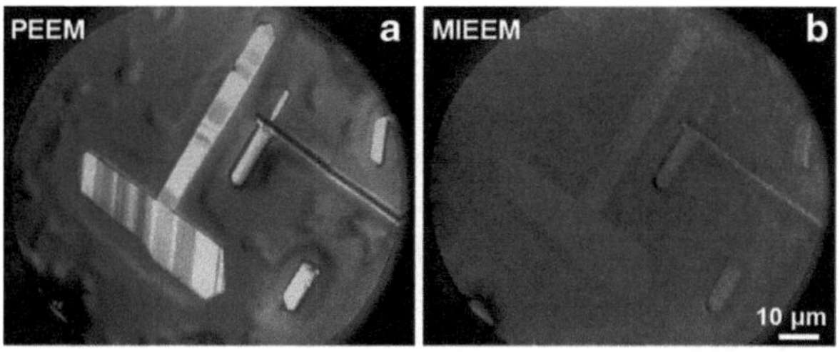

Fig. 6.13 SrO segregation from a $SrTiO_3(100)$ surface heated in air at 1300 °C for 120 h. (**a**) UVPEEM image, (**b**) MIIEM image. Field of view 100 μm. The image acquisition time is 0.5 s and 50 s, respectively. Contrast outside the SrO crystals seen in (**a**) is not visible in (**b**) which is sensitive only to the topmost layer and therefore attributed to subsurface features. The comparison of the two images taking contrast, image acquisition time and signal/noise ratio into account illustrates why MIEEM had only a short lifetime. Reproduced with permission from Ref. [207]. Copyright 2001 Elsevier

In all the studies discussed up to now doping was made by ion implantation, which produces damage. In addition, the resulting concentration depth profile and lateral distribution depends on annealing conditions and ion energy, which makes a quantitative analysis difficult because of the large IMFP of the slow electrons. This problem was overcome by producing B and P-doped Si regions on a lithographically patterned surface via CVD using CH_2Cl_2, B_2H_6, and PH_3 precursors. In this case a well-defined doping distribution was achieved, however with a height difference between the doped region and its surrounding. The goal was to measure the electric field at the horizontal p-n junction by imaging with secondary electrons with various energies, excited by synchrotron light, the doping edge sharpness depending on the deflection of the electrons by the lateral field. Taking into account the electric field distortion caused by the height difference at this edge required modeling of the influence of these two fields on the angular distribution of the electrons and the effect of the contrast aperture [204]. Summarizing the efforts to understand doping contrast formation in cathode lens microscopy and to apply the contrast to quantitative analysis, it is evident that this problem is more complex than originally thought and that more work is still necessary, in particular concerning the influence of overlayers.

Turning to other materials used in electronics, $SrTiO_3$ is used in sensors. Its surface composition is orientation dependent. This dependence was studied with secondary electron and O 1s, Sr 3d, and Ti 2p core electron XPEEM on Nb-doped polycrystalline samples annealed at 800 °C. Comparison with high energy electron backscattering diffraction showed a clear correlation between surface abundance of the elements and grain orientation [205]. In another study the work function of the various grains was measured from images around the photoemission threshold [206]. Annealing at much higher temperatures forms islands on the surface (Fig. 6.13) which is detrimental to sensor performance. This was studied on the (100) surface of a crystal annealed in air at 1300 °C with UVPEEM, energy-resolved MIEEM, metastable impact electron spectroscopy, and XPS studies, which identified the islands as SrO [207, 208].

In a Pt/TiO_2/Pt memristor local defect formation upon electroforming was studied with XPEEM and μXPS, which showed transformation of the amorphous TiO_2 into anatas at the defect [209]. An electronic device in operation, an organic field effect transistor, was also imaged with UVPEEM; the local potential in the channel was measured via the shift of the energy distribution of the photoelectrons from small regions [210].

Though not strictly an electronic device, solar cells will be mentioned here briefly too. One study looked at the distribution of Cd and S of the CdS layer on the thin film Cu(In,Ga)Se_2 absorber layer, which is deposited before the deposition of the n type ZnO window, as a function of thickness and annealing. Cd 4d, S 2p,In 4d, and Se 3d core level XPEEM, XSEM, and UVPEEM indicated preferential CdS growth on cation-terminated surfaces [211]. A more fundamental study using an epitaxial $CuInSe_2$(112) layer on GaAs(111) was significantly limited by up to 1 μm high arc-over-prone and shadow-casting protrusions caused by $Cu_{2-x}Se$ segregation. Nevertheless LEED, work function, and chemical characterization via core level XPEEM and XPS was possible and allowed to follow the surface changes as a function of the thickness of ex situ deposited ZnO films [212].

Finally an application to a vacuum electronics problem should be mentioned yet: a study of the high current density, long lifetime oxide cathodes, in particular scandate cathodes, used in microwave power electronics. Kordesch et al. [213–216] have devoted considerable effort to the understanding of the processes producing the high current density, using TEEM, PEEM, and electron spectroscopy in the study of 200 nm thick oxide films on polycrystalline W. These efforts resulted in a model in which the Sc_2O_3-W interface barrier plays a dominant role and, in cathodes mixed with BaO, also the barrier between Sc_2O_3 and the BaO layer segregated to or deposited on top of the Sc_2O_3 layer. The loss of strong emission was found to be caused by the loss of BaO and Sc_2O_3. The remaining poor emission was attributed to three-dimensional $BaWO_4$ and Ba_2WO_3, based on Raman spectroscopy [217].

Device inspection. With increasing miniaturization of devices semiconductor and magnetic recording industry needs rapid ("high throughput") detection of defects down to the 10 nm range. The parallel image acquisition and strong electron–matter interaction makes cathode lens microscopy superior to the serial acquisition and weaker electron–matter interaction of SEM. Instruments, which have been developed for this purpose and some results obtained with them are described in Sect. 3.1.2. An application example is the inspection of master templates for nanoimprint lithography. A defect in the master copy will be transferred to copies of it and to the nanoimprints made of them, which makes defect inspection of the master copy very important. MEM studies of a periodic linear grating with 100 nm periodicity have shown that at a S/N ratio of 3 a spacing error of 3 nm can be measured with an acquisition time of 10 ms per $30 \times 30\ \mu m^2$ image. This makes MEM a viable inspection tool [218]. In another application of MEM in inspection, stacking faults in homoepitaxial SiC(0001) layers have been imaged by two contrast mechanisms: topographic contrast from the intersection of the stacking fault with the surface and charging of its submerged parts by illumination with above-band gap light [219].

Another interesting inspection method has been suggested for the detection of defects in multilayer mask blanks for extreme ultraviolet (EUV) lithography, which uses light with 13.5 nm wavelength and allows fabricating structures 50 nm or less in size. The mask is a Mo/Si multilayer filter reflective at 13.5 nm, covered with an absorbing pattern. Any defect in or on this multilayer changes the phase relations in the multilayer and causes shifts of the maxima and minima of the wave field, which alters the resist pattern produced by the illumination. Exactly this phenomenon can be used to image defects with PEEM when excited with light of the same energy ("actinic inspection") at normal incidence: incident and reflected wave (at the multilayer/substrate interface) form antinodes or nodes of the electric field on the surface, which enhance or reduce electron emission. In this manner deeply buried defects can be imaged although the electrons are emitted only from the top few nm. The experimental set-up and demonstration experiments with different designed buried defects are described in a series of publications [220–227]. Here only an example, which illustrates the possibilities and limitations, will be mentioned. The defects were 4 nm high stripes varying in width from 35 to 1000 nm, buried below a 40 period Mo/Si multilayer with 6.9 nm period thickness. All stripes could be clearly imaged but the width of the images changed only by a factor of about 8 from the narrowest of the widest stripe. This is due to the spread of the wave field from the defect through the about 280 nm thick layer to the surface. Thus size determination is possible only by modeling the wave propagation in the multilayer [226]. More recent actinic EUV mask inspection tools, which use optical imaging with zone plates [228] or Schwarzschild optics [229] have replaced cathode lens UVPEEM. With Schwarzschild optics test structures 1 nm high and 33 nm wide have been reproducibly detected.

6.4 Biology

The application of UVPEM to biological problems dates back as far as 1972 when Griffith et al. published the first low magnification images of thin sections of rat epididymis cooled to 80–100 K, excited with H_2 light illumination. At that time it was not clear whether or not biological samples could be imaged, where the electrons originated and what determined the contrast. A few years later Grund et al. produced with Engel's PEEM instrument already excellent images of dried biological samples, which showed features down to 160 nm in diameter [230]. By the early 1980s Griffith's group became very productive and published impressive images of a wide variety of biological samples and methodical aspects. Attempts with LEEM imaging were made too [231]. The early work has been reviewed by Griffith [232] and by Birrell et al. [233]. Here only the work with the second generation of PEEM instruments will be discussed.

In the early 1990s TEM cryomicroscopy of ultrathin sections with its superior resolution had made UVPEEM studies of biological materials obsolete. Now synchrotron radiation opened up a new field for PEEM, chemical characterization via core level absorption contrast. De Stasio et al. started this effort with Tonner's PEEM [234] with a study of neuron cells grown on a gold substrate, initially without chemical resolution, using the secondary electrons for imaging. Later by taking absorption spectra from micron-size regions in the image they succeeded in chemical analysis of samples with locally varying concentrations of Al, Co, Zn, and other elements due to preferential uptake of the staining material in different parts of the cells. Over the years experimental procedures such as sputtering, ashing of the samples, and charging suppression were improved, also image analysis by taking the ratio of images at and below the absorption peak or image stacks covering the energy region. As a result quantitative distribution maps could be made. Highlights in the evolution are the studies of the medically important distribution and chemical state of boron, which is used in neutron capture therapy, in human tissue [235] and of gadolinium, which also is promising in brain cancer therapy [236, 237]. The experimental aspects of these later studies have been summarized by Frazer [238].

Inasmuch as the results of XPEEM studies of thin biological objects are dominated by the bulk, other imaging methods, such as core electron loss TEM or spectroscopic XSTM, are more suited for this purpose and have more or less replaced chemical XPEEM. Another application of PEEM, in this case UVPEEM has become available with tunable lasers, in particular with free electron lasers (FELs). By changing the frequency of the light the ionization threshold of biological objects can be determined using Fowler fits to the measured photoelectron emission intensities. This has been done with spontaneous FEL emission for melanosomes, isolated from human hair [239–241], from retinal epithelium cells and from sepia [242]. The melanosomes were supported on Si wafers on which they formed micron size and larger clusters, whose emission was measured. The ionization potential, which is a measure of the oxidation potential, of red hair melanosomes (pheomelanosomes) was found to be 0.6 eV lower than that of black hair melanosomes (eumelanosomes) in agreement with the higher sensitivity of red-haired humans to UV radiation. In another FELPEEM study the ionization potential of fibrinogen was measured. This protein molecule plays an important role in blood coagulation and is about 50 nm long with two globular regions at its ends. A single molecule was imaged. Strong photoemission occurred from its ends with an ionization threshold of 4.9–5.0 eV as obtained from different fits of the intensity as function of photon energy [243]. Efforts to image individual biological objects with MEM have revived again with the attempt to image tobacco mosaic viruses [244] and plans for DNA sequencing in an aberration corrected LEEM [245].

While chemical imaging of biological objects can be done better with other methods than with XPEEM, XPEEM has one unique application, imaging the orientation of dichroitic materials by making use of the search light effect provided by its $\boldsymbol{E}$ vector in XNLDPEEM. This produces polarization-dependent image contrast (PIC) in NEXAFS spectromicroscopy in addition to chemical contrast, well known from XMLDPEEM of antiferromagnetic materials. Gilbert (formerly de

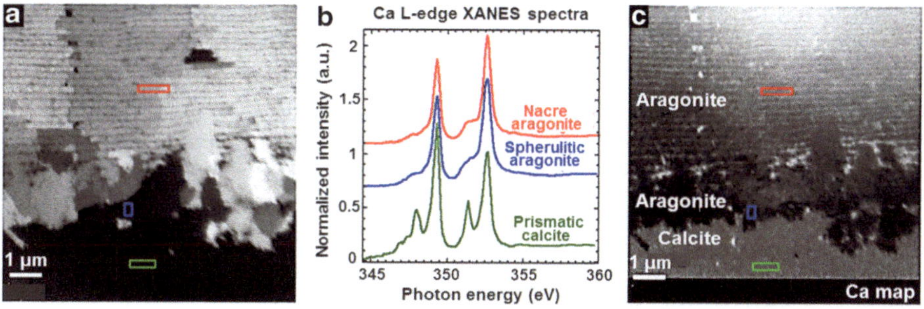

Fig. 6.14 XPEEM analysis of the abalone shell. (**a**) Linear dichroism image obtained from the ratio of the C 1s π^* to σ^* images of $CaCO_3$. (**b**) shows the Ca L edge µXANES spectra of the regions marked in (**a**) and (**c**). (**c**) Ca polytype image obtained by dividing an image taken at 351.5 eV by a pre-adsorption edge image at 345 eV. The absorption peak at 351.5 in (**b**) is characteristic for calcite so that the calcite region appears *brighter* in (**c**) than in (**a**). The fact that the *blue* region is *dark* in (**c**) shows that the transition from calcite to the ordered aragonite platelets occurs via a disordered aragonite region. Adapted with permission from Ref. [248]. Copyright 2008 American Chemical Society

Stasio) et al. introduced it into biology with studies of nacre, the mixed organic-mineral layer, which is found in the shells of many mollusks such as abalone [246]. The structure, formation mechanism ("biomineralization"), and mechanical properties of nacre have been subject of many studies, to which XNLDPEEM ("PIC-XPEEM") brought new light and stimulated also studies of other organic-mineral system such as the sea urchin teeth. The remarkable amount of work, which has been done in this field up to 2012, including purely chemical imaging studies, has been reviewed by Gilbert [247] so that here only a few general points will be made.

Most synchrotron radiation facilities provide elliptical-polarized light with the limiting cases of parallel and perpendicular linear polarized light so that $\boldsymbol{E}$ can be rotated about the incident beam from parallel to perpendicular to the object plane, allowing partial orientation determination. Usually the C $1s \rightarrow \pi^*$ resonance transition or its ratio to the $1s \rightarrow \sigma^*$ transition is used for PIC-XPEEM but similar contrast is also obtained at the O K edge. Figure 6.14 shows as example a study of abalone nacre [248]. In (a) an image with well-oriented nacre platelets on the top and disordered material at the bottom is shown. That both regions are Ca-containing is demonstrated by the Ca L edge ratio image in (c). The characteristic local spectra shown in (b) differentiate between the two polymorphs of $CaCO_3$, calcite, and aragonite. The c-axis-oriented aragonite platelets in nacre are separated by an amino acid peptide layer. It is believed to be porous, allowing epitaxial growth within each platelet stack, but different stacks are misaligned azimuthally. The combination of organic and crystalline layers is believed to be responsible for the high strength of nacre. Many of these XPEEM studies have been combined with TEM, SEM, µ-X-ray diffraction, nano-secondary ion mass spectroscopy, and polarized light microscopy. This and the collaboration between many scientific disciplines have given deep insight in the structure of these organic–mineral systems and their formation processes and promises to remain a fruitful field for XPEEM. Another possible application field of XNLDPEEM is the study of organic molecules, which form chains with preferred orientation as illustrated by a study of collagen fibrils [249]. To complete this section on biological applications a study of a fossil plant, the bark of a coal swamp forest tree, should be mentioned. In this case PIC allowed distinction between calcite and organic matter in the specimen [250].

6.5 A Multimethod Case Study

In most cathode lens microscopy studies a considerable amount of information on the specimen is available at the beginning such as crystal structure, chemical composition, magnetic properties and the experiment aims at the understanding of changes in these properties, such as phase transitions, growth, reactions, response to fields or to currents. There are, however, situations in which little is known about the specimen and the full potential of cathode lens microscopy, that is all imaging, diffraction, and spectroscopy modes are needed for its characterization and understanding.

An example of such a situation will be discussed now as an illustration of the possibilities of SPELEEM. It is the material, which forms long thin crystals ("nanowires") with unusual epitaxial orientations when Fe is deposited onto a ZnS(100) surface at temperatures between 300 and 400 °C. High resolution transmission electron microscopy (HRTEM), transmission electron diffraction, and energy-dispersive X-ray spectroscopy in the HRTEM instrument of these (ex situ prepared) wires had previously identified them as Fe although resistivity measurements gave a 2.4 times larger value than that of bulk Fe, which could be explained by a variety of reasons [251].

In an attempt to understand the nature of these wires their growth was studied in situ with LEEM and LEED [252]. It became immediately evident that these wires showed a wide variety of shapes and orientations, two of which are shown in Fig. 6.15a, depending upon deposition and annealing conditions. Also their complex LEED pattern (Fig. 6.15b) was incompatible with Fe. The LEED pattern was dominated by facet spots, i.e., spots that do move with electron energy, in addition to weaker stationary spots, and could not be identified without further knowledge about the crystal. In contrast, the dark crystals seen in the figure were clearly identified as one of the Fe crystal orientations reported in the previous study. In order to help in the identification of the wires, growth was repeated in a SPELEEM instrument, which allowed chemical and magnetic characterization.

The chemical analysis in the SPELEEM instrument showed that the delay of crystal nucleation seen in LEEM was due to the formation of a (Fe,Zn)S reaction layer. Fe 3p and S 2p μXPS of the Fe crystals (Fig. 6 in [252]) showed that they were covered with a sulfur adsorption layer, compatible with the superstructure LEED pattern but the corresponding spectra of the wires (Fig. 6.16a, b) clearly identified them as a Fe–S compound. The fitting of their Fe 3p spectrum required a spin-up/spin-down ratio different from that of the Fe crystals and bulk Fe.

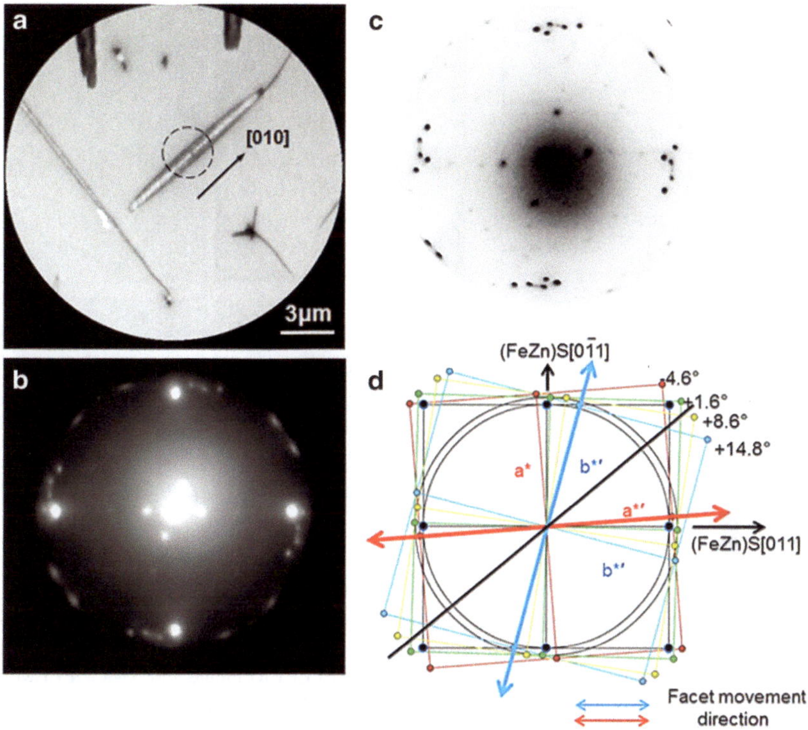

Fig. 6.15 LEEM image taken with 10 eV electrons (**a**) and μLEED pattern (**b**) from the *marked* area of the crystal in center of the image, taken with 17 eV electrons. The second bright crystal ("nanowire") produces the same LEED pattern, the *black* crystals on the top of the image a Fe(100) pattern with a c(2 × 2) superstructure. (**c**) inverted μLEED pattern from the same region as in (**b**), taken with 20 eV electrons. The four *inner facet spots* in the center have moved compared to (**b**) but the four spots near the substrate spots not. The spot doublet at the edge is also from facets. (**d**) Sketch of the two-dimensional reciprocal lattice with spot assignments. The *thick red* and *blue double arrows* indicate the facet spot movement directions, which are symmetric with respect to the *thick black line* along the long axis of the crystal. The four lattices, marked by *thin lines* are also pairwise (*red–blue*, *yellow–green*) symmetric with respect to the long axis. The angles give the rotation of the four lattices with respect to the [11] direction of the two-dimensional reciprocal lattice ([100] direction in the substrate). The *black spots* and *thin lines* through them indicate the substrate lattice. Adapted from Ref. [252]. Permission of the Royal Society of Chemistry

For the fitting of their S 2p spectrum two doublets characteristic for S in the volume and on the surface were needed. Insufficient knowledge of the angular distribution of the Fe and S photoelectron, which determines their transmission through the system, and of the IMFP precluded a precise determination of the chemical composition so that it could only be concluded that the wires contained a comparable amount of Fe and S. Fe 2p μXAS (Fig. 6.16c) showed that the wires contained an amount of Fe comparable to that of the Fe crystals and much more than the (Fe,Zn)S reaction layer, which forms before the crystals. Again, here the unknown difference in the IMFP between the Fe and Fe–S crystals does not allow a quantitative conclusion.

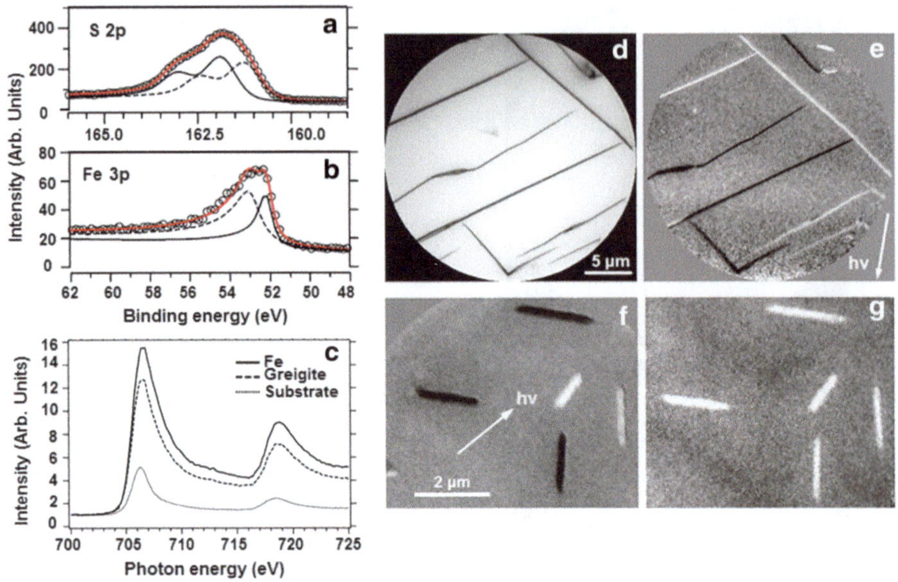

Fig. 6.16 (**a**) S 2p µXPS spectrum, (**b**) Fe 3p µXPS spectrum, both excited with 260 eV photons, of a typical "nanowire" crystal and their fits. (**c**) Fe 2p µXAS spectra of an Fe crystal, of a nanowire and of the substrate covered by reaction layer. (**d**) and (**e**) LEEM image and XMCDPEEM image of a group of particularly long "nanowires". (**f**) and (**g**) XMCDPEEM and Fe 3p XPEEM image after heating crystals with smaller aspect ratio to 450 °C. Adapted from Ref. [252]. Permission of the Royal Society of Chemistry

Next XMCDPEEM was used to clarify the magnetic state of the wires. Figure 6.16d, e show a LEEM and the corresponding XMCDPEEM image of a group of particularly thin crystals in two equivalent orientation, which are in the magnetic single domain state due to shape anisotropy. For the quantitative analysis of the magnetization wider crystals were used as shown in the XMCDPEEM image of Fig. 6.16f, which was taken after heating to 450 °C. At this temperature there was still clear magnetic contrast although the length of the crystals was decreasing due to reaction with the ZnS substrate. Quantitative analysis of the XMCD contrast showed a decrease of only 10 % between room temperature and 450 °C, a clear indication that the Curie temperature of this Fe–S compound is above this temperature.

Equipped with this information on the approximate chemical composition and lower limit of the Curie temperature the analysis of the LEED pattern could be attempted. The Fe–S phase diagram shows many structurally distinct phases of which only two are known to be ferrimagnetic, pyrrhotite (Fe_7S_8), and greigite (Fe_3S_4). Pyrrhotite, like most other Fe–S compounds, has a complex crystal structure, which is not compatible with the LEED pattern. Tetragonal mackinawite (FeS) and cubic Fe_3S_4 are the only Fe–S compounds with a square two-dimensional reciprocal lattice, the tetragonal FeS when the c-plane is parallel to the substrate, the cubic Fe_3S_4 with the {100} plane parallel to the substrate. They have very similar two-dimensional reciprocal lattice distances: for FeS the (10) distance is 0.272 $Å^{-1}$, for Fe_3S_4 it is 0.286 $Å^{-1}$. The LEED pattern of the crystal, which is shown in Fig. 6.15c, taken at a somewhat higher energy, shows its stationary spots as quadruplets close to the (10) spots of the (Fe,Zn)S layer, whose lattice constant is exactly known. The four quadruplet spots around these spots form nearly square lattices (Fig. 6.15d), whose reciprocal lattice distances can be determined with high accuracy by referencing them to those of the (Fe,Zn)S layer. Using these spots for reference, the stationary spots indicate two lattices with $a^* = 0.286$ $Å^{-1}$, $b^* = 0.270$ $Å^{-1}$ and two with $a^{*\prime} = 0.280$ $Å^{-1}$, $b^{*\prime} = 0.277$ $Å^{-1}$. Although the error bar of these measured reciprocal lattice distances was believed so small, that FeS could be excluded, additional information was desired to support this conclusion. This was supplied by XMCDPEEM results, which showed that the crystals are magnetic with a Curie temperature above 450 °C while mackinawite is known to be nonmagnetic. With all this information the wires could be identified as greigite (Fe_3S_4).

In addition to this material analysis there is much more information in the LEED pattern and in the LEEM images, in particular in the movies taken during the growth of the crystals. The movement of the facet spots reveals unusual facets and twinning. The LEEM images show a variety of growth shapes: growth started from facets, regions terminated with the (100) surface developed only later and the crystals had a fine-grained substructure. Description of these features is beyond the scope of this section, which just wants to demonstrate that cathode lens microscopy, used with its full potential, is useful not only for the study of material properties but also for material analysis.

References

1. Novoselov, K.S., Geim, A.K., Morozov, S.V., Jiang, D., Zhang, Y., Dubonos, S.V., Grigorieva, I.V., Firsov, A.A.: Electric field effect in atomically thin carbon films. Science **306**, 666–669 (2004)
2. Berger, C., Song, Z., Li, T., Li, X., Ogbazghi, A.Y., Feng, R., Dai, Z., Marchenkov, A.N., Conrad, E.H., First, P.N., de Heer, W.A.: Ultrathin epitaxial graphite: 2D electron gas properties and a route toward graphene-based nanoelectronics. J. Phys. Chem. B **108**, 19912–19916 (2004)
3. Warner, J.H., Schäffel, F., Rümmeli, M.H., Bachmatiuk, A. (eds.): Graphene: Fundamentals and Emergent Applications. Elsevier Inc., Amsterdam (2013)
4. Batzill, M.: The surface science of graphene: metal interfaces, CVD synthesis, nanoribbons, chemical modifications, and defects. Surf. Sci. Rep. **67**, 83–115 (2012)
5. Weiss, N.O., Zhou, H., Liao, L., Liu, Y., Jiang, S., Huang, Y., Duan, X.: Graphene: an emerging electronic material. Adv. Mater. **24**, 5782–5825 (2012)
6. Avouris, P., Dimitrakopoulos, C.: Graphene: synthesis and applications. Mater Today **15**, 86–97 (2012)
7. Lu, W., Soukiassian, P., Boeckl, J. (eds.): Graphene: fundamentals and functionalities. MRS Bull. **37**, 1119–1321 (2012)
8. Singh, V., Joung, D., Zhai, L., Das, S., Khondaker, S.I., Seal, S.: Graphene based materials: past, present and future. Prog. Mater. Sci. **56**, 1178–1271 (2011)

9. Hibino, H., Kageshima, H., Maeda, F., Nagase, M., Kobayashi, Y., Yamaguchi, H.: Microscopic thickness determination of thin graphite films formed on SiC from quantized oscillation in reflectivity of low-energy electrons. Phys. Rev. B **77**, 075413, 7 pages (2008)
10. Knox, K.R., Wang, S., Morgante, A., Cvetko, D., Locatelli, A., Mentes, T.O., Niño, M.A., Kim, P., Osgood Jr., R.M.: Spectromicroscopy of single and multilayer graphene supported by a weakly interacting substrate. Phys. Rev. B **78**, 201408, 4 pages (2008)
11. Man, K.L., Altman, M.S.: Low energy electron microscopy and photoemission electron microscopy investigation of graphene. J. Phys. Condens. Matter **24**, 314209, 20 pages (2012)
12. Sutter, P., Sutter, E.: Microscopy of graphene growth, processing, and properties. Adv. Funct. Mater. **23**, 2617–2634 (2013)
13. Hwang, C., Park, C.-H., Siegel, D.A., Fedorov, A.V., Louie, S.G., Lanzara, A.: Direct measurement of quantum phases in graphene via photoemission spectroscopy. Phys. Rev. B **84**, 125422, 10 pages (2011)
14. Liu, Y., Bian, G., Miller, T., Chiang, T.-C.: Visualizing electronic chirality and berry phases in graphene systems using photoemission with circularly polarized light. Phys. Rev. Lett. **107**, 166803, 5 pages (2011)
15. Knox, K.R., Locatelli, A., Yilmaz, M.B., Cvetko, D., Menteş, T.O., Niño, M.A., Kim, P., Morgante, A., Osgood, R.M.: Making angle-resolved photoemission measurements on corrugated monolayer crystals: suspended exfoliated single-crystal graphene. Phys. Rev. B **84**, 115401, 12 pages (2011)
16. Barrett, N., Winkler, K., Krömker, B., Conrad, E.H.: Laboratory-based real and reciprocals pace imaging of the electronic structure of few layer graphene on SiC(0001) using photoelectron emission microscopy. Ultramicroscopy **30**, 94–100 (2013)
17. Krasovskii, E.E.: Augmented-plane-wave approach to scattering of Bloch electrons by an interface. Phys. Rev. B **70**, 245322, 11 pages (2004)
18. Feenstra, R.M., Widom, M.: Low-energy electron reflectivity from graphene: first-principles computations and approximate models. Ultramicroscopy **130**, 101–108 (2013)
19. Feenstra, R.M., Srivastava, N., Gao, Q., Widom, M., Diaconescu, B., Ohta, T., Kellogg, G.L., Robinson, J.T., Vlassiouk, I.V.: Low-energy electron reflectivity from graphene. Phys. Rev. B **87**, 041406, 4 pages (2013)
20. Srivastava, N., Gao, Q., Widom, M., Feenstra, R.M., Nie, S., McCarty, K.F., Vlassiouk, I.V.: Low-energy electron reflectivity of graphene on copper and other substrates. Phys. Rev. B **87**, 245414, 19 pages (2013)
21. Zheng, J., Wang, Y., Wang, L., Quhe, R., Ni, Z., Mei, W.-N., Gao, Z., Yu, D., Shi, J., Lu, J.: Interfacial properties of bilayer and trilayer graphene on metal substrates. Sci. Rep. **3**, 2081, 11 pages (2013)
22. Busse, C., Lazic, P., Djemour, R., Coraux, J., Gerber, T., Atodiresei, N., Caciuc, V., Brako, R., N'Diaye, A.T., Blügel, S., Zegenhagen, J., Michely, T.: Graphene on Ir(111): physisorption with chemical modulation. Phys. Rev. Lett. **107**, 036101, 4 pages (2011)
23. Menteş, T.O., Locatelli, A.: Angle-resolved X-ray photoemission electron microscopy. J. Electron Spectrosc. Relat. Phenom. **185**, 323–329 (2012)
24. Hibino, H., Tanabe, S., Mizuno, S., Kageshima, H.: Growth and electronic transport properties of epitaxial graphene on SiC. J. Phys. D. Appl. Phys. **45**, 154008, 12 pages (2012)
25. Kageshima, H., Hibino, H., Tanabe, S.: The physics of epitaxial graphene on SiC(0001). J. Phys. Condens. Matter **24**, 314215, 11 pages (2012)
26. Tromp, R.M., Hannon, J.B.: Thermodynamics and kinetics of graphene growth on SiC(0001). Phys. Rev. Lett. **102**, 106104, 4 pages (2009)
27. Pavlovska, A., Bauer, E., Smith, D.J.: In situ studies of the role of excess Ga on the growth morphology of thin GaN layers. Surf. Sci. **496**, 160–178 (2002)
28. Hannon, J.B., Tromp, R.M.: Pit formation during graphene synthesis on SiC(0001): in situ electron microscopy. Phys. Rev. B **77**, 241404, 4 pages (2008)
29. Ouerghi, A., Ridene, M., Mathieu, C., Gogneau, N., Belkhou, R.: From nanographene to monolayer graphene on 6H-SiC(0001) substrate. Appl. Phys. Lett. **102**, 253108 (2013)

30. Tanaka, S., Morita, K., Hibino, H.: Anisotropic layer-by-layer growth of graphene on vicinal SiC(0001) surfaces. Phys. Rev. B **81**, 041406, 4 pages (2010)
31. Ji, S.-H., Hannon, J.B., Tromp, R.M., Perebeinos, V., Tersoff, J., Ross, F.M.: Atomic-scale transport in epitaxial graphene. Nat. Mater. **11**, 114–119 (2012)
32. Fukidome, H., Kawai, Y., Fromm, F., Kotsugi, M., Handa, H., Ide, T., Ohkouchi, T., Miyashita, H., Enta, Y., Kinoshita, T., Seyller, T., Suemitsu, M.: Precise control of epitaxy of graphene by microfabricating SiC substrate. Appl. Phys. Lett. **101**, 041605 (2012)
33. Hicks, J., Conrad, E.H.: Graphene investigated by synchrotron radiation. MRS Bull. **37**, 1203–1213 (2012)
34. Johansson, L.I., Watcharinyanon, S., Zakharov, A.A., Iakimov, T., Yakimova, R., Virojanadara, C.: Stacking of adjacent graphene layers grown on C-face SiC. Phys. Rev. B **84**, 125405, 8 pages (2011)
35. Johansson, L.I., Xia, C., Hassan, J.U., Iakimov, T., Zakharov, A.A., Watcharinyanon, S., Yakimova, R., Janzén, E., Virojanadara, C.: Is the registry between adjacent graphene layers grown on C-face SiC different compared to that on Si-face SiC. Crystals **3**, 1–13 (2013)
36. Mathieu, C., Barrett, N., Rault, J., Mi, Y.Y., Zhang, B., de Heer, W.A., Berger, C., Conrad, E. H., Renault, O.: Microscopic correlation between chemical and electronic states in epitaxial graphene on SiC(0001). Phys. Rev. B **83**, 235436, 11 pages (2011)
37. He, G., Srivastava, N., Feenstra, R.M.: Formation of graphene on SiC(000) surfaces in disilane and neon environments. J. Vac. Sci. Technol. B **30**, 04E102, 6 pages (2012)
38. Srivastava, N., He, G., Luxmi, Feenstra, R.M.: Interface structure of graphene on SiC(0001). Phys. Rev. B **85**, 041404, 5 pages (2012)
39. Srivastava, N., He, G., Feenstra, R.M.: Graphene on carbon-face SiC{0001} surfaces formed in a disilane environment. Mater. Sci. Forum **609**, 717–720 (2012)
40. Ohta, T., Beechem, T.E., Robinson, J.T., Kellogg, G.L.: Long-range atomic ordering and variable interlayer interactions in two overlapping graphene lattices with stacking misorientations. Phys. Rev. B **85**, 075415, 7 pages (2012)
41. Ohta, T., Robinson, J.T., Feibelman, P.J., Bostwick, A., Rotenberg, E., Beechem, T.E.: Evidence for interlayer coupling and moire periodic potentials in twisted bilayer graphene. Phys. Rev. Lett. **109**, 186807, 6 pages (2012)
42. Ouerghi, A., Belkhou, R., Marangolo, M., Silly, M.G., El Moussaoui, S., Eddrief, M., Largeau, L., Portail, M., Sirotti, F.: Structural coherency of epitaxial graphene on 3C-SiC (111) epilayers on Si(111). Appl. Phys. Lett. **97**, 161905, 3 pages (2010)
43. Ouerghi, A., Ridene, M., Balan, A., Belkhou, R., Barbier, A., Gogneau, N., Portail, M., Michon, A., Latil, S., Jegou, P., Shukla, A.: Sharp interface in epitaxial graphene layers on *3C*-SiC(100)/Si(100) wafers. Phys. Rev. B **83**, 205429, 5 pages (2011)
44. Fukidome, H., Kawai, Y., Handa, H., Hibino, H., Miyashita, H., Kotsugi, M., Ohkochi, T., Jung, M.-H., Suemitsu, T., Kinoshita, T., Otsuji, T., Suemitsu, M.: Site-selective epitaxy of graphene on Si wafers. Proc. IEEE **101**, 1557–1566 (2013)
45. Coletti, C., Emtsev, K.V., Zakharov, A.A., Ouisse, T., Chaussende, D., Starke, U.: Large area quasi-free standing monolayer graphene on 3C-SiC(111). Appl. Phys. Lett. **99**, 081904, 3 pages (2011)
46. Yazdi, G.R., Vasiliauskas, R., Iakimov, T., Zakharov, A., Syväjärvi, M., Yakimova, R.: Growth of large area monolayer graphene on 3C-SiC and a comparison with other SiC polytypes. Carbon **57**, 477–484 (2013)
47. Watcharinyanon, S., Johansson, L.I., Zakharov, A.A., Virojanadara, C.: Studies of Li intercalation of hydrogenated graphene on SiC(0001). Surf. Sci. **606**, 401–406 (2012)
48. Walter, A.L., Jeon, K.-J., Bostwick, A., Speck, F., Ostler, M., Seyller, T., Moreschini, L., Kim, Y.S., Chang, Y.J., Horn, K., Rotenberg, E.: Highly p -doped epitaxial graphene obtained by fluorine intercalation. Appl. Phys. Lett. **98**, 184102, 3 pages (2011)
49. Zakharov, A.A., Virojanadara, C., Watcharinyanon, S., Yakimova, R., Johansson, L.I.: Nanoscale 3-D (*E, kx, ky*) band structure imaging on graphene and intercalated graphene. IBM J. Res. Dev. **55**, Paper 6, 6 pages (2011)

50. Ristein, J., Mammadov, S., Seyller, T.: Origin of doping in quasi-free-standing graphene on silicon carbide. Phys. Rev. Lett. **108**, 246104, 5 pages (2012)
51. Starke, U., Forti, S., Emtsev, K.V., Coletti, C.: Engineering the electronic structure of epitaxial graphene by transfer doping and atomic intercalation. MRS Bull. **37**, 1177–1186 (2012)
52. Riedl, C., Coletti, C., Iwasaki, T., Zakharov, A.A., Starke, U.: Quasi-free-standing epitaxial graphene on SiC obtained by hydrogen intercalation. Phys. Rev. Lett. **103**, 246804, 4 pages (2009)
53. Virojanadara, C., Zakharov, A.A., Yakimova, R., Johansson, L.I.: Buffer layer free large area bi-layer graphene on SiC(0001). Surf. Sci. **604**, L4–L7 (2010)
54. Virojanadara, C., Yakimova, R., Zakharov, A.A., Johansson, L.I.: Large homogeneous mono-/bi-layer graphene on 6H-SiC(0001) and buffer layer elimination. J. Phys. D. Appl. Phys. **43**, 374010, 13 pages (2010)
55. Watcharinyanon, S., Virojanadara, C., Osiecki, J.R., Zakharov, A.A., Yakimova, R., Uhrberg, R.I.G., Johansson, L.I.: Hydrogen intercalation of graphene grown on 6H-SiC (0001). Surf. Sci. **605**, 1662–1668 (2011)
56. Forti, S., Emtsev, K.V., Coletti, C., Zakharov, A.A., Riedl, C., Starke, U.: Large-area homogeneous quasifree standing epitaxial graphene on SiC(0001): electronic and structural characterization. Phys. Rev. B **84**, 125449, 10 pages (2011)
57. Virojanadara, C., Watcharinyanon, S., Zakharov, A.A., Johansson, L.I.: Epitaxial graphene on 6H-SiC and Li intercalation. Phys. Rev. B **82**, 205402, 6 pages (2010)
58. Virojanadara, C., Zakharov, A.A., Watcharinyanon, S., Yakimova, R., Johansson, L.I.: A low-energy electron microscopy and x-ray photo-emission electron microscopy study of Li intercalated into graphene on SiC(0001). New J. Phys. **12**, 125015, 9 pages (2010)
59. Xia, C., Watcharinyanon, S., Zakharov, A.A., Johansson, L.I., Yakimova, R., Virojanadara, C.: Detailed studies of Na intercalation on furnace-grown graphene on 6H-SiC(0001). Surf. Sci. **613**, 88–94 (2013)
60. Oida, S., McFeely, F.R., Hannon, J.B., Tromp, R.M., Copel, M., Chen, Z., Sun, Y., Farmer, D.B., Yurkas, J.: Decoupling graphene from SiC(0001) via oxidation. Phys. Rev. B **82**, 041411, 4 pages (2010)
61. Mathieu, C., Lalmi, B., Menteş, T.O., Pallecchi, E., Locatelli, A., Latil, S., Belkhou, R., Ouerghi, A.: Effect of oxygen adsorption on the local properties of epitaxial graphene on SiC (0001). Phys. Rev. B **86**, 035435, 5 pages (2012)
62. Xia, C., Watcharinyanon, S., Zakharov, A.A., Yakimova, R., Hultman, L., Johansson, L.I., Virojanadara, C.: Si intercalation/deintercalation of graphene on 6H-SiC(0001). Phys. Rev. B **85**, 045418, 7 pages (2012)
63. Wang, F., Shepperd, K., Hicks, J., Nevius, M.S., Tinkey, H., Tejeda, A., Taleb-Ibrahimi, A., Bertran, F., Le Fèvre, P., Torrance, D.B., First, P.N., de Heer, W.A., Zakharov, A.A., Conrad, E.H.: Silicon intercalation into the graphene-SiC interface. Phys. Rev. B **85**, 165449, 6 pages (2012)
64. Emtsev, K.V., Zakharov, A.A., Coletti, C., Forti, S., Starke, U.: Ambipolar doping in quasifree epitaxial graphene on SiC(0001) controlled by Ge intercalation. Phys. Rev. B **84**, 125423, 6 pages (2011)
65. van Bommel, A.J., Crombeen, J.E., van Tooren, A.: LEED and AUGER electron observations of the SiC (0001) surface. Surf. Sci. **48**, 463–472 (1975)
66. Bartelt, N.C., McCarty, K.F.: Graphene growth on metal surfaces. MRS Bull. **37**, 1158–1165 (2012)
67. Olsen, T., Thygesen, K.S.: Random phase approximation applied to solids, molecules, and graphene-metal interfaces: from van der Waals to covalent bonding. Phys. Rev. B **87**, 075111, 13 pages (2013)
68. Alfè, D., Pozzo, M., Miniussi, E., Günther, S., Lacovig, P., Lizzit, S., Larciprete, R., Santos Burgos, B., Menteş, T.O., Locatelli, A., Baraldi, A.: Fine tuning of graphene-metal adhesion by surface alloying. Sci. Rep. **3**, 2430, 6 pages (2013)

69. Jin, L., Fu, Q., Zhang, H., Mu, R., Zhang, Y., Tan, D., Bao, X.: Tailoring the growth of graphene on Ru(0001) via engineering of the substrate surface. J. Phys. Chem. C **116**, 2988–2993 (2012)
70. Sutter, P., Ciobanu, C.V., Sutter, E.: Real-time microscopy of graphene growth on epitaxial metal films: role of template thickness and strain. Small **8**, 2250–2257 (2012)
71. Locatelli, A., Wang, C., Africh, C., Stojić, N., Menteş, T.O., Comelli, G., Binggeli, N.: Temperature-driven reversible rippling and bonding of a graphene superlattice. ACS Nano **7**, 6955–6963 (2013)
72. Miniussi, E., Pozzo, M., Baraldi, A., Vesselli, E., Zhan, R.R., Comelli, G., Menteş, T.O., Niño, M.A., Locatelli, A., Lizzit, S., Alfè, D.: Thermal stability of corrugated epitaxial graphene grown on Re(0001). Phys. Rev. Lett. **106**, 216101, 4 pages (2011)
73. Gao, L., Ren, W., Xu, H., Jin, L., Wang, Z., Ma, T., Ma, L.-P., Zhang, Z., Fu, Q., Peng, L.-M., Bao, X., Cheng, H.-M.: Repeated growth and bubbling transfer of graphene with millimetre-size single-crystal grains using platinum. Nat. Commun. **3**, 1702, 7 pages (2012)
74. Murata, Y., Nie, S., Ebnonnasir, A., Starodub, E., Kappes, B.B., McCarty, K.F., Ciobanu, C. V., Kodambaka, S.: Growth structure and work function of bilayer graphene on Pd(111). Phys. Rev. B **85**, 205443, 6 pages (2012)
75. Odahara, G., Otani, S., Oshima, C., Suzuki, M., Yasue, T., Koshikawa, T.: In-situ Observation of Graphene Growth on Ni(111). In: Proc. Atomic Level Characterizations for New Materials and Devices (ALC'11), Seoul, Korea 2011, pp. 268–271. JSPS 141 Activity Report (2012)
76. Odahara, G., Otani, S., Oshima, C., Suzuki, M., Yasue, T., Koshikawa, T.: In-situ observation of graphene growth on Ni(111). Surf. Sci. **605**, 1095–1098 (2011)
77. Odahara, G., Otani, S., Oshima, C., Suzuki, M., Yasue, T., Koshikawa, T.: Macroscopic single-domain graphene sheet on Ni(111). Surf. Interface Anal. **43**, 1491–1493 (2011)
78. Addou, R., Dahal, A., Sutter, P., Batzill, M.: Monolayer graphene growth on Ni(111) by low temperature chemical vapor deposition. Appl. Phys. Lett. **100**, 021601, 3 pages (2012)
79. Dahal, A., Addou, R., Sutter, P., Batzill, M.: Graphene monolayer rotation on Ni(111) facilitates bilayer graphene growth. Appl. Phys. Lett. **100**, 241602, 4 pages (2012)
80. Odahara, G., Hibino, H., Nakayama, N., Shimbata, T., Oshima, C., Otani, S., Suzuki, M., Yasue, T., Koshikawa, T.: Macroscopic single-domain graphene growth on polycrystalline nickel surface. Appl. Phys. Express **5**, 035501, 3 pages (2012)
81. Kanzaki, K., Hibino, H., Makimoto, T.: Graphene layer formation on polycrystalline nickel grown by chemical vapor deposition. Jpn. J. Appl. Phys. **52**, 035103, 8 pages (2013)
82. Vinogradov, N.A., Zakharov, A.A., Kocevski, V., Rusz, J., Simonov, K.A., Eriksson, O., Mikkelsen, A., Lundgren, E., Vinogradov, A.S., Mårtensson, N., Preobrajenski, A.B.: Formation and structure of graphene waves on Fe(110). Phys. Rev. Lett. **109**, 026101, 5 pages (2012)
83. Bae, S., Kim, H., Lee, Y., Xu, X., Park, J.-S., Zheng, Y., Balakrishnan, J., Lei, T., Kim, H.R., Song, Y.I., Kim, Y.-J., Kim, K.S., Özyilmaz, B., Ahn, J.-H., Hong, B.H., Iijima, S.: Roll-to-roll production of 30-inch graphene films for transparent electrodes. Nat. Nanotechnol. **5**, 574–578 (2010)
84. Ogawa, Y., Hu, B., Orofeo, C.M., Tsuji, M., Ikeda, K., Mizuno, S., Hibino, H., Ago, H.: Domain structure and boundary in single-layer graphene grown on Cu(111) and Cu(100) films. J. Phys. Chem. Lett. **3**, 219–226 (2012)
85. Orofeo, C.M., Hibino, H., Kawahara, K., Ogawa, Y., Tsuji, M., Ikeda, K., Mizuno, S., Ago, H.: Influence of Cu metal on the domain structure and carrier mobility in single-layer graphene. Carbon **50**, 2189–2196 (2012)
86. Kraus, J., Böcklein, S., Reichelt, R., Günther, S., Santos, B., Menteş, T.O., Locatelli, A.: Towards the perfect graphene membrane? - Improvement and limits during formation of high quality graphene grown on Cu-foils. Carbon **64**, 377–390 (2013)
87. Wofford, J.M., Starodub, E., Walter, A.L., Nie, S., Bostwick, A., Bartelt, N.C., Thürmer, K., Rotenberg, E., McCarty, K.F., Dubon, O.D.: Extraordinary epitaxial alignment of graphene islands on Au(111). New J. Phys. **14**, 053008, 10 pages (2012)

88. Nie, S., Wofford, J.M., Bartelt, N.C., Dubon, O.D., McCarty, K.F.: Origin of the mosaicity in graphene grown on Cu(111). Phys. Rev. B **84**, 155425, 7 pages (2011)
89. Jin, L., Fu, Q., Mu, R., Tan, D., Bao, X.: Pb intercalation underneath a graphene layer on Ru (0001) and its effect on graphene oxidation. Phys. Chem. Chem. Phys. **13**, 16655–16660 (2011)
90. Jin, L., Fu, Q., Yang, Y., Bao, X.: A comparative study of intercalation mechanism at graphene/Ru(0001) interface. Surf. Sci. **617**, 81–86 (2013)
91. Cui, Y., Gao, J., Jin, L., Zhao, J., Tan, D., Fu, Q., Bao, X.: An exchange intercalation mechanism for the formation of a two-dimensional Si structure underneath graphene. Nano Res. **5**, 352–360 (2012)
92. Lizzit, S., Larciprete, R., Lacovig, P., Dalmiglio, M., Orlando, F., Baraldi, A., Gammelgaard, L., Barreto, L., Bianchi, M., Perkins, E., Hofmann, P.: Transfer-free electrical insulation of epitaxial graphene from its metal substrate. Nano Lett. **12**, 4503–4507 (2012)
93. Cui, Y., Fu, Q., Zhang, H., Tan, D., Bao, X.: Dynamic characterization of graphene growth and etching by oxygen on Ru(0001) by photoemission electron microscopy. J. Phys. Chem. C **113**, 20365–20370 (2009)
94. Starodub, E., Bartelt, N.C., McCarty, K.F.: Viable thermionic emission from graphene-covered metals. Appl. Phys. Lett. **100**, 181604, 4 pages (2012)
95. Schmidt, O., Fecher, G.H., Hwu, Y., Schönhense, G.: The spatial distribution of non-linear effects in multi-photon photoemission from metallic adsorbates on Si(111). Surf. Sci. **482–485**, 687–692 (2001)
96. Schmidt, O., Bauer, M., Wiemann, C., Porath, R., Scharte, M., Andreyev, O., Schönhense, G., M, A.: Time-resolved two photon photoemission electron microscopy. Appl. Phys. B **74**, 223–227 (2002)
97. Fecher, G.H., Schmidt, O., Hwu, Y., Schönhense, G.: Multiphoton photoemission electron microscopy using femtosecond laser radiation. J. Electr. Spectrosc. Relat. Phenom. **126**, 77–87 (2002)
98. Cinchetti, M., Oelsner, A., Fecher, G.H., Elmers, H.J., Schönhense, G.: Observation of Cu surface inhomogeneities by multiphoton photoemission spectromicroscopy. Appl. Phys. Lett. **83**, 1503–1506 (2003)
99. Cinchetti, M., Schönhense, G.: Two-photon photoemission spectromicroscopy of noble metal clusters on surfaces studied using time-of-flight photoemission electron microscopy. J. Phys. Condens. Matter **17**, S1319–S1328 (2005)
100. Cinchetti, M., Gloskovskii, A., Nepjiko, S.A., Schönhense, G., Rochholz, H., Kreiter, M.: Photoemission electron microscopy as a tool for the investigation of optical near fields. Phys. Rev. Lett. **95**, 047601, 10 pages (2005)
101. Kubo, A., Onda, K., Petek, H., Sun, Z., Jung, Y.S., Kim, H.K.: Femtosecond imaging of surface plasmon dynamics in a nano-structured silver film. Nano Lett. **5**, 1123–1127 (2005)
102. Petek, H., Ogawa, S.: Femtosecond time-resolved two-photon photoemission studies of electron dynamics in metals. Prog. Surf. Sci. **56**, 239–310 (1997)
103. Kubo, A., Jung, Y.S., Kim, H.K., Petek, H.: Femtosecond microscopy of localized and propagating surface plasmons in silver gratings. J. Phys. B **40**, S259–S272 (2007)
104. Kubo, A., Pontius, N., Petek, H.: Femtosecond microscopy of surface plasmon polariton wave packet evolution at the silver/vacuum interface. Nano Lett. **7**, 470–475 (2007)
105. Zhang, L., Kubo, A., Wang, L., Petek, H., Seideman, T.: Imaging of surface plasmon polariton fields excited at a nanometer-scale slit. Phys. Rev. B **84**, 245442, 10 pages (2011)
106. Lemke, C., Leißner, T., Jauernik, S., Klick, A., Fiutowski, J., Kjelstrup-Hansen, J., Rubahn, H.-G., Bauer, M.: Mapping surface plasmon polariton propagation via counter-propagating light pulses. Opt. Express **20**, 12877, 8 pages (2012)
107. Leißner, T., Thilsing-Hansen, K., Lemke, C., Jauernik, S., Kjelstrup-Hansen, J., Bauer, M., Rubahn, H.-G.: Surface plasmon polariton emission prompted by organic nanofibers on thin gold films. Plasmonics **7**, 253–260 (2012)
108. Lange, J., Bayer, D., Rohmer, M., Wiemann, C., Gaier, O., Aeschlimann, M., Bauer, M.: Probing femtosecond plasmon dynamics with nanometer resolution. In: Andrews, D.L., Nunzi, J.-M., Ostendorf, A. (eds.) Nanophotonics 2006. Proc. SPIE, vol. 6195, pp. Z-1–Z-9. SPIE, Bellingham, WA (2006)

109. Bauer, M., Wiemann, C., Lange, J., Bayer, D., Rohmer, M., Aeschlimann, M.: Phase propagation of localized surface plasmons probed by time-resolved photoemission electron microscopy. Appl. Phys. A **88**, 473–480 (2007)
110. Bayer, D., Wiemann, C., Gaier, O., Bauer, M., Aeschlimann, M.: Time-resolved 2PPE and time-resolved PEEM as a probe of LSP's in silver nanoparticles. J. Nanomater. **2008** (249514), 11 pages (2008)
111. Wiemann, C., Bayer, D., Rohmer, M., Aeschlimann, M., Bauer, M.: Local 2PPE-yield enchancement in a defined periodic silver nanodisk array. Surf. Sci. **601**, 4712–4721 (2007)
112. Chelaru, L.I., Horn-von Hoegen, M., Thien, D., Meyer zu Heringdorf, F.-J.: Fringe fields in nonlinear photoemission microscopy. Phys. Rev. B **73**, 115416, 5 pages (2006)
113. Chelaru, L.I., Meyer zu Heringdorf, F.-J.: In situ monitoring of surface plasmons in single-crystalline Ag-nanowires. Surf. Sci. **601**, 4541–4545 (2007)
114. Meyer zu Heringdorf, F.-J., Chelaru, L.I., Möllenbeck, S., Thien, D., Horn-von Hoegen, M.: Femtosecond photoemission microscopy. Surf. Sci. **601**, 4700–4705 (2007)
115. Schmidt, Th., Slezak, J., Heun, S., Diaz, J., Blyth, R.R., Delaunay, R., Cocco, D., Prince, K. C., Bauer, E., Coreno, M.: Optical layout of a beamline for photoemission microscopy. J. Synchrotron Rad. **6**, 957–963 (1999)
116. Kirschbaum, P., Buckanie, N.M., Meyer zu Heringdorf, F.-J.: Impact of C60 adsorption on surface plasmon polaritons on self-assembled Ag(111) islands on Si(111). Plasmonics **7**, 229–233 (2012)
117. Buckanie, N.M., Kirschbaum, P., Sindermann, S., Meyer zu Heringdorf, F.-J.: Interaction of light and surface plasmon polaritons in Ag Islands studied by nonlinear photoemission microscopy. Ultramicroscopy **130**, 49–53 (2013)
118. Gloskovskii, A., Valdaitsev, D., Nepijko, S.A., Schönhense, G., Rethfeld, B.: Coexisting electron emission mechanisms in small metal particles observed in fs-laser excited PEEM. Surf. Sci. **601**, 4706–4713 (2007)
119. Gloskovskii, A., Valdaitsev, D.A., Cinchetti, M., Nepijko, S.A., Lange, J., Aeschlimann, M., Bauer, M., Klimenkov, M., Viduta, L.V., Tomchuk, P.M., Schönhense, G.: Electron emission from films of Ag and Au nanoparticles excited by a femtosecond pump-probe laser. Phys. Rev. B **77**, 195427, 11 pages (2008)
120. Munzinger, M., Wiemann, C., Rohmer, M., Guo, L., Aeschlimann, M., Bauer, M.: The lateral photoemission distribution from a defined cluster/substrate system as probed by photoemission electron microscopy. New J. Phys. **7**, 68, 17 pages (2005)
121. Rohmer, M., Wiemann, C., Munzinger, M., Guo, L., Aeschlimann, M., Bauer, M.: Local correlation of photoemission electron microscopy and STM at a defined cluster substrate system. Appl. Phys. A **82**, 87–93 (2006)
122. Georgiev, N., Martinotti, D., Ernst, H.-J.: Enhanced photoemission from localized inhomogeneities on Cu(001) characterized by laser assisted photoemission electron microscopy and low-energy electron microscopy. Phys. Rev. B **75**, 085430, 6 pages (2007)
123. Peppernick, S.J., Joly, A.G., Beck, K.M., Hess, W.P.: Plasmonic field enhancement of individual nanoparticles by correlated scanning and photoemission electron microscopy. J. Chem. Phys. **134**, 034507, 7 pages (2011)
124. Douillard, L., Charra, F., Fiorini, C., Adam, P.M., Bachelot, R., Kostcheev, S., Lerondel, G., Lamy de la Chapelle, M., Royer, P.: Optical properties of metal nanoparticles as probed by photoemission electron microscopy. J. Appl. Phys. **101**, 083518, 5 pages (2007)
125. Douillard, L., Charra, F., Korczak, Z., Bachelot, R., Kostcheev, S., Lerondel, G., Adam, P.-M., Royer, P.: Short range plasmon resonators probed by photoemission electron microscopy. Nano Lett. **8**, 935–940 (2008)
126. Hrelescu, C., Sau, T.K., Rogach, A.L., Jäckel, F., Laurent, G., Douillard, L., Charra, F.: Selective excitation of individual plasmonic hotspots at the tips of single gold nanostars. Nano Lett. **11**, 402–407 (2011)

127. Awada, C., Barbillon, G., Charra, F., Douillard, L., Greffet, J.-J.: Experimental study of hot spots in gold/glass nanocomposite films by photoemission electron microscopy. Phys. Rev. B **85**, 045438, 6 pages (2012)
128. Könenkamp, R., Word, R.C., Fitzgerald, J., Nadarajah, A., Saliba, S.: Controlled spatial switching and routing of surface plasmons in designed single-crystalline gold nanostructures. Appl. Phys. Lett. **101**, 141114, 4 pages (2012)
129. Word, R.C., Dornan, T., Könenkamp, R.: Photoemission from localized surface plasmons in fractal metal nanostructures. Appl. Phys. Lett. **96**, 251110, 3 pages (2010)
130. Word, R.C., Fitzgerald, J., Könenkamp, R.: Photoelectron emission control with polarized light in plasmonic metal random structures. Appl. Phys. Lett. **99**, 041106, 3 pages (2011)
131. Word, R.C., Fitzgerald, J.P.S., Könenkamp, R.: Electron emission in the near-field of surface plasmons. Surf. Sci. **607**, 148–152 (2013)
132. Schertz, F., Schmelzeisen, M., Mohammadi, R., Kreiter, M., Elmers, H.-J., Schönhense, G.: Near field of strongly coupled plasmons: uncovering dark modes. Nano Lett. **12**, 1885–1890 (2012)
133. Schertz, F., Schmelzeisen, M., Kreiter, M., Elmers, H.-J., Schönhense, G.: Field emission of electrons generated by the near field of strongly coupled plasmons. Phys. Rev. Lett. **108**, 237602, 5 pages (2012)
134. Hutter, E., Fendler, J.H., Roy, D.: Surface plasmon resonance studies of gold and silver nanoparticles linked to gold and silver substrates by 2-aminoethanethiol and 1,6-hexanedithiol. J. Phys. Chem. B **105**, 11159–11168 (2001)
135. Brixner, T., Gerber, G.: Femtosecond polarization pulse shaping. Opt. Lett. **26**, 557–559 (2001)
136. Brixner, T., Krampert, G., Niklaus, P., Gerber, G.: Generation and characterization of polarization-shaped femtosecond laser pulses. Appl. Phys. B **74**, S133–S144 (2002)
137. Aeschlimann, M., Bauer, M., Bayer, D., Brixner, T., Garcıa de Abajo, F.J., Pfeiffer, W., Rohmer, M., Spindler, C., Steeb, F.: Adaptive subwavelength control of nano-optical fields. Nature **446**, 301–304 (2007)
138. Aeschlimann, M., Bauer, M., Bayer, D., Brixner, T., Cunovic, S., Dimler, F., Fischer, A., Pfeifferd, W., Rohmer, M., Schneider, C., Steeb, F., Strüber, C., Voronine, D.V.: Spatiotemporal control of nanooptical excitations. Proc. Natl. Acad. Sci. U. S. A. **107**, 5329–5333 (2010)
139. Melchior, P., Bayer, D., Schneider, C., Fischer, A., Rohmer, M., Pfeiffer, W., Aeschlimann, M.: Optical near-field interference in the excitation of a bowtie nanoantenna. Phys. Rev. B **83**, 235407, 7 pages (2011)
140. Aeschlimann, M., Bauer, M., Bayer, D., Brixner, T., Cunovic, S., Fischer, A., Melchior, P., Pfeiffer, W., Rohmer, M., Schneider, C., Strüber, C., Tuchscherer, P., Voronine, D.V.: Optimal open-loop near-field control of plasmonic nanostructures. New J. Phys. **14**, 033030, 10 pages (2012)
141. Aeschlimann, M., Brixner, T., Cunovic, S., Fischer, A., Melchior, P., Pfeiffer, W., Rohmer, M., Schneider, C., Strüber, C., Tuchscherer, P., Voronine, D.V.: Nano-optical control of hot-spot field superenhancement on a corrugated silver surface. IEEE J. Sel. Top. Quant. Electron. **18**, 275–282 (2012)
142. Tuchscherer, P., Rewitz, C., Voronine, D.V., Garcıa de Abajo, F.J., Pfeiffer, W., Brixner, T.: Analytic coherent control of plasmon propagation in nanostructures. Opt. Express **17**, 14235–14259 (2009)
143. Brixner, T., García de Abajo, F.J., Schneider, J., Spindler, C., Pfeiffer, W.: Ultrafast adaptive optical near-field control. Phys. Rev. B **73**, 125437, 11 pages (2006)
144. Aeschlimann, M., Brixner, T., Fischer, A., Kramer, C., Melchior, P., Pfeiffer, W., Schneider, C., Strüber, C., Tuchscherer, P., Voronine, D.V.: Coherent two-dimensional nanoscopy. Science **333**, 1723–1725 (2011)
145. Stockman, M., Kling, M.F., Kleineberg, U., Krausz, F.: Attosecond nanoplasmonic-field microscope. Nat. Photon. **1**, 539–544 (2007)
146. Buckanie, N.M., Göhre, J., Zhou, P., von der Linde, D., Horn-von Hoegen, M., Meyer zu Heringdorf, F.-J.: Space charge effects in photoemission electron microscopy using amplified femtosecond laser pulses. J. Phys. Condens. Matter **21**, 314003, 7 pages (2009)

147. Douillard, L., Charra, F.: High-resolution mapping of plasmonic modes: photoemission and scanning tunnelling luminescence microscopies. J. Phys. D. Appl. Phys. **44**, 464002, 13 pages (2011)
148. Douillard, L., Charra, F.: Photoemission electron microscopy, a tool for plasmonics. J. Electr. Spectrosc. Relat. Phenom. **189**, 24–29 (2013)
149. Halas, N.J., Lal, S., Chang, W.-S., Link, S., Nordlander, P.: Plasmons in strongly coupled metallic nanostructures. Chem. Rev. **111**, 3913–3961 (2011)
150. Stockman, M.I.: Nanoplasmonics: past, present, and glimpse into future. Opt. Express **19**, 22029–22106 (2011)
151. Pelton, M., Bryant, G.: Introduction to Metal-Nanoparticle Plasmonics. Willey, Hoboken, NJ (2013)
152. Shvets, G., Tsukerman, I. (eds.): Plasmonics and Plasmonic Metamaterials: Analysis and Applications. World Scientific Publ, Singapore (2012)
153. Enoch, S., Bonod, N. (eds.): Plasmonics: From Basics to Advanced Topics. Springer, Berlin (2012)
154. De Stasio, G., Frazera, B.H., Gilbert, B., Richter, K.L., Valley, J.W.: Compensation of charging in X-PEEM: a successful test on mineral inclusions in 4.4 Ga old zircon. Ultramicroscopy **98**, 57–62 (2003)
155. Vantelon, D., Belkhou, R., Bihannic, I., Michot, L.J., Montargès-Pelletier, E., Robert, J.L.: An XPEEM study of structural cation distribution in swelling clays. I. Synthetic trioctahedral smectites. Phys. Chem. Mineral. **36**, 593–602 (2009)
156. Acres, R.G., Harmer, S.L., Beattie, D.A.: Synchrotron PEEM and ToF-SIMS study of oxidized heterogeneous pentlandite, pyrrhotite and chalcopyrite. J. Synchrotron Radiat. **17**, 606–615 (2010)
157. Chandra, A.P., Gerson, A.R.: The mechanisms of pyrite oxidation and leaching: a fundamental perspective. Surf. Sci. Rep. **65**, 293–315 (2010)
158. Marsh, S.P., Imam, M.A., Rath, B.B., Pande, C.S.: On the kinetics of shrinking grains. Acta Metall. Mater. **41**, 297–304 (1993)
159. Xiong, G., Joly, A.G., Beck, K.M., Hess, W.P., Cai, M., Langford, S.C., Dickinson, J.T.: In situ photoelectron emission microscopy of a thermally induced martensitic transformation in a CuZnAl shape memory alloy. Appl. Phys. Lett. **88**, 091910, 3 pages (2006)
160. Cai, M., Langford, S.C., Dickinson, J.T., Xiong, G., Droubay, T.C., Joly, A.G., Beck, K.M., Hess, W.P.: An in situ study of the martensitic transformation in shape memory alloys using photoemission electron microscopy. J. Nucl. Mater. **361**, 306–312 (2007)
161. Dobbins, T., Abrecht, M., Uprety, Y., Moore, K.: An x-ray photoemission electron microscopy study of the formation of Ti-Al phases in 4 mol% $TiCl_3$ catalyzed $NaAlH_4$ during high energy ball milling. Nanotechnology **20**, 204014, 9 pages (2009)
162. Montei, E.L., Kordesch, M.E.: Detection of tribochemical reactions using photoelectron emission microscopy. J. Vac. Sci. Technol. A **14**, 1352–1356 (1996)
163. Montei, E.L., Kordesch, M.E.: High resolution electron energy loss spectroscopy and photoelectron emission microscopy study of Fomblin Y on molybdenum surfaces. J. Vac. Sci. Technol. A **15**, 1173–1178 (1997)
164. Anders, S., Stammler, T., Bhatia, C.S., Stöhr, J., Fong, W., Chen, C.-Y., Bogy, D.B.: Study of hard disc and slider surfaces using x-ray photoemission elecrron microscopy and near edge x-ray absorption fine structure spectroscopy. Proc. Mat. Res. Soc. Symp. **517**, 415–420 (1998). Materials Research Society
165. Anders, S., Stammler, T., Fong, W., Bogy, D.B., Bhatia, C.S., Stöhr, J.: Investigation of slider surfaces after wear using photoemission electron microscopy. J. Vac. Sci. Technol. A **17**, 2731–2736 (1999)
166. Anders, S., Stammler, T., Fong, W., Chen, C.-Y., Bogy, D.B., Bhatia, C.S., Stöhr, J.: Study of tribochemical processes on hard disks using photoemission electron microscopy. J. Tribol. **121**, 961–967 (1999)
167. Pereira, G., Lachenwitzer, A., Munoz-Paniagua, D., Kasrai, M., Norton, P.R., Abrecht, M., Gilbert, P.U.P.A.: The role of the cation in antiwear films formed from ZDDP on 52100 steel. Tribol. Lett. **23**, 109–119 (2006)
168. Wang, C., Kordesch, M.: The morphology of carbon films and surfaces studied by photoemission electron microscopy. Ultramicroscopy **36**, 154–163 (1991)

169. Garcia, A., Wang, C., Kordesch, M.E.: Controlled deposition and lateral growth of an ordered monolayer of carbon on Mo(100) observed in situ. Appl. Phys. Lett. **61**, 2984–2986 (1992)
170. Garcia, A., Kordesch, M.E.: Surface reaction-diffusion fronts observed with photoelectron emission microscopy during carbon deposition on Mo(310). J. Vac. Sci. Technol. A **13**, 1396–1399 (1995)
171. Wang, C., Shovlin, J.D., Kordesch, M.E.: Real-time, in situ photoelectron emission microscopy observation of CVD diamond oxidation and dissolution on molybdenum. Diam. Relat. Mater. **3**, 1066–1071 (1994)
172. Ziethen, C., Schmidt, O., Marx, G.K.L., Schönhense, G., Frömter, R., Gilles, J., Kirschner, J., Schneider, C.M., Gröning, O.: Orbital mapping of carbon thin films by XANES-spectromicroscopy. J. Electron Spectrosc. Relat. Phenom. **107**, 261–271 (2000)
173. Ziethen, C., Wegelin, F., Schöhense, G., Ohr, R., Neuhäuser, M., Hilgers, H.: Soft X-ray photoelectron microscopy used for the characterization of diamond, a-C and CN_x, thin films. Diam. Relat. Mater. **11**, 1068–1073 (2002)
174. Kono, S., Saito, T., Kang, S.H., Jung, W.Y., Kim, B.Y., Kawata, H., Goto, T., Kakefud, Y., Yeom, H.W.: Band diagram for chemical vapor deposition diamond surface conductive layer: presence of downward band bending due to shallow acceptors. Surf. Sci. **604**, 1148–1165 (2010)
175. Shovlin, J.D., Kordesch, M.E.: Electron emission from chemical vapor deposited diamond and dielectric breakdown. Appl. Phys. Lett. **65**, 863–865 (1994)
176. Shovlin, J.D., Kordesch, M.E., Dunham, D., Tonner, B.P., Engel, W.: Synchrotron radiation photoelectron emission microscopy of chemical-vapor-deposited diamond electron emitters. J. Vac. Sci. Technol. A **13**, 1111–1115 (1995)
177. Cui, J.B., Ristein, J., Stammler, M., Janischowsky, K., Kleber, G., Ley, L.: Hydrogen termination and electron emission from CVD diamond surfaces: a combined secondary electron emission, photoelectron emission microscopy, photoelectron yield, and field emission study. Diam. Relat. Mater. **9**, 1143–1147 (2000)
178. Köck, F.A.M., Garguilo, J.M., Nemanich, R.J.: Direct correlation of surface morphology with electron emission sites for intrinsic nanocrystalline diamond films. Diam. Relat. Mater. **13**, 1022–1025 (2004)
179. Koeck, F.A.M., Garguilo, J.M., Nemanich, R.J.: On the thermionic emission from nitrogen-doped diamond films with respect to energy conversion. Diam. Relat. Mater. **13**, 2052–2055 (2004)
180. Garguilo, J.M., Koeck, F.A.M., Nemanich, R.J., Xiao, X.C., Carlisle, J.A., Auciello, O.: Thermionic field emission from nanocrystalline diamond-coated silicon tip arrays. Phys. Rev. B **72**, 165404, 6 pages (2005)
181. Koeck, F.A.M., Nemanich, R.J.: Sulfur doped nanocrystalline diamond films as field enhancement based termionic emitters and their role in energy conversion. Diam. Relat. Mater. **14**, 2051–2054 (2005)
182. Köck, F.A.M., Garguilo, J.M., Nemanich, R.J.: Field enhanced thermionic electron emission from sulfur doped nanocrystalline diamond films. Diam. Relat. Mater. **14**, 704–708 (2005)
183. Kataoka, M., Zhu, C., Koeck, F.A.M., Nemanich, R.J.: Thermionic electron emission from nitrogen-doped homoepitaxial diamond. Diam. Relat. Mater. **19**, 110–113 (2010)
184. Koeck, F.A.M., Obraztsov, A.N., Nemanich, R.J.: Electron emission microscopy of nano-crystal graphitic films as high current density electron sources. Diam. Relat. Mater. **15**, 875–879 (2006)
185. Shang, N., Papakonstantinou, P., Wang, P., Zakharov, A., Palnitkar, U., Lin, I.-N., Chu, M., Stamboulis, A.: Self-assembled growth, microstructure, and field-emission high-performance of ultrathin diamond nanorods. ACS Nano **3**, 1032–1038 (2009)
186. Zeman, M.C., Fulton, C.C., Lucovsky, G., Nemanich, R.J., Yang, W.-C.: Thermal stability of TiO_2, ZrO_2, or HfO_2 on Si(100) by photoelectron emission microscopy. J. Appl. Phys. **99**, 023519, 6 pages (2006)
187. Döscher, H., Lilienkamp, G., Iskra, P., Kazempoor, M., Daum, W.: Thermal stability of thin ZrO_2 films prepared by a sol-gel process on Si(001) substrates. J. Vac. Sci. Technol. B **28**, C5B5–C5B11 (2010)

188. Hoffmann, P., Mikalo, R.P., Schmeisser, D.: PEEM - a spectromicroscopic tool for mc-Si surface evaluation. Sol. Energ. Mat. Sol. C **72**, 517–523 (2002)
189. Mikkelsen, A., Hilner, E., Andersen, J.N., Ghatnekar-Nilsson, S., Montelius, L., Zakharov, A. A.: Low temperature Ga surface diffusion from focused ion beam milled grooves. Nanotechnology **20**, 325304, 7 pages (2009)
190. Batabyal, R., Patra, S., Roy, A., Roy, S., Bischoff, L., Dev, B.N.: Estimation of diffusion coefficient by photoemission electron microscopy in ion-implanted nanostructures. Appl. Surf. Sci. **256**, 536–540 (2009)
191. Batabyal, R., Mahato, J.C., Roy, A., Roy, S., Bischoff, L., Dev, B.N.: Lateral straggling and its influence on lateral diffusion in implantation with a focused ion beam. Nucl. Instrum. Meth. B **269**, 856–860 (2011)
192. Driesel, W., Bethge, H.: Investigation of implantation structures in semiconductors using a photoelectron-emission microscope. In: Hennig, K. (ed.) Proc. Conf. Energy Pulse Modification of Semiconductors and Related Materials, vol. 1, pp. 595–599. Akademie der Wissenschaften der DDR, Dresden (1984)
193. Giesen, M., Phaneuf, R.J., Williams, E.D., Einstein, T.L., Ibach, H.: Characterization of p-n junctions and surface-states on silicon devices by photoemission electron microscopy. Appl. Phys. A **64**, 423–430 (1997)
194. Ballarotto, V.W., Siegrist, K., Phaneuf, R.J., Williams, E.D., Mogren, S.: PEEM imaging of dopant contrast in Si(001). Surf. Sci. **461**, L570–L574 (2000)
195. Ballarotto, V.W., Siegrist, K., Phaneuf, R.J., Williams, E.D., Yang, W.-C., Nemanich, R.J.: Photon energy dependence of contrast in photoelectron emission microscopy of Si devices. Appl. Phys. Lett. **78**, 3547–3549 (2001)
196. Ballarotto, V.W., Siegrist, K., Phaneuf, R.J., Williams, E.D.: Model for doping-induced contrast in photoelectron emission microscopy. J. Appl. Phys. **91**, 469–475 (2002)
197. Frank, L., Müllerová, I., Valdaitsev, D.A., Gloskovskii, A., Nepijko, S.A., Elmers, H.-J., Schönhense, G.: The origin of contrast in the imaging of doped areas in silicon by slow electrons. J. Appl. Phys. **100**, 093712, 5 pages (2006)
198. Hovorka, M., Mika, F., Mikulık, P., Frank, L.: Profiling n-type dopants in silicon. Mat. Trans. JIM **51**, 237–242 (2010)
199. Frank, L., Mika, F., Hovorka, M., Valdaitsev, D., Schönhense, G., Müllerova, I.: Dopant contrast in semiconductors as interpretation challenge at imaging by electrons. Mat. Trans. JIM **48**, 936–939 (2007)
200. Walker, C.G.H., Zaggout, F., El-Gomati, M.M.: The role of oxygen in secondary electron contrast in doped semiconductors using low voltage scanning electron microscopy. J. Appl. Phys. **104**, 123713, 6 pages (2008)
201. Fukidome, H., Yoshimura, M., Ueda, K., Guo, F.Z., Kinoshita, T., Kobayashi, K.: Variation of contrast of H/*pn*-Si(100) imaged with different emission electron microscopies. Proc. Conf. ISSS-4: e-J. Surf. Sci. Nanotech. **4**, 539–543 (2006). The Surface Science Society of Japan
202. Barrett, N., Zagonel, L.F., Renault, O., Bailly, A.: Spatially resolved, energy-filtered imaging of core level and valence band photoemission of highly p and n doped silicon patterns. J. Phys. Condens. Matter **21**, 314015, 7 pages (2009)
203. de la Peña, F., Barrett, N., Zagonel, L.F., Walls, M., Renault, O.: Full field chemical imaging of buried native sub-oxide layers on doped silicon patterns. Surf. Sci. **604**, 1628–1636 (2010)
204. Lavayssière, M., Escher, M., Renault, O., Mariolle, D., Barrett, N.: Electrical and physical topography in energy-filtered photoelectron emission microscopy of two-dimensional silicon pn junctions. J. Electron Spectrosc. Relat. Phenom. **186**, 30–38 (2013)
205. Zagonel, L.F., Barrett, N., Renault, O., Bailly, A., Bäurer, M., Hoffmann, M., Shih, S.-J., Cockayne, D.: Orientation-dependent surface composition of in situ annealed strontium titanate. Surf. Interface Anal. **40**, 1709–1712 (2008)
206. Zagonel, L.F., Bäurer, M., Bailly, A., Renault, O., Hoffmann, M., Shih, S.-J., Cockayne, D., Barrett, N.: Orientation-dependent work function of in situ annealed strontium titanate. J. Phys. Condens. Matter **21**, 314013, 5 pages (2009)

207. Wei, H., Beuermann, L., Helmbold, J., Borchardt, G., Kempter, V., Lilienkamp, G., Maus-Friedrichs, W.: Study of SrO segregation on $SrTiO_3$(100) surfaces. J. Eur. Ceram. Soc. **21**, 1677–1680 (2001)
208. Wei, H., Maus-Friedrichs, W., Lilienkamp, G., Kempter, V., Helmbold, J., Gömann, K., Borchard, G.: Surface structure of oxygen annealed donor doped $SrTiO_3$(100) single crystals studied with spectroscopic electron microscopy. J. Electroceram. **8**, 221–228 (2002)
209. Strachan, J.P., Yang, J.J., Münstermann, R., Scholl, A., Medeiros-Ribeiro, G., Stewart, D.R., Williams, R.S.: Structural and chemical characterization of TiO_2 memristive devices by spatially-resolved NEXAFS. Nanotechnology **20**, 485701, 6 pages (2009)
210. Müller, K., Burkov, Y., Schmeißer, D.: Photo-emission-electron-microscopy for characterization of an operating organic electronic device. Thin Solid Films **495**, 219–223 (2006)
211. Sadewasser, S., Bremsteller, W., Plake, T., Kaufmann, C.A., Pettenkofer, C.: Microscopic investigation of the CdS buffer layer growth on Cu(In, Ga)Se2 absorbers. J. Vac. Sci. Technol. B **26**, 901–903 (2008)
212. Pettenkofer, C., Hofmann, A., Bremsteller, W., Lehmann, C., Kelleter, F.: Photoelectron spectromicroscopy at chalcopyrite films. Ultramicroscopy **119**, 102–105 (2012)
213. Vaughn, J.M., Kordesch, M.: Photoelectron- and thermionic-emission microscopy of barium/scandium thin films on tungsten. Proc. Mat. Res. Soc. Symp. **1088**, W03–W09 (2008). Materials Research Society
214. Kordesch, M.E., Vaughn, J.M., Wan, C., Jamison, K.D.: Model scandate cathodes investigated by thermionic-emission microscopy. J. Vac. Sci. Technol. B **29**, 04E102, 7 pages (2011)
215. Vaughn, J.M., Wan, C., Jamison, K.D., Kordesch, M.E.: Thermionic electron emission microscopy of metal-oxide multilayers on tungsten. IBM J. Res. Develop. **55**, 14, 6 pages (2011)
216. Wan, C., Vaughn, J.M., Sadowski, J.T., Kordesch, M.E.: Scandium oxide coated polycrystalline tungsten studied using emission microscopy and photoelectron spectroscopy. Ultramicroscopy **119**, 106–110 (2012)
217. Wan, C., Kordesch, M.E.: Tungstate formation in a model scandate thermionic cathode. J. Vac. Sci. Technol. B **31**, 011210, 11 pages (2013)
218. Shimakura, T., Hasegawa, M., Suzuki, H., Ohta, H.: Inspection technique for nanoimprint template with mirror electron microscopy. In: Montgomery, M.W., Maurer, W. (eds.) Proc. of SPIE, Photomask Technology, 2010, vol. 7823, pp. Q-1–Q-6. SPIE, Bellingham, WA (2010)
219. Hasegawa, M., Ohno, T.: Non-destructive observation of in-grown stacking faults in 4H-SiC epitaxial layer using mirror electron microscope. J. Appl. Phys. **110**, 073507, 6 pages (2011)
220. Neuhäusler, U., Oelsner, A., Slieh, J., Brzeska, M., Wonisch, A., Westerwalbesloh, T., Brückl, H., Schicketanz, M., Weber, N., Escher, M., Merkel, M., Schönhense, G., Kleineberg, U., Heinzmann, U.: High-resolution actinic defect inspection for extreme ultraviolet lithography multilayer mask blanks by photoemission electron microscopy. Appl. Phys. Lett. **88**, 053113, 3 pages (2006)
221. Neuhäusler, U., Lin, J., Oelsner, A., Schicketanz, M., Valdaitsev, D., Slieh, J., Weber, N., Brzeska, M., Wonisch, A., Westerwalbesloh, T., Brechling, A., Escher, M., Merkel, M., Schönhense, G., Kleineberg, U., Heinzmann, U.: A new approach for actinic defect inspection of EUVL multilayer mask blanks: standing wave photoemission electron microscopy. Microelectron. Eng. **83**, 680–683 (2006)
222. Lin, J., Neuhaeusler, U., Slieh, J., Brechling, A., Kleineberg, U., Heinzmann, U., Oelsner, A., Valdaitsev, D., Schoenhense, G., Weber, N., Escher, M., Merkel, M.: Actinic extreme ultraviolet lithography mask blank defect inspection by photoemission electron microscopy. J. Vac. Sci. Technol. B **24**, 2631–2635 (2006)
223. Kleineberg, U., Lin, J., Neuhaeusler, U., Slieh, J., Heinzmann, U., Weber, N., Escher, M., Merkel, M., Oelsner, A., Valsaitsev, D., Schoenhense, G.: Actinic EUVL mask blank defect inspection by EUV photoelectron microscopy. In: Lercel, M.J. (ed.) Emerging Lithographic Technologies X. Proc. of SPIE, vol. 6151, 615120. SPIE, Bellingham, WA, 11 pages (2006)

224. Lin, J., Neuhaeusler, U., Slieh, J., Brechling, A., Heinzmann, U., Weber, N., Escher, M., Merkel, M., Oelsner, A., Valdaitsev, D., Schoenhense, G., Quesnel, E., Kleineberg, U.: Actinic inspection of EUVL mask blank defects by photoemission electron microscopy: effect of inspection wavelength variation. Microelectron. Eng. **84**, 1011–1014 (2007)
225. Maul, J., Lin, J., Oelsner, A., Valdaitsev, D., Weber, N., Escher, M., Merkel, M., Seitz, H., Heinzmann, U., Kleineberg, U., Schönhense, G.: Phase defect inspection of multilayer masks for 13.5 nm optical lithography using PEEM in a standing wave mode. Surf. Sci. **601**, 4758–4763 (2007)
226. Lin, J., Weber, N., Maul, J., Hendel, S., Rott, K., Merkel, M., Schoenhense, G., Kleineberg, U.: At-wavelength inspection of sub-40 nm defects in extreme ultraviolet lithography mask blank by photoemission electron microscopy. Opt. Lett. **32**, 1875–1877 (2007)
227. Lin, J., Weber, N., Escher, M., Maul, J., Han, H.-S., Merkel, M., Wurm, S., Schönhense, G., Kleineberg, U.: Three-dimensional characterization of extreme ultraviolet mask blank defects by interference contrast photoemission electron microscopy. Opt. Express **16**, 15343–15352 (2008)
228. Goldberg, K.A., Mochi, I., Benk, M., Allezy, A.P., Dickinson, M.R., Cork, C.W., Zehm, D., Macdougall, J.B., Anderson, E., Salmassi, F., Chao, W.L., Vytla, V.K., Gullikson, E.M., DePonte, J.C., Jones, M.S.G., Van Camp, D., Gamsby, J.F., Ghiorso, W.B., Huang, H., Cork, W., Martin, E., Van Every, E., Acome, E., Milanovic, V., Delano, R., Naulleau, P.P., Rekawa, S.B.: Commissioning an EUV mask microscope for lithography generations reaching 8 nm. In: Naulleau, P.P. (ed.) Extreme Ultraviolet (EUV) Lithography IV. Proc. of SPIE, vol. 8679, 867919. SPIE, Bellingham, WA, 10 pages (2013)
229. Tchikoulaeva, A., Miyai, H., Suzuki, T., Takehisa, K., Kusunose, H., Yamane, T., Terasawa, T., Watanabe, H., Inoue, S., Mori, I.: EUV actinic blank inspection: from prototype to production. In: Naulleau, P.P. (ed.) Extreme Ultraviolet (EUV) Lithography IV. Proc. of SPIE, vol. 8679, 86790I. SPIE, Bellingham, WA, 6 pages (2013)
230. Grund, S., Engel, W., Teufel, P.: Photoelektronen-Emissionsmikroskop und Immunofluoreszenz. J. Ultrastruct. Res. **50**, 284–288 (1975)
231. Griffith, O.H., Hedberg, K.K., Desloge, D., Rempfer, G.F.: Low-energy electron microscopy (LEEM) and mirror electron microscopy (MEM) of biological specimens: preliminary results with a novel beam separator system. J. Microsc. **168**, 249–258 (1992)
232. Griffith, O.H.: Photoelectron microscopy - applications to biological surfaces. Appl. Surf. Sci. **26**, 265–279 (1986)
233. Birrell, G.B., Hedberg, K.K., Habliston, D.L., Griffith, O.H.: Biological applications of photoelectron imaging: a practical perspective. Ultramicroscopy **36**, 235–251 (1991)
234. De Stasio, G., Koranda, S.F., Tonner, P., Harp, G.R., Merkanti, D., Ciotti, M.T., Margaritondo, G.: X-ray secondary-emission microscopy (XSEM) of neurons. Europhys. Lett. **19**, 655–659 (1992)
235. Gilbert, B., Perfetti, L., Fauchoux, O., Redondo, J., Baudat, P.-A., Andres, R., Neumann, M., Steen, S., Gabel, D., Mercanti, D., Ciotti, M.T., Perfetti, P., Margaritondo, G., De Stasio, G.: Spectromicroscopy of boron in human glioblastomas following administration of $Na_2B_{12}H_{11}SH$. Phys. Rev. E. **62**, 1110–1118 (2000)
236. De Stasio, G., Casalbore, P., Pallini, R., Gilbert, B., Sanita, F., Ciotti, M.T., Rosi, G., Festinesi, A., Larocca, L.M., Rinelli, A., Perret, D., Mogk, D.W., Perfetti, P., Mehta, M.P., Mercanti, D.: Gadolinium in human glioblastoma cells for gadolinium neutron capture therapy. Cancer Res. **61**, 4272–4277 (2001)
237. De Stasio, G., Rajesh, D., Ford, J.M., Daniels, M.J., Erhardt, R.J., Frazer, B.H., Tyliszczak, T., Gilles, M.K., Conhaim, R.L., Howard, S.P., Fowler, J.F., Esteve, F., Mehta, M.P.: Motexafin-gadoliniumtaken up in vitro by at least 90% of glioblastoma cell nuclei. Clin. Cancer Res. **12**, 206–213 (2006)
238. Frazer, B.H., Girasole, M., Wiese, L.M., Franz, T., De Stasio, G.: Spectromicroscope for the photoelectron imaging of nanostructures with x-rays (SPHINX): performance in biology, medicine and geology. Ultramicroscopy **99**, 87–94 (2004)

239. Samokhvalov, A., Garguilo, J., Yang, W.-C., Edwards, G.S., Nemanich, R.J., Simon, J.D.: Photoionization threshold of eumelanosomes determined using uv free electron laser-photoelectron emission microscopy. J. Phys. Chem. B **108**, 16334–16338 (2004)
240. Samokhvalov, A., Hong, L., Liu, Y., Garguilo, J., Nemanich, R.J., Edwards, G.S., Simon, J. D.: Oxidation potentials of human eumelanosomes and pheomelanosomes. Photochem. Photobiol. **81**, 145–148 (2005)
241. Ye, T., Hong, L., Garguilo, J., Pawlak, A., Edwards, G.S., Nemanich, R.J., Sarna, T., Simon, J.D.: Photoionization thresholds of melanins obtained from free electron laser-photoelectron emission microscopy. Femtosecond transient absorption spectroscopy and electron paramagnetic resonance measurements of oxygen photoconsumption. Photochem. Photobiol. **82**, 733–737 (2006)
242. Garguilo, J., Hong, L., Edwards, G.S., Nemanich, R.J., Simon, J.D.: The surface oxidation potential of melanosomes measured by free electron laser-photoelectron emission microscopy. Photochem. Photobiol. **83**, 692–697 (2007)
243. Kong, X., Rowe, J.E., Nemanich, R.J.: Single molecule measurements with photoelectron emission microscopy. J. Vac. Sci. Technol. B **26**, 1461–1465 (2008)
244. Pereiro, J., Sadowski, J.T., Lin, B., Panagopoulos, C., Bozović, I.: Low-energy electron microscope study of tobacco mosaic viruses. J. Nanosci. Lett. **2**, 29, 6 pages (2012)
245. Mankos, M., Shadman, K.: A monochromatic, aberration-corrected, dual-beam low energy electron microscope. Ultramicroscopy **130**, 13–28 (2013)
246. Metzler, R.A., Abrecht, M., Olabisi, R.M., Ariosa, D., Johnson, C.J., Frazer, B.H., Coppersmith, S.N., Gilbert, P.U.P.A.: Architecture of columnar nacre, and implications for its formation mechanism. Phys. Rev. Lett. **98**, 268102, 4 pages (2007)
247. Gilbert, P.U.P.A.: Polarization-dependent imaging contrast (PIC) mapping reveals nanocrystal orientation patterns in carbonate biominerals. J. Electron Spectrosc. Relat. Phenom. **185**, 395–405 (2012)
248. Gilbert, P.U.P.A., Metzler, R.A., Zhou, D., Scholl, A., Doran, A., Young, A., Kunz, M., Tamura, N., Coppersmith, S.N.: Gradual ordering in red abalone nacre. J. Am. Chem. Soc. **130**, 17519–17527 (2008)
249. Lam, R.S.K., Metzler, R.A., Gilbert, P.U.P.A., Beniash, E.: Anisotropy of chemical bonds in collagen molecules studied by x-ray absorption near-edge structure (XANES) spectroscopy. ACS Chem. Biol. **7**, 476–480 (2012)
250. Boyce, C.K., Abrecht, M., Zhou, D., Gilbert, P.U.P.A.: X-ray photoelectron emission spectromicroscopic analysis of arborescent lycopsid cell wall composition and Carboniferous coal ball preservation. Int. J. Coal Geol. **83**, 146–153 (2010)
251. Lok, S.K., Chan, S.K., Wong, G.K.L., Sou, I.K.: MBE-grown Fe nanowires on a ZnS(100) surface. J. Cryst. Growth **311**, 2208–2211 (2009)
252. Bauer, E., Man, K.L., Pavlovska, A., Locatelli, A., Menteş, T.O., Niño, M.A., Altman, M.S.: Fe_3S_4 (greigite) formation by vapor–solid reaction. J. Mater. Chem. A **2**, 1903–1913 (2014)

Chapter 7
Magnetic Imaging

Abstract This chapter covers the study of thin films, nanostructures, and surfaces of bulk ferromagnetic and other domain-forming materials such as antiferromagnetics, ferroelectrics, and multiferroics. Several methods are used in these studies: spin-polarized low energy electron microscopy (SPLEEM), X-ray magnetic circular dichroism photoemission electron microscopy (XMCDPEEM), X-ray magnetic linear dichroism photoemission electron microscopy (XMLDPEEM), UV magnetic circular dichroism photoemission electron microscopy (UVMCDPEEM), spin-resolved ultraviolet photoemission electron microscopy (UVPEEM), and mirror electron microscopy (MEM). In the spirit of this book the emphasis is on imaging. For a detailed discussion of the magnetic aspects the reader is referred to review papers which will be cited in the context of the various applications and to the comprehensive book on magnetism by Stöhr and Siegmann (Stöhr and Siegmann, Magnetism. Springer, Berlin, 2006).

7.1 Introduction

The problems discussed in this chapter are not only of fundamental interest because of the richness of the physical phenomena involved but also important for practical applications in sensing and storing devices, which has been the major driving force for their study. With the exception of ferroelectrics, the spin, magnetic moment, and magnetization distribution can and has been studied with the methods mentioned above as function of material, system geometry, chemical structure, temperature, applied fields, and other parameters. In connection with switching devices the response to field and current pulses has attracted special attention.

The study of the magnetization distribution with cathode lens microscopy has been preceded by another method using slow electrons for imaging, scanning electron microscopy with polarization analysis (SEMPA), which will be briefly discussed in Chap. 8. SPLEEM was stimulated by one of the most active SEMPA researchers,

E. Bauer, *Surface Microscopy with Low Energy Electrons*, 409
DOI 10.1007/978-1-4939-0935-3_7, © Springer Science+Business Media New York 2014

Dan Pierce, and developed in Clausthal in a cooperation of Bauer's group and Poppa with IBM support by replacing the field emission gun in the first LEEM instrument with a spin-polarized electron gun. The first SPLEEM images were presented in 1991 [2]. The first XMCDPEEM images were obtained in 1993 by Stöhr et al. with Tonner's PEEM [3]. Stöhr et al. also did the first XMLDPEEM study using the PEEM2 instrument at ALS in 1999 [4]. In 2000 Marx et al. demonstrated magnetic contrast in UVMLDPEEM with a homebuilt PEEM [5]. Magnetic field imaging with MEM had already been introduced in the early years of cathode lens microscopy as mentioned in Chap. 1. Since these pioneering studies so much work has been done with increasingly improved instruments that only part of it can be included in this book. The reader is referred to some of the numerous review papers for most of the work covered in them. SPLEEM studies prior to 2010 are reviewed in [6–8] while for XMCDPEEM studies many reviews on various aspects of micromagnetism have been published [9–21]. Only a general overview illustrated by examples will be given here with emphasis on more recent work.

Beforehand some comments on the experimental methods have to be made. With the exception of SPLEEM the basics of the various imaging methods have already been discussed briefly in Chap. 2 so that we can limit ourselves here to SPLEEM. In SPLEEM a spin-polarized electron beam, characterized by its polarization vector $\boldsymbol{P}$, is incident normal to the surface of a ferromagnetic system, whose magnetization $\boldsymbol{M}$ determines the quantization direction. $\boldsymbol{M}$ results from the difference between the number of electrons with magnetic moment $\boldsymbol{m}$ parallel and antiparallel to $\boldsymbol{M}$ in the majority and minority bands in Fig. 2.23c, respectively. The arrows in this figure show the direction of the spin, whose direction is opposite to that of the magnetic moment ($\boldsymbol{m} = -g\mu_B \boldsymbol{s}$), a difference which is not taken very seriously in much of the literature. Electron–electron interactions produce the spin-dependent exchange-correlation potential V_{xc}. The incident electron experiences this potential so that electrons with spin parallel and antiparallel to those of the "crystal" electrons experience different interaction potentials $V_{xc}^{\uparrow\uparrow} \equiv V_{xc}^{\uparrow}$ and $V_{xc}^{\uparrow\downarrow} \equiv V_{xc}^{\downarrow}$. Accordingly, electrons with $\boldsymbol{P}$ parallel and antiparallel to $\boldsymbol{M}$ are scattered differently. This leads to different intensities $I_{\uparrow\uparrow} \equiv I_{\uparrow}$, $I_{\uparrow\downarrow} \equiv I_{\downarrow}$ of the backscattered beams with opposite polarization. In monolayers the intensity difference is proportional to the magnetization, otherwise it is an average weighted by the damping caused by inelastic scattering. Normalized to the sum of the intensities, this difference is the so-called exchange scattering asymmetry

$$A = \frac{I_{\uparrow} - I_{\downarrow}}{I_{\uparrow} + I_{\downarrow}}. \tag{7.1}$$

A magnetic image is thus obtained by taking two images and displaying A as illustrated in Fig. 7.1 [22] by one of the first SPLEEM images of a thin film. When $\boldsymbol{P}$ is not parallel to $\boldsymbol{M}$ as is frequently the case, $A \sim \boldsymbol{P} \bullet \boldsymbol{M}$.

The intensities can be calculated in principle with the dynamical LEED theory as shown by Feder [23]. This has been done actually for a number of ferromagnetic

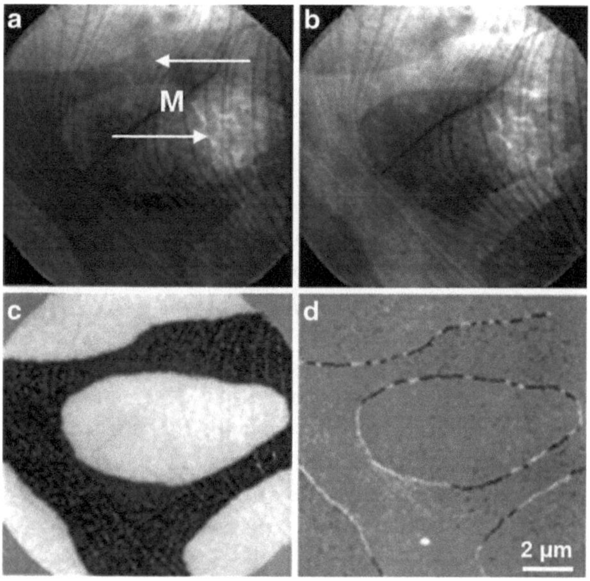

Fig. 7.1 Spin-polarized LEEM image formation, illustrated with a 6 monolayer thick Co film on W(110). (**a**) and (**b**) show images taken with ***P*** parallel and antiparallel to ***M***, (**c**) the difference image between (**a**) and (**b**), (**d**) the difference image between images taken with ***P*** perpendicular to ***M*** in opposite directions. In (**c**) the domains are seen, in (**d**) the domain walls. The "*snake*" in (**a**) and (**b**) is a crack in the channel plate, the *lines* are monatomic steps. Adapted with permission from Ref. [22]. Copyright 1996 Elsevier

systems in connection with spin-polarized LEED (SPLEED) studies (for a review see [24], for more recent work [25]). SPLEED studies are usually done at energies significantly higher than LEEM studies. No calculations have been made which could help in the interpretation of the magnetic contrast in SPLEEM. Nevertheless the basic ideas will be briefly presented. The incident electron is considered to interact with the electron system in the crystal described by the spin-dependent charge densities $\rho^{\uparrow,\downarrow}(\boldsymbol{r})$ via the spin-dependent exchange correlation potential

$$V_{\mathrm{xc}}^{\uparrow,\downarrow}(\boldsymbol{r}) = -\left(2 + \frac{1-\eta^2}{\eta}\ln\frac{|1+\eta|}{|1-\eta|}\right)\left(3\rho^{\uparrow,\downarrow}(\boldsymbol{r})/8\pi\right)^{1/3}, \tag{7.2}$$

where $\eta = k/k_{\mathrm{F}}$ and k, k_{F} are the wave number of the incident electron and the Fermi wave number, respectively. This Slater type local spin-density potential was originally derived for a homogeneous electron gas, which is obviously very different from the strong electron density variations in a crystal. These are later taken into account in various potentials called "Local Spin Density Approximation" which were further improved by taking the gradient of the densities into account ("Generalized Gradient Approximation") Exchange and correlation derive from spin-dependent *pair*-densities. Why the potentials just mentioned, which use *one-particle* densities $\rho^{\uparrow,\downarrow}(\boldsymbol{r})$ can describe exchange and correlation so well has been studied carefully in Ref. [26].

SPLEED calculations use potentials of this kind which describe the ground state of the system with a spin-split band structure $E^{\uparrow,\downarrow}(\boldsymbol{k})$. This does not include the incident electron, which is in an excited state and is taken into account with a largely constant, $\boldsymbol{k}$-independent complex energy correction called self-energy. The imaginary part describes the damping of the excited state due to excitations of electrons from the ground state. When such calculations are not available, which is usually the case, the experimentalist uses the ground state data, for example the ground state band structure $E^{\uparrow,\downarrow}(\boldsymbol{k})$ for the interpretation of the measurements.

The assumptions on which the energy dependence (the η-dependent part in Eq. (7.2)) of the exchange correlation potential is based are inappropriate for a crystal (L. Fritsche, private communication). Here the energy dependence is a consequence of the fact that at low energies the electron does not penetrate much into the core region of the atoms, which is described by the Hartree potential $V_{\mathrm{H}}(\boldsymbol{r})$. It interacts mainly in the interstitial region where $V_{\mathrm{xc}}^{\uparrow,\downarrow}(\boldsymbol{r})$ is not small compared to $V_{\mathrm{H}}(\boldsymbol{r})$ and where the difference $V_{\mathrm{xc}}^{\uparrow}(\boldsymbol{r}) - V_{\mathrm{xc}}^{\downarrow}(\boldsymbol{r})$ is sizeable. This is a consequence of the fact that the densities $\rho^{\uparrow,\downarrow}(\boldsymbol{r})$ of the occupied 3d-type band states have their largest amplitude in the outer regions of the atoms. With increasing energy the electron penetrates more and more into the atomic core where $V_{\mathrm{H}}(\boldsymbol{r})$ increasingly dominates the total potential $V(\boldsymbol{r}) = V_{\mathrm{H}}(\boldsymbol{r}) + V_{\mathrm{xc}}^{\uparrow,\downarrow}(\boldsymbol{r})$ and where the difference $V_{\mathrm{xc}}^{\uparrow}(\boldsymbol{r}) - V_{\mathrm{xc}}^{\downarrow}(\boldsymbol{r})$ practically vanishes. Spin-orbit coupling, which increases with decreasing $\boldsymbol{r}$, has no influence in specular reflection at normal incidence below the threshold energy of excitation of the first diffracted wave. There is presently no theory formulating this simple picture in a rigorous manner so that no quantitative expression for the energy dependence can be given. In any case, the decrease of the sensitivity to the magnetization and the wish to exclude spin-orbit effect contributions to the scattering asymmetry limits SPLEEM to energies typically below 20 eV.

Up to now we have considered only elastic processes. Inelastic processes play an important role in SPLEEM, at least in 3d metals because of the high density of unoccupied states just above the Fermi level and the small energy spacing between unoccupied and occupied states, which allow excitations at the low energies used in SPLEEM. As the number of these states is larger in the minority spin bands than in the majority spin bands (see Fig. 2.23c), energy losses of minority spin electrons are correspondingly more probable. The inelastic mean free path (IMFP) is therefore spin-dependent. In the energy range of SPLEEM there is only one calculation of spin-dependent mean free paths of the incident electrons, based on a realistic model of 3d metals [27]. It shows that an electron, whose spin is parallel to a minority spin in the metal, has a high probability for a transition into an unoccupied minority spin band with ejection of an electron from the majority spin band ("spin flip transition," see inset in Fig. 7.2a [27]). For an electron with spin parallel to the majority spin this probability is smaller because of the unfavorable band occupation for this spin flip process. As a consequence the minority spin electrons have shorter life times $\tau_{\downarrow}$

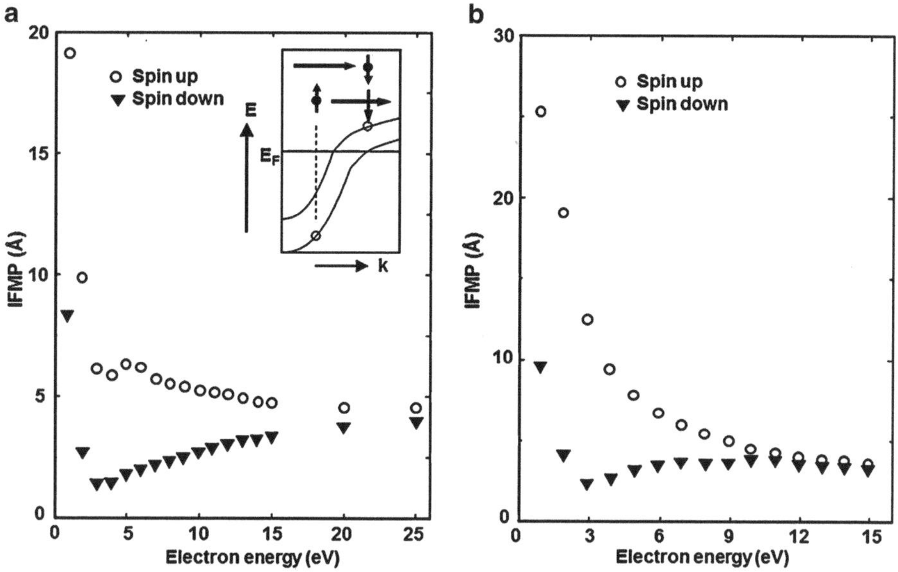

Fig. 7.2 Theoretical spin-dependent inelastic mean free paths above the Fermi level of Fe (**a**) and Ni (**b**). *Open circles*: spin up, *triangles*: spin down. The *inset* in (**a**) shows the dominant loss process in Fe: minority electrons in, majority electrons out. The vacuum level of the (100) surfaces of these metals is at 4.6 eV (Fe) and 5.2 eV (Ni). Modified from Ref. [27]

than the majority spin electrons $\tau_{\uparrow}$, the more so, the larger the difference in the band occupations. The same is true for the spin-dependent IMFPs $l_{\uparrow,\downarrow} = v\tau_{\uparrow,\downarrow}$, where the velocity is determined by the dispersion of the energy bands, $v = \hbar^{-1}\nabla_k E(\boldsymbol{k})$, which is approximately independent of spin. The results of these calculations for Fe and Ni are shown in Fig. 7.2 [27]. $l_{\downarrow}$ is clearly smaller than $l_{\uparrow}$, both values are very small, less so in Ni than in Fe because of the lower number of holes on the minority band. In the SPLEEM energy range the energy dependence is small and $l_{\downarrow} \approx l_{\uparrow}$ already at 20 eV (10 eV) above the vacuum level of Fe (Ni). Figure 7.2 is an illustration of how the IMFP decreases and its spin-dependence increases with increasing number of unoccupied minority d states from Ni to Fe.

Up to now we have assumed that the interaction with the spins in the crystal does not change the direction of the spin of the elastically scattered electrons but only determines the spin-dependent reflected current and its spin-dependent attenuation by inelastic scattering. However even elementary physics tells us that the magnetic moment connected with the spin precesses around the direction of a magnetic field. This is true for the internal field $\boldsymbol{M}$ in the crystal as it is for the magnetic fields used in the spin manipulators discussed in Chap. 3. The correct description of how the direction of the electron spin changes in a magnetic field has been given already 1956 by Tolhoek [28]. The interaction of the polarization $\boldsymbol{P}$ with $\boldsymbol{M}$ causes a precession of $\boldsymbol{P}$ about $\boldsymbol{M}$ and a rotation towards $\boldsymbol{M}$. This was first experimentally demonstrated in 1999 [29] in the transmission of a spin-polarized beam through a Co film and has since been confirmed and studied in more detail in other systems as reviewed in Ref. [30] together with the basic physics. Strong changes have not only been observed in transmission but also in reflection. For example, upon specular reflection of a transverse-polarized beam with ***P*** perpendicular to ***M***, incident at 45° on a 32 monolayer (4.6 nm) thick (100)-oriented Fe film on Ag(100), a complete reversal of the polarization was found at the energy at which a diffracted beam parallel to the surface was excited [31]. Under SPLEEM conditions (normal incidence, normal reflection) such changes will also occur except when ***P*** and ***M*** are perpendicular to the surface and no spin reversal, for example at the boundaries or at defects, occurs. Even if the polarization of the reflected beam were different from that of the incident beam it would not be seen in SPLEEM, which detects only the intensity but not the polarization. Furthermore the interaction length in the magnetic field is very short so that the direction of ***P*** of the reflected beam differs very little from that of the incident beam except in special cases such as the one described in [31]. Therefore the precession of ***P*** about ***M*** and a rotation towards ***M*** is neglected in SPLEEM. However in spin current phenomena at high current densities in magnetic materials, in particular in the spin transfer torque, it plays an important role. For a review of this field see Ref. [32].

Finally some comments on the interpretation of SPLEEM results should be made. SPLEEM, which uses the (00) LEED beam suffers from the same limitations as LEED does at the very low energies used in LEEM (see Chap. 2). The shape of the surface potential and the self-energy of the excited state (the incident electron) are in addition spin-dependent. In particular the imaginary part of the self-energy and its energy dependence pose problems. All these problems are taken care of in the theory with model assumptions, partially adjusted to experiment. In addition the magnetic moments vary in the crystal, generally as a function of distance from the surface, not only in magnitude, but in noncollinear systems also in direction. For all these reasons quantitative SPLEEM image analysis has to rely on simple models such as optical analogues.

7.2 Ferromagnetic Films

In this section we discuss studies of thin magnetic systems with large lateral dimensions so that their magnetic properties are not influenced by size and shape effects. We start with single layers, i.e., layers consisting of only one magnetic material, including alloys and the influence of nonmagnetic overlayers. Then bilayers consisting of two different magnetic materials will be discussed. The next step in complication are trilayers or sandwiches consisting of nonmagnetic and magnetic layers between two magnetic layers, which then leads to multilayers consisting of alternating nonmagnetic and magnetic layers. Finally some compound layers will be briefly discussed. All these aspects have been studied with many other methods as reviewed in [33].

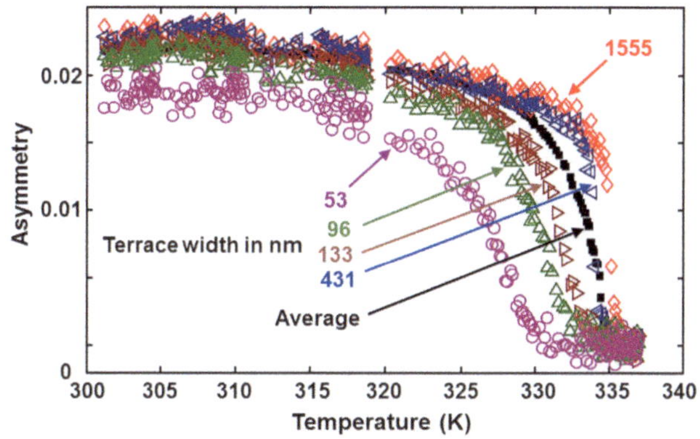

Fig. 7.3 Exchange asymmetry of a Fe monolayer measured at 3.5 eV versus temperature from sample regions with different average terrace widths. The average asymmetry curve taken from the whole image is also shown. The tails above T_C increase with decreasing terrace width and are, therefore, attributed to finite size rounding. The Fe layer was grown on a Au bilayer and covered with a Au monolayer in order to increase T_C above room temperature. Adapted with permission from Ref. [36]. Copyright 2008 by the American Physical Society

7.2.1 *Single Layers*

In single layers, ranging from the monolayer to films many monolayers thick, several aspects of magnetism have been studied with SPLEEM and XMCDPEEM: magnetic anisotropy, domain and domain wall structure, spin reorientation transitions, and magnetic phase transitions. Most of the quantitative information comes, however, from various laterally averaging magnetic measurement techniques cited in the review of magnetism in ultrathin film structures by Vaz [34]. For monolayers, in which the exchange asymmetry A_{ex} is proportional to the magnetization, also SPLEED gives quantitative information [35]. However unless the terrace width of the substrate is large as in the studies cited in [35], lateral resolution is needed in order to avoid misinterpretation. This is illustrated in Fig. 7.3 [36], which shows the temperature dependence of P_0A_{ex} for various terrace widths (P_0 polarization of the incident beam). Determining critical behavior from narrow terraces without taking finite size effects into account or averaging over the whole surface would give completely wrong results. SPLEEM allows selecting wide terraces and thus avoids the shortcomings of lateral averaging studies. It also shows the magnetization fluctuations in the critical region.

In the study of the magnetic microstructure, which will be discussed in the following, SPLEEM and XMCDPEEM are complementary: SPLEEM allows to measure all three components of the magnetization $\boldsymbol{M}$ with high image acquisition rate (subseconds) but contains no chemical information and relies on strong elastic backscattering, which usually requires single-crystalline or at least highly oriented samples. It gives also structural information via LEED. XMCDPEEM is chemically sensitive but slow (many seconds) and contains no structural information.

Its strength is in systems of two or more magnetic components because it allows determination of the magnetization of the various components. The intensity I_{XMCD} of the emitted electrons is proportional to $\boldsymbol{L}_P \cdot \boldsymbol{M}$ (Eq. 2.20), where $\boldsymbol{L}_P$ is the angular momentum of the incident photon beam. In most instruments the glancing angle of incidence of the photon beam is less than 30° so that XMCDPEEM is sampling mainly the in-plane magnetization $\boldsymbol{M}_{\|}$. The out-of-plane component $\boldsymbol{M}_{\perp}$ can be separated from $\boldsymbol{M}_{\|}$ by sample rotation so that $\boldsymbol{M}_{\|}$ is perpendicular to $\boldsymbol{L}_P$.

The film/substrate systems most studied with various imaging and nonimaging methods are Co and Fe on W(110). Here only a few examples of older SPLEEM studies will be mentioned and work published after the review [8] briefly discussed. Already the first SPLEEM studies of Co films on W(110) as a function of thickness showed that the magnetization direction tilted with increasing thickness stepwise into the film plane without significant change of the domain size and shape. The tilting into the plane was attributed to the thickness-dependent competition between shape anisotropy and perpendicular anisotropy [37]. When Au was deposited before Co the magnetization was completely in-plane up to 4 Au monolayers at all Co coverages. At higher Au coverages the magnetization was perpendicular to the film up to a Au coverage-dependent Co thickness and then turned in-plane, starting with domain wall broadening, continuing with magnetization switching and ending in Neél wall formation. This study, in which the potential of fast image acquisition in SPLEEM was fully used, provided detailed insight into one of the processes of how the transition of one spin orientation to another one occurs [38]. Previous studies of this so-called spin reorientation transition (SRT) with SEMPA and later studies with SPLEEM, XMCDPEEM, and SEMPA, to which references can be found in [6–8, 34, 39], have shown several paths via which SRTs can occur. FeCo alloy films on a (111)-oriented Au film on W(110) are an example. In these films the initially nearly perpendicular magnetization tilts continuously into the plane, with steadily decreasing size of stripe-shaped domains, until at about 2.7 monolayers suddenly large, predominantly in-plane magnetized domains nucleate [40]. A comparison between these two examples, Co and Fe, can be found in [7].

In addition to these out-of-plane to in-plane SRTs there are also frequently in-plane SRTs with increasing thickness, an example being Fe films on two Au monolayers on W(110) [41]. The average exchange asymmetry does not change monotonically with thickness. At 2 monolayers $\boldsymbol{M}$ jumps abruptly from the $[1\bar{1}0]$ direction to the [001] direction, then rotates again towards the $[1\bar{1}0]$ direction, by an amount varying from domain to domain by as much as 30°, with a maximum around 7 monolayer, and then rotates back towards [001] again. In addition to this lateral noncollinear magnetization simultaneously weak, thickness-dependent internal structure appears within the domains suggesting also perpendicular noncollinearity. Noncollinear magnetization in 1–5 monolayer thick Fe films on W(110) has been demonstrated later clearly by nuclear resonant scattering of synchrotron radiation [42, 43]. A long-known in-plane SRT occurs in Fe films on the clean W(110) surface at room temperature at about 35 monolayers, where $\boldsymbol{M}$ rotates from the $[1\bar{1}0]$ to the [001] direction. Recently a XMCDPEEM study

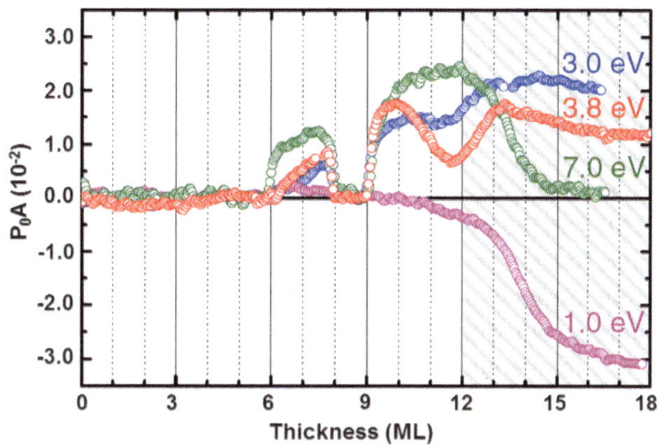

Fig. 7.4 Exchange asymmetry of Fe films on W(111) as a function of thickness, measured at different energies during deposition. In the hatched region a multidomain domain state is observed. Reprinted with permission from Ref. [45]. Copyright 2013 by the American Physical Society

[44] showed that this SRT is first order like the first transition in [41]. In this case an interesting domain boundary faceting with triangular morphology was observed, which was attributed to the competition between short- and long-range forces.

LEED studies which accompanied the SPLEEM studies of Fe films on 2 Au monolayers [41] showed that the changes in the magnetization were intimately connected with structural changes. The strong influence of structure is even more evident when the substrate imposes an unusual structure on the film. An example is Fe on W(111) [45], a very open surface. During uninterrupted deposition Fe grows pseudomorphic but highly strained up to 12 monolayers; then the film breaks up into small, probably bcc, Fe crystals. Concomitant with this structural change the original single-domain state breaks up into small domains, with the magnetization remaining in-plane. At lower coverages a magnetic signal appears at room temperature first at 6 monolayers, corresponding in atomic density to about two close-packed Fe layers, disappears at 8 monolayers and reappears again at 9 monolayers, from where it rises sharply. Between 6 and 8 monolayers the Curie temperature was found to be 305 K, beyond 9 monolayers it is significantly higher. The exchange asymmetry depends strongly on electron energy due to quantum size effects. The A_{ex} dependence shown in Fig. 7.4 [45] for Fe is in strong contrast to that in Co films, in which A_{ex} increases monotonically with increasing thickness after the onset of magnetization at about 6 pseudomorphic monolayers corresponding to 2 slightly compressed and buckled fcc(111) or hcp(0001) monolayers [46]. The differences between Fe and Co are undoubtedly due to a large extent to the fact that Fe has several polymorphs (bcc, fcc, and intermediate structures) and has a tendency to noncollinear magnetism, while Co has only two structures, hcp and fcc, which differ only in layer stacking and less than 0.4 % in (0001)/(111) intralayer atomic distances. The strong influence of the structure on the magnetic properties is particularly evident in the much studied Fe films on Cu(100) [47–50].

It should be noted, however, that complications may also occur with hcp and fcc films. For example, after forming initially a pseudomorphic layer, Co film on W(100) coalesce into thin elongated crystals with the $(11\overline{2}0)$ plane parallel to the substrate, which grow preferentially normal to the surface instead of forming a continuous layer. A 10 monolayer thick film agglomerates already at 400 K into large faceted crystals, which do not produce the strong (00) beam needed for SPLEEM [51]. This phenomenon is a general feature for film/substrate combinations within a narrow range of the lattice constant ratios [52].

Possibly because of the complications with Fe films most SPLEEM and XMCDPEEM work on single layers has been done on Co and Ni films. In particular, Ni films on Cu(100) have been studied extensively because of its unusual SRT from in-plane to out-of-plane with increasing thickness, which is reviewed in [8]. In more recent work the disordering of a perpendicularly magnetized Ni film near the Curie temperature was studied [53]. Ni films on Cu(100) have since become one of the testing grounds for the understanding of the physical processes involved in UVMCDPEEM [54–56]. Here only some work on Co films will be briefly discussed. An interesting SRT was discovered with SPLEEM in Co films on Ru (0001), in which the spin changes from in-plane to out-of-plane and to in-plane again from 1 to 3 monolayers, a phenomenon attributed to the competition between magnetoelastic, surface and interface anisotropies [57]. In Co films on the Au(455) surface, which is faceted into (233) and (677) facets, the magnetization of the step-induced stripe domains was found to rotate from in-plane to out-of-plane in a complex manner [58]. Just as interlayers between substrate and magnetic film have a strong influence on the magnetization by modifying the interface anisotropy, H_2 adsorption on a Co bilayer on Ru(0001) causes a reversible out-of-plane to in-plane SRT, which was explained by a purely electronic effect, the hybridization between hydrogen and Co orbitals [59]. The influence of overlayers on the magnetization direction is more complicated because it involves both electronic and magnetoelastic effects caused by the misfit. The first SPLEEM studies of how overlayers change the direction of $\boldsymbol{M}$ [60] showed that Au and Cu overlayers on Co films on W(110) rotate $\boldsymbol{M}$ from in-plane towards out-of-plane in manner depending upon Co, Au, Cu coverage, and deposition conditions. These changes were explained by changes in the surface anisotropy constants and by the growth modes of the overlayers [60]. A much more detailed SPLEEM study of the influence of Cu, Ag, and Au overlayers on Co films on Ru(0001), combined with first principle calculations, allowed to untangle the electronic and magnetoelastic contributions to the anisotropy constants [61]. An unusual overlayer is graphene. Co deposited *on* graphene on Ir(111) intercalates (see Chap. 6) when annealed above 530 K. In contrast to a Co film on Ir(111) without graphene, in which the SRT occurs at about 6 monolayers so that a 11 monolayer thick film is in-plane magnetized, an intercalated layer is still predominantly out-of-plane magnetized at the same thickness and has its SRT between 12 and 13 monolayers. This shift of the SRT to larger thickness was attributed to a significant increase of the surface anisotropy constant by the graphene overlayer [62]. In another context, spin injection into semiconductors, attempts to grow Co films on pentacene unsurprisingly failed because Co displaced pentacene even at 196 K and grew directly at the oxidized SiO(100) substrate, whether bare or Cu covered, as evident in Co XMCDPEEM and C 1s XPEEM [63].

Concluding this section on single films a few studies of alloy films should be mentioned. A XMCDPEEM study of epitaxial $CoPt_3$ films on $WSe_2(0001)$ found a SRT from perpendicular to in-plane magnetization at about 6 monolayers, using the ratio of Co L_2 and L_3 images [64]. Another interesting XMDPEEM study looked at the reversible antiferromagnetic–ferromagnetic transition in epitaxial FeRh films on MgO as a function of temperature, using the XMCD at the Fe L_3 edge. Nucleation occurred locally within several ten degrees, giving macroscopically the impression of a continuous transition with hysteresis [65]. The domain structure of an amorphous TbCo alloy film on glass, which has perpendicular magnetization and is of interest for optical magnetization switching, has been the subject of a study with energy-resolved valence band MCDPEEM. Two-photon photoemission and—after cesiation—single photon photoemission with circular polarized frequency-doubled Ti:sapphire laser pulses at normal incidence was used. The contrast was attributed mainly to the Co 3d bands [66].

7.2.2 *Quantum Well Effects*

Quantum well effects in the reflectivity of thin nonmagnetic films have been known for a long time (see Sect. 2.3.6). In SPLEEM they were first reported by Poppa et al. [67] in a brief study of the influence of Cu overlayers on the magnetic contrast from an epitaxial Co film on W(110), which showed that the magnetic contrast inverted with every second Cu monolayer. These contrast oscillations were studied later in more detail with Cu films on Co films on Cu(100) and attributed to quantum size effect (QSE) oscillations in the Cu film with a spin-dependent reflectivity at the Cu/Co interface [68]. The first SPLEEM studies of spin-dependent QSE *in* a magnetic film were made by Wurm [69], who measured the spin-dependent oscillations of the reflectivity $R = I/I_0$ of Co films on W(110) in monolayer steps up to 12 monolayers from 0 to 40 eV and extracted the A_{xc} oscillations from them. An example of his results is shown in Fig. 7.5 [22]. Attempts to determine the spin splitting of the band structure from the data were noise-limited so that only the spin-averaged band structure could be obtained in good agreement with theory. He also calculated the figure of merit (FOM) $Q = A_{ex}^2 R$ of these Co layers for use as a spin detector and found an optimum value of 0.02 [69], which is orders of magnitude higher than conventional spin detectors (see also [70]). His results stimulated the first and only SPLEED calculation at normal incidence at very low energies, based on band structure calculations and a realistic surface potential barrier, which achieved good agreement with experiment, using an empirical energy-dependent but spin-independent imaginary part of the self-energy [71].

Of course, the spin-dependent band structure above the vacuum level can be extracted directly from experiment making use of the QSE oscillation conditions (Eq. 2.25a), in which k is now spin-dependent. This was done first successfully for Fe films on W(110) [72, 73] from large regions with constant thickness so that the

Fig. 7.5 Quantum size oscillations of the specular reflectivity (**a**) and exchange asymmetry (**b**) of a 5 monolayer thick Co film on W(110) as a function of electron energy. Reproduced with permission from Ref. [22]. Copyright 1996 Elsevier

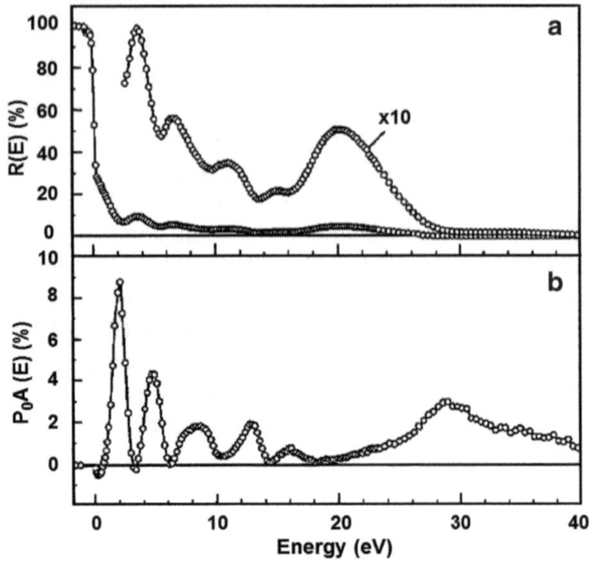

oscillations could be measured at well-defined thickness. In earlier [69] and later studies [74, 75] the films were grown at room temperature, at which usually three levels grow simultaneously so that the measurements average over several thicknesses. With averaging absent, a Q value of 0.05 was obtained, assuming a polarization $P_0 = 0.2$ of the incident beam, compared to $Q = 0.0225$ for a film grown at room temperature. A later study [74] found a smaller Q value for Fe but obtained for Co the same value (0.02) found earlier [69], possibly another sign of the sensitivity of the structure of Fe layers to the growth conditions. The band structure derived from Fe films grown directly on W(110) [72] was precisely reproduced also from films grown on a pseudomorphic Cu monolayer on the W (110) surface but only by an upward shift of 1.1 eV. The Cu layer acted as the reflecting barrier without a phase shift [75].

The QSE not only allows determining the band structure but also another interesting quantity, the spin-dependent inelastic mean free paths $l_\uparrow$, $l_\downarrow$. The QSE reflectivity oscillations are strongly damped due to electronic excitations as discussed in Sect. 7.1, in particular those of minority electrons. $l_\uparrow$ and $l_\downarrow$ can be extracted from the experimental data by considering the thin film as an absorbing Fabry-Perot interference layer [76]. The resulting values confirm the theoretical predictions of short inelastic mean free paths [27], giving values of $l_\uparrow(l_\downarrow)$ decreasing from 8(6) Å at 5 eV to 5(4) Å at 16 eV [77]. These values are slightly larger than the theoretical values but within the scatter of isolated values derived from photoemission. The spin-averaged IMFP can also be obtained from the attenuation of the W 4f photoelectrons from the substrate as a function of thickness, however, with large errors because their peaks are superimposed on the much larger secondary electron background at low energies [76]. These numbers should not be confused with the sampling depth via secondary electron emission in XMCDPEEM, 17, 25, and 25 Å for Fe, Co, Ni [78] or 17, 17 Å for Fe, Co [79], determined from the increase (decrease) of the overlayer (substrate) XMCD signal with increasing thickness.

Finally two studies of the connection between quantum well states in nonmagnetic materials and magnetism should be mentioned yet. One looked at the influence of quantum well states in Au films on Ru(0001) on the magnetic properties of Co films grown on these Au films [80]. The Curie temperature and the magnetic anisotropy of the Co film on the 2 monolayer thick Au wetting layer was found to be different from that on the thicker regions of the Au film but was not influenced by the quantum well states in the Au film. Another study was concerned with QSE oscillations in MgO films on a Fe(100) single crystal which play an important role in magnetic Fe/MgO/Fe tunnel junctions. Interestingly, the oscillations of the reflectivity with $\boldsymbol{P}//\boldsymbol{M}$ and $\boldsymbol{P}//-\boldsymbol{M}$ had different amplitudes but were in-phase, indicating spin-dependent reflectivity at the MgO/Fe interface without phase change [81, 82].

7.2.3 Bilayers

In bilayers, that is layers consisting of two magnetic layers, a new phenomenon is encountered, the coupling and competition between the magnetizations in the two layers. All combinations between Fe, Co, and Ni, grown on Cu(100), have been studied, mostly with XMCDPEEM, making use of the chemical discrimination, but also with SPLEEM, making use of the structural information. In the SPLEEM studies film combinations with various constant thicknesses were used, in the XMVDPEEM studies film thickness wedges, rotated 90° with respect to each other, which allowed collecting results from many thickness combinations. The early XMCDPEEM work in Kuch's group, which not only determined the magnetic state of the film pairs but also separated spin and orbital moments, has been reviewed repeatedly [9, 11, 13]. Here we illustrate some bilayer phenomena with the work done at ALS and some SPLEEM results.

The most studied system has been and still is Fe/Ni/Cu(100), mainly because the Ni film on Cu(100) has a somewhat temperature-dependent in-plane to out-of-plane transition via a canted state, which provides a substrate with different magnetization directions for Fe [83–88]. A SPLEEM study found a critical Fe thickness of 2.7 monolayers, at which out-of-plane and canted magnetizations of the film pair rotated in-plane in a spiral-like motion, which was attributed to a strong Fe/Ni interface magnetic anisotropy [83]. The SRT at 2.7 monolayers was also observed with XMCDPEEM on a wedge film, which nicely revealed the decrease of the stripe domain width with increasing Fe thickness [84]. Subsequent work looked in more detail at the out-of-plane to in-plane SRT of Fe on a 5.4 monolayer thick Ni film, which has in-plane magnetization before Fe deposition, and also at the reverse transition with increasing Fe thickness as a function temperature. This allowed determination of the magnetic phase diagram of this particular system [85].

A magnetic field, slightly tilted a few degrees away from the film plane, was found to produce a bubble phase close to the SRT region, which increases from 3.2 to 4.1 Fe monolayers when the Ni film thickness is increased from 6 to 12 monolayers. Upon heating to 370 K the bubbles reverted back into stripes, indicating that the bubble phase is metastable in the absence of a magnetic field. Imaging with Fe 2p and Ni 2p XMCDPEEM showed that the Fe/Ni bilayer behaved as a single magnetic layer due to the strong coupling of the magnetization between the Fe and Ni films [86].

A quite different phenomenon, the existence of the Dzyaloshinskii-Moriya interaction on Neél walls was also discovered in Fe/Ni bilayers on Cu(100) [87]. This interaction is briefly and precisely explained in Ref. [88]. It is a spin-orbit interaction effect and comes into play whenever the inversion symmetry is broken, for example at interfaces, and was discovered originally in spin-polarized STM studies of a Mn monolayer on W(110). In its presence the energy of the magnetic system becomes

$$E_{\mathrm{H}} + E_{\mathrm{DM}} = \sum_{i,j} J_{i,j} \boldsymbol{S}_i \cdot \boldsymbol{S}_j + \sum_{i,j} \boldsymbol{D}_{i,j} \cdot \left(\boldsymbol{S}_i \times \boldsymbol{S}_j\right), \tag{7.3}$$

where E_H is the Heisenberg exchange interaction, which is symmetric in $\boldsymbol{S}_i$ and $\boldsymbol{S}_j$, E_{DM} is the Dzyaloshinskii-Moriya interaction energy, which is antisymmetric in $\boldsymbol{S}_i$ and $\boldsymbol{S}_j$, and $\boldsymbol{D}_{i,j}$ is the Dzyaloshinskii vector. To Eq. (7.3) the various anisotropy energies, characterized by anisotropy constants K_i, still have to be added. Minimization of the resulting total energy can lead to very complex magnetization structures. In domain walls the Dzyaloshinskii-Moriya interaction introduces opposite rotational directions (chiralities), which differ in energy, contrary to exchange interactions, in which both chiralities have the same energy. In their study of domain walls of stripe domains in Fe/Ni bilayers with perpendicular magnetization Chen et al. [87] did not find the Bloch walls expected for this configuration but Neél walls with fixed chirality: the $\boldsymbol{S}_i$ in-plane components in the walls between neighboring spin-up and spin-down domains were always pointing from the spin-up to the spin-down direction (right-handed); in the inverted film sequence (Ni/Fe) the handedness inverted. Accompanying Monte Carlo simulations and energy considerations brought considerable insight into the conditions which will lead to the formation of monochiral domain walls.

The inverse bilayer system, Ni/Fe/Cu(100), has attracted much less attention, possibly because of the complex thickness dependence of the structure and magnetic properties of Fe on Cu(100). In a Cu-capped bilayer a paramagnetic (PM) to ferromagnetic (FM) transition and a SRT were found with increasing Ni thickness. The PM-FM Ni transition thickness decreased monotonically with Fe thickness as long as Fe was ferromagnetic (0–4 monolayers) and rose thereafter to a constant value in the thickness range in which Fe is antiferromagnetic (5–11 monolayers). The Ni SRT thickness behaved in the opposite way. The results were interpreted in terms of the various anisotropies, of which the in-plane magnetic anisotropy of the Ni/Fe interface was ascribed a critical role [89]. In a study of a Fe/Co bilayer on Cu

(100) Fe showed the same in-plane domain structure as Co and was ferromagnetic up to 4 monolayers. Thicker Fe films were antiferromagnetic, covered with a ferromagnetic monolayer [79]. In the Ni/Co bilayer the in-plane to perpendicular SRT of the Ni layer was found by MOKE to increase with increasing Co thickness linearly but with different slopes below and above 1 monolayer Co, which was explained by the relative magnitude of the Ni/Co, Cu/Cu, and Ni/Cu interface anisotropy constants [90].

7.2.4 Trilayers

Here we consider first trilayers ("sandwiches") consisting of two magnetic layers separated by a nonmagnetic layer. The magnetic coupling between the layers is of interest in spin valves and magnetic tunnel junctions. Subsequently we discuss trilayers consisting of three magnetic materials. The first studies of such sandwiches were made with SPLEEM using the same magnetic material for top and bottom layer and Cu for the spacer layer. SPLEEM is particularly useful for these studies because the signal from the bottom layer is already strongly suppressed by 2 monolayers in the top layer. Imaging before deposition of the top layer therefore allows correlation between the domain structures of the two layers. Already the first results found an interesting feature, the so so-called biquadratic coupling, in which the magnetizations in the two layers are rotated by 90° with respect to each other [22]. In subsequent more detailed studies of Co/M/Co trilayers (M = Cu, Au) as a function of top layer and spacer thickness [91, 92] a wide variety of coupling phenomena was fund as a function of thickness: ferromagnetic (FC), antiferromagnetic (AFC), biquadractic, and intermediate coupling and was explained in terms of the interlayer exchange energy. For example, on a 6 monolayer thick Au spacer on a 7 monolayer thick Co bottom layer on W(110) the magnetization of the top Co layer tilts initially out of plane then moves back into plane and ends up at 8 monolayers in-plane but rotated up to about 60° off the direction in the bottom layer [91]. As a function of Au layer thickness the coupling between 7 monolayer thick Co films changes from FC to AFC to FC with increasing thickness (Fig. 7.6) [91]. This is a consequence of the fact that the trilayer is a quantum well system and that the coupling between the two sides depends upon the spin-dependent reflectivities of the spin-up and spin-down electrons and the well width. The biquadratic coupling on the other hand has been attributed to various deviations from the ideal quantum well. A detailed description of the various aspects of interlayer coupling can be found in [92]. Cu spacers produced quite different coupling [92], a consequence of the different electronic structure as discussed in [93] and the microstructure of the Cu layer.

In contrast to these SPLEEM studies of trilayers with hcp(0001) and fcc(111) orientation on W(110), later XMCDPEEM studies concentrated on (100)-oriented layers on Cu(100). They used crossed wedge structures and focused on magnetic phase transitions in Co/Cu/Ni [94], Fe/Cu/Ni [95] and Co/Cu/Fe/Ni [84].

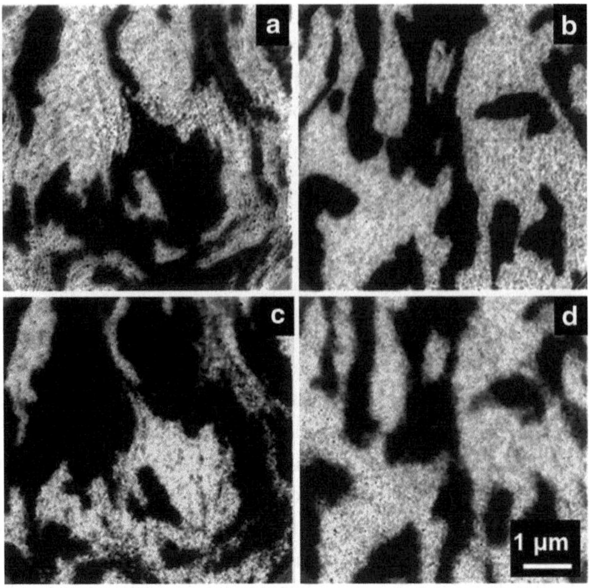

Fig. 7.6 SPLEEM images of the interlayer exchange coupling between two 7 monolayer thick Co films through an *n* monolayer thick Au film. *Left side* (**a**, **c**): $n = 5$, antiferromagnetic coupling; *right side* (**b**, **d**): $n = 6$, ferromagnetic coupling. (**a**, **b**) Co bottom layer, (**c**, **d**) Co top layer. Substrate: W(110). Adapted with permission from Ref. [91]. Copyright 1999 by the American Physical Society

Because of the larger sampling depth of XMCDPEEM the magnetization in bottom and top layer could be determined quasi-simultaneously via their respective 2p edges. In the Co/Cu/Ni trilayer the spacer thickness was kept constant at 2 monolayers. Three PM to FM phase transitions were observed as a function of Co and Ni thickness, which could be explained by Monte Carlo simulations with different Co and Ni exchange interaction constants J_{Co}, J_{Ni} [94]. In the Fe/Cu/Ni trilayer Fe and Cu layers were grown as crossed wedges on a thick Ni layer with perpendicular magnetization in the single domain state. Clear oscillations between FC and AFC interlayer coupling were observed, similar to the earlier SPLEEM studies, with single domain states also in the Fe layer. No coupling occurred in between the FC and AFC coupled thicknesses as evidenced by a multidomain state [95]. The influence of the interlayer coupling on the SRT was studied in the Co/Cu/Fe trilayer under conditions of zero coupling, FC, and AFC coupling with the result that increasing interlayer coupling acts in a similar manner as decreasing perpendicular magnetic anisotropy. This could be explained by the competition between exchange interaction, magnetic anisotropy, dipolar interaction, and Zeeman energy [84]. Finally a Fe/Ni/Cu/Ni sandwich structure was used to study the influence of a magnetic field on the SRT in the Fe/Ni/Cu bilayer mentioned earlier [84]. The perpendicular magnetic field of the Ni base layer coupled through the wedge-shaped Cu layer into the Fe/Ni bilayer and produced a break-up of the stripe phase into a bubble phase. The transition point depended on the effective field acting on the Fe/Ni layer, i.e. the Cu film thickness and on the perpendicular magnetic anisotropy of the Fe/Ni layer, i.e. on its thickness [96].

A quite different aspect of the magnetic interlayer coupling, the influence of magnetostatic interactions caused by the stray fields of domain walls, was studied in $Fe_{19}Ni_{81}$/Cu/Co trilayers on GaAs at three Cu spacer thicknesses. Micromagnetic simulations showed that the observed preferred domain wall configurations in the magnetic soft $Fe_{19}Ni_{81}$ layer were due to the reduction of the magnetostatic energy originating from the fringe field of the domain walls in the magnetically harder FM layer [97]. Domain wall interactions have also been studied in Co/Al_2O_3/$Fe_{20}Ni_{80}$ trilayer on a Si(111) surface which was miscut to introduce a uniaxial magnetic anisotropy. Close correlations between the walls were found [98]. Co-permalloy sandwiches with Cu and Al_2O_3 spacers were subject of time-resolved XMCDPEEM studies using the pump-probe technique (Sect. 3.2.9), which aimed at the understanding of the switching mechanism upon applied magnetic field pulses [99–104]. They were grown on oxidized flat Si(100) surfaces and on miscut, step-bunched Si(111) surfaces with several terrace widths The results of these studies are reviewed in [104]. Some of them are: (1) the stray field of domain walls in the hard Co layer, which produces an internal bias field, considerably reduces the nucleation barrier of domains in the soft permalloy layer, (2) the domain wall energy acts against domain expansion and plays an important role in switching (magnetization reversal), (3) the switching depends also strongly on the topography as seen in the dependence on terrace width on the miscut Si(111) surfaces: on wide terraces separated by high steps nucleation is dominating but domain wall propagation (domain growth) is strongly pinned, while on narrow terraces separated by low steps domain wall propagation dominates over domain nucleation.

The studies of trilayers, in which all layers are ferromagnetic, were concerned with the influence of the thickness and layer sequence on the magnetic properties of the individual layers and of the complete system. In wedge-shaped Ni/Co/Ni trilayers with fixed Co spacers, 0.7–1.8 monolayers thick, oscillations of the SRT thickness and of the magnetization direction were found as a function of the thickness of the bottom Ni layer whose Curie temperature was increased by the coupling to the Ni/Co layer on top. Quantum well effects were suggested as the cause of the oscillations [90]. In the Co/Fe/Ni trilayer the Ni and Co layers show AFC coupling between 4.5 and 6.5 Fe monolayers and FC coupling between 6.5 and 9 Fe monolayers. At 5 Fe monolayers the FM-PM phase diagram was similar to that of the Co/Cu/Ni trilayer discussed above [94]. More detailed studies of the trilayers with the inverse sequence, Ni/Fe/Co(fcc) [79, 105], with crossed Fe and Ni wedges on top of 10 monolayer Co give a clearer picture of the SRTs. With increasing Fe thickness the coupling oscillates with rather sharp transitions from FC to AFC to FC to AFC to FC. In the thickness region in which Fe is antiferromagnetic (5–11 monolayers) Fe is ferromagnetically ordered not only at the interface with Co but also at the interface with Ni as long as Ni is above the Curie temperature $T_{C,Ni}$. The oscillatory coupling between the Co and Ni layers caused also oscillations of $T_{C,Ni}$ at low Ni thicknesses, with maxima at the coupling maxima [79]. In the SRT region of Ni the interlayer coupling of the in-plane components of $\boldsymbol{M}_{Ni}$ to $\boldsymbol{M}_{Co}$ causes a continuous rotation of $\boldsymbol{M}_{Ni}$ towards the layer's normal. The Ni thickness at which this SRT occurs oscillates with the Fe thickness due to the interlayer coupling, with maxima coinciding with the coupling maxima [105].

Concluding this section on trilayers, the XMCDPEEM studies with its element-sensitive imaging capabilities give deep insight in the variety of magnetic phenomena caused by interlayer coupling, interface anisotropies, magnetostatic, and other effects. Only a rough sketch of these phenomena and their interpretation could be given here, the many details can be found in the original literature and some of the reviews mentioned earlier.

7.2.5 Multilayers

Multilayers are used in magnetic tunnel junctions [106] and other magnetic tunneling devices but have rarely been studied with XMCDPEEM although its atomic selectivity would be very useful for their structure and magnetic properties. One of the components of some tunnel junction devices, superlattices with perpendicular magnetic anisotropy, however, has been studied with SPLEEM. These superlattices are also used in magnetic memory and logic devices which are based on the motion of domain walls caused by the spin torque [107] exerted on them by a spin-polarized current, such as the Racetrack Memory [108]. Several superlattices have been studied such as Co/Pt, Co/Pd, and Co/Ni, the last having attracted most of the attention because both metals are ferromagnetic and thus provide the largest magnetization density. (111)-oriented Co/Ni superlattices were predicted and experimentally confirmed to have a very large perpendicular magnetic anisotropy already in the early 1990s (for a review of the early work see [109]) and have become the favorites in research and applications (for references to later work see [110]). For this reason they have been selected for a SPLEEM study of the evolution of the spin orientation and domain structure with the number of superlattice periods and the layer sequence for several layer pairs (Ni_xCo_y) [111, 112]. Based on the existing knowledge of the optimum combinations, Ni_2Co, Ni_3Co, and $CoNi_2$ superlattices were grown on W(110) and the evolution of the magnetic domain structure continuously recorded during growth, which is possible in the SPLEEM instrument with the high polarization, high brightness spin gun described in Sect. 3.2.8. W(110) was used as a substrate because after an initial pseudomorphic layer and a few slightly distorted layers Co and Ni grow on it in (111) orientation with little distortions. Both Ni and Co on W(110) have a pronounced magnetic in-plane anisotropy so that the influence of the substrate on the anisotropy can be followed during film growth. Figure 7.7 [112] illustrates the evolution of the domain structure with images taken at the completion of the various layers for the Ni_2Co sequence. The perpendicular magnetization does not increase continuously with thickness but oscillates with decreasing amplitude. This is a consequence of the competition between interface anisotropy, which dominates in each Ni_2Co pair and favors perpendicular magnetization, and the shape anisotropy, which favors

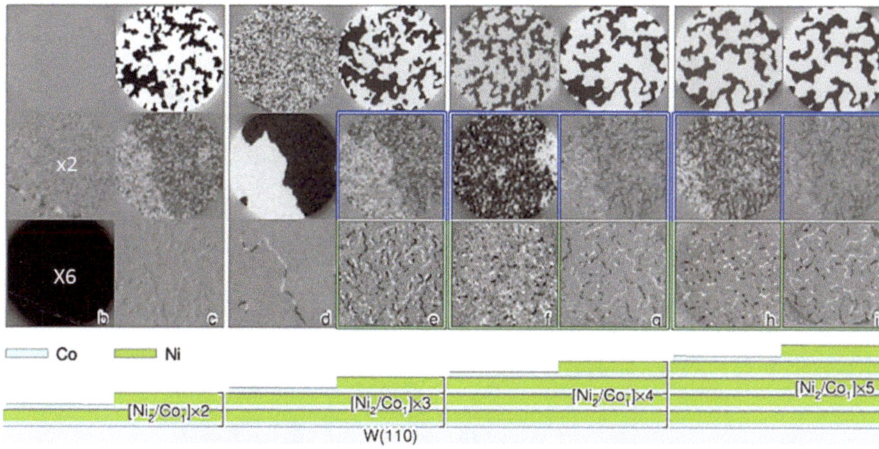

Fig. 7.7 SPLEEM images taken at the selected growth stages indicated at the *bottom*. From *top to bottom*: Images taken with the polarization vector in the [110], $[1\bar{1}0]$ and [001] directions of W (110) showing the three magnetization components. The contrast of the images marked *blue* and *green* in the second and third row is enhanced by a factor of 4 and 8, respectively. Because of pronounced quantum size effects during growth the energy was varied between 0.9 and 2.5 eV to optimize the asymmetry (contrast). Reproduced with permission from Ref. [112]. © IOP Publishing 2013. All rights reserved

in-plane magnetization and is large whenever a Co layer is added because of the strong in-plane magnetization of Co. This competition could be modeled with Monte Carlo simulations using effective anisotropy constants fitted to the experimental data [113]. Superlattices with the inverse layer sequence ($CoNi_2$) developed complete perpendicular magnetization faster, Ni_3Co superlattices somewhat slower, in agreement with theory and laterally averaging measurements. High magnification images show interesting features in the deposition sequence-dependent Neél domain wall structure in the perpendicular magnetized state. A more detailed analysis requires the knowledge of the IMFP in Ni, which is significantly larger than that in Fe so that SPLEEM is expected to average over several CoNi pairs as indicated in some of the in-plane images at nearly perpendicular magnetization. The results already obtained show that SPLEEM can produce valuable details not accessible to laterally averaging studies.

7.2.6 Compound Layers

There are a number of ferromagnetic compound layers which have been studied, but mostly as nanostructures or in connection with exchange coupling with antiferromagnetic single crystals and films, where they will be discussed. Here only continuous films will be considered. One study was concerned with the nature of the

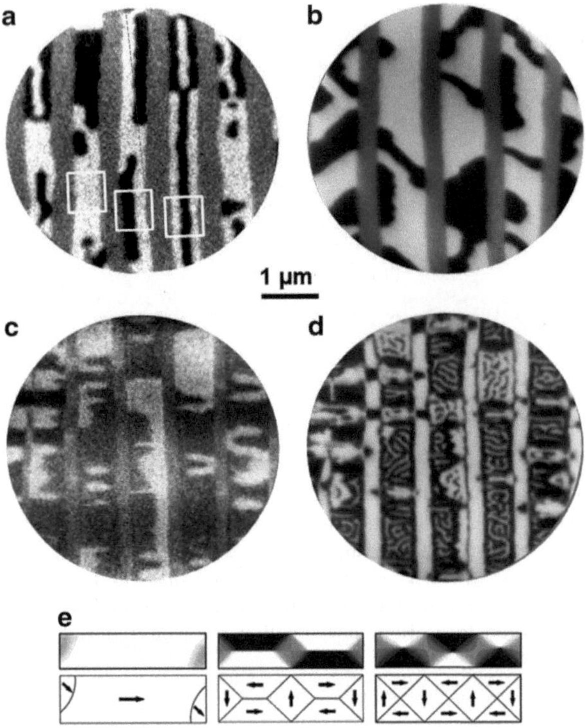

Fig. 7.8 XMCDPEEM images of MnAs films on GaAs(100). (**a**) 180 nm, (**b–d**) 300 m thick films. Images taken at the Mn L_3 edge (**a–c**). (**c, d**) are from a MnAs film coated with one monolayer Fe (**d**) taken at the Fe L_3 edge. All images at room temperature; (**e**) shows schematically the domain structures of the regions marked in (**a**) seen from the top with XMCDPEEM (*top row*) and in cross section as derived from micromagnetic simulations (*bottom row*). (**a, b**) Reproduced from Ref. [123] with permission from the American Institute of Physics, © 2007. (**c, d**) R. Belkhou, E. Bauer unpublished; (**e**) Reproduced with permission from Ref. [122]. Copyright 2006 Elsevier

ferromagnetic phase in dilute magnetic semiconductors, in particular in Mn- and Cr-doped Ge films. Although hysteresis curves showed ferromagnetism, no XMCD signal was observed. Chemical imaging with XPEEM in the absorption mode showed significant segregation of Mn and Cr in form of oxides. In this example PEEM could not clarify the nature of their ferromagnetic phase [114]. XMCDPEEM studies of half-metallic $La_{1-x}Sr_xMnO_3$ (LSMO) films ($x = 0.2$–0.4) on stepped $SrTiO_2(100)$ surfaces found in thin films step-induced uniaxial magnetic anisotropy. With increasing thickness biaxial magneto-crystalline anisotropy developed [115, 116].

An interesting ferromagnetic compound is MnAs. Epitaxial MnAs layers on GeAs show a complicated transition from the low temperature ferromagnetic α-state to the high temperature β-state, which has been a controversy for some time. XMCDPEEM studies have contributed significantly to the clarification of this transition and XMLDPEEM temporarily to some confusion [117–125]. Figure 7.8 [122, 123] shows some examples of MnAs films on GaAs and of a Fe overlayer on a

MnAs film. The black-white stripes are the ferromagnetic α-phase consisting of domains with opposite in-plane magnetization perpendicular to the stripes, the gray stripes are the β-phase. The relative stripe width depends upon temperature, the stripe period on film thickness. Interpretation of the domain pattern has to keep in mind that XMCDPEEM sees only a small fraction of the thickness of the films shown in Fig.7.8. Only micromagnetic simulations, inspired by the domain structure of thin rectangular magnetic nanostructures, which will be discussed in Sect. 7.5, gave an explanation [121, 122], which for the simple structures marked in Fig. 7.8a are shown in Fig. 7.8e. The more complex pattern in Fig. 7.8b was more elusive [123]. Fe overlayers change not only the domain structure of the MnAs film but also produce a still unexplained complex domain structure in the Fe film (Fig. 7.8c, d) (R. Belkhou, E. Bauer, unpublished).

The β-phase has been predicted by theory to be antiferromagnetic [126] and XMLDPEEM studies also seemed to support this prediction [123]. Subsequent laterally averaging XMLD studies (R. Belkhou, private communication) found a very weak XMLD signal in the α-phase and an even weaker one in the β-phase. However magneto-optic Kerr effect studies of films with various thicknesses as a function of temperature did not show a reproducible exchange bias shift of the hysteresis curve, which is considered to be a criterion for AFM-FM coupling [124] and (E. Bauer, unpublished). Possibly the XMLD signal is not caused by antiferromagnetism but rather by strain in the epitaxial film.

7.3 Bulk Magnetic Materials

7.3.1 Ferromagnetic Materials

Little work has been done on bulk magnetic materials with cathode lens electron microscopy after the early work discussed in Chap. 1. The first SPLEEM image of a magnetic surface, Co(0001), was made only to demonstrate the capabilities of this method [2]. Some systematic SPLEEM work was done much later on another surface. UVPEEM of magnetic surfaces, using the stray field of the domain walls for contrast was used only once to image the domain structure of permanent magnet, $Nd_2Fe_{14}B$ [127], and analyzed in some detail later [128]. The first XMCDPEEM images of a bulk material, stainless steel, looked at the influence of annealing and quenching on the domain structure [129] but no systematic studies were made until much later. The reasons for this inactivity are that the scientific problems described before and in the following sections were more interesting and also that the resolution in stray field UVPEEM is not better than that of traditional magnetic imaging methods. Also, it has to be kept in mind that the methods used here see only the tip of the iceberg and what is below the water may be quite different. The (0001) surface of Co is a good example. The easy magnetic axis in the bulk of Co is perpendicular to this surface but what surface microscopy sees are the stray field-minimizing closure domains, which look quite different from the domain structure in the bulk. Nevertheless a number of studies of bulk surfaces were made.

One series of experiments was concerned with the magnetic domain structure of bcc-FeNi lamella in an iron meteorite. It showed in XMLDPEEM an unusual magnetization direction, which was attributed to an extremely slow formation process [130]. An XMCDPEEM study of the interface between the bcc α- and the fcc β-phase of FeNi, whose composition was determined by Fe and Ni K-edge absorption PEEM [131], showed a head-to-head domain wall between the two phases. This was attributed to segregation of an ordered $L1_0$-FeNi lamella between the two phases [132]. XMCDPEEM studies of (100)-oriented epitaxial $L1_0$-FeNi films in the context of this work showed an interesting domain structure: wide regions with $\boldsymbol{M}$ in-plane parallel [011], alternating with narrow regions with an out-of-plane $\boldsymbol{M}$ component, separated by lines with strong out-out-of $\boldsymbol{M}$, which were attributed to Bloch walls [133].

More mundane studies were concerned with the magnetic domain structure of a Fe-3% Si(110) sheet, which is important for the eddy current losses in this soft magnetic material. XMCDPEEM was used in all studies. The first one looked at the temperature dependence of the domain structure up to the Curie temperature. Interestingly, the so-called Lancet domains, which strongly affect the losses, reappear in the same size and position upon cooling but with opposite magnetization [134]. The second study was concerned with the elimination of local strain by annealing. Focused laser irradiation was used to produce strain-induced fine, regular μm-wide stripe domain patterns, which were stable up to about 450 °C and could be annealed out completely only above 675 °C [135]. In the third study the influence of nonmagnetic inclusions on the changes of the domain structure during magnetization and demagnetization was followed and correlated with the hysteresis curve. The inclusions were simulated by large and small holes in the center of the disc samples [136]. These three studies are a nice example of the application of XMCDPEEM to problems of immediate technological importance.

Another example of technological relevance is the study of Nd-Fe-B sintered magnets with absorption edge XPEEM for chemical analysis and XMCDPEEM for the determination of the magnetic domain structure. In these magnets the ferromagnetic $Nd_2Fe_{14}B$ crystals are imbedded in a nonmagnetic Nd-rich phase and the relationship between their microstructures is important for the coercitivity of these permanent magnets. In a first study it was found that after thermal demagnetization the grains were in a multi-domain state while after DC-field demagnetization they were in a single domain state with $\boldsymbol{M}$ pointing opposite to the applied field, which was attributed to a lower coercivity in the surface region [137]. A second study looked at the temperature dependence of the domain structure of a Dy-doped magnet of the same type. Dy increases the thermal stability and strongly increases the coercivity. Chemical imaging with Dy M_5 absorption edge XPEEM showed inhomogeneous Dy distribution. In XMCDPEEM at the Fe L_3 edge a window frame specimen holder (Fig. 3.63a) was used in order to minimize the magnetic stray field. The results showed that Dy significantly reduced the number of domains with reversed magnetization and the growth of the reversed domains perpendicular to the c axis during heating, which was attributed to Dy-enhanced magnetic anisotropy. The temperature dependence of the domain structure gave a detailed picture of the thermal demagnetization process [138].

Some bulk half-metallic compounds, i.e. materials with complete spin polarization at the Fermi level, have also been studied: the Heusler compounds $Co_2Cr_{0.6}Fe_{0.4}Al$ [139] and Co_2FeSi [140] and magnetite (Fe_3O_4) [141, 142]. These materials are of interest for a variety of magneto-electronic applications. In the first study polycrystalline samples were prepared from arc-melted mixtures of the elements and imaged with Fe and Co XMCDPEEM. The images showed identical μm size domains, indicating ferromagnetic coupling between Fe and Co, with a fine superimposed ripple structure. Selective oxidation of Cr on the surface was pointed out as a problem for the use of the surface study in the interpretation of the bulk properties and of some surface properties. In the second study [140] both a polycrystalline sample, prepared by arc melting like the first sample and a (110)-oriented single crystal were used after removing the oxide layer by sputter-annealing cycles. The polycrystalline sample showed a micro-magnetic ripple structure similar to that in the first study. On the single crystal surface the domain patterns varied locally. In some regions the domains were small and indicated out-of-plane magnetization, in other regions they were up to several 10 μm large with four in-plane magnetization directions and 180° Bloch walls.

The studies of magnetite were concerned with the relation between structure and magnetism, clearly a domain of SPLEEM. The sample was the (100) surface of a synthetic crystal. The first study [141] looked at the Verwey transition, a first-order transition from cubic to monoclinic structure upon cooling below about 120 K, the second one [142] in more detail at the domain structure at room temperature. In the cubic phase at room temperature large domains with the magnetization pointing along the [011] or the $[01\overline{1}]$ axis were observed in contrast to the easy $\langle 111 \rangle$ directions in the bulk. The distortions in the monoclinic phase cause a ridge-valley topography of the surface along one of the in-plane $\langle 100 \rangle$ directions, which is visible in LEEM in form of stripes, and determines now also the direction of the magnetization. This direction is the c axis of the monoclinic magnetite, which is also the easy axis in the bulk. The second study [142], covering a larger area, revealed a complex closure domain structure with regions of both domain orientations. Many of the domain walls showed substantial curvature and were interpreted in terms of Neél-capped bulk Bloch walls and Bloch lines.

7.3.2 *Antiferromagnetic Materials*

In contrast to ferromagnetic surfaces, in which stray field minimization leads to closure domain formation, which hides the magnetization in the interior of the crystal, antiferromagnetic surfaces have no stray field so that surface imaging allows to a certain extent also a view of the interior. The only antiferromagnetic

(AFM) material, which has been studied with cathode electron microscopy in detail is NiO. NiO is extremely well-suited for these studies because of its cubic rock salt crystal structure, cleavage along (100) and convenient Neél temperature of $T_N = 523$ K. It is at the same time challenging because upon cooling below T_N the magnetostriction caused by the AFM ordering leads to contraction along the $\langle 111 \rangle$ directions, resulting in the formation of four $\{111\}$ twin domains. In the $\{111\}$ Ni planes the spins are aligned along the three $\langle 11\overline{2} \rangle$ directions with opposite spin direction in the neighboring $\{111\}$ Ni planes. Thus there are a total of 12 spin domains. A nice pictorial presentation of the twin and spin domains as well as of the domain walls between them can be found in [143]. The contraction along the $\langle 111 \rangle$ directions also changes the coordination of the O atoms between the Ni layers making it slightly rhombohedral along the $\langle 111 \rangle$ directions. This asymmetry of the charge distribution causes an X-ray linear dichroism at the O K edge, so that in addition to the XMCD at the Ni L_2 edge, which is sensitive to the spin domains, XLD at the O K edge can be used for imaging the twin domains (Sect. 2.2.6.2).

This understanding has developed over several studies since the early work, in which epitaxial NiO(100) films were used and XMLD contrast was mainly deduced from the temperature dependence of the contrast and of the fine structure of the Ni L_2 absorption spectrum because the domains were below the resolution limit of the PEEM used [144, 4]. The transition to cleaved single crystals and instrumental improvements over the years has revealed the full complexity of this AFM system. Important steps were the discovery of contrast at the O K edge [145] and of the possibility of imaging the twin domains with LEEM without spin polarization [146]. The comparative O K edge and Ni L_2 edge XLDPEEM studies [143, 145, 147] clearly showed that the contrast at the O K edge was a pure, nonmagnetic XLD ("searchlight") contrast (Sect. 2.2.6.2) and imaged only the twin domain structure, while imaging at the Ni L_2 revealed the spin domain structure via XMCDPEEM. Image contrast at the O K edge could also be obtained with circular polarized light due to the fact that this is a coherent combination of s- and p-polarized light with a $\pi/2$ phase shift [147]. The domain images obtained with LEEM (Fig. 7.9a [146]) were taken with the four (0, ½) spots, which had been found already in early LEED studies and were interpreted in terms of magnetic contrast caused by exchange scattering [148]. In the meantime it has become clear that the exchange interaction between spins is much too weak to cause spin-dependent scattering of slow electrons but that it is rather the spin-dependent exchange-correlation potential, which causes spin-sensitive scattering (Sect. 7.1). Inasmuch as in an antiferromagnet both spin-orientations are present in equal numbers, no effect is expected even with polarized electrons. The fact that the LEEM images show the same domain distribution as the O K edge XLDPEEM images [146] strongly suggest the same basic origin, strain. Although the strain in the crystal is small, on the surface it can cause significant atomic displacements leading to the well-pronounced $p(2 \times 1)$ structure. Another subject of controversy was the imaging of the twin domain structure with UVPEEM. References to this and other studies of NiO can be found in [143, 146, 147].

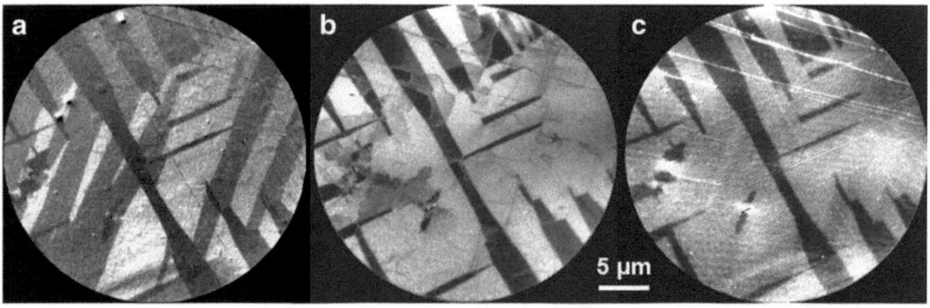

Fig. 7.9 Domain images on NiO(100). (**a**) Twin domain LEEM image taken with a (1/2,0) beam with 24 eV electrons, (**b**) Spin domain XMLDPEEM image taken at the Ni L_2 edge, (**c**) Twin domain XLDPEEM image taken at the O K edge. Adapted with permission from Ref. [146]. Copyright 2011 by the American Physical Society and Courtesy of R. Belkhou

7.4 Ferromagnet–Antiferromagnet Interfaces

The coupling between the spins in the ferromagnet and the spins in the antiferromagnet plays an important role in spin valves and magnetic tunnel junctions, in which one layer is free to change its spin orientation (free layer) under the influence of a magnetic field or a spin current, while in the other layer the spin direction is fixed. This fixing (pinning) can be achieved by the so-called exchange coupling between FM and AFM layer. The physical mechanism of this coupling has been the subject of many combined XMCDPEEM and XMLDPEEM studies starting in 2000 with the system Co/LaFeO3 [149]. AFM domains in thin epitaxial $LaFeO_3$ films on a $SrTiO_3$(100) bicrystal had been imaged previously with XMLD [150].

$LaFeO_3$ is orthorhombic with $a \approx b$, $c \approx \sqrt{2}a$ and a very high Neél temperature (740 K). A comparison of the FM and AFM domain images in the polycrystalline FM Co layer and the single-crystalline $LaFeO_3$ showed that the spin in the Co layer was aligned parallel or antiparallel to the in-plane projections of the AFM axes, which are tilted 45° with respect to the (100) surface of $LaFeO_3$. After a magnetic field was applied along these projections, the hysteresis curves extracted from the contrast showed a clear shift of the magnetization direction in the Fe domains, which were coupled to $LaFeO_3$ domains with AFM axis projections in this direction. This local "exchange" bias was attributed to uncompensated spins in the individual AFM domains, which were frozen in after growth. In subsequent studies $LaFeO_3$ films were grown on different substrates, which produced different domain structures, and Fe films on top of them. Also Co on Ni-diluted $LaFeO_3$ films was studied. The results, reviewed in [151] together with the early work, give a significantly more complex picture of the cause of the exchange bias effect.

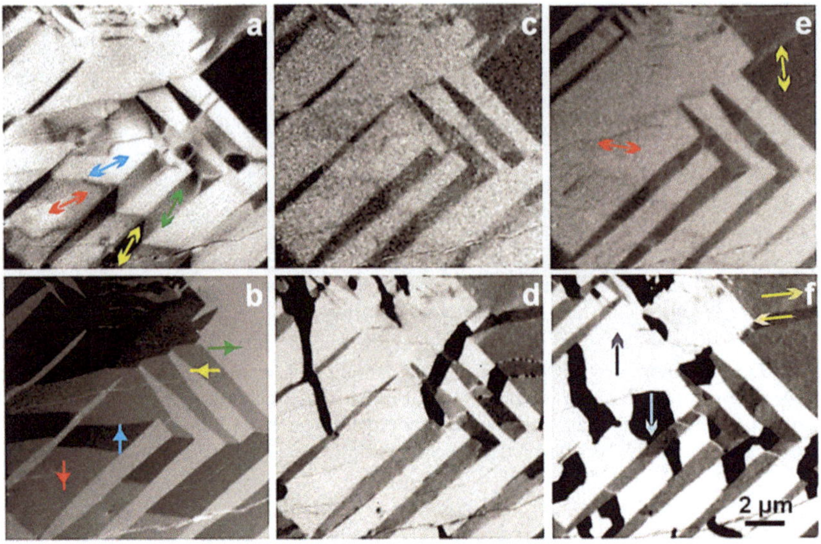

Fig. 7.10 NiO(100) surface. (**a**, **c**, **e**) Ni L_2 XMLDPEEM images, (**b**) O K XLDPEEM image, (**d**, **f**) Co L_2 XMCDPEEM images of a 3 nm thick Co layer on the NiO(100) surface. (**a**) and (**b**) show the spin and twin domain structure before Co deposition, respectively, (**c**, **d**) the domain structure in NiO and Co in the as-prepared state, (**e**, **f**) the corresponding structures after annealing above the Neél temperature and cooling to room temperature. The *arrows* indicate the spin orientation. Annealing caused some domain growth in NiO and significant domain changes in the Co film but the relative spin orientation in Co and NiO was retained. Adapted with permission from Ref. [154]. Copyright 2009 by the American Physical Society and Courtesy of H. Ohldag

The difficulties in relating structural and magnetic aspects encountered in the epitaxial but fine-grained $LaFeO_3$ do not exist on the cleaved and annealed NiO (100) surface as illustrated at the first study of exchange coupling with Co on this surface [152]. Annealing is essential otherwise the strain resulting from cleaving produces a surface layer with different domain structure (R. Belkhou, private communication). Therefore most of the later work was done with Co and Fe on this surface although there have been some studies also on the reverse system, NiO on epitaxial Fe on MgO [153]. Here only two of the many studies of Co on NiO will be briefly mentioned. Building on earlier work [145, 152] a study of the correlation of crystallographic and magnetic domains at Co/NiO(001) interfaces [154] compared the spin distribution in a 3 nm thick Co film, in the NiO substrate and in the interface before and after annealing 25 K above the Neél temperature. The images (Fig. 7.10) [154]) showed that the in-plane components of the Co and Ni spins are aligned perpendicular to each other, both in the as-prepared state and in the annealed state. In the as-prepared state no Ni $L_{3,2}$ edge XMCD signal could be detected but after annealing, which formed a thin $CoNiO_x$ interface alloy a small Ni XMCD signal was seen, indicating uncompensated Ni spins. This shows that uncompensated spins are not the cause of the perpendicular spin alignment in Co and NiO, contrary to earlier opinion. The formation of the $CoNiO_x$ interface layer caused a reorientation of the Ni spins from the 112 directions to the 110 directions. A further complication was discovered when the ***E*** vector of the linear polarized light parallel to the surface (π polarization) was rotated around the surface normal, which revealed a canting of the Ni spins. This canting was attributed to uncompensated Ni spins at the interface, which are oriented parallel to the Co spins [155]. A critical reexamination of the theory of X-ray magnetic linear dichroism revealed that the analysis in previous work with the methods described in [1] and briefly discussed in Sect. 2.2.6.2 is insufficient in a cubic crystal field and that angular distribution measurements are needed for a complete analysis [155]. This requires specimen rotation, which is increasingly incorporated in PEEM instruments.

The most complicated NiO/FM metal system studied is the system CoNiOFe on a Ag(100) surface as a function of NiO thickness [156]. The Co/NiO interface showed perpendicular coupling between Co and Ni spins and uncompensated Ni spins independent of NiO thickness while at the Fe/NiO interface no uncompensated Ni spins were observed and the coupling to the Fe spins changed from perpendicular to parallel with increasing NiO film thickness. FM/AFM interface studies have also been performed on pure oxide systems and analyzed in terms of the anisotropic XMLD theory. NiO films on Fe_3O_4(100), (110), and (111) single surfaces are prime examples of the importance of the interface orientation for the magnetic coupling between AF and AFM systems [157]. The coupling at the (111) and (110) surfaces was found to be collinear, at the (100) surface perpendicular. The density of uncompensated spins at the (111) and (100) interfaces was significantly lower ($\approx$15 %) than on the (110) interface and thus not related to the type of coupling. On the (110) surface it is so large that clear Ni edge XMCD images with identical contrast as in the Fe edge XMCD image could be obtained, indicating strong alignment of the uncompensated Ni spins at the interface. On the (111) and (100) surfaces, on which all easy axes are tilted relative to the surface, the situation is more complicated as analyzed in detail in [157].

There have been numerous other studies of the FM-AFM couples mentioned here and of systems with other AFM materials such as FeMn, of trilayers, superlattices, and even of tunnel junction film stacks, but in view of the large amount of material the examples cited have to suffice here. Information on them can be found dispersed in some of the review papers cited at the beginning of this chapter. A lesson learned from the past work is that for a full understanding of the interface coupling not only the polarization must be rotated about the beam direction but also the specimen about its normal so that the angular dependence of the XLD effect can be explored as much as possible. While the glancing angle of incidence of the photon beam is fixed by the instrument design at values ranging from 15° to about 40°—or at 90° when illuminated through the objective lens—the azimuthal angle can be changed by incorporation of a UHV-compatible rotating platform. In this manner the understanding of the interface coupling should improve significantly, in particular by comparative studies of the influence of interface orientation, surface modification, and thin film microstructure.

7.5 Magnetic Nanostructures

7.5.1 Introduction

When the lateral dimensions of a ferromagnetic thin film system become small the shape of the pattern determines to a large extent the magnetization distribution by minimizing the lateral stray field. A closure domain structure with domain walls forms similar to the closure domains on a bulk ferromagnet, whose easy axis in the interior is not parallel to the surface. In smaller patterns a closed magnetization loop with a vortex in the center can form and very small patterns may be in the single domain state. Much interesting physics happens in these dimensions, which has stimulated a large amount of XMCDPEEM work and some SPLEEM studies. However the main impetus for this effort comes from the practical applications of small FM structures in technology. Accordingly, there are already many reviews and several books, which cover various aspects of the physics and technology of magnetic nanostructures, of which only a few recent ones are cited here [158–163] in addition to those mentioned already earlier [9–21].

Because of this interest the study of magnetic nanostructures has become one of the most important applications of cathode lens electron microscopy. To be more precise, it is more appropriate to talk about microstructures and submicron structures because most of the work has been done in the μm and many-hundred nm range. Below about 100 nm the internal magnetic structure of the features, which is generally the subject of the study reaches the resolution limit. In the true nm range the samples are in the single domain or in the superparamagnetic state though interparticle interaction phenomena still make their study interesting. However most of the understanding has been obtained from studies of structures with dimensions above several 100 nm to several 10 μm.

The preparation of well-defined structures is usually done by electron beam lithography so that most of the work uses ex-situ prepared samples and XMCDPEEM for imaging because the structures are generally polycrystalline and frequently covered with a corrosion-inhibiting layer, which excludes SPLEEM for imaging. In-situ preparation of defined structures suitable for SPLEEM is limited by nature and can be influenced only to a limited extent by the experimentalist so that very little work with SPLEEM has been done. The subject of the early studies was the static domain structure as a function of shape and dimension. The desire to understand how this structure changes under the influence of a magnetic field soon led to investigations of the domain structure after a magnetic field pulse was applied, to pump-probe experiments and finally to imaging in an applied magnetic field. The discovery of the spin current torque and its promising applications has led to intense XMCDPEEM efforts to understand and use it. Nanostructure arrays have been studied because of their importance in device physics and technology. At the limits of the imaging methods real nanoparticles were explored.

7.5.2 Static Domain Structure

Already the first patterned "nanostructures," permalloy squares with 20 μm edge lengths on silicon, showed in XMCDPEEM the theoretically predicted closure structure with a vortex in the center [164]. A subsequent study demonstrated the usefulness of XMCDPEEM in a more complicate case, the influence of exchange coupling in rings and square frames of a FeNi/NiO/FeNi trilayer on the magnetization distribution, depending upon the orientation of the frame relative to the exchange bias field [165]. A comparison of square and circular permalloy structures showed that in the square patterns the vortex structure could be imaged down to 1 μm edge length. In the circular discs the closed magnetic flux structure was visible down to 200 nm but the vortex core could not be seen although in magnetic force microscopy (MFM) it was clearly visible [166]. This limitation is not only a question of resolution but also of the geometry of the instrument: the photon beam probes the magnetization at a glancing angle of incidence of 16° so that the magnetization normal to the surface in the vortex core contributes little to the intensity.

A study aiming at the question to what extent micromagnetic simulations can predict the domain structure of rectangular patterns, which had not been exposed to a magnetic field, was answered qualitatively positively. Significant differences were attributed to preparation conditions, which did not produce the field-free ("virgin") configuration assumed in the simulations, and to defects [167]. Starting in 2003 the study of rings become popular based on the idea that the handedness of the magnetization in the ring could be switched by inverting the current through a wire in its center perpendicular to the ring, thus providing the 0 and 1 for computing [168]. These studies were later extended to other patterns and to the influence of thickness, ring width, and ring defects [169, 170]. Examples of such patterns are shown in Fig. 7.11 [170]. For a review of the ring studies see [171, 172]. A considerable amount of work has been done since, not only on rings but also on other structures. Only a few examples can be given here in order to illustrate the applications of cathode lens microscopy in this field and the variety of phenomena observed.

While in most of the earlier work permalloy was used because of its low magnetocrystalline anisotropy, more recent work increasingly concentrated on more complex materials with significant magnetic anisotropy, either of magnetocrystalline or extrinsic character, e.g., step-induced. For example rings of epitaxial semimetals, Fe_3O_4 on MgO [173] and a Heusler alloy on Cr/MgO [174] have been studied, which showed a strong influence of the magnetocrystalline anisotropy on the domain structure and the domain walls. In long Ni-rich NiPd rectangles on oxide-covered Si, with aspect ratios of 5:1 and 10:1, complicated domain structures were observed, with pronounced lateral anisotropy. In thicker films parallel SEMPA studies revealed a significant out-of-plane component of the magnetization. These anisotropies were attributed to strain effects [175]. Epitaxial $La_{0.7}Sr_{0.3}MnO_3$ (LSMO) squares, rectangles, and triangles on $SrTiO_3$ were studied with the specific goal to determine the influence of the magnetic anisotropy on the spin configuration. The results showed in

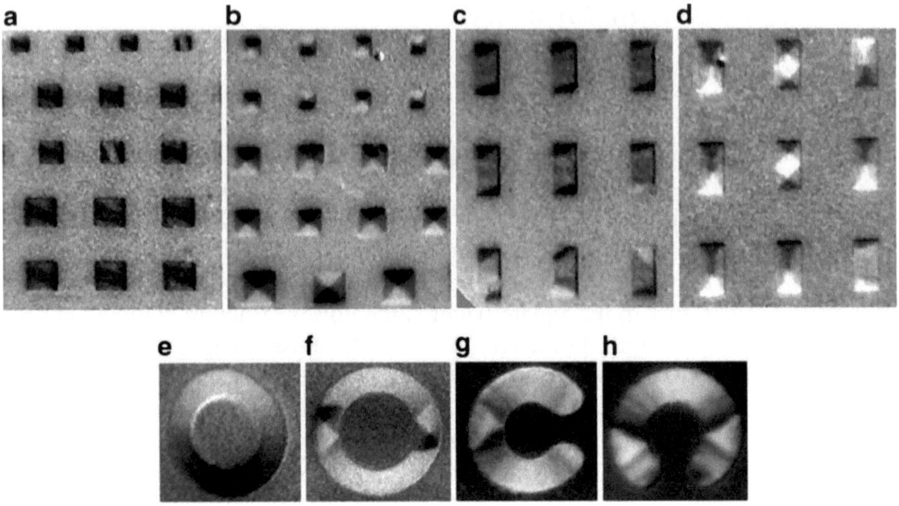

Fig. 7.11 Domain structures of *square*, *rectangular* and *circular* magnetic patterns. XMCDPEEM images. (**a–d**, **g**, **h**) Co, (**e**, **f**) Permalloy. (**a**, **b**) 400, 600 and 800 nm *squares*, (**c**, **d**) 400 nm wide *rectangles* with aspect ratio 2:1. (**a**, **c**) 10 nm thick, (**b**, **d**) 20 nm thick. (**e**, **f**) 30 nm thick *rings* with 2400 nm outside diameter and 350 nm width. (**g**, **h**) 20 nm thick *open C* patterns with 1600 nm outside diameter and 400 nm width. (**a–d**) show the influence of size, shape and thickness resulting in domain states called c, s (**a**, **c**), vortex (**b**), vortex and diamond (**d**). (**e**) is called vortex state, (**f**) onion state, which has two vortex domain walls. Adapted from Ref. [170] with permission by John Wiley & Sons

thin films (15 nm) strong distortions of the domain structures caused by the effective magnetocrystalline anisotropy but no influence in thick films (50 nm) [176].

A strong influence of the magnetocrystalline anisotropy was also found in a comparison of trilayer patterns with differing anisotropies, permalloy/Cr/FeCo, and Co/Cr/Fe, consisting of an epitaxial bottom layer, a wedge-shaped Cr interlayer and a polycrystalline top layer. Squares with a soft top layer (permalloy) and 12 μm edge length showed the oscillating FM-AFM interlayer coupling with increasing spacer thickness known from continuous films, but the simple vortex structure of noninteracting permalloy squares occurred only in FM coupling regions. In contrast, in the magnetically harder film couple (Co, Fe) vortex structures formed: at the thinnest spacer (1 nm) a distorted AFM-coupled vortex structure was observed, with the magnetization tilted towards the easy direction in Fe; at 2 nm—a distorted vortex with biquadratic (90°) coupling. At a spacer thickness of 2.5 nm, which is too large to influence the magnetization in the Co layer, a regular vortex structure appeared [177]. In 2–4 μm large Co/Cu/Py/Cu(001) trilayer disks also a loss of the vortex structure was observed at a Cu thickness which causes AFM coupling [178]. These two experiments show that for practical devices, which consist of several magnetic layers, some of them with considerable magnetic anisotropy, the domain configurations may be very different from those found in single layer patterns.

Another interesting question is how reduced lateral dimensions influence the domain structure in ferromagnetic/antiferromagnetic films. This was studied with periodic epitaxial $LaFeO_3/La_{0.7}Sr_{0.3}MnO_3$ zigzag patterns on $SrTiO_3(001)$ with line widths varying from 2 μm down to 200 nm, a masterpiece of specimen preparation [179]. The magnetic ordering temperature of the AFM layer is higher than that of the FM layer so that the AFM domain structure can be studied below and above the ordering temperature of the FM layer. Combined XMLDPEEM and XMCDPEEM measurements showed that below a critical width, which depends on the line orientation with respect to the substrate, the exchange coupling favors collinear alignment of the spins in AFM and FM layers in contrast to the perpendicular alignment above this thickness. For the interesting details see Ref. [179], where also the work leading up to this study is cited.

All studies mentioned up to now made use of specimens, which were prepared ex situ by electron beam lithography, focused ion beam milling or, as in the last example, by more elaborate methods. These allow producing patterns with well-defined dimensions. There have also been some studies in which the samples were prepared in situ in the microscope, most of them making use of the fact that many materials tend to form at high temperatures three-dimensional crystals after an initial wetting layer on the substrate (Stranski-Krastanov growth mode, see Sect. 5.3.1.1). Structures with desired shape can be obtained by either deposition at high temperature or by annealing films deposited at low temperature. This approach was used first in ex situ preparation for XMCDPEEM studies of μm size Fe crystals on Mo(110). The observed complex domain patterns could be explained by three-dimensional micromagnetic simulations, which showed that the magnetization depends more on the particle size and thickness than on the details of its shape [180]. In situ preparation is necessary for SPLEEM studies. This was done by annealing or growth of Fe films on W(110) at 600–720 K [181, 182]. Annealing of continuous films causes a strain-induced breakup into crystals with a wide variety of shapes, depending on film thickness, annealing temperature, and surface topography. This allows production of (a) small crystals whose domain structure is mainly determined by exchange and stray field energy or (b) large flat crystals in which the surface and magnetoelastic energy produces a single domain state with the magnetization pointing in the $[1\bar{1}0]$ direction, the easy direction in Fe films on W(110) below 50-nm thickness or (c) narrow Fe wires along the [001] direction in which shape anisotropy dominates the magnetization. Figure 7.12 [170] illustrates one of the intermediate structures with variously shaped closure domains and single domain regions. Long thin wire arrays with wire widths less than 100 nm can be obtained by annealing of thick films [181].

In situ preparation by high temperature growth has also been used in an XMCDPEEM study of Fe_3O_4 microcrystals on Ru(0001). In this study reactive Fe deposition in oxygen at 900 K was used to grow Fe_3O_4 microcrystals of various thickness on the FeO wetting layer. The thickness was determined by QSE LEED intensity oscillations. XMCDPEEM showed that crystals at thin as 1 nm were ferrimagnetic up to 520 K [183]. A final example for the use of in situ preparation

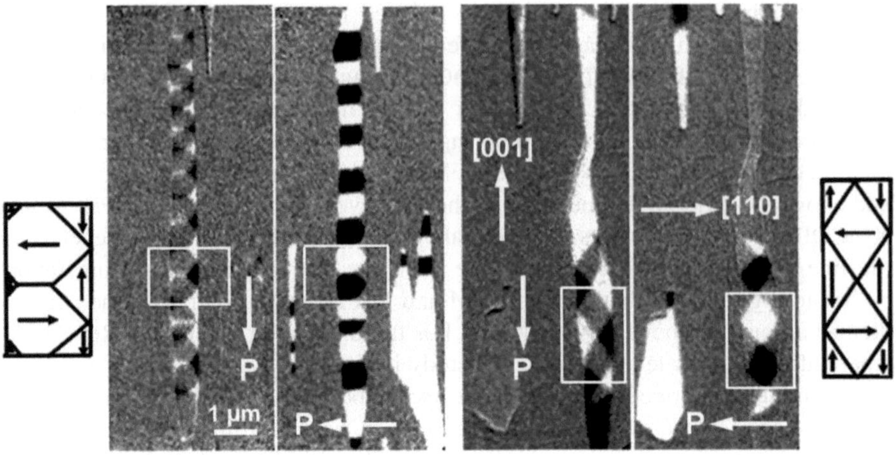

Fig. 7.12 SPLEEM images of domain structures of epitaxial Fe ribbon crystals on W(110) produced by annealing a 5-monolayer thick Fe film at 650 K. The image pairs were taken with the polarization vector of the electrons parallel to the [001] and $[1\bar{1}0]$ directions. The magnetization distribution in the *marked* regions is indicated on the *sides*. Adapted from Ref. [170] with permission by John Wiley & Sons

are FePd nanowires, which were obtained by first growing self-organized Pd stripe nanopatterns on partially oxygen-covered W(110) at high temperature followed by deposition of up to 4 Fe monolayers at 500 K. At a wire width of 30 nm and period of 50 nm XMCDPEEM showed that the magnetization was not along the wires as expected on the basis of shape anisotropy but perpendicular to it [184].

7.5.3 Field and Current Influence

For practical applications of nanostructures it is important to understand the influence of applied magnetic fields and of currents flowing through them on the domain structure, in particular on the domain walls. Various aspects of this understanding can be obtained with several methods: imaging in a magnetic field allows to measure characteristic magnetic properties such as coercivity or magnetic susceptibility on a submicroscopic scale, imaging after application of magnetic field or current pulses gives information on the response of the domain structure and domain walls to this disturbance and pump-probe experiments allow to follow the dynamics of the processes involved in this response.

Imaging in magnetic fields normal to the specimen of sample is relatively straightforward with a high Ampere-turn coil with a high permeability core below the specimen, which has to be well-aligned to the optical axis. This has been done with a continuous film already in the early years of magnetic imaging in SPLEEM [185]. Imaging in fields parallel to the specimen surface is more problematic because of the beam deflection in the absence of perfect shielding but has been achieved on macroscopic samples with magnetic window frame specimens (Sect. 3.2.9) and on nanostructures with well-designed shielding. Imaging *after* application of a magnetic field or current pulse shows the state into which the system has relaxed and gives only partial information on its state in the field or current. The pump-probe method, first applied to continuous films [186], is now widely used in synchrotron radiation facilities for the study of nanostructures. In the following some examples of the application of these methods will be given, starting with in-field studies.

A well-designed magnetic yoke was used to image the magnetization reversal of an exchange-coupled permalloy/FeMn bilayer on a Cu-covered Si(110) surface. In-plane magnetic fields up to 33 mT could be applied without reducing the resolution below the resolution limit of the nanoESCA instrument, revealing the complex domain structures caused by the coupling. The zero crossings of local hysteresis curves allowed determining local variations of the exchange bias [187]. Using the same magnetization specimen holder, hysteresis curves from Fe nanoparticles with edge lengths as small 18 nm could be measured with XMCDPEEM in the same instrument, although the particles, whose size and shape was determined with SEM, appeared much larger due to resolution limitations [188]. With an AC magnetic field applied to the specimen, synchronized with the image detection of the instrument, the magnetic AC susceptibility of 3 μm diameter Fe discs could be measured as a function of temperature by choosing the film thickness such that the Curie temperature was below the agglomeration temperature of the film [188]. These examples illustrate the possibilities of in-field measurements.

Most work up to now has been done, however, by imaging *after* field or current pulses, starting with a XMCDPEEM study of 30 nm thick Co triangles on Si with edge lengths of about 30 μm. The images showed interesting changes in the domain structure along the "hysteresis curve," which depended upon the direction of the magnetization [189]. A more recent example is a XMCDPEEM study of the switching of submicron U- and H-shaped permalloy patterns, planned for domain wall "seasaw" memory applications, by magnetic fields. In this experiment the switching probability and reliability as a function of the applied field was determined by using large arrays of these patterns [190].

By far the most important application of imaging after pulsing is its use with *current* pulses in thin wires to study the movement and blocking of domain walls. These studies are driven by the development of magnetic domain wall memory (IBM-Hitachi Racetrack, NEC magnetic domain wall motion cell), switching and detection technology. For spin transfer torque-driven domain wall motion to occur high current densities are necessary. This can be achieved with acceptable power consumption only in thin submicron wires. The first current-induced domain wall propagation studies in a wire with MFM were soon followed by XMCDPEEM studies of 28 nm thick permalloy zigzag wires with widths between 100 nm and 2 μm, lengths up to 100 μm and current densities of to about 11×10^{11} A/m^2 in 10 μs pulses The domain walls, initially located at the kinks of the wires, started to

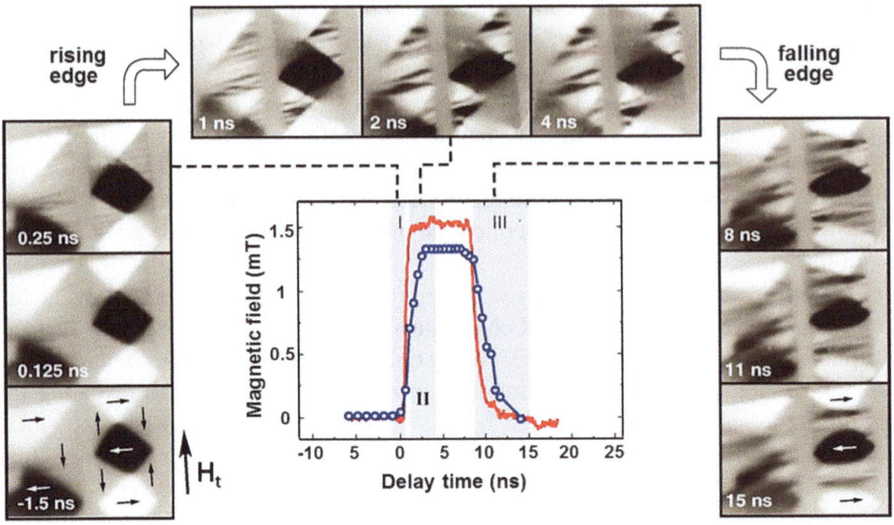

Fig. 7.13 Pump-probe imaging of the evolution of the magnetic domain structure in 40 nm thick rectangular permalloy patterns, the smaller one having the dimensions 20 μm × 80 μm, in an in-plane magnetic field pulse along the long direction of the pattern. The *red* and *blue* curves show the shape of the field pulse shape, measured in two different ways, the *arrows* in the images the direction of the local magnetization. The processes occurring in the various regions (I–III) are described in [194]. Reproduced with permission from Ref. [194]. Copyright 2005 Elsevier

move at about 8×10^{11} A/m^2. Single vortex walls, which did not transform into extended or multiple vortex walls moved fastest. With decreasing wire width vortex wall nucleation decreased, resulting in better single vortex wall propagation [191]. A more recent XMCDPEEM study of 5 nm NiFe/8 nm Cu/7 nm Co spin-valve trilayer zigzag wires, 200 and 300 nm wide, determined the wall displacement as a function of pulse duration (3–200 ns) and current density (2×10^{11}–4.2×10^{11} A/m^2) and found a wide spread of the velocities from 0 up to 600 m/s and of the wall displacements ranging from 0 to about 4 μm. This spread was attributed to a large number of pinning sites resulting from the dipolar interaction of the NiFe layer with magnetic inhomogeneities in the Co layer [192]. More information on this subject can be found in [18].

The first application of the pump-probe method to nanostructures was a study of the vortex movement in 20 nm thick, μm size FeCo patterns upon application of 200 ps magnetic field pulses produced in a Cu waveguide as described in Sect. 3.2.9. It found that the vortex performed a precessional motion whose handedness was determined by the direction of the vortex and was much faster than expected from the static susceptibility [193]. Studies on larger microstructures show that not only the vortex core moves but that the whole magnetization distribution changes in them. This is illustrated in Fig. 7.13 [194] for two rectangular permalloy patterns, which shows the evolution of the domain structure during and after the application of an in-plane pulse with a maximum field of about 2 mT. In this particular

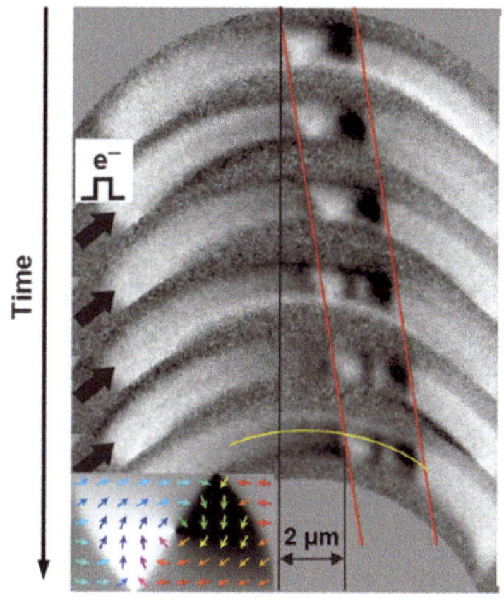

Fig. 7.14 Vortex domain wall motion caused by current pulses with 100 ps rise time injected from the left into a permalloy wire, imaged with XMCDPEEM. The fourth pulse nucleates a second vortex wall but the wall velocity of 130 m/s, obtained from pulse duration (3 ns) and wall displacement/ pulse (400 nm) did not change. The *inset* shows the magnetization distribution in the initial vortex wall. Reproduced from Ref. [196] with permission from the American Institute of Physics, © 2010

experiment the 48 ps light pulses had a repetition time of 176 ns, which required image accumulation over 10^8 pulses for each delay time, a value typical for pump-probe experiments. The duration of the field pulse as measured by the pulser output and on the coplanar waveguide is indicated in the center of the figure [194]. The magnetocrystalline anisotropy has a strong influence on the response of the magnetization distribution as demonstrated in a comparison of epitaxial $10 \times 10\ \mu m^2$ permalloy and Fe patterns: in the permalloy pattern the magnetization can rotate freely while in the Fe pattern it is pinned to the four easy axes. In this experiment electron beam deflection gating has been used to control the image acquisition period between the photon and field pulses, allowing complete relaxation of the magnetization after the disturbance caused by the pulses [195]. In trilayer systems such as encountered in spin valves the response of the magnetization distribution is more complex due to the different response of the layers involved to the applied field, due to the interlayer coupling and due to the magnetocrystalline anisotropy as illustrated in a pump-probe study of Co/Cr/Fe and FeCo/Cr/FeNi microstructures [177].

Pump-probe measurements are of particular importance in nanowire studies because they give detailed insight in the domain wall motion processes under the influence of the spin current. For example, one study, which used a short 40 nm thick, 1 μm wide curved permalloy wire, 3 ns current pulses with ultrashort rise time (<100 ps) and a peak current density of about $4 \times 10^{12}\ A/m^2$ found a high domain wall velocity independent of the complexity of the domain wall—contrary to the earlier observations mentioned before—which was attributed to the additional spin torque connected with the fast rise of the current pulse. Figure 7.14

shows the domain wall motion with the number of pulses [196]. A study of 400 nm wide 5 nm CoFeB/8 nm Cu/7 nm Co zigzag wires with 90° corners brought out another important factor for domain wall motion and pinning: the Oersted field associated with the current which is perpendicular to the current direction: it caused a tilt of the magnetization direction as large as 75° at a current density of 7×10^{11} A/m^2 [197]. A final example is an experiment with perpendicularly magnetized nanowires, the most promising configuration for spin torque devices. Twenty 500 nm wide, 10 μm long 3 nm Pt/0.6 nm Co wires in parallel connection to a bipolar pulsed voltage source were studied at two current densities (7.7×10^{11} and 1.3×10^{12} A/m^2). The first part of the pulse, 120 or 160 ns long, was used for the XMCDPEEM measurements in time steps of 5–20 ns, the second for restoring the original domain wall position. The results showed that during the 120 ns 7.7×10^{11} A/m^2 pulse the domain wall displacement was linear, corresponding to a domain wall velocity of 7 m/s, with very little delay time relative to the pulse onset. This small domain wall inertia was attributed to a narrow domain wall width and the Rashba interaction caused by the Pt/Co inversion asymmetry [198].

7.5.4 Nanodots and Nanostructure Arrays

The applications discussed up to now were above the resolution limit of the instruments used and the structures were sufficiently separated so that no interaction occurred between them. Now we briefly consider the few studies of structures below or at the practical resolution limit of the instruments used in them. A SPLEEM study looked at the relative stability of single domain (SD) and vortex (V) states in small epitaxial Co nanodots grown on the Co wetting layer on Ru (0001) above 700 K. Depending upon Co dose (up to 3 monolayers) and growth temperature (700–860 K) the Co particles were 20–800 nm wide and 1–4 nm thick. The SD states could be switched to V states by annealing and the V states to SD states by in-plane magnetic field pulses, in qualitative agreement with a state phase diagram obtained from micromagnetic calculations [199].

The magnetic properties of particles below the resolution limit of XMCDPEEM imaging were studied by combining chemical-sensitive PEEM imaging with XMCD of selected particles [200]. Fe particles ranging in size from 6 to 25 nm were prepared in an arc cluster ion source operating at a noble gas pressure around 20 mbar, deposited on 13 nm thick polycrystalline Co films on silicon and transferred to the PEEM connected to the cluster source. The dimensions of the particles were determined with AFM and SEM and their crystal structure and shape was known from other studies to be bbc and bounded by {100} planes with the edges cut off by {110} planes. In PEEM the particles were identified by the ratio of Fe L_3 edge to pre-edge absorption images and spectra, their magnetization was obtained by acquiring their XMCD spectra, adding up many images and spectra for sufficient signal to noise ratio. Careful studies showed that a qualitative analysis of the electronic and magnetic properties was possibly down to a particle size of 6 nm and that they were comparable to those of bulk Fe [200]. A detailed comparison of the magnetization direction $\boldsymbol{M}$ in the Fe particles with that in the Co substrate showed a particle size-dependent misalignment between the two direction ranging from 0° in the smallest particles to 90° in the largest particles. Numerical calculations, assuming that the particles are exchange coupled to the ferromagnetic substrate, explained this phenomenon by parallel magnetization at the interface and continuous rotation to perpendicular magnetization with increasing thickness, forming a spin spiral. Because of their fixed shape small and thus thin particles have parallel $\boldsymbol{M}$ while large, thick particles show perpendicular $\boldsymbol{M}$ because of the limited sampling depth [201]. These two studies are a good illustration of what can be achieved at the limits of cathode lens microscopy in combination with complementary methods. Another study of such nanocubes was mentioned already before in connection with local hysteresis measurements [188]. In this case there was no ferromagnetic support and the particles had to be cooled below the blocking temperature to stabilize the direction of their magnetization.

Nanopattern arrays are of interest because of the competing interactions between their constituents as reviewed in [202]. They can be prepared by several methods: (1) by electron beam lithography from continuous films, (2) by growth on self-organized or self-assembled surfaces or (3) by sequential deposition of a ferromagnetic material with a nonferromagnetic material in such a manner that closely spaced but nonpercolating ferromagnetic particles are formed. Here first some examples using the second two methods will be discussed. As seen in Sect. 5.3.1.2 and described in detail in [203] Ge and Si deposition can lead to a variety of self-organized patterns. One of them, which showed a lateral feature distance of about 500 nm after coating with 25 monolayer of Au, has been used as substrate for the formation of self-organized Co/Au and Au/Co/Au arrays by grazing incidence deposition of Co. XMCDPEEM, laterally averaging XMCD and EXAFS studies of the resulting array showed significant differences from corresponding observations in continuous flat film structures (Sect. 7.2), in particular that the Co films are in-plane oriented at nearly all thicknesses due to magnetostatic coupling between the dots and that out-of-plane magnetization is seen only at low temperature (100 K), which was attributed to an increase of the orbital magnetic moment [204].

Self-assembly of spherical particles is a method, which produces periodic nanostructure arrays. 20 and 40 nm thick permalloy layers were deposited at normal incidence onto Ta-covered arrays of SiO_2 spheres with 100, 330, and 800 nm diameter. This produced a period array of closely packed permalloy caps, whose magnetization distribution was studied with XMCDPEEM in the remanent state and also in an in-plane magnetic field. Isolated caps revealed vortex states but the interaction of 100 and 330 nm diameter caps with their neighbors caused the chirality to alternate and to produce frustration in the chirality. The interaction between 800 nm diameter caps is too weak to cause any correlation between the chiralities, which allows forming close-packed vortex arrays. In the magnetic field sweeps through the hysteresis curve vortex annihilation and nucleation occurred at significantly higher field values than in closely packed arrays of planar disks, allowing better stabilization of the vortex remanent state [205].

An example for the third method is the $Co_{80}Fe_{20}$ nanoparticle ensemble embedded in Al_2O_3 [206]. Multilayers are produced by alternating deposition of small amounts of the alloy and of 3 nm alumina. The alloy does not wet the alumina so that it grows in the Weber-Volmer mechanism, i.e., it forms small particles, which are immobilized by the subsequent alumina deposition. The mean size d and distance s can be controlled by the deposition temperature and by the deposited thickness t, e.g., for $t = 0.9$ nm, $d = 2.8$ nm and $s \approx 6$ nm. Variation of t changes s and thus the dipolar coupling between the particles, which allows formation of a variety of magnetic states such as the superferromagnetic state in which the spins of the superparamagnetic particles are coupled by the dipolar interaction between them. An example is a [1.3 nm $Co_{80}Fe_{20}$/3 nm $Al_2O_3]_{10}$ film. When grown in an applied field $\boldsymbol{H}_g$ it shows in XMCDPEEM in the demagnetized state lamellar domains aligned along the easy axis determined by $\boldsymbol{H}_g$. It can be transformed into a homogeneously magnetized state by a field as low as 0.2 mT, which makes these films interesting for applications requiring high permeability [206].

Electron beam lithography, combined with focused ion beam milling has been used mainly to produce two-dimensional artificial spin ice, a subject which has received increasing interest because it is an excellent model system for the study of some fundamental physics problems such as magnetic monopoles and Dirac strings, a subject extensively reviewed in [207]. Reference [202] also gives a good description of spin ice systems so that here only a few examples of XMCDPEEM studies will be mentioned. They were made with kagomé lattices, whose elementary units consist of six elongated islands arranged in a hexagonal pattern, separated by a small distance at their ends (nodes). Their aspect ratio and size is chosen so that they are single domain dipoles. In a single building block the six dipoles are oriented such that they form a right- or left-handed chirality pattern in the lowest energy configuration, but when joined to other units in a kagomé pattern this is no longer possible: the resulting system is "frustrated." A good introduction to the increasing number of possible dipole arrangements and resulting degree of frustration with increasing number of joined units can be found in [208].

In an ideal kagomé spin ice the spins are assumed to interact only via nearest neighbor exchange coupling, which are short range, while in the dipolar kagomé ice the interactions are dipolar (magnetostatic), which are long range. A study of a lattice consisting of 342 Co dipoles with $470 \times 70 \times 10\ \mathrm{nm}^3$ (length × width × thickness) with 500 nm separation at the nodes actually showed that the long range interactions lift the ground state degeneracy of the ideal spin ice caused by the multitude of configurations. In the *ground* state, achieved by careful demagnetization only 0.6 % of the nodes showed in XMCDPEEM a wrong configuration but a detailed analysis showed significant deviations of the spin correlations between different dipoles from those of the ideal spin ice model [209]. *Excited* states can be produced by applying in-plane fields. Such studies were performed with a huge ($50 \times 50\ \mu\mathrm{m}^2$) kagomé of $470 \times 160 \times 20\ \mathrm{nm}^3$ permalloy dipoles with 500 nm dipole-dipole center distance, an impressive feat of specimen preparation. Initially a field was applied high enough to orient all dipoles in one direction. Subsequent lower fields in the opposite direction induced locally dipole-coupled dipole strings with the reversed orientation ("Dirac strings") terminated by magnetic monopoles (nodes at which the magnetization of all three dipoles either point towards the node or away from it). With increasing field the strings proliferate until all dipoles are oriented in the same direction. Counting the number of reversed dipole via their XMCDPEEM contrast the hysteresis curve and other parameters characterizing the spin ice response can be measured [210, 211].

Concluding this section, the examples shown illustrate that cathode lens microscopy not only makes important contributions to the understanding of technical devices and their improvement but also to the understanding of fundamental problems. For this reason the study of magnetic nanostructures has become a major, steadily growing field in research with PEEM and LEEM.

7.6 Ferroelectrics/Multiferroics

7.6.1 Ferroelectrics

The surfaces of ferroelectric crystals were studied with cathode lens microscopy already long before UHV instruments were available. Spivak et al. imaged already 1959 ferroelectric domains on a $BaTiO_3$ (BTO) single crystal with MEM [212] and thereafter several groups studied this crystal and other ferroelectrics such as $PbZrTiO_3$ (PZT), $Gd_2(MoO_4)_3$ (GMO), and triglycine sulfate (TGS) with Mayer's MEM instrument and the JEOL MEM (see Chap. 1) References to this early work can be found in [213].

Except for a study of field induced ferroelectric electron emission from TGS [214] interest in ferroelectrics did not revive until the development of ferroelectric random access memory started. In these memories positive and negative polarized micropatterns (P^+, P^-) store the information. Cathode lens microscopy became an important tool for the understanding of these patterns and their properties. The first study used the tunable spontaneous emission of a UV FEL (see Sect. 3.2.7) to image patterns with opposite polarity on the (0001) surface of $LiNbO_3$ (LNO). Domain contrast appeared at photon energies above 4.6 eV in the negative poled regions and dominated at least up to 6.2 eV, at which energy emission became also stronger from the positive poled regions. A polarity-patterned 200 nm thick tetragonal (111)-oriented PZT film on Pt showed similar differences between the two oppositely poled regions, with the emission difference irreversibly disappearing upon heating above the Curie temperature T_C (about 300 °C). The domain contrast was attributed in both cases to differences in the adsorbate dipole layers on the uncleaned surface, resulting in electron affinity differences [215].

The influence of adsorbates on the contrast was clearly demonstrated in a later study combining O 1s and C 1s XPS with MEM-LEEM transition measurements before and after in situ cleaning by heating in oxygen, which completely eliminated the C 1s signal. In this study the sample was a (100)-oriented PZT film on a conducting $SrRuO_3$ layer on a $SrTiO_3(100)$ single crystal. Although the surface potential difference between P^+ and P^- regions increased only from 50 to 90 meV upon cleaning, a significant increase in contrast was seen, mainly caused by the stronger cleaning of the P^- regions [216]. Screening of the surface charge by water adsorption has been clearly demonstrated by measuring the surface potential difference between P^+ and P^- regions via the MEM-LEEM transition. A much stronger effect was observed by screening with photo-generated carriers upon UV illumination, which is connected with significant delay time [217]. The factors, which determine the contrast are obviously difficult to pin down. In this connection the results of a study of the temperature dependence of the domain contrast on the tetragonal $BaTiO_3$(001), imaged with 4.28 eV light from a pulsed laser should be mentioned. On this surface three domains occur (P^+, P^-, and an in-plane polarized domain P_i). Above T_C there is still domain contrast, but the contrast between P^- and P_i, which is due to the work function difference, inverted and continuously decreased with temperature. The contrast above T_C has been attributed to near-surface domains with in- and out-of-plane tetragonality, which do not need to be ferroelectric, on top of the cubic paraelectric bulk [218]. This example shows that contrast on ferroelectric surfaces is not necessarily purely due to polarization. A comparison of different imaging methods MEM, PFM (piezoresponse force microscopy), absorption XPEEM (i.e., secondary electron imaging) and other imaging methods should help to untangle the various contributions to the contrast [219]. A review of the results obtained up to 2013 and of the methods with which they were obtained illustrates these problems and the present understanding of this field [220].

7.6.2 Multiferroics

Multiferroic or magnetoelectric materials, which combine magnetic and electric order, have become one of the most exciting fields in solid state physics at the beginning of this century. For this reason there are already many reviews of the bulk aspects of these materials of which only the most recent ones are cited here for background information [221–224]. These materials are also attracting increasing attention in cathode lens microscopy and some of the work done to date will be briefly sketched here. Multiferroic behavior can either be an inherent property of a material such as $BiFeO_3$ (BFO) or can be engineered by coupling of a ferroelectric material with a magnetically ordered material. As their study requires chemical and magnetic information all cathode lens microscopy work has been done up to now with XPEEM.

We first consider studies of multiferroic materials. The only one whose ordering temperature is above room temperature is $BiFeO_3$, which is antiferromagnetic with a Neél temperature T_N of 643 K and a ferroelectric Curie temperature T_C of 1143 K. As no large single crystals are available most of the work has been done with epitaxial films. In the first study 600 nm thick epitaxial films were grown on a conducting $SrRuO_3$ layer on a $SrTiO_3$(100) surface. The film surface was imaged with piezoelectric force microscopy (PFM) and XLDPEEM at the Fe L_3 edge in

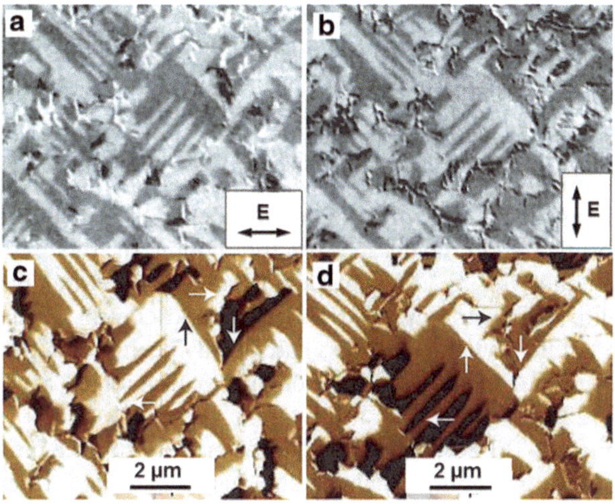

Fig. 7.15 XLDPEEM images (**a**, **b**) and piezoelectric force microscopy (PFM) images (**c**, **d**) of a (100)-oriented epitaxial $BiFeO_3$ layer at room temperature. In (**b**) and (**d**) the sample was rotated by 90°. The contrast change from (**a**) to (**b**) resulting in the 90° change of the ***E*** vector with respect to the sample is an indication of antiferromagnetism. The *arrows* in (**c**) and (**d**) show the directions of the in-plane component of ferroelectric polarization. Reproduced from Ref. [225] with permission from Nature Publishing Group (UK)

order to determine the correlation between ferroelectric and antiferromagnetic (AFM) domain structure before and after domain switching by electrical poling of the ferroelectric domains with the PFM tip. As Fig. 7.15 [225] shows for two different X-ray polarizations, there is perfect agreement between the domains. That the XLD contrast was not simply caused by the ferroelectric asymmetric charge distribution but also by antiferromagnetism was confirmed by XLDPEEM imaging at various temperatures up to and above the T_N. The antiferromagnetic order parameter obtained from the XLD values showed a clear decrease up to T_N above which it was less than half of the room temperature value. This showed that most of the XLD contrast was due to antiferromagnetism (XMLD) and that the correlated switching seen with PFM and XLDPEEM was due to the switching of the AFM domains with the ferroelectric domains [225].

A later very detailed XLDPEEM study as a function of film thickness showed that the magnetic structure of BFO depended strongly on film thickness due to the strain caused by the misfit between film and substrate, which leads to significant crystal distortions. While thick films ($\approx$1 μm) were similar in their magnetic structure to the bulk, having six nearly equivalent magnetization directions ($\langle 1\bar{1}0 \rangle$ and $\langle 11\bar{2} \rangle$) in the plane normal to the $\langle 111 \rangle$ polarization direction, thin films ($\approx$20–200 nm) developed a strong preference for $\langle 11\bar{2} \rangle$ directions perpendicular to the 111 polarization direction and pointing out of the surface. This is an important aspect for the exchange coupling between FM and BFO layers [226]. In further studies of epitaxial $BiFeO_3$ films it was found that the high strain produced an intimate mixture of a highly strained rhombohedral (R) and a tetragonal (T) phase of $BiFeO_3$, which produced a clear XMCD signal, i.e., indicated a net magnetic moment of the mixture, in contrast to the negligible and zero XMCD signal of the pure R and T phases, respectively. XMCDPEEM of a 2.5 nm thick $Co_{.90}Fe_{.10}$ film on top of this layer showed rectangular domains, coincident with the topography of the layer. Based on AFM and PFM imaging these domains could be attributed to the R phase. Angular-dependent studies indicated that moments in the CoFe film are magnetically coupled to the moment in the R-phase of the BFO layer and electric poling of this ferroelectric layer with voltages around 10 V by PFM tip scanning allowed to switch the magnetization in the FeCo film, demonstrating clearly electric-field control of magnetism [227]. A detailed study of the switching process had already been shown previously that the average magnetization direction in the ferromagnetic film rotates by 90° upon electric field switching [228].

The first study of a $BiFeO_3$ single crystal was made with a 0.8 mm long, 0.36 mm wide, and 0.05 mm thick multidomain crystal. In this case the ferroelectric domains were imaged with XLDPEEM at the O 1s edge similar to the method used to measure the twin domains in NiO discussed earlier. Several domains were seen, which were attributed to the various $\langle 111 \rangle$ polarization directions. Imaging of the AFM domains with Fe L_2 XMLDPEEM was hampered by artifacts but locally several AFM domains inside larger ferroelectric domains could be identified. Obviously working with such small insulating crystals is difficult [229].

Ferromagnetic-ferroelectric coupling does not necessarily require that the ferroelectric partner is magnetically ordered. Important is the close proximity and contact between the two partners. This was first discovered in MFM studies of specially prepared mixed $CoFe_2O_4$–$BiFeO_3$ (CFO–BFO) films [230, 231]. The two oxides are immiscible and when co-deposited at high temperature they phase separate. Co-deposition of a 35:65 mixture on a $SrTiO_3(100)$ surface covered with a conductive $SrRuO_3$ layer results in a self-organized array of rectangular ferrimagnetic CFO columns with about 100 nm edge length in the ferroelectric BFO matrix. When the magnetization direction of the CFO columns was set by a high magnetic field, the magnetization could be inverted by poling with voltages of less than ± 20 V, demonstrating clearly electric magnetization switching. Later XPEEM and XMCDPEEM studies at the Fe and Co edges showed that the spin in both Co and Fe magnetic sublattices was completely reversed in about 50 % of the pillars by the magnetoelectric coupling between BFO and CFO [232].

Magnetoelectric coupling between ferromagnetic oxides such as CFO and NFO ($NiFe_2O_4$) with $BaTiO_3$ (BTO), which is only ferroelectric but not antiferromagnetic, so that exchange coupling can be excluded, should be easier to analyze. Combined Fe L_3 XMCDPEEM, Fe L_2 XLDPEEM, and Ti L_3 XLDPEEM studies of about 30 nm thick epitaxial CFO of NFO films on $BaTiO_3(001)$ have indeed given considerable insight in the correlation between the ferroelectric domain structure of BTO, the strain, and the magnetization in the films, whose average magnetic domain size was about 300 nm. The surface ferroelectric domain structure of BTO is clearly transferred to the ferromagnetic films via the strong elastic interaction between film and substrate and the strain in the film has a strong influence on the magnetic anisotropy. This strain-mediated coupling at the ferroelectric–ferromagnetic interface thus produces an artificial multiferroic system consisting of two nonmultiferroic components [233].

As a final example of the application of XMCDPEEM a study of strain-mediated magnetocaloric effects in epitaxial ferromagnetic films on BTO should be mentioned. $La_{0.7}Ca_{0.3}MnO_3$ (LCMO) shows a weak magnetocaloric effect at the transition from the ferromagnetic to the paramagnetic state. LCMO films on BTO show a sharp jump of the magnetization near the rhombohedral-orthorhombic phase transition of BTO at about 200 K, which is connected with a strong entropy change, causing a giant magnetocaloric effect. XMCDPEEM was used to study how this magnetization change occurs by imaging the magnetic domain structure at the surface above and below the transition. A complex pattern of many randomly sized and distributed submicron domains was observed. Its analysis showed that it was not a change of the magnetization of the domains but of the area fraction of magnetic domains which caused the change of the macroscopically measured magnetization [234].

Concluding this section, it is clear that cathode lens microscopy is well suited for the study of multiferroics thanks to the information it can provide on chemical composition, strain, and ferroelectric, ferromagnetic, and antiferromagnetic ordering. From a broader point of view, cathode lens electron microscopy, in particular the various XPEEM modes, has already contributed invaluable information to the understanding of magnetic phenomena and is expected to do so even more with instrument improvements such as higher spatial resolution and more versatile specimen manipulation (azimuthal rotation, magnetic field application, liquid He cooling, etc.).

References

1. Stöhr, J., Siegmann, H.C.: Magnetism. Springer, Berlin (2006)
2. Altman, M.S., Hurst, I., Marx, G., Pinkvos, H., Poppa, H., Bauer, E.: Spin polarized low energy electron microscopy of surface magnetic structure. In: Suzuki, Y., et al. (eds.) MRS Symp. Proc., Philadelphia 1991, pp. 125–132. MRS Society, Philadelphia, PA (1991)
3. Stöhr, J., Wu, Y., Hermsmeier, B.D., Samant, M.G., Harp, G.R., Koranda, S., Dunham, D., Tonner, B.P.: Element-specific magnetic microscopy with circularly polarized x-rays. Science **259**, 658–661 (1993)
4. Stöhr, J., Scholl, A., Regan, T.J., Anders, S., Lüning, J., Scheinfein, M.R., Padmore, H.A., White, R.L.: Images of the antiferromagnetic structure of a NiO(100) surface by means of x-ray magnetic linear dichroism spectromicroscopy. Phys. Rev. Lett. **83**, 1862–1865 (1999)
5. Marx, G.K.L., Elmers, H.J., Schönhense, G.: Magneto-optical linear dichroism in threshold photoemission electron microscopy of polycrystalline Fe films. Phys. Rev. Lett. **84**, 5888–5891 (2000)
6. Bauer, E.: SPLEEM. In: Hopster, H., Oepen, H.P. (eds.) Magnetic Microscopy of Nanostructures, pp. 111–136. Springer, Berlin (2005)
7. Bauer, E.: Spin-polarized low energy electron microscopy (SPLEEM). In: Zhu, Y. (ed.) Novel Techniques for Characterizing Magnetic Materials, pp. 361–379. Kluwer Academic Publ, Boston, MA (2005)

8. Rougemaille, N., Schmid, A.K.: Magnetic imaging with spin-polarized low-energy electron microscopy. Eur. Phys. J. Appl. Phys. **50**, 20101, 19 pages (2010)
9. Kuch, W.: X-ray magnetic circular dichroism for quantitative element-resolved magnetic microscopy. Phys. Scripta **T109**, 89–95 (2004)
10. Schönhense, G.: Surface magnetism studied by photoelectron spectromicroscopy with high spatial and time resolution. J. Electron Spectrosc. Relat. Phenom. **137**, 769–783 (2004)
11. Kuch, W.: Imaging magnetic microspectroscopy. In: Hopster, H., Oepen, H.P. (eds.) Magnetic Microscopy of Nanostructures, pp. 1–28. Springer, Berlin (2005)
12. Scholl, A., Ohldag, H., Nolting, F., Anders, S., Stöhr, J.: Study of ferromagnet-antiferromagnet interfaces using x-ray PEEM. In: Hopster, H., Oepen, H.P. (eds.) Magnetic Microscopy of Nanostructures, pp. 29–50. Springer, Berlin (2005)
13. Kuch, W.: Magnetic imaging. In: Beaurepaire, E., Bulou, H., Scheurer, F., Kappler, J.-P. (eds.) Magnetism: A Synchrotron Radiation Approach, vol. 697, pp. 275–320. Springer, Berlin (2006)
14. Feng, J., Scholl, A.: Photoemission electron microscopy (PEEM). In: Hawkes, P.W., Spence, J.C.H. (eds.) Science of Microscopy, vol. 1, pp. 657–695. Springer, New York, NY (2007)
15. Locatelli, A., Bauer, E.: Recent advances in chemical and magnetic imaging of surfaces and interfaces by XPEEM. J. Phys. Condens. Matter **20**, 82202–82224 (2008) (and references therein)
16. Schneider, C.M., Krug, I., Müller, M., Matthes, F., Kaiser, A., Wiemann, C., Cramma, S., Elmers, H.-J., Wegelin, F., Krasyuk, A., Nepijko, S.A., Schönhense, G.: Investigating spintronics thin film systems with synchrotron radiation. Radiat. Phys. Chem. **78**, S5–S10 (2009)
17. Nolting, F.: Magnetic imaging with x-rays. In: Beaurepaire, E., Bulou, H., Scheurer, F., Jean-Paul, K. (eds.) Magnetism and Synchrotron Radiation. New Trends. Springer Proceedings in Physics, vol. 133, pp. 345–366. Springer, Berlin (2010)
18. Kläui, M.: Domain wall spin structures and dynamics probed by synchrotron techniques. In: Beaurepaire, E., Bulou, H., Scheurer, F., Jean-Paul, K. (eds.) Magnetism and Synchrotron Radiation. New Trends. Springer Proceedings in Physics, vol. 133, pp. 367–384. Springer, Berlin (2010)
19. Quitmann, C., Raabe, J., Puzic, A., Kuepper, K.: Dynamics of mesoscopic magnetic objects. In: Beaurepaire, E., Bulou, H., Scheurer, F., Jean-Paul, K. (eds.) Magnetism and Synchrotron Radiation New Trends. Springer Proceedings in Physics, vol. 133, pp. 385–405. Springer, Berlin (2010)
20. Cheng, X.M., Keavney, D.J.: Studies of nanomagnetism using synchrotron-based x-ray photoemission electron microscopy (X-PEEM). Rep. Prog. Phys. **75**, 026501, 29 pages (2012)
21. Kinoshita, T., Arai, K., Fukumoto, K., Ohkochi, T., Kotsugi, M., Guo, F., Muro, T., Nakamura, T., Osawa, H., Matsushita, T., Okuda, T.: Observation of micro-magnetic structures by synchrotron radiation photoelectron emission microscopy. J. Phys. Soc. Jpn. **82**, 021005, 24 pages (2013)
22. Bauer, E., Duden, T., Pinkvos, H., Poppa, H., Wurm, K.: LEEM studies of the microstructure and magnetic domain structure of ultrathin films. J. Magn. Magn. Mater. **156**, 1–6 (1996)
23. Feder, R., Kirschner, J.: Spin-polarized low energy electron diffraction: theory, experiment and analysis of results from W(001)(1x1). Surf. Sci. **103**, 75–102 (1981)
24. Feder, R.: Spin-polarised low-energy electron diffraction. J. Phys. C Solid State Phys. **14**, 2049–2091 (1981)
25. Samarin, S.N., Williams, J.F., Sergeant, A.D., Artamonov, O.M., Gollisch, H., Feder, R.: Spin-dependent reflection of very-low-energy electrons from W(110). Phys. Rev. B **76**, 125402, 8 pages (2007)
26. Fritsche, L., Koller, J.: Potentials in density functional theory and the importance of sum rules. J. Solid State Chem. **176**, 652–670 (2003)

27. Hong, J., Mills, D.L.: Spin dependence of the inelastic electron mean free path in Fe and Ni: explicit calculations and implications. Phys. Rev. B **62**, 5589–5600 (2000)
28. Tolhoek, H.A.: Electron polarization, theory and experiment. Rev. Mod. Phys. **28**, 277–298 (1956)
29. Weber, W., Oberli, D., Riesen, S., Siegmann, H.C.: The electron analogue to the Faraday rotation. New J. Phys. **9**, 01949, 7 pages (1999)
30. Dey, P., Weber, W.: Electron-spin motion: a new tool to study ferromagnetic films and surfaces. J. Phys. Condens. Matter **23**(473201), 1–19 (2011)
31. Hallal, A., Berdot, T., Dey, P., Bismaths, L.T., Joly, L., Bourzami, A., Scheurer, F., Bulou, H., Henk, J., Alouani, M., Weber, W.: Ultimate limit of electron-spin precession upon reflection in ferromagnetic films. Phys. Rev. Lett. **107**, 087203, 5 pages (2011)
32. Maekawa, S., Adachi, H., Uchida, K., Ieda, J., Saitoh, E.: Spin current: experimental and theoretical aspects. J. Phys. Soc. Jpn. **82**, 102002, 23 pages (2013)
33. Jensen, P.J., Bennemann, K.H.: Magnetic structure of films: dependence on anisotropy and atomic morphology. Surf. Sci. Rep. **61**, 129–199 (2006)
34. Vaz, C.A.F., Bland, J.A.C., Lauhoff, G.: Magnetism in ultrathin film structures. Rep. Prog. Phys. **71**, 056501, 78 pages (2008)
35. Elmers, H.-J.: Spin-polarized low-energy electron diffraction. In: Kronmüller, H., Parkin, S. (eds.) Handbook of Magnetism and Advanced Magnetic Materials, pp. 1444–1469. John Wiley & Sons, Chichester (2007)
36. Zdyb, R., Bauer, E.: Coexistence of ferromagnetism and paramagnetism in a ferromagnetic monolayer. Phys. Rev. Lett. **100**, 155704, 4 pages (2008)
37. Duden, T., Bauer, E.: Magnetization wrinkle in thin ferromagnetic films. Phys. Rev. Lett. **77**, 2308–2311 (1996)
38. Duden, T., Bauer, E.: Magnetic domain structure and spin reorientation transition in the system Co/Au/W(110). MRS Symp. Proc. **475**, 283–288 (1997)
39. Stickler, D., Frömter, R., Stillrich, H., Menk, C., Oepen, H.P., Gutt, C., Streit-Nierobisch, S., Stadler, L.-M., Grübel, G., Tieg, C., Yakhou-Harris, F.: Domain size in systems with canted magnetization. Phys. Rev. B **84**, 104412, 7 pages (2011)
40. Zdyb, R., Bauer, E.: Magnetic domain structure and spin reorientation transition in ultrathin Fe-Co alloy films. Phys. Rev. B **67**, 134420, 8 pages (2003)
41. Zdyb, R., Menteş, T.O., Locatelli, A., Niño, M.A., Bauer, E.: Magnetization and structure of ultrathin Fe films. Phys. Rev. B **80**, 184425, 9 pages (2009)
42. Ślęzak, T., Ślęzak, M., Zajac, M., Freindl, K., Kozioł-Rachwał, A., Matlak, K., Spiridis, N., Wilgocka-Ślęzak, D., Partyka-Jankowska, E., Rennhofer, M., Chumakov, A.I., Stankov, S., Rüffer, R., Korecki, J.: Noncollinear magnetization structure at the thickness-driven spin-reorientation transition in epitaxial Fe films on W(110). Phys. Rev. Lett. **105**, 027206, 4 pages (2010)
43. Ślęzak, M., Ślęzak, T., Freindl, K., Karás, W., Spiridis, N., Zając, M., Chumakov, A.I., Stankov, S., Rüffer, R., Korecki, J.: Perpendicular magnetic anisotropy and noncollinear magnetic structure in ultrathin Fe films on W(110). Phys. Rev. B **87**, 134411, 8 pages (2013)
44. Vescovo, E., Menteş, T.O., Sadowski, T., Ablett, J.M., Niño, M.A., Locatelli, A.: Domain faceting in an in-plane magnetic reorientation transition. Phys. Rev. B **82**, 184405, 4 pages (2010)
45. Wu, Q., Zdyb, R., Bauer, E., Altman, M.S.: Growth, magnetism and ferromagnetic thickness gap in Fe films on the W(111) surface. Phys. Rev. B **87**, 104410, 9 pages (2013)
46. Man, K.L., Zdyb, R., Feng, Y.I., Chan, T., Altman, M.S., Bauer, E.: Growth morphology, structure and magnetism of ultrathin Co films on W(111). Phys. Rev. B **67**, 184402, 11 pages (2003)
47. Man, K.L., Altman, M.S., Poppa, H.: Spin polarized low energy electron microscopy investigations of magnetic transitions in Fe/Cu(100). Surf. Sci. **480**, 163–172 (2001)
48. Man, K.L., Ling, W.L., Paik, S.Y., Poppa, H., Altman, M.S., Qiu, Z.Q.: Modification of initial growth and magnetism in Fe/Cu(100). Phys. Rev. B **65**, 024409, 11 pages (2001)

49. Biedermann, A., Tscheliessnig, R., Schmid, M., Varga, P.: Local atomic structure of ultrathin Fe films grown on Cu(100). Appl. Phys. A **78**, 807–816 (2004)
50. Wuttig, M., Liu, X.: Ultrathin Metal Films: Magnetic and Structural Properties. Springer Tracts in Modern Physics, vol. 206. Springer, Berlin (2004). Chapter 6
51. Wormeester, H., Hüger, E., Bauer, E.: Growth and electronic structure of thin epitaxial Pd and Co films on W(100). Phys. Rev. B **54**, 17108–17117 (1996)
52. Hüger, E., Wormeester, H., Bauer, E.: Hexagonal metal modifications and thin-film ferromagnetism. Surf. Sci. **438**, 185–190 (1999)
53. Kuch, W., Fukumoto, K., Wang, J., Nolting, F., Quitmann, C., Ramsvik, T.: Thermal melting of magnetic stripe domains. Phys. Rev. B **83**, 172406, 4 pages (2011)
54. Nakagawa, T., Watanabe, K., Matsumoto, Y., Yokoyama, T.: Magnetic circular dichroism photoemission electron microscopy using laser and threshold photoemission. J. Phys. Condens. Matter **21**, 314010, 6 pages (2009)
55. Kronseder, M., Minár, J., Braun, J., Günther, S., Woltersdorf, G., Ebert, H., Back, C.H.: Threshold photoemission magnetic circular dichroism of perpendicularly magnetized Ni films on Cu(001): theory and experiment. Phys. Rev. B **83**, 132404, 4 pages (2011)
56. Nakagawa, T., Yokoyama, T.: Laser induced threshold photoemission magnetic circular dichroism and its application to photoelectron microscope. J. Electron Spectrosc. Relat. Phenom. **185**, 356–364 (2012)
57. El Gabaly, F., Gallego, S., Munoz, C., Szunyogh, L., Weinberger, P., Klein, C., Schmid, A.K., McCarty, K.F., de la Figuera, J.: Imaging spin-reorientation transitions in consecutive atomic Co layers on Ru(0001). Phys. Rev. Lett. **96**, 147202, 4 pages (2006)
58. Gao, J.H., Girard, Y., Repain, V., Tejeda, A., Belkhou, R., Rougemaille, N., Chacon, C., Rodary, G., Rousset, S.: Spin reorientation transition and magnetic domain structure of Co ultrathin films grown on a faceted Au(455) surface. Phys. Rev. B **77**, 134429, 8 pages (2008)
59. Santos, B., Galego, S., Mascaraque, A., McCarty, K.F., Quesada, A., N'Diaye, A.T., Schmid, A.K., de la Figuera, J.: Hydrogen-induced reversible spin-reorientation transition and magnetic stripe domain phase in bilayer Co on Ru(0001). Phys. Rev. B **85**, 134409, 6 pages (2012)
60. Duden, T., Bauer, E.: Influence of Au and Cu overlayers on the magnetic structure of Co films on W(110). Phys. Rev. B **59**, 468–473 (1999)
61. El Gabaly, F., McCarty, K.F., Schmid, A.K., de la Figuera, J., Munoz, M.C., Szunyogh, L., Weinberger, P., Gallego, S.: Noble metal capping effects on the spin-reorientation transitions of Co/Ru(0001). New J. Phys. **10**, 073024, 22 pages (2008)
62. Rougemaille, N., N'Diaye, A.T., Coraux, J., Vo-Van, C., Fruchart, O., Schmid, A.K.: Perpendicular magnetic anisotropy of cobalt films intercalated under graphene. Appl. Phys. Lett. **101**, 142403, 3 pages (2012)
63. Wei, D.H., Wang, C.-H., Chang, H.-C., Chan, Y.-L., Lee, C.-H., Hsu, Y.-J.: Direct imaging and spectral identification of the interfaces in organic semiconductor-ferromagnet heterojunction. Appl. Phys. Lett. **101**, 141605, 4 pages (2012)
64. Albrecht, M., Maret, M., Maier, A., Treubel, F., Riedlinger, B., Mazur, U., Schatz, G., Anders, S.: Perpendicular magnetic anisotropy in $CoPt_3(111)$ films grown on a low energy surface at room temperature. J. Appl. Phys. **91**, 8153–8155 (2002)
65. Baldasseroni, C., Bordel, C., Gray, A.X., Kaiser, A.M., Kronast, F., Herrero-Albillos, J., Schneider, C.M., Fadley, C.S., Hellman, F.: Temperature-driven nucleation of ferromagnetic domains in FeRh thin films. Appl. Phys. Lett. **100**, 262401, 5 pages (2012)
66. Melchior, P., Rollinger, M., Thielen, P., Alebrand, S., Bierbrauer, U., Schneider, C., Gottwald, M., Hehn, M., Mangin, S., Cinchetti, M., Aeschlimann, M.: Energy-resolved magnetic domain imaging in TbCo alloys by valence band photoemission magnetic circular dichroism. Phys. Rev. B **88**, 104415, 6 pages (2013)
67. Poppa, H., Pinkvos, H., Wurm, K., Bauer, E.: Spin polarized low energy electron microscopy (SPLEEM) of single and combined layers of Co, Cu, and Pd on W(110). In: Jonker, B.T., Chambers, S.A., Farrow, R.F.C., Chappert, C., Clarke, R., de Jonge, W.J.N., Egami, T., Grunberg, P., Krishnan, K.M., Marinero, E.E., Rau, C., Tsunashima, S. (eds.) Magnetic

Ultrathin Films, Multilayers and Surfaces: Interfaces and Characterization Symposium, San Francisco, USA, pp. 219–226. Material Research Society, Philadelphia, PA (1993)
68. Wu, Y.Z., Schmid, A.K., Altman, M.S., Jin, X.F., Qiu, Z.Q.: Spin-dependent fabry-perot interference from a Cu thin film grown on fcc Co(001). Phys. Rev. Lett. **94**, 027201, 4 pages (2005)
69. Wurm, K.: Spin-Polarisierte LEEM-Untersuchungen An Dunnen Kobalt-Epitaxieschichten Auf W(110). M.S. thesis, TU Clausthal (1994)
70. Bauer, E.: Polarized electrons in low energy electron microscopy. In: Hatanaka, K., Nakano, T., Imai, K., Ejiri, H. (eds.) 14th Intern. Spin Physics Symp. SPIN 2000 Osaka, Japan 2000, pp. 965–975. AIP Conf. Proc. (2001)
71. Scheunemann, T., Feder, R., Henk, J., Bauer, E., Duden, T., Pinkvos, H., Poppa, H., Wurm, K.: Quantum-well resonance in ferromagnetic Co films. Solid State Commun. **104**, 787–792 (1997)
72. Zdyb, R., Bauer, E.: Spin-resolved unoccupied electronic band structure from quantum size oscillations in the reflectivity of slow electrons from ultrathin ferromagnetic crystals. Phys. Rev. Lett. **88**, 166403, 4 pages (2002)
73. Zdyb, R., Bauer, E.: Spin dependent quantum size effects in the electron reflectivity of ultrathin ferromagnetic crystals. Surf. Rev. Lett. **9**, 1485–1491 (2002)
74. Graf, J., Jozwiak, C., Schmid, A.K., Hussain, Z., Lanzara, A.: Mapping the spin-dependent electron reflectivity of Fe and Co ferromagnetic films. Phys. Rev. B **71**, 144429, 5 pages (2005)
75. Wu, Q., Altman, M.S.: Spin polarized low energy electron microscopy of quantum well resonances in Fe films on the Cu-covered W(110) surface. Ultramicroscopy **130**, 109–114 (2013)
76. Zdyb, R., Menteş, T.O., Locatelli, A., Niño, M.A., Bauer, E.: Inelastic mean free path from reflectivity of slow electrons. Phys. Rev. B **87**, 075436, 5 pages (2013)
77. Zdyb, R., Bauer, E.: Spin-resolved inelastic mean free path of slow electrons in Fe. J. Phys. Condens. Matter **25**, 272201, 4 pages (2013)
78. Nakajima, R., Stöhr, J., Idzerda, Y.U.: Electron-yield saturation effects in L-edge x-ray magnetic circular dichroism spectra of Fe, Co, and Ni. Phys. Rev. B **59**, 6421–6429 (1999)
79. Wu, Y.Z., Won, C., Scholl, A., Doran, A., Toyoma, F., Jin, X.F., Smith, N.V., Qiu, Z.Q.: Interfacial magnetism of fcc Fe and the effect of the oscillatory interlayer coupling on the Ni magnetic properties in Ni/Fe/Co/Cu (100). Phys. Rev. B **65**, 214417, 7 pages (2002)
80. Choi, J., Wu, J., El Gabaly, F., Schmid, A.K., Hwang, C., Qiu, Z.Q.: Quantum well states in Au/Ru(0001) and their effect on the magnetic properties of a Co overlayer. New J. Phys. **11**, 043016 (2009)
81. Wu, Y.Z., Schmid, A.K., Qiu, Z.Q.: Spin-dependent quantum interference from epitaxial MgO thin films on Fe(001). Phys. Rev. Lett. **97**, 217205, 4 pages (2006)
82. Wu, Y.Z., Qiu, Z.Q., Schmid, A.K.: Spin-dependent quantum interference from MgO thin films grown on Fe(001). J. Magn. Magn. Mater. **310**, 1629–1631 (2007)
83. Ramchal, R., Schmid, A.K., Farle, M., Poppa, H.: Spiral-like continuous spin-reorientation transition of Fe/Ni bilayers on Cu(100). Phys. Rev. B **69**, 214401, 6 pages (2004)
84. Wu, Y.Z., Won, C., Scholl, A., Doran, A., Zhao, H.W., Jin, X.F., Qiu, Z.Q.: Magnetic stripe domains in coupled magnetic sandwiches. Phys. Rev. Lett. **93**, 117205, 4 pages (2004)
85. Won, C., Wu, Y.Z., Choi, J., Kim, W., Scholl, A., Doran, A., Owens, T., Wu, J., Jin, X.F., Zhao, H.W., Qiu, Z.Q.: Magnetic stripe melting at the spin reorientation transition in Fe/Ni/Cu(001). Phys. Rev. B **71**, 224429, 5 pages (2005)
86. Choi, J., Wu, J., Won, C., Wu, Y.Z., Scholl, A., Doran, A., Owens, T., Qiu, Z.Q.: Magnetic bubble domain phase at the spin reorientation transition of ultrathin Fe/Ni/Cu(001) Film. Phys. Rev. Lett. **98**, 207205, 4 pages (2007)
87. Chen, G., Zhu, J., Quesada, A., Li, J., N'Diaye, A.T., Huo, Y., Ma, T.P., Chen, Y., Kwon, H.Y., Won, C., Qiu, Z.Q., Schmid, A.K., Wu, Y.Z.: Novel chiral magnetic domain wall structure in FeO/Ni/Cu(001) films. Phys. Rev. Lett. **110**, 177204, 5 pages (2013)

88. Heide, M., Bihlmayer, G., Blügel, S.: Dzyaloshinskii-Moriya interaction accounting for the orientation of magnetic domains in ultrathin films: Fe/W(110). Phys. Rev. B **78**, 140403, 4 pages (2008)
89. Zhao, H.W., Won, C., Wu, Y.Z., Scholl, A., Doran, A., Qiu, Z.Q.: Magnetic phase transition and spin-reorientation transition of Cu/Ni/Fe/Cu(001) studied by photoemission electron microscopy. Phys. Rev. B **70**, 024423, 6 pages (2004)
90. Won, C., Wu, Y.Z., Kurahashi, N., Law, K.T., Zhao, H.W., Scholl, A., Doran, A., Qiu, Z.Q.: Evidence of the oscillatory magnetic anisotropy in Ni/Co/Ni/Cu(100). Phys. Rev. B **67**, 174425, 6 pages (2003)
91. Duden, T., Bauer, E.: Biquadratic exchange in ferromagnetic/nonferromagnetic sandwiches: a spin-polarized low energy electron microscopy study. Phys. Rev. B **59**, 474–479 (1999)
92. Duden, T., Bauer, E.: Exchange coupling in Co/Cu/Co sandwiches studied by spin-polarized low energy electron microscopy. J. Magn. Magn. Mater. **191**, 301–312 (1999)
93. Stiles, M.D.: Interlayer magnetic coupling. In: Bland, J.A.C., Heinrich, B. (eds.) Ultrathin Magnetic Structures, pp. 99–142. Springer, Berlin (2005)
94. Won, C., Wu, Y.Z., Scholl, A., Doran, A., Kurahashi, N., Zhao, H.W., Qiu, Z.Q.: Magnetic phase transition in Co/Cu/Ni/Cu(100) and Co/Fe/Ni/Cu(100). Phys. Rev. Lett. **91**, 147202, 4 pages (2003)
95. Choi, H.J., Ling, W.L., Scholl, A., Wolfe, J.H., Bovensiepen, U., Toyama, F., Qiu, Z.Q.: Spin reorientation transition of Fe films in magnetically coupled Fe/Cu/Ni/Cu(001). Phys. Rev. B **66**, 014409, 5 pages (2002)
96. Wu, J., Choi, J., Won, C., Wu, Y.Z., Scholl, A., Doran, A., Hwang, C., Qiu, Z.Q.: Stripe-to-bubble transition of magnetic domains at the spin reorientation of (Fe/Ni)/Cu/Ni/Cu(001). Phys. Rev. B **79**, 014429, 6 pages (2009)
97. Kurde, J., Miguel, J., Bayer, D., Sánchez-Barriga, J., Kronast, F., Aeschlimann, M., Dürr, H.A., Kuch, W.: Magnetostatic coupling of 90° domain walls in $Fe_{19}Ni_{81}$/Cu/Co trilayers. New J. Phys. **13**, 033015, 13 pages (2011)
98. Vogel, J., Cherifi, S., Pizzini, S., Romanens, F., Camarero, J., Petroff, F., Heun, S., Locatelli, A.: Layer-resolved imaging of domain wall interactions in magnetic tunnel junction-like trilayers. J. Phys. Condens. Matter **19**, 476204, 9 pages (2007)
99. Kuch, W., Vogel, J., Camarero, J., Fukumoto, K., Pennec, Y., Pizzini, S., Bonfim, M., Kirschner, J.: Exploring spin valve magnetization reversal dynamics with temporal, spatial and layer resolution: influence of domain-wall energy. Appl. Phys. Lett. **85**, 440–442 (2004)
100. Vogel, J., Kuch, W., Camarero, J., Fukumoto, K., Pennec, Y., Pizzini, S., Bonfim, M., Petroff, F., Fontaine, A., Kirschner, J.: Interplay between magnetic anisotropy and interlayer coupling in nanosecond magnetization reversal of spin-valve trilayers. Phys. Rev. B **71**, 060404(R), 4 pages (2005)
101. Vogel, J., Kuch, W., Hertel, R., Camarero, J., Fukumoto, K., Romanens, F., Pizzini, S., Bonfim, M., Petroff, F., Fontaine, A., Kirschner, J.: Influence of domain wall interactions on nanosecond switching in magnetic tunnel junctions. Phys. Rev. B **72**, 220402, 4 pages (2005)
102. Fukumoto, K., Kuch, W., Vogel, J., Romanens, F., Pizzini, S., Camarero, J., Bonfim, M., Kirschner, J.: Dynamics of magnetic domain wall motion after nucleation: dependence on the wall energy. Phys. Rev. Lett. **96**, 097204, 4 pages (2006)
103. Romanens, F., Vogel, J., Kuch, W., Fukumoto, K., Camarero, J., Pizzini, S., Bonfim, M., Petroff, F.: Influence of topography and Co domain walls on the magnetization reversal of the FeNi layer in FeNi/Al_2O_3/Co magnetic tunnel junctions. Phys. Rev. B **74**, 184419, 9 pages (2006)
104. Vogel, J., Kuch, W., Fukumoto, K., Romanens, F., Pizzini, S., Camarero, J.: Domain wall dynamics and interlayer interactions in magnetic trilayer systems studied by XMCD-PEEM. Appl. Phys. A **92**, 505–510 (2008)
105. Won, C., Wu, Y.Z., Zhao, H.W., Scholl, A., Doran, A., Qiu, Z.Q.: Effect of the interlayer coupling on the Ni spin reorientation in Ni/Fe/Co/Cu(100). Phys. Rev. B **68**, 052404, 4 pages (2003)

106. Ikeda, S., Sato, H., Yamanouchi, M., Gan, H., Miura, K., Mizinuma, K., Kanai, S., Fukami, S., Matsukura, F., Kasai, N., Ohno, H.: Recent progress of perpendicular anisotropy magnetic tunnel junctions for nonvolatile VLSI. SPIN **2**, 1240003, 12 pages (2012)
107. Beach, G.S.D., Tsoi, M., Erskine, J.L.: Current-induced domain wall motion. J. Magn. Magn. Mater. **320**, 1272–1281 (2008)
108. Parkin, S.S.P., Hayashi, M., Thomas, L.: Magnetic domain-wall racetrack memory. Science **320**, 190–194 (2008)
109. Johnson, M.T., Bloemen, P.J.H., den Broeder, F.J.A., de Vries, J.J.: Magnetic anisotropy in metallic multilayers. Rep. Prog. Phys. **59**, 1409–1458 (1996)
110. Gimbert, F., Calmels, L.: First-principles investigation of the magnetic anisotropy and magnetic properties of Co/Ni(111) superlattices. Phys. Rev. B **86**, 184407, 10 pages (2012)
111. Suzuki, M., Yasue, T., Koshikawa, T., Bauer, E.: Magnetic structure of Co/Ni thin film Magnetic structure of Co/Ni thin films on W(110) studied by high brightness and highly spin-polarized LEEM. In: Proc. 8th Int. Symp. on Atomic Level Characterizations for New Materials and Devices (ALC'11), Seoul, Korea 2011 (2011)
112. Suzuki, M., Kudo, K., Kojima, K., Yasue, T., Akutsu, N., Agerico Diño, W., Kasai, H., Bauer, E., Koshikawa, T.: Magnetic domain patterns on strong perpendicular magnetization of Co/Ni multilayers as spintronics materials: I. Dynamic observations. J. Phys. Condens. Matter **25**, 406001, 8 pages (2013)
113. Kudo, K., Suzuki, M., Kojima, K., Yasue, T., Akutsu, N., Diño, W.A., Kasai, H., Bauer, E., Koshikawa, T.: Magnetic domain patterns on strong perpendicular magnetization of Co/Ni multilayers as spintronics materials: II. Numerical simulations. J. Phys. Condens. Matter **25**, 395005, 6 pages (2013)
114. Gambardella, P., Claude, L., Rusponi, S., Franke, K.J., Brune, H., Raabe, J., Nolting, F., Bencok, P., Hanbicki, A.T., Jonker, B.T., Grazioli, C., Veronese, M., Carbone, C.: Surface characterization of MnxGe1 − x and CryMnxGe1 − x − y dilute magnetic semiconductors. Phys. Rev. B **75**, 125211, 7 pages (2007)
115. Taniuchi, T., Kumigashira, H., Oshima, M., Wakita, T., Yokoya, T., Kubota, M., Ono, K., Akinaga, H., Lippmaa, M., Kawasaki, M., Koinuma, H.: Observation of step-induced magnetic domain formation in $La_{1-x}Sr_xMnO_3$ thin films by photoelectron emission microscopy. Appl. Phys. Lett. **89**, 112505, 3 pages (2006)
116. Taniuchi, T., Yasuhara, R., Kumigashira, H., Kubota, M., Okazaki, H., Wakita, T., Yokoya, T., Ono, K., Oshima, M., Lippmaa, M., Kawasaki, M., Koinuma, H.: Thickness dependence of magnetic domain formation in $La_{0.6}Sr_{0.4}MnO_3$ epitaxial thin films studied by XMCD-PEEM. Surf. Sci. **601**, 4690–4693 (2007)
117. Bauer, E., Cherifi, S., Daeweritz, L., Kaestner, M., Heun, S., Locatelli, A.: Low-energy electron microscopy/x-ray magnetic circular dichroism photoemission electron microscopy study of epitaxial MnAs on GaAs. J. Vac. Sci. Technol. B **20**, 2539–2542 (2002)
118. Däweritz, L., Herrmann, C., Mohanty, J., Hesjedal, T., Ploog, K.H., Bauer, E., Locatelli, A., Cherifi, S., Belkhou, R., Pavlovska, A., Heun, S.: Tailoring of the structural and magnetic properties of MnAs films grown on GaAs: strain and annealing effects. J. Vac. Sci. Technol. B **23**, 1759–1768 (2005)
119. Däweritz, L., Kolovos-Velliantis, D., Trampet, A., Herrmann, C., Ploog, K.H., Bauer, E., Locatelli, A., Cherifi, S., Heun, S.: Orientation and interface effects on the structural and magnetic properties of MnAs-onGaAs hybrid structures. J. Phys. IV France **132**, 159–162 (2006)
120. Engel-Herbert, R., Locatelli, A., Cherifi, S., Schaadt, D.M., Mohanty, J., Ploog, K.H., Bauer, E., Belkhou, R., Heun, S., Pavlovska, A., Leo, T., Hesjedal, T.: Investigation of magnetically coupled ferromagnetic stripe arrays. Appl. Phys. A **84**, 231–236 (2006)
121. Engel-Herbert, R., Hesjedal, T., Mohanty, J., Schaadt, D.M., Ploog, K.H.: Magnetization reversal in MnAs films: magnetic force microscopy, SQUID magnetometry, and micromagnetic simulations. Phys. Rev. B **73**, 104441, 7 pages (2006)

122. Engel-Herbert, R., Schaadt, D.M., Cherifi, S., Bauer, E., Belkhou, R., Locatelli, A., Heun, S., Pavlovska, A., Mohanty, J., Ploog, K.H., Hesjedal, T.: The nature of charged zig-zag domains in MnAs thin films. J. Magn. Magn. Mater. **305**, 457–463 (2006)
123. Bauer, E., Belkhou, R., Cherifi, S., Locatelli, A., Pavlovska, A., Rougemaille, N.: Magnetostructure of MnAs on GaAs revisited. J. Vac. Sci. Technol. B **25**, 1470–1475 (2007)
124. Bauer, E., Belkhou, R., Maccherozzi, F., El Moussaoui, S., Rougemaille, N., Jalochowski, M., Akinaga, H., Takano, F.: Epitaxial MnAs on GaAs(100):A X-ray Magnetic Dichroism PEEM and LEEM study. In: Proc. 6th Int. Symp. on Atomic Level Characterizations for New Materials and Devices (ALC'07), Kanasawa, Japan 2007, pp. 87–91. JSPS 141 Activity Report (2007)
125. Engel-Herbert, R., Hesjedal, T.: Micromagnetic analysis of unusual, V-shaped domain transitions in MnAs nanowires. J. Magn. Magn. Mater. **323**, 1840–1845 (2011)
126. Niranjan, M.K., Sahu, B.R., Kleinman, L.: Density functional determination of the magnetic state of b-MnAs. Phys. Rev. B **70**, 80406(R), 4 pages (2004)
127. Mundschau, M., Romanowicz, J., Wang, J.Y., Sun, D.L., Chen, H.C.: Imaging of ferromagnetic domains using photoelectrons: photoelectron emission microscopy of neodymium-iron-boron ($Nd_2Fe_{14}B$). J. Vac. Sci. Technol. B **14**, 3126–3130 (1996)
128. Nepijko, S.A., Mundschau, M., Schönhense, G.: Photoemission electron microscopy of neodymium-iron-boron ($Nd_2Fe_{14}B$). Appl. Phys. A **86**, 515–519 (2007)
129. Lilienkamp, G., Koziol, C., Schmidt, T., Bauer, E.: Cathode lens spectromicroscopy with a low-energy electron microscope. In: Thieme, J., Schmahl, G., Rudolph, D., Umbach, E. (eds.) X-ray Microscopy and Spectromicroscopy, pp. III-25–III-34. Springer, Berlin (1998)
130. Kotsugi, M., Guo, F.Z., Taniguchi, M., Ishimatsu, N., Maruyama, H.: Easy-axis rotation in meteoritic iron probed by photoelectron emission microscope (PEEM). Surf. Sci. **601**, 4326–4328 (2007)
131. Kotsugi, M., Wakita, T., Taniuchi, T., Ono, K., Suzuki, M., Kawamura, N., Takagaki, M., Taniguchi, M., Kobayashi, K., Oshima, M., Ishimatsu, N., Maruyama, H.: Local electronic structure analysis using a photoelectron emission microscope (PEEM) with hard X-ray. e-J. Surf. Sci. Nanotech. **4**, 490–493 (2006)
132. Kotsugi, M., Mitsumata, C., Maruyama, H., Wakita, T., Taniuchi, T., Ono, K., Suzuki, M., Kawamura, N., Ishimatsu, N., Oshima, M., Watanabe, Y., Taniguchi, M.: Novel magnetic domain structure in iron meteorite induced by the presence of $L1_0$-FeNi. Appl. Phys. Express **3**, 013001, 3 pages (2010)
133. Kotsugi, M., Mizuguchi, M., Sekiya, S., Ohkouchi, T., Kojima, T., Takanashi, K., Watanabe, Y.: Determination of local magnetic moment in $L1_0$-FeNi using photoelectron emission microscopy (PEEM). J. Phys.: Conf. Ser. **266**, 012095, 6 pages (2011)
134. Hashimoto, M., Iwata, K., Suzuki, M., Ueda, M., Matuoka, Y., Kotsugi, M., Kinoshita, T., Watanabe, Y., Tanaka, K., Yasue, T., Koshikawa, T.: Temperature dependence of magnetic domain structure of Fe-3 % Si(110) by XMCD-PEEM. In: Proc. 7th Int. Symp. on Atomic Level Characterizations for New Materials and Devices (ALC'09), Maui, Hawaii 2009, pp. 626–628. JSPS 141 Activity Report (2009)
135. Suzuki, M., Iwata, K., Hashimoto, M., Ueda, M., Matuoka, Y., Yasue, T., Kotsugi, M., Ohkochi, T., Kinoshita, T., Watanabe, Y., Tanaka, K., Koshikawa, T.: XMCD-PEEM study on stabilization of magnetic domain structure of Fe-3 % Si(110). In: Proc. 7th Int Symp. on Atomic Level Characterizations for New Materials and Devices (ALC'09), Maui, Hawaii 2009, pp. 622–625. JSPS 141 Activity Report (2009)
136. Hashimoto, M., Iwata, K., Suzuki, M., Kotsugi, M., Ohkochi, T., Kinoshita, T., Tanaka, K., Yasue, T., Koshikawa, T.: Magnetic domain structure of Fe-3 % Si(110) by external magnetic field by XMCD-PEEM. In: Proc. 8th Int. Symp. on Atomic Level Characterizations for New Materials and Devices (ALC'11), Seoul, Korea 2011, 4 pp
137. Yamamoto, S., Yonemura, M., Wakita, T., Fukumoto, K., Nakamura, T., Kinoshita, T., Watanabe, Y., Guo, F.Z., Sato, M., Terai, T., Kakeshita, T.: Magnetic-domain structure

analysis of Nd-Fe-B sintered magnets. Using XMCD-PEEM technique. Mater. Trans. **49**, 2354–2359 (2008)
138. Yamaguchi, R., Terashima, K., Fukumoto, K., Takada, Y., Kotsugi, M., Miyata, Y., Mima, K., Komori, S., Itoda, S., Nakatsu, Y., Yano, M., Miyamoto, N., Nakamura, T., Kinoshita, T., Watanabe, Y., Manabe, A., Suga, S., Imada, S.: An XMCD-PEEM study on magnetized Dy-doped Nd-Fe-B permanent magnets. IBM J. Res. Dev. **55**, 4, Paper 12, 5 pages (2011)
139. Felser, C., Heitkamp, B., Kronast, F., Schmitz, D., Cramm, S., Dürr, H.A., Elmers, H.-J., Fecher, G.H., Wurmehl, S., Block, T., Valdaitsev, D., Nepijko, S.A., Gloskovskii, A., Jakob, G., Schönhense, G., Eberhardt, W.: Investigation of a novel material for magnetoelectronics: $Co_2Cr_{0.6}Fe_{0.4}Al$. J. Phys. Condens. Matter **15**, 7019–7027 (2003)
140. Gloskovskii, A., Barth, J., Balke, B., Fecher, G.H., Felser, C., Kronast, F., Ovsyannikov, R., Dürr, H., Eberhard, W., Schönhense, G.: A spatially resolved investigation of the local, micro-magnetic domain structure of single and polycrystalline Co_2FeSi. J. Phys. D. Appl. Phys. **40**, 1570–1575 (2007)
141. de la Figuera, J., Novotny, Z., Setvin, M., Liu, T., Mao, Z., Chen, G., N'Diaye, A.T., Schmid, M., Diebold, U., Schmid, A.K., Parkinson, G.S.: Real-space imaging of the Verwey transition at the (100) surface of magnetite. Phys. Rev. B **88**, 161410(R), 5 pages (2013)
142. de la Figuera, J., Vergara, L., N'Diaye, A.T., Quesada, A., Schmid, A.K.: Micromagnetism in (001) magnetite by spin-polarized low-energy electron microscopy. Ultramicroscopy **130**, 77–81 (2013)
143. Arai, K., Okuda, T., Tanaka, A., Kotsugi, M., Fukumoto, K., Ohkochi, T., Nakamura, T., Matsushita, T., Muro, T., Oura, M., Senba, Y., Ohashi, H., Kakizaki, A., Mitsumata, C., Kinoshita, T.: Three-dimensional spin orientation in antiferromagnetic domain walls of NiO studied by x-ray magnetic linear dichroism photoemission electron microscopy. Phys. Rev. B **85**, 104418, 12 pages (2012)
144. Spanke, D., Solinus, V., Knabben, D., Hillebrecht, F.U., Ciccacci, F., Gregoratti, L., Marsi, M.: Evidence for in-plane antiferromagnetic domains in ultrathin NiO films. Phys. Rev. B **58**, 5201–5204 (1998)
145. Kinoshita, T., Wakita, T., Sun, H.L., Tohyama, T., Harasawa, A., Kiwata, H., Hillebrecht, F.U., Ono, K., Matsushima, T., Oshima, M., Ueno, N., Okuda, T.: Antiferromagnetic domain structure imaging of cleaved NiO(100) surface using nonmagnetic linear dichroism at O K edge: essential effect of antiferromagnetic crystal distortion. J. Phys. Soc. Jpn. **73**, 2932–2935 (2004)
146. Menon, K.S.R., Mandal, S., Das, J., Menteş, T.O., Niño, M.A., Locatelli, A., Belkhou, R.: Surface antiferromagnetic domain imaging using low-energy unpolarized electrons. Phys. Rev. B **84**, 132402, 5 pages (2011)
147. Arai, K., Okuda, T., Tanaka, A., Kotsugi, M., Fukumoto, K., Ohkochi, T., Guo, F., Nakamura, T., Matsushita, T., Muro, T., Oura, M., Senba, Y., Ohashi, H., Kakizaki, A., Kinoshita, T.: Direct observation of twin domains of NiO(100) by x-ray linear dichroism at the O K edge using photoemission electron microscopy. Phys. Rev. B **85**, 174401, 6 pages (2012)
148. Palmberg, P.W., De Wames, R.E., Vredevoe, L.A., Wolfram, T.: Coherent exchange scattering of low energy electrons by antiferromagnetic crystals. J. Appl. Phys. **40**, 1158–1163 (1969)
149. Nolting, F., Scholl, A., Stöhr, J., Seo, J.W., Fompeyrine, J., Siegwart, H., Locquet, J.-P., Anders, S., Lüning, J., Fullerton, E.E., Toney, M.F., Scheinfein, M.R., Padmore, H.A.: Direct observation of the alignment of ferromagnetic spins by antiferromagnetic spins. Nature **405**, 767–769 (2000)
150. Scholl, A., Stohr, J., Luning, J., Seo, J.W., Fompeyrine, J., Siegwart, H., Locquet, J.-P., Nolting, F., Anders, S., Fullerton, E.E., Scheinfein, M.R., Padmore, H.A.: Observation of antiferromagnetic domains in epitaxial thin films. Science **287**, 1014–1016 (2000)

151. Seo, J.W., Fullerton, E.E., Nolting, F., Scholl, A., Fompeyrine, J., Locquet, J.-P.: Antiferromagnetic $LaFeO_3$ thin films and their effect on exchange bias. J. Phys. Condens. Matter **20**, 264014, 10 pages (2008)
152. Ohldag, H., Scholl, A., Nolting, F., Anders, S., Hillebrecht, F.U., Stöhr, J.: Spin reorientation at the antiferromagnetic NiO(001) surface in response to an adjacent ferromagnet. Phys. Rev. Lett. **86**, 2878–2881 (2001)
153. Finazzi, M., Brambilla, A., Biagioni, P., Graf, J., Gweon, G.-H., Scholl, A., Lanzara, A., Duo, L.: Interface coupling transition in a thin epitaxial antiferromagnetic film interacting with a ferromagnetic substrate. Phys. Rev. Lett. **97**, 097202, 4 pages (2006)
154. Ohldag, H., van der Laan, G., Arenholz, E.: Correlation of crystallographic and magnetic domains at Co/NiO(001) interfaces. Phys. Rev. B **79**, 052403, 4 pages (2009)
155. van der Laan, G., Telling, N.D., Potenza, A., Dhesi, S.S., Arenholz, E.: Anisotropic x-ray magnetic linear dichroism and spectromicroscopy of interfacial Co/NiO(001). Phys. Rev. B **83**, 064409, 9 pages (2011)
156. Wu, J., Choi, J., Scholl, A., Doran, A., Arenholz, E., Wu, Y.Z., Won, C., Hwang, C., Qiu, Z.Q.: Element-specific study of the anomalous magnetic interlayer coupling across NiO spacer layer in Co/NiO/Fe/Ag(001) using XMCD and XMLD. Phys. Rev. B **80**, 012409, 4 pages (2009)
157. Krug, I.P., Hillebrecht, F.U., Haverkort, M.W., Tanaka, A., Tjeng, L.H., Gomonay, H., Fraile-Rodríguez, A., Nolting, F., Cramm, S., Schneider, C.M.: Impact of interface orientation on magnetic coupling in highly ordered systems: a case study of the low-indexed Fe_3O_4/NiO interfaces. Phys. Rev. B **78**, 064427, 14 pages (2008)
158. Kronmüller, H., Parkin, S. (eds.): Handbook of Magnetism and Advanced Magnetic Materials. Micromagnetism, vol. 2. Wiley, Chichester (2007)
159. Bedanta, S., Kleemann, W.: Supermagnetism. J. Phys. D. Appl. Phys. **42**, 013001, 28 pages (2009)
160. Bennemann, K.: Magnetic nanostructures. J. Phys. Condens. Matter **22**, 243201, 39 pages (2010)
161. Boulle, O., Malinowski, G., Kläui, M.: Current-induced domain wall motion in nanoscale ferromagnetic elements. Mater. Sci. Eng. R **72**, 159–187 (2011)
162. Nasirpouri, F., Nogaret, A. (eds.): Nanomagnetism and Spintronics: Fabrication, Materials, Characterization and Applications. World Scientific, Singapore (2011)
163. Zabel, H., Farle, M. (eds.): Magnetic Nanostructures. Springer, Heidelberg (2013)
164. Swiech, W., Fecher, G.H., Ziethen, C., Schmidt, O., Schönhense, G., Grzelakowski, K., Schneider, C.M., Frömter, R., Oepen, H.P., Kirschner, J.: Recent progress in photoemission microscopy with emphasis on chemical and magnetic sensitivity. J. Electron Spectrosc. Relat. Phenom. **84**, 171–188 (1997)
165. Schneider, C.M., de Haas, O., Muschiol, U., Cramer, N., Oelsner, A., Klais, M., Schmidt, O., Fecher, G.H., Jark, W., Schonhense, G.: Photoemission microscopy from magnetically coupled thin-film systems. J. Magn. Magn. Mater. **233**, 14–20 (2001)
166. Locatelli, A., Cherifi, S., Heun, S., Marsi, M., Ono, K., Pavlovska, A., Bauer, E.: X-ray magnetic circular dichroism imaging in a low energy electron microscope. Surf. Rev. Lett. **9**, 171–176 (2002)
167. Cherifi, S., Hertel, R., Kirschner, J., Wang, H., Belkhou, R., Locatelli, A., Heun, S., Pavlovska, A., Bauer, E.: Virgin domain structures in mesoscopic Co patterns: comparison between simulation and experiment. J. Appl. Phys. **98**, 043901, 6 pages (2005)
168. Kläui, M., Vaz, C.A.F., Bland, J.A.C., Monchesky, T.L., Unguris, J., Bauer, E., Cherifi, S., Heun, S., Locatelli, A., Heyderman, L.J., Cui, Z.: Direct observation of spin configurations and classification of switching processes in mesoscopic ferromagnetic rings. Phys. Rev. B **68**, 134426, 5 pages (2003)
169. Zdyb, R., Pavlovska, A., Locatelli, A., Heun, S., Cherifi, S., Belkhou, R., Bauer, E.: Imaging low-dimensional magnetism with slow electrons. Appl. Surf. Sci. **249**, 38–44 (2005)

170. Bauer, E., Belkhou, R., Cherifi, S., Hertel, R., Heun, S., Locatelli, A., Pavlovska, A., Zdyb, R., Agarwal, N., Wang, H.: Microscopy of mesoscopic ferromagnetic systems with slow electrons. Surf. Interface Anal. **38**, 1622–1627 (2006)
171. Vaz, C.A.F., Hayward, T.J., Llandro, J., Schackert, F., Morecroft, D., Bland, J.A.C., Kläui, M., Laufenberg, M., Backes, D., Rüdiger, U., Castaño, F.J., Ross, C.A., Heyderman, L.J., Nolting, F., Locatelli, A., Faini, G., Cherifi, S., Wernsdorfer, W.: Ferromagnetic nanorings. J. Phys. Condens. Matter **19**, 255207, 14 pages (2007)
172. Laufenberg, M., Kläui, M., Backes, D., Bührer, W., Ehrke, H., Bedau, D., Rüdiger, U., Nolting, F., Heyderman, L.J., Cherifi, S., Locatelli, A., Belkhou, R., Heun, S., Vaz, C.A.F., Bland, J.A.C., Kasama, T., Dunin-Borkowski, R.E., Pavlovska, A., Bauer, E.: Domain wall spin structures in 3d metal ferromagnetic nanostructures. In: Haug, R. (ed.) Advances in Solid State Physics, vol. 46, pp. 281–293. Springer, Berlin (2008)
173. Fonin, M., Hartung, C., Rüdiger, U., Backes, D., Heyderman, L., Nolting, F., Fraile Rodrıguez, A., Kläui, M.: Formation of magnetic domains and domain walls in epitaxial Fe_3O_4(100) elements. J. Appl. Phys. **109**, 07D315, 4 pages (2011)
174. Vaz, C.A.F., Rhensius, J., Heidler, J., Wohlhüter, P., Bisig, A., Kömer, H.S., Menteş, T.O., Locatelli, A., Le Guyader, L., Nolting, F., Graf, T., Felser, C., Heyderman, L.J., Kläui, M.: Spin configurations in $Co_2FeAl_{0.4}Si_{0.6}$ Heusler alloy thin film elements. Appl. Phys. Lett. **99**, 182510, 3 pages (2011)
175. Chauleau, J.-Y., McMorran, B.J., Belkhou, R., Bergeard, N., Menteş, T.O., Niño, M.A., Locatelli, A., Unguris, J., Rohart, S., Miltat, J., Thiaville, A.: Magnetization textures in NiPd nanostructures. Phys. Rev. B **84**, 094416, 8 pages (2011)
176. Wohlhüter, P., Rhensius, J., Vaz, C.A.F., Heidler, J., Körner, H.S., Bisig, A., Foerster, M., Méchin, L., Gaucher, F., Locatelli, A., Niño, M.A., El Moussaoui, S., Nolting, F., Goering, E., Heyderman, L.J., Kläui, M.: The effect of magnetic anisotropy on the spin configurations of patterned $La_{0.7}Sr_{0.3}MnO_3$ elements. J. Phys. Condens. Matter **25**, 176004, 6 pages (2013)
177. Schneider, C.M., Kaiser, A., Wiemann, C., Tieg, C., Cramm, S.: Photoemission microscopy study of picosecond magnetodynamics in spin-valve-type thin film elements. J. Electron Spectrosc. Relat. Phenom. **181**, 159–163 (2010)
178. Wu, J., Carlton, D., Oelker, E., Park, J.S., Jin, E., Arenholz, E., Scholl, A., Hwang, C., Bokor, J., Qiu, Z.Q.: Switching a magnetic vortex by interlayer coupling in epitaxially grown Co/Cu/Py/Cu(001) trilayer disks. J. Phys. Condens. Matter **22**, 342001, 4 pages (2010)
179. Folven, E., Scholl, A., Young, A., Retterer, S.T., Boschker, J.E., Tybell, T., Takamura, Y., Grepstad, J.K.: Effects of nanostructuring and substrate symmetry on antiferromagnetic domain structure in $LaFeO_3$ thin films. Phys. Rev. B **84**, 220410, 5 pages (2011)
180. Hertel, R., Fruchart, O., Cherifi, S., Jubert, P.-O., Heun, S., Locatelli, A., Kirschner, J.: Three-dimensional magnetic-flux-closure patterns in mesoscopic Fe islands. Phys. Rev. B **72**, 214409, 11 pages (2005)
181. Zdyb, R., Pavlovska, A., Jałochowski, M., Bauer, E.: Self-organized Fe nanostructures on W (110). Surf. Sci. **600**, 1586–1591 (2006)
182. Rougemaille, N., Schmid, A.K.: Self-organization and magnetic domain microstructure of Fe nanowire arrays. J. Appl. Phys. **99**, 08S502, 3 pages (2006)
183. Monti, M., Santos, B., Mascaraque, A., Rodrıguez de la Fuente, O., Niño, M.A., Menteş, T.O., Locatelli, A., McCarty, K.F., Marco, J.F., de la Figuera, J.: Magnetism in nanometer-thick magnetite. Phys. Rev. B **85**, 020404(R), 5 pages (2012)
184. Menteş, T.O., Locatelli, A., Aballe, L., Niño, M.A., Bauer, E.: Growth of magnetic nanowires on self-organized stripe templates:Fe on Pd-O/W(110). Ultramicroscopy **130**, 82–86 (2013)
185. Poppa, H., Tober, E.D., Schmid, A.K.: In situ observation of magnetic domain pattern evolution in applied fields by spin-polarized low energy electron microscopy. J. Appl. Phys. **91**, 6932–6934 (2002)
186. Vogel, J., Kuch, W., Bonfim, M., Camarero, J., Pennec, Y., Offi, F., Fukumoto, K., Kirschner, J., Fontaine, A., Pizzini, S.: Time-resolved magnetic domain imaging by x-ray photoemission electron microscopy. Appl. Phys. Lett. **82**, 2299–2301 (2003)

187. Kronast, F., Schlichting, J., Radu, F., Mishra, S., Noll, T., Dürr, H.A.: Spin resolved photoemission microscopy and magnetic imaging in applied magnetic fields. Surf. Interface Anal. **42**, 1532–1536 (2010)
188. Sandig, O., Herrero-Albillos, J., Römer, F.M., Friedenberger, N., Kurde, J., Noll, T., Farle, M., Kronast, F.: Imaging magnetic responses of nanomagnets by XPEEM. J. Electron Spectrosc. Relat. Phenom. **185**, 365–370 (2012)
189. Imada, S., Suga, S., Kuch, W., Kirschner, J.: Magnetic microspectroscopy by a combination of XMCD and PEEM. Surf. Rev. Lett. **9**, 877–881 (2002)
190. Ohshima, N., Numata, H., Fukami, S., Nagahara, K., Suzuki, T., Ishiwata, N., Fukumoto, K., Kinoshita, T., Ono, T.: Magnetic configuration of submicron-sized magnetic patterns in domain wall motion memory. J. Appl. Phys. **107**, 103912, 9 pages (2010)
191. Kläui, M., Laufenberg, M., Heyne, L., Backes, D., Rüdiger, U., Vaz, C.A.F., Bland, J.A.C., Heyderman, L.J., Cherifi, S., Locatelli, A., Mentes, T.O., Aballe, L.: Current-induced vortex nucleation and annihilation in vortex domain walls. Appl. Phys. Lett. **88**, 232507, 3 pages (2006)
192. Uhlíř, V., Pizzini, S., Rougemaille, N., Novotný, J., Cros, V., Jiménez, E., Faini, G., Heyne, L., Sirotti, F., Tieg, C., Bendounan, A., Maccherozzi, F., Belkhou, R., Grollier, J., Anane, A., Vogel, J.: Current-induced motion and pinning of domain walls in spin-valve nanowires studied by XMCD-PEEM. Phys. Rev. B **81**, 224418, 10 pages (2010)
193. Choe, S.-B., Acremann, Y., Scholl, A., Bauer, A., Doran, A., Stöhr, J., Padmore, H.A.: Vortex core-driven magnetization dynamics. Science **304**, 420–422 (2004)
194. Schneider, C.M., Kuksov, A., Krasyuk, A., Oelsner, A., Nepijko, S.A., Schönhense, G.: Time-resolved X-ray photoemission electron microscopy: imaging magnetodynamics on the 100 ps scale and below. J. Electron. Spectr. Relat. Phenom. **144–147**, 967–971 (2005)
195. Kaiser, A., Wiemann, C., Cramm, S., Schneider, C.M.: Influence of magnetocrystalline anisotropy on the magnetization dynamics of magnetic microstructures. J. Phys. Condens. Matter **21**, 314008, 6 pages (2009)
196. Heyne, L., Rhensius, J., Bisig, A., Krzyk, S., Punke, P., Kläui, M., Heyderman, L.J., Le Guyader, L., Nolting, F.: Direct observation of high velocity current induced domain wall motion. Appl. Phys. Lett. **96**, 032504, 3 pages (2010)
197. Uhlíř, V., Vogel, J., Rougemaille, N., Fruchart, O., Ishaque, Z., Cros, V., Camarero, J., Cezar, J.C., Sirotti, F., Pizzini, S.: Current-induced domain wall motion and magnetization dynamics in CoFeB/Cu/Co nanostripes. J. Phys. Condens. Matter **24**, 024213, 9 pages (2012)
198. Vogel, J., Bonfim, M., Rougemaille, N., Boulle, O., Miron, I.M., Auffret, S., Rodmacq, B., Gaudin, G., Cezar, J.C., Sirotti, F., Pizzini, S.: Direct observation of massless domain wall dynamics in nanostripes with perpendicular magnetic anisotropy. Phys. Rev. Lett. **108**, 247202, 5 pages (2012)
199. Ding, H.F., Schmid, A.K., Keavney, D.J., Li, D., Cheng, R., Pearson, J.E., Fradin, F.Y., Bader, S.D.: Selective growth of Co nanoislands on an oxygen-patterned Ru(0001) surface. Phys. Rev. B **72**, 035413, 5 pages (2005)
200. Fraile Rodríguez, A., Kleibert, A., Bansmann, J., Nolting, F.: Probing single magnetic nanoparticles by polarization-dependent soft x-ray absorption spectromicroscopy. J. Phys. D. Appl. Phys. **43**, 474006, 8 pages (2010)
201. Fraile Rodríguez, A., Kleibert, A., Bansmann, J., Voitkans, A., Heyderman, L.J., Nolting, F.: Size-dependent spin structures in iron nanoparticles. Phys. Rev. Lett. **104**, 127201, 4 pages (2010)
202. Hellwig, O., Heydermann, L.J., Petracic, O., Zabel, H.: Competing interactions in patterned and self-assembled magnetic nanostructures. In: Zabel, H., Farle, M. (eds.) Magnetic Nanostructures, pp. 189–234. Springer, Heidelberg (2013)
203. Teichert, C., Bean, J.C., Lagally, M.G.: Self-organized nanostructures in $Si_{1-x}Ge_x$ films on Si (001). Appl. Phys. A **67**, 675–685 (1998)
204. Gridneva, L., Persson, A., Niño, M.Á., Camarero, J., de Miguel, J.J., Miranda, R., Hofer, C., Teichert, C., Bobek, T., Locatelli, A., Heun, S., Carlsson, S., Arvanitis, D.: Experimental

investigation of the spin reorientation of Co/Au based magnetic nanodot arrays. Phys. Rev. B **77**, 104425, 9 pages (2008)
205. Streubel, R., Makarov, D., Kronast, F., Kravchuk, V., Albrecht, M., Schmidt, O.G.: Magnetic vortices on closely packed spherically curved surfaces. Phys. Rev. B **85**, 174429, 4 pages (2012)
206. Bedanta, S., Eimüller, T., Kleemann, W., Rhensius, J., Stromberg, F., Amaladass, E., Cardoso, S., Freitas, P.P.: Overcoming the dipolar disorder in dense CoFe nanoparticle ensembles: superferromagnetism. Phys. Rev. Lett. **98**, 176601, 4 pages (2007)
207. Nisoli, C., Moessner, R., Schiffer, P.: Colloquium: artificial spin ice: designing and imaging magnetic frustration. Rev. Mod. Phys. **85**, 1473–1490 (2013)
208. Mengotti, E., Heyderman, L.J., Fraile Rodríguez, A., Bisig, A., Le Guyader, L., Nolting, F., Braun, H.B.: Building blocks of an artificial kagome spin ice: photoemission electron microscopy of arrays of ferromagnetic islands. Phys. Rev. B **78**, 144402, 7 pages (2008)
209. Rougemaille, N., Montaigne, F., Canals, B., Duluard, A., Lacour, D., Hehn, M., Belkhou, R., Fruchart, O., El Moussaoui, S., Bendounan, A., Maccherozzi, F.: Artificial kagome arrays of nanomagnets: a frozen dipolar spin ice. Phys. Rev. Lett. **106**, 057209, 4 pages (2011)
210. Mengotti, E., Heyderman, L.J., Fraile Rodríguez, A., Nolting, F., Hügli, R.V., Braun, H.-B.: Real-space observation of emergent magnetic monopoles and associated Dirac strings in artificial kagome spin ice. Nat. Phys. **7**, 68–74 (2011)
211. Hügli, R.V., Duff, G., O'Conchuir, B., Mengotti, E., Heyderman, L.J., Fraile Rodríguez, A., Nolting, F., Braun, H.B.: Emergent magnetic monopoles, disorder, and avalanches in artificial kagome spin ice. J. Appl. Phys. **111**, 07E103, 5 pages (2012)
212. Spivak, G.V., Igras, É., Pryamkova, I.A., Zheludov, I.S.: Detection of the domain structure in barium titanate with an electron reflector. Sov. Phys. Cryst. **4**, 115–117 (1959)
213. Luk'yanov, A.E., Spivak, G.V., Gvozdover, R.S.: Mirror electron microscopy. Sov. Phys. Usp. **16**, 529–552 (1973)
214. Klais, M., Averty, D., Gundel, H.W., Schönhense, G.: Visualization of field induced ferroelectric electron emission from TGS using emission electron microscopy. Appl. Phys. A **78**, 67–72 (2004)
215. Yang, W.C., Rodriguez, B.J., Gruverman, A., Nemanich, R.J.: Photo electron emission microscopy of polarity-patterned materials. J. Phys. Condens. Matter **17**, S1415–S1426 (2005)
216. Krug, I., Barrett, N., Petraru, A., Locatelli, A., Mentes, T.O., Niño, M.A., Rahmanizadeh, K., Bihlmayer, G., Schneider, C.M.: Extrinsic screening of ferroelectric domains in Pb $(Zr_{0.48}Ti_{0.52})O_3$. Appl. Phys. Lett. **97**, 222903, 3 pages (2010)
217. Wang, J.L., Vilquin, B., Barrett, N.: Screening of ferroelectric domains on BaTiO3(001) surface by ultraviolet photo-induced charge and dissociative water adsorption. Appl. Phys. Lett. **101**, 092902, 3 pages (2012)
218. Höfer, A., Fechner, M., Duncker, K., Hölzer, M., Mertig, I., Widdra, W.: Persistence of surface domain structures for a bulk ferroelectric above T_C. Phys. Rev. Lett. **108**, 087602 (2012)
219. Cherifi, S., Hertel, R., Fusil, S., Béa, H., Bouzehouane, K., Allibe, J., Bibes, M., Barthélémy, A.: Imaging ferroelectric domains in multiferroics using a low-energy electron microscope in the mirror operation mode. Phys. Status Solidi (RRL) **4**, 22–24 (2010)
220. Barrett, N., Rault, J.E., Wang, J.L., Mathieu, C., Locatelli, A., Mentes, T.O., Niño, M.A., Fusil, S., Bibes, M., Barthelemy, A., Sando, D., Ren, W., Prosandeev, S., Bellaiche, L., Vilquin, B., Petraru, A., Krug, I.P., Schneider, C.M.: Full field electron spectromicroscopy applied to ferroelectric materials. J. Appl. Phys. **113**, 187217, 13 pages (2013)
221. Wang, K.F., Liu, J.-M., Ren, Z.F.: Multiferroicity: the coupling between magnetic and polarization orders. Adv. Phys. **58**, 321–448 (2009)
222. Pyatakov, A.P., Zvezdin, A.K.: Magnetoelectric and multiferroic media. Physics - Uspekhi **55**, 557–581 (2012)

223. Vaz, C.A.F.: Electric field control of magnetism in multiferroic heterostructures. J. Phys. Condens. Matter **24**, 333201, 29 pages (2012)
224. Kleemann, W., Binek, C.: Multiferroic and Magnetoelectric Materials. In: Zabel, H., Farle, M. (eds.) Magnetic Nanostructures, pp. 163–187. Springer, Heidelberg (2013)
225. Zhao, T., Scholl, A., Zavaliche, F., Lee, K., Barry, M., Doran, A., Cruz, M.P., Chu, Y.H., Edeker, C., Spaldin, N.A., Das, R.R., Kim, D.M., Baek, S.H., Eom, C.B., Ramesh, R.: Electrical control of antiferromagnetic domains in multiferroic $BiFeO_3$ films at room temperature. Nat. Mat. **5**, 823–829 (2006)
226. Holcomb, M.B., Martin, L.W., Scholl, A., He, Q., Yu, P., Yang, C.-H., Yang, S.Y., Glans, P.-A., Valvidares, M., Huijben, M., Kortright, J.B., Guo, J., Chu, Y.-H., Ramesh, R.: Probing the evolution of antiferromagnetism in multiferroics. Phys. Rev. B **81**, 134406, 6 pages (2010)
227. He, Q., Chu, Y.-H., Heron, J.T., Yang, S.Y., Liang, W.I., Kuo, C.Y., Lin, H.J., Yu, P., Liang, C.W., Zeches, R.J., Kuo, W.C., Juang, J.Y., Chen, C.T., Arenholz, E., Scholl, A., Ramesh, R.: Electrically controllable spontaneous magnetism in nanoscale mixed phase multiferroics. Nat. Commun. **2** (Article number: 225), 5 pages (2011)
228. Chu, Y.-H., Martin, L.W., Holcomb, M.B., Gajek, M., Han, S.-J., He, Q., Balke, N., Yang, C.-H., Lee, D., Hu, W., Zhan, Q., Yang, P.-L., Fraile-Rodrígez, A., Scholl, A., Wang, S.X., Ramesh, R.: Electric-field control of local ferromagnetism using a magnetoelectric multiferroic. Nat. Mater. **7**, 478–482 (2008)
229. Moubah, R., Elzo, M., El Moussaoui, S., Colson, D., Jaouen, N., Belkhou, R., Viret, M.: Direct imaging of both ferroelectric and antiferromagnetic domains in multiferroic $BiFeO_3$ single crystal using x-ray photoemission electron microscopy. Appl. Phys. Lett. **100**, 042406, 4 pages (2012)
230. Zavaliche, F., Zheng, H., Mohaddes-Ardabili, L., Yang, S.Y., Zhan, Q., Shafer, P., Reilly, E., Chopdekar, R., Jia, Y., Wright, P., Schlom, D.G., Suzuki, Y., Ramesh, R.: Electric field-induced magnetization switching in epitaxial columnar nanostructures. Nano Lett. **5**, 1793–1796 (2005)
231. Zheng, R., Gu, H., Xu, B., Fung, K.K., Zhang, X., Ringer, S.P.: Self-assembly and self-orientation of truncated octahedral magnetite nanocrystals. Adv. Mater. **18**, 2418–2421 (2006)
232. Zhao, T., Scholl, A., Zavaliche, F., Zheng, H., Barry, M., Doran, A., Lee, K., Cruz, M.P., Ramesh, R.: Nanoscale x-ray magnetic circular dichroism probing of electric-fieldinduced magnetic switching in multiferroic nanostructures. Appl. Phys. Lett. **90**, 123104, 3 pages (2007)
233. Chopdekar, R.V., Malik, V.K., Fraile Rodríguez, A., Le Guyader, L., Takamura, Y., Scholl, A., Stender, D., Schneider, C.W., Bernhard, C., Nolting, F., Heyderman, L.J.: Spatially resolved strain-imprinted magnetic states in an artificial multiferroic. Phys. Rev. B **86**, 014408, 9 pages (2012)
234. Moya, X., Hueso, L.E., Maccherozzi, F., Tovstolytkin, A.I., Podyalovskii, D.I., Ducati, C., Phillips, L.C., Ghidini, M., Hovorka, O., Berger, A., Vickers, M.E., Defay, E., Dhesi, S.S., Mathur, N.D.: Giant and reversible extrinsic magnetocaloric effects in $La_{0.7}Ca_{0.3}MnO_3$ films due to strain. Nat. Mater. **12**, 52–58 (2013)

Chapter 8
Other Surface Imaging Methods with Electrons

Abstract There are many other surface imaging methods, which are competing or complementary to cathode lens microscopy with slow electrons. Scanning probe microscopy in various imaging modes extends the resolution down to the atomic level and is frequently combined with the imaging methods discussed in this book. Here we make only a short comparison with imaging methods using reflected or emitted electrons, independent of energy and experimental setup.

8.1 Scanning Low Energy Electron Microscopy

Scanning low energy electron microscopy (SLEEM), in its construction and application more properly called very low energy scanning electron microscopy, at first sight seems to be most closely related to LEEM as discussed in this book because it uses a cathode lens to decelerate the incident beam focused by the magnetic objective lens in a scanning electron microscope (SEM). The deceleration is achieved by putting the sample at high negative potential and mounting a detector at ground potential with a hole in the center for incident beam passage between magnetic objective lens and sample.

The detector is usually a YAG scintillator single crystal, which is connected to the outside with a light pipe. It integrates over all reflected and emitted electrons but not the specular reflected beam from a flat surface perpendicular to the incident beam, which passes through the bore. Thus at very low energies, at which there is only a specular reflected beam, imaging of single crystal surfaces is not possible. However most surfaces studied in a SEM are polycrystalline and have some roughness, so that imaging at all energies is possible in general. These considerations are illustrated by the first SLEEM images taken under UHV conditions (Fig. 8.1) [1]. The sample is an assembly of flat and three-dimensional Pb crystals deposited in situ on an oxidized Si surface. With some exceptions, which are

E. Bauer, *Surface Microscopy with Low Energy Electrons*, DOI 10.1007/978-1-4939-0935-3_8, © Springer Science+Business Media New York 2014

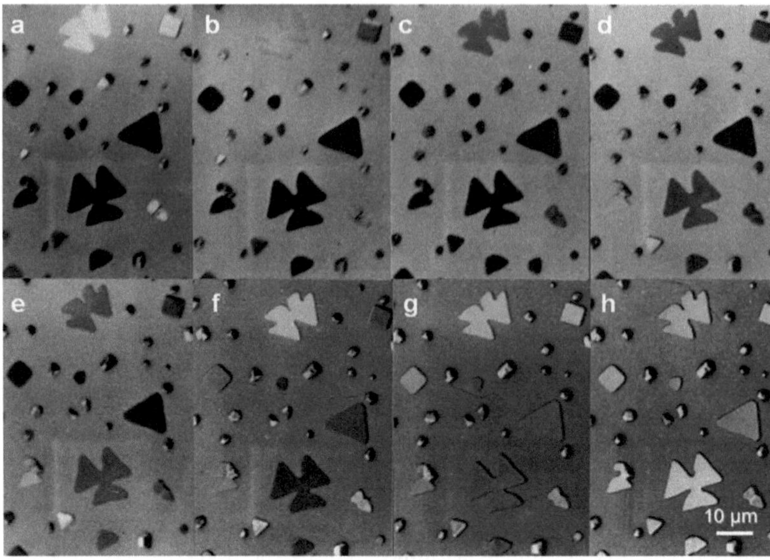

Fig. 8.1 SLEEM images of Pb crystals on an oxidized Si surface deposited in situ in a JEOL JAMP 30 UHV SEM, modified for SLEEM operation. Electron energy (**a–f**) in eV: 6.5, 7.5, 10.5, 16, 18, 22, 29, 34.5. With increasing energy differently oriented crystals develop different contrast due to the appearance of diffracted beams. Reproduced from Ref. [1] with permission from John Wiley & Sons

probably due to beam deflection by local charging, the flat crystals near the optical axis are dark on the brighter background caused by the diffuse scattering from the amorphous SiO_x surface. Above 16–18 eV, when diffracted beams appear, differently oriented crystals develop contrast, e.g., between the (100)-oriented squares and the (111)-oriented triangles. Interestingly, there are also contrast differences between (111)-oriented crystals, which could be due to quantum size effects. In any case, these images show that at low energies real LEEM contrast is possible. With increasing energy, however, the contrast becomes increasingly dominated by secondary electron emission, i.e., by the work function and other parameters, such as the electronic structure.

Despite the limited usefulness for LEEM studies, SEM with a cathode lens has many important applications in materials research because of the possibility of contrast control via landing energy control without dramatic loss of resolution with decreasing energy. Calculations for a typical experimental setup have shown that the spot size on the specimen increases from 0.8 nm at 10 kV only to 3.5 nm and 7 nm at 10 eV and 1 eV, respectively, when using a field emission gun. This makes SLEEM interesting for samples, which scatter diffusely and therefore do not produce LEEM contrast. Some recent reviews by Frank and Müllerová, who have developed this method, highlight the wide application range of cathode lens SEM and give also details on its theory and design [2–4].

8.2 Scanning Low Energy Electron Diffraction Microscopy

Scanning low energy electron diffraction microscopy is a method, which can be incorporated into a standard UHV scanning microscope by adding a detector consisting of a LEED optics, channel plate image intensifier, and position-sensitive detector (PSD). It was developed in the 1980s by Ichinokawa et al. [5] but used only for a short period [6] because it could not compete with LEEM. It is included here in the list of other surface imaging methods with slow electrons because the instrument developments, which have happened since, make this method a useful complement to standard SEM without requiring a cathode lens. The original instrument [5] was a modified Hitachi S-800 UHV SEM with a cold field emission gun, which could be operated from 30 keV down to 100 eV after modification. At a working distance of 3 mm from the objective lens it produced a spot size of 10 nm (100 nm) at an energy of 1 keV (100 eV). The spot was scanned across the surface and the LEED pattern recorded on the PSD. An electronic window was set around a selected LEED spot and its intensity stored for subsequent display with 256×256 pixels, thus producing a dark field LEEM image of the surface. The image acquisition time was limited by the count rate of the total area of the PSD and the size of the electronic window. Depending upon the desired signal-to-noise ratio 10 min to 1 h was needed per image. The spatial resolution of the detector was comparable to the spot size on the sample.

The advantage of this method compared to SLEEM is that the sample is in field-free space and that the LEEM image is formed by a selected LEED spot, while in SLEEM the sample is in a field and the image is formed by many LEED spots and secondary electrons. Disadvantages are the longer image acquisition times and the poorer resolution. With improved PSDs, better data acquisition, and possible improvements in the electron optics, this method nevertheless might be useful for SEM studies which require information on the crystal structure of the surface.

It should be mentioned that the scanning LEED microscope was preceded by a scanning medium energy electron diffraction (MEED) microscope with similar but less sophisticated design [7]. The electron optics was simpler, and the LEED detector was a movable small Spiraltron electron multiplayer behind a four-grid LEED system for the measurement of the intensity of selected diffracted beams. The long focal length of the objective lens and the small glancing angle of incidence limited resolution and caused strong foreshortening of the image.

8.3 Reflection Electron Microscopy

Reflection electron microscopy (REM) is nearly as old as emission electron microscopy. The mode of operation, which was later used in diffraction imaging, was introduced in 1940 by von Borries [8]. Different from the scanning diffraction imaging modes described in Sect. 8.2, in which the sample is mounted after the objective lens, in REM the sample sits in the objective lens of a transmission electron microscope, with its surface and the illuminating beam tilted so that the

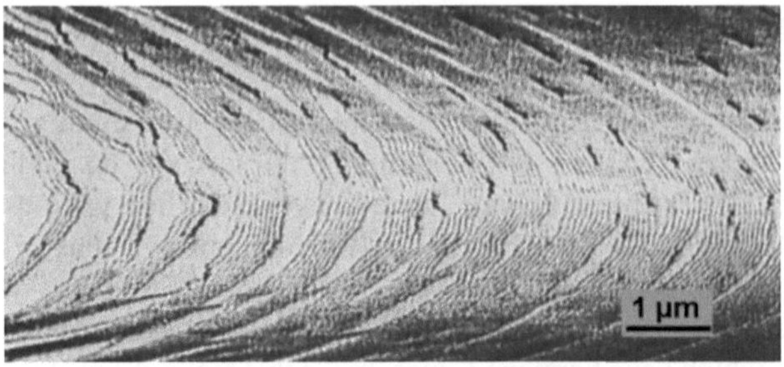

Fig. 8.2 REM image of the vicinals of a (111) plane on a spherical Pt crystal. The (111) plane is to the left. The surface consists of bands of monatomic steps and atomically flat (111) regions, which disappear with increasing deviation from the (111) plane (*increasing curvature*). The random distribution of double steps (*black regions*) and the roughness of the steps suggest slight contamination. Reproduced with permission from Ref. [9]. Copyright 1983 Elsevier

reflection high energy electron diffraction (RHEED) pattern passes through the objective lens. Similar to LEEM the contrast aperture in the back focal plane, where the primary RHEED pattern is located, selects one of the diffracted beams for imaging. Similar to LEEM by removing this aperture the RHEED pattern in the back focal plane can be imaged. Different from LEEM, which mostly uses the specular reflected beam, REM uses one of the higher order beams. Also different from LEEM, in which the diffracted beams are described in terms of the two-dimensional reciprocal lattice, in REM a three-dimensional description is used because the diffracted beams come from deeper layers in the crystal at the usual diffraction angles. The angle of incidence is usually 1°–3° and the angle of reflection more than 5° depending upon electron energy and diffracted beam chosen for imaging. Because the surface is nearly parallel to the electron beam the object distance varies along the crystal in the beam direction, which causes a different magnification $M_{\|}$ in this direction from that ($M_{\perp}$) normal to it, with $M_{\perp}/M_{\|}$ ranging from 10 to 70 depending upon tilt angle and position of the specimen with respect to the lens ("foreshortening"). The foreshortening can be seen in the strong curvature of the monatomic step bunches and in the decreasing step distance towards the bottom and top of the typical REM image shown in Fig. 8.2 [9]. The defocus connected with the different distance limits the focused distance along the beam, but at high magnifications the defocusing within the field of view is acceptable.

The foreshortening problem has been a least partially solved by Fert et al. [10] with a cylinder lens, which increases $M_{\|}$ and decreases $M_{\perp}$, a method which interestingly was used later by only one other group. Another approach is numerical distortion correction of the image recorded in the camera, but this is limited by the pixel size of the detector. A third method is to reduce the foreshortening by tilting the detector such that the image distance varies with the object distance on the sample surface. In this method, low distortion REM ("LODREM") [11], two tilt controls are needed for the detector, one against the optical axis for foreshortening control and the other around the optical axis to follow the image rotation of the magnetic objective lens when the magnification is changed. The intensity on the detector is reduced due to the small angle of incidence of the beam, and its surface has to be very smooth to avoid shadows, but the results reported are better than those obtained with numerical image processing.

REM has been a popular method for surface imaging in the last decades of the twentieth century, originally mainly for imaging the topography and starting around 1980, when UHV instruments became commercially available, also for real surface structure imaging. This popularity is reflected in the number of review papers [12–14] and a book on this subject [15]. In the early twenty-first century the interest in REM declined, possibly because of the commercial availability of LEEM instruments, which do not suffer from foreshortening, but one instrument is still used successfully for surface studies as evident from a review in 2012 [16] and a 2013 original publication [17]. With aberration-corrected transmission electron microscopes (TEMs), now increasingly available, REM might find some resurrection for surface problems requiring better resolution than LEEM, whose resolution is not expected to get better than 1 nm. Resolutions in the direction perpendicular to the beam direction as fine as 0.3 nm have been reported in the uncorrected TEM systems used for REM already in 1993 [18]. REM in aberration-corrected TEM instruments with samples in UHV environment should achieve still better resolution, possibly atomic resolution, with less effort or luck. Combination with well-designed quadrupole lenses should also reduce the foreshortening sufficiently so that sub-nm resolution images with little distortion in both directions can be achieved.

These considerations also apply to scanning REM (SREM) which also has been developed in the 1980s when a UHV scanning TEM (STEM) became commercially available. Compared to REM it has the advantage that the focus can be maintained over the imaged area by changing the lens current with the position of the beam along the optical axis ("dynamic focusing"). This goes at the expense of image acquisition rate because of the inductance of the lens. Also image interpretation becomes more difficult because the RHEED pattern is no longer a spot but a convergent beam pattern (Kikuchi pattern) due to the convergent illumination in the spot on the sample. However, the vastly improved image quality makes this mode of operation very attractive. SREM has the additional advantage over REM that it can be easily combined with secondary electron and Auger electron imaging (see the review [19]). Nevertheless SREM disappeared from research even earlier than REM.

In both REM and SREM the size and access to the sample are very limited because of its location in the objective lens. The desire to image larger samples and to have access to it, for example for deposition from several evaporators, has led, also in the early 1980s, to the development of microprobe high energy electron diffraction microscopy. In this method, a scanning electron microscope column is mounted on a UHV chamber, which accommodates not only the detector for the RHEED pattern, from which the diffraction spot used for image formation is selected, but also several other analysis methods for complementary characterization of the specimen over which the beam is scanned. These instruments reached over the years a high level of sophistication [20–22] but did not achieve the high resolution as in-lens SREMs because of the larger working distance needed for the various characterization methods. This may be the reason why little work was done with these instruments after 2000.

8.4 Secondary and Auger Electron Microscopy

As mentioned in the preceding section, secondary and Auger electron imaging is possible with the specimen in the lens field. These imaging modes can also be used in cathode lens electron microscopes. In LEEM instruments this is possible with an additional power supply and a beam separator, which allows different energies in the illumination and imaging section of the instrument. For Auger electron imaging in addition a band-pass energy filter is needed. In practice these imaging modes are not used because the combination of LEEM, MEM, and UVPEEM gives more information than secondary electron imaging. Auger electron imaging actually has been used before undulator beamlines at synchrotron radiation facilities became available, which allows much more flexible chemical imaging with XPEEM. Furthermore, samples with pronounced three-dimensional structure are poorly suited for these instruments because of discharge problems, field distortion, and in most PEEM illumination setups by shadow effects.

For all these reasons field-free post-lens scanning secondary electron microscopy (SEM) and (scanning) Auger electron microscopy (AEM) have developed into an important field in materials analysis, in particular for chemical analysis, once UHV scanning electron illumination systems became available. These were attached to UHV chambers, equipped with well-developed secondary electron and Auger electron detectors [23]. The first applications to surface studies [24, 25] were so convincing that dedicated instruments were developed commercially, called microprobes, with specified resolutions of 3 nm and 8 nm in secondary electron and Auger electron imaging, respectively, in one of the instruments (JAMP-9500F field emission Auger microprobe). AEM played and still plays an important role in applied science for chemical surface characterization. SEM has been used in the 1990s successfully in semiconductor surface physics ([26] and earlier work), similar to LEEM and REM, in particular in growth studies. Another important application was the study of micron-sized particles, which are too large to be suitable for LEEM, under UHV conditions. An example is the study of the thermal disordering near the melting point and of the melting of metals (Pb, In, Sn, Bi, Tl) under UHV conditions [27, 28]. Near the melting point two phenomena can occur on a surface: surface roughening and surface melting, which can be imaged via the associated work function change. Figure 8.3a, b [28] illustrates this for an In crystal at 5.5 K and 1 K, respectively, below the melting point. The dark regions are already surface-molten and encroach between 5.5 and 1 K onto the flat equilibrium faces and their surface-roughened vicinals. Figure 8.3c shows the various faces of this crystal and the boundaries of the surface-molten regions for various temperatures (because the crystal is tetragonal not all faces with the same indices (hkl) are equivalent). Careful intensity analysis of the original micrographs allows delineating the phase boundaries between flat, rough, and surface-molten regions as a function of temperature as shown in Fig. 8.3d for the [011] crystal zone [28]. Studies of the growth shape of such crystals have revealed an unusual temperature dependence and a strong influence of reactive gases such as H_2O in the low 10^{-6} mbar

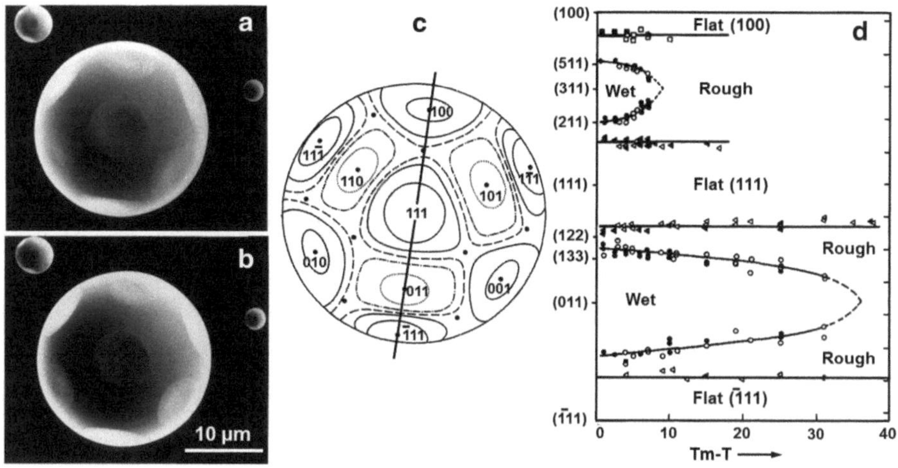

Fig. 8.3 Equilibrium shape of a spherical indium crystal near the melting point. (**a**, **b**) SEM images taken at 5.5 and 1 K below the melting point. The boundaries between the quasi-liquid layer and the major surfaces are shown in (**c**): *dotted line* at 31 K below the melting point, *dash-dotted line* at 15 K, *dashed line* at 9 K, and *full line* at 4 K below the melting point. (**d**) Surface-phase diagram along the [011] zone, indicated by the *thick line* in (**c**). Wet means covered by quasi-liquid layer. The boundaries were obtained from images during heating and cooling. Adapted with permission from Ref. [28]. Copyright 1994 Elsevier

range—compared to the clean growth in the low 10^{-10} Torr range—on the growth shape [29]. Such studies demonstrate that UHV-SEM complements LEEM nicely where LEEM is not applicable.

8.5 Scanning Electron Microscopy with Spin Analysis

Scanning electron microscopy with spin analysis (SEMPA), also called spin-SEM, differs from the SEM instruments used in the previous sections in that it has a detector, which analyzes the spin orientation of the emitted secondary electrons. Several spin analyzer ("polarimeter") types are used, all of which rely on the asymmetry of the spin-orbit scattering at heavy atoms such as Au or W: classical Mott polarimeters, operating at 40–60 keV; mini-Mott polarimeters, working at 20–30 keV; and low energy diffuse scattering detectors and low energy diffraction detectors, which have optimum operation energies around 100 eV. Information on these detectors can be found in several reviews [30–33]. More recent developments are reported in [34] and a later review [35]. SEMPA has some advantages compared to SPLEEM: the sample is in field-free space, can be amorphous or polycrystalline, and can have considerable topography. Its disadvantages are the nonparallel image acquisition and the low spin sensitivity of the spin detectors, which together lead to much longer image acquisition times then in SPLEEM. The sample is usually

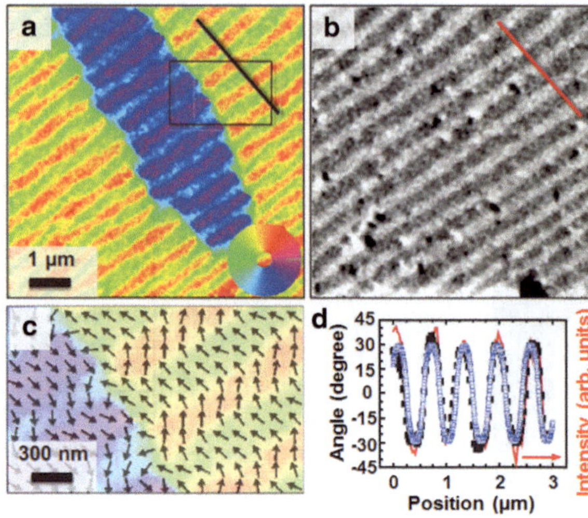

Fig. 8.4 Exchange coupling between ferromagnetic and multiferroic films. (**a**) SEMPA image of the ferromagnetic CoFe film, with the color wheel indicating the magnetization direction. (**b**) Electron backscattering image from the multiferroic $BeFeO_3$ film below the CoFe film. (**c**) Micromagnetic simulation of the spin distribution in the *rectangle* marked in (**a**). (**d**) Scans along the *lines* marked in (**a**) (*black*) and in (**b**) (*red*) and results of the micromagnetic simulations (*blue*). Reproduced with permission from Ref. [36]. Copyright 2013 by the American Physical Society

mounted in a large chamber, which can accept other analysis systems, such as an Auger system for chemical analysis, a fluorescent screen for RHEED or an electron backscattering diffraction system, windows for magneto-optical Kerr effect measurements, and other components. All these possibilities, together with more flexibility in the sample mounting (degrees of freedom, heating, cooling, and application of fields), make SEMPA a more flexible spin-sensitive imaging system than SPLEEM. An illustration is the system described in [35], which has many of these features, including liquid He cooling, and has reached a resolution of 3 nm with aberration-corrected electron optics.

The importance of combining SEMPA with complementary methods is illustrated in Fig. 8.4 [36]. The sample is a non-magnetostrictive 2.5 nm thick amorphous $Co_{0.9}Fe_{0.1}$ film on a thick (>10 nm) epitaxial multiferroic (ferroelectric-antiferromagnetic) $BiFeO_3$ layer, which had been grown on a $SrRuO_3$ buffered (110)-oriented $DyScO_3$ substrate. Figure 8.4a shows a SEMPA image of the CoFe film, and Fig. 8.4b shows a simultaneously acquired backscattered electron image, which images the domain structure of the $BiFeO_3$ layer via electron channeling contrast. The one-to-one correlation between the nearly completely in-plane magnetized CoFe domains and the $BiFeO_3$ domains disappears in CoFe films thicker than 20 nm or when a 2 nm thin fully strained $SrTiO_3$ layer is inserted between CoFe and $BiFeO_3$ layer. This shows that the coupling is localized at the interface and is not strain-induced but rather by the direct interaction between CoFe and magnetic order in the $BiFeO_3$ domains, i.e., by exchange coupling.

A final comparison between SEMPA, SPLEEM, and XMCDPEEM shows that each method has its strengths. SEMPA has the best resolution and is universally applicable, and SPLEEM is the fastest method and gives structural information via LEED, while XMCDPEEM gives chemical information. SEMPA is the oldest method and best developed. SPLEEM still has some way to go by combining a high-polarization spin gun, aberration correction, and liquid He cooling in one and the same system. With a high-polarization spin gun it should also be able to study samples, which do not produce a strong diffracted beam, probably faster than SEMPA, eliminating one of the main restrictions of SPLEEM. Also with the improvement of magnetization sample holders and the addition of a chamber for complementary measurements some of the other advantages of SEMPA can be overcome.

8.6 Scanning Photoemission Microscopy

In scanning photoemission microscopy (SPEM) the sample is scanned across the finely focused photon beam and the image is formed by sequential detection of the emitted photoelectrons. The first imaging of this type was done in 1989 by Rotermund et al. with UV light from a deuterium lamp which was focused by a Schwarzschild optics into a 0.5 μm diameter spot on the sample, and all emitted electrons were detected with a channeltron to form an image via work function contrast [37]. This method was soon displaced by full-field PEEM (see Sect. 5.2.1.2). Scanning UVPEEM was revived again by Munakata et al. in 2002 in more sophisticated form as time-resolved photoemission microspectroscopy with fs-VUV laser light [38]. Frequency upconversion of 100 fs Ti:sapphire laser light gave photon energies up to 8.9 eV. A Schwarzschild optics produced a diffraction-limited spot size of 0.2 μm. A time-of-flight spectrometer allowed an energy resolution of 30 meV, which makes this instrument useful for imaging local energy and work function variations within the practical resolution limit of 0.5 μm. An application example is the study of the lateral inhomogeneity of unoccupied states in Pb phthalocyanine films [39].

By far the most important application of SPEM is however in synchrotron radiation-based photoemission microscopy, beginning around 1990, using imaging with Schwarzschild optics and zone plates. XSPEM is now used in most synchrotron radiation facilities and has reached a high level of sophistication as described in two reviews [40, 41]. Zone plate instruments work now routinely with spatial resolutions of better than 200 nm, and with special plates individual 50 nm structures have been imaged (M. Kiskinova, private communication), approaching the resolution of XPEEM. An advantage over XPEEM is the faster acquisition of spectra from very small areas. In XPEEM fast spectra acquisition via imaging of the dispersive plane from areas selected with the field-limiting aperture is also possible, but selection of features below 1 μm in diameter in this manner is difficult because the photon beam is focused into an area of typically 50 μm^2 or larger

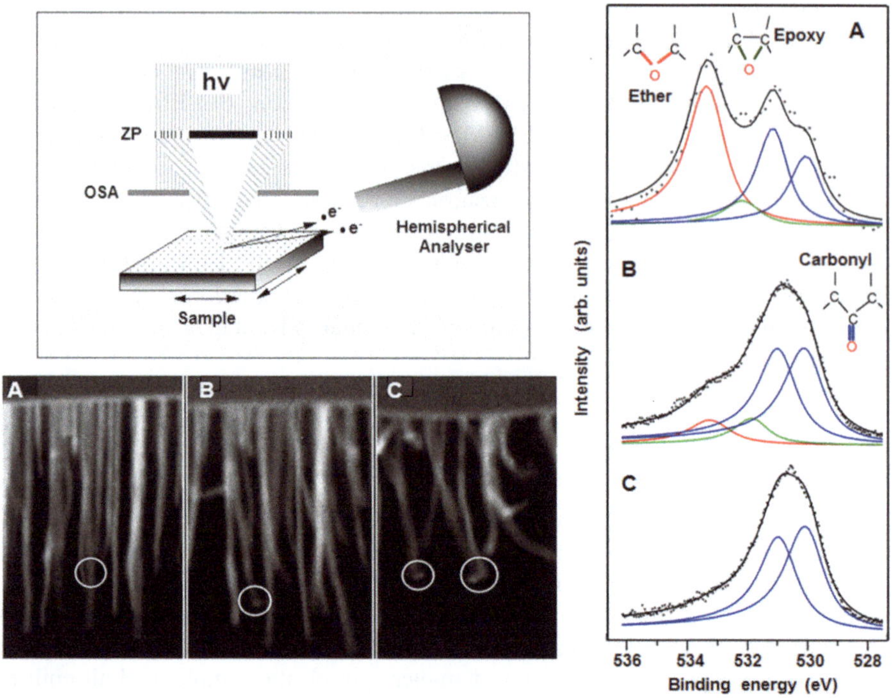

Fig. 8.5 Scanning photoemission microscopy. The *upper left* shows schematically the experimental setup of a zone plate microscope. The incident monochromatic photon beam is focused by the outer parts of the zone plate ZP onto the specimen. Diffraction orders higher than one are eliminated by the order-sorting aperture, OSA. The sample is scanned under the focused beam, and the emitted electrons are analyzed by a hemispherical analyzer. The XSPEM images (*lower left*) show multi-wall carbon nanotubes at different oxidation states and the spectra (on the *right*) their corresponding oxygen K 1s multiplet structure. For details see text. Courtesy of M. Kiskinova except for the *upper left* part which is reproduced with permission from Ref. [41]. Copyright 2009 Elsevier

compared to several 10^{-2} μm^2 in XSPEM. In XPEEM spectra from smaller areas than in XSPEM can be obtained from images taken as a function of energy in small steps and setting windows in the resulting image stack to extract the intensity as a function of energy, however with significantly longer acquisition time than local spectra with XSPEM.

The main advantage of XSPEM compared to XPEEM is that the sample is in the field-free space. This allows microspectroscopy of crystals in the 10 μm size range with sharp edges and corners, such as indium zinc oxide pyramids [42] which would cause strong field distortions and probably discharges in XPEEM. Another advantage is that the sample does not need a support. This is illustrated in Fig. 8.5 [courtesy of M. Kiskinova and [41]] which shows the changes of the morphology and chemical composition of multi-walled carbon nanotubes, supported at one end on Si, during reaction with oxygen. The images were taken with C 1s photoelectrons with about 10 min acquisition time, and the spectra are O 1s spectra from the regions marked in the images, which were acquired twice faster. The initial state at room temperature (A) is stable even at high oxygen doses, but exposure at 200 °C to doses of 2.7×10^{17} (B) and 5.4×10^{17} (C) oxygen atoms causes strong oxidation evident in the decreasing length of the nanotubes. Simultaneously several intermediate oxygen functional groups with binding energies of about 533.5 (ether, red), 532 (epoxy, green), 531, and 530 eV (carbonyl, blue) appear with increasing oxidation [41, 43]. Another advantage of XSPEM is the higher acceptable pressure, which is limited only by the mean free path of the photoelectrons in the surrounding gas, while in XPEEM the pressure is limited to the middle 10^{-5} mbar range to avoid high voltage discharges. Thus XSPEM, when viewed as microspectroscopy, is in many cases complementary to XPEEM, i.e., to spectromicroscopy.

8.7 Concluding Remarks

The preceding sections have shown that some of the other surface imaging methods such as SEMPA and XSPEM have some advantages for specific studies compared to the full-field cathode lens imaging methods described in this book. The combination of a wide variety of complementary methods (LEEM, μLEED, MEM, UVPEEM, XPEEM in the secondary electron and photoelectron mode, μXPS, μXPD, various dichroism PEEM modes (MCD, MLD, NLD)) gives full-field cathode lens microscopy a unique position in the field of surface analysis. Microstructure, crystal structure, electronic structure, chemical composition and boding, magnetic domain structure, work function, and other surface properties can be obtained in this manner. In addition, there is sufficient access to the sample during investigation that it can be exposed to gas or vapor beams so that the microscope becomes a versatile system for experimentation, not only a system for imaging of externally prepared samples. These possibilities are enhanced in large specimen chambers, which allow attachment of multiple evaporators, gas sources, a mass spectrometer, and other accessories. This makes these instruments not simple imaging systems but small-scale laboratories for the study of the properties of and processes on/in surfaces, thin films, and nanostructures.

References

1. Frank, L., Müllerová, I., Faulian, K., Bauer, E.: The scanning low-energy electron microscope: first attainment of diffraction contrast in the scanning electron microscope. Scanning **21**, 1–13 (1999)
2. Frank, L., Hovorka, M., Konvalina, I., Mikmeková, Š., Müllerová, I.: Very low energy scanning electron microscopy. Nucl. Instrum. Meth. Phys. Res. A **645**, 46–54 (2011)

3. Müllerová, L., Hovorka, M., Mika, F., Mikmeková, E., Mikmeková, Š., Pokorná, Z., Frank, L.: Very low energy scanning electron microscopy in nanotechnology. Int. J. Nanotechnol. **9**, 695–716 (2012)
4. Frank, L., Hovorka, M., Mikmeková, Š., Mikmeková, E., Müllerová, I., Pokorná, Z.: Scanning electron microscopy with samples in an electric field. Materials **5**, 2731–2756 (2012)
5. Ichinokawa, T., Ishikawa, Y., Kemmmochi, M., Ikeda, N., Hosokawa, Y.: Low energy scanning electron microscopy combined with low energy electron diffraction. Surf. Sci. **176**, 397–414 (1986)
6. Ichinokawa, T., Hamaguchi, I., Hibino, M.: Surface defects of MoS_2 crystals observed by scanning LEED microscopy. Surf. Sci. Lett. **231**, L189–L195 (1990)
7. Cowley, J.M., Albain, J.L., Hembree, G.G., Højlund-Nielsen, P.E., Koch, F.A., Landry, J.D., Shuman, H.: System for reflection electron microscopy and electron diffraction at intermediate energies. Rev. Sci. Instrum. **46**, 826–829 (1975)
8. von Borries, B.: Sublichtmikroskopische Auflösungen bei der Abbildung von Oberflächen im Übermikroskop. Z. Physik **116**, 370–378 (1940)
9. Hsu, T.: Reflection electron microscopy (REM) of vicinal surfaces of fcc metals. Ultramicroscopy **11**, 167–172 (1983)
10. Fert, C.: Obervation directe des surfaces métalliques par réflexion. In: Sjöstrand, F.S., Rhodin, J.A.G. (eds.) Proceedings of the Conference Electron Microscopy, Stockholm 1956, pp. 8–12. Academic, New York, NY (1957)
11. Müller, P., Métois, J.J.: Low distortion reflection electron microscopy for surface studies. Surf. Sci. **599**, 187–195 (2005)
12. Hsu, T.: Technique of reflection electron microscopy. Microsc. Res. Techn. **20**, 318–332 (1992)
13. Wang, Z.L.: Electron reflection, diffraction and imaging of bulk crystal surfaces in TEM and STEM. Rep. Prog. Phys. **56**, 997–1065 (1993)
14. Yagi, K.: Reflection electron microscopy: studies of surface structures and surface dynamic processes. Surf. Sci. Rep. **17**, 305–362 (1993)
15. Wang, Z.L.: Reflection electron microscopy and spectroscopy for surface analysis. Cambridge University Press, Cambridge (1996)
16. Latyshev, A.V.: Formation of surface patterns observed by reflection electron microscopy. In: Dehm, G., Howe, J.M., Zweck, J. (eds.) In Situ Electron Microscopy: Applications in Physics, pp. 99–122. Chemistry and Material Science, Weinheim, Wiley-VCH Verlag (2012)
17. Rogilo, D.I., Fedina, L.I., Kosolobov, S.S., Ranguelov, B.S., Latyshev, A.: Critical terrace width for two-dimensional nucleation during Si growth on Si(111)-(7x7) surface. Phys. Rev. Lett. **111**, 036105, 4 pages (2013)
18. Cowley, J.M., Liu, J.: Contrast and resolution in REM, SEM and SAM. Surf. Sci. **298**, 456–467 (1993)
19. Liu, J., Cowley, J.M.: Scanning reflection electron microscopy and associated techniques for surface studies. Ultramicroscopy **48**, 381–416 (1993)
20. Isu, T., Watanabe, A., Hata, M., Katayama, Y.: In-situ microscopic observation of GaAs surfaces during molecular beam epitaxy and metalorganic molecular beam epitaxy by scanning microprobe reflection high energy electron diffraction. J. Cryst. Growth **100**, 433–438 (1990)
21. Watanabe, H., Ichikawa, M.: Development of a multifunctional surface analysis system based on a nanometer scale scanning electron beam: combination of ultrahigh vacuum-scanning electron microscopy, scanning reflection electron microscopy, Auger electron spectroscopy, and x-ray photoelectron spectroscopy. Rev. Sci. Instrum. **67**, 4185–4190 (1996)
22. Maruno, S., Nakahara, H., Fujita, S., Watanabe, H., Kusumi, Y., Ichikawa, M.: A combined apparatus of scanning reflection electron microscope and scanning tunneling microscope. Rev. Sci. Instrum. **68**, 116–119 (1997)

23. Venables, J.A.: An UHV SEM for in-situ deposition and surface studies. In: Venables, J.A. (ed.) Developments in Electron Microscopy and Analysis, pp. 23–26. Academic, London (1976)
24. Hartig, K., Janssen, A.P., Venables, J.A.: Nucleation and growth in the system Ag/Mo(100): a comparison of UHV-SEM and AES/LEED observations. Surf. Sci. **74**, 69–78 (1978)
25. Venables, J.A., Janssen, A.P., Akhter, P., Derrien, J., Harland, C.J.: Surface studies in a UHV field emission gun scanning electron microscope. J. Microsc. **118**, 351–365 (1980)
26. Homma, Y., Finnie, P., Uwaha, M.: Morphological instability of atomic steps observed on Si (111) surfaces. Surf. Sci. **492**, 125–136 (2001)
27. Pavlovska, A., Faulian, K., Bauer, E.: Surface roughening and surface melting in the high temperature equilibrium shape of small Pb crystals. Surf. Sci. **221**, 233–243 (1989)
28. Pavlovska, A., Dobrev, D., Bauer, E.: Orientation dependence of the quasi-liquid layer on tin and indium crystals. Surf. Sci. **314**, 341–352 (1994)
29. Pavlovska, A., Dobrev, D., Bauer, E.: Facet growth of spherical lead crystals. Surf. Sci. **326**, 101–112 (1995)
30. Unguris, J.: Scanning electron microscopy with polarization analysis (SEMPA) and its applications. In: de Graef, M., Zyu, Y. (eds.) Magnetic Imaging and its Applications to Materials 2001, pp. 167–193. Academic, San Diego, CA (2001)
31. Allenspach, R.: Spin-polarized scanning electron microscopy. In: Zhu, Y. (ed.) Modern Techniques for Characterizing Magnetic Materials, pp. 327–359. Kluwer Academic Publishers, Boston, MA (2005)
32. Oepen, H.P., Hopster, H.: SEMPA studies of thin films, structures and exchange coupled layers. In: Hopster, H., Oepen, H.P. (eds.) Magnetic Microscopy of Nanostructures, pp. 137–167. Springer, Berlin (2005)
33. Oepen, H.P., Frömter, R.: Scanning electron microscopy with polarisation analysis. In: Kronmüller, H., Parkin, S. (eds.) Handbook of Magnetism and Advanced Magnetic Materials, vol. 3, pp. 1488–1509. John Wiley & Sons, Chichester (2007)
34. Frömter, R., Hankemeier, S., Oepen, H.P., Kirschner, J.: Optimizing a low-energy electron diffraction spin-polarization analyzer for imaging of magnetic surface structures. Rev. Sci. Instrum. **82**, 033704, 11 pages (2011)
35. Koike, K.: Spin-polarized scanning electron microscopy. Microscopy **62**, 177–191 (2013)
36. Trassin, M., Clarkson, J.D., Bowden, S.R., Liu, J., Heron, J., Paull, R.J., Arenholz, E., Pierce, D.T., Unguris, J.: Interfacial coupling in multiferroic/ferromagnet heterostructures. Phys. Rev. B **87**, 134426, 6 pages (2013)
37. Rotermund, H.H., Ertl, G., Sesselmann, W.: Scanning photoemission microscopy of surfaces. Surf. Sci. **217**, L383–L390 (1989)
38. Munakata, T., Masuda, T., Ueno, N., Abdureyim, A., Sonoda, Y.: Time-resolved photoemission microspectroscopy based on fs-VUV laser light. Surf. Sci. **507–510**, 434–440 (2002)
39. Yamamoto, R., Yamamoto, I., Mikamori, M., Yamada, T., Miyakubo, K., Munakata, T.: Lateral inhomogeneity of unoccupied states for PbPc films. Surf. Sci. **605**, 982–986 (2011)
40. Günther, S., Kaulich, B., Gregoratti, L., Kiskinova, M.: Photoelectron microscopy and applications in surface and materials science. Prog. Surf. Sci. **70**, 187–260 (2002)
41. Barinov, A., Dudin, P., Gregoratti, L., Locatelli, A., Menteş, T.O., Niño, M.A., Kiskinova, M.: Synchrotron-based photoelectron microscopy. Nucl. Instrum. Meth. Phys. Res. A **601**, 195–202 (2009)
42. Bartolome, J., Maestre, D., Amati, M., Cremades, A., Piqueras, J.: Indium zinc oxide pyramids with pinholes and nanopipes. J. Phys. Chem. C **115**, 8354–8360 (2011)
43. Barinov, A., Gregoratti, L., Dudin, P., La Rosa, S., Kiskinova, M.: Imaging and spectroscopy of multiwalled carbon nanotubes during oxidation: defects and oxygen bonding. Adv. Mater. **21**, 1916–1920 (2009)

Index

E. Bauer, *Surface Microscopy with Low Energy Electrons*,
DOI 10.1007/978-1-4939-0935-3, © Springer Science+Business Media New York 2014

D

J

K

L

S

MIX
Papier aus verantwortungsvollen Quellen
Paper from responsible sources
FSC® C105338

If you have any concerns about our products,
you can contact us on
ProductSafety@springernature.com

In case Publisher is established outside the EU,
the EU authorized representative is:
Springer Nature Customer Service Center GmbH
Europaplatz 3, 69115 Heidelberg, Germany

Printed by Libri Plureos GmbH
in Hamburg, Germany